KINESIOLOGY
for Occupational Therapy

SECOND EDITION

KINESIOLOGY
for Occupational Therapy

SECOND EDITION

MELINDA F. RYBSKI, PhD, OTR/L

The School of Allied Health Medical Professions
Occupational Therapy Division
The Ohio State University
Columbus, Ohio

SLACK
INCORPORATED

www.slackbooks.com

ISBN 978-1-55642-916-3

Kinesiology for Occupational Therapy Second Edition Instructor's Manual is also available from SLACK Incorporated. Don't miss this important companion to *Kinesiology for Occupational Therapy Second Edition*. To obtain the Instructor's Manual, please visit http://www.efacultylounge.com

Melinda F. Rybski and Linda M. Martin have no financial or proprietary interest in the materials presented herein.

The procedures and practices described in this publication should be implemented in a manner consistent with the professional standards set for the circumstances that apply in each specific situation. Every effort has been made to confirm the accuracy of the information presented and to correctly relate generally accepted practices. The authors, editors, and publisher cannot accept responsibility for errors or exclusions or for the outcome of the material presented herein. There is no expressed or implied warranty of this book or information imparted by it. Care has been taken to ensure that drug selection and dosages are in accordance with currently accepted/recommended practice. Off-label uses of drugs may be discussed. Due to continuing research, changes in government policy and regulations, and various effects of drug reactions and interactions, it is recommended that the reader carefully review all materials and literature provided for each drug, especially those that are new or not frequently used. Some drugs or devices in this publication have clearance for use in a restricted research setting by the Food and Drug and Administration or FDA. Each professional should determine the FDA status of any drug or device prior to use in their practice.

Any review or mention of specific companies or products is not intended as an endorsement by the author or publisher.

SLACK Incorporated uses a review process to evaluate submitted material. Prior to publication, educators or clinicians provide important feedback on the content that we publish. We welcome feedback on this work.

Published by: SLACK Incorporated
                 6900 Grove Road
                 Thorofare, NJ 08086 USA
                 Telephone: 856-848-1000
                 Fax: 856-848-6091
                 www.slackbooks.com

Contact SLACK Incorporated for more information about other books in this field or about the availability of our books from distributors outside the United States.

Library of Congress Cataloging-in-Publication Data
Rybski, Melinda.
Kinesiology for occupational therapy / Melinda F. Rybski. -- 2nd ed.
p. ; cm.
Includes bibliographical references and index.
ISBN 978-1-55642-916-3 (pbk. : alk. paper) 1. Kinesiology. 2. Human mechanics. 3. Musculoskeletal system. 4. Occupational therapy. I. Title.
  [DNLM: 1. Kinesiology, Applied. 2. Movement--physiology. 3. Musculoskeletal Physiological Phenomena. 4. Occupational Therapy--methods.
WE 103]
  QP303.R93 2012
  612.7'6--dc23   2011021074

For permission to reprint material in another publication, contact SLACK Incorporated. Authorization to photocopy items for internal, personal, or academic use is granted by SLACK Incorporated provided that the appropriate fee is paid directly to Copyright Clearance Center. Prior to photocopying items, please contact the Copyright Clearance Center at 222 Rosewood Drive, Danvers, MA 01923 USA; phone: 978-750-8400; website: www.copyright.com; email: info@copyright.com

Printed in the United States of America.

Last digit is print number: 10 9 8 7 6 5 4 3 2 1

Contents

Kinesiology for Occupational Therapy Second Edition Instructor's Manual is also available from SLACK Incorporated. Don't miss this important companion to *Kinesiology for Occupational Therapy Second Edition*. To obtain the Instructor's Manual, please visit http://www.efacultylounge.com

ACKNOWLEDGMENTS

I wish to thank Sharon Flinn and Lori DeMott for our regular discussions about the content of this textbook and for their valuable contributions. Thanks to Kim Szucs for her expertise and helpful resources for the chapter on the shoulder. Thanks also to Sandra Rogers, who commended the first edition and was recruited to help with content for this second edition.

I would like to thank Brien Cummings and John Bond for their patience and encouragement throughout the writing of this second edition. Thanks always to Amy McShane, who encouraged me to write this text initially. The occupational therapy faculty members at The Ohio State University, my teachers, as well as my peers have been steadfast in their confidence in my abilities. My parents taught me the wonder of life and the fun of curiosity. Most of all, I want to thank Tom, Katherine, and Greg for their patience and understanding, and I appreciate Ivy, who was my constant companion as I wrote this text.

ABOUT THE AUTHOR

Melinda Fritts Rybski, PhD, OTR/L, is a faculty member of the Occupational Therapy Division at The Ohio State University. Dr. Rybski graduated from The Ohio State University College of Allied Medical Professions with a bachelor of science degree in 1979 and a Master's degree in 1987. She completed her PhD in 2010 with a degree from The Ohio State University, focusing on quantitative research, evaluation, and measurement. She has taught occupational therapy students for 25 years, with a primary focus on kinesiology, physical disabilities, and the practice of occupational therapy and Level I fieldwork. She has also taught or assisted with other occupational therapy courses, including Application of Neurodevelopmental Construct, Practice of Occupational Therapy in the Hospital Setting, Interpersonal Dynamics and Task-Oriented Groups in Occupational Therapy, Occupational Therapy in Mental Health, Introduction to Alternative Research Methodologies, and Critical Phases of Life.

Her clinical experience includes inpatient rehabilitation and acute, outpatient, and long-term care. She is currently serving on the Ohio Occupational Therapy Board, is a book reviewer for several publishing companies and for the *American Journal of Occupational Therapy*, and is section editor for the *Health and Interprofessional Practice Journal.*

CONTRIBUTORS

The author extends her heartfelt gratitude to the contributors, whose feedback and expertise were invaluable in the creation of this second edition.

Lori DeMott, OTR/L, CHT (Chapters 3, 4)
Ohio Therapy Institute
Ohio Orthopedic Center of Excellence
Columbus, Ohio

Sharon Flinn, PhD, OTR/L, CHT, CVE (Chapters 3, 4)
The Ohio State University
Columbus, Ohio

Linda M. Martin, PhD, OTR/L, FAOTA (Chapter 2)
Florida Gulf Coast University
Department of Occupational Therapy
Fort Myers, Florida

Sandra Rogers, PhD, OTR/L (Chapter 1)
Pacific University
School of Occupational Therapy
Hillsboro, Ohio

Kim Szucs, PhD, OTR/L (Chapter 5)
The Ohio State University
Columbus, Ohio

Preface

This book is written for occupational therapists and occupational therapy students. The purpose of this book is to explore and explain how movement occurs from a musculoskeletal orientation. This text does not discuss the influence and contribution of the sensory system, nervous systems, volition, or cognition on the production of movement, although these are clearly vital parts of movement.

This text includes descriptions of how joints, muscles, and bones all interact to produce movement. General information about muscles and assessment of strength, as well as joints and assessment of joint motion, are contained in two chapters that will elucidate this idea of movement. There are six chapters devoted to how movement is produced at each joint (shoulder, elbow, wrist, hand, lower extremity, and posture). Being able to visualize the internal mechanisms of joint movement and to accurately assess observable joint characteristics is an important part of understanding movement.

In order to understand how movement is produced, kinesiology concepts are explained with regard to forces acting on the body and how these forces influence not only movement but ultimately our intervention with clients.

Because this book is written for occupational therapists, the first chapter briefly explains concepts particularly related to the profession of occupational therapy. Terminology is defined according to *Occupational Therapy Practice Framework: Domain and Process* as well as *International Classification of Functioning, Disability, and Health* (ICF) terminology.

Once one understands how movement is produced and how to assess strength and joint motion, the next logical step is to learn about appropriate intervention. The last two chapters are devoted to two intervention frames of reference used in occupational therapy. Included in each of these chapters are goals relative to areas of focus that include the theoretic principles that underlie intervention. It is important to be able to clearly articulate to our clients, their families, other professionals, and third-party payers what we are doing and why.

While this textbook focuses on a very small part of the occupational therapy domain (musculoskeletal client factors), it is imperative for the occupational therapist to remain true to occupational therapy values of client-centered, holistic, and systems-oriented practice. Include the client and family in the entire intervention process, which will ensure better treatment outcomes and improved client satisfaction.

Instructor's materials include class activities, discussion questions, and learning tasks.

Melinda F. Rybski, PhD, OTR/L

SECTION I

Foundations and Assessment

1

OCCUPATIONAL THERAPY CONCEPTS

Occupational therapy is directed toward "supporting health and participation in life through engagement in occupation" (American Occupational Therapy Association [AOTA], 2008, p. 626). This statement reflects the philosophy and values of the profession, which include a holistic, client-centered, occupation-based, and systems-oriented approach focused on participation and health (Cole, 2010).

CONCEPTUAL FOUNDATIONS

The *Occupational Therapy Practice Framework: Domain and Process* (OTPF) (AOTA, 2008) is an official document of the American Occupational Therapy Association providing constructs that define and guide the practice of occupational therapy. The OTPF is based on the core values and beliefs of the profession and uses much of the language of the *International Classification of Functioning, Disability, and Health* (ICF) from the World Health Organization (WHO).

Systems-Oriented

The ICF, developed by the WHO as a part of a "family" of classifications for application to various aspects of health, was written to "provide a unified and standard language and framework for the description of health and health-related states" (WHO, 2001, p. 3). This classification system and conceptual model can provide a uniform language to describe health and health-related conditions and a conceptual model to visualize the relationships of functioning and disability with contextual factors.

The OTPF also provides consistent language and concepts that can be used by internal and external audiences to clearly express the role of occupational therapy contribution to health promotion and participation in occupation (AOTA, 2008). The ICF and OTPF documents reflect a shift from a disease perspective to one related to health. Health and wellness involves individuals, organizations, and societies and is seen as an active process of making choices for an optimal state of physical, mental, and social well-being. The OTPF is aligned with global health trends emphasizing health and wellness as well as a growing awareness of the need to provide opportunities and resources for success in activity participation (occupational justice).

Taking a systems-oriented approach, larger contexts of intervention are recognized. With an expanded view of who the client can be, intervention is directed not only toward those clients who may already have activity limitations and participation restrictions, but also toward those who may be at risk for health conditions or toward the population as a whole.

A systems-approach recognizes the variety of contexts in which intervention occurs. The ICF considers the physical, social, and attitudinal environment in which people live and includes detailed descriptions of environmental factors that include products and technology; natural environment and human-made changes to the environment; support and relationships; attitudes; and services, systems, and policies. The OTPF defines contexts as cultural, personal, temporal, virtual, physical, and social.

Benefits of systems-oriented frameworks may "influence universal design, public education and legislation, permit

Rybski MF.
Kinesiology for Occupational Therapy, Second Edition (pp 3-12)
© 2012 SLACK Incorporated

comparison across patients, studies, countries and clinical services, populations, predict health care system usage and costs and provide evidence for social policies and laws" (Jette, Norweg, & Haley, 2008, p. 964). Determining the health of populations and prevalence of health outcomes in terms of health-care needs and effectiveness of health-care systems can serve public health purposes. Further, the use of a standard language can help in policy development in the areas of social security, employment, transportation, and access to technology (Unstun, 2002). As Madden, Choi, and Sykes (2003) state, "the use of a common framework, with its common definitions and classifications, thus helps to produce meaningful information for decision making and policy development and increases the likelihood of improved outcomes for people with disabilities" (p. 676).

Systems-oriented approaches in occupational therapy (also called overarching frames of reference [Dunn, 2000], conceptual models [Reed & Sanderson, 1999], or occupation-based frameworks [Baum, Christensen, & Haugen, 2005]) include relationships between the person, environment, and occupational performance with the focus on occupation (Cole, 2010). The Canadian Model of Occupational Performance focuses on the relationships between Person-Environment-Occupation (PEO) and is an example of a conceptual umbrella on which intervention can be based. Other occupation-based models are Occupational Behavior (Reilly, 1969), Model of Human Occupation (Kielhofner & Burke, 1980), Occupational Adaptation (Schkade & Schultz, 1992a, 1992b), Ecology of Human Performance (Dunn, Brown, & McGuigan, 1994), and Person, Environment, Occupation, Performance (Christiansen, 1994).

Holistic and Client-Centered

The ICF framework is described as a biopsychosocial model that integrates aspects of the more traditional medical model with the social model advocated by the disability community (Crimmins & Seeman, 2004; WHO, 2001). The ICF model integrates the need to cure and prevent disease (medical model) with the goal of increasing participation in daily life (social model) (Iezzoni & Freedman, 2008). The ICF model describes "the situation of each person within an array of health or health related domains...made within the context of environmental and personal factors" (WHO, 2001, p. 8). The ICF framework describes functioning and disability as "a dynamic interaction between health conditions and contextual factors" (WHO, 2001, p. 8).

The OTPF also reflects a holistic understanding of the client. Just like the ICF and the biopsychosocial model, the practice of occupational therapy involves intervention that is remedial (medical model) and social (social model). Kielhofner and Burke (1977) identified paradigm shifts that have occurred in the history of occupational therapy. The first paradigm shift was from a focus on occupation to a mechanistic or reductionistic model. This occurred between 1940 and 1970 and was a result of greater alignment with the medical model. From this shift, three models emerged: kinesiology (including the biomechanical and rehabilitation approaches), psychoanalytic (psychodynamic), and sensory integrative (neuroscience, motor control). Since the 1980s, occupational therapy has moved toward a focus on occupation itself and broader models (such as ICF model and occupation-based models) and away from the primary emphasis on physical, sensory, psychological, emotional, or cognitive components of function.

The profession's early roots in humanism and pragmatism are evident in the OTPF with the "dedication to the betterment of the human condition and the right of each person to respect, dignity and a meaningful and productive role in society" (Cole, 2010, p. 78). Client-centered practice is driven by respect for the client and caregivers and for the choices they make for their lives. Intervention is individualized based on active participation of the client in determining goals with clients assuming ultimate responsibility for decisions about occupations they wish to resume. The therapist collaborates with the client to solve occupational performance issues.

In providing services that are client-centered, occupational therapists use many different types of reasoning, which may include procedural, interactive, pragmatic, conditional, and narrative thinking. The practice of occupational therapy requires the use of scientific and objective knowledge used in procedural reasoning; the understanding of the illness experience based on the subjective reality of each individual client used in interactive reasoning; and the use of conditional reasoning that integrates objective and subjective information with contextual factors. This is the melding of information from the person, environment, and occupation.

In a client-centered practice, the activities, roles, and tasks of the person are considered as are systems and services that can support the person. The client is an active participant in the intervention process and assumes responsibility for his or her care. The therapist collaborates with the client in establishing treatment priorities and provides education, information, and resources in the community to help clients develop skills and behaviors that prevent disabilities and promote healthy lifestyles (Law, 1998).

In a study by Neistadt (1995), 99% of occupational therapists who were surveyed reported that they routinely identify clients' priorities for treatment, although more formal means of assessment would ensure that all clients were helped to delineate their goals. Northern and colleagues (1995) found that therapists did involve patients and families in the goal-setting process, although there were some discrepancies in the verbal preparation of client and family for intervention and potential outcomes; in attempts to elicit client concerns; and in the level of collaboration to establish treatment goals. Occupational therapists involve their clients in the treatment process. A client-centered approach has been shown to result in shorter hospital stays, better goal attainment, and improved client satisfaction (McAndrew, McDermott, Vizakovitch, Warunek, & Holm, 1999).

Occupation-Based

Occupation is used to organize and define occupational therapy's domain of concern (AOTA, 1995a). The unique focus on occupation is a distinguishing feature of our profession (Rogers, 2007). It is the client's designation of the meaning and importance of each occupation or activity that is the focus of intervention. This is consistent with the occupational therapy views of the relationship between occupation and health and that people are occupational beings (AOTA, 2008, p. 625). By using these everyday activities, or occupations, increased functional performance that is meaningful to the client is the outcome of occupational therapy intervention. Purposefulness helps to organize while meaningfulness of activities motivates clients (Trombly, 1995).

Moyers (1999) lists nine principles of occupation that guide occupational therapy interventions:

1. Occupations act as the therapeutic change agent to remediate/restore impaired abilities or capacities in the performance components

2. Occupations facilitate transfer of performance component skills to multiple contexts

3. Occupations are selected to enhance motivation for making change

4. Occupations promote self-exploration and identification of values or interests

5. Chosen therapeutic occupations start with the current capacity of the client

6. Occupations create opportunities to practice skills

7. Occupations are selected to support the most appropriate intervention approach

8. Active engagement in occupations produces feedback that successively grades performance

9. Successful occupational experiences are necessary for achieving goals (pp. 270-272).

These principles of occupation apply to all models, frames of references, and interventions relevant to occupational therapy.

DOMAIN OF OCCUPATIONAL THERAPY

The OTPF identifies six specific domains of occupational therapy practice. The first, areas of occupation, includes many of the same items that are classified as activities and participation in the ICF model. Areas of occupation include activities of daily living (ADLs), instrumental activities of daily living (IADLs), rest and sleep, education, work, play, leisure, and social participation. Improvement or enhancement of occupational performance is often a desired outcome of intervention. Occupations will be defined by the client as part of an occupational profile in which the therapist gains an understanding of the client's history, interests, values, and priorities that forms the basis of intervention.

The remaining five aspects of the occupational therapy domain are factors that may influence the client's ability to successfully engage in occupations and participate in health-promoting activities. These include *client factors* (body functions, body structures, values, beliefs, and spirituality); *performance skills* (sensory and perceptual skills, motor and praxis skills, emotional regulation skills, cognitive skills, and communication and social skills); *performance patterns* (habits, routines, roles, and rituals); *context and environment* (cultural,

physical, personal, social, temporal, and virtual); and *activity demands* (objects, space, social demands, sequencing, and timing; required actions, required body functions, and required body structures. All of the aspects of the occupational therapy domain interact to influence the client's performance.

Contextual factors are those related to the physical environment, cultural and social systems, simulation of environmental conditions, and spiritual aspects of being (AOTA, 2002). This definition of contextual factors is very similar to that used in the ICF. Environmental factors are those in the natural environment and in the human-made environment and include social attitudes, customs, rules, practices, institutions, and other individuals. Personal factors are those components that are not part of the health condition, including age, race, gender, educational background, experiences, personality and character style, aptitudes, other health conditions, fitness, lifestyle, habits, upbringing, coping styles, social background, profession, and past and current experience (WHO, 2001).

Occupational therapists also conceptualize occupational performance as being a function of activity demands and client factors. Activity demands are those variables directly related to purposeful activities in which the client engages. Whether the activity demands are too great for the client depends on the client's abilities and performance skills. Abilities are related to learning and involve cognition and social-emotional and physical factors. Performance skills are observable and relate to successful participation in activities.

OCCUPATIONAL THERAPY PROCESS

The occupational therapy domain and occupational therapy process "are inextricably linked" (AOTA, 2008, p. 627). The occupational therapy process includes evaluation, intervention, and outcomes. These are shown in Table 1-1.

Evaluation

The OTPF describes assessment as a process that involves two steps. Step one is assessment of the occupational profile of the client in order to gather information about the client's interests, values, needs, and goals. By starting the assessment process with the occupational profile, this client-centered focus can be incorporated throughout the treatment process. This assessment stage is where the therapist and client begin the collaborative process of therapy.

Therapeutic rapport is established with the client as the therapist uses his or her own unique characteristics of personality, style, perceptions, and judgments as part of the therapeutic process (AOTA, 2008). The intention is to understand the perspective of the client, with these particular limitations, within the specific context and environment. ~~Clients are considered the experts regarding their own situation and methods for problem solving.~~ Evaluation does not always have to start with the occupational profile nor is assessment in this area ever completed. ~~Ongoing collaboration between the therapist and client continuously determines if the client's needs and goals are being addressed.~~

Table 1-1

OVERVIEW OF OCCUPATIONAL THERAPY PROCESS

OCCUPATIONAL PROFILE

Who is the client?

Why is the client seeking services?

What areas of occupation are affected?
- ADL
- IADL
- Work
- Play
- Leisure
- Social participation

Contexts
- Life experiences
- Values
- Interests
- Previous patterns of engagement
- Meanings of patterns

Client's occupational history

Client's priorities and targeted outcomes

INTERVENTION PLAN

Plan that includes:
- Objective and measurable goals with timeframe
- Theory and evidence
- Create or promote (health promotion)
- Establish or restore (remediation/ (biomechanical/NDT/cognitive-perceptual)
- Maintain
- Modify (rehabilitation: adaptation/ compensation
- Prevent (disability prevention)

Mechanisms for delivery:
- Who will deliver.... role delineation
- Types of intervention
- Frequency, duration
- Outcome measures
- D/c needs and plans
- Recommendations or referrals

OUTCOME MEASURES

Select outcome measures
- Occupational performance
- Client satisfaction
- Adaptation
- Health and wellness
- Prevention
- Quality of life

ANALYSIS OF OCCUPATIONAL PERFORMANCE

- Performance skills (motor, process, communication/interaction)
- Performance patterns (habits, routines, roles)
- Factors (context, activity demands, client)

TYPE OF INTERVENTION

Therapeutic use of self

Use of occupations or activities
- Occupation-based
- Purposeful activity
- Preparatory

Consultation

Education

MEASURE AND USE OUTCOMES

- Compare goal achievement to targeted outcomes
- Assess outcome results

#2 In step two, an analysis of the client's occupational performance is done. Occupational performance is the interaction of the client, context, and activity that enables successful engagement in the areas of occupation. Areas of occupation, performance skills, patterns, contexts, activity demands, and specific client factors are considered as to how these might enhance or hinder engagement in desired occupations. Step two is more focused on the component factors leading to the occupational performance participation restrictions identified in step one. The client's actual performance may be observed in the context in which it normally occurs so that performance skills and patterns can be clearly seen.

Intervention

Assessment and intervention in occupational therapy are based on the theoretic understanding of the problems that are presented by the client. How the therapist views the restrictions in participation influences how he or she would assess and provide treatment.

Occupational therapy interventions include occupation-based intervention, purposeful activity, and preparatory methods (Table 1-2). Additional interventions include consultation, education, and advocacy. These last three interventions are less germane to this textbook so will not be discussed in detail.

Table 1-2

COMPARISON OF OCCUPATIONAL THERAPY REMEDIATION INTERVENTION CONTINUA

OT PRACTICE FRAMEWORK	*PEDRETTI*	*FISHER*
Preparatory Methods	**Adjunctive Methods**	**Exercise**
Prepares client for occupational performance. Includes sensory input, physical agent modalities, orthotics/splinting, exercise.	"Preliminary to the use of purposeful activities and that prepare the client for occupational performance." Include exercise, orthotics, sensory stimulation, physical agent modalities, ROM, inhibition/facilitation techniques. Necessary to provide structural stability, prevent deformities, provide rest for a part, to increase function in client factors (body structure/function).	Rote exercise and practice activities. A purpose but no goal. Little meaning.
Purposeful Activity	**Enabling Methods**	**Contrived Occupation**
Goal directed behaviors or activities within therapeutically designed context that leads to occupation. Examples: practice slicing vegetables, role play to manage anger.	Intermediate activities. Practice specific motor patterns. Train in perceptual and cognitive skills. Practice sensorimotor skills. Examples: inclined sanding boards, cone stacking, puzzles, fastening boards, work simulators, pegboards, computer programs. Lack client-centered goals.	Exercise with added purpose or occupation with contrived component. Little meaning. Examples: exercise embedded in an activity; pounding nails into a board to pretend to build a birdhouse.
	Purposeful Activities	**Therapeutic Occupation**
	Identifiable goal and frequently areas of occupation are addressed. Skill generalization greater than for adjunctive or enabling methods.	Client actively participates in areas of occupation. Client sees task as purposeful and meaningful. Real objects used in context. Still focused on remediation.
Occupation-Based Activity	**Occupation-Based Intervention**	**Adaptive or Compensatory Occupation**
Actual occupation part of their own context and meeting their goals. Examples: grocery shopping, dressing without assistance.	Includes information about client needs, wants, goals, expectations. Most beneficial to client and most challenging. Activities in appropriate context and match client goals. Goal directed, meaningful.	Active participant in areas of occupation that they choose. Focus on improved performance in areas of occupation May include assistive devices, teaching alternative or compensatory strategies, modification of environment.

Adapted from Fisher, A. G. (1998). Uniting practice and theory in an occupational framework: 1998 Eleanor Clarke Slagle lecture. *Am J Occup Ther, 52*(7), 509-521; Pedretti, L. W. (1996). *Occupational therapy: Practice skills for physical dysfunction* (4th ed.). St Louis, MO: C.V. Mosby; and American Occupational Therapy Association (2002). Occupational therapy practice framework: Domain and process. *Am J Occup Ther, 56*(6), 609-639.

Occupation-Based Intervention

Occupation-based activities are client-centered activities, collaboratively chosen by the client and therapist, that are meaningful and relevant to the client in the expected environment. The actual task is done in the same context as is typical that meets the client's goal. This is the most beneficial and most challenging level of intervention. Fisher (1998) called this adaptive or compensatory occupation, which comprised active participation in chosen occupations but also included using assistive devices, teaching alternative methods, or modifying the environment as goals of intervention. The intervention is focused on improved occupational performance and is not directed toward remediation of impairments.

Purposeful Activity

Purposeful activity is defined as "goal directed behaviors or activities within therapeutically designed contexts that lead to occupation" (AOTA, 2008, p. 674). Examples of purposeful activity would be to practice slicing vegetables for a salad or to practice transfers in and out of a bathtub.

Pedretti (1996) and Fisher (1998) developed models that further delineated this level of intervention. In Pedretti's model, purposeful activities have an identifiable goal, and frequently areas of occupation are addressed. The focus of purposeful activities is on skill generalization. Enabling methods are intermediate activities for the purpose of practicing specific motor, perceptual, or cognitive skills. Examples of enabling methods would be computer programs to increase attention or having a client replicate a pegboard design. Enabling methods do not necessarily include client-centered goals.

Fisher (1998) divides the OTPF category of purposeful activity into therapeutic occupation and contrived occupation. The client participates in areas of occupations, in tasks that are seen as purposeful and meaningful, when engaged in the stage of therapeutic occupation. Real objects are used in natural environments but the focus is on remediation of particular skills. Contrived occupation is exercise with added purpose or occupation with a contrived component. There may be a purpose or a goal, but it originated with the practitioner so it is less meaningful to the client. The focus is on remediation of impairments, and the objects and the potential purpose or meaning is contrived. Exercise may be embedded into an activity such as reaching into the cupboard to simulate getting dishes with the intention of increasing range of motion. Or another variation might be occupation with a contrived component where the objects are real but the task has no purpose. For example, pounding nails into a board to pretend to build a birdhouse. The purpose and meaning are contrived.

Occupational therapy intervention that is primarily in the therapeutic occupation or compensatory adaptive models is more client-centered and provides meaningful and functional outcomes for the client.

Preparatory Methods

Not all clients who are seen by occupational therapists are ready to participate in occupations or purposeful activities. Preparatory methods are used to prepare the client for occupational performance and may include physical agent modalities, orthotics/splinting, or exercise as examples. Pedretti (1996)

calls these adjunctive methods, which also include range of motion, inhibition or facilitation techniques, and sensory stimulation. Fisher (1998) calls rote exercise and practice activities exercise. These are activities done for a purpose but with little meaning.

The activities originate with the therapist and not the client, with the focus on remediation of impairments at the client factor level (see Table 1-2).

Occupation-based activities engage the client in the intervention process and have meaning and relevance to them. Infusing occupation into purposeful activities and preparatory methods is part of the creative challenge therapists may face. Barriers to occupation-based intervention may include limited institutional support with the expectation that preparatory methods alone will enable successful integration into roles once the client is discharged. Reimbursement is not always straightforward, so there is a need to justify treatment (Rogers, 2007).

Having clients actively participating in choosing priorities and setting goals will engage the client meaningfully in the intervention process and make the intervention more occupation-based (Deshaies, Bauer, & Berro, 2001). Give clients choices of activities. Identify clients with similar interests, and arrange occupation-based groups so that socialization and peer mentoring can facilitate the intervention, and the intervention can be more fun and meaningful. Use the facility to its fullest potential: have the client use the vending machines; go to activity rooms; walk on the hospital grounds and community areas. Go on outings to homes, job sites, and schools or to other places that have meaning for the client (Deshaies et al., 2001; Rogers, 2007).

Intervention Approaches

Systems-based or occupation-based models such as the PEO provide an overarching theory about the relationship between occupation, the person, and the environment. How is this theory used in practice? Theories applied to individual clients in specific practice situations are considered frames of reference, practice models (Kielhofner, 2009; Reed & Sanderson, 1999), or intervention approaches. Using frames of reference, practitioners link the "concrete particular with the abstract general" (Mattingly & Fleming, 1994).

Intervention is guided by both the occupation-based models and by frames of reference.

Trombly (1993) calls this "layers of occupational functioning," where all parts of domains and roles need to be considered in treatment, including tasks, activities, abilities, and capacities. Knowing that the client wishes to resume a homemaking role would also entail assessment of the ability to prepare meals, perform specific tasks related to meal preparation, and have the necessary physical and cognitive capacities to perform specific activities.

Frames of reference are not necessarily occupation based. These models of practice were developed as guidelines to address specific disability areas. Several different practice models may be used simultaneously to address different limitations. Initially, restorative/restoration approaches may be used with the client to improve limitations at the body structure

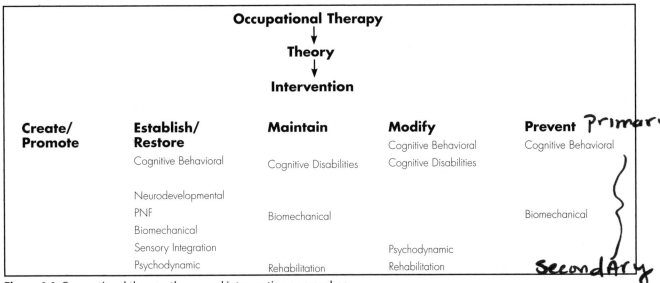

Figure 1-1. Occupational therapy theory and intervention approaches.

and body function levels. If further progress is not made or if the client wants to see more immediate results, compensation/adaptation approaches may be used in conjunction with other approaches or alone.

A frame of reference matches the client concerns with the abstract theory, so it is specific to the individual and specific areas of practice. Crepeau and Schell (2003) stated that "major theories in occupational therapy differ in purpose, scope, complexity, extent of development and validation through research, and usefulness in practice" (p. 204). Some occupational therapy theories are useful in one area of practice and not another.

A frame of reference is "a set of interrelated, internally consistent concepts, definitions and postulates derived from or compatible with empirical data [and] provides a systematic description of or prescription for particular designs of the environment for the purpose of facilitating evaluation and effecting changes relative to a specified part of the profession's domain of concern" (Mosey, 1986, p. 12). The concepts and relationships are drawn from evidence to explain possible causes of factors contributing to function or dysfunction (Tufano, 2010).

Each frame of reference has specific assumptions about the causes of activity and participation restrictions and provides guidelines for intervention. The frame of reference provides a framework on which to explain problems in occupational performance, to select assessment methods relevant to the problems, and to develop an intervention plan based on evidence. Regardless of the frame of reference used, referring to an overarching occupation-based model will ensure that all parts of the person-environment-occupation relationship are included in the client-centered goals.

Five different intervention approaches are identified in the OTPF. They include create/promote (health promotion), establish/restore (remediation, restoration), maintain, modify (compensation, adaptation), and prevent (disability prevention) (AOTA, 2008; Moyers, 1999). Figure 1-1 illustrates the relationship between theory, intervention approaches, and some common frames of reference used in occupational

therapy. The dark ink indicates the primary function of the frame of reference while the grey ink is a secondary function. For example, the biomechanical frame of reference is primarily a remediation approach but some of the strategies can also be considered to fall within the maintenance or prevention approaches.

Remediation: Biomechanical

The biomechanical approach is a remediation (establish/restore) approach. The intention of this approach is to change the underlying structures that are limited by disease, trauma, or overuse with the understanding that these gains will lead to improvement in occupational performance. The biomechanical approach focuses on "the intersection of motion and occupational performance" (Kielhofner, 2009, p. 70). The focus is on remediation of limitations in tissue integrity, structural support, range of motion, strength, coordination, and endurance.

Therapeutic occupation as defined by Fisher (1998) is considered a remediation approach. Remediation approaches are selected when there is an expectation for reduction in the limitations in client factors that influence performance in areas of occupation. It may involve learning new skills, slowing the decline in abilities, or maintaining or improving quality of life (Moyers, 1999). Examples of remediation techniques might be the use of enabling activities, sensorimotor techniques, graded exercises, physical agent modalities, or manual techniques (Moyers, 1999). Arts, crafts, games, sports, exercise, and daily activities may also be used to improve the function of a specific body structure or function, and each of these therapeutic methods would be tailored to each individual to fit the capacities and goals of that person.

If the remediation is to be considered successful and to be considered occupational therapy, the intervention will need to occur in a natural context. Moyers (1999) adds "simply expecting improvements in impairments to automatically produce change in level of disablement without addressing performance in occupations within the intervention plan is inappropriate" (p. 276). Intervention that concentrates on client factors must also be contextually meaningful and occur in a natural environment to avoid being rote exercise or contrived occupation. The biomechanical approach used in occupational therapy intervention is the remediation approach discussed in this text, because this approach is most beneficial in intervention for clients with decreased strength, endurance, and range of motion.

Compensate, Adapt: Rehabilitation

Compensation and adaptation (modify) approaches are often referred to as the rehabilitation approach (Pedretti, 1996), and this approach concentrates not on the client factors but instead on performance in areas of occupation. An assumption of this approach is that the performance of the activity or task that the person sees as important is the focus of the intervention. Techniques used in this approach include changing the task, altering the task method, adapting the task object, changing the context, educating the family and caregiver, and adapting the environment. Disability prevention may be considered part of this approach in that education of the client and caregivers is vital in preventing further disease or disability, especially in those who already have impairments or limitations.

Commonly, the biomechanical and rehabilitation approaches are seen as a continuum of intervention for physical limitations, and the process is fluid and not linear. The focus in the rehabilitation approach is on the client's remaining strengths. Tasks and task objects are adapted to these remaining strengths, and new methods of completing tasks are taught to clients to promote independence in performance areas, increase client satisfaction, and enhance participation.

The compensation and adaptation intervention is used when there is little expectation for change in client factors and subsequent performance skills. This approach may be selected when there is limited time for intervention or when the client or family prefers this approach to remedial techniques, seeing compensation and adaptation as providing more immediate success in performance of the areas of occupation (Moyers, 1999).

The rehabilitation approach is discussed in this text as an intervention strategy for use with those clients with movement impairments preventing full participation in occupations for which remediation is not possible. Both biomechanical and rehabilitation interventions are used extensively in occupational therapy and require knowledge of the structure and function of the body, knowledge about specific diagnostic categories and procedural reasoning, and knowledge about specific individual clients and their activities, roles, and values.

Create, Promote (Health Promotion)

This approach clearly reflects the emphasis on health in the practice of occupational therapy. Health promotion services are directed toward the general population and do not assume that there is any disability or dysfunction present (although health promotion can and does occur with clients with limitations, too). Teaching the well elderly to eat well and to arrange their environment for greater mobility and safety is an example of a health promotion activity in people without limitations. Inclusion of client hobbies and interests in intervention plans is often a way health is promoted for those with limitations.

Maintain

Maintenance approaches are used to support the client's current level of functioning. Without intervention at this level, it is assumed that performance would decrease, occupational needs would not be met, or both, thereby affecting health and quality of life (AOTA, 2008). This may be an appropriate approach to use for people with degenerative diseases where preserving the current level of functioning is a successful outcome.

Prevent (Disability Prevention)

The disability prevention approach is designed to address clients who are at risk for developing occupational performance problems. This approach may be used with clients with and without a disability (AOTA, 2008). Backpack awareness programs for school-aged children are directed toward children without disabilities but with continual poor posture due to inappropriately worn backpacks, who could develop dysfunction and pain. Secondary consequences of an injury or disease would be another area of prevention intervention. A person with a spinal cord injury may have areas of insensate skin as a secondary result of the spinal cord injury. Disability prevention intervention would address the necessity of pressure relief, awareness of areas lacking sensation, and methods of skin inspection to prevent the development of decubiti ulcers.

Outcomes

Occupational therapy intervention is designed to "improve the occupational performance of persons who lack the ability to perform an action or activity considered necessary for their everyday lives" (AOTA, 1995b, p. 1019), as well as to achieve the outcomes of prevention of injury or disability, promotion of health, and quality of life (Moyers, 1999). Outcomes in occupational therapy include occupational performance; adaptation; health and wellness; participation; prevention; quality of life; role competence; self-advocacy; and occupational justice. Outcomes are the end-result of occupational therapy intervention and involve the conceptual inter-relationships between health, participation, and engagement in occupation. Outcomes are used to measure progress and adjust goals and interventions and are interwoven throughout the occupational therapy process.

OVERVIEW OF THIS TEXT

This text covers in detail only a small part of the domain of occupational therapy. By focusing only on the study of movement (kinesiology), client factors, performance skills, and context and environmental aspects of the occupational therapy domain are the focus as these directly relate to movement. Detailed explanations of the evaluation of these areas are included in each chapter. Within each chapter, evaluation of the person via the occupational profile will remind the reader to stay true to the conceptual foundations of occupational therapy, which value client-centered, holistic, and occupation-based intervention collaboratively completed with the client. While seemingly reductionistic with the focus on muscles and joints, the relationship of the person, environment, and occupation are always part of the treatment process.

Only a few intervention approaches are addressed in this text. A thorough explanation of the concepts and methods used in the biomechanical remediation approach and rehabilitation adaptation/compensation approaches are presented in the last two chapters. Seen as a continuum, if remediation efforts for increasing structural limitations are unsuccessful or have plateaued, the next logical step is to adapt or compensate for the limitations. Disability prevention and maintenance approaches are also discussed in the intervention chapters.

This text is comprised of three distinct sections. Section I is designed to provide background information, knowledge, and facts that will be applied to later sections. The chapters in this foundational section pertain to basic concepts about movement, factors that influence range of motion, and factors that influence strength. Terminology is explained that will be used throughout the text, and the relationships between body structures and functional movement are introduced. Assessment of joint movement and strength is included in this section. This section is provided to enable greater understanding of body structure and function as a basis of movement necessary for performance in areas of occupation.

Section II discusses how the general principles introduced in Section I apply to specific joints. How is joint motion produced at the wrist? What bones and muscles are involved? Is this a stable joint? These chapters that focus on specific joints relate structural anatomy to functional motion. Readers are encouraged to palpate the anatomical structures (muscles and bones) and to see and feel how these structures work together to produce joint movement. Most joints are actually made up of multiple articulations; each articulation contributes to the overall motion produced at that joint, so these articulations are discussed separately and as contributors to joint motion. Not every structure of the joint is included or discussed in detail; only those structures that most influence movement or stability are included. A summary chart describing joint movements, the plane and axis in which the movement occurs, normal limiting factors in movement production, the specific "end feel" per motion, and the muscles producing the motion is provided for each joint to recapitulate information presented in the chapter.

Detailed descriptions of muscles are provided so that students can clearly see how muscle orientation affects muscle action. It is not always clear how one muscle can have multiple and seemingly incongruent muscle actions without seeing how a muscle acts at different joints. A clear understanding of normal joint movement is essential to occupational therapy practice and is the basis of the biomechanical intervention approach. Knowing how a normal joint should move is the essence of assessment. Specific factors that are relevant to the assessment of individual joints are discussed, and specific joint pathology is provided as an overview of possible limitations in function related to specific structures and disease processes or trauma.

Section III of this text provides two different intervention strategies that could be used if there are activity limitations or participation restrictions due primarily to musculoskeletal or motor impairments. Remediation via the biomechanical approach and adaptation/compensation via the rehabilitation approach are the primary intervention strategies discussed. For each approach, intervention principles are provided and explained, as are examples of goal statements and specific methods that could be used to implement the intervention selected.

SUMMARY

1. Occupational therapy is a systems-oriented, holistic, client-centered practice that uses occupation to organize and define occupational therapy's domain of concern.

2. Occupational therapy uses occupation, or the everyday things that people do, as the basis of intervention.

3. The occupational therapy process involves evaluation intervention and outcomes.

4. The occupational therapy process is inextricably linked to the occupational therapy domain.

5. Intervention may be occupation-based and use purposeful activities or preparatory methods.

6. Intervention approaches include remediation, compensate/adaptation, health promotion, maintenance, or disability prevention.

7. Outcomes include occupational performance; adaptation; health and wellness; participation and prevention; quality of life; role competence; self-advocacy; and occupational justice.

8. This text focuses on movement; client factors, performance skills, and context and environmental aspects of the occupational therapy domain; biomechanical remediation, rehabilitation adaptation/compensation, and disability prevention and maintenance approaches.

REFERENCES

American Occupational Therapy Association (1995a). Position paper: Occupation. *Am J Occup Ther, 49*(10), 1015-1018.

American Occupational Therapy Association (1995b). Position paper: Occupational performance: Occupational therapy's definition of function. *Am J Occup Ther, 49*(10), 1019-1020.

American Occupational Therapy Association (2002). Occupational therapy practice framework. *Am J Occup Ther, 56*(6), 609-639.

American Occupational Therapy Association (2008). Occupational therapy practice framework: Domain and Process, second edition. *Am J Occup Ther, 62*, 625-683.

Baum, C., Christensen, C., & Haugen, J. B. (2005). *Occupational therapy: Performance, participation, and well-being.* Thorofare, NJ: SLACK Incorporated.

Christiansen, C. (1994). *Ways of living: Self-care strategies for special needs.* Rockville, MD: American Occupational Therapy Association.

Cole, M. B. (2010). Occupational therapy theory development and organization. In K. Sladyk, K. Jacobs, & N. MacRae (Eds.), *Occupational therapy essentials for clinical competence* (pp. 75-86). Thorofare, NJ: SLACK Incorporated.

Crepeau, E. B., & Schell, B. A. B. (2003). Theory and practice in occupational therapy. In E. B. Crepeau, E. S. Cohn, & B. A. B. Schell (Eds.), *Willard & Spackman's occupational therapy* (10th ed.) (pp. 203-208). Philadelphia, PA: Lippincott, Williams & Wilkins.

Crimmins, E. M., & Seeman, T. E. (2004). Integrating biology into the study of health disparities. *Population and Development Review, 30*, 89-107.

Deshaies, L. D., Bauer, E. R., & Berro, M. (2001). Occupation-based treatment in physical disabilities rehabilitation. *Occupational Therapy Practice,* July 2, 13-18.

Dunn, W. (2000). *Best practice occupational therapy: in community service with children and families.* Thorofare, NJ: SLACK Incorporated.

Dunn, W., Brown, C., & McGuigan, A. (1994). The ecology of human performance: A framework for considering the effect of context. *Am J Occup Ther, 48*(7), 595-607.

Fisher, A. G. (1998). Uniting practice and theory in an occupational framework. *Am J Occup Ther, 52*(7), 509-521.

Iezzoni, L., & Freedman, V. A. (2008). Turning the disability tide: The importance of definitions. *Journal of the American Medical Association, 299*(3), 332-334.

Jette, A. M., Norweg, A., & Haley, S. M. (2008). Achieving meaningful measurements of ICF concepts. *Disability and Rehabilitation, 30*(12-13), 963-969.

Kielhofner, G. (2009). *Conceptual foundations of occupational therapy practice* (4th ed.). Philadelphia, PA: F. A. Davis Company.

Kielhofner, G., & Burke, J. P. (1980). A model of human occupation: part 1: conceptual framework and content. *Am J Occup Ther, 34*(9), 572-581.

Kielhofner, G., & Burke, J. P. (1977). Occupational therapy after 60 years: An account changing identity and knowledge. *Am J Occup Ther, 31*(10), 675-689.

Law, M. (1998). *Client-centered occupational therapy.* Thorofare, NJ: SLACK Incorporated.

Madden, R., Choi, C., & Sykes, C. (2003). The ICF as a framework for national data: The introduction of ICF into Australian data dictionaries. *Disability and Rehabilitation, 25*(11-12), 676-682.

Mattingly, C., & Fleming, M. H. (1994). *Clinical reasoning: Forms of inquiry in a therapeutic practice.* Philadelphia, PA: F. A. Davis.

McAndrew, E., McDermott, S., Vizakovitch, S., Warunek, M., & Holm, M. B. (1999). Therapist and patient perceptions of the occupational therapy goal-setting process: A pilot study. *Physical and Occupational Therapy in Geriatrics, 17*(1), 55-63.

Mosey, A. C. (1986). *Psychosocial components of occupational therapy.* New York, N. Y.: Raven Press.

Moyers, P. A. (1999). Guide to occupational therapy practice. *Am J Occup Ther, 53,* 247-322.

Neistadt, M. (1995). Methods of assessing clients' priorities: A survey of adult physical dysfunction settings. *Am J Occup Ther, 49*(5), 428-436.

Northern, J. G., Rust, D. M., Nelson, C. E., & Watts, J. H. (1995). Involvement of adult rehabilitation patients in setting occupational therapy goals. *Am J Occup Ther, 49*(3), 214-220.

Pedretti, L. W. (1996). *Occupational therapy: Practice skills for physical dysfunction.* St. Louis, MO: Mosby.

Reed, K. L., & Sanderson, S. N. (1999). *Concepts of occupational therapy.* Philadelphia, PA: Lippincott, Williams & Wilkins.

Reilly, M. (1969). The educational process. *Am J Occup Ther, 23,* 299-307.

Rogers, S. (2007). Occupation-based intervention in medical-based settings. *Occupational Therapy Practice.* Retrieved from http://www.aota.org/Pubs/ OTP/1997-2007/Features/2007/f0827072.aspx; April 23, 2009.

Schkade, J. K., & Schultz, S. (1992a). Occupational adaptation: Toward a holistic approach for contemporary practice: Part I. *Am J Occup Ther, 46*(9), 829-837.

Schkade, J. K., & Schultz, S. (1992b). Occupational adaptation: toward a holistic approach for contemporary practice: Part II. *Am J Occup Ther, 46*(10), 917-926.

Trombly, C. A. (1995). Occupation: Purposefulness and meaningfulness as therapeutic mechanisms. *Am J Occup Ther, 49*(10), 960-970.

Trombly, C. (1993). Anticipating the future: Assessment of occupational function. *Am J Occup Ther, 47*(3), 253-257.

Tufano, R. (2010). Occupational therapy theory use in the process of evaluation and intervention. In K. Sladyk, K. Jacobs, & N. MacRae (Eds.), *Occupational therapy essentials for clinical competence* (pp. 87-97). Thorofare, NJ: SLACK Incorporated.

Unstun, T. B. (2002). Towards a common language for functioning, disability and heath: ICF. Retrieved from http://www.who.int/classifications/icf/site/icftemplate.cfm?myurl=beginners.html&mytitle=Beginners%27s%Guide; May 5, 2008.

World Health Organization (2001). *International classification of functioning, disability and health* (ICF). Geneva, Switzerland: World Health Organization.

ADDITIONAL RESOURCES

American Occupational Therapy Association (1993). Core values and attitudes of occupational therapy practice. *Am J Occup Ther, 47*(12), 1085-1086.

American Occupational Therapy Association (1994a). Uniform terminology for occupational therapy (3rd ed.). *Am J Occup Ther, 48*(11), 1047-1054.

American Occupational Therapy Association (1994b). The philosophical base of occupational therapy. *Am J Occup Ther, 49*(10), 1026.

American Occupational Therapy Association (1997). Statement: Fundamental concepts of occupational therapy: Occupation, purposeful activity and function. *Am J Occup Ther, 51*(10), 864-866.

Arilotta, A. E. (2003). Performance in the areas of occupation: The impact of the environment. *Physical Disabilities Special Interest Section Quarterly, 26*(December), 1-3.

Bonzani, P. J. (2003). Cumulative trauma disorders: An occupation-based perspective. *Physical Disabilities Special Interest Section Quarterly, 27*(6), 1-3.

Chisholm, D., Dolhi, C., & Schreiber, J. (2000). Creating occupation-based opportunities in a medical model clinical practice setting. *O.T. Practice,* 1-7.

Coppola, S. (2003). An introduction to practice with older adults using the occupational therapy practice framework: Domain and process. *Gerontology Special Interest Section Quarterly, 26*(March), 1-4.

Dutton, R. (1995). *Clinical reasoning in physical disabilities.* Baltimore, MD: Williams and Wilkins.

Kielhofner, G. (1983). *Health through occupations: Theory and practice in occupational therapy.* Philadelphia, PA: F. A. Davis Co.

Kielhofner, G. (1992). *Conceptual foundations of occupational therapy.* Philadelphia, PA: F. A. Davis Co.

Killian, A. (2006). Making occupation-based practice a reality: Part 1. *Administration & Management Special Interest Section Quarterly, 22*(2), 1-4.

Killian, A. A. (2006). Making occupation-based practice a reality: Part 2. *Administration & Management Special Interest Section Quarterly, 22*(3), 1-4.

Marrelli, T. M., & Krulish, L. H. (1999). *Home care therapy: Quality, documentation and reimbursement.* Boca Grande, FL: Marrelli and Associates, Inc.

McGovern-Denk, M., Levine, M., & Casey, P. (2005). Approaching occupation with the person with amyotrophic lateral sclerosis. *Physical Disabilities Special Interest Section Quarterly, 28*(4), 1-4.

Neistadt, M. E., & Seymour, S. G. (1995). Treatment activity preferences of occupational therapists in adult physical dysfunction settings. *Am J Occup Ther, 49*(5), 437-443.

Paris, C., & Boyle, D. C. (2005). Leadership in practice: Implementing the practice framework in a rehabilitation setting. *Administration & Management Special Interest Section Quarterly, 21*(1), 1-4.

Sabata, D. (2004). Home and community health context:: What does the environment mean to occupation? *Home and Community Health Special Interest Section Quarterly,* June, 1-4.

Trombly, C. A. (Ed.). (1995). *Occupational therapy for physical dysfunction* (4th ed.). Baltimore, MD: Williams and Wilkins.

2

KINESIOLOGY CONCEPTS

Melinda F. Rybski, PhD, OTR/L and Linda M. Martin, PhD, OTR/L, FAOTA

It is important to take a moment to appreciate the complexity of movement and the integration of functions that are necessary at many levels to enable successful participation in everyday activities. A person wants to move. This desire to move sets off a series of physiologic responses to produce efficient volitional movement:

- Spinal reflexes that underlie reciprocal movements need to be able to respond quickly to stimuli.

- Sensory afferent tracts need to be intact to receive sensory input through intact skin and deep receptors in order to guide movement.

- Perceptual processes must be intact in order to interpret the sensory information.

- Motor efferent tracts need to be intact to send messages to muscle fibers.

- The brain stem must mediate tonic patterned responses of trunk and limb in relation to the head.

- Basal ganglia facilitate adjustments of head in relation to body and aid in balance and automatic motor plans.

- The cerebellum is concerned with speed, smoothness, force, and accuracy of movement.

- The association cortex plans movement.

- The motor cortex executes movement and controls recruitment of motor units.

- Movement is completed due to normal elasticity in tissues, contractile capacity of muscle fibers, freely moveable joints in which the motion is occurring, and stabilization of other joints.

- The person needs motivation to move volitionally, which is reflective of mood, interest, and emotion.

- Cognition is required to maintain interest, use of prior motor learning and experiences, and use of judgment, concentration, and memory (Trombly, 1995).

A more simplified definition of movement is the observable behavioral response that will result in the displacement of one or more limb segments of the body. This definition includes aspects of movement that can be described by watching a person move, as well as those factors that influence movement that are not visible.

THE SCIENCE OF MOVEMENT

The science of movement, kinesiology, is the study of the forces and of the active and passive structures that are involved in human movement. By understanding these forces and their impact on the body, occupational therapy intervention can be directed toward the optimal occupational performance in our clients and also in preventing injury by recommending appropriate positioning of the body and location of objects in relation to the body.

According to Gench, Hinson, and Harvey (1995), a further refinement in the science of movement includes two subsections of kinesiology—anatomy and mechanics. Figure 2-1 illustrates the relationships between the sciences associated with movement. Anatomy is the understanding of the production of movement by muscles of the body. Mechanics is the study of forces and motion; when applied to the living human body,

Rybski MF.
Kinesiology for Occupational Therapy, Second Edition (pp 13-36)
© 2012 SLACK Incorporated

Figure 2-1. The science of movement.

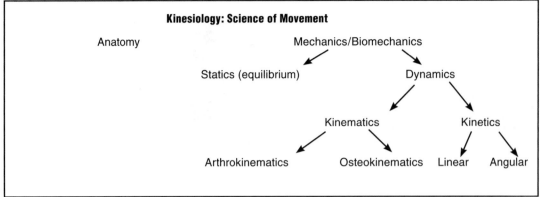

it is most often referred to as biomechanics. Biomechanics is the application of mechanics to the analysis of biological and physiological systems. Biomechanics deals with forces related to the body and their effect on body movement, size, shape, and structure. Biomechanics can be further subdivided into statics and dynamics.

Statics

Statics examines systems not moving or moving at constant speed (equilibrium). An understanding of equilibrium and statics is helpful in identifying stresses on anatomical structures (Smith, Weiss, & Lehmkuhl, 1996). Equilibrium has to do with two or more forces that enable a body or object to remain at rest or to move at a constant speed.

Forces act in pairs, and there will be no movement or equilibrium without the pairing of forces. If a force is greater than the resistance offered by an object, then movement will occur. If a force is not applied to an object that is moving, then the object will continue to move at a constant speed unless acted upon by another force. The definition of equilibrium clarifies this summation of forces: the sum of all forces that are applied to the object must be zero for the object to be in equilibrium.

Multiple forces act upon the body at all times. Forces are internal, external, observable, and indiscernible. It is important to realize, for example, that gravity exerts a constant force on all objects. This force of gravity has an effect on both equilibrium and on movement. Gravity is a constant force that we accommodate without conscious thought. Gravity can and frequently does provide forces for movement, and this force is directed vertically downward at an acceleration rate of 32 feet/second/second (9.8 meters/second²). Gravity affects the stability of the body as well as movements of the extremities, neck, and head. We adjust to the weight of our arm as we use it in different positions just as we adjust easily, and without a loss of stability, when we use tools or pick up heavy objects. Other external forces include wind or air resistance, water, friction, other people, and objects.

Internal forces act on the body but arise from sources within the body. These forces serve to counteract forces (both internal and external) as well as produce movement. Forces produced internally are the result of the contraction of muscles, the passive resistance of ligaments and bones, or the elastic properties of various tissues including muscles, tendons,

ligaments, fascia, and skin. Internal forces provide movement and stability. Stability of the joints depends upon bony architecture (as in the elbow), ligamental support (as in the wrist), and tendon and muscle tension (as in the shoulder).

Dynamics and Kinematics

Dynamics is concerned with movement. Within dynamics, kinematics is the science of motion of bodies in space over time and is concerned with issues of displacement, velocity, and acceleration. Kinematics is the study of the amount and direction of movement, speed and acceleration, and determination of joint angles (Radomski & Latham, 2008). Considerations of how fast, how high, and how far can be thought of as the "observational geometry of motion" (Durward, Baer, & Rowe, 1999). To understand normal human movement, one must first understand basic laws of mechanics onto which principles of static and dynamic biomechanics may be overlaid.

Kinetics

Forces acting in and on the body that produce stability or mobility are the focus of study in kinetics. This study includes consideration of internal forces (such as body mass and muscular forces applied to move the body) and external forces (such as gravity and forces impacting on the body).

Studying kinetics is helpful in identifying which muscles are the strongest, as well as which aspects of movement are most prone to injury (Hamill & Knutzen, 1995). An occupational therapist uses this knowledge of forces to plan interventions to minimize deforming positions and detrimental stresses on joints to prevent further disability. Remediation of limitations of structures of the neuromusculoskeletal system to elongate tissues and to strengthen muscles is based on an understanding of how forces affect these soft tissues. Remediation of these impairments is an important step toward increasing performance in meaningful occupations.

FORCE AND MOTION

Human motion is the result of complex interaction of force and motion, following the physical laws and anatomic capability of the body. Forces acting in pairs occur often in the body in order to produce movements of great variety and strength. Often, more than one force is acting on a joint at a time. These efforts may mobilize or stabilize joints, as required by the activity. For example, the flexor carpi radialis (FCR) and the extensor carpi radialis brevis (ECRB) cross the wrist near the axis of motion, the FCR on the volar side and the ECRB on the dorsal side of the joint. If the FCR contracts, there is flexion of the wrist. If the ECRB contracts, there is extension of the wrist. If both muscles exert a force concurrently, there is movement in the direction of the pull of the fibers. Because each of these muscles effectively cancels the action of the other (flexion and extension), there is no movement. This type of muscle action produces stabilization of the joint. This can be seen when carrying a heavy suitcase; the FCR and ECRB work together to stabilize the wrist against the distracting forces exerted by the weight. At other times, two or more muscles must act together to produce movements that are not possible by the single action of either muscle alone, such as the synergistic actions that produce wrist ulnar and radial deviation.

The shoulder complex is a good example of a very mobile structure that is capable of a great variety and range of movements. The glenohumeral joint, however, is not very stable. The joint serves to mobilize the arm rather than provide stabilization. The stability of the shoulder is accomplished by the muscles in and around the scapula and humeral head. In contrast, the elbow is a much more stable joint due to the congruence of the bones that make up the joint, but it lacks the variety of movement (Flowers, 1998).

The study of mechanics centers on the concepts of force and motion and the laws of mechanics developed by Newton. Newton's First Law, known as the Law of Inertia, is at the heart of equilibrium. It states that a body will remain at rest or will continue in motion unless acted upon by an external force that brings about a change in the existing state. Momentum (p) is the amount of force required to change the inertial state and is equal to the mass times the velocity of an object, or p=mv. An object at rest has no observable momentum because its velocity is zero (actually, its velocity is the same as that of the spin of the earth, but that has no practical application here).

Theoretically, its inertial state can be changed by any additional force; however, observable movement will be dependent on the amount of force exerted.

An object in motion has observable momentum; the larger it is and faster it is going, the more momentum it has. Anyone knows that a head-on collision between a car and a dump truck, both traveling the same speed when impact occurs, will result in significantly greater damage to the car than to the dump truck, not to mention the human costs. This is because the truck has greater momentum due to its greater mass.

Inertia is also illustrated in attempting to push a wheelchair: the hardest part of pushing a wheelchair is when you first push it because you need to overcome inertia to accelerate the chair. Once the wheelchair is set in motion, it will continue through momentum until another force stops the motion. The other force may be another object or it can be friction (a product of gravity and surface features) and air resistance, which is admittedly minimal in this case. Greene and Roberts (1999) provided this example of Newton's First Law:

Bernice Richards has quadriplegia and minimal control of the muscles balancing her trunk. She directs her electric wheelchair through the main thoroughfare of a shopping mall, and a child darts in front of her. As she stops the chair, Bernice continues moving forward, stopped only by her chest and pelvic seatbelts. In Bernice's situation, when she stopped quickly, she and her wheelchair traveled together. An outside force (friction) slowed the chair, but Bernice's body continued moving forward. The force of friction on the chair did not affect Bernice's body. A separate, outside force, provided by the seatbelt, stopped her body from continuing forward (p. 32).

Newton's First Law can be interpreted in several ways:

1. If an object is at rest and no external forces act on it, the object remains at rest. If you don't push a person in a wheelchair and no other forces act on the wheelchair, the wheelchair will not move.

2. If an object is in motion and no external forces act on it, the object will continue moving at a constant velocity in a straight line. If gravity and other external forces didn't act on a wheelchair moving on a straight path with no changes in horizontal alignment, then the wheelchair could continue indefinitely.

3. If an object is at rest and external forces do act on it, the object remains at rest only if the resultant of the external forces is zero (the net external force is zero). If a person pushes on a wheelchair and the weight of the person (plus the chair) is equal to the force applied, the resultant of the two forces will be zero, and no movement will occur. If the person pushing on the wheelchair applies a force greater than that of the weight of the person (and chair), then movement occurs.

4. If an object is in motion and external forces do act on it, the object will continue moving at constant velocity in a straight line only if the net external force is zero (McGinnis, 1999).

While Newton's First Law discusses how forces act in pairs and how externally applied forces bring about changes in inertia, the magnitude of externally applied forces affects motion according to Newton's Second Law, the Law of Acceleration. Acceleration is what happens when inertia is overcome and movement of the object begins—it is the rate of change in velocity, or displacement/time/time.

Newton's Second Law states that acceleration of a body is proportional to the magnitude of the resultant forces on it and is inversely proportional to the mass of the body. The formula that expresses this law is F=ma, or a=F/m. This means that the change in motion (acceleration or deceleration) is a direct result of the net force exerted, occurs in the direction of the applied force, and is inversely related to the mass of the object moved. Conversely, the amount of force required to accelerate an object depends on the amount of weight of the object and how fast you want it to move. It takes more force from the pitcher's arm to pitch a 90 mph fast ball than to throw a 50 mph pitch. If you push a wheelchair with a great force, there is faster acceleration than if you pushed it with minimal force. If you push a wheelchair in which a large person is sitting, you will get less acceleration than if the person in the wheelchair has less mass, with force application remaining constant in

Figure 2-2. Effects of applied forces. (Adapted from Hamill, J., & Knutzen, K. M. (2003). *Biomechanical basis of human movement.* Philadelphia, PA: Lippincott Williams & Wilkins.)

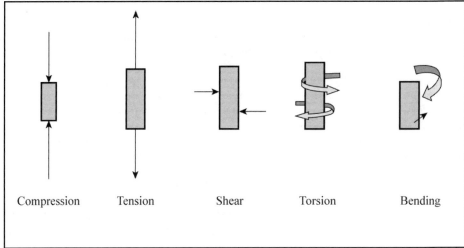

Compression   Tension   Shear   Torsion   Bending

both cases. Similarly, it takes a stronger initial muscle contraction to move your arm quickly than it does to move your arm slowly. In the clinic, this has implications for patients recovering from muscle or tendon injuries—rapid motion is avoided until late in the treatment regimen because the initial amount of force needed by a muscle to accelerate the part rapidly may exceed the amount of strain the healing tissues can withstand without re-injury.

Newton's Third Law of Action/Reaction states that "for every action, there is an equal and opposite reaction." This is interpreted to mean that forces are equalized (sum to zero). This means that whenever two objects are pressed together, they exert an equal and opposite force on each other. It also means that when a force is exerted by one object (or body) on another, its effects are distributed equally on both. If both objects are moveable, the result will be motion in opposite directions. When you jump from a rowboat, you jump in one direction, while the rowboat is pushed in the opposite direction. The force applied during the push-off is directed equally to both your body and the boat. When you row a rowboat, the oar on the water pushes back but the boat moves forward. Rockets can be used as another example of Newton's Third Law in that the thrust of gases and fuel goes in one direction while the rocket goes in the other.

When the interacting object is immobile, the results are less obvious. For example, when you are standing, your feet exert a force on the ground while, at the same time, the ground is exerting the same amount of resistance force to your feet. Pressure is the result of this phenomenon. The amount of pressure at the contacting surfaces is dependent upon the sum of the combined forces divided by the total area over which the forces are applied, or $P = F/A$.

The Effects of Forces

If the surfaces in the above example are not capable of withstanding the combined forces, stress to the material is the result. Stress is a measure of the force intensity that results from pressure. Stress is not visible. These applied forces or loads lead to changes in the size and shape of the object or material, and this is called deformation.

Deformation can be temporary or permanent, depending on the characteristics of the material and the magnitude of the force. The amount of deformation depends upon the load and the ability of the material or object to resist the load. Walking in sand at the beach involves shifting/compression of the sand until it reaches a point it can support your weight. In the clinic, when a person sits in a wheelchair, the tissues of the buttocks are compressed by the person's weight on the seat of the chair. Temporary deformation of tissues causes strain by occluding blood vessels and decreasing circulation to the area, particularly over bony prominences. This can lead to decubitus ulcers if the pressure is not relieved before tissue damage has occurred. To minimize the compression forces on skin surfaces, one needs to either decrease the force or increase the area of skin contact to decrease the pressure. Because changing the patient's weight is not immediately feasible, this is typically accomplished by distributing the body weight over a larger area using special cushions. Another example of this is that by making wider, longer splints, there is less likelihood of skin breakdown and more comfort for the client (Fess & Philips, 1987).

There are different modes of force transmission, which results in different types of deformation of structures. The amount of deformation depends on the type of force, the amount of force, and the type of structure.

Linear forces produce tension or compression (Figure 2-2). Compression forces press bones together and are produced by muscles, weight bearing on a limb, gravity, or external loading along length of bone. Compression pushes parts together, and the direction of forces is at right angles to the surfaces. Compression stress is what is commonly called pressure. Tension elongates tissues and pulls bones apart. Maximum stress is perpendicular to the plane of the applied load, usually by the pull of the muscle. Sprains, strains, and shin splints are usually the result of tension forces (Hamill & Knutzen, 2003). In both compression and tension, the force is applied in a perpendicular direction. Compression and tension are often called normal or axial forces.

Shear forces result in deformation internally in an angular direction and when the forces are parallel to the surfaces in contact. High shear effects can be avoided when fabricating splints by rounding sharp edges (Fess & Philips, 1987). Shear forces are also produced with the auto-recline feature of some electric wheelchairs. Torsion is a twisting force that creates a shear stress over an entire bone.

When a force is applied to a part with no support of the structure, the result is bending. One side of the structure becomes convex due to tensile forces acting upon the part, and the other side of the structure is concave where compressive forces are present (Hamill & Knutzen, 2003). The idea of bending forces is important when designing adaptive equipment and splints, because often there are three or four points of force used (Fess & Philips, 1987).

Friction

Friction is a special type of resistance force that is also derived in part from Newton's Third Law—it is largely dependent upon the net forces focused on two contacting surfaces. Friction can provide a stabilizing force that can retard movement or contribute to instability. Friction is a resistance force to smooth movement (Lafferty, 1992).

Skipping a stone on ice goes far because there is little friction. A house of cards stays up because there is friction between the cards, and they stay in place as long as there is a steep angle. Hovercraft use a cushion of air that holds the boat away from the water so friction is reduced, enabling the boat to travel at high speeds (Lafferty, 1992). How much friction limits a movement depends on how firmly the two surfaces are pressed together, as well as the nature of the materials that are in contact and their effects on each other (Galley & Forster, 1987).

To change the degree of friction between two surfaces, it is necessary to change the forces that hold the surfaces together or to change the nature of surfaces in contact (Galley & Forster, 1987). This idea is used in ball-bearing feeding orthoses (mobile arm supports), where friction is reduced by rolling steel balls, which enable movement by weak clients. Oil and lubricants also reduce friction, as illustrated in automobiles. Synovial joints in the human body have synovial fluid, which reduces friction during joint movements. The least amount of friction in synovial joints occurs when the joints are moving because the synovial fluid lubricates and smoothes the movement of the articulating surfaces. The loss of smooth, friction-reduced movement is a problem encountered in rheumatoid arthritis, where there is decreased gliding due to damage to synovial tissues.

Decreasing friction is important when designing splints and in adapted equipment, because poorly fitting devices, improper joint alignment, or inefficient fastenings can result in skin irritation, blisters, and skin breakdown (Fess & Philips, 1987).

Newton's laws interweave concepts of force and motion, but to understand the whole, we must look more carefully at each of these concepts.

Force

The study of linear forces involves the use of vectors. Forces have both magnitude and direction, which can be represented either graphically or mathematically. Graphic representations are used throughout this book to illustrate the application and results of forces acting on particular objects to produce specific motions. Graphic representation is done via vectors that take into consideration the qualities of magnitude and direction. Analysis of forces can be done by showing the combined effect of one or more forces in the same plane (co-planar) and on the same point (concurrent), which is called composition of forces. Or it may be necessary to replace a single force by two or more equivalent forces, which is called the resolution of forces (LeVeau, 1992).

Force Systems

Forces may be grouped to form different types of force systems. Linear force systems occur along the same action line. These forces produce tension or compression. Any system of forces acting in the same plane but along different action lines are called co-planar and are parallel force systems.

Concurrent force systems occur when all of the forces meet in one point. The forces do not act along the same line but do act on the same point (McGinnis, 1999). The sternal and clavicular heads of the pectoralis major muscle are anatomical examples of a concurrent force system. A third type of force system is a general force system, which occurs when all of the forces are in the same plane but are not covered in the above categories.

When parallel or general force systems are applied to the same object, the object reacts as a lever. Levers have unique and special properties that themselves require study in order to understand how forces applied to them affect motion of the lever. This topic will be explored further under the section on motion.

Composition of Forces

It is generally accepted that the simplest system to analyze is a linear force system. The parallel forces are co-planar (same plane), concurrent (same object), and co-linear, or acting along the same line (Figure 2-3).

In Figure 2-3, force #1 is acting on the same object in the same direction as force #2 along the same line, so the forces may be added together to obtain the resultant force. Only in the composition of parallel forces is the resultant the same as simple addition of the magnitude of the forces. In the second part of the illustration, the two forces are acting along the same line but in opposite directions. The resultant force is actually the simple difference in the two forces (LeVeau, 1992). Whenever the vectors are even slightly off parallel, however, these are concurrent force systems, and the resultant force is always less than their combined magnitude.

Polygons and parallelograms are used to graphically represent the composition of forces (Figure 2-4). In Figure 2-4, vectors A and B are exerting force on the same object, but in

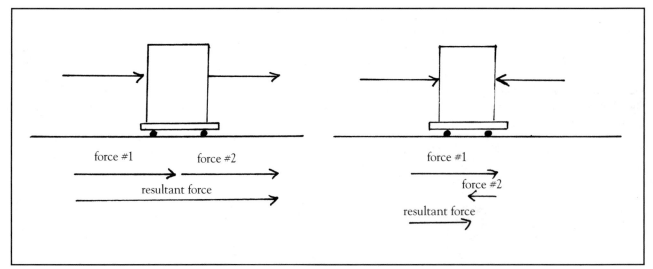

Figure 2-3. Composition of forces.

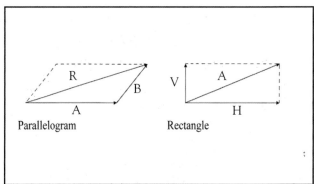

Figure 2-4. Vectors representing the composition and resolution of forces.

different directions. Their vector representations are aligned so that the tail of one vector is at the point of the other. A parallelogram is constructed by replicating the angle formed; the resultant R bisects the parallelogram, beginning at the tail of the first vector and ending at the point of the second.

Resolution of Forces

Resolution of forces is the opposite process described in the composition of forces to determine the combined effect of force. Any force can be broken down into its horizontal and vertical components. By constructing a rectangle around the original vector, with the original vector dividing the rectangle, one can find the horizontal and vertical components of the force. This exercise illustrates the relative amount of force that is directed vertically and horizontally by the vector. The rectangular shape, depicted in the second part of Figure 2-4 shows how vector A can be broken down into its horizontal (H) and vertical (V) components by constructing a rectangle around the vector, with the vector bisecting opposite right angles. In this illustration, there is approximately 33% as much horizontal force as there is vertical force.

The clinical relevance of this can be seen in Figure 2-5. When traction is positioned to extend the finger joint, the angle of application of the force is important. In the illustration, the angle of application is not perpendicular to the

bone. Analysis of the horizontal and vertical components of the force will reveal that some of the force is being directed horizontally along the longitudinal axis of the bone, which will result in linear motion (the sling will slip distally along the finger); less than full force is vertical (rotary), and the force will actually rotate the finger at the joint.

Motion

Kinematics looks at aspects of movement like acceleration, velocity, and associated forces with respect to time (Loth & Wadsworth, 1998). The description of movement can represent displacement of a body's position in space or a change in position relative to time or velocity (Loth & Wadsworth, 1998). The motion of bodies in space may be concerned with a single point in space (i.e., center of gravity), the position of several segments (i.e., the upper extremity), or the position of a single joint or motion (Hamill & Knutzen, 1995). It may represent linear or angular movements. Kinematics is not concerned with the causes of motion but rather with the results.

Linear or Translatory Movement

In linear or translatory movement, all parts of an object or person move the same distance in the same direction at the same time (Gench et al., 1995). If you push a book across

Vector =
a Quantity → magnitude
→ direction

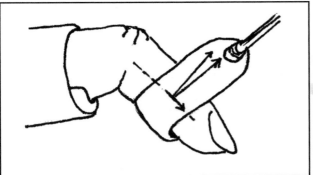

Figure 2-5. Efficient application of force. Note that the angle of application of the traction allows horizontal force, the effect of which will be to slip the sling distally on the finger.

a table, that is translatory movement—all pieces of the book traveled the same distance along a parallel path at the same time. A person traveling in a car is moving with translatory motion because all parts of that person are moving in the same direction at the same velocity. A figure skater gliding across the ice in a static position or a bicyclist coasting on a flat section of road are examples of linear movement. Protraction and retraction of the scapula are examples of linear motion.

Imagine a straight line connecting two points of an object. If the line between the two points maintains the same orientation as the object moves, then this would demonstrate linear movement. If both points move in parallel straight lines, this movement is rectilinear; if both points move in parallel but not straight lines, the motion is said to be curvilinear (McGinnis, 1999).

Angular Movement

Movement that occurs around an axis or a pivot point is angular movement (Smith et al., 1996). Angular kinetics depends on the amount of force applied and the distance of force from axis of movement. Angular velocity and acceleration can be treated as vectors in the same way that displacement and velocity are vectors in linear motion.

In angular motion, different segments of the same object do not move through the same distance (Hamill & Knutzen, 1995) because different segments are at varying distances from the axis. All parts of the object, however, move through the same angle in the same direction at the same time (Watkins, 1999), and the path of motion at each point along the lever is an ever-widening arc. An example is swinging on a bar: the whole body rotates around one point on the bar but the feet travel a greater distance than the arm because the feet are further from the axis (Pedretti, 2001). The lever doesn't have to be a rigid bar; it can be any object to which the application of force will cause rotation around an axis. Everyday examples of angular motion can be seen in a swinging door, a steering wheel, or a turning doorknob. A seesaw is a common example of a lever and a parallel force system, where both children on the seesaw exert the force of their weight.

A special case of parallel force system is called a force couple, where the parallel forces are equal in magnitude but opposite in direction. This particular force arrangement is powerful in causing rotation of the lever; all force is directed to that purpose. This can be visualized by imagining a steering wheel on a car where both hands on the steering wheel (one pushing up, the other pulling down) combine to turn the steering wheel in one direction (Galley & Forster, 1987). Force couples in the body consist of muscle co-contraction to stabilize a joint or by one muscle group contracting concentrically as the opposing group acts eccentrically to produce joint movement. Pathology to force couples can result in imbalance, instability, and loss of smooth coordinated movement (Magee, 2008).

Levers

Angular motion requires a lever. A lever can be any rigid object, usually depicted as a rigid bar with a fulcrum, a point around which the object will rotate when force is applied. Some objects, such as balls, do not have a fixed fulcrum, but rather an axis of rotation that becomes evident when the ball is spun. In the human body, the long bones are levers, and the fulcrum, the axis of motion, is at the axis of the joint.

Torque

Levers have the effect of magnifying forces applied to them. When forces are directed perpendicular to the lever at a distance from the fulcrum, the resulting turning force, or torque, is the product of the amount of force multiplied by the moment arm, the distance from the fulcrum. When the force is directed at something other than a 90° angle, the moment arm is shorter. The length of the moment arm is the distance between the fulcrum and the line of action of the force, not the point of application of the force itself (Figure 2-6). The force (F) is being applied to the lever at an oblique angle, while the resistance force (R) is applied perpendicularly. The moment arm is the perpendicular distance from the line of action of each force to the axis of motion. FA represents the force arm, while RA represents the resistance arm.

An everyday example of torque is when you use a wrench. Grasping the wrench at the end maximizes the moment arm and torque. If the wrench was grasped in the middle, it would require nearly twice the force to use the tool (Hamill & Knutzen, 2003).

Clinical examples of torque can be seen in the ease with which a short reacher is used as compared to a long one: the longer reacher requires more force because there is greater distance between the object and the person's joint axis of movement. Keeping a client close to you when performing a pivot transfer is a principle of good body mechanics because the client provides less torque when positioned closer to the therapist, so less effort is needed (Radomski & Latham, 2008). If you want to increase elbow strength, initial intervention may apply the resistance force at the mid-forearm because this creates less torque and gradually moves the resistance toward the wrist, which increases the moment arm and thereby increases torque while keeping the force constant (Fess & Philips, 1987).

External resistance encountered by the body includes the weight of casts, braces, plates of food, dumbbells, crutches, or exercise equipment, and these are often applied distally. The muscle force required to overcome the resistance must be taken into account with the understanding that the maximum resistance torque of the weight occurs when the segment

Figure 2-6. Application of force determines moment arm.

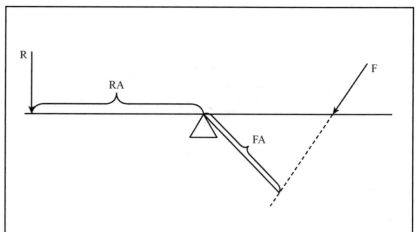

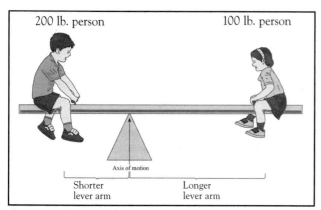

Figure 2-7. First-class lever.

or extremity is horizontal because the line of action of the resistance (the line of gravity) is always vertical, perpendicular to the ground, producing the maximum resistance moment arm (Smith et al., 1996).

Torque is also important in manual muscle testing. When testing elbow flexion, resistance is applied at the wrist rather than the middle of the forearm. The torque produced by the patient is the same, but the force applied by the therapist at the forearm is about half the amount if applied at the wrist due to a shorter resistance arm (Smith et al., 1996). If force is applied distally, forces might be small but there would be relatively large torques that the muscles must match (Smith et al., 1996).

Leverage

The principle of levers states that a lever of any class will balance if the product of the effort force (F) and its moment arm (FA) are equal to the product of the resistance (R) and its moment arm (RA), or F•FA=R•RA. The first part of the formula represents the torque of effort, and the second half represents the torque of resistance. When these elements are equal, no acceleration of the lever occurs. When they are out of balance, acceleration of the lever occurs. Leverage in the body is a function of the angle of pull of the muscle at its bony attachment. The optimal leverage can be achieved when a muscle pulls in a perpendicular direction to the axis of the bone, but this rarely occurs within the normal range of motion. The leverage of the muscle changes from one joint position to another because the angle of pull relative to the bone changes.

Turning force (torque) can be illustrated by opening a door. If you push or pull at the center of the door, you will not be able to move the door as effectively as if you push the door at a point as far from the hinges as possible. The hinge acts as an axis, and when you push far from the hinges, you increase the force arm (Hamill & Knutzen, 1995). In canoeing, torque is applied by the person using the oar to cause it to rotate. In golf or tennis, torque is applied to the club or racquet to swing these pieces of equipment (McGinnis, 1999).

First-Class Levers

There are three types of levers. A first-class lever occurs when forces are exerted on opposite sides of each other with the axis or fulcrum in between them. One can gain either force or speed, depending upon distance from the axis to the resistance and the distance from the axis to the effort force. Greater force can be achieved if the axis is closer to the resistance force, whereas greater speed is accomplished with the axis closer to the effort force. On a seesaw or teeter-totter, the heavier child sits closer to the axis so that his greater mass can be overcome by the smaller gravitational force that his lighter playmate exerts. The lighter child's movement further from the axis produces a longer lever arm so less effort is required to move the seesaw due to greater leverage (Figure 2-7).

The same idea that a longer lever produces increased force can be seen in Figure 2-8. The muscles depicted in the figure are the same size, cross section, and fiber type, crossing the same joint but at varying distances from the joint axis. Muscle B has a lever arm that is twice as long as the lever arm of muscle A, so muscle B has greater leverage and can produce more force to move the bone.

Common examples of first-class levers are a seesaw, scissors (actually a double first-class lever), and splints. Figure 2-9 shows several examples of first-class levers. When using scissors, the effort comes from the muscles closing the scissors, the resistance is what you are cutting, and the fulcrum is the screw holding the scissors together. Examples in the body include the atlanto-occipital joint in which the head is balanced by the neck extensors; the intervertebral joints in sitting or standing where the trunk is balanced by the erector spinae acting on

1st = E Fu Res (advantage for effort or speed)
2nd = FRE (large masses w/ little effort)
3rd FER (more effort)

Effort = Force
Resistance = weight

Fulcrum = Axis

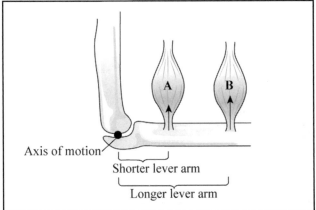

Axis of motion
Shorter lever arm
Longer lever arm

Figure 2-8. Lever arm and muscle force.

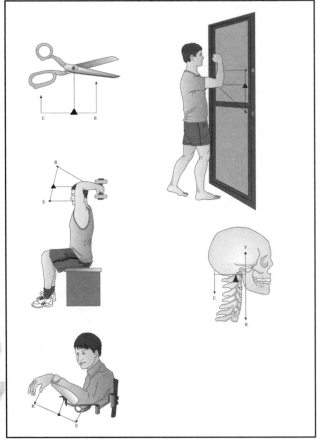

Figure 2-9. Examples of first-class levers.

the vertebral axis; and the triceps muscle acting on the ulna. Clinical examples of first-class levers are splints and the mobile arm support (or balanced forearm orthosis).

Second-Class Levers

A second-class lever occurs when a large amount of weight is supported or moved by a smaller force; a second-class lever is considered a force magnifier (Lafferty, 1992). Archimedes is reputed to have said in 200 B.C., "give me a fulcrum on which to rest and I will move the earth!" Second-class levers are able to move large masses with little force. Common examples are a wheelbarrow, a nutcracker, or a faucet, and some of these are shown in Figure 2-10.

In a second-class lever, the weight or resistance is in the middle with the effort force and axis on either side. There are few examples of second-class levers in the body, and those that are identified are seen as controversial. Standing on one's toes is an example in the body where the metatarsophalangeal joint of the big toe is seen as the axis, the muscles inserting into the heel are the effort forces, and the weight of the body acting through the ankle is the resistance force. Other examples are the splenus muscle action in extending the head at the atlanto-occipital joint (Lafferty, 1992). The forearm and brachialis are also identified as a second-class lever because the center of gravity of the forearm is between the elbow and the insertion of the brachialis muscle (Gench et al., 1995).

Third-Class Levers

Third-class levers are the most common in the body. In third-class levers, the effort force is central with the resistance force and the axis on either side. Third-class levers permit speed or movement of a small weight for a long distance and are considered force reducers (Lafferty, 1992), because the effort force is greater than the resistance or load. The muscle forces must always be greater than the force of resistance, which sacrifices mechanical advantage to produce wide ranges of motion and high-velocity movements (Greene & Roberts, 1999). A broom, fishing pole, tweezers, chopsticks, and a crab's pincer are examples of third-class levers (Figure 2-11).

Anatomically, small muscular forces can produce large movements of long bony segments. The deltoid, extensor carpi radialis, and iliopsoas muscles are examples of forces interact-

ing with bone in third-class lever configurations in the body. Biceps brachii is another example (see Figure 2-10), and an animated website demonstrates the actions of levers at http://www.enchantedlearning.com/physics/machines/Levers.shtml.

Line of Action

When force is applied to a lever, the angle of application is of great importance.

In the human body, the number of active muscle fibers determines the magnitude of muscle force (Roberts & Falkenburg, 1992), while the direction of pull establishes the line of action, which then determines the force moment arm. The direction of pull is typically the direction of the alignment of the muscle fibers and can be seen as a line (the action line or line of action) that is parallel to the fibers and intersects the attachment of the muscle. The muscle force multiplied by the moment arm is the torque, or turning force, that is the resultant force available to move the bone.

Because the action line of the muscle is fixed by the anatomic attachments, the muscle often pulls in a line other than in the direction of movement, and some of the efficiency of the muscle is reduced (LeVeau, 1992) (Figure 2-12). In this instance, the resolution of its vector can illustrate the relative amount of force that is directed horizontally and vertically. The horizontal component applies a compressing force to the joint, providing stability, while the vertical component applies rotational force to move the bone. As we have seen, any force

Figure 2-10. Examples of second-class levers.

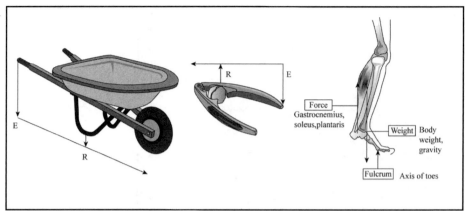

Figure 2-11. Examples of third-class levers.

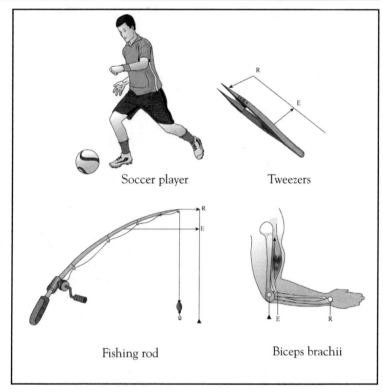

Soccer player   Tweezers

Fishing rod   Biceps brachii

can be resolved into its horizontal and vertical components, and when a force is directed to a lever in any direction other than perpendicular, some of the force is directed linearly along the lever rather than as rotational force.

The effectiveness of the muscle as a stabilizer or mobilizer can be estimated based on comparisons of the lengths of horizontal and vertical vectors as well as the angle of muscle pull. The resolved force vector is depicted as a diagonal line located between, and at right angles to, the two component forces. The tips of the three lines can be connected to form a rectangle. If the stabilizing line is longer than the rotary or angular line, then this muscle is more effective in stabilization in this position than it is in causing rotation or movement. The angle of pull is important in considering the effectiveness of a muscle as a stabilizer or as a mobilizer. Gench and colleagues (1995) state that "the farther from the perpendicular the angle of insertion becomes, the weaker the muscle will be

as a joint mover" (p. 35). In the example of the levator scapula muscle, the line of pull of the muscle is longer for the vertical component than it is for the horizontal component. Because of this, the levator scapulae muscle functions more effectively as an elevator of the scapula than it does as an adductor.

Another example is elbow flexion. When the elbow is extended, the force is directed primarily in a parallel direction along the length of the bone, making the initiation of movement difficult. Once flexion begins, the angle of insertion of the muscle increases so more of the muscular force is directed toward moving the joint than toward stabilization. At the end of flexion, the rotary (movement) force diminishes, and the parallel forces increase. Biceps are strongest at 90° of elbow flexion when the combined effects of biceps, brachialis, and brachioradialis are directed toward development of joint tension with a line of muscle pull perpendicular to the bony axis (Hamill & Knutzen, 2003). By analyzing the components of

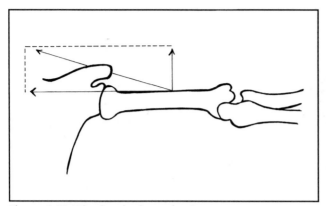

Figure 2-12. Wheel and axle mechanics.

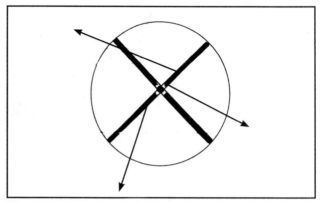

Figure 2-13. Wheel and axle mechanics.

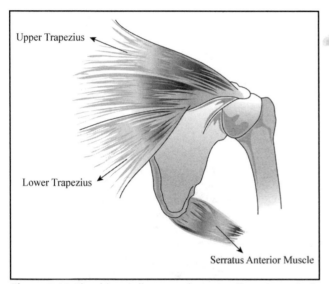

Figure 2-14. Shoulder girdle upward rotation (lateral rotation).

a muscle action, one can determine the angle of pull of the muscle, function of the muscle, and the relative strength of that muscle in any position of the muscle fibers. This information is relevant to prevention of injury and to intervention with the intention of increasing muscle strength by suggesting optimal positioning of the muscle during activities.

The ideas of force resolution and an understanding of the relationship of stabilizing and rotary components can also be applied to splint effectiveness. If forces are applied perpendicular to the segment being mobilized, the translational forces are lessened so there is less unwanted tension on the joint surfaces, and the splint is more effective. As the client improves, adjustments to the splint are necessary to maintain the 90° force application (Fess & Philips, 1987).

Wheel and Axle Mechanics

The wheel and axle is one configuration of levers that can be second or third class. The center of rotation of the wheel is the axis, and the radian of the circle is the moment arm. The axle and the rim are where the application of force occurs. With a steering wheel, a second-class lever, force is applied to the rim of the wheel in order to overcome the resistance, which is focused at the steering column. The effect of wheel

and axle mechanics is such that if a wheel is turned by a small force at its rim, a larger force can be generated at the axle. This is the idea behind waterwheels, round door handles, and water faucets (Lafferty, 1992). In the case of a bicycle tire (a third-class lever), the force is applied on the gear through the chain (turned by the feet), and the resistance (weight) is focused through the central core of the wheel onto the point where the tire meets the road.

The rotary movements of the scapula can be compared to the turning of a wheel (Figure 2-13). If one pulls on vector A, the wheel will turn clockwise. Likewise, if one pulls on vector B, the wheel will turn counterclockwise. The next drawing (Figure 2-14) shows a few of the muscles of the scapular region. Contraction of the upper trapezius muscle will produce an upward rotation of the scapula. Contraction of the serratus anterior will cause a lateral movement of the inferior angle of the scapula. Lower trapezius will cause the scapula to downwardly rotate. The resultant force is shoulder girdle upward rotation of the glenoid fossa (or lateral rotation of the inferior angle). Combined actions occur with many of the scapular and shoulder muscles producing different movements.

Mechanical Advantage

Levers differ in their capacity to balance and overcome resistance. This is known as mechanical advantage. Mechanical advantage is ratio of the force arm to resistance arm, or the force multiplied by its moment arm divided by the resistance multiplied by its moment arm: MA=FA/RA.

In second-class levers, the FA is always greater than the RA, so second-class levers can have significant mechanical advantage (always more than 1). Conversely, in third-class levers, the FA is less than the RA. In third-class levers, there is minimal mechanical advantage (always less than 1), and third-class levers have the least capacity to overcome resistance. First-class levers can vary in the amount of mechanical advantage, depending on the location of the axis in relation to the resistance force and effort force arms. In the human body, muscles that attach further from the joint generally have greater mechanical advantage than do muscles that attach near the joint, but this will vary at position in the range of motion (Muscolino, 2006).

Gench and colleagues (1995) point out that it may seem paradoxical that the levers least found in the body are the levers that are capable of the greatest mechanical advantage.

Figure 2-15. Bending and forces.

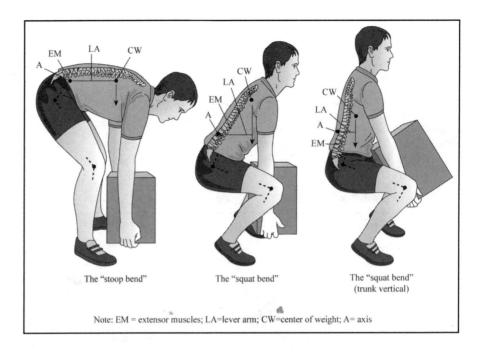

The "stoop bend"   The "squat bend"   The "squat bend" (trunk vertical)

Note: EM = extensor muscles; LA=lever arm; CW=center of weight; A= axis

They add that this would be true if we used our bodies to move great amounts of weight slowly. However, our bodies are more often used in activities requiring movement of smaller weights or moving quickly, and this is consistent with an abundance of third-class lever systems.

While it may seem that the predominance of third-class levers would make the human body perform inefficiently and ineffectively, there are distinct benefits to this arrangement. In a third-class lever, while the magnitude of the force needed to move the joint is great and therefore inefficient, the movement produced distally is through a much greater arc than that produced proximally. "In fact, the shorter the lever arm of the effort force (the closer the effort force is applied to the axis), the greater is the movement of the distal end of the lever" (Norkin & Levangie, 1992). This greater movement and speed is not true in a second-class lever where there is mechanical advantage and efficiency in terms of force output, but relatively little gain in movement distally (Norkin & Levangie, 1992).

Norkin and Levangie summarize these concepts:

- When a muscle is the effort force and the FA is smaller than the RA, the magnitude of the effort force must be large, but the expenditure of muscle force is offset by gains in speed and distance of the distal portions of the segment. This is true for all third-class levers and for first-class levers in which RA>FA.

- When the external force is the effort force and the FA is larger than the RA, the magnitude of the effort can be small as compared to the resistance, but relatively little is gained in speed and distance. This is true for all second-class levers and for first-class levers in which FA>RA (1992, p. 40).

These concepts are inherent in the logic behind the joint protection technique to use the strongest joints possible for activities and for body mechanics principles that advocate keeping objects closer to the body, not only for greater stability but also so that less force is needed to lift the objects, as in transfers.

Figure 2-15 illustrates how changing the length of the lever arm in lifting can reduce the risk of injury. In the first figure, the "stoop bend" creates a long lever arm for the force of the weight of the trunk (lumbosacral joint) and upper body, requiring contraction of the extensor muscles of the back. This increases the risk of back strain. In the "squat bend," the back is straight, decreasing the lever arm of the weight of the trunk and upper body, decreasing back strain. The "squat bend" reduces the lever arm even further, reducing the forces acting on the back, reducing the risk of injury.

Momentum

Angular motion is present when levers are in action, and it has distinct differences from linear motion in the development of momentum. Linear momentum is the product of the mass of an object and its velocity. In angular systems, however, momentum is the mass times its distance from the fulcrum times velocity. This means that rotary motion produces powerful momentum. Linear momentum builds linearly: doubling the mass (at the same velocity) will result in doubling the momentum. Angular momentum builds exponentially: doubling the mass (at the same velocity) will result in quadrupling the momentum because of the distance of the squared moment arm.

Increasing the moment arm is also a very efficient way to build momentum. The pitcher goes through his wind-up before releasing the ball in order to build velocity; the side-arm throw increases the moment arm from the axis (the shoulder) to the ball. The result is powerful momentum to put the fastball over the plate. The baseball bat has a thinner shaft at the grip and a widened end, bringing the center of gravity of the bat towards the distal end—this allows more momentum to be developed as the batter swings, enabling him to put the ball out of the park.

f RA>FA then Effort is great but Speed+distance is gained
f FA>RA then effort is small compared to resistance, but little is gained in terms of speed+distance

Kinesiology Concepts   25

In the clinic, momentum can be a helpful or a hazardous thing. In transfers, we have the patient rock back and forth before standing, in part because the forward momentum on the count of "three" will help him achieve the standing position. However, if the patient is tall and the therapist short, the higher center of gravity of the patient can result in too much momentum, the shorter therapist may not have sufficient leverage to control it, and a fall can occur. Patients on crutches have to be especially cautious because of the danger of building too much momentum during crutch-walking and experiencing a fall.

Pulleys

Pulleys act to change the direction of force. In the gym, the pulley allows you to push forward against the rope handle while the weight is moved up. Pulleys exist in the body in the form of bony prominences or ligaments; these structures control the direction of pull of muscles through controlling placement of their tendons.

Pulleys also have the capability of increasing mechanical advantage. They do this by changing the line of action of a force, thereby increasing the moment arm. An example of this is seen at the knee. The angle of pull of the quadriceps muscles is altered by the patella as it moves on the condylar groove of the femur (Hamill & Knutzen, 1995) in the action of knee extension. From a flexed position, the tendon passes over the patella and changes direction down to attach to the tibia. As the knee is extended, the patella rides over the condyles of the femur and pushes the quadriceps tendon further from the axis of motion at the knee, thereby increasing the force moment arm of the muscle and increasing its ability to overcome resistance. Other examples of pulleys are the peroneus longus muscle action around the ankle (Figure 2-16) and the extensor pollicis longus around Lister's tubercle.

MOTIONS OF BONES AND JOINTS

The study of kinematics also extends to the study of how motion in the human body occurs. Osteokinematics describes observable movement of bones, while arthrokinematics describes the movement of articulating surfaces in relation to the direction of movement of the extremities.

Osteokinematics

Osteokinematics is the description of the movement of the bones, focusing on the angular changes between the bones forming the joints. Osteokinematic motion is referred to as physiologic or classical movement, and this movement differs from joint to joint due to the bony alignments and soft tissues around the joint (Lippert, 2006). Osteokinematics describes visible motion.

The movements of bones have been classified using planes and axes as determinants of motion (see Figure 2-15). Planes and axes provide a three-dimensional system of recording the descriptions of movements in space of specific points on the body.

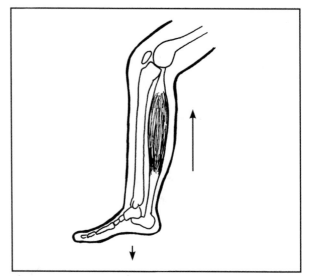

Figure 2-16. Example of a pulley.

The location of movement occurs in three planes that are at right angles to each other:

1. The first plane is the sagittal plane, which corresponds to the sagittal suture in the skull. This is a vertical plane that divides the body into right and left halves. Other names for this plane are the anterior-posterior plane or the YZ plane.

2. The second plane is the frontal/coronal plane, which corresponds to the coronal suture in the skull. This, too, is a vertical plane that divides the body into front and back portions. Alternative names for this plane are lateral plane or XY plane.

3. The transverse plane is a horizontal plane that separates the body into upper and lower (cranial-caudal) portions. The transverse plane is also referred to as the XZ or horizontal plane.

When these planes divide the two hemispheres precisely in the center, they are referred to as cardinal planes. The center of gravity is the point at which the three cardinal planes intersect.

Planes and Axes

The direction of movement is determined by the axis, which can be thought of as the line around which movement takes place. Axes are related to planes and are perpendicular to them. Three axes divide the three planes into four quadrants (Figure 2-17).

1. The anteroposterior or sagittal axis extends horizontally from front to back. The sagittal axis determines the direction of movement in the frontal plane (Table 2-1). Jumping jacks are an example of movement of the arms and legs in a frontal plane with a sagittal axis.

2. The frontal/coronal axis extends horizontally from side to side. With the sagittal plane, the frontal axis determines the direction of flexion and extension, as well as thumb abduction (see Table 2-1). Marching like a soldier is an example of movement in the sagittal plane with a frontal axis.

Linear momentum = mass × velocity
ngular = mass × distance from axis × velocity

Table 2-1

SUMMARY OF MOTIONS IN EACH PLANE AND AXIS

MOTIONS IN A FRONTAL/CORONAL PLANE AROUND A ANTERIOPOSTERIOR/SAGITTAL AXIS

Shoulder: Abduction and adduction

Wrist: Radial and ulnar deviation

Thumb: Extension

Hip: Abduction and adduction

Foot: Eversion and inversion

MOTIONS IN A SAGITTAL PLANE AROUND A FRONTAL/CORONAL AXIS

Shoulder: Flexion and extension

Hip: Flexion and extension

Knee: Flexion and extension

Ankle: Dorsi and plantar flexion

Thumb: Abduction

MOTIONS IN A HORIZONTAL PLANE AROUND A LONGITUDINAL/VERTICAL AXIS

Shoulder: Internal and external rotation

Forearm: Supination and pronation

Hip: Internal and external rotation

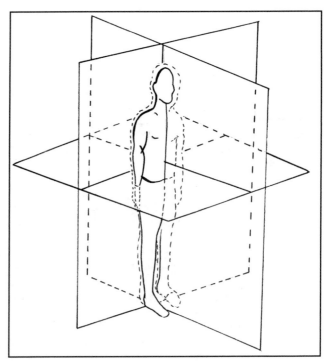

Figure 2-17. Planes and axes.

3. A vertical line in a cranial-caudal direction is the longitudinal/vertical axis, which intersects the transverse/horizontal plane. Movements that occur in this transverse plane/longitudinal axis are rotational (see Table 2-1).

Most joint axes are offset from the anatomic planes, and many joint axes in the body move during movement of the extremity. This is an important concept to remember when measuring joint range of motion, so that alignment of the goniometer is correct and the measurement therefore reliable.

The importance of understanding planes and axes is in the ability of this three-dimensional system to describe movements clearly and consistently so the movement can be understood explicitly. Anatomic concepts are based on this system, and this is the basis for describing these movements. Clear understanding of planes and axes is critical in conducting biomechanical assessments of muscle strength and joint range of motion.

Arthrokinematics

Arthrokinematics describes the relative motion of the articular surfaces and is part of the movement that cannot be observed. The study of arthrokinematic movement is closely related to mechanical and engineering principles. Motions of

bone surfaces within the joint are a combination of rolling, sliding (gliding), and/or spinning (MacConaill, 1957).

Rolling is analogous to a ball rolling on a table or a rocking chair in motion, where each point on one surface contacts a new point on the other surface (Figure 2-18). Characteristics of one bone rolling on another include the following:

- The surfaces are incongruent.
- New points on one surface meet new points on the opposing surface.
- Rolling results in angular motion of the bone.
- Rolling is always in the same direction as the angulating bone motion whether the bone is convex or concave.
- If it occurs alone, rolling causes compression of the surfaces on the side to which the bone is angulating and separation on the other side. Passive stretching done using bone angulation alone may cause stressful compressive forces to portions of the joint surface, potentially leading to joint damage.
- In normally functioning joints, pure rolling does not occur alone but in combination with joint sliding and spinning. Some muscles may function to cause the slide with normal active motion (Kisner & Colby, 1990; Nicholas & Hershman, 1990).

Slide is another type of motion of the bone surfaces within the joint. The gliding movement can be visualized as seeing a box move down an incline with translatory motion of one surface without rotation. An example in the body is the movement of the humeral head on the glenoid fossa during full abduction. Characteristics of slide include the following:

- For pure slide, the surfaces must be congruent, either flat or curved.

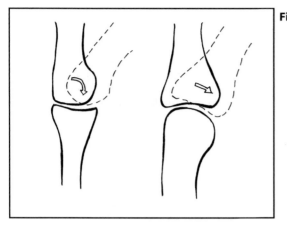

Figure 2-18. Rolling in the same direction as movement.

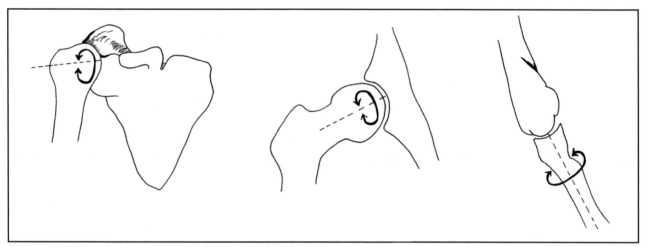

Figure 2-19. Examples of spin movements.

- The same point on one surface comes into contact with new points of the opposing surface.

- Pure sliding does not occur in joints because the surfaces are not completely congruent.

- The direction in which sliding occurs depends on whether the moving surface is convex or concave. Sliding is in the opposite direction of the angular movement of the bone if the moving joint surface is convex. Sliding is in the same direction as the angular movement of the bone if the moving surface is concave. This is the basis for determining the direction of the mobilizing force when joint mobilization gliding techniques are used (Kisner & Colby, 1990).

Spin occurs when one bone rotates around a stationary longitudinal axis. The same point on the moving surface creates a circle as the bone spins. An example of spin is a top as it rotates around an axis perpendicular to the surface on which it spins or a car tire that loses traction on ice.

Combined roll-sliding occurs naturally in joints as the body segment is moved. The more congruent the surfaces are, the more sliding of one bony segment on another occurs. The more incongruent the surfaces are between two bones, the more rolling occurs.

If there is rotation of a segment about a stationary axis, this is called spin. The same point on the moving surface creates an arc of a circle as the bone spins. Spinning rarely occurs in joints but in combination with rolling and sliding. Flexion and extension of the hip and shoulder are examples of where spin occurs in the body (Kisner & Colby, 1990) (Figure 2-19).

These motions of rolling, spinning, and sliding provide large amounts of motion at the joints using small articulating surfaces, and these motions contribute additional motion between joint surfaces.

Accessory Motions

Accessory motion is defined as motion that accompanies active motion and is necessary for normal motion, but cannot be isolated voluntarily (Thomas, 1997). Rolling, sliding, and spin are accessory motions (sometimes referred to as joint play). Joint mobilization techniques are used to restore or maintain joint play within a joint as well as in the treatment of stiffness, hypomobility, and pain. Joint play is the distensibility of the joint capsule or the "give" that occurs when joints are passively moved. The sliding component is used in joint mobilization to restore joint play and reverse joint hypomobility, whereas rolling is not used because it causes joint compression (Kisner & Colby, 1990). Spinning is also used where it is a normal accessory motion of the joint.

The term *accessory motion* also refers to motion in related joints that causes a joint axis to move, usually in a curved pattern, as the joint motion occurs. This can be seen in the scapula and clavicle as the humerus is flexed or abducted. As the humerus is raised, the scapula rotates and may protract, while the clavicle moves forward and elevates. This movement of the scapula and clavicle occurs without conscious control, and full humeral flexion or abduction would not be possible without this component or accessory motion. It is this changing of the position of joint axis during movement that creates problems for proper fitting of orthotic devices (such as splints) and in reliable measure of joint angles (as in goniometric assessment).

Accessory movements are essential for pain-free joint movement. For example, if distal movement of the head of the humerus on the glenoid fossa did not occur, then the greater tuberosity would hit the acromion process leading to injury, pain, and loss of range of motion. This could result in an impingement syndrome that could develop into a "frozen shoulder" (adhesive capsulitis). If there is a loss of joint play accompanied by pain, there will be joint dysfunction (Smith et al., 1996). If this occurs, then the joint is not free to move nor are muscles free to move them, and this can lead to deterioration in the joints as well as muscle weakness (Smith et al., 1996).

GROSS MOVEMENTS OF BONY SEGMENTS

The movement of bony levers is called swing (Kisner & Colby, 1990). The amount of swing is measured in degrees of motion using a goniometer. The movements that are considered to occur in plane are pure motions: flexion, extension, abduction, adduction, and rotation. Other motion classifications, such as circumduction, opposition, inversion, and eversion, are actually combinations of the pure motions.

Many of the movements in the human body are a combination of linear and angular movements. General motion is most easily understood when one considers walking or running where there is linear motion of the head and trunk and angular motion of the arms and legs. A person bicycling is demonstrating linear motion of the head, trunk, and arms, but angular motion of the legs to cycle and propel the bicycle. To review, differences between angular and linear (translatory) movements are as follows:

- In angular motion, there is movement in a circular path around a central point. In translatory motion, movement starts in one place and ends in another.

- In angular movement, the object changes orientation during the movement. In translatory movement, the object remains in the original orientation.

- In angular motion, two points moving in a segment around a circle move at different velocities. In translatory movement, two points in a line move at the same speed (Greene & Roberts, 1999, p. 47).

Table 2-2 provides a summary of terminology related to movements of the joints. Clinically, we also differentiate between active motion, that produced by contraction of the patient's muscles, and passive motion, produced by the therapist or other device applied to the patient's extremities. Ballistic motion is dependent on momentum; in the body, it is started by a quick firing of a muscle or muscles; the segment continues moving through the range as a result of momentum and is stopped by an opposing force such as an antagonistic muscle, joint constraints, or an external object.

Joint Classifications

Functional descriptions of the movement capability of joints are one type of joint classification. A synarthrodial joint is one that is immovable (such as cranial and facial bones). An amphiarthrodial joint is slightly movable, as seen in some intervertebral articulations. Diarthrodial joints exist between two bones separated by a distinct space and are freely moveable.

Diarthrodial joints are characterized by two separated bony surfaces covered with a layer of hyaline cartilage or fibrocartilage and the joint cavity enclosed by a fibrous capsule and ligaments. The articular (hyaline) cartilage, which covers the bones, helps in additional load transmission, stability, improved joint fit, protection of joint edges, and lubrication. The capsule protects the joint, creates the interarticular portion of the joint, and helps to sustain some of the load applied to the joint. The ligaments, which are integral to the capsule, provide additional support.

Structural classifications have also been developed to define the composition of the tissue of the joint. Fibrous joints are made up of fibrous connective tissue and permit little or no movement. Cartilaginous joints have continuous cartilage connecting the articulating surfaces with moderate but limited movement. Synovial joints are more complex: they are diarthrodial and include additional structures such as a joint capsule with a synovial lining and synovial fluid. A summary of joint classifications and examples are provided in Table 2-3.

Diarthrodial joint surfaces have been compared to geometric shapes and their movements likened to mechanical devices. Using this classification system, there are three major types of joints:

1. Uniaxial: Movement in one axis (one degree of freedom).

2. Biaxial: Movement in two axes (two degrees of freedom, usually flexion/extension and abduction/adduction).

3. Triaxial/multiaxial joints: Movement in all axes, including all movements.

Uniaxial Joints

Uniaxial joints include joints that work like a hinge and are aptly called hinge joints or ginglymus joints (Figure 2-20). These joints permit flexion and extension, which are angular motions. Examples of hinge joints are the humeroradial and humeroulnar joints (commonly known as the elbow joint). The interphalangeal joints of the fingers are also ginglymus joints.

Another type of uniaxial joint is a trochoid or pivot joint (Figure 2-21). This is where an arch-shaped process fits around a peg-like process. The axis of movement is longitudinal so

Table 2-2

SUMMARY OF TERMINOLOGY RELATED TO JOINT MOVEMENTS

Flexion: The bending of a part (like the elbow) so that surfaces (usually anterior) come closer together. In flexion, the angle of the joint decreases as the joint is moved. Example: when you use your shoulder to reach overhead to retrieve an object from a high shelf.

Extension: When the angle at the joint increases as the joint is moved. Example: when the elbow is inserted into the armhole of a shirt. Movement of the shoulder when the arm goes behind the body in the coronal plane has been termed both *extension* and *hyperextension*, depending upon the source. For simplicity, this text will assert that shoulder extension occurs from directly overhead to placement at the side of the body.

Abduction: Movement away from a defined line, such as the center of the body. In the hand, midline is the third digit and in the foot, the second digit. Example of abduction of the shoulder: when you raise your arm to comb the back of your hair. A special case of abduction is *shoulder abduction in the plane of the scapula*. This motion has several alternative names including scapular plane elevation, elevation in the scapular plane, and "scaption." This motion is approximately "30-40 degrees anterior to the frontal plane" (Smith et al., 1996). The motion is midway between forward shoulder flexion and abduction of the humerus in the plane of the scapula (Nordin & Frankel, 1989). This is a good position to test the motion of the glenohumeral joint since the capsule is in a loose packed position, so one is less likely to impinge on the coracoacromial structures (Smith et al., 1996).

Adduction is movement toward a defined line, such as the center of the body. When you are sitting with your arms held at the sides of your body, your shoulders are in adduction.

Horizontal adduction occurs at the shoulder joint. Example: when the shoulder is in 90 degrees of flexion, and the arm is moved in a direction toward the midline or toward the anterior.

Internal (medial) rotation occurs at the shoulder and at the hip. This is when there is a transverse rotation oriented toward the anterior side of the body. Example: when you place the back of your hand on your vertebral column.

External (lateral) rotation: When there is a transverse rotation oriented toward the posterior side of the body. Example: when you brush your hair.

Scapular rotation is described in terms of the direction of movement of either the inferior angle of the scapula or movement of the glenoid fossa of the scapula. Medial (downward) rotation of the scapula occurs when the inferior angle of the scapula moves toward the midline and movement of the glenoid fossa moves in a downward or caudal direction.

Lateral (upward) rotation: When the inferior angle moves away from the midline and movement of the glenoid fossa is in an upward or cranial direction.

Additional terms are also used to describe joint positions and movements.

anterior	palmar/volar
posterior	dorsal
medial	ulnar
lateral	radial

Pronation: when the palm is directed downward. Pronation can actually be considered internal rotation of the forearm. Example: when you are sanding a board.

Supination: the position of the hand in anatomical position where the palm is facing upward. Again, this movement can actually be considered external rotation of the forearm.

Table 2-3

SUMMARY OF JOINT CLASSIFICATIONS

JOINT TYPE	CHARACTERISTICS	TYPES	EXAMPLES
Synarthroses (fibrous)	• Immovable joints • Connective tissue or hyaline cartilage	Sutura Schindylesis Gomphosis Syndemosis	Skull Sphenoid Ethmoid bones Teeth insertion into mandible and maxilla Between ulna and radius Between tibia and fibula
Amphiarthrosis (cartilaginous)	• Slight movement	Synchondrosis Symphysis	Epiphyses of long bones in children Costal cartilage Pelvic bone articulation
Diarthrosis (synovial)	• Most numerous in body • Freely moveable • Specific characteristics	Hinge Condyloid Saddle Ball and socket Pivot Gliding	Distal interphalangeal joints of fingers Radiocarpal joint Thumb carpometacarpal joint Shoulder Hip Radioulnar joints Sternocostal joints

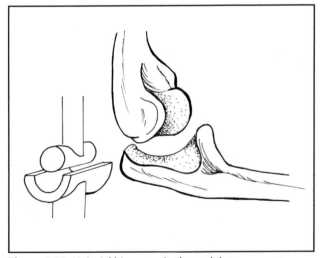

Figure 2-20. Uniaxial hinge or ginglymus joint.

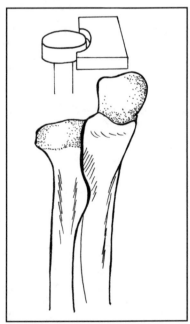

Figure 2-21. Pivot or trochoid joint.

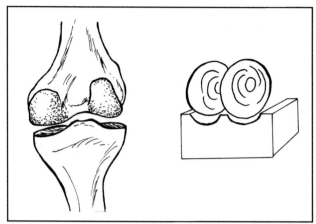

Figure 2-22. Condyloid or ellipsoid joint.

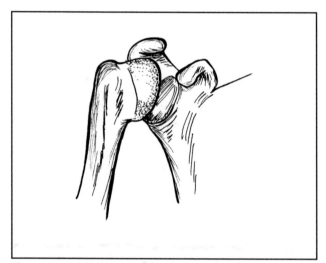

Figure 2-24. Ball and socket joint.

the movement is rotary. The atlanto-axial joint in the cervical spine is a trochoid joint, with the dens, projecting from the second vertebra, providing the axis around which the first vertebra rotates. The proximal and distal radioulnar joints, which are responsible for pronation and supination of the forearm, are also trochoid joints.

Biaxial Joints

There are two types of biaxial joints. In the first, a condyloid or ellipsoid joint (Figure 2-22), one oval condyle fits into an elliptical socket. There is flexion and extension in one plane and abduction and adduction in the other; circumduction at these joints is accomplished through a combination of these motions. The wrist joint (radiocarpal joint) and the metacarpophalangeal joints of the hand are examples.

Saddle joints are the second type of biaxial joints (Figure 2-23). In a saddle joint, each joint is convex in one plane and concave in the other, and these surfaces fit together. A saddle-shaped bone fits into a socket that is concave-convex-concave in shape. There is movement in two planes as with condyloid joints, which are flexion/extension in one plane and abduction/adduction in the other. The carpometacarpal joint of the

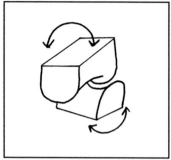

Figure 2-23. Saddle joint.

thumb is the most universally accepted saddle joint using this classification, although the sternoclavicular joint of the shoulder complex is also considered a saddle joint according to some resources (Konin, 1999).

Multiaxial Joints

The shoulder (glenohumeral joint) and the hip are considered ball and socket joints, a type of multiaxial or triaxial joint (Figure 2-24). All movements are possible, including flexion/extension, abduction/adduction, rotation, and combined movements, such as circumduction. These joints have three axes perpendicular to each other. Independent movement is possible about each of these axes. These types of joints require many muscles for control and stabilization, and muscles are placed in the proximal portions of the limbs to be closer to the center of gravity, thereby limiting the amount of movement required (Brand & Hollister, 1999).

In addition to ball and socket joints, there are joints that are capable of a gliding motion that can occur in all planes to some slight degree. These are gliding or planar joints, which are also considered multiaxial joints. While discussed here as multiaxial or triaxial, some sources refer to gliding joints as nonaxial joints, capable only of linear motion. Slight gliding motion occurs in the vertebrocostal, sternocostal, and acromioclavicular joints, as well as between the carpal bones of the wrist.

Some authors contend that joints are not flat, cylindrical, conical, or spherical (Kisner & Colby, 1990; Roberts & Falkenburg, 1992). Instead, joints can be classified as ovoid/oval or sellar joints (Figure 2-25).

Ovoid joints are joint surfaces that are egg-shaped and may be nearly planar or nearly spherical in shape. Ovoid joints have a convex and concave surface on the articulating bones. In an ovoid joint, the radius of the curvature varies from point to point (Smith et al., 1996). Some joints have a convex and concave surface on each of the articulating bones and are called sellar joints. Examples of sellar joints are the carpometacarpal joint of the thumb, the elbow, the sternoclavicular joint, and the talocrural joints.

Concave/Convex Relationships

The gliding movements of the articulating surfaces of the bones demonstrate convex-concave relationships; there are two principles relative to this convex-concave relationship that determine the movement between joints:

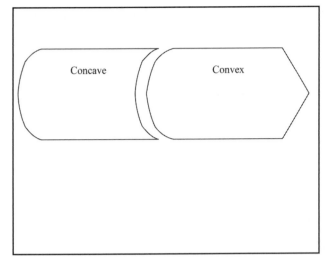

Figure 2-25. Ovoid joints.

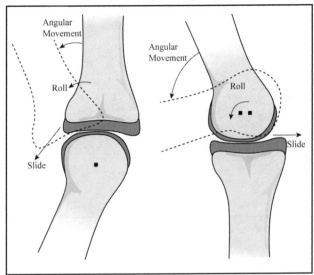

Figure 2-26. Convex/concave rules.

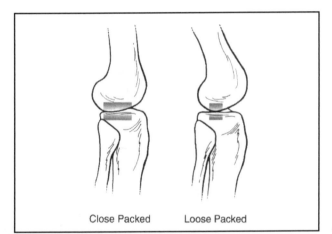

Figure 2-27. Close-packed and loose-packed joint positions.

1. When the bone with the convex surface moves on the bone with concavity, then the convex surface moves in the opposite direction to the bone segment. An example of the concave-convex principles could be when abduction of the shoulder occurs. There is a downward rotation of the humeral head on the glenoid fossa when the humerus moves up.

2. When the bone with concavity moves on the convex surface, the concave articulation moves in the same direction as the bone segment (Figure 2-26) (Kaltenborn, 1980). Passive motion of the glenohumeral joint produces rolling of the convex humeral head and downward gliding on the scapula's concave glenoid fossa so that the motions of the distal and proximal humerus are reciprocally opposite during glenohumeral movement.

For example, during shoulder abduction, when the humerus moves up, the head of the humerus slides inferiorly. Likewise, during adduction, as the arm comes to rest at the side of the body, the humeral head slides upward or superiorly (Shafer, 1997).

To better understand these two rules, one joint surface is mobile, and the other is stable. A stationary convex bone and a concave moving bone glide in the same direction as the angular motion (convex stationary + concave moving = same). With a concave stationary bone and a convex moving bone, the gliding occurs opposite to the direction of angular motion (concave stationary = convex moving = opposite). Figure 2-25 illustrates these relationships.

Joint Congruency

Joint surfaces match each other perfectly in only one position—the close-packed position (Figure 2-27). In a close-packed position, there is maximum surface contact between the bones. The attaching ligaments are at a point that is the farthest apart, and these structures are under tension. In addition, the capsular structures are taut. The joint mechanisms are compressed and are difficult to distract (separate) because of these factors. Further passive movement of the joint is not possible. All other positions of the joint are called open or loose-packed, where the joint surfaces do not fit perfectly. Loose-packed positions permit additional or accessory motions. Accessory motions are necessary for pain-free movement but cannot be performed voluntarily by the client.

Knowledge of close-packed and loose-packed positions is used in the clinic in treatments such as splinting of the hand. The close-packed position of the metacarpophalangeal (MCP) joint is near full flexion; in this position, the collateral ligaments of the joint are taut (Chase, 2002). If the patient needs to have the hand immobilized during a period of healing, the splint is constructed so that the joint is maintained in this flexed position. That way, when recovery begins, the joint will still be able to be moved through the entire range. If the joint had been allowed to remain in a loose-packed position, the collateral ligaments would have been able to shorten and become stiff, thereby limiting any potential future motion.

The loose-packed position allows joint mobilization to be performed in order to reduce joint stiffness and increase range. Joint mobilization utilizes accessory motions to increase joint play in an effort to promote motion in the joint.

KINEMATIC CHAINS

Kinematic chains are several joints that unite successive segments. The distal segments have higher degrees of freedom, which enables the body to transform stereotyped angular motion of joints into efficient curvilinear motion (Smith et al., 1996). Smith and colleagues indicate that the large number of degrees of freedom in kinematic chains is an advantage to maintaining function because one can compensate for loss of motion at one joint by substituting combined motions of other joints, but this requires more energy.

Kinematic chains can be open or closed. In open kinematic chains, the distal end is moveable and may be placed in a wide array of positions. When muscles contract, the proximal skeleton remains relatively still while the force of the muscle moves the less stabilized, distal extremity. In closed kinematic chains, the distal end of the segment is fixed, so movement is seen closer to the core of the body.

Imagine an orchestra conductor: the wave of his baton represents an open kinematic chain, while the sway of his body in time to the music represents a closed kinematic chain. Consideration of kinematic chains also applies to splint fabrication. Consideration of the passive motion between all of the joints along the kinematic chain will prevent damage to normal joints and produce forces directed toward the less mobile joints (Fess & Philips, 1987). Evaluation and intervention often need to go beyond the client's localization of pain and tenderness to structures distal along the kinematic chain (Nadler, 2004).

BALANCE AND STABILITY

The stability of one's body is dependent upon the relationship of the center of gravity with the base of support. The center of gravity is a point of the body at which the entire weight of the body is concentrated. This is the balance point at which the vertical and horizontal planes meet (Greene & Roberts, 1999). In an upright position, the center of gravity in humans is generally accepted to be at sacral level 1 or 2 (S1 or S2), although the precise center of gravity for each person depends on the proportions of that individual. Specific centers of gravity for each body segment have been determined as well.

The position of the body affects the center of gravity so that changes in movement produce changes in the center of gravity. Stability is maintained as long as the line of gravity is maintained within the base of support. Try this exercise: put an object on the floor in front of you a few feet away from a wall; put your heels up against the wall and then try to pick up the object while your feet remain against the wall. What happens? You lose your balance. Why? Not because the weight of the object changes, not because the force of gravity

has changed, and not because your mass has changed. What has changed is the base of support relative to your center of gravity. Your center of gravity was no longer within your base of support, and, being unable to make postural adjustments, balance and stability were lost. Had you not been standing against the wall, your hips would have moved posteriorly as you bent forward, thereby maintaining your center of gravity over the base of support.

Postural control enables the body to remain in equilibrium or in balance. Balance is maintained by automatic postural adjustments (equilibrium reactions) that serve to keep the center of gravity over the base of support, to provide a wide base and lower center of gravity (protective extension reactions), and to move the base to keep it under a moving center of gravity.

The larger the base of support, the greater the stability. Consider a toddler who is learning to walk and the wide base of support that the toddler uses to maintain balance. This concept is further elucidated when considering the use of proper body mechanics in transfer and mobility activities with clients. Widening your base of support so that your vertical gravity line falls within a wider base of support and maintaining your center of gravity close to the client's center of gravity provide greater stability and safety in transfer.

Crutches, canes, and walkers are excellent means of improving mobility by increasing stability; these devices provide a broadened base of support. Clients need to learn to shift their center of gravity so that it lies within the newly broadened base of support when using these devices. In addition to having a wider base of support, if the line of gravity falls at the center of the base of support, this, too, provides greater stability.

Individual client variables also affect balance and stability. Greater mass increases stability. If one considers a football player crouched at the beginning of a play, this position affords much in terms of stability. Not only is there a great deal of mass, but the crouched position also permits a lower center of gravity and a wide base of support, which helps to stabilize the player. Changes in posture will cause the center of gravity to move.

Another individual variable is level of skill. While we have automatic adjustments to postural displacement, we also can train ourselves to have better balance. Gymnasts, for example, have a high level of learned balance and are skilled in maintaining equilibrium and in the use of spatial judgments. Age is also an individual variable, and both static and dynamic balance are most efficient in young adults and middle-aged people. The elderly are especially at risk for falls due to problems with balance and mobility secondary to medical conditions, fear of falling, weakness, and environmental hazards (Thompson & Floyd, 1994).

Maintaining balance depends on the following:

- Adequate functioning central nervous system
- Musculoskeletal system
- Adequate vision
- Vestibular function
- Proprioceptive efficiency
- Tactile input, especially hands and feet

- Integration of the central nervous system with sensory stimuli
- Visuo-spatial perception
- Effective muscle tone that adapts to changes
- Muscle strength and endurance
- Joint flexibility (Galley & Forster, 1987)

The importance of visual, vestibular, tactile, and proprioceptive sensory input can be clearly seen in the maintenance of balance. The vestibular system is vital in detecting motion and the position of the head in space in relation to gravity. This system influences muscle tone, especially in the antigravity extensor muscles, and helps to maintain stable visual perception when the person or environment moves. Proprioception is important in informing the body where the head is in relation to body. The muscle spindles, Golgi tendon organs, and proprioceptors are instrumental in informing the body about the current position of joints, whether the joint is static or moving, the range and duration of movement, the velocity and acceleration of body segments, the pressure and tension in joint structures, and the relative lengths of muscles. Tactile input relays information regarding the focus of pressure on the soles of the feet, which is indicative of the location of the line of gravity within the base of support. Vision tells us the relation of the head to the object. The role of vision cannot be overemphasized because vision is the most far-reaching sensory system and strongly dominates our perception of the environment and our adaptation to it.

Other factors that influence balance are hearing, psychological factors (fear of falling, anxiety, inattention, and mental confusion), environmental factors (illumination, surface characteristics, clothing and footwear, and noise), or movement (Galley & Forster, 1987; Thompson & Floyd, 1994).

When posture is disturbed, rescue maneuvers may be done by the person to prevent a fall. The person may sway to correct the imbalance, stagger, or perform sweeping motions to prevent the fall. Vertigo may be associated with damaged vestibular mechanisms, so it is important to determine when loss of balance occurs, in which positions, and under what circumstances. A person may also experience dizziness or giddiness and other symptoms such as vomiting, nausea, pallor, sweating, prostration, or hypotension (Galley & Forster, 1987). Postural hypotension occurs when a person moves from lying to sitting or standing and experiences a drop in systemic systolic blood pressure of 20 mm Hg from what was measured when lying down (Galley & Forster, 1987).

Clinical practice involves the application of kinesiology principles in countless ways. A clinical example of the leverage principle can be seen in the contraindication of performing passive stretching after fractures, surgery, or joint pathology when force is applied to the distal end of the bone. The force on the lever arm has greater mechanical advantage over the injured part, and the force can be amplified 10 to 20 times (Smith et al., 1996). Re-injury could occur.

Mobilization or passive movement of joint surfaces following normal accessory motion is indicated in pathological conditions to relieve pain and restore movement, but one must consider the following:

- Direction and force must follow normal arthrokinematics of the joint.
- Magnitude of the force must be controlled and gentle.
- Motions of the joint surfaces are small; use very short lever arms with force applied at the base (head) of the bone forming the joints.

If motion is not maintained, the effects of prolonged immobilization (due to splinting, casts, or other devices) and inactivity due to these limiting factors, injury, or paralysis lead to changes in tissues and are often the focus of remedial intervention. Adaptive shortening and adaptive lengthening can occur because of the normal cell replacement cycle in the absence of demand on tissues, as in paralysis (Cummings, Crutchfield, & Barnes, 1983). Stiffness can also result from the scarring process that occurs during normal wound-healing. The challenge to the therapist to achieve and preserve motion as a means to enable full participation in occupation requires sound clinical reasoning based on knowledge of the underlying science.

SUMMARY

- Movement is complex and involves all aspects of the body.
- Different parts of science study different aspects of movement. Anatomy looks at the production of movement by the muscles of the body. Biomechanics looks at forces relative to the body.
- Dynamics is concerned with movement and is subdivided into kinematics and kinetics. Kinematics describes motion and looks at the result of movement, whereas kinetics studies the forces and causes of motion.
- Newton's Laws of Physics (inertia, acceleration, and action/reaction) apply to the body as well as to other physical objects.
- Movements of the joints are based on the direction of movement in planes around an axis. There are three cardinal planes and three axes used as references when describing motion.
- Movement is angular, linear, or general motion, which is a combination of angular and linear.
- Joints are classified in terms of the motion that is produced, the structure of the joint, or the shape of the bones. Joints are often described as being hinge, pivot, condyloid, saddle, ball and socket, or gliding.
- Accessory motion also occurs at joint surfaces, resulting in rolling, sliding, or spinning motions of bones, which provide additional motion at joints.
- Gravity is a constant force acting on the body. Other external forces include friction, wind, water, other people, or objects. These are forces outside of the body.
- Internal forces are those from within the body that produce movement or counteract other forces. Internal forces are due to muscles, ligaments, and bones.

- Forces acting on the body produce deformation of structures, which is measured by strain.

- Forces can be combined to determine the overall effect of all forces on an object. Composition of forces can be linear or can be represented by polygons and parallelograms.

- Forces can also be resolved into component parts. This is helpful in determining the line of pull of a muscle. By looking at how the muscle pulls on the bone during a muscle action, conclusions about the effectiveness of the muscle in terms of stabilization or mobilization can be made.

- In linear force systems, forces act on the body as first-, second-, or third-class levers. While third-class levers are the most common in the body, they are not the type of lever that provides the greatest mechanical advantage. While this seems contradictory, the predominance of third-class levers allows for greater speed and mobility than would levers with greater mechanical advantage.

REFERENCES

Brand, P. W., & Hollister, A. M. (1999). *Clinical mechanics of the hand* (3rd ed.). St. Louis, MO: Mosby.

Chase, R. W. (2002). Anatomy and kinesiology of the hand. In T. M. Skirven, A. L. Osterman, J. Fedorczyk, et al (Eds.), *Hunter, Mackin, & Callahan's rehabilitation of the hand and upper extremity* (vol. 1, 5th ed.). St. Louis, MO: Mosby.

Cummings, G., Crutchfield, C., & Barnes, M. (1983). *Soft tissue changes in contractures.* Orthopedic Physical Therapy Series (vol. 1). Atlanta, GA: Stokesville Publishing.

Durward, B. R., Baer, G. D., & Rowe, P. J. (1999). *Functional human movement: Measurement and analysis.* Oxford: Butterworth Heinemann.

Fess, E. E., & Philips, C. A. (1987). *Hand splinting: Principles and methods.* St. Louis: C.V. Mosby Company.

Flowers, K. (1998). Shoulder anatomy and biomechanics. Paper presented at the Ohio Occupational Therapy Association. Columbus, OH.

Galley, P. M., & Forster, A. L. (1987). *Human movement: An introductory text for physiotherapy students* (2nd ed.). New York, NY: Churchill-Livingstone.

Gench, B. E., Hinson, M. M., & Harvey, P. T. (1995). *Anatomical kinesiology.* Dubuque, IA: Eddie Bowers Publishing, Inc.

Greene, D. P., & Roberts, S. L. (1999). *Kinesiology: Movement in the context of activity.* St. Louis, MO: Mosby.

Hamill, J., & Knutzen, K. M. (1995). *Biomechanical basis of human movement.* Baltimore, MD: Williams & Wilkins.

Hamill, J., & Knutzen, K. M. (2003). *Biomechanical basis of human movement.* Philadelphia, PA: Lippincott Williams & Wilkins.

Kaltenborn, F. M. (1980). *Manual therapy for the extremity joints.* Oslo: Bokhandel.

Kisner, C., & Colby, L. A. (1990). *Therapeutic exercise: Foundations and techniques* (2nd ed.). Philadelphia, PA: F. A. Davis.

Konin, J. G. (1999). *Practical kinesiology for the physical therapist assistant.* Thorofare, NJ: SLACK Incorporated.

Lafferty, P. (1992). *Force and motion.* New York, NY: Darling Kindersley Inc.

LeVeau, B. F. (1992). *Williams and Lissner's biomechanics of human motion* (3rd ed.). Philadelphia, PA: W. B. Saunders Co.

Lippert, L. S. (2006). *Clinical kinesiology and anatomy.* Philadelphia, PA: F. A. Davis Company.

Loth, T., & Wadsworth, C. T. (1998). *Orthopedic review for physical therapists.* St. Louis, MO: C. V. Mosby Co.

MacConaill, M. A. (1957). Movement of bones and joints: The significance of shape. *Journal of Bone and Joint Surgery, 35B,* 290-297.

Magee, D. J. (2008). *Orthopedic physical assessment.* Philadelphia, PA: W. B. Saunders Co.

McGinnis, P. M. (1999). *Biomechanics of sport and exercise.* Champaign, IL: Human Kinetics.

Muscolino, J. E. (2006). *Kinesiology: The skeletal system and muscle function.* St. Louis, MO: Mosby.

Nadler, S. F. (2004). Injury in a throwing athlete: Understanding the kinematic chain. *American Journal of Physical Medicine and Rehabilitation, 83*(1), 79.

Nicholas, J. A., & Hershman, E. B. (1990). *The upper extremity in sports medicine.* St. Louis, MO: C. V. Mosby Co.

Norkin, C. C., & Levangie, P. K. (1992). *Joint structure and function: A comprehensive analysis* (2nd ed.). Philadelphia, PA: F. A. Davis Co.

Pedretti, L. W. (2001). *Occupational therapy: Practice skills for physical dysfunction.* St. Louis, MO: Mosby.

Radomski, M. V., & Latham, C. A. T. (Eds.). (2008). *Occupational therapy for physical dysfunction.* Philadelphia, PA: Wolters Kluwer/Lippincott, Williams & Wilkins.

Roberts, S., & Falkenburg, S. A. (1992). *Biomechanics: Problem solving for functional activity.* St. Louis, MO: C. V. Mosby Co.

Shafer, R. C. (1997). Shoulder girdle trauma: Monograph 16. Retrieved from http://www.chiro.org/ACAPress.

Smith, L. K, Weiss, E. L., & Lehmkuhl, L. D. (1996). *Brunnstrom's clinical kinesiology* (5th ed.). Philadelphia, PA: F. A. Davis Co.

Thomas, C. L. (Ed.). (1997). *Taber's cyclopedic medical dictionary* (18th ed.). Philadelphia, PA: F. A. Davis.

Thompson, C. W., & Floyd, R. T. (1994). *Manual of structural kinesiology* (12th ed.). St. Louis, MO: C. V. Mosby Co.

Trombly, C. A. (Ed.). (1995). *Occupational therapy for physical dysfunction* (4th ed.). Baltimore, MD: Williams & Wilkins.

Watkins, J. (1999). *Structure and function of the musculoskeletal system.* New York, NY: Human Kinetics.

ADDITIONAL RESOURCES

Basmajian, J. V., & DeLuca, C. J. (1985). *Muscles alive* (5th ed.). Baltimore, MD: Williams & Wilkins.

Burstein, A. H., & Wright, T. M. (1994). *Fundamentals of orthopaedic biomechanics.* Baltimore, MD: Williams & Wilkins.

Esch, D. L. (1989). *Analysis of human motion.* Minneapolis, MN: University of Minnesota Press.

Esch, D. L. (1989) *Musculoskeletal function: An anatomy and kinesiology laboratory manual.* Minneapolis, MN: University of Minnesota.

Frankel, V. H., & Burstein, A. (1970). *Orthopaedic biomechanics: The application of engineering to musculoskeletal system.* Philadelphia, PA: Lea and Febiger.

Jenkins, D. B. (1998). *Hollingshead's functional anatomy of the limbs and back* (7th ed.). Philadelphia, PA: W. B. Saunders Co.

Kendall, F. P., & McCreary, E. K. (1983). *Muscles: Testing and function* (3rd ed.). Baltimore, MD: Williams & Wilkins.

Lehrman, R. L. (1998). *Physics the easy way* (3rd ed.). Hauppauge, NY: Barron's Educational Series, Inc.

MacKenna, B. R., & Callender, R. (1990). *Illustrated physiology* (5th ed.). New York, NY: Churchill Livingstone.

Magee, D. J. (1992). *Orthopedic physical assessment.* Philadelphia, PA: W. B. Saunders Co.

Marieb, E. N. (1998). *Human anatomy and physiology.* Glenview, IL: Addison Wesley Longman Publishers.

Nordin, M., & Frankel, V. H. (1989). *Basic biomechanics of the musculoskeletal system* (2nd ed.). Philadelphia, PA: Lea and Febiger.

Perr, A. (1998). Elements of seating and wheeled mobility intervention. *OT Practice,* 16-24.

Rasch, P. J., & Burke, R. K. (1978). *Kinesiology and applied anatomy: The science of human movement.* Philadelphia, PA: Lea and Febiger.

Shankar, K. (1999). *Exercise prescription*. Philadelphia, PA: Hanley & Belfus, Inc.

Snell, M. A. (1999). Guidelines for safely transporting wheelchair users. *OT Practice*, 35-38.

Spaulding, S. J. Biomechanics of prehension. *Am J Occup Ther, 43*(5), 302-306.

Zelenka, J. P., Floren, A. E., & Jordan, J. J. (1966). Minimal forces to move patients. *Am J Occup Ther, 50*(5), 354-361.

3

RANGE OF MOTION

Range of motion (ROM) is the amount of movement that occurs at a joint and can be defined as the measurement of motion available (or the arc of motion available) at a joint or through which the joint passes, resulting from the joint structure and surrounding soft tissue (Norkin & White, 1985; Pedretti, 1996; Trombly, 1995). Joint function is influenced by the structure of the joint, externally applied forces, and internal forces. Joint motion depends on many variables, which may include the restraining effects of ligaments and muscles crossing the joint, skin and other soft tissues, the bulk of tissue in adjacent segments, age, and gender. In addition, measurement of range of motion involves methodological factors such as accurate recording, instrumentation, and the type of testing done (Figure 3-1).

FACTORS INFLUENCING RANGE OF MOTION

Client Factors

Client factors are person-level factors such as genetics, gender, age, and lifestyle choices affecting health as it relates to joint movement.

Genetics

Individual subject factors can vary due to genetic predispositions for greater motion (hypermobility) as is sometimes seen in hyperextension at the elbow. There may be less motion (hypomobility), which may happen when there is soft tissue tightness or contractures that limit full motion.

Different activities put different stresses on joints, which may change the amount of motion that occurs. Gymnasts and cheerleaders, for example, may have greater wrist extension due to repeated handstands or placing body weight on extended wrists. Pianists may have more finger abduction and extension due to years of playing and reaching for keys at the far ends of the keyboard. Musicians of stringed instruments often need full and prolonged finger abduction and flexion to move into different positions along the neck of the instrument and to apply pressure on the strings. Individual variability needs to be considered in assessment when the range of motion is different from the expected or normative values. Questions asked about activities that are done currently and in the past may help explain variances in flexibility and would be valuable information gathered from the occupational profile.

Health Status

Decreased motion occurs for many reasons. The overall health of the client is important in determining the amount of motion that occurs. Joint disease or injury, edema, pain, skin tightness or scarring, muscle or tendon shortening due to immobilization, muscle weakness or muscle hypertrophy, muscle tone abnormalities, and excess adipose tissue are all possible reasons for decreased range of motion.

The consequences of decreased motion are many. Supporting structures may become loose, creating joints that are unstable and painful. The joint structures may be insufficient to hold the joint in stable and functional positions during activities. Ligaments and muscles may be stretched and, combined with the effects of gravity, can lead to subluxation and instability

Rybski MF.
Kinesiology for Occupational Therapy, Second Edition (pp 37-64)
© 2012 SLACK Incorporated

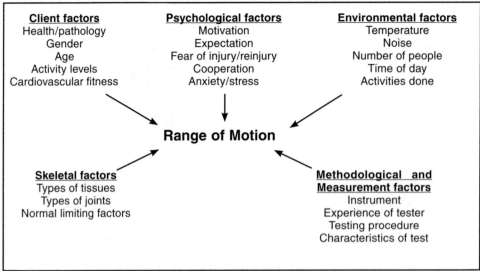

Figure 3-1. Factors influencing range of motion.

of the joint. This is often seen in the shoulders of clients with hemiplegia where the shoulder and scapular muscles are inactive or diminished and the weight of the arm plus gravity pull the humeral head out of the glenoid fossa.

Inactivity of the muscles can lead to contractures due to muscle imbalance and the inability to perform normal activities. Muscles and tendons can lose their tensile strength. Scar tissue adhesions, which occur secondarily to chronic inflammation with fibrotic changes, can tear when the muscle is stretched. As muscles lose their normal flexibility, changes occur in the length-tension relationships. The muscles are no longer capable of producing peak tension, which can lead to pain and decreased strength.

Fatigue due to lack of sleep, pain, and emotional distress may prevent a client from participating fully in the therapeutic process. Fatigue may also influence the client's ability to communicate effectively and may interfere with concentration, memory, or attention to tasks. In addition, fatigue affects the physiologic properties of muscles and joints, limiting sustained, purposeful movement during activities.

Age and Gender

Age and gender are subject factors that also influence the amount of motion that occurs at joints. Normal age-related changes in bone, cartilage, tendons, and joints that affect ROM include the following:

- Loss of tensile strength and mobility in collagen tissue occurs due to dehydration, increased density, and cross-linkage of fibers
- Decreased ease of movement of tissues is due to decreased elasticity
- Decreased ease of movement may be due to progressive diminution of hyaluronic acid, which helps to regulate the viscosity of tissues (Bonder & Bello-Haas, 2009).

The concept that normal ROM may vary with age and gender with resultant decreased range of motion has generally been supported in the literature and in clinical practice. However, in a study of lower extremity range of motion, Roach and Miles found that normative values of hip and knee range

of motion may not be representative of the population. In their study, there were differences between the oldest and youngest subjects, but the differences were small and of limited clinical importance. They concluded that "any substantial loss of joint mobility should be viewed as abnormal and not attributable to ageing" (1991, p. 644).

Normal age-related changes in joints can be offset somewhat by the types and level of activities in which one engages throughout life. Understanding all of the occupations in which the client engages throughout the lifespan will provide a more accurate picture of the health of the person. This is as true for the 35-year-old construction worker as it is for the retired 75-year-old grandfather who enjoys golf. Understanding desired occupations gives the therapist insight into the activity level and motivation as well as information about joint function.

Age-related changes occur symmetrically unless accompanied by other impairments, with the greatest changes occurring in the spine. Women generally have more flexible joints with greater range of motion than men, and this is true throughout life.

Cultural Variations

Few studies have been done that indicate trends in regards to culture, race, or ethnicity (Van Deusen & Brunt, 1997), although positioning in childhood and customary postures may be attributed to variability of lower extremity joint motion and related to culture (Demeter, Andersson, & Smith, 1996).

Biocultural variations in the musculoskeletal system have been identified by Jarvis (2000), which might affect the amount of joint movement available to a client. From this study, the following variations were found:

- There is greater torsion of the proximal end of the right humerus in Whites, but it is more symmetrical in Blacks.
- Long bones are significantly longer in Blacks than Whites.
- Bone density varies by race with Blacks having the densest bones, then Chinese, Japanese, Inuit, and American Whites.

- Curvature of long bones varies with Native Americans having more anteriorly convex femurs while Blacks generally have straight femurs.
- Length of the ulna and radius varies. The ulna and radius are of equal length in Swedes 61% of the time while only 16% of the time in Chinese. The ulna is longer than the radius in Swedes 16% of the time, and it is longer in 48% of the Chinese. The radius is longer in Swedes 23% of the time, but only 10% of the time in Chinese.
- Palmaris longus is absent in 12% to 20% of Whites, 2% to 12% of Native Americans, 5% of Blacks, and in 3% of Asians (Jarvis, 2000, p. 634).

Psychological and Psychosocial Factors

Psychological and psychosocial factors also influence range of motion. Client participation is a vital component to active movement and is related to the person's emotional state at the time the movement is requested. Understanding the client's perspective about the meaning of the limitations and potential losses of occupation is an important part of learning about his or her illness experience. Fear of injury or re-injury will prevent a client from engaging in activities that are perceived to be potentially painful. By not moving, the part is immobilized, and soft tissue changes will result. Anxiety and stress can cause muscle guarding, tightness in tissues, or inactivity.

In providing instructions for movement or engagement in activities, cognition is a variable as well. If the client cannot understand what movement or activity you are requesting, whether due to anxiety, depression, or inability to cognitively attend to the task, efforts at evaluation will be invalid, and intervention will be unsuccessful.

Environmental Factors

Temperature, level of noise, and the number of people in the room affect not only the comfort of the client, but the client's ability to attend to the task and the physiological readiness of muscles to respond. The time of day has a bearing on the amount of motion possible, as can be seen in clients with rheumatoid arthritis who are stiff in the mornings and able to move more comfortably later in the day. Less motion may be possible later in the day for some clients (for example, multiple sclerosis patients) due to fatigue.

Skeletal Factors

Range of motion is facilitated by bony shape and characteristics of the tissues around the joint. The shape of the bones allows certain types of movements but not others. The tissues comprising the joint provide stability and flexibility for joint movement. Muscolino states that "joints allow movement, muscles create movement, and ligaments and joint capsules limit motion" (2006, p. 161).

Types of Tissue

The amount of movement that occurs in a joint is based on the types of tissue that make up the joint. There are four major types of tissue in the body: epithelial, connective, muscle, and nervous (Table 3-1).

Epithelial Tissue

Epithelium lines both the outside of an organ or organism (i.e., skin) and the inside cavities and lumen of bodies (i.e., blood vessels or the lining of the stomach). Epithelial tissue is characterized by a continuous surface of cells with few interruptions or gaps between adjacent cells. Several of the body's organs are primarily epithelial tissue, and examples include the lungs, kidneys, and liver. Many glands are also formed by epithelial tissue.

There are several types of epithelial tissue, and they have various functions. Epithelial cells protect underlying tissue from mechanical injury, harmful chemicals, pathogens, and excessive water loss. Many of the sensory receptors of the eyes, skin, ears, nose, and tongue are made of specialized epithelial tissues and are important in the reception of sensory information. Enzymes, hormones, and lubricating fluids are specialized epithelial tissue as are endocrine and exocrine glands with secretory functions. Epithelial cells lining the small intestine absorb nutrients, and epithelial tissue in the kidney excretes waste products. Because epithelial tissue lines the walls of capillaries and the lungs, it helps to promote diffusion of gases, liquids, and nutrients (Currey, 2005; King, 2010).

Connective Tissue

Connective tissue is the "passive" element of the musculoskeletal system and includes fibrous tissues, cartilage, and bone. Some other specialized categories of connective tissue include lymphoid tissue and blood. Connective tissues provide the pathways for nutrients and waste disposal. In addition, connective tissues provide semi-permeable barriers to pathogens and serve a protective and immunological function by providing a refuge for phagocytes, mast cells, and fibroblasts. Connective tissue also provides information to the central nervous system about internal and external forces acting on the body (Smith, 2005).

Fibers of connective tissue are made of varying degrees of collagenous or elastic types of tissues. Collagen, the most abundant protein in the body, offers tensile stiffness and strength, while elastic fibers permit flexibility. There are three types of collagen tissues. The different types of collagen can be differentiated by the protein composition of the fiber. Type I collagen is found primarily in ligaments, tendons, fascia, and fibrous capsules and constitutes about 90% of the total collagen in the body. These thick fibers do not elongate much when stretched. Type II collagen is found in cartilage and intervertebral disks serving to resist pressure on joints. Collagen in arteries, the liver, and the spleen is Type III collagen (Cooper, 2007).

There are three types of connective tissue: bones, fibrous tissues, and cartilage.

Bones

Osseous connective tissue is the hardest of the connective tissue types. Bones are highly vascular and can self-repair. Bones can alter their properties and configuration based on stresses applied to the body. Adaptive remodeling occurs as a result, and calcium is laid down in response to stress (Muscolino, 2006; Snyder, Conner, & Lorenz, 2007). Bones

Table 3-1

TYPES AND CHARACTERISTICS OF TISSUES

TISSUE TYPE	TYPES	SUBCATEGORIES	FUNCTION/DESCRIPTION
Epithelial			• Forms the epidermis of the skin • Surface layer of mucous and serous membranes • Serves function of protection, absorption, secretion • Specialized functions of movement of substances through ducts, production of germ cells, and reception of stimuli
Connective	Bone		• Provides axial and appendicular skeleton • Protection for vital structures
	Fibrous tissue	Dense fibrous Unorganized/irregular • Fascia • Dermis • Periosteum • Capsules of the organs Organized/regular • Aponeurosis • Tendons • Ligaments Loose fibrous (areolar) • Adipose tissue • Submucosa • Fascia (superficial and deep) • Liver • Bone marrow	• Fibers are either collagen or elastic types • Bind or strengthen organs and muscles • Compartmentalize tissues • Transport nutrients • Help with immunological defense connective tissue function within muscle • Provides gross structure to muscle • Serves as conduit for blood vessels and nerves • Generates passive tension • Recoils after stretch • Conveys contractile force to tendon
	Cartilage	• Hyaline • White fibrocartilage • Yellow or elastic fibrocartilage	• Distribute joint loads over a wide area • Decrease stress of joint surfaces • Allow movement of opposing joint surfaces with minimal friction and wear

(continued)

will thicken in response to stress. However, in response to excessive demands, bone spurs or arthritic (degenerative joint disease or osteoarthritis) changes may occur (Muscolino, 2006). The purpose of the skeletal system is to protect internal organs, to provide a rigid lever system, and to present muscle attachment sites. Bones achieve this by providing a firm struc-

tural support to facilitate muscle action and body movement.

Bone shape determines how stable the joint will be and the types of movement that occur at a joint. If the bony surfaces are congruent and provide a tight fit, then the joint will have articular stability. A saddle-shaped bone that fits into a socket and is concave-convex-concave in shape (biaxial saddle joint)

Table 3-1 (continued)

TYPES AND CHARACTERISTICS OF TISSUES

TISSUE TYPE	TYPES	SUBCATEGORIES	FUNCTION/DESCRIPTION
Muscle	Voluntary/ striated/ skeletal		• Skeletal muscles pull on bones and act as lever systems
	Involuntary/ unstriated/ smooth		• Smooth muscle • Regulates the flow of blood in the arteries • Helps with peristalsis • Expels urine from urinary bladder • Aids in childbirth • Regulates the flow of air through the lungs
	Cardiac muscle		• Cardiac muscles pump blood throughout the heart
Nerves	Efferent Afferent	• Central nervous system • Peripheral nervous system	• Electrical and chemical impulses transmit sensory information to brain, muscles, spinal cord

will permit distinctly different movements than a convex bone that fits into the concave part of another bone, only allowing motion in only one plane (as in a uniaxial hinge joint). The convex/concave surfaces are also associated with accessory motions that are possible only with passive movement (Trew & Everett, 2005).

Fibrous Connective Tissue

There are two types fibrous connective tissue: loose fibrous connective tissue and dense connective tissue. Dense fibrous connective tissue is further subdivided into dense organized/regular tissues and dense unorganized/irregular fibrous tissues.

Loose connective tissue has the same types of fibers that dense connective tissue has (collagenous, elastic, and reticular tissue types) but also has a relatively large proportion of ground substance so these tissues lack the reinforcing structure found in dense connective tissue. Loose connective tissue, because of the larger proportion of gel-like ground substance, is easily distorted. However, when sufficiently distorted, loose connective tissue will resist further deformation due to the strength of the collagen fibers (King, 2010).

Types of loose connective tissue include areolar, adipose, and reticular tissues. Areolar loose connective tissue is charac-

terized by a loose network of intercellular material, abundant blood vessels, and significant empty space with predominantly collagenous fibers. Areolar tissues surround blood vessels and nerves, and epithelium rests on a layer of areolar tissue. It is also a component of mucus membranes found in the digestive, respiratory, reproductive, and urinary systems (Goss, 1973). Areolar tissue holds organs in place by filling in the spaces between organs, cushions and protects organs and blood vessels, attaches epithelial tissue to other tissues, acts as a reservoir for water and nutrients, and provides space for waste release from other tissues. Adipose tissue is dominated by fat cells (adipocytes), which serve to pad and insulate the body and to regulate body temperature. Reticular loose connective tissue forms the structure of organs (liver, pancreas) and lymph nodes (King, 2010). Made up of primarily type II collagen fibers that provide scaffolding for other cells (such as bone marrow and superficial and deep fascia), loose connective tissue is the focus of myofascial intervention techniques (Donatelli & Wooden, 2010).

Dense connective tissue is made up of fibers that are thicker and in closer contact than loose connective tissue with either parallel or latticed arrangements (Smith, 2005). Dense connective tissue is composed primarily of type I collagen fibers, creating strong connections to bone and muscles. These tissues are supplied sparingly with blood vessels.

Dense connective tissue is considered organized/regular if all of the fibers are aligned in a single direction (such as tendons, ligaments, and aponeurosis), giving tensile strength in that direction.

Tendons attach skeletal muscles to bones for the purpose of transmitting the pulling force of a muscle to its bony attachment, thereby creating movement (Muscolino, 2006). Considered inelastic and comprised of 80% collagen fibers, which are arranged parallel to direction of force application of muscle, tendons can withstand high tensile forces and exhibit viscoelastic behavior in response to loading (Cooper, 2007; Hamill & Knutzen, 2003). This makes tendons strong enough to sustain tensile forces from muscle contractions but also flexible enough to change the direction of muscle pull (Nordin & Frankel, 2001). In addition to attaching muscle to bone, transmitting tensile loads, producing movement, or maintaining posture, tendons also position the muscle belly to be at an optimal distance from the joint. Further, the myotendinous junction, where the collagen fibers of the tendon and myofibrils of the muscle join, tends to act as a dynamic restraint to muscle actions (Nordin & Frankel, 2001). Tendons are a factor in joint stability because tendons become taut when muscles are contracted. Aponeuroses are considered within this tendon description, because they are tendons that are flat rather than rope-like (Muscolino, 2006).

Ligaments connect bones to bones or two or more cartilages or structures together, thus strengthening and stabilizing joints during rest and movement. Ligaments also support various organs, including the uterus, bladder, liver, and diaphragm, and help maintain the shape of the breasts (King, 2010). The deeper surfaces of ligaments may form part of the synovial sheath of the joint, and superficial surfaces may blend with surrounding connective tissue. Ligaments may be found as thickenings in the wall of the joint capsule (such as in the glenohumeral ligaments), lie outside of the joint (as in collateral ligaments), or be inside the joint (as in the cruciate ligaments) (Goss, 1973; Hamill & Knutzen, 2003; Nordin & Frankel, 2001).

Ligaments and joint capsules enhance the mechanical stability of joint, help to guide joint motion, serve to prevent excessive motion, and act as static restraints to joint movement (Nordin & Frankel, 2001). Because a ligament can only stretch 6% of its length before rupture, joints where ligaments are the major means of supporting a joint are not very stable (Konin, 1999).

Ligaments are more flexible and contain more elastic fibers than tendons. Unlike tendons, ligaments gradually lengthen when under tension. Athletes, gymnasts, dancers, and martial artists, who engage in activities that apply frequent or constant stretch forces to ligaments, gradually increase the range of motion available at their joints. The danger in this is when joints become too lax (hyperlaxity), because this can lead to joint weakness and possible future dislocations or subluxations. In normal joints, excessive movements such as hyperextension or hyperflexion may be restricted by ligaments, thereby limiting movement to certain directions only.

Ligaments respond to loads by becoming stiffer and stronger over time so there is diminished strength of ligaments when a joint is immobilized. Because ligaments are so important in the stabilization of joints, they are also highly susceptible to injury. Ligaments must withstand a great deal of stress in daily activities, but because they have a relatively low blood supply, injuries can take up to 2 years to heal (Konin, 1999). Tensile injury to a ligament is a sprain that is rated "1" (partial tear), "2" (tear with some loss of stability), or "3" (complete tear with loss of stability) in severity. Because ligaments stabilize, control, and limit joint motion, any injury to a ligament will influence joint movement (Hamill & Knutzen, 2003). Injured ligaments tend to be less flexible and more prone to repeated injury or rupture under stressful conditions.

Dense irregular connective tissue includes joint capsules, periosteum, the dermis of skin, white layer of the eyeball, labrum of the shoulder, menisci in the knee, and fascial sheaths in which the fibers are arranged randomly in all directions. This fiber arrangement enables these structures to resist tensile stresses in many directions, while tendons can only withstand unidirectional tension forces (Hamill & Knutzen, 2003). There is a higher proportion of ground substance than in the dense regular/organized connective tissues, and there is increased vascularity (Donatelli & Wooden, 2010; Goss, 1973). Because of more ground substance, dense irregular connective tissue reduces friction, acts as a shock absorber, and provides greater joint nutrition. These structures also increase joint congruity, which can enhance joint stability.

A specialized type of dense fibrous irregular connective tissue is periosteum, which is a double-layered membrane covering the outer surfaces of bone. Periosteum includes cells important in repairing and forming bone tissue and houses blood vessels, lymph vessels, and nerve fibers. It is a site of attachment for ligaments and tendons, and because there are nerve fibers, when a tendon is severely strained or torn, the periosteum can become inflamed and painful.

Cartilage

Cartilage is a nonvascular structure without a nutrient supply. This means that if there is injury to this structure, it does not heal well. It is comprised of primarily type II collagen fibers, which are thinner and more flexible than the collagen fibers found in other types of connective tissue. Cartilage is found between bones of cartilaginous joints, at the caps of joints, and in the intra-articular disks.

Cartilage is strong, flexible, and acts as a shock absorber for the joint. Cartilage is able to resist shear forces, and it deforms quickly to low or moderate loads and responds stiffly to rapid loads (Hamill & Knutzen, 2003). Cartilage also helps to distribute joint loads over a wide area to decrease stress of joint surfaces and to allow movement of opposing joint surfaces with minimal friction and wear (Muscolino, 2006; Nordin & Frankel, 2001; Nyland, 2006). While stiff and strong under tensile forces, cartilage buckles under compression and is subject to wear during one's lifetime.

There are three types of cartilage: elastic, fibrocartilage, and hyaline. Elastic cartilage is found in the epiglottis, laryngeal cartilage, and the walls of the Eustachian tubes, external ear, and auditory canal. Elastic cartilage contains collagen and elastin fibers so that elastic cartilage has the firmness of cartilage plus great flexibility. Consider how flexible your ear is when passively moved; this is another example of elastic cartilage.

Fibrocartilage has tensile strength and the ability to withstand high pressure due to a great density of fibrous collagen fibers. It is the toughest form of cartilaginous tissue. It is found in the intervertebral disks, some articular surfaces, pubic symphyses, joint capsules, and when articular cartilage melds with a tendon or ligament. Fibrocartilage, also called menisci or disks, contains some elastic components to allow for accommodation of pressure, friction, and shear forces (Kisner & Colby, 2007). Because of the location of fibrocartilage in the joint, it helps to improve the fit between articulating bones. Interarticular fibrocartilage (seen in the sternoclavicular, acromioclavicular, wrist, and knee joints) serves to minimize the distance between joint surfaces, increase the depths of the articular surfaces, ease gliding movements, moderate the effects of great pressure, and lessen the intensity of the shocks to which the parts may be subjected (Muscolino, 2006; Nordin & Frankel, 2001; Nyland, 2006). Circumferential fibrocartilage surrounding the margins of the labrum of the hip and of the shoulder joints deepens the articular cavities and protects their edges.

Hyaline cartilage is the most common type of cartilaginous tissue. It is found on the articular surfaces of the peripheral joints, the sternal ends of the ribs, the nasal septum, larynx, bronchi, and tracheal rings. Hyaline cartilage provides slightly flexible support and reduces friction within joints by providing a smooth surface over which bones can glide.

Muscles

There are three kinds of muscles in the body. The first type, voluntary/striated/skeletal muscles, appears striped, attaches primarily to the skeleton, and have contractions that are usually rapid and intermittent. The second type, involuntary/unstriated/smooth muscles, is found in areas where movement occurs without conscious thought, such as in the stomach, intestines, and blood vessels. Smooth muscles contract slowly and rhythmically, as is seen in peristalsis. Cardiac muscle is the third type of muscle and is a specialized heart muscle that has characteristics of both smooth and skeletal muscles. Muscles have good supply, enabling nutrition and waste removal. Nerves within the fiber and fascia carry both motor and sensory input to the nervous system. The type of muscle discussed in this text will be voluntary muscles as these are the muscles that produce human movement necessary for everyday tasks.

Muscles contain skeletal muscle tissue and a fibrous fascial connective tissue, collectively called the myofascial unit. The muscle fibers are the force generators for movement. The fascia transfers forces from the muscle contraction to the bone and provides a structural framework for the muscle. Fascia encloses the entire muscle (epimysium), groups of muscle cells within the muscle (perimysium), and each individual muscle cell (endomysium). The fascia then continues at both ends to create tendons that attach the muscle to bone (Muscolino, 2006). Examples of muscle attaching to bone via tendons would be the long head of the triceps, hamstring muscles, or the flexor carpi radialis muscle. Muscle-tendon attachments are the most common means of connecting muscle with bone. Muscles can

also attach directly to the bone (trapezius muscle and coracobrachialis muscle) or attach via an aponeurosis (a broad and flat tendon) as can be seen in the abdominal muscles, latissimus dorsi, or palmaris longus muscle.

Characteristics of muscle and how force develops in muscle fibers and connective tissue to produce movement are discussed in detail in Chapter 4.

Nerves

The nervous system is comprised of neurons, or specialized cells consisting of cylindrical bundles of tissue that originate from the brain and spinal cord, and branch repeatedly to innervate and coordinate the actions of every part of the body. Sensory neurons transmit signals to inform the central nervous system about the status of the body and external environment, and motor neurons connect the nervous system to muscles or other organs in the body. Neurons are distinctive because they communicate with other cells via synapses, which are electrical or chemical signals through membrane-to-membrane junctions.

Types of Joints

The shapes of the bones and the way in which the structures in a joint join together determine the movement at that joint. The bones of the humeroradial and humeroulnar (elbow) joints only permit the active movements of flexion and extension due to the shape of the bony surfaces. The radioulnar joint, which has movement around the bony axis, allows pronation and supination. Biaxial and triaxial joints permit active movement in two and three different planes and axes. Passive accessory movement (such as roll, spin, and glide) is dependent upon the shapes of the bones and the direction of movement. When observing or assessing the movement of the joint, consideration of the type of joint determines the movement you expect to see.

Methodological Factors

Methodological variables affect the precision of measurement, the accuracy and stability of results, and the worth of the test to functional activities. These variables include how the testing is done, the method of testing, the instruments being used, and the knowledge and experience of the tester. Methodology is concerned with the particular assessment chosen, the domain of concern, and the person conducting the assessment.

The purpose of an assessment in general is as follows:

- To support effective clinical reasoning.

- To define the nature and scope of clinical problems.

- To provide baselines against which to monitor progress.

- To summarize changes that occur as a result of therapy and to define areas of treatment requiring revision based on changes in the client.

- To allow peers and managers to critically evaluate the effectiveness of interventions and to develop directions for quality improvements.

- To aid in the decision-making process regarding allocation of health-care resources.

- To aid in the classification of different kinds of client groups and in justification for the need for ongoing service provision.

- To aid in the determination of a functional outcome, determination of discharge plans, and potentials for work, play, and self-care within the person's own context.

- To provide an opportunity to establish rapport with the client and as an aid in communication with other professionals.

- To provide a means of tracking improvements, which is often motivational for the client (Asher, 1996; Hinojosa & Kramer, 1998).

There are ethical considerations in choosing and using assessment tools. Therapists are consumers of tests and measurement instruments and, as such, are accountable to clients for the choices made. It is the therapist's responsibility to elicit the client's best performance on any assessment by being familiar with the test in advance, by maintaining an impartial and scientific attitude during testing, by establishing a therapeutic and positive rapport with the client, and by providing an optimal environment in which to give the test (Demeter, Andersson, & Smith, 1996).

As consumers of assessment tools, there are characteristics of the particular test that will provide the clinical information necessary for intervention given this particular client, in a particular setting or context, and for a particular functional limitation. Minimally, you should be familiar with the test's validity and reliability to determine if this tool is assessing the domain of interest consistently. How well the tool discriminates changes within the individual client and between two clients is important in documenting changes that have occurred because of intervention. If the test has norms, knowing on what population (age, gender, diagnosis, etc.) the norms are based will inform you how well the results of the test will apply to your particular client. Because an important outcome of occupational therapy is resumption of roles and engagement in areas of occupation, an assessment must be related to functional performance. What is the level of performance that is being assessed: occupational performance, contexts, habits/patterns, skills, or client factors (range of motion, sensation, etc). Length of stay often determines the depth of assessment. If a client is only going to be in your facility for 2 days, a range of motion screen is perhaps a more appropriate assessment of joint motion than goniometric measurement of each joint. Cultural concerns, literacy, and language issues should also be considered during evaluation in order to elicit the best possible response from the client. Other test characteristics are whether the test is easy to administer, the time requirements to administer the test, and the cost of administration (Table 3-2).

The decision to use a standardized or nonstandardized assessment is also a consideration. Nonstandardized assessments would include observation, interviews, and questionnaires that rely heavily on experience and clinical reasoning (Leubben & Royeen, 2005). A nonstandardized test hasn't undergone rigorous development or analysis. Many activities of daily living tools that are "home grown" or developed specifically for a particular hospital or department are nonstandardized assessments. Nonstandardized assessments may require less time to learn to use the assessment, the administration of the test may be more flexible and take less time, and the tool may be less costly to administer than standardized tools (Lorch & Herge, 2007).

Standardized assessments might also include interview and observation, but tests that measure specific behavior, performance abilities, attitudes, skills, and traits are also options. Standardized assessments have established reliability and validity based on extensive research (Lorch & Herge, 2007). Standardized assessments lend credibility to observations, provide consistent and factual information, enable evaluation of progress and outcomes, and support evidenced-based evaluation and outcomes measurement (Lorch & Herge, 2007). The advantages and disadvantages of using standardized and nonstandardized assessments are summarized in Table 3-3.

Reliability and Validity of Assessments

Accurate measurements are essential to the development of intervention plans based on the goals and needs of the client, the limitations in performance, and the expected outcome. Accurate measurements are also used to document effective remediation that will ensure reimbursement of therapy services and become part of the legal record of these services. Accurate measurements are crucial in providing research data for treatment efficacy.

It is important that the evaluations used yield results that truly represent the client's function (validity) and that yield similar results through time and between testers (reliability). Multiple studies (Awan, Smith, & Boon, 2002; Ellis & Bruton, 2002; Gajdosik & Bohannon, 1987; LaStayo & Wheeler, 1994; Rothstein, Roy, Wolf, & Scalzitti, 2005; Sabari, Maltzev, Lubarsky, Liszkay, & Homel, 1998; Somers, Hanson, Kedzierski, Nestor, & Quinlivan, 1997) have been done to determine the reliability and validity of range of motion assessments, but results vary based on the part being measured, measurement error due to mass of body segments, inconsistent operational definitions, variety in instruments used and placement of tools, and variability in identification of body landmarks.

Validity

Different types of validity are shown in Table 3-4. Range of motion evaluations have what's known as logical validity, in which the performance is clearly defined and the therapist directly observes the behavior. In a range of motion assessment, the movement itself is being tested, not the ability to move (which would entail aspects of the central nervous system as well as muscle strength), the length of the muscle, or the flexibility of the structures. Range of motion assessments also have content validity because testing is based on knowledge of anatomy for proper placement of the goniometer, identification of bony landmarks, and the axis of rotation.

Difficulties in validity arise in using consistent explicit definitions of range of motion. While normative data are available for the motion at each joint, the norms are not sensitive enough to reflect differences based on age, gender, occupation,

Table 3-2

CONSIDERATIONS IN SELECTING AN ASSESSMENT

PRINCIPLE	CONSIDERATIONS
Objectivity	Validity, intra-tester reliability, accuracy, relevance preferred
Consistency	Inter-tester reliability; small range of normal human variability preferred
Fairness	Relationship between numerical impairment rating and true alteration of health status and physical function
Accuracy	Precision, specificity, sensitivity; potential to discriminate, intra-individual and inter-individual alterations in health status
Relevance	Correlation between assessment technique and alteration of health status from "normal"; requires comparison to normal functioning
Convenience	Test must be easy to perform, preferably with techniques known to all potential evaluators; limited educational requirements; limited time required for assessment
Cost	Minimum desirable by parties requesting assessments but not at the price of significantly increased variability; shortest possible time commitment of expert evaluators

Reprinted with permission from Demeter, S., Andersson, G., Smith, B. J., & Smith, G. M. (1996). *Disability evaluation.* St. Louis, MO: C. V. Mosby.

or sociocultural considerations. However, in a study comparing radiographs (x-rays) with goniometric measurements, there was found to be a high correlation between these two types of measurements, which indicates that this assessment has concurrent or congruent validity, which is a criterion referenced measure. The goniometer is considered the gold standard by which other tools of joint measurement are compared (Flinn, Latham, & Podolski, 2008; Palmer & Epler, 1998). Validity can be improved by controlling for outside variables, maturation, sensitivity to the test, and consideration of environmental variables.

Reliability

Consistent results are also important. When a test is given to a client one time and the same test is given to the same client again by a different tester, this is known as inter-rater reliability (Table 3-5). When the same tester assesses the same client at different times, this is called intra-rater reliability. Intra-rater reliability is consistently higher than inter-rater reliability (Boone, 1978; Flinn, Latham, & Podolski, 2008; Hamilton & Lachenbruch, 1969; Hellebrandt, Duvall, & Moore, 1949), and active range of motion measurements have higher reliability than passive measurements (Sabari et al., 1998). Table 3-5 summarizes some of the types of reliability.

While some studies suggest that if reliability has been established among testers, there can be a high degree of accuracy, other authors recommend that examiners with little experience may want to take several measurements and record the mean of the values to increase reliability.

Somers and colleagues (1997) suggest that inter-rater reliability may be more dependent on training than on experience, and uniform training may be more of a determinant of inter-rater reliability.

Error estimates for range of motion vary according to the particular joints tested. Awan, Smith, and Boon (2002) determined that intra-rater reliability for internal and external rotation of the shoulder was good (r = 0.58-0.71) and inter-rater reliability was fair to good (r = 0.41-0.66). Rothstein and colleagues (2005) reported intra-rater and inter-rater reliability values for specific joint motions, and the intraclass correlation coefficient was used to assess the consistency of measurements made by multiple observers measuring the same quantity (Table 3-6).

In measuring range of motion, there is equal reliability between readings of different goniometers for determining joint angles and between goniometers with different scale increments (Bear-Lehman & Abreu, 1989). Generally, ±5 degrees is considered the standard error of measurement for range of motion. Reliability can be improved by using consistent and well-defined testing positions and anatomic landmarks to align the goniometer.

Limitations in range of motion may be due to tissue changes (when passive range of motion is less than normative values) or muscle weakness (where active range of motion is less than passive). An understanding of why the joint is limited will direct the intervention. If the joint is limited due to shortened tissues, the goal would be to stretch the tissues to enable full or functional range of motion. If the limitations are due to

Table 3-3

COMPARISON OF STANDARDIZED AND NONSTANDARDIZED ASSESSMENTS

NONSTANDARDIZED	STANDARDIZED
More sensitive to individual client needs and environmental factors	Research evidence is available to read about the assessment and to support assessment development and use
May be administered within the context of dynamic assessment	Allows for critique of the assessment's level of quality—validity, reliability and appropriateness for client
May be more sensitive to identifying changes in areas of occupational performance	Provides consistency in methods of administering and scoring
Training time to learn to use the assessment is often less than with standardized assessments	Provides a statistical basis to quantify client's performance results in comparisons to other clients (norms)
Administration may be more flexible and take less time	Administration techniques and score are more objective, resulting in findings that should be consistent from examiner to examiner
May be less costly to administer, requiring fewer materials, forms, or equipment	Results may be used for outcomes management purposes
Incorporates therapist's clinical reasoning and experience for interpretation	Results may be seen as a more objective measurement by other professionals, clients, and third-party payers

Adapted from Lorch, A., & Herge, E. A. (2007). Using standardized assessments in practice. *Occupational Therapy Practice*, May 28, 17-22.

edema, pain, or spasticity, the goal of treatment would be to minimize these problems and then reassess joint movement. Long-standing limitations or contractures that are not able to be remediated would best be treated using compensatory or adaptation strategies.

ASSESSMENT OF RANGE OF MOTION

Range of motion limitations affect the person's ability to engage in occupations and can result in physiologic changes in muscles and tissues. When a part is immobilized or normal movement does not occur, there may be a loss of muscle fibers, changes in the length and number of muscle sarcomeres, ligamental and tendon weakness, disorganized synthesis of new collagen, and shortening of muscle fibers. Disruptions in joint movement affect the synovial fluid, synovial membranes, and articular cartilage, further affecting joint movement and increasing the circumference of the joint. Increasing range of motion is important physiologically to prevent muscle imbalance, minimize discomfort, avert skin breakdown, avoid hygiene problems, and enable people to care for the client or to enable the client to move actively.

Assessment of active range of motion has also been predictive of function in clients who have experienced a stroke. Active range of motion, particularly of the shoulder and

Table 3-4

TYPES OF VALIDITY

FACE VALIDITY

From appearance, and without statistical proof, the test items appear to address the purpose of the test and the variables to be measured; subjective, logical judgment by the author or experts on the topic.

CONTENT VALIDITY

The items on a test represent sufficient, representative samples of the domain or construct being examined; requires selection of a specific aspect of the behavioral domain evaluated; designed to measure the level of mastery of a particular content domain.

CRITERION-RELATED VALIDITY

The ability to determine performance on one test based on the performance on another test; the degree of agreement between two tests; determines the accuracy of prediction that is obtained by squaring the validity coefficient (r'), which tells us how much variance the predictor variable is able to explain.

CONCURRENT OR CONGRUENT VALIDITY

The extent of agreement between two simultaneous measures of the same behaviors or traits.

PREDICTIVE VALIDITY

The extent of agreement between the current test results and a future assessment; used to make predictions about future behavior.

CONSTRUCT VALIDITY

Based on the theoretical framework to test the "goodness-of-fit" between the theorized construct and resulting data from a test; a gauge of the ability of a test to measure a trait or hypothesis that is not observable.

CONVERGENT VALIDITY

The test results should correlate highly with another measure of the same variable or construct; infer degree of agreement measuring the same trait with two different tests of the same trait or construct.

DISCRIMINANT VALIDITY

The test results should not correlate with another measure of the same variable or construct; infer degree of disagreement measuring the same trait with two different tests of the same trait or construct.

FACTORIAL VALIDITY

The identification of interrelated behaviors, abilities, or functions in an individual that contribute to collective abilities or functions; correlation of the test with other group to define the common traits it measures.

Reprinted with permission from Hinojosa, J. & Kramer, P. (1998). *Occupational therapy evaluation: Obtaining and interpreting data*. Bethesda, MD: American Occupational Therapy Association.

middle finger, taken at 1 month post-stroke was predictive of functional recovery 3 months later (Beebe & Land, 2009).

Range of motion can be measured passively and actively with different information gleaned from each procedure. For clarification, the following abbreviations and terminology are defined:

- Passive range of motion (PROM): Arc of motion through which a joint passes when moved by an outside force.

- Active range of motion (AROM): Arc of motion through which a joint passes when moved by muscles acting on a joint.

- Active assistive range of motion (AAROM): Arc of motion through which a joint passes when moved initially by muscles then completed by an outside force.

- "Functional ROM": Amount of motion necessary to perform essential activities of daily living (ADLs) tasks

---

Table 3-5

TYPES OF RELIABILITY

TEST-RETEST

A measure of test score stability on the same version of the test repeated over two occasions; based on a correlation of the scores obtained during each of two administrations; useful for tools monitoring change over time.

INTRARATER RELIABILITY

Consistency in measurement and scoring by the evaluator when two test results from two similar situations are correlated.

INTERRATER OR INTEROBSERVER RELIABILITY

The degree of agreement between the scores from two raters following observation and rating of the same subject; correlations of .85 or higher are expected to compare the objective competency between two raters of the same testing condition.

ALTERNATE OR PARALLEL FORMS

Useful when multiple, equivalent forms of the same test are needed; particularly useful when one's response to the earlier test items can easily be recalled and influence the responses on the second test after a lapse of time (alternate); while the forms contain different questions, similar items on each test are expected to have item equality, making the tests equal at a given point in time (parallel); correlation should be >.80.

INTERNAL CONSISTENCY

Reflects the degree of homogeneity between test items; items measure the same construct; correlation should be >.80.

SPLIT HALF

The results from performance on the first half of the test are correlated with the second half or the scores on all even-numbered items are compared to odd-numbered ones.

COVARIANCE PROCEDURES

The average of all split-half tests; expressed as KR20 or KR21; the consistency of response between all items on a test; referred to as interitem consistency.

Reprinted with permission from Hinojosa, J. & Kramer, P. (1998). *Occupational therapy evaluation: Obtaining and interpreting data.* Bethesda, MD: American Occupational Therapy Association.

---

without adaptations or equipment. Another definition is the "minimum motion necessary to comfortably and effectively perform ADL" (Vasen, Lacey, Keith, & Shaffer, 1995, p. 291).

- Torque range of motion (TROM): The external force (generally 200 g) is applied to the joint by an orthotic gauge or torque device to achieve objective and precise passive measurements (Flinn & DeMott, 2010).
- Total active range of motion (TAM): Total active motion is computed using AROM from MCP, PIP, DIP of one digit minus the extension lag for that joint (Flinn & DeMott, 2010).
- Total passive motion (TPM): TAM and TPM are useful in identifying true progress in intervention and can identify

a patient who is making improvements in flexion at the expense of losing extension (Flinn & DeMott, 2010).

The assessment of range of motion seems like a simple assessment of the movement of the joints. But an assessment of the mobility of the joint also includes an assessment of the physiological stability of the joint and its structural integrity. Many conditions limit range of motion and joint stability (such as edema, adhesions, shortened tissues, decreased strength), and the assessment of joint motion may lead to the administration of other physical assessments. The range of motion assessment would include understanding the client and his or her experiences, observing the way the client moves, palpating joint structures, and measuring range of motion.

Table 3-6

INTRA-RATER AND INTRA-RATER RELIABILITY FOR JOINT MOTIONS

Joint	Motion	INTRA-RATER RELIABILITY		INTER-RATER RELIABILITY	
		ICC	N	ICC	N
Shoulder	Flexion	0.98	100	0.88	50
	Extension	0.94	100	0.27	50
	Abduction	0.98	100	0.85	50
	Horizontal abduction	0.91	100	0.29	50
	Horizontal adduction	0.96	100	0.37	50
		0.98	100	0.90	50
		0.94	100	0.48	50
Elbow	Flexion	0.95*	24	0.93*	12
	Extension	0.95*	24	0.94*	12
Knee	Flexion	0.98*	24	0.85*	12
	Extension	0.95*	24	0.70*	12
Ankle	Subtalar neutral	0.77	100	0.25	50
	Inversion	0.74**	100	0.32**	50
	Eversion	0.75**	100	0.17**	50
	Dorsiflexion	0.90	100	0.50	50
	Plantarflexion	0.86	100	0.72	50

*ICC values calculated using a less conservative form of the ICC than other values in the table

** These measurements were not referenced back to the subtalar position

Reprinted with permission from Rothstein, J. M., Roy, S. H., Wolf, S. L., & Scalzitti, D. A. (2005). *The rehabilitation specialist's handbook* (3rd ed.). Philadelphia, PA: F. A. Davis Company.

Occupational Profile

The occupational profile includes information about the client's medical and occupational histories, patterns and habits in daily life, interest and needs, roles, and valued occupations (American Occupational Therapy Association, 2008). Have the client talk with you about previous conditions and how that may influence the current problem. Discuss the current problems the client is having, what caused the limitation, how long it has lasted, and under what conditions the problem seems to get better or worse. Careful consideration of the client's concerns helps to establish rapport, understand the client's perspective about the limitation, identify underlying factors causing the limitation, and suggest intervention strategies that are based on meaningful client occupations and roles.

Observation and Palpation

Much information can be obtained by watching the client move. When the client comes in, do you observe inequality in the levels of the shoulders or hips? You are looking for symmetry between the limbs. Does the client use compensatory movements in activities such as elevating the scapula rather than flexing the shoulder? Note the rhythm of movement and whether the motions seem coordinated. You can observe whether the client seems to be guarding or protecting a body part or whether the affected limb is used at all or elicits a pain response when it is used. Observe for scars and wounds, and note any deformities. Observe the overall body posture.

Gentle movement of the client's limbs will tell you if movement produces pain, if the temperature of the skin is hot or cold, and if movement is blocked by normal limiting structures. Visually, you can observe areas of edematous tissues, and palpation will help determine the type of edema.

Range of Motion Screening

Joint range of motion can be assessed by means of a goniometer or by screening techniques. Screening tools are used as a rapid method of assessing multiple joint movements or combined movements to determine if more in-depth and standardized assessments are needed. While visual estimation is often used clinically, this method is unreliable and deemed "not acceptable" (Demeter, Andersson, & Smith, 1996) for a complete joint assessment, but is often used as a screening mechanism.

Clinical range of motion screens ask the client to move in a specified way, or the therapist passively moves the client. These can be done actively by the client, or the therapist moves the client into the positions. Table 3-7 provides active range of motion screening tests for the upper and lower extremities.

Functional motion tests or motion inventories often use observation to determine decreased motion or use of compensatory motions when achieving combined motions or simulated tasks. This might include tasks such as the following:

- Placing an object behind the patient and having him or her look for it and retrieve the object while sitting and then standing

- Having the client reach overhead for an object

- Having the client don a button-up shirt or blouse

- Having the client pick up items off of the floor (can be done in sitting or standing)

- Observing the client comb/brush his or her hair

- Having the client cross his or her legs to don a pair of pants (Gutman & Schonfeld, 2003, p. 120)

Generally, a functional motion screening tool uses active motion and gives information about patterns of movement and about the client's willingness to move. Descriptions that are more detailed may minimize this problem but may also be more difficult for the client to understand. Scoring of functional motion tests often involves use of the following nominal scales:

n = completes normally

s = completes with substitution

I = initiates movement but cannot complete

u = unable to perform

NT = not tested

Percentages are also used to indicate what part of range is limited, such as the following:

0 limitation = full ROM

minimal limitation = 1% to 33% ROM

moderate = 33% to 66% ROM

maximal = 66% to 100% (Tan, 1998)

Some screening tools combine functional motion test items, such as picking up an object or turning a doorknob, with active range of motion screens of isolated joint motions (such as touching the top of the head for shoulder flexion). These tests may then be followed by simulated tasks such as using a hammer or screwdriver, throwing a ball, or threading a needle.

A disadvantage of functional motion screening tools is the variability of movements that can occur that satisfy the request for movement. For example, if asked to touch the top of the head, the humerus can be abducted or flexed with the elbow and wrist flexed to reach the hand to the head, or the elbow can be flexed with the neck and trunk flexed to reach the head to the hand.

Instrumentation

If deficits have been noted or a more detailed measurement of joint motion is desired, goniometric measurements are made. Goniometry is a commonly used measurement of joint motion. The word *goniometer* comes from two Greek words: "gonia," which means angle, and "metron," which means measure (Norkin & White, 1985). A goniometer is a tool that is used to determine the amount of motion available at a specific joint.

There are various types of goniometers, but the most widely used is the universal goniometer. Universal goniometers have "good to excellent" reliability (Palmer & Epler, 1998). These are simple, uniplanar protractors commonly used in clinical practice. The goniometer has either 360-degree or 180-degree scales (or both), with two arms used to measure the motion. One arm follows the movement, and the other aligns with a stationary part of the body. These can be full-circle goniometers or half-moon (in the shape of half of a circle).

Gravity-dependent goniometers use gravity's effect on pointers and fluid levels. The device is strapped onto a distal leg segment with the proximal limb positioned vertically or horizontally. The pointer or bubble is read at the end of range of motion. This type of goniometer does not need to be aligned with skeletal landmarks and makes the measurement of passive range of motion easier. Disadvantages are that the gravity-dependent goniometer is bulkier, more expensive, not as readily available, and more difficult to use on small joints and for rotational movements than the universal goniometer.

An inclinometer is a pendulum-based device that has a 360-degree scale protractor with a counter-weighted pointer. Electronic versions are available, and this type of goniometer is especially good for measurements of spinal range of motion. A fluid goniometer (bubble goniometer or hydrogoniometer) also has a 360degrees scale with fluid in a tube with small air bubbles. An electrogoniometer is a potentiometer in which movement causes changes in resistance and voltage that can be calibrated to represent range of motion. This tool is expensive and is often used in research (Tan, 1998).

Other methods of measuring joint motion include use of tape measures, photometric (video-based) motion analysis systems, radiographs, photocopies, free hand drawings or tracings, infrared light sources, light-emitting diodes (or reflectors), or laser lights, which have good validity and reliability but are expensive (Tan, 1998). The type of goniometer chosen reflects the level of accuracy required, resources available, and convenience.

Active Movement

It is important to distinguish whether the limitations in range of motion occur during active or passive range of motion. Combining the results of range of motion and of contractile and non-contractile tissue assessments allows the therapist

Table 3-7

ACTIVE RANGE OF MOTION SCREEN

POSITION OF CLIENT	MOTION BEING TESTED	INSTRUCTIONS TO CLIENT
Sitting	Scapular elevation*	Shrug shoulder toward ears and release
	Scapular retraction*	Squeeze shoulder blades back as if they are touching
	Shoulder abduction *	Raise right arm out to side and then as high up as possible; repeat with left arm
	Shoulder internal rotation*	Place hands behind the back as though fastening a bra or tucking in a shirt
	Shoulder abduction and lateral rotation	Reach behind your head and touch the opposite shoulder blade
		Place hands behind your neck, and push your elbows back (posteriorly)
	Shoulder adduction and medial rotation	Reach to your opposite shoulder
		Touch your shoulder blade on the opposite side
		Place both hands behind your back as high as possible
	Shoulder flexion and extension	Raise your arms in front of your body and up as high as you can
	Elbow flexion and extension	Bend and straighten your elbows
	Supination and pronation	With elbows bent (to 90°) and with arm at your side, turn your palm up and down
	Wrist flexion and extension	Bend your wrist up and down
	Radial and ulnar deviation	With palms down, move your wrist toward and away from the midline
	Finger flexion and extension	Make a fist and then open up your fingers
	Finger abduction and adduction	Spread your fingers apart and bring them together
	Thumb flexion and extension	Bend your thumb across your palm and then out to the side
	Neck flexion and extension	Place your chin on your chest and then tilt your head back
	Neck rotation	Turn your head to the right, left, and then in a circle
	Hip flexion*	Ask the patient to march in place while seated
	Hip flexion, abduction, and lateral rotation	Uncross thighs and place the side of your foot on your opposite knee
	Ankle inversion	Turn your foot in
	Ankle eversion	Turn your foot out
	Hip abduction and adduction	Spread legs apart and bring them back together
Supine	Hip flexion and knee flexion	Bend hips and knees

(continued)

Table 3-7 (continued)

ACTIVE RANGE OF MOTION SCREEN

POSITION OF CLIENT	MOTION BEING TESTED	INSTRUCTIONS TO CLIENT
Standing	Trunk flexion	Bend forward and reach for toes with knees straight
	Trunk extension	Bend backward while I stand beside you
	Trunk lateral bending	Lean to the left, then right while I stabilize your pelvis
	Trunk rotation	Turn to the right, then left while I stabilize your pelvis
	Ankle plantarflexion and toe extension	Stand on your tiptoes
	Ankle dorsiflexion	Stand on your heels

Reprinted with permission from Palmer, M. L., & Epler, M. E. (1998). *Fundamentals of musculoskeletal assessment techniques* (2nd ed.). Philadelphia, PA: J. B. Lippincott.

* Gutman, S. A., & Schonfeld, A. B. (2003). *Screening adult neurological populations.* Bethesda, MD: American Occupational Therapy Press.

to determine what tissue or structure is affected. Treatment can then be directed to a specific tissue site. Optimally, active range of motion assessments are completed first. If active range of motion is less than passive range of motion, this may be due to muscle weakness, pain, paralysis, spasm, tight or shortened tissues, changes in the length-tension relationships of the tissues, or modified neuromuscular factors or changes in the joint-muscle interaction (Magee, 2002).

When using a goniometer to measure active range of motion, the testing position of the client and the goniometer placement are given for each joint. In consideration of the client's comfort, the client is seated in an armless chair that provides back support, although some tests can be done in the supine or standing positions. To ensure the most accurate results, make sure the client is comfortable and relaxed. Having established rapport during the initial interview, it is important to reassure the client that you are aware of motions that may be uncomfortable and to address any client concerns about being touched (due to fear of pain, cultural beliefs, etc.) prior to range of motion testing initiation.

Tell the client exactly what you will be doing and why. State the purpose of the range of motion simply, such as "this test will provide information about the way your joints move and this tool, a goniometer, will measure the arcs of motion of your joints." Then, have the client assume the starting position for the test, which is usually 0 degrees in the anatomical position except for rotary movements. Stabilize the joint proximal to the joint being measured, which will help to minimize substitution movements, increase accuracy, and decrease client fatigue.

In aligning the goniometer to the joint, you need to consider the goniometer arm placement and identification of the joint axis. The goniometer is essentially a protractor used to measure joint angles. The axis of the goniometer is the rivet. This is aligned to the axis of movement of the joint. Often, this is a specific bony landmark (such as the acromion of the shoulder). Palpation is necessary to identify bony skeletal landmarks used as reference points when measuring the arc of motion. Because identification of joint axes involves bony landmarks, for the most precise range of motion measurements, uncovering the joint will be necessary to palpate structures, avoid interference with motion, and for alignment of the goniometer axis.

Because joint movements are often a combination of movements at several joints simultaneously, the joint axis may appear to shift during movement. This occurs most notably during shoulder flexion and abduction. It is important in these cases to maintain the original orientation of the goniometer axis in spite of the shift in joint axis due to movement.

The two arms of the goniometer are then aligned to body segments. One arm of the goniometer (stationary arm) lies parallel to the longitudinal axis of a proximal segment or will point to a distal bony prominence. The other goniometer arm (considered the movable arm) follows the motion and lies parallel to the longitudinal axis of the moving distal joint or points to a distal bony prominence. To make sense of this, just remember you are measuring a part of the body that moves: one part of the goniometer has to be a reference (stationary arm); the other part has to follow the movement (moveable arm).

Demonstrate the motions you want the client to do first and how you would use the goniometer to measure the joint. Use as many sensory modalities as necessary to clearly convey to the client how you want him or her to move. You can verbally describe the motion, demonstrate the motion yourself so they can visually see the motion, and then have the clients demonstrate the motion themselves so they can kinesthetically and proprioceptively feel the motion. You will have already asked if the client has any previous fractures or fused joints, which you note on the recording form so you know in which joints to expect limitations. If performing passive range of motion, hold the part securely but gently, and do not force any joints to move. During the joint movement, be aware of the client's comfort, and note any motions that produce pain. Record the number of degrees of motion at the initial and final positions.

Passive Movement

Passive range of motion gives the examiner information about the integrity of the articular surfaces and the extensibility of the joint capsule, ligaments, bursa, cartilage, nerves, nerve sheaths, and muscles. Passive movement is analogous to anatomical movements (as compared to active movements, which are seen to be physiologic movements).

Passive range of motion is tested in the same way that active range of motion is tested with the exception that the therapist is moving the part, rather than the client moving the part through volitional movement. Normally PROM is greater than AROM due to accessory motions and joint play. Passive joint assessment is able to assess the "tensegrity" of a joint, which is a term denoting both tension and tissue integrity that occurs in normal structures. Passive range of motion assessment provides information about the status of the joint by evaluating normal limiting factors and patterns of limitation or restriction. Passive range of motion may be considered more objective because the motion is controlled by the examiner and is free of control by the subject (Demeter, Andersson, & Smith, 1996).

Normal Limiting Factors

Normal limiting factors of joint movement can be felt when passively moving a part to the extreme ends of range of motion. This might be when a joint is in a close-packed position and the articular surfaces are in contact. The limited ligamental extensibility, due to the relative inelasticity of white fibrous collagen, restricts some motions, while limited muscle and tendon extensibility restricts other motions. In some joints, the normal limiting factor is due to the apposition of soft tissues. Structures that limit the amount of motion at a specific joint have a characteristic feel that is felt as resistance to further movement. This is known as end feel. End feel has been defined as the "subjective assessment of the quality of the feel when slight pressure is applied at the end of a joint's passive range of motion" (Lippert, 2006).

While there are some variations in the terminology used to describe an end feel, generally it is accepted that there are six types of end feel: three are descriptions of normal limiting factors and three are pathological impediments to movement (Table 3-8).

Further motion in a joint with soft end feel is due to compression of soft tissue. This is a normal limiting factor and is seen in knee flexion when the soft tissue of the posterior leg contacts the posterior thigh (Norkin & White, 1985). Firm normal end feel occurs when there is tension in the way the joint feels but also a slight "give" in the structures, comparable

to pulling a rubber band or a strip of leather, depending upon the structure limiting the motion.

Normal firm end feel can occur due to muscles, capsules, or ligaments. A firm muscular end feel occurs with hip flexion with the knee straight because there is passive elastic tension in the hamstring muscles. Extension of the metacarpophalangeal joints of the fingers creates tension in the anterior capsule and is an example of firm capsular end feel. Firm ligamental end feel is demonstrated in forearm supination with tension occurring in the palmar radioulnar ligament of the inferior radioulnar joint, interosseous membrane, and oblique cord. The third type of normal end feel is hard, in which there is an abrupt hard sensation at the extreme end of range of motion. This is when a bone contacts another bone and can be felt in elbow extension (Cyriax, 1982).

Abnormal limitations in range of motion may be due to any of the following:

- Destruction of bone and cartilage (osteoarthritis or rheumatoid arthritis)
- Bone fractures
- Foreign body in joint (Bankart lesion in shoulder; tearing of meniscus in knee)
- Tearing or displacement of intracapsular structures
- Adhesions or scar tissue
- Muscle atrophy or hypertrophy
- Muscle tear, rupture, or denervation
- Pain
- Psychological factors
- Edema
- Neurological impairment (Trew & Everett, 2005)

These abnormal limitations would manifest as pathological end feels. Slight overpressure at the end of the range of motion that reveals soft end feel in a joint that normally had a firm or hard end feel may be indicative of edema or synovitis of that joint. Other conditions may present with end feels different than the normal limiting factors or may present with the normal type of end feel, but the limitation occurs earlier in the range of motion than in a normal joint. A hard end feel may occur in the cervical spine due to osteophytes, or empty end feel may occur due to acute bursitis or a tumor. Empty end feel is due to a lack of mechanical limitation of joint range of motion and is where joint motion is limited by pain and complete disruption of soft tissue constraints (Lippert, 2006). If there is an abnormal hard end feel and limited but pain-free movement, this may be due to osteoarthritis where osteophytes restrict movement but are not compressing nerves or sensitive structures (Magee, 2002).

Other abnormal end feels include the following:

- Spasm that is felt as a "vibrant twang" (Marieb, 1998, p. 8; Cyriax, 1982) and is the result of a prolonged muscle contraction in response to circulatory and metabolic changes (Kisner & Colby, 2007; Lippert, 2006).
- Capsular end feel, which is similar to tissue stretch, but occurs earlier in the range of motion. Hard capsular end feel is more in chronic conditions or capsular abnormalities,

Table 3-8

PHYSIOLOGIC AND PATHOLOGIC END FEEL

	PHYSIOLOGIC (NORMAL)		PATHOLOGIC	
Type of End Feel	**Description of End Feel**	**Joint Examples**	**Description of End Feel**	**Possible Reasons**
Soft	Soft tissue	Knee flexion Elbow flexion CMC thumb flexion	• Occurs sooner or later in range of motion • Occurs in a joint normally firm or hard • Feels boggy	• Synovitis • Edema
Firm	• Muscular stretch • Capsular stretch • Ligamentous stretch	Scapular elevation Scapular depression Shoulder abduction Shoulder adduction Shoulder internal rotation Shoulder external rotation Shoulder extension Shoulder horizontal adduction Shoulder horizontal abduction Forearm supination Wrist flexion Wrist extension Wrist radial deviation Wrist ulnar deviation Metacarpophalangeal extension Carpometacarpal extension Thumb extension Interphalangeal extension Distal interphalangeal flexion distal Interphalangeal extension Hip flexion	• Occurs sooner or later in range of motion • In joints, normally soft or hard	• Increased tone • Shortening of capsules, muscles or ligaments

(continued)

and it comes on abruptly after smooth, friction-free movement. Soft capsular end feel is more common in acute conditions with stiffness occurring early in the range of motion and increasing to end of range of motion. Soft capsular end feel presents as soft or boggy and may be the result of synovitis or soft-tissue edema. Major injuries to ligaments and capsule may cause soft end feel (Magee, 2002).

• Springy block is when a joint rebounds at the end of range of motion due to internal articular derangement that may

Table 3-8 (continued)

PHYSIOLOGIC AND PATHOLOGIC END FEEL

	PHYSIOLOGIC (NORMAL)		PATHOLOGIC	
Type of End Feel	Description of End Feel	Joint Examples	Description of End Feel	Possible Reasons
Hard	Bone against bone	Elbow extension Forearm pronation PIP flexion MCP flexion of thumb	• Occurs sooner or later in range of motion • In joints, normally soft or firm • Bony grating or block felt	• Osteoarthritis • Chondromalacia • Loose bodies in a joint • Myositis ossificans • Fractures
Empty			• Range of motion never reached due to pain • No resistance felt	• Acute joint inflammation • Bursitis • Abscesses • Fractures • Psychogenic

Adapted from Norkin, C., & White, D. (1985). *Measurement of joint motion: a guide to goniometry.* Philadelphia, PA: F. A. Davis Co.

be indicative of intra-articular blocks, such as torn meniscus or articular cartilage.

- Muscle guarding is an involuntary muscle contraction in response to acute pain (Kisner & Colby, 2002; Lippert, 2006).

Patterns of Limitation

Joint play is the "give" or distensibility (or extensibility) of the joint capsule and soft tissues that can only be obtained passively. It is usually a motion less than 4 mm that must be tested in the loose-packed (resting) position because this is the position in which the joint is under the least amount of stress. The joint capsule works at its greatest capacity in this position, and there is minimal congruency between articular surfaces and the joint capsule. The ligaments are in a position of greatest laxity, which allows joint distraction via rolling, spinning, and sliding. The close-packed position should be avoided because, in this position, the two joint surfaces fit together perfectly, ligaments and tendons are maximally tight, and the joint is under maximal tension.

Assessing joint gliding requires an understanding of the convex-concave rule to successfully replicate the accessory motion. The direction of glide is in the same direction as the moving segment in joints that are concave, and the direction of glide is in an opposite direction to the bony segment in a convex joint.

Assessing the inert structures of the joint requires a passive stretch of these tissues to evaluate pain, laxity, or limitations in range of motion. Normally, passive stretching or compression is painless but if there is injury, immobilization, or inflammation resulting in abnormal tissues, the passive movement may be painful. Ligaments that are injured often result in a particular pattern of limitation that is found to be in a specific pattern in each particular joint. Noncapsular patterns are also possible. These are limitations that exist but do not correspond to the classic capsular patterns described for each joint (Magee, 2002). Local restriction (i.e., ligamentous adhesion), bursitis, internal derangement (as seen in knees, ankles, and the elbow), intracapsular fragments, extra-articular lesions, and subluxations are possible reasons for noncapsular patterns (Magee, 2002). Capsular patterns are listed in Table 3-9. The value of assessing the inert structures is to identify the structures causing the limitation.

Pain

The assessment of a client's pain is an integral part of the evaluation process. Pain will invalidate many physical assessments because the client will be unable or unwilling to move

Table 3-9

CAPSULAR PATTERNS OF THE JOINTS

JOINT	CAPSULAR PATTERN (IN ORDER OF LIMITATION)
Glenohumeral	• Lateral rotation, abduction, medial rotation
Sternoclavicular	• Pain at extreme range of movement
Acromioclavicular	• Pain at extreme range of movement
Humeroulnar	• Flexion, extension
Radiohumeral	• Flexion is more limited than extension
	• Supination and pronation are equally limited
Proximal radioulnar	• Supination, pronation
Distal radioulnar	• Pain at extremes of rotation
	• Pronation and supination are mildly limited at the distal radioulnar joint
Wrist	• Flexion and extension equally limited
Fingers	• Abduction is more limited than adduction of the thumb CMC
	• Flexion is more limited than extension of MCP and IP joints
Thoracic spine	• Side flexion and rotation equally limited, extension
Lumbar spine	• If a left facet is limited, forward bending produced a deviation to the left
	• Side bending right is limited while side bending left is unrestricted; rotation left is limited while rotation right is unrestricted
Cervical Spine	• If left facet is limited, forward bending produces some deviation on the left; side bending right is unrestricted while side bending left is comparatively unrestricted; rotation right is comparatively unrestricted and rotation right is most limited.
Hip	• Flexion, abduction, medial rotation (order varies)
Knee	• Flexion is more limited than extension
Foot and toes	• Extension more limited than flexion

through the total range of motion. More importantly, pain may interfere with the client's ability to engage in desired occupations. For example, limitations in activities of daily living might be seen in avoidance or slowed performance of grooming or showering tasks if the movement elicits pain. Pain can interfere with sleep, which can affect cognitive abilities such as attention and memory recall. The client may be preoccupied with thoughts of pain so is unable to engage in leisure activities or socialize with friends. The client may be unable to resume work or parenting roles following injury or trauma. Pain is a significant reason for lost work days and disability and adds considerably to family and marital discord (Solet, 2008).

Pain is directly related to the well-being of those with physical disabilities. The client may present with a flat affect and express feelings of loss of self-esteem, helplessness or hopelessness, and frustration (Fisher, Emerson, Firpo, Ptak, Wonn, & Bartolacci, 2007; Reed, 2001). The client may appear agitated or lethargic and may feel grief over lost mastery of roles and tasks. Depression, anxiety, and fear are common sequelae of chronic or long duration pain that directly affect the person's quality of life.

The assessment of pain includes understanding the painful experience that the individual client is experiencing and the direct impact pain has on activity. It is a highly personal experience, and individuals turn to their social environment for validation and meaning of their pain behavior and symptoms. The subjective pain experience is affected by cultural, historical, environmental, and social factors.

Expectations, manifestations, and management of pain are embedded in cultural context, and beliefs of illness guide individuals to specific types of intervention (Crepeau & Schell, 2003; Solet, 2008). Pain experiences have different meanings in different cultures. Different names are given to diseases in various parts of the world, and symptoms have different meanings in different cultures. For example, epilepsy is considered a shameful disease among some Greeks, while Ugandans believe that it is contagious and untreatable. Some Mexican-Americans feel that epilepsy is a reflection of physical imbalance, while Mennonites believe is it a sign of favor from God (Jarvis, 2000).

The experience of pain is subjective, so self-report is the most valid measure of the experience (Jarvis, 2000; Magee, 2002; Solet, 2008). During the patient interview, be aware of any nonverbal indicators of pain such as grimaces, frequent position changes, and other nonverbal cues. Monitoring vital signs is also a key to understanding pain levels during a musculoskeletal assessment. The Pain Behavior Checklist was developed to classify and record pain behaviors during client interviews (Dirks, Wunder, Kinsman, McElhinny, & Jones, 1993).

There are various ways to have clients tell you about their pain. There are verbal description scales where clients are asked to describe their pain by referring to a list of adjectives. Numeric rating scales are used, and usually clients are asked to rate their pain on a 0-10 scale, where 0 indicates no pain and 10 is indicative of maximal pain. The Borg Scale (Borg, 1982) has clients rate their pain on a 0 to 10 scale, and the scale items are operationalized so that a 1 indicates very weak pain, 2 is weak pain, 3 is moderate, and so on. Alterations in movement are expected with pain levels of 6 or greater (Palmer & Epler, 1998). A modification of this rating system is the Pain Disability Index in which the client is asked to rate his or her level of pain on a 0 to 10 scale in the areas of family/home responsibilities, recreation, social activities, occupation (job), sexual behavior, and life support activities (eating, sleeping, breathing) (Pollard, 1984). The Functional Status Index also looks at pain in relation to activity performance (Jette, 1980). Clients can also be asked to keep a pain diary to record pain daily on a numeric scale during ADLs, and they might also include medication, alcohol use, and emotional responses during the same period.

Client descriptions of pain may suggest structures that are the source of the pain. Table 3-10 lists descriptions of pain and possible causes and indications about the progression of the disability.

A visual analog scale is often used to visually represent the client's pain intensity. The client can either indicate his or her level of pain on a line (Cline, Herman, Shaw, & Morton, 1992), choose from six faces, or indicate his or her pain intensity on a thermometer (Brodie, Burnett, Walker, & Lydes-Reid, 1990). These scales may include numbers, drawings, and/or adjectives to further measure pain. There are no norms for this scale, and it cannot be used to compare different clients but is helpful to monitor the same client over time.

The McGill Pain Questionnaire (Melzak, 1967) combines several of these methods of gathering information about a client's pain. First, there is a list of 20 categories of adjectives to describe pain. Columns 1 to 10 describe qualities of pain (e.g., throbbing, pinching, dull), and columns 11 to 15 describe affective qualities of pain (e.g., tiring, blinding). Column 16 uses evaluative words about the overall intensity of pain (such as annoying, miserable, intense), and columns 17 to 20 include miscellaneous descriptors (spreading, squeezing, agonizing). One disadvantage to such an extensive list of descriptors is that not everyone you treat will be able to discriminate or will be able to define all of the terms presented. The second part of the McGill Pain Questionnaire asks that the client indicate on a drawing of the body where the pain is and what type of pain is experienced. Finally, the client is asked to describe if and how the pain changes with time.

There may be pain with active or passive movement. Pain with active range of motion but not with passive range of motion is most likely to be due to a muscle or tendon problem. This may be characterized by pain and limitation or excessive movement in some directions but not others. If only one movement is painful, this may suggest a sprain of a single ligament, and often there is greater pain at the extreme end of range. Pain with passive range of motion is likely due to tight joint structures, ligamental injury, cartilage injury, or inflammation. A capsular pattern or capsulitis might be present if there is pain when the joint is passively moved in nearly all directions. If there is limitation in motion because of pain, and pain is present during distraction of joint surfaces but not compression, this is likely due to stretching of ligaments or the joint capsule. If there is pain during compression of joint surfaces that is relieved by distraction, this is probably due to thinning or loss of cartilage, inflammation within the joint, or a surface abnormality (Magee, 2002).

Recording of Range of Motion Results

While the method of performing a range of motion assessment is standardized, there are different methods of recording range of motion results. There are three systems used in clinical practice: the 360 degrees system, the 180 degrees system, and the STFR system. The 180 degrees system is more commonly used (i.e., American Academy of Orthopaedic Surgeons) and accepted in clinical practice. In this system, the neutral zero or starting position is with the body in anatomical position with the zero position toward the feet. The body is in the plane in which the motion is to occur, with the axis of the joint acting as the arc of motion (Pedretti, 1996).

In the 360 degrees system, the zero starting position is overhead with the arc of motion related to a full circle. In both systems, some motions do not readily lend themselves to movements around a semicircle or a full circle. These are generally rotational movements such as pronation, supination, and internal and external rotation, radial and ulnar deviation, and thumb carpometacarpal flexion and extension. To illustrate the difference in recording between these two systems, consider the motion of shoulder flexion. The arc of motion for shoulder flexion is generally accepted to be 180 degrees, which would be recorded as 0-180 degrees in the 180 degrees system. A separate movement of shoulder hyperextension (movement from the arm held at the side then moved in a posterior direction) would be recorded as 0-60 degrees. In the 360 degrees system, this same arc of motion would include flexion and the additional movement of hyperextension. The total movement would be recorded as 0 to 240 degrees. It is clear that identification of the system of recording is vital to clear communication and documentation of range of motion results.

Another method of recording is the SFTR system (Gerhardt, 1983). In this system, the measurement of range of motion is completed as in either the 180 degrees or 360 degrees systems. The recording of the results is related directly to the cardinal planes, with "S" indicating the sagittal plane, "F" the frontal, "T" the transverse plane, and "R" indicative of rotational movements. Neutral zero is the anatomical position with three

Table 3-10

DESCRIPTIONS OF PAIN AND POSSIBLE CAUSES

DESCRIPTION OF PAIN	POSSIBLE CAUSES
Intensity, duration, frequency increasing	Condition worsening
Constant pain	Suggestive of chemical irritation tumors, visceral lesions
Periodic, occasional pain	May be activity- or position-dependent; likely to be mechanical, related to movement and stress
Episodic pain	Related to specific activities
Morning stiffness that improves with activity	Chronic inflammation and edema
Pain and aching progresses through day	Increased congestion at a joint
Pain at rest	Acute inflammation
Pain worse at beginning of activity	Acute inflammation
Pain not affected by rest or activity	Bone pain
	Organic or systemic disorders (e.g., cancer)
Intractable pain at night	May indicate serious pathology (e.g., tumor)
Pain at night	Peripheral entrapment neuropathies (carpal tunnel, thoracic outlet syndromes)
Sharp pain	Nerve injury
Burning pain	Nerve injury
Deep, boring, localized pain	Bone injury, trauma, or disease
Diffuse, aching, poorly localized pain	Vascular pain
Pain that is hard to localize, dull, aching	Muscle pain
Severe chronic or aching pain inconsistent with injury or pathology	Somatic pain
Acute pain	Muscle strain
	Tendonitis
	Contusions
	Ligamental injuries
Chronic pain	Fibromyalgia
	Chronic fatigue syndrome
	Rheumatoid arthritis
	Low back pain
Psychogenic or idiopathic pain	Unknown cause
	Malingering
	Munchausen syndrome

numbers used to record the motion rather than a ROM. The first number indicates movements away from the body such as abduction, flexion, or external rotation; the second number indicates the starting position (usually zero); and the third number represents movements toward the body, such as extension, adduction, and internal rotation. An SFTR recording of the shoulder movement of full flexion and hyperextension would be 180 degrees-0-60 degrees to indicate full flexion, starting position, and hyperextension. An elbow that can be hyperextended would be recorded as S 10-0-150 degrees.

Usually, in clinical practice, the recording of range of motion uses the 180 degrees system. The range of motion results for shoulder flexion would be 0-180 degrees, and the "0" signifies the starting position of full extension while the "180" is indicative of the motion overhead into full forward flexion. This one range of motion recording of two numbers describes both flexion and extension because the movement began in full extension and moved to full flexion. Other motions where one set of range of motion values describes two motions are shoulder abduction and adduction, elbow and knee flexion and extension, and finger flexion and extension (PIP, DIP, MCP). Separate measurements are required for shoulder internal rotation, shoulder external rotation, pronation, supination, wrist flexion, wrist extension, ulnar deviation, radial deviation, hip adduction, hip abduction, hip flexion, hip extension, hip internal rotation, hip external rotation, dorsiflexion, plantarflexion, inversion, and eversion.

Regardless of whether the 360 degrees or 180 degrees system is used, it is important to record both numbers, as these signify the starting and the end positions. Limitations will be readily apparent in the designation of these values. If normal shoulder range of motion is 0-180 degrees, a value of 15-180 degrees would indicate a deficit in the ability to assume the zero starting position or an inability to assume full extension. Similarly, 0-165 degrees would indicate a 15 degrees deficit in shoulder flexion. Familiarity with normative values is helpful in determining deficits in ROM. The standard error of measurement for ROM values is ±5 degrees (Radomski & Latham, 2008).

However, lack of achievement of the norms for a given motion does not necessarily signify this as an intervention goal, because consideration of client goals and limitations in functional activities are the determinants of treatment planning as well as considering how imbalances in joint function might contribute to joint deformities.

The use of negative recordings leads to reporting of unclear data (Bear-Lehman & Abreu, 1989; Trombly, 1995). For example, if the client lacked 20 degrees of shoulder flexion, a negative recording would be -20 degrees. This lack of 20 degrees does not indicate the total range of motion that occurred and can be easily misunderstood. Fused joints have the same starting and end points, and results should state "fused at x _."

Interpretation of Range of Motion Results

To interpret range of motion, one needs to compare the client's values with established norms, to compare uninvolved with involved sides (if applicable), and/or to compare the client range of motion values with the functional range of motion limits required for most activities. It is not the range of motion value per se that is of the greatest concern for the occupational therapist but how limitations in range of motion will impact that client's ability to engage in meaningful occupations.

The client's range of motion values give an indication of how the client's joint motion compares with normative values. In interpreting the results, it is important to recognize that these norms are not specific in regard to age or gender and that there is variability in what is considered "normal" based on environmental, occupational, methodological, individual, and skeletal factors (Table 3-11).

Often, it is suggested that if one extremity is impaired and the other is not, comparing the two sides is a way to interpret range of motion differences. However, Günal and colleagues (1996) noted there are very significant differences between right and left side and that the contralateral, normal side "may not always be a reliable control in the evaluation of restriction of motion of a joint" (p. 1401).

Full or "within normal limits" (WNL) values for range of motion are not necessary for many daily tasks. Shampooing or combing hair requires only 115 degrees of active range of motion, while the norms for shoulder flexion are 0-180 degrees. Table 3-12 compares normative range of motion values with functional range of motion for selected activities.

Precautions and Contraindications

Measurement of range of motion is not always indicated or safe for clients. Measurement of active range of motion may be a safer method, because there is voluntary cooperation. Active motion assesses the physiologic movement of the joint as well as the degree of coordination, the level of consciousness, the length of the attention span, indications about pain, the ability to follow directions, and the skill in performance of functional activities. An unintended effect of AROM is that of self-limitation of end ranges. The client may limit the amount of active movement for which he or she is capable due to pain or fear of pain, which may limit the usefulness of the movement to joint health.

Precautions and contraindications to range of motion assessment are included in Table 3-13.

SUMMARY

- Range of motion is the arch of motion that occurs at a joint and is often measured with a tool called a goniometer.

- Individual factors such as genetic disposition, types of preferred activities, overall health of the person, age, and gender influence the amount of ROM at a joint. Anxiety and stress also influence joint mobility.

- The amount of movement that occurs in a joint is based on the types of tissue that make up the joint. Not only is the joint structure important for movement, but movement is also important to the joint structure for cartilage and bone nutrition and growth, and adequate functioning of ligaments and tendons.

- Structures limit the amount of motion at a joint, and these structures have a characteristic resistance called end feel. End feel can be firm, soft, or hard in normal joints, and these end feels also occur in joints with pathology. Other pathological end feels are springy blocks, spasms, muscle guarding, and muscle spasticity.

- Variations in the measurement of ROM occur due to different methods, devices, means of recording, and other variables that influence the reliability and validity of goniometric measurement. Goniometry measurements generally are seen to have better intra-rater reliability than inter-rater, and measurement of the upper extrem-

Table 3-11

COMPARISON OF RANGE OF MOTION NORMATIVE VALUES

	Wiechec and Krusen	Dorinson and Wagner	JAMA	Daniels and Worthingham	Esch and Lepley	Gerhardt and Russe	Boone and Azen	AAOS	CMA	Clarke
SHOULDER										
Flexion	180	180	150	90	170	170	167	180	170	130
Extension	45	45	40	50	60	50	62	60	30	80
Abduction	180	180	150	90	170	170	184	180	170	180
Internal rotation	90	90	40	90	80	80	69	70	60	90
External rotation	90	90	90	90	90	90	104	90	80	40
Horizontal abduction							45			
Horizontal adduction						135	140	135		
ELBOW										
Flexion	135	145	150	160	150	150	143	150	135	150
Pronation	90	80	80	90	90	80	76	80	75	50
Supination	90	70	80	90	90	90	82	80	85	90
WRIST										
Flexion	60	80	70	90	90	60	76	80	70	80
Extension	55	55	60	90	70	50	75	70	65	70
Radial deviation	35	20	20	25	20	20	22	20	20	15
Ulnar deviation	75	40	30	65	30	30	36	30	40	30
HIP										
Flexion	120	125	100	125	130	125	122	120	110	120
Extension	45	50	30	15	45	15	10	30	30	20

(continued)

Table 3-11 (continued)

COMPARISON OF RANGE OF MOTION NORMATIVE VALUES

	Wiechec and Krusen	Dorinson and Wagner	JAMA	Daniels and Worthingham	Esch and Lepley	Gerhardt and Russe	Boone and Azen	AAOS	CMA	Clarke
HIP										
Abduction	45	45	40	45	45	45	46	45	50	55
Adduction		20	20	0	15	15	27	30	30	45
Internal rotation		30	40	45	33	45	47	45	35	20
External rotation		50	50	45	36	45	47	45	50	45
KNEE										
Flexion	135	140	120	130	135	130	143	135	135	145
ANKLE										
Plantarflexion	55	45	40	45	65	45	56	50	50	50
Dorsiflexion	30	20	20		10	20	13	20	15	15
Inversion		50	30		30	40	37	35	35	
Eversion		20	20		15	20	26	15	20	

Reprinted with permission from Baxter, R. (1998). *Pocket guide to musculoskeletal assessment*. Philadelphia: WB Saunders.

Table 3-12

FUNCTIONAL RANGE OF MOTION FOR SPECIFIC ACTIVITIES

FUNCTIONAL MOVEMENT	ACTIVITY	FUNCTIONAL ROM IN DEGREES	NORMATIVE ROM IN DEGREES
Shoulder flexion	Shampoo or comb hair	115	0-180
Shoulder extension/ hyperextension	Reaching for surface when moving from stand to sit	20	0-60
Shoulder abduction	Placing a barrette in hair	80	0-180
Shoulder adduction	Washing contralateral side of the body	25	180-0
Shoulder internal rotation	Fastening a bra	80	0-90
Shoulder external rotation	Reaching behind body for shirtsleeve when dressing	80	0-90
Elbow flexion	Washing face; bringing hand to mouth	150	0-150
Elbow extension	Reaching to don/doff LE clothes	Less than 0	150-0
Forearm supination	Brushing hair or teeth	15	0-90
Forearm pronation	Turning a key in a lock	15	0-90
Wrist extension	Pushing up from support surface	25	0-70
Wrist flexion	Pulling up on a car door handle to open it	30	0-80
Ulnar/radial deviation	Open a door; wipe down a countertop	30	0-30/0-20

Adapted from Gutman, S. A., & Schonfeld, A. B. (2003). *Screening adult neurological populations.* Bethesda, MD: American Occupational Therapy Press.

ity is a more reliable measurement than lower extremity. Using consistent test procedures with the same client under the same environmental conditions enhances reliability and validity.

REFERENCES

American Occupational Therapy Association. (2008). Occupational therapy practice framework: Domain and process, second edition. *Am J Occup Ther, 62,* 625-683.

Asher, I. E. (1996). *Occupational therapy assessment tools: An annotated index* (2nd ed.). Bethesda, MD: American Occupational Therapy Association.

Awan, R., Smith, J., & Boon, A. J. (2002). Measuring shoulder internal rotation range of motion: A comparison of 3 techniques. *Archives of Physical Medicine and Rehabilitation, 83,* 1229-1234.

Bear-Lehman, J., & Abreu, B. C. (1989). Evaluating the hand: Issues in reliability and validity. *Physical Therapy, 69*(12), 1025-1033.

Beebe, J. A., & Land, C. E. (2009). Active range of motion predicts upper extremity function 3 months after stroke. *Stroke, 40,* 1772-1779.

Bonder, B. R., & Bello-Haas, V. D. (2009). *Functional performance in older adults* (3rd ed.). Philadelphia, PA: F. A. Davis Company.

Boone, D. C. (1978). Reliability of goniometric measurements. *Physical Therapy, 58,* 1255.

Borg, G. A. V. (1982). Psychophysical bases of perceived exertion. *Medical Science Sports Exercise, 12,* 377-381.

Brodie, D. J., Burnett, J. V., Walker, J. M., & Lydes-Reid, D. (1990). Evaluation of low back pain by patient questionnaires and therapist assessment. *Journal of Orthopaedic Sports Physical Therapy, 11,* 519-529.

Cline, M. E., Herman, J., Shaw, E. R., & Morton, R. D. (1992). Standardization of the Visual Analogue Scale. *Nursing Research, 41,* 378-380.

Cooper, C. (Ed.). (2007). *Fundamentals of hand therapy: Clinical reasoning and treatment guidelines for common diagnoses of the upper extremity.* St. Louis, MO: Mosby Elsevier.

Crepeau, E. B., & Schell, B. A. B. (2003). Theory and practice in occupational therapy. In E. B. Crepeau, E. S. Cohn, & B. A. B. Schell (Eds.), *Willard & Spackman's occupational therapy* (10th ed.). Philadelphia, PA: Lippincott, Williams & Wilkins.

Currey, C. J. (2005). Part II: Epithelial Tissue, from http://medinfo.ufl.edu/pa/chuck/summer/handouts/epi.htm; Accessed April 23, 2009.

Cyriax, J. (1982). *Textbook of orthopaedic medicine.* London: Bailler Tindal.

Table 3-13

PRECAUTIONS AND CONTRAINDICATIONS TO RANGE OF MOTION ASSESSMENT

CONTRAINDICATIONS

Dislocations

Unhealed fracture

Myositis ossificans

Immediately following surgery to tendons,
ligaments, muscles, joint capsule, or skin

PRECAUTIONS

Infected joints

Inflamed joints

Client on pain medication

Client on muscle relaxants

Marked osteoporosis

Hypermobile joints

Subluxed joints

Hemophiliacs

Regions of hematomas (especially elbow, hip, knee)

Bony ankylosis

Demeter, S. L., Andersson, G., & Smith, G. M. (1996). *Disability evaluation.* St. Louis, MO: Mosby.

Dirks, J. F., Wunder, J., Kinsman, R., McElhinny, J., & Jones, N. F. (1993). A Pain Rating Scale and a Pain Behavior Checklist for clinical use: development, norms, and the consistency score. *Psychotherapy and Psychosomatics, 59*(1), 41-49.

Donatelli, R. A., & Wooden, M. J. (2010). *Orthopaedic physical therapy* (4th ed.). St. Louis, MO: Churchill Livingstone Elsevier.

Ellis, B., & Bruton, A. (2002). A study to compare the reliability of composite finger flexion with goniometry for measurement of range of motion in the hand. *Clinical Rehabilitation, 16,* 562-570.

Fisher, G. S., Emerson, L., Firpo, C., Ptak, J., Wonn, J., & Bartolacci, G. (2007). Chronic pain and occupation: An exploration of the lived experience. *American Journal of Occupational Therapy, 61*(3), 290-302.

Flinn, N. A., Latham, C. A. T., & Podolski, C. R. (2008). Assessing abilities and capacities: range of motion, strength, and endurance. In M. V. Radomski & C. A. T. Latham (Eds.), *Occupational therapy for physical dysfunction* (6th ed., pp. 91-185). Philadelphia, PA: Wolters Kluwer/Lippincott, Williams and Wilkins.

Flinn, S., & DeMott, L. (2010). Dysfunction, evaluation, and treatment of the wrist and hand. In R. A. Donatelli & M. J. Wooden (Eds.), *Orthopaedic physical therapy* (pp. 267-312). St. Louis, MO: Churchill Livingstone Elsevier.

Gajdosik, R. L., & Bohannon, R. L. (1987). Clinical measurement of range of motion: Review of goniometry emphasizing reliability and validity. *Physical Therapy, 67*(12), 1867-1872.

Gerhardt, J. J. (1983). Clinical measurements of joint motion and position in the neutral-zero method and SFTR recording: Basic principles. *International Journal of Rehabilitation Medicine, 5,* 161.

Goss, C. M. (1973). *Anatomy of the human body* (29th ed.). Philadelphia, PA: Lea & Febiger.

Gutman, S. A. & Schonfeld, A. B. (2003). *Screening adult neurological populations.* Bethesda, MD: American Occupational Therapy Press.

Günal, I., Köse, N., Erdogan, O., Goktürk, E., & Seber, S. (1996). Normal range of motion of the joints of the upper extremity in male subjects with special reference to side. *Journal of Bone and Joint Surgery, 78,* 1401-1444.

Hamill, J., & Knutzen, K. M. (2003). *Biomechanical basis of human movement.* Philadelphia, PA: Lippincott, Williams & Wilkins.

Hamilton, G. F., & Lachenbruch, P. A. (1969). Reliability of goniometers in assessing finger joint angle. *Physical Therapy, 49,* 465.

Hellebrandt, R. A., Duvall, E. N., & Moore, M. L. (1949). The measurement of joint motion: Part 3: Reliability of goniometry. *Physical Therapy Review, 29,* 302-307.

Jarvis, C. (2000). *Physical examination and health assessment.* Philadelphia, PA: W. B. Saunders.

Jette, A. M. (1980). Functional capacity evaluation: an empirical approach. *Archives of Physical Medicine and Rehabilitation, 61,* 86-89.

King, D. G. (2010). Histology at the Southern Illinois University School of Medicine 2010, from http://www.siumed.edu/~dking2/index.htm; Accessed March 3, 2010.

Kisner, C., & Colby, L. A. (2007). *Therapeutic exercise: Foundations and techniques* (5th ed.). Philadelphia, PA: F. A. Davis Company.

Konin, J. G. (1999). *Practical kinesiology for the physical therapy assistant.* Thorofare, NJ: SLACK Incorporated.

LaStayo, P. C., & Wheeler, D. L. (1994). Reliability of passive wrist flexion and extension goniometric measurements: A multicenter study. *Physical Therapy, 74,* 162-176.

Leubben, A. J., & Royeen, C. B. (2005). Nonstandardized testing. In J. Hinojosa, P. Kramer, & P. Crist (Eds.), *Evaluation: Obtaining and interpreting data* (2nd ed., pp. 125-146). Bethesda, MD: American Occupational Therapy Association.

Lippert, L. S. (2006). *Clinical kinesiology and anatomy.* Philadelphia, PA: F. A. Davis Company.

Lorch, A., & Herge, E. A. (2007). Using standardized assessments in practice. *Occupational Therapy Practice,* May 28, 17-22.

Magee, D. J. (2002). *Orthopedic physical assessment.* Philadelphia, PA: Saunders.

Melzak, R. (1967). The McGill pain questionnaire: Major properties and scoring. *Pain, 1,* 277-299.

Muscolino, J. E. (2006). *Kinesiology: The skeletal system and muscle function.* St. Louis, MO: Mosby.

Nordin, M., & Frankel, V. H. (2001). *Basic biomechanics of the musculoskeletal system.* Philadelphia, PA: Lippincott, Williams & Wilkins.

Norkin, C. C. & White, D. J. (1985). *Measurement of joint motion: A guide to goniometry.* Philadelphia, PA: F.A. Davis Co.

Nyland, J. (2006). *Clinical decisions in therapeutic exercise: Planning and implementation.* Upper Saddle River, N.J.: Pearson Prentice Hall.

Palmer, M. L., & Epler, M. E. (1998). *Fundamentals of musculoskeletal assessment techniques* (2nd ed.). Philadelphia, PA: Lippincott.

Pedretti, L. W. (1996). *Occupational therapy: Practice skills for physical dysfunction.* St. Louis, MO: Mosby.

Pollard, C. A. (1984). Preliminary validity study of the Pain Disability Index. *Perceptual Motor Skills, 59*(3), 974.

Radomski, M. V. & Latham, C. A. T. (2008). *Occupational therapy for physical dysfunction.* Philadelphia, PA: Lippincott, Williams & Wilkins.

Reed, K. L. (2001). *Quick reference to occupational therapy* (2nd ed.). Gaithersburg, MD: Aspen Publishing.

Roach, K. E. & Miles, T. P. (1991). Normal hip and knee active range of motion: The relationship to age. *Physical Therapy 71*(9), 656-665.

Rothstein, J. M., Roy, S. H., Wolf, S. L., & Scalzitti, D. A. (2005). *The rehabilitation specialist's handbook* (3rd ed.). Philadelphia, PA: F. A. Davis Company.

Sabari, J. S., Maltzev, I., Lubarsky, D., Liszkay, E., & Homel, P. (1998). Goniometric assessment of shoulder range of motion: Comparison of testing in supine and sitting positions. *Archives of Physical Medicine and Rehabilitation, 79*(6), 647-651.

Smith, J. (2005). *Structural bodywork.* St. Louis, MO: Elsevier Churchill Livingstone.

Snyder, D. C., Conner, L. A., & Lorenz, G. (2007). *Kinesiology foundations for OTAs.* Clifton Park, N.Y.: Thomson Delmar Learning.

Solet, J. M. (2008). Optimizing personal and social adaptation. In M. V. Radomski & C. A. T. Latham (Eds.), *Occupational therapy for physical dysfunction* (6th ed., pp. 924-950). Philadelphia, PA: Wolters Kluwer/Lippincott Williams & Wilkins.

Somers, D. L., Hanson, J. A., Kedzierski, C. M., Nestor, K. L., & Quinlivan, K. Y. (1997). The influence of experience on the reliability of goniometric and visual measurement of forefoot position. *Journal of Orthopaedic Sports Physical Therapy, 25,* 192-202.

Tan, J. C. (1998). *Practical manual of physical medicine and rehabilitation.* St. Louis, MO: C.V. Mosby.

Trew, M., & Everett, T. (Eds.). (2005). *Human movement: An introductory text.* London: Elsevier Churchill Livingstone.

Trombly, C. A. (Ed.) (1995). *Occupational therapy for physical dysfunction* (4th ed.). Baltimore, MD: Williams & Wilkins.

Van Deusen, J., & Brunt, D. (1997). *Assessment in occupational therapy and physical therapy.* Philadelphia, PA: W. B. Saunders Co.

Vasen, A. P., Lacey, S. H., Keith, M. W., & Shaffer, J. W. (1995). Functional range of motion of the elbow. *The Journal of Hand Surgery, 20A*(2): 288-292.

ADDITIONAL RESOURCES

Baxter, R. (1998). *Pocket guide to musculoskeletal assessment.* Philadelphia, PA: W. B. Saunders Co.

Christiansen, C., & Baum, C. (1997). *Occupational therapy: Enabling function and well being* (2nd ed.). Thorofare, NJ: SLACK Incorporated.

Durward, B. R., Baer, G. D., & Rowe, P. J. (1999). *Functional human movement: Measurement and analysis.* Oxford: Butterworth Heinemann.

Dutton, R. (1995). *Clinical reasoning in physical disabilities.* Baltimore, MD: Williams & Wilkins.

Esch, D. L. (1989). *Analysis of human motion.* Minneapolis, MN: University of Minnesota.

Falkenstein, N., & Weiss-Lessard, S. (1999). *Hand rehabilitation: A quick reference guide and review.* St. Louis, MO, Mosby Publisher.

Flinn, N. A., Jackson, J., et al. (2008). Optimizing abilities and capacities: Range of motion, strength and endurance. In: M. V. Radomski & C. A. T. Latham, *Occupational therapy for physical dysfunction* (pp. 573-597). Philadelphia, PA: Wolters Kluwer/Lippincott Williams & Wilkins.

Galley, P. M., & Forster, A. L. (1987). *Human movement: An introductory text for physiotherapy students* (2nd ed.). New York, NY: Churchill-Livingstone.

Gench, B. E., Hinson, M. M., & Harvey, P. T. (1995). *Anatomical kinesiology.* Dubuque, IA: Eddie Bowers Publishing, Inc.

Greene, D. P., & Roberts, S. L. (1999). *Kinesiology: Movement in the context of activity.* St. Louis, MO: C. V. Mosby.

Hall, C. M., & Brady, L. T. (1999). *Therapeutic exercise: Moving toward function.* Philadelphia, PA: Lippincott Williams & Wilkins.

Hamill, J., & Knutzen, K. M. (2003). *Biomechanical basis of human movement.* Philadelphia, PA: Lippincott Williams & Wilkins.

Hayes, C. (Ed.). (1980). *Sample forms for occupational therapy.* Rockville, MD: American Occupational Therapy Association.

Jarvis, C. (2000). *Physical examination and health assessment.* Philadelphia, PA: W. B. Saunders Co.

Magee, D. J. (2002). *Orthopedic physical assessment.* Philadelphia, PA: W. B. Saunders Publisher.

Melvin, J. L. (1982). *Rheumatic disease: Occupational therapy and rehabilitation* (2nd ed.). Philadelphia, PA: F. A. Davis Co.

Norkin, C. C., & Levangie, P. K. (1992). *Joint structure and function: A comprehensive analysis* (2nd ed.). Philadelphia, PA: F. A. Davis Co.

Nyland, J. (2006). *Clinical decisions in therapeutic exercise: Planning and implementation.* Upper Saddle River, N. Y.: Pearson Prentice Hall.

Oatis, C. A. (2004). *Kinesiology: The mechanics and pathomechanics of human movement.* Philadelphia, PA: Lippincott Williams & Wilkins.

Prosser, R., & Conolly, W. B. (2003). *Rehabilitation of the hand and upper limb.* London: Butterworth Heinemann.

Shankar, K. (1999). *Exercise prescription.* Philadelphia, PA: Hanley & Belfus, Inc.

Van Ost, L. (2000). *Goniometry: An interactive tutorial.* Thorofare, NJ: SLACK Incorporated.

Zimmerman, M. E. (1969). The functional motion test as an evaluation tool for patients. *Am J Occup Ther, 23*(1), 49-56.

4

FACTORS INFLUENCING STRENGTH

Strength can be defined simply as a muscular force or power, which is a vague, unscientific definition. A more elaborate definition might be a force or torque produced by a muscle during maximal voluntary contraction. The differentiation of strength, power, and endurance is necessary at this point. Strength and power are not the same. Strength is a force that is directly related to the amount of tension a muscle can produce. Power is the product of force and velocity (Hamill & Knutzen, 1995) or the amount of work per unit of time. Power requires timing and coordination. Endurance is the ability to maintain a force over time or for a set number of contractions or repetitions. Whether muscles contract for strength, power, or for endurance, the development of active force comes from the muscles, and many factors contribute to this development of force in the muscles (Figure 4-1).

CLIENT FACTORS

Changes in muscle strength occur due to changes in muscle, the nervous system, and immunological changes as well as individual client factors such as activity levels, health status, nutrition, age and gender, cognitive status, and individual responses to stress and pain.

Age and Gender

Developmentally, both genders develop the greatest strength capacities from birth through adolescence, with peak strength between 20 and 30 years of age. Both genders experience a decrease in strength with increasing age due to deterioration of muscle mass, decreased muscle fiber size and number, increases in connective tissue and fat, decreased vascularization, altera-

tions in capacity to generate force, and decreased respiratory capacity of the muscle.

Age

Muscle and strength changes occur throughout the lifespan. At birth, muscle accounts for about 25% of the body weight, and the total number of muscle fibers is established prior to or early during infancy. In puberty, there is a rapid acceleration in muscle fiber size and mass, and marked differences in strength levels develop between boys and girls. Muscle mass in women peaks between 16 and 20 years of age and between 18 and 25 for men. By the third decade, strength declines between 8% and 10% per decade through the fifth or sixth decades (Kisner & Colby, 2007).

Just how muscle strength decreases with age has been attributed to a variety of changes in the muscle. Muscle function changes due to muscle mass is one variable. Tjoi and Kaneko (2007) found that the decline in muscle force and shortening velocity was attributed to declining muscle function rather than a decrease in cross-sectional area of muscles, which was true in a study by Runnels and colleagues in which decreases in both upper and lower extremity muscle function in the elderly occurred, independent of changes in muscle mass (Runnels, Bemben, Anderson, & Bemben, 2005). In contrast to this finding, other studies indicate that muscle decline is due to decreased percentages of muscle fibers and loss of muscle mass (Doherty, 2001; Gaur, Shenoy, & Sandhu, 2007). Dalbo and colleagues identify the loss of type II muscle fibers and an increase in intramuscular fat storage as a change occurring with age in which 20% to 30% of lean body mass is lost between the third and eighth decades of life (Dalbo, Roberts, Lockwood, Tucker, Kreider, & Kerksick, 2009).

Muscle fiber composition is another factor influencing muscle function with age. There is a shift in fiber composition,

Rybski MF.
Kinesiology for Occupational Therapy, Second Edition (pp 65-94)
© 2012 SLACK Incorporated

Figure 4-1. Summary of factors influencing strength.

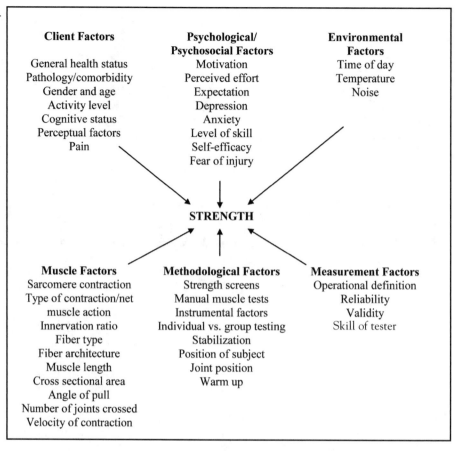

Client Factors

General health status
Pathology/comorbidity
Gender and age
Activity level
Cognitive status
Perceptual factors
Pain

**Psychological/
Psychosocial Factors**
Motivation
Perceived effort
Expectation
Depression
Anxiety
Level of skill
Self-efficacy
Fear of injury

**Environmental
Factors**
Time of day
Temperature
Noise

STRENGTH

Muscle Factors
Sarcomere contraction
Type of contraction/net
muscle action
Innervation ratio
Fiber type
Fiber architecture
Muscle length
Cross sectional area
Angle of pull
Number of joints crossed
Velocity of contraction

Methodological Factors
Strength screens
Manual muscle tests
Instrumental factors
Individual vs. group testing
Stabilization
Position of subject
Joint position
Warm up

Measurement Factors
Operational definition
Reliability
Validity
Skill of tester

resulting in a greater percentage of type I fibers enabling more prolonged contraction of muscle fibers as compared to type II fibers, which are involved in high-force, short-duration contractions (Dalbo et al., 2009). Anderson, Erzis, and Kryger suggest that, in the elderly, there are greater percentages of fibers that are a cross between type I and type II fiber types (Anderson, Erzis, & Kryger, 1999). Changes in muscle fiber types may affect the types of activities done and level of endurance in the elderly.

The decrease in type II fibers may be related to age-related loss of motor neurons, which occurs at a rate of 1% of the total number lost per year (Rice, 2000). Coupled with decreases in motor axon conduction velocity, decreased rates of protein synthesis, and increased muscle fatigability, muscle force generation is decreased (Bello-Haas, 2009). In addition, due to decreased strength, greater recruitment of other muscles is necessary to perform activities as a compensatory mechanism, which may lead to instability and structural deformities (Gaur et al., 2007).

Gender

Men, with bigger bones and larger muscles, are stronger than women. Generally, women are 40% to 50% weaker in the upper body and 20% to 30% weaker in the lower body (Tan, 1998). Women may experience strength losses earlier than men, but, overall, age-associated decreases in strength are similar when controlling for muscle mass. The rate of isokinetic strength loss for women is 2% per decade, while it is 12% per decade for men (Frontera, Hughes, Fielding, Fiatarone, Evans,

& Roubenoff, 2000; Hughes, Frontera, Wood, Evans, Dallal, Roubenoff et al., 2001). Older subjects demonstrated a greater rate of decline in strength. Men may experience greater losses of total muscle mass, but there are greater declines in muscle quality in older women (Doherty, 2001).

Many of the studies that compare strength by gender also compare strength based on the type of muscle action. Overall, isometric performance is the least affected type of muscle action (Runnels et al., 2005) and concentric the most (Lindle, Metter, Lynch, Fleg, Fozard, Tobin et al., 1997; Lynch, Metter, Lindle, Fozard, Tobin, Roy et al., 1999; Pousson, Lepers, & Hoecke, 2001). Muscle quality is affected by age and gender, but the magnitude of this effect depends on the muscle group studied and the type of muscle action (Lynch et al., 1999).

Individual Differences

Individual differences may be due to genetic factors, engagement in specific activities, lifestyle choices, cultural beliefs, gender roles, overall health and nutrition, and any pre-existing co-morbidities. Inheritable factors may result in differences in tissue anatomy, resulting in variations in rates of atrophy, the development of hypertrophy, and the physical capacity of tissues. Individuals also vary in their tissue physiology, affecting rates of healing and remodeling potential in one's response to chronic loading. This may be genetically determined or as the result of engaging in specific activities. Large variations occur in each gender due to diet, exercise, and level of activity.

While changes in muscle function are age-related, the amount of decreased strength is a function of training and regular exercise (Gaur et al., 2007; Rantanen, Guralnik, Sakar-Rantala, Leveille, Simonsick, & Ling, 1999). Decreased physical activity is associated with increased disability and loss of muscle strength, so continuation of valued occupations throughout life for our clients is a vital part of occupational therapy intervention.

Pain

Pain is a subject-related variable important in strength assessment. Pain is a multidimensional phenomenon with cognitive and affective components as well as physiologic properties that also have culturally determined values. How one perceives pain and reacts to it is part of a culturally sanctioned role that contributes to a person's ability to function in his or her environment. When assessing strength, indicators of the client's willingness to endure discomfort or pain can be observed. Pain invalidates the assessment of strength if the person is unable to provide a maximal voluntary contraction of the muscle being tested, so observation of discomfort and pain is important in the assessment of strength.

Muscle pain may be due to delayed-onset muscle soreness that occurs 1 or 2 days after injury. It may be possible to localize areas of palpable tenderness, and the client may experience loss of range of motion and muscle stiffness. This type of pain is also characterized by reduced blood flow to the area, restricting oxygen to muscle tissues. Acute muscle soreness may occur immediately after exercise especially if the activity required isometric contractions. The client may experience a burning sensation (Konin, 1999).

Cognitive Status and Perceptual Factors

When assessing a client's strength, the ability to follow directions, plan and execute the motor action, and the ability to pay attention are all factors influencing the test and can be observed. It is important to differentiate whether the client cannot complete the test motion due to motor planning problems (apraxia) or due to cognitive impairments. Language may be another reason a client is unable to follow directions, or there may be pathological reasons for lack of communication (as in aphasia).

PSYCHOLOGICAL/PSYCHOSOCIAL FACTORS

Often, clients want to perform well on tests of strength, and the link between muscle force and function seems clear. Motivation is related to accuracy in muscle testing and depends on the benefits of performance to occupation, the ability to understand what is expected, fear of pain or injury, and anxiety about movement. Occasionally, true effort may not be demonstrated, which may be due to perceived benefits of secondary gains (such as discontinuation of work), parental or other roles, or due to depression.

ENVIRONMENTAL FACTORS

Controlling environmental variables such as noise and number of people may facilitate better performance on tests of strength with clients. Temperature has a direct bearing on muscle activity. Muscle function is most efficient at around 101°F (38.5°C). A rise in muscle temperature causes an increase in conduction velocity and frequency of stimulation, thereby increasing the muscle force. Increased enzyme activity results and is related to greater efficiency in muscle contractions. Heat increases the elasticity of the collagen in the passive components (Nyland, 2006). These factors combined produce increased tension and a stronger muscular effort by the client. Additional environmental factors that might influence maximal muscle output are air quality, tasks with repetitive motion or excessive force, static exertion, awkward postures, and mechanical stress.

MUSCULAR FACTORS

Muscles contribute to the production of skeletal movement, which is the primary focus of this chapter. However, muscles also assist in joint stability, maintenance of posture, support of visceral organs, protection of internal tissues from injury, continuation of pressure within body cavities, contribution to the maintenance of body temperature, and control of swallowing and bowel and bladder functions (Hamill & Knutzen, 2003).

Characteristics of Muscles

Of particular interest in the study of movement is the ability of the muscular system to stabilize and support the body and to allow movement. Muscle actions generate tension that is then transferred to bone. Muscles contribute significantly to joint stabilization when the muscle tension is generated and applied across joints via tendons. This is especially true in the shoulder and knee joints (Hamill & Knutzen, 2003). The tension developed by muscles applies a compression force to the joints, enhancing stability. Muscles could also pull segments apart and create instability, depending upon the line of pull of the muscle and the direction of movement.

Skeletal muscles are different from smooth and cardiac muscles in that each skeletal muscle receives a branch from a nerve cell called an alpha motor neuron, which is part of the somatic nervous system. These alpha motor neurons signal the muscle fibers to contract. Contraction of muscle fibers and not the muscle as a whole enable fine gradations of force to be produced by these muscles. Cardiac and smooth muscles are innervated by the autonomic nervous system, and muscle contraction is not determined directly by a single message from the nerve nor under voluntary control (Irion, 2000).

Irion states that "muscles act as a transducer to convert chemical energy into mechanical energy producing the force necessary to provide movement, support, and other mechanical functions" (2000, p. 206). This being true, it is reasonable that skeletal muscles are the main energy-consuming tissues in the body. However, of the energy produced during a muscle contraction, only 20% of that energy is used to produce movement; the rest is lost as heat (Hamill & Knutzen, 2003).

Figure 4-2. Organization of muscle fibers.

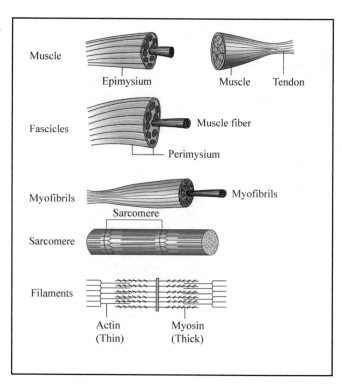

Muscles are also the most abundant tissue in the body, making up 40% to 45% of the total body weight. While it is obvious that muscles contribute to joint movement and strength, muscles also add protection to the skeleton by distributing loads and acting as shock absorbers.

Skeletal muscles have four general characteristics. Contractility is the ability to produce tension between the ends of two bones to exert a pull, as when a muscle contracts. Irritability (excitability) is the ability of the muscle to respond to stimuli and transmit impulses. A muscle is distensible (or extensible) because it can be lengthened or stretched by a force outside the muscle itself, which is used therapeutically to increase range of motion (ROM). Elasticity describes the ability of a muscle to recoil from a distended stretch. Not all parts of the muscle have each of these characteristics; some parts of the muscle contain the contractile elements, and other parts contain the elastic components (one of two major parts of connective tissue with the other part being collagen).

Skeletal muscle contains muscle tissue and fibrous fascial connective tissue. The fascia wraps around muscle. The fascia also may extend beyond the muscle to create tendons, which attach to bones. The major component of fascia is collagen with only small amounts of elastin. The collagen provides strength to the fascia so that muscle contraction forces can be transferred to the bone (Muscolino, 2006).

Structural Unit

The structural unit of skeletal muscle is the muscle fiber (Figure 4-2). Fibers range in thickness and length. Fibers are organized into various-sized bundles called fascicles; there may be as many as 200 muscle fibers in one fascicle. Fascicles are encased in dense connective tissue (perimysium), which creates pathways for nerves and blood vessels. While each muscle is covered externally by epimysium, each muscle fiber is encased in a membrane (endomysium), which carries capillaries and nerves that innervate and nourish each muscle fiber. The connective tissue in the perimysium and epimysium gives the muscle the ability to be stretched. Perimysium is the focus of flexibility training because it can be stretched, which enables muscle elongation (Hamill & Knutzen, 2003). The function of the epimysium is to transfer muscular tension to tendons and then to bone.

Each muscle fiber is composed of a large number of delicate strands called myofibrils. Each muscle fiber is filled with 80% myofibrils, and the remaining 20% is made up of mitochondria, sarcoplasm, reticulum, and T tubules (Hamill & Knutzen, 2003). Myofibrils are made up of filaments. One unit consists of thin, light bands of myofilaments and thick, dark bands and is called a sarcomere, which is the basic contractile unit of the muscle. The dark, thick bands are made up of myosin (tropomyosin) proteins, and the light, thin bands are actin proteins. The bands of light and dark myofilaments are what give skeletal muscles the characteristically striped (or striated) appearance. These two proteins, actin and myosin, create muscle fiber contraction due to the creeping action of actin protein along the thicker myosin protein.

Sarcomere Contraction

The sliding filament theory of muscle fiber contraction describes this creeping action of the thick and thin filaments as they increase or decrease their degree of overlap. Figure 4-3 illustrates the movement of the actin and myosin myofilaments. Once a muscle is stimulated to contract via neu-

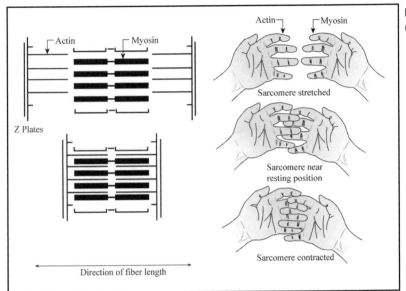

Figure 4-3. Myofilament movement of actin and myosin.

rochemical stimulation, calcium is released, which causes the sliding of actin to the center of the sarcomere, which is likened to "pulling on a rope hand over hand" (McPhee, 1987). This stimulation of single muscle cells occurs in an "all-or-none" manner where the cells either respond completely or they don't respond at all. The sarcomere, myofibril, and muscle fiber are all innervated by a single motor neuron that carries the message to either contract or not. This contraction does not extend to the entire muscle, because there are many motor units in the entire muscle. This all or none property permits partial contraction of a muscle to create the right amount of force required for the muscle action.

If the stimulus is adequate, the myosin protein moves along the actin, forming cross-bridges between the head of the myosin and the actin filaments. Each cross-bridge acts independently. It is the simultaneous sliding of many filaments that creates a change in length and force in the muscle, which is proportional to the number of cross-bridges formed. The maximum number of cross-bridges between the actin and myosin proteins and the maximum contractile force in sarcomeres occurs when the full length of the actin at each end of sarcomere is in contract with myosin. This occurs when the muscle is in a lengthened position. The action of the myofibril cross-bridging is the active, contractile element of the muscle (Oatis, 2004).

Active Components of Muscle Contraction

Voluntary contraction of a skeletal muscle requires the interaction of the contractile and elastic components of the muscle, intact nervous system input, and sufficient circulation and nutrition provided by the vascular system. The motor units and sensory receptors provide critical information about the status of the muscle and the extent of the demand requiring a motor response. The vascular tissues provide the energy for muscle contraction. While the emphasis of this chapter is on the contractile tissues to produce skeletal movement, rec-

ognizing the importance of the neural and vascular structures is crucial to consider in treatment of muscular limitations.

The contractile structures include the muscle body, musculotendinous junction, and the tendon. A muscle contraction is a shortening of the fibers toward the center. This creates a pulling force on the bones attached to the muscle, ultimately resulting in movement of the body part (Muscolino, 2006). Contractile tissue has both passive and active components.

Passive Components of Muscle Contraction

Two noncontractile (or inert) components of the muscle serve to absorb, transmit, and store energy (Hamill & Knutzen, 2003). The first is the series elastic components, which are found primarily in tendons (85%) with the remaining found in actin-myosin cross-bridges. When the active contractile components shorten, this stretches the series elastic components, which act like a spring to maintain the tendon-muscle length and slows down the forces generated (Hamill & Knutzen, 2003). Of the two passive components, the series elastic components are more important in the production of muscle tension.

The elastic components not only prevent overstretching of the muscle but also ensure that the contractile elements return to their resting length after a muscle contraction. The elastic components keep the muscle in readiness for contraction and ensure that muscle tension is produced and transmitted smoothly during contraction (Nordin & Frankel, 2001).

The second passive element, the parallel elastic component, is found in the sarcolemma, epimysium, perimysium, and endomysium. These parallel components are important when a passive muscle is being elongated (when the contractile elements are active, the parallel elastic components do not function) (Kisner & Colby, 2007). As the passive muscle is lengthened, the parallel component offers an opposing force and prevents the contractile elements from being torn apart by external forces.

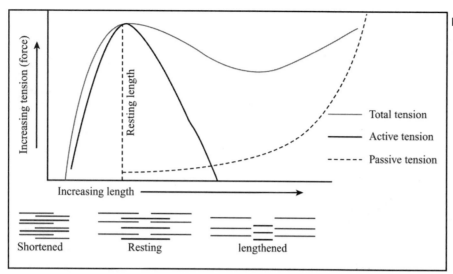

Figure 4-4. Length-tension curve.

Stretching a muscle elongates both passive structures, which generates a springy resistance in the muscle. Without a muscle contraction, the resistance to stretch increases as the stretch force continues due to the elastic recoil of the parallel elastic components and the slowing action of the series elastic components. This is passive muscle tension.

When a muscle is stretched prior to testing, it will have greater tension because the elastic energy stored in the non-contractile elastic elements is converted to kinetic energy when stretched (Kisner & Colby, 2007). This passive tension helps prevent maximal elongation, and the viscosity of the stretched fibers helps protect muscles from being damaged by quick, forceful stretch due to prolonging the application of force to allow a more gradual elongation (Neumann, 2002).

Length-Tension Curve

The active and passive tension produced by muscle fibers follows a specific pattern represented by the length-tension curve in Figure 4-4. The curve represents the relationship between the length of the sarcomere and the tension that it generates at that length (Muscolino, 2006). It is valid for isometric contractions and for lengthening muscles (Irion, 2000). The solid thick line represents active tension where the strength of the sarcomere's contractions is greatest at the resting length because the greatest number of cross-bridges can be formed. While the active contractile contribution to muscle tension peaks in midrange of the stretch, passive components make an increasing contribution to force after midrange. As the tissues continue to be lengthened, active tension decreases, but passive tension increases. The passive tension force is generated when the sarcomere is stretched. When the muscle is stretched, the overlap area between actin and myosin decreases, and the number of cross-bridges is less, resulting in a decrease in tension in the active components due to minimal overlap of actin and myosin. Continual stretch would result in tearing and failure of the sarcomere. The overall tension developed by a muscle is due to both the active (contractile) and passive (connective tissue) elements. Total tension is represented by the thin line and represents the total pulling force of the sarcomere (Muscolino, 2006).

The length-tension relationship is based not only on the histological overlap of actin and myosin but is also related to biomechanical and neurophysiologic factors.

Biomechanically, the angle of muscle insertion into bone dictates where the greatest tension development will occur. The greatest moment arm for muscle force is when forces are perpendicular to the lever arm, which is seen in length-tension curve at the apex of the active tension curve. Neurophysiologically, joint mechanoreceptors respond to muscle actions at joints. The least amount of power due to firing of the mechanoreceptors is at the beginning and end of the range of motion, which is consistent with the active tension depicted in the length-tension curve (Cooper, 2007).

Muscle Actions

Skeletal muscles as a whole are capable of responding with great variances of force, time, and speed rather than the all-or-none responses of muscle cells. This is due to the nerve-muscle functional unit called the motor unit, which is comprised of a single motor neuron and all of the muscle fibers innervated by it. Motor units affect the functioning of a muscle in:

- The number of muscle fibers, which affects the magnitude of the response.

- The diameter of the axon, which determines the conduction velocity.

- The number of motor units firing at any one time, which affects the total response of the muscle.

- The frequency of the motor unit firing (Nordin & Frankel, 2001).

The number of muscle fibers in a motor unit relates to the degree of control required of that muscle. Small muscles that perform fine movements may have less than a dozen muscle fibers so that the neuron controls only a few muscle fibers for greater control and precision. Large muscles performing grosser movements may have several thousand fibers in one motor unit. Fewer fibers mean more neural control per motor unit. An example of this is the gastrocnemius muscle, which may have 1720 fibers per motor unit while the first lumbrical may have only 110 fibers per motor unit (Guyton, 1991). This innervation ratio has a direct bearing on the skill level achieved by the muscle.

The size of the motor unit is also related to muscle function. Smaller motor units produce contractions of a smaller number of muscle fibers, resulting in actions that are more precise. Larger motor units produce contractions of greater numbers of muscle fibers so these produce larger, more powerful contractions. Each muscle generally has a mixture of small and large motor units (Muscolino, 2006).

Recruiting more motor units to respond to resistance is another way muscles can generate gradations of force. Fibers of a motor unit are not contiguous but are interspersed throughout the muscle with other motor units. When stimulated, all fibers in the motor unit respond, but the muscle as a whole will not unless additional motor units are recruited.

A muscle as a whole contracts as larger numbers of motor units are added to contraction, which is called the graded strength principle. Graded muscle responses occur in two ways: by increasing the rapidity of stimulation, which produces a wave summation, and by recruiting more motor units to produce a multiple motor unit summation.

Wave summation or temporal summation occurs when more than one stimulus is received and the second contraction is induced before the muscle fully relaxes after the first contraction. Because the muscle was already partially contracted, the tension produced by the second contraction causes more shortening than the first by actually "summing" the contractions (Konin, 1999). If the stimulus is applied at a constant rate, the muscle is stimulated at faster rates, and relaxation time decreases. The summation becomes greater until a smooth, sustained contraction occurs, called tetanus. Because continuous, prolonged contraction cannot continue indefinitely, prolonged tetanus leads to muscle fatigue. The repetitive twitching of all recruited motor units of a muscle in an asynchronous manner ensures that every motor unit isn't recruited at the same time, reducing the potential for fatigue while enabling smooth movement to occur (Nordin & Frankel, 2001).

Multiple motor unit summation serves to control the force of muscle contractions more precisely by neural activation of increasingly larger numbers of motor units (Konin, 1999). If weak and precise movements are required, few motor units are activated. When many motor units are stimulated, the muscle contracts more forcefully. The smallest motor units (with the fewest muscle fibers) are controlled by the most excitable motor neurons and are activated first. Larger motor units, activated by less excitable neurons, are activated only when stronger contractions are needed. This illustrates the size principle of recruitment, where small motor units are activated initially, and as the force of contraction increases, gradually larger motor neurons and larger numbers of motor units are required. Small motor neurons generally innervate the types of muscle fibers (type I/slow-twitch) that are required for sustained muscle actions and that fatigue slowly. This ensures weak but sustained postural contraction. By recruiting motor units controlling larger numbers of fibers, more cross-bridges are formed so more tension is produced. Smooth movements occur because different motor units fire asynchronously, so that while one motor unit is firing, another is relaxing, ensur-

ing that even weak contractions are smooth. While the size principle of motor unit recruitment has been demonstrated in some studies, it has not been demonstrated in human muscle (Nyland, 2006).

The Law of Parsimony states that the nervous system tends to activate the fewest muscles or muscle fibers to control a given joint action (Neumann, 2002). This means that smaller, one-joint muscles are activated first, because they require less energy. If greater power is required, larger motor units are recruited, more motor units are added, and increased stimulation of motor units occurs to respond to the force. In this way, unnecessary fatigue is avoided, and yet the necessary muscle force is generated to meet the resistance.

Muscle Fiber Types

There are three types of fibers based on varied physiologic features: Type I fibers, Type IIB fibers, and Type IIA muscle fibers. Each motor unit will have only one type of fiber (Table 4-1).

Type I/tonic (antigravity-postural)/red/slow-twitch oxidative (SO) are muscle fibers with a small diameter that are slow to fatigue. These muscle fibers have a good capillary supply so there less build-up of lactate and metabolic waste. Tonic muscle fibers atrophy almost immediately when immobilized after injury because they depend on oxygen for metabolism (Cooper, 2007). Postural or tonic muscle fibers tend to become tight and hypertonic with pathology and develop contractures (Magee, 2008).

Type I fibers are recruited early but respond with weak contractions due to a small number of fibers activated. Type I fibers are generally involved in maintaining posture against gravity and are generally more deeply located and more medial. Examples of postural muscles are upper trapezius, levator scapulae, pectoralis major (upper part), pectoralis minor, scalenes, erector spinae, quadratus laborum, tensor fascia latae, hamstrings, rectus femoris, short hip adductors, gastrocnemius, soleus, piriformis, iliopsoas, and tibialis posterior muscles.

Type IIB fibers/phasic/white/fast-twitch glycolic (FG) are basically anaerobic because the capillary supply is not as abundant as Type I fibers. Type IIB fibers have large diameters and are recruited later to produce more powerful contractions of the muscle. Type IIB fibers are usually more numerous, producing sharp bursts of energy that enable quick postural changes or skilled movements. While Type IIB fibers contract more readily than Type I, these fibers also fatigue more readily.

Generally, Type IIB muscles are more superficial and laterally located, are often longer muscles, and cross more than one joint. Phasic muscle fibers tend to become weak and inhibited with pathology (Magee, 2008). Examples of phasic muscles are tibialis anterior, trapezius (mid and lower), latissimus dorsi, rhomboids, serratus anterior, rectus abdominus, vastus medialis and lateralis, gluteus maximus, gluteus medius, gluteus minimus, and peroneals.

Type IIA fast-twitch oxidative glycolic (FOG) fibers tend to be intermediate in physiologic characteristics as compared to Type I and Type IIB fibers. FOG fibers have fast conduction rates and are able to engage in both aerobic and anaerobic muscle activities.

Table 4-1

DIFFERENTIATION OF MUSCLE FIBER TYPES

OTHER NAMES	FAST TWITCH GLYCOLIC (FG) (LIGHT WORK) WHITE TYPE I	FAST TWITCH OXIDATIVE (FOG) INTERMEDIATE FAST TWITCH RED TYPE II A	SLOW TWITCH OXIDATIVE (SO) (HEAVY WORK MUSCLES) SLOW TWITCH RED TYPE I
Diameter	Large	Intermediate	Small
Color	White	Red	Red
Capillary supply	Sparse	Dense	Dense
Speed of contraction	Fast	Fast	Slow
Rate of fatigue	Fast	Intermediate	Slow
Motor unit size	Large	Intermediate to large	Small
Axon conduction velocity	Fast	Fast	Slow

Adapted from Norkin, C. C., & Levangie, P. K. (1992). *Joint structure and function: A comprehensive analysis.* (2nd ed.). Philadelphia: F. A. Davis Co.

The ratio of muscle fiber types varies in each individual and is thought to be genetically determined. The ratio of muscle fiber types varies per each muscle, too. It is for this reason that some individuals, with presumed genetically pre-established ratios of specific fiber types, become sprinters (with a predominance of fast-twitch fibers) or marathon runners (with a predominance of slow-twitch fibers), each with a different percentage of fiber type in each muscle.

In terms of tension development, Type IIB muscles are capable of the greatest tension, followed by Type IIA and finally by Type I fibers. In general, fast-twitch muscles are more affected by immobilization and disuse, which has implications for treatment. In addition, the fast-twitch muscles of the lower extremity (especially extensors) are more affected than muscles in the upper extremity. The differentiation of muscles by fiber type is an important consideration when developing intervention that focuses on strengthening muscles, because consideration of intensity and duration of muscle contraction will be variables directly linked to fiber type.

Fiber Architecture (Muscle Morphology)

Muscle fibers are arranged either in a parallel or an oblique manner within the muscle. Parallel or series fibers are longer muscles that are capable of producing greater ROM. Parallel fibers can be subdivided further (Table 4-2). Strap fibers are long and thin, running the entire length of the muscle, as is seen in the sartorius muscle. Fusiform fibers are spindle-shaped or wider at the middle, tapering off at each end (i.e., brachialis muscle or biceps muscle). Rhomboid fibers are four-sided, short, fibers arranged in a flat, rectangular, or square shape (i.e., rhomboid muscle, pronator quadratus muscle). Triangular fibers are flat and fan-shaped, with fibers coming from narrow attachment at one end and broad attachment at the other (i.e., pectoralis major muscle) (Snyder, Conner, & Lorenz, 2007).

The parallel muscle fibers are arranged parallel to the line of pull in a longitudinal manner. The fiber length is greater

Table 4-2

FIBER ARCHITECTURE

PARALLEL OR SERIES FIBERS

Type of Fiber	Example
Strap	Sartorius
Fusiform	Brachialis
Rhomboid	Rhomboid
Triangular	Pronator
	Quadratus
	Pectoralis
	Major

PENNATE/OBLIQUE FIBERS

Type of Fiber	Example
Unipennate	Tibialis posterior
	Flexor pollicis longus
	Semimembranosus
	Extensor digitorum
	longus
Bipennate	Rectus femoris
	Gastrocnemius
	Solus
	Vastus medialis
	Vastus lateralis
Multipennate	Deltoid
	Gluteus maximus
	Subscapularis

than the tendon length, so these muscles have the potential for shortening through greater distances. Contraction occurs through the maximal distance allowed by the length of the muscle fibers so one can achieve maximum range of motion but with limited power.

Oblique fibers can be seen in the tibialis posterior, flexor pollicis longus, and semimembranosus or extensor digitorum longus muscles. These muscles have a series of short fibers attaching diagonally along the length of a central tendon. These are unipennate fibers, which look like a one-sided feather. Bipennate oblique fibers obliquely attach to both sides of a central tendon (i.e., rectus femoris, gastrocnemius, solus, vastus medialis, vastus lateralis muscles). Many tendons with oblique fibers are multipennate fibers and include deltoid, gluteus maximus, and subscapularis muscles.

The oblique arrangement of the fibers makes a diagonal force to line of pull, and the muscle cannot shorten as much as parallel muscles can. However, due to the greater number of muscle fibers in the same area, oblique fibers are capable of greater power. Therefore, the arrangement of the fibers influences whether the muscle is primarily functioning for greater movement and mobility or whether it is more functional for strength and stability. Parallel muscle fibers produce greater range of motion, and oblique fibers can generate greater force. As a comparison, biceps (with a parallel fiber arrangement) are capable of greater range of motion than is possible with

the triceps muscle (an oblique fiber arrangement), which is capable of larger force production.

Angle of Pull

A muscle's action is dependent on its line of pull relative to the joint that it crosses. Lines of pull of muscle fibers can be in a cardinal plane, in an oblique plane, in more than one line of pull, and can cross more than one joint. Brachialis is an example of a muscle with all fibers oriented parallel in the sagittal plane. There is only one muscle action, which is flexion of the forearm at the elbow. Coracobrachialis has muscle action in an oblique plane (or combination of sagittal and frontal planes). The motion produced is flexion in the sagittal plane and adduction in the frontal, and the arm is pulled diagonally in an anterior and medial direction. The deltoid muscle has fibers that run in anterior, posterior, and medial directions so there are three lines of pull of the muscle fibers producing different muscle actions in each direction. Multi-joint muscles have lines of pull on two or more joints. Flexor carpi ulnaris crosses the elbow with one line of pull in a cardinal plane and one in an oblique plane so that this muscle can flex the forearm at the elbow in the sagittal plane and flex the wrist in an ulnar direction at the wrist in the sagittal and frontal planes (Muscolino, 2006).

Figure 4-5. Effects of different angles of pull of muscle fibers.

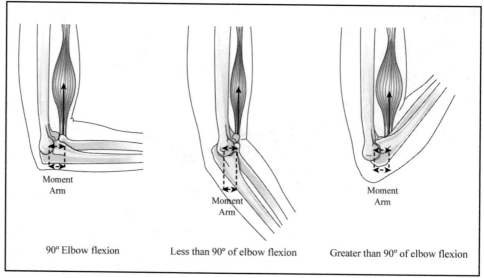

Moment Arm

90° Elbow flexion

Moment Arm

Less than 90° of elbow flexion

Moment Arm

Greater than 90° of elbow flexion

If the line of pull to the joint changes, the muscle action changes. For example, the clavicular head of pectoralis major is an adductor of the shoulder until the arm is adducted to more than 100 degrees and the clavicular head becomes an abductor. Clinical implications of this are that if the supraspinatus and deltoid become weak, pectoralis major can contribute to adduction (Muscolino, 2006).

When a muscle contracts, it creates a force that causes the body segment in which it inserts to rotate around an axis of the joint. This turning effect, or torque, is the product of the muscle force and the perpendicular distance between the axis of rotation and the muscle force (Gench, Hinson, & Harvey, 1995; Neumann, 2002). Strength is the product of muscle force and the distance between the muscle's line of force and the axis of rotation. This length of the moment arm (or torque) changes through range of motion. This is why muscle strength is greater in some positions and why, during a manual muscle test, minimal force can elicit maximal strength (Neumann, 2002).

The optimal angle of muscle pull occurs when the muscle is pulling at a 90 degrees angle or perpendicular to the bony segment. It is at this point that all of the muscle force is acting to rotate the segment, and no force is used to distract or stabilize the limb (Gench et al., 1995). If the muscle is pulling at an optimal length or is perpendicular to the bony segment, then this will produce a stronger contraction.

The angles of pull and length-tension relationships interact to produce this force. Generally, there is a decrease in force production in the extreme outer and inner ranges, with greater force produced in the middle range of motion. As a muscle begins contracting through its full range of shortening, it begins in a weakened condition, gradually becomes stronger, and then it approaches its shortest length and becomes weakened again. The force resolution of the biceps muscle demonstrates that, mechanically, the biceps muscle reaches peak strength at 90 degrees of elbow flexion or in the middle of full range of motion (Figure 4-5). There is a larger moment

arm, and 100% of the muscle force is rotary movement. If the elbow is in full extension, the angle of the biceps insertion is small, and there is a long stabilizing component that makes the biceps muscle less effective or forceful in this position. In elbow flexion greater than 90 degrees (or stated another way, when the angle of muscle attachment is less than 90 degrees), the resolution of forces yields an angular and a dislocating force that runs along the bone and away from the joint. These forces decrease the effectiveness of the biceps muscles as an elbow flexor because the forces are now directed toward dislocating the joint, not in force production (Gench et al., 1995).

Types of Muscle Contraction

When muscle fibers change in length, different forces are generated. In isotonic muscle contractions, internal forces result in movement of the joint, which may include lengthening (eccentric) or shortening (concentric) (Figure 4-6).

Many texts refer to concentric and eccentric contractions as isotonic contractions, although some texts define muscle contractions not only in terms of tension/energy/length but also in terms of the leverage effects at the joint (Richards, Olson, & Palmiter-Thomas, 1996). By strict definition, isotonic means "equal tension," and because the muscle force changes through range of motion, muscle tension must also change. Norkin and Levangie (1992) add that "the tension generated in a muscle cannot be controlled or kept constant" and that the concept of equal or constant tension in a muscle is "unphysiologic" (p. 107). So in the strictest sense, isotonic contractions do not exist in the production of the joint motion. In general use, the term "isotonic" refers to a description of muscle length and not tension, and isotonic contractions may be one of two types: concentric or eccentric.

Concentric contractions occur when the internal force produced by a muscle is greater than the external force or resistance that produces a shortening of the muscle. An example of this is when one is walking up stairs. The quadriceps is

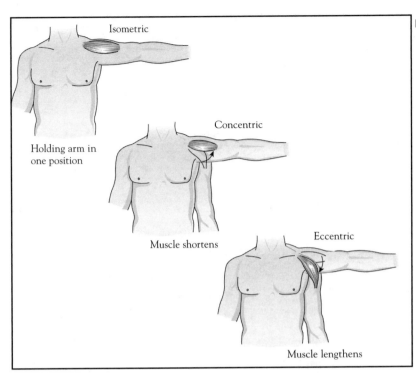

Figure 4-6. Types of muscle contractions.

Isometric

Holding arm in
one position

Concentric

Muscle shortens

Eccentric

Muscle lengthens

demonstrating concentric contractions in the extending knee. Concentric muscle contractions produce the least force output as compared with eccentric or isometric contractions.

Eccentric contractions produce a lengthening of the muscle as a whole because the internal force produced by the muscle is less than the external force or resistance.

The antagonist muscle can eccentrically contract and lengthen, which serves as a braking action to motion. The antagonistic muscle can also relax and lengthen to allow movement.

Eccentric muscle contractions can develop the same force output as other types of muscle contractions at the level of the sarcomere but with fewer muscle fibers activated. This type of contraction is seen as more efficient and uses less oxygen consumption (Hamill & Knutzen, 2003). Eccentric contractions help to regulate movements caused by external forces, such as gravity, and can be seen when one lowers an object onto a table or one eases into a chair. Eccentric contractions act as a brake to decelerate motion of the joint as seen in the quadriceps muscles when one descends stairs. Muscle tension is less than gravity pulling the body down but sufficient to allow controlled movement.

Eccentric contractions occur in every movement in the direction of gravity. In eccentric contractions, active muscles are those that are antagonists of the same movement when it is made against gravity. Hamill and Knutzen (2003) state that "even the weakest individual may be able to perform controlled lowering of body part or small weight but not able to hold or raise the weight" (p. 73), so strengthening activities and exercise programs might start by using eccentric muscle contractions.

Eccentric muscle contractions are thought to be related to delayed-onset muscle soreness (DOMS), which occurs 24 to 48 hours after exercise. This occurs because cross-bridges of myosin stay attached to the active sites while the resistance lowers, resulting in "tearing" away of cross-bridges when lowering against heavy resistance. Muscle swelling occurs, caused by damage to muscle and shifts in muscle length (Cooper, 2007; Donatelli & Wooden, 2010).

Isometric contractions of muscles happen when the internal force generated by the contractile elements of the muscle does not exceed the external force of resistance. The tension produced against resistance is in equilibrium, and there will be no change in the external muscle length, no motion, and no mechanical work. The portion of the myosin filament that pulls the actin toward center of sarcomere is equal to resistive force (Cooper, 2007). There is static physiologic work being done (energy is expended), but no joint (mechanical) work is done. It is important to point out that all dynamic work involves an initial static (isometric) phase as the tension in the muscle develops. The types of muscle contractions are summarized in Table 4-3.

Isometric contractions enable muscles to act in a restraining or holding action. Muscles acting as stabilizers and some synergists produce tension equal to the resistance that it needs to overcome. An example of an isometric contraction is a contraction in which a constant external load is lifted at the extreme ends of motion. When you grasp the handle of a coffee cup, there is muscle tension but no movement. The sustained contraction helps to hold the cup.

In terms of force production, eccentric muscle contractions produce the greatest force followed by isometric. Eccentric muscle contractions produce two to three times the amount of force that a concentric contraction can produce (Donatelli & Wooden, 2010).

Table 4-3

TYPES OF MUSCLE CONTRACTIONS

Static muscle work	Isometric contraction	No joint movement Joint tension
Dynamic muscle work	Concentric contraction	Internal muscle force > external force Muscle shortens
	Eccentric contraction	Internal muscle force Muscle lengthens < external force
	Isokinetic contraction	Constant velocity of joint motion Maximal muscle moment produced
	Isoinertial contraction	Constant external load Submaximal muscle moment produced

Length of Muscle

The tension developed within a muscle depends upon the initial length of the muscle. A muscle is capable of generating maximal force at or near resting length because a maximal number of cross-bridges of actin and myosin can be formed. Moderate tension is produced when the muscle is lengthened, and minimal tension is possible in a shortened or contracted muscle. In a shortened state, the maximum number of possible cross-bridges has already been formed, and no additional tension can be produced.

In the shortened state, tension in muscle is equal to the tension in the series elastic components. When the muscle is lengthened, passive tension is generated. As the tension-developing characteristics of active components diminish with muscle elongation, tension in the total muscle increases due to the passive elements in muscle. As the series elastic component is stretched and tension develops in the tendon and cross-bridges, significant tension in the parallel component occurs as the connective tissue offers resistance to stretch. At extreme lengths, tension is almost exclusively elastic or passive tension (Hamill & Knutzen, 2003). This factor has implications not only for strengthening programs but also for stretching of muscles and soft tissues.

Location of Muscle and Axis

Muscles that move a part usually do not lie over that part but are often proximal to the part moved. The distance of the origin and the insertion of the muscle to the joint axis have an influence on the type of movement produced. If the distance from the insertion to the joint axis is greater than the distance from the origin, this is considered a shunt muscle. Shunt muscles tend to have the line of pull along a bone, so the muscle tends to pull bones together, creating a more translatory effect, and the muscle acts as a stabilizer. Examples of shunt muscles are the sternocleidomastoid muscle and the brachioradialis muscle where the origin is near the axis and the insertion is far.

A spurt muscle is one where the origin is farther from the joint axis and the insertion nearer. Spurt muscles have their line of pull across bones so that there is a larger rotary component to the movement produced. While shunt muscles act to stabilize joints, spurt muscles help to overcome inertia and produce rapid movements throughout a wide range of motion. Two-joint muscles may possess spurt characteristics at one joint and shunt at the other joint. Spurt and shunt classifications are based on 19th century engineering terms, with spurt defined as a force that "provides energy to impel a body into motion or keep it in motion," which basically is a restatement of Newton's Second Law (Gench et al., 1995).

The muscle's location or line of action in regard to the location of the joint axis determines what motion the muscle will perform. Muscles crossing anterior to joint axis in the upper extremity, trunk, and hip are flexors, while muscles located posterior to the axis are extensors. Muscles located laterally and medially are abductors and adductors (Gench et al., 1995; Nordin & Frankel, 2001). By applying knowledge about anatomy and muscle fiber arrangement, one can determine optimal angles of pull and locations of muscles to use muscles most advantageously in everyday occupations.

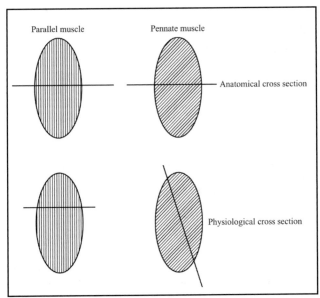

Figure 4-7. Anatomic and physiologic cross-sections.

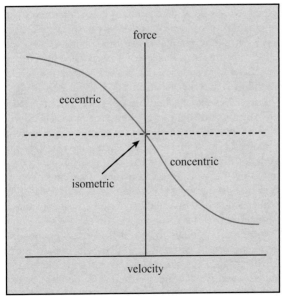

Figure 4-8. Force-velocity curve.

Cross-Section of the Muscle

Larger muscles with larger cross-sectional areas are capable of producing greater strength. This is what is known as anatomical cross-section, which is the cross-section made at right angles to the longitudinal axis of the muscle at the widest point (Hamill & Knutzen, 2003) (Figure 4-7). Physiologic cross-section is defined as "the sum total of all of the cross sections of fibers in the muscle, measuring the area perpendicular to the direction of the fibers" (Hamill & Knutzen, 2003). The physiologic cross-section is a slice that passes through all of the fibers in the muscle (Oatis, 2004).

In parallel fibers, the anatomic and physiologic cross-sections are the same, whereas in pennate fibers, the physiologic cross-section is greater than the anatomic cross-section. Because pennate muscles generally have more muscle fibers and shorter fibers that run diagonally into the tendon, pennate muscles have a much larger physiologic cross-section than parallel muscles, which enables greater force generation. The physiologic cross-sectional area will give an indication of the maximum tensile force the muscle is capable of producing (Marieb, 1998). The physiologic cross-sectional area is directly proportion to the maximum tension that can be generated, because it represents the sum of the cross-sectional areas of all of the muscle fibers within the muscle. Maximum muscle tension is not simply proportional to muscle mass (Lieber, 2002).

Velocity of Contraction

While muscle force is proportional to the physiologic cross-sectional area of the muscle, muscle velocity is proportion to muscle fiber length (Donatelli & Wooden, 2010). There is a direct relationship between the amount of force generated and velocity. As the velocity increases, the force decreases. The length-tension relationship describes muscle behavior at a constant length, but the force-velocity relationship involves

movement (Lieber, 2002). This inverse relationship is shown in Figure 4-8. This is due to neuromuscular recruitment patterns of both Type I and II fibers, which are activated together at lower speeds. However, as speed increases, there are fewer Type I fibers recruited, and eventually these become inactive. At very high velocities, smaller and smaller fiber populations are recruited. Simply stated, this relationship indicates that high-velocity movement corresponds to low muscle force, and low-velocity movement corresponds to high muscle force: you can move a light load more quickly than a heavy load.

This follows the same pattern as seen with tension production: eccentric muscle contractions, capable of the greatest force production, are strongest at slower velocities.

The greatest speed of shortening is when the external load is zero. As the load increases, the muscle shortens more slowly. Isometric contractions occur when the velocity is zero. Concentric contractions, the least in force production, occur when the velocity is greater than zero or at higher velocities. In concentric contractions, velocity is increased at the expense of a decrease in force. The maximum force is at zero velocity because a large number of cross-bridges are attached, and maximum velocity is achieved at the lightest load. As the velocity of the muscle shortening (concentric) increases, fewer cross-bridges are formed, so force production is negligible (Hamill & Knutzen, 2003).

If a concentric muscle action (shortening) is preceded by an eccentric (lengthening) action, the resulting concentric action is capable of generating greater force. This pre-stretch condition changes a muscle's characteristics by increasing the tension through storage of potential elastic energy in series elastic component. When a muscle is stretched, small changes occur in the muscle and tendon length and in the maximum accumulation of stored energy. When concentric muscle action follows, there is an enhanced recoil effect adding to the force output. Stored elastic energy in parallel components in connective tissue also contributes to high-force output at initial portions of concentric action as the tissue returns to a resting length. The parallel elastic contribution drops off as the muscle continues to shorten. If the stretch is held too long before shortening occurs, stored elastic energy is converted to heat and lost (Hamill & Knutzen, 2003; Irion, 2000).

The force-velocity relationship does not indicate that the muscle cannot generate a strong force at a fast speed, because maximum strength can be generated either by recruitment of more motor units or by increased muscle length. The force-time relationship specifies that the force generated by a muscle is proportional to the contraction time. The longer the contraction time, the greater the force that is developed because this enables changes in length and in recruitment, allowing time for tension to be produced by the contractile elements and transmitted through the parallel elastic components to the tendon (Nordin & Frankel, 2001). Slow, steady contractions will produce the greatest force, which is important to remember when instructing clients how to perform activities using the most of their strength.

Number of Joints Crossed by the Muscle

Norkin and Levangie (1992) call muscles that cross more than one joint "economic" because they are able to produce motion at more than one joint (p. 116). However, one-joint muscles are recruited first for single joint motions, because additional muscle fibers or motor units may be required to prevent unwanted motions of a two-joint muscle.

Two-joint or multi-joint muscles are able to be stretched over a greater range of motion. A multi-joint muscle is most effective when shortened at one end and lengthened at the other, so that the muscle works most effectively at one joint while being disadvantaged at the other. Manual muscle tests are designed to put one end of multi-joint muscles "on slack" or lengthened so that the strength of the contracting muscle can be tested.

Active Insufficiency

Active insufficiency occurs when a multi-joint muscle is unable to exert enough tension to shorten sufficiently to complete full range of motion in both joints simultaneously. This occurs because of a decrease in myosin-actin cross-bridges because a shortened muscle is composed of shortened sarcomeres, and therefore fewer cross-bridges can be formed. The agonist muscle cannot shorten any further because no further force can be produced. An example is the long head of the biceps. The long head of the biceps attaches to the superior portion of the glenoid, and the short head attaches to the coracoid. The biceps can assist the deltoid with shoulder flexion with the elbow in extension. When the elbow is fully flexed and the forearm is supinated, there is limited ability of the biceps to assist with shoulder flexion because it is actively insufficient. Other examples of active insufficiency occur with the triceps as it assists the posterior deltoid with glenohumeral extension, rectus femoris with hip and knee flexion, or the hamstrings as hip and knee extensors (Smith, Weiss, & Lehmkuhl, 1996; Snyder et al., 2007; Trew & Everett, 2005). A practical example of this is if you are being attacked, grab the

attacker's wrist and maximally flex it to reduce the attacker's ability to produce force in the hands (Konin, 1999).

Passive Insufficiency

Passive insufficiency occurs when an antagonist muscle cannot be elongated any further without damage. The antagonist muscle is passively stretched, and passive tension reaches the limit of extensibility. The passive tension that develops may be significant enough to cause joint movement, as in the tenodesis action of the wrist and fingers. The finger extensors become passively insufficient as they lengthen over the wrist during wrist flexion, causing passive tension to produce finger extension. The opposite occurs when the finger flexors develop passive tension when the wrist is extended. Passive insufficiency is usually felt as an uncomfortable or painful sensation and can be felt when one tries to simultaneously flex the wrist and fingers at the same time.

Passive insufficiency is related to the full stretch of the antagonist muscle when the agonist muscle is in active insufficiency, disallowing complete and full range of motion in both joints simultaneously (Snyder et al., 2007). Both active and passive insufficiency result in incomplete movement of some or all of the joints crossed by the muscles. Muscles that become actively insufficient due to excessive shortening are likely to develop passive insufficiency in the opposite or antagonistic muscle group (Nordin & Frankel, 2001). The force generated by the position of the multi-joint muscles can be seen by noting that the greatest isometric grip strength is achieved when the wrist is in slight extension. When the wrist is in full flexion, there is a combination of active insufficiency of the long finger flexors and passive insufficiency of the antagonistic long finger extensors (Smith et al., 1996).

McGinnis (1999) proposes this self-experiment relative to muscle length and insufficiency:

Consider the muscles that flex or extend your fingers. The finger flexor muscles are located in your anterior forearm, whereas the finger extensor muscles are located in your posterior forearm. Their tendons cross the wrist joint, the carpometacarpal joints, the metacarpophalangeal joints, and the interphalangeal joints. Flex your fingers, and grip a pencil as tightly as you can. Now, flex your wrist as far as you can. You may notice that you cannot flex your wrist as far with your fingers flexed as you can without your fingers flexed. You may also notice that your grip on the pencil weakened as you flexed your wrist (try to pull the pencil out in both positions). If you push on your hand and cause it to flex further, the pencil may even fall out of your grip. The finger flexor muscles were unable to produce much tension. The finger extensor tendons were lengthened at each joint they cross and were stretched by the extensor muscles beyond 160% of their resting length. Passive tension created by the stretching of the connective tissue resulted in an extension of the fingers when you pushed your hand further into wrist flexion (p. 272).

METHODOLOGICAL FACTORS

In testing the strength of muscles, you are assessing impaired muscle performance that might indicate muscle strain, neurologic injury (peripheral nerve or nerve root dysfunction), general weakness due to muscle imbalance of agonist/antagonist or due to immobilization, de-conditioning or reduced force or torque production, altered length-tension relationships, or the presence of pain (Hall & Brady, 2005). Methods for muscle strength testing include strength screens, standardized manual muscle tests, functional motion tests, or observation of performance in daily activities. Handheld instruments may be used but often the external force applied in the test is the force of the examiner. Palpation of muscles during testing ensures the evaluation of the correct muscle and adds to the test's validity.

Muscle Strength Screens

An occupation-based assessment would start first at assessing activities of daily living. Occupational therapists often use daily activities to assess performance and function because the task performance involves more than muscle strength. Some inferences about muscle function can be made by observing and analyzing muscle performance during specific tasks that combine key muscle groups. This type of assessment enables the assessment of other factors that may also be interfering with successful performance in tasks, such as balance or cognition, as well as the dynamic interaction of muscles and observation of the complex interactions of musculoskeletal, neuromuscular, cardiopulmonary, and integumentary systems (Reese, 2005). Characteristics of the task itself are considered, such as the dynamics, complexity, context, and meaning of specific tasks to the individual client (Reese, 2005). If the client is observed to have difficulty performing these tasks, a muscle screen or standardized muscle test would then be done.

Screening for muscle weakness is appropriate for conditions where muscle weakness is not the primary symptom or the primary limiting factor in performance of daily activities. Screening tools are designed to provide a general estimate of strength, and if deficits are indicated, further discrete testing may be done. Often, these quick muscle strength screens place clients in positions of convenience rather than specific positions (such as against gravity or in gravity-eliminated positions) (Reese, 2005). Muscle screens can be as simple as asking a client to hold a position while the therapist applies resistance to the muscle. An example would be to have the client shrug the shoulders, hold that position, and you apply resistance downward for a test of the strength of the scapular elevators.

Functional motion tests vary in what they assess and how the assessment is done. Some tests look at the active movement of a client as an assessment of joint range of motion and active, volitional movement patterns. The Zimmerman Functional Motion test screens for strength, range of motion, and coordination for the upper extremity (Zimmerman, 1969). In some tests, the part being tested is passively placed, and the client is asked to hold the part in that position while resistance is applied. In other tests, active movement into the test position is required. Some tests describe first and second positions designed to take into account the effect of gravity on the part during testing, while others do not. Functional tasks might include asking the client to rise from a chair, ascend steps without assistance, and move from a seated to a standing position as an assessment of lower extremity strength and placing a garment in a closet, drinking from a glass, or squeezing toothpaste as an assessment of upper extremity strength and grasp.

Gutman and Schonfeld (2003) describe functional muscle strength testing as the amount of resistance a joint can sustain during movement. They associate muscle groups with specific activities and the functional muscle grade required to perform the activity. For example, drinking from a glass requires F+ functional muscle grades of elbow flexors, while pulling up knee-high socks requires P+ strength of elbow extensors. This can be a helpful guide when deciding intervention strategies for desired activities based on muscle strength.

Manual Muscle Testing

More discrete testing of muscle strength is actually a measurement of impairment rather than function (Van Deusen & Brunt, 1997). Tan (1998) describes the potential uses of strength testing:

For physical medicine and rehabilitation

- Prescribing treatment and establishing a baseline
- Monitoring progress
- Evaluating level of impairment

Medical-legal

- Test effort consistency
- Determine maximal exertion
- Assess disability

Industry

- Provide ergonomic and rehabilitation guidelines
- Screen for job placement and return to work
- Compliance with ADA (p. 54)

Other purposes for strength testing are to determine how muscle weakness limits performance; to prevent deformities caused by muscle imbalance; to establish the need for assistive or technological devices; and to aid in the selection of appropriate activities. Results are useful in establishing differential diagnoses and prognoses.

Strength testing can be done in various ways. Manual testing would involve the active contraction of the muscle by the client with the application of resistance by the tester. Several muscles can be tested as a group that produces a particular motion. For example, the anterior deltoid, coracobrachialis, pectoralis major, and biceps muscles might be tested as a group of muscles that produce shoulder flexion. Individual muscles can be isolated and tested separately.

Manual muscle testing by definition is "the application of resistance by the tester or by the force of gravity to the voluntary maximum contraction of the patient's muscle" (Jenkins, 1998). This can be done by means of isotonic, isokinetic, or isometric force application.

Isotonic Testing

In isotonic testing, muscle strength is tested by using a constant external force, often applied by weights or machines. Isotonic dynamometers, which are force-measuring devices, use a constant weight or resistance to measure muscle contractions at an accommodating speed and at an accommodating resistance. An example of a commercially available isotonic dynamometer at an accommodating speed would be the Ariel Computerized Exercise System (Ariel Dynamics Worldwide, Trabuco Canyon, CA). Some disadvantages to this type of muscle testing is that the repetition maximum, determined largely by trial and error, is determined for each muscle group and may be overtaxing for some clients. The testing can be time-consuming, and the equipment may not be portable (Tan, 1998).

Isokinetic Testing

Isokinetic resistance is provided by machines, which provide the resistance through a specific range of motion at a constant velocity. This provides a peak torque at one point in the range of motion, and this measurement has high reliability. Disadvantages are portability and high cost of the equipment (Tan, 1998). Isokinetic instruments include Cybex (Cybex International, Medway, MA), and active isokinetic tools include Biodex (Biodex Medical Systems, Inc., Shirley, N.Y.) and Kin-Com (Kin-Com, Harrison, TN).

Isometric Testing

Resisted isometric movements are done to determine the function of the contractile tissues. The muscle is usually tested in a position of optimum length, so maximum force can be generated. In most of the manual muscle testing done in clinical settings, isometric muscle testing is done. Muscles generate force against an immovable resistance (either the therapist or a handheld dynamometer), so muscle length remains the same throughout the test. By testing using isometric muscle contractions, variability in muscle length and velocity of joint motion are eliminated. Because little or no equipment is needed, isometric manual muscle tests or handheld dynamometry (HHD) are portable and inexpensive (Magee, 2008).

Isometric resistive tests can be used to assess strength and for the provocation of pain. Cyriax (1982) recommends specific test positions to evoke varying levels of pain indicative of underlying dysfunction. For example, resisting shoulder abduction may produce pain during muscle contraction, which may be indicative of a lesion within a muscle belly. If the pain occurs once the contraction is completed, this may indicate a lesion within a tendon. Strong and painless resisted contractions are normal. If the resisted action resulted in a strong contraction that was painful, this might be a minor muscle dysfunction. Gross trauma or partial rupture of muscle or tendon may occur when there is a weak muscle contraction accompanied by a painful response. If the contraction is weak and pain-free, this might indicate a muscle or tendon lesion or a neurological dysfunction (Magee, 2008).

Tester Requirements

Those performing manual muscle tests need a thorough knowledge of the location and directional line of pull of muscles and of the structural anatomy of the parts being tested. This is especially true when palpating muscles to determine muscle grades of zero and trace. Knowledge of muscles with the same innervation can help in identifying patterns of weakness or those that may be related to specific diagnostic categories. The function of participating muscles (synergist, prime mover, etc.) is important, as is familiarity with patterns of substitu-

tion. Experience and adherence to standardized positioning and stabilization required in the specific muscle tests enables more accurate test results. Sensitivity to differences in normal muscle contours and joint laxity is invaluable to the assessment of muscles. The effects of fatigue and of sensory loss on the ability to move also cannot be disregarded. Manual muscle testing requires attention to detail, knowledge of normal movement and anatomy, and time.

Manual muscle tests might need to be modified when testing older clients due to pain, joint deformities, or limitations in endurance and flexibility, and extra caution is advised in the presence of osteoporosis (Bonder & Bello-Haas, 2009).

Make Versus Break Tests

In manual muscle testing advocated by Daniels and Worthingham (Daniels & Worthingham, 1986), motions using agonists and synergists are the focus of assessment. Group muscle testing is seen as more functional than individual muscle testing, and the assessment occurs through the full test range of motion. The patient performs a concentric contraction or holds the test position at the end of the available range of motion while the muscle is in a shortened position. Not as objective as individualized muscle testing, group strength testing can help to identify where in the range weakness exists (Palmer & Epler, 1998).

Testing at the end of the range with the muscle or muscles providing the isometric hold is referred to as the break test or method. Manual muscle testing follows muscle length-tension relationships and joint mechanics in the positioning of the client for assessment. One-joint muscles often have external force applied at the end of range, whereas two-joint muscles often have the point of maximum resistance at or near mid-range. In the break test, the primary movers of the joint are positioned so that they have mechanical advantage or when the muscle or muscles are at resting length or slightly more than resting length. In this position, the prime movers are at their strongest and the actions of the synergistic muscles are reduced (Radomski & Latham, 2008).

Kendall and McCreary (1983) propose methods for testing specific, individual muscles rather than testing muscle groups. This type of manual muscle testing requires a greater knowledge of anatomy and kinesiology. While seen as more accurate, this type of testing is also more time-consuming to perform and fatiguing for the clients. The test is performed while the client contracts isometrically with the segment aligned in the direction of the muscle fibers in a midrange position. The client is then asked to "hold" this position against resistance. Similarly, the Medical Research Council Scale uses resistance as an isometric hold at the end of the test range, and this group advocates using numbers rather than words for muscle grades (for example, a good muscle grade is a 4/5).

Resistance that is applied throughout the test range in a direction opposite to the muscle's rotary component is known as make test. This method is done while the muscle is performing a concentric contraction (Palmer & Epler, 1998). The make test is especially useful for quantifying muscle grades F+ to N (Radomski & Latham, 2008). The make test is the preferred method when using handheld dynamometers (Reese, 2005).

Table 4-4

MANUAL MUSCLE TEST PROCEDURE (BREAK TEST AGAINST GRAVITY)

1. Ensure that the client is comfortable, warm, rested.
2. Arrange the environment to ensure that it is quiet, well lit, and warm.
3. Explain what you will be doing, how the test will be done, the purpose of the test, how the results will be used, and who will see the results.
4. Place the client (or have the client move) into the described test position.
5. Provide fixation (stabilization, support, or counterpressure) as specified in the test procedure.
6. Tell the patient to "hold this position and don't let me move you."
7. Apply resistance slowly and for 4 to 5 seconds, paying attention to signs of exertion or discomfort.
8. Palpate the contracting muscle.
9. Determine the muscle grade.
10. Record the results and any additional information from observations during the test.

Test Procedure

Making sure the client is comfortable, warm, and rested is important not only to elicit optimal muscle contractions by controlling environmental factors, but is vital to establishment of therapist-client trust and the development of the therapeutic relationship. Be aware of cultural and personal boundaries when performing assessments because you will be touching clients, sometimes asking them to expose parts of their body so that skeletal landmarks can be identified and palpation of structures can occur.

The test is standardized so that the test can be given in the same manner every time and by different testers. The test protocols specify the starting position of the client, where stabilization is needed, instructions to give to the client, and the direction of the applied resistance. Two test positions are given: one against gravity and one with gravity eliminated. Common substitution movements and where to palpate specific muscles are also given to ensure testing of the correct muscle or muscle action.

Patient positioning in the specified positions is essential to prevent substitution movements and to ensure that the appropriate muscle is being tested. In general, the distal joint segment is placed in relation to pull of gravity, and the proximal segment of the joint is stabilized. The origin of the muscle is fixed so that a maximal contraction against the insertion can occur. Inadequate stabilization may cause an underestimation of strength because the client will be less able to produce optimal force. Be aware, too, of your own body mechanics when performing a manual muscle test. Position yourself so that you can use your muscles in the most mechanically advantaged position while still maintaining the test protocols. The proper positioning of the patient in the specific tests not only elicits the strongest muscle contraction from the client but also optimizes the examiners' body mechanics when proper stabilization and application of resistance are applied.

Controlling for environmental factors can elicit the client's best response. So, make sure the client is comfortable, warm, and rested, and make sure the room is quiet, well-lit, and warm. Reassure the client that the test should not cause pain, and explain to the client what you will be doing and why. Explain the purpose of the test, how it will be performed, who will see the results, and what the results will be used for. Simply tell the client that this is a test to assess the strength of muscles and that you will be positioning them in specific ways to test specific muscles. You will also be applying some resistance to that muscle to test its strength, but the resistance will be applied slowly.

Tell the client to inform you if he or she is uncomfortable at any time during the testing. This is important for the rapport, trust, and therapeutic relationship but also can help to identify at what point in the range limitations or pain may occur.

When applying resistance in the against-gravity position, apply the resistance slowly. This will enable recruitment of larger and more motor units to respond to the force you are applying. The subjective part of the manual muscle test is when resistance is applied. Nicholas and colleagues studied whether it was the duration or amount of force that was applied that enabled the clinician to make the muscle grade determination. In this study, it was the clinician's perceptions of both the time required to move the limb through certain ranges of motion and the average force that was involved in the mental process of the evaluation (Nicholas, Sapega, Kraus, & Webb, 1978).

While resistance is often applied at end of the gravity-resisted range of motion (which is frequently not the strongest point in range of motion for muscle being tested), this provides a consistent point for application of resistance and may prevent overestimating strength (Reese, 2005). Table 4-4 outlines the steps of the break test against gravity test.

Table 4-5

FACTORS RELATED TO MUSCLE GRADES

Related to evidence of contraction	Zero = 0: or 0/5	No muscle contraction felt.
	Trace = T: or 1/5	Evidence of contractility on palpation. No joint motion.
Related to gravity	Poor = P: or 2/5	Movement with gravity eliminated but not against gravity or complete ROM with gravity eliminated.
	Fair = F: or 3/5	Can raise part against gravity or complete ROM against gravity.
Related to resistance	Good = G: or 4/5	Can raise part against outside resistance and against gravity or complete ROM against gravity with some resistance.
	Normal = N: or 5/5	Can overcome a greater amount of resistance than a good muscle or complete ROM against gravity with full resistance.

Muscle Grades

The definition of manual muscle testing that was provided earlier identifies three variables that are used in determining the grade of the muscle: voluntary maximum muscle contraction, gravity, and resistance. Table 4-5 summarizes these three factors. Muscle grades can be identified as a number (1 to 5) or as a letter to identify zero (0), trace (T), poor (P), fair (F), good (G), and normal (N).

Contractility is the first variable, and the client's ability to volitionally contract a muscle determines the muscle grades of zero (0) and trace (T). Contractility is determined by palpation of the muscle. When palpating a muscle, move slowly, avoid excessive pressure, and focus on what you are feeling (Biel, 2001). You need to be sure you are locating the correct structure and if you feel tension as the muscle contracts.

If the client is unable to contract a muscle and no muscle tension is felt with palpation, the muscle grade is zero (0). Palpation is best performed by using the middle finger, which is the most sensitive finger to identify muscle tension. Every muscle tested must be palpated to be sure the correct muscle is being tested and to minimize substitution actions by synergists or other muscles. If the client is able to contract by volitional control, then the muscle grade would be at least a trace or 1/5.

Gravity, the second variable, affects muscle's ability to contract by adding an external force. Muscles that can contract volitionally in a gravity-eliminated situation would be considered a poor muscle (or 2/5). If the person can move the extremity in a position against gravity, the muscle grade is fair or 3/5. Manual muscle testing positions are against gravity (to test for F, G, N muscles) and gravity-eliminated (to test for 0, T, and P muscles). In a gravity-eliminated muscle test, the best muscle grade possible is a poor, because a fair muscle is one that can contract against gravity.

The third variable, resistance, is used to differentiate the good muscle (4/5) from the normal muscle (5/5). The application of manual resistance is seen as a subjective variable, which may account for the poor reliability and validity of muscle grading above the fair muscle grade. The amount of resistance is dependent upon the muscle being tested. Less resistance should be applied to muscles with a predominance of type II fibers or those of parallel arrangement because they are less likely to be as strong as those that are type I or pennate. Less resistance should be applied to smaller muscle such as the hand muscles as compared to larger shoulder and hip muscles. Consideration of the individual characteristics of the client, such as age and gender, also affect the amount of resistance used.

Generally, if a client has G to N muscle strength and good to normal endurance, the person will be able to perform all work, play, and self-care tasks without undue fatigue. Those clients with F+ strength will have low endurance for tasks, will fatigue more easily, and may need frequent rests. A client with F muscle strength can move against gravity and perform light tasks with little or no resistance and with decreased endur-

Table 4-6

COMPARISON OF GRAVITY-RESISTED MANUAL MUSCLE TEST GRADING CRITERIA

LOVETT AND DANIELS AND WORTHINGHAM	KENDALL AND MCCREARY	MEDICAL RESEARCH COUNCIL
N (Normal): Subject completes range of motion against gravity, against maximal resistance.	100%: Subject moves into and holds test position against gravity, against maximal resistance.	5
G+ (Good Plus): Subject completes range of motion against gravity, against nearly maximal resistance.		4+
G (Good): Subject completes range of motion against gravity, against moderate resistance.	80%: Subject moves into and holds test position against gravity, against less than maximal resistance.	4
G- (Good Minus): Subject completes range of motion against gravity, against less than moderate resistance.		4-
F+ (Fair Plus): Subject completes range of motion against gravity, against minimal resistance.		3+
F (Fair): Subject completes range of motion against gravity with no manual resistance.	50%: Subject moves into and holds a test position against gravity.	3
F- (Fair Minus): Subject does not complete range against gravity but does complete more than half the range.		3-
P+ (Poor Plus): Subject initiates range of motion against gravity or completes range with gravity minimized against slight resistance.		2+
P (Poor): Subject completes range of motion with gravity minimized.	20%: Subject moves through small motion with gravity minimized.	2
P- (Poor Minus): Subject does not complete range of motion with gravity minimized.		2-
T (Trace): Subject's muscle can be palpated, but there is no joint motion.	5%: Contraction is palpable with no joint motion.	1
O (Zero): Subject exhibits no palpable contraction.	0%: No contraction is palpable.	0

Reprinted with permission from Palmer, M. L. & Epler, M. (1998). *Fundamentals of musculoskeletal assessment*. 2nd ed. Philadelphia: Lippincott Williams & Wilkins.

ance. A muscle grade of P is considered below the functional range, but the client can perform some light activities of daily living with adaptive equipment and assistance.

Muscle grades can be further defined by adding a "+" or "-" to further quantify the muscle grade (Table 4-6). Muscle grades

of G- (4-) or below are considered weak and, therefore, a possible focus of remediation (Flinn, Latham, & Podolski, 2008). Others consider muscle grades of G and below as the point at which to consider intervention (Reese, 2005).

Instruments for Muscle Strength

Handheld dynamometry (HHD) devices can be easily used. Handheld dynamometers are a portable force-measuring device that the examiner holds while the client exerts maximal force. Some handheld dynamometers that are commercially available include Lafayette hand dynamometer, Lafayette pediatric hand dynamometer, Nicholas manual muscle tester (Lafayette Manual Muscle Test System, Lafayette, IN), Jamar dynamometer, Tech power-track II, and MicroFet dynamometer.

Using a HHD is noninvasive, relatively quick, inexpensive, and applicable to a wide range of applications (Reese, 2005). As with manual muscle testing, the clients need to be able to follow directions in order to assume the test positions and maintain muscle contractions against gravity. Strength testing done with dynamometers is generally regarded as more reliable and valid than with manual resistance. Using a handheld dynamometer for strength testing yielded a greater than 0.84 test-retest value when performed by an experienced therapist (Trombly, 1995).

Isometric dynamometers for assessment of limb muscle groups measure peak and average force, reaction time, rate of motor recruitment of motor units, and maximal voluntary exertion and fatigue (Tan, 1998). In this test, the muscle length is held constant, which is a preferred method in conditions where joint motion causes pain. The disadvantage to this type of testing using strain gauges is that the movements are not well-correlated with real-life dynamic activities (Tan, 1998). The Baltimore Therapeutic Equipment (BTE) device has both isometric and isotonic testing capacities.

When considering whether to use HHD or a manual muscle test, consider the client's overall strength. If the client has significant weakness, muscle strength is best assessed using manual muscle testing or functional motion testing, which are more sensitive to low levels of muscle strength. If your client has a muscle grade of good or better, the HHD is a better choice of an assessment because manual muscle testing does not clearly discriminate between higher muscle grades (Reese, 2005).

Manual muscle testing is appropriate for those clients who are able to voluntarily contract muscles, which excludes those with central nervous system dysfunction and those with tone problems or stereotypic, automatic movements. Manual muscle testing is often not performed with children younger than 3 years, and the elderly may not be able to assume and/or tolerate some testing positions. Manual muscle testing is contraindicated for those with dislocations. Extra care and attention in testing should be taken for clients with cardiovascular and chronic respiratory diseases (i.e., COPD, COLD), as well as those clients just recovering from abdominal surgery. It is possible to tear muscle fibers if excess application of resistance is applied, and muscle cramping can occur if resistance is applied for too long.

MEASUREMENT FACTORS

Manual muscle testing, while used extensively clinically, has not demonstrated consistent reliability and validity. The break test is considered inaccurate in testing strength greater than fair muscle grades (or 3/5) (Sisto & Dyson-Hudson, 2007) and is insensitive to change (Van Deusen & Brunt, 1997). There is a tendency to overestimate the normalcy of muscle strength, and manual muscle testing has been seen as inadequate in determining functional capacity (Tan, 1998). However, there is no other method for evaluating muscle weakness that has the reliability, validity, and ease of use of manual muscle testing (Hall, Lewith, Brien, & Little, 2008). Manual muscle testing is seen as a valid and reliable method to measure muscle strength (Herbison, Isaac, Cohne, & Ditunno, 1996; Marx, Bombadier, & Wright, 1999; Swartz, Cohen, Herbison, & Shah, 1992).

Manual Muscle Testing Reliability

Manual muscle testing has been determined to have inter-rater reliability within one muscle grade 60% to 75% of the time (Frese, Brown, & Norton, 1987; Williams, 1956), and others have found that results do not vary more than one half of a muscle grade (Palmer & Epler, 1998; Van Deusen & Brunt, 1997). Further, there is approximately 50% complete agreement on muscle grades, including the addition of + and -, 66% agreement within + or -, and 90% agreement within one full grade (Iddings, Smith, & Spencer, 1961; Palmer & Epler, 1998). Brandsma and colleagues found that there was a range of 0.71 to 0.96 for intra-rater reliability and 0.72-0.93 for inter-rater reliability when testing intrinsic hand muscles (Brandsma, Schreuders, Birke, Piefer, & Oostendorp, 1985). Savic and colleagues found overall agreement of assigning muscle grades was 82% on the right and 84% on the left with strongest agreement for zero muscle grades and the weakest for fair muscles (Savic, Bergstrom, Frankel, Jamous, & Jones, 2007).

There is poor inter-rater reliability in muscle grades below fair (Beasley, 1961; Frese et al., 1987; Palmer & Epler, 1998; Rothstein, Roy, Wolf, & Scalzitti, 2005; Tan, 1998), but Florence and colleagues found that reliability ranged from 0.80 to 0.99, with the muscle grades for the gravity-eliminated positions (P,O,T) having the highest reliability values.

Muscle grades of G and N are seen as subjective and inaccurate (Stuberg & Metcalf, 1988) because they involve the application of force by the examiner, which will vary from person to person and by age and gender of the examiner. To overcome this problem, the use of handheld dynamometers might be considered for these muscle grades. HHD can achieve ICC values ranging from 0.91 to 0.99 for intra-rater, inter-rater, intra-session, and inter-session comparisons (Lu, Hsu, Chang, & Chen, 2007). HHDs are seen to be reliable and valid (Kolber & Cleland, 2005; Li, Jasiewicz, Middleton, Condie, Barriskill, Hebnes, et al., 2006). Clinically, there are fewer adaptations that need to be made to activities or exercises at the G and N range, and the discrepancies in accuracy do not limit functional performance as much as with lower muscle grades.

It is vital to follow procedures, provide clear directions, demonstrate and explain movement, and passively move the client to facilitate understanding of required motions. Following the standardized procedures produces reliable results even if the tester is inexperienced (Pollard, Lakay, Tucker, Watson, & Bablis, 2005), but more experience leads to greater reliability. Other factors that need to be considered to improve reliability include the following:

- Cooperation of the client
- Experience of the tester
- Tone of voice of tester
- Ambient temperature
- Temperature of limb
- Distractions to patient and to tester or other environmental conditions
- Posture
- Fatigue
- Operational definitions of muscle grades (Trombly, 1995).

Manual Muscle Testing Validity

It is generally accepted that manual muscle testing has face and content validity, because the testing is based on anatomic and physiologic structures (Palmer & Epler, 1998; Lamb, 1985; Payton, 1979), but there is little credence placed on the ability to generalize the results to immediate and future behavior of patients (Palmer & Epler, 1998). Validity is improved by palpating each muscle being tested, stabilizing proximal segments during testing, and preventing substitution movements and muscle actions.

GRASP AND PINCH ASSESSMENT

Grip strength is important in many of the activities in which we engage every day. The assessment of grip strength was found to be a good predictor of overall strength (Bohannan, 1998), of whole body strength and disability (Bassey & Harries, 1993; Davis, Ross, Preston, Nevitt, & Wasnich, 1998; Giampaoli, Ferrucci, Cechi, LoNoce, Poce, & Dima, 1999), and of mortality (Bassey & Harries, 1993; Milne & Maule, 1984; Philips, 1990), and midlife grip strength is predictive of functional limitations and disability 25 years later (Rantanen, Guralnik, Foley, Masaki, Leveille, & Curb, 1999). Strength testing can assess nutrition (Harries, 1985) and the risk of mortality in people with acute illness (Philips, 1990), can be a prognostic factor (Sunderland, Tinson, Bradley, & Hewer, 1989), is associated with hospital outcome, and can be used to identify ICU-acquired paresis (Ali, O'Brien, Hoffmann, Phillips, Garland, Finely et al., 2008).

The loss of muscle strength, including grip, is associated with the development of physical disability in diabetes (Park, Goodpaster, Strotmeyer, deRekeneire, Harris, & Swartz, 2006; Redmond, Bain, Laslett, & McNeil, 2009; Sayer, Dennison, Syddall, Gilbody, Philips, & Cooper, 2005). In addition, Redmond and colleagues state that there is an inter-relationship between hands, obesity, and physical functioning in women (Redmond et al., 2009), and decreased hand grip and key pinch power have been found to be decreased in people with type 2 diabetes mellitus (Cetinus, Buyukbese, Uzel, Ekerbicer, & Karaoguz, 2005).

Factors affecting grip and pinch include gender (Balogun, Akomolafe, & Amusa, 1991; Crosby, Wehbe, & Mawr, 1994; Mathiowetz, Volland, Weber, Dowe, & Rogers, 1985; Mathiowetz, Wiemer, & Federman, 1986; Petersen, Petrick, Connor, & Conklin, 1989), age (Crosby, Wehbe, & Mawr, 1994; Mathiowetz, Volland, Weber, Dowe, & Rogers, 1985; Mathiowetz, Wiemer, & Federman, 1986; Petersen, Petrick, Connor, & Conklin, 1989), hand dominance (Crosby, Wehbe, & Mawr, 1994; Josty, Tyler, Shewell, & Roberts, 1997; Mathiowetz, Volland, Weber, Dowe, & Rogers, 1985; Petersen, Petrick, Connor, & Conklin, 1989), occupation (Josty, Tyler, Shewell, & Roberts, 1997), body weight and height (Balogun, et al., 1991; Chau, et al., 1997; Peolsson, Hedulnd, & Obert, 2001), position of wrist, shoulder, and elbow (Balogun, Akomolafe, & Amusa, 1991; Chau, Petry, Bourgkard, Huguenin, Remy, & Andre, 1997; Crosby, Wehbe, & Mawr, 1994; O'Driscoll, Horii, Ness, Callahan, Richards, & Kai-Nan, 1992; Peolsson, Hedulnd, & Obert, 2001; Pryce, 1980; Su, Lin, Chien, Cheng, & Sung, 1994), interest and cooperation of client, experience and tone of tester (Johannson, Kent, & Shepard, 1983), posture, fatigue, ability to understand directions, and ability to understand various grades and test positions (Kendall, McCreary, Provance, Rogers, & Romani, 2005).

As with body strength, age and gender are factors in the assessment of grasp and pinch. Men are stronger than women, and this is true throughout adult life. Men have higher grasp values in all postures and joint angles (Gutierrez & Shechtman, 2003). While men are stronger, some studies suggest that women are more dexterous (Durward, Baer, & Rowe, 1999; Richards et al., 1996; Rice, Leonard, & Carter, 1998). There are also significant differences between men and women for grip endurance, which is the number of seconds one can sustain maximal grip or the level of strength output after repetitive or sustained contraction for a predetermined time (Wallström & Nordenskiöld, 2001).

For both genders, grip strength increases curvilinearly until age 20, peaks between 19 and 50 years of age, then declines (Schechtman, Mann, Justis, & Tomita, 2004). With females, the strength increases until age 13 then remains constant until 20 to 29 years of age. Males demonstrate linear increases before peaking at ages 30 to 39 years, after which decline in strength occurs (Durward et al., 1999). Shiffman found a significant effect of aging on functional performance with statistically significant differences with age on grip strength, prehension, and time to perform tasks (Durward et al., 1999). Desrosiers, Bravo, Hebert, and Dutil (1995) state that "grip strength in persons aged 60 years and older varies negatively and curvilinearly with age and that the loss seems more marked across older subjects" (p. 641). These authors propose that reductions in the number and size of muscle fibers (especially fast-twitch fibers), decreased ability to achieve maximum tension levels, as well as normal age-related changes in the nervous system and vascular and circulatory systems account for decreased grip strength with increased age.

Grip scores may be meaningless when tested in cognitively impaired clients as seen in a study comparing grip strength between minimally impaired, visually impaired, motor impaired, and cognitively impaired clients (Schechtman et al., 2004). Age and gender are not the only determining factors of grip strength in the frail elderly.

Hand dominance differences in strength are discrepant. Some studies indicate no statistically significant differences in grip strength between right and left hands (Bear-Lehman & Abreu, 1989; Mathiowetz et al., 1985; Peolsson et al., 2001; Richards et al., 1996), while others found that the dominant hand is 10% to 13% stronger than the nondominant hand (Crosby et al., 1994; Desrosiers et al., 1995; Petersen, Petrick, Connor, & Conklin, 1989; Richards et al., 1996; Schechtman et al., 2004). Many studies resulting in normative values for grip strength reflect the greater grip strength values for the preferred hand.

Grip Strength

The measurement of grip strength has been the focus of many studies that consider the position of the shoulder, elbow, forearm, and wrist; type and complexity of directions; position of the body (sitting, standing); instruments used to measure grasp; terminology; and protocols of testing. Table 4-7 lists some of the studies done on grip strength.

Grip strength is measured isometrically, but, interestingly, most activities are done dynamically. An assessment of grasp and pinch is only a small part of the evaluation of the client's ability to use and manipulate objects with the hands. Observation of the use of the hands during activities is also needed to understand the client's occupational performance.

A grip force of 9 kg (or 20 pounds) has been considered necessary and functional for most activities of daily living (Nalebuff & Philips, 1990; Philips, 1990), although Rice, Leonard, and Carter (1998) found that it took less than 20 pounds of grip force to open and manipulate common objects, such as an aerosol sprays can, medicine bottles, or dual-pinch safety squeeze bottles. Grasp values are considered abnormal when associated with functional limitation and/or if value is ±3 standard deviations from normative values.

Normative values based on the mean of three trials by gender, age, and hand preference are listed in Table 4-8. The results from various instruments (Jamar, Rolyan, Grippit, Dexter, and BTE-Primus models) can be compared to these norms established by Mathiowetz and colleagues (1985) (Bellace, Healy, Bryon, & Hohman, 2000; Mathiowetz, 2002; Schechtman, Davenport, Malcolm, & Nabavi, 2003; Wallström & Nordenskiöld, 2001). To ensure the most accurate results, the same model should be used consistently with any one client.

The American Society of Hand Therapists suggests that grip be measured with the client seated in a straight chair with the feet flat on the floor. The shoulder should be adducted against the body in neutral rotation with the elbow flexed to 90 degrees and the forearm in neutral rotation (Fess, 1992). Other studies support this positioning during grasp and pinch testing (Richards et al., 1996). Grip strength tests are only valid and reliable when the client is exerting maximal voluntary effort and are not a test of sincerity of effort (Gutierrez & Shechtman, 2003).

Several instruments are used in clinical practice to evaluate grasp and pinch. The Jamar dynamometer is considered the gold standard for grip measurement with ICC of 0.98 to 0.99 (Lindstrom-Hazel, Kratt, & Bix, 2009; Mathiowetz, Weber, Volland, & Kashman, 1984; Peolsson et al., 2001). The Jamar handheld dynamometer is a sealed hydraulic strain gauge system that measures force exerted on the device in pounds or kilograms. Calibration accuracy is important to validity, and calibration accuracy of the Jamar dynamometer is ±3% to 5% (Richards et al., 1996). This means that if one scored 50 pounds using the Jamar dynamometer, the actual value is somewhere between 47.5 and 52.5 pounds.

Using the second handle position yielded high inter-rater reliability with a correlation coefficient of 0.97 or above for all tests; test-retest reliability using the mean of three tests was most consistent with a correlation coefficient of 0.80 or better. Lowest correlations occurred when only one trial was done. The Jamar dynamometer is considered the most precise of the instruments used to measure grasp (Desrosiers et al., 1995).

The Jamar dynamometer measures strength in pounds with norms based on age and gender. The procedure for use is as follows:

1. Adjust the handle to fit the patient's hand size and to allow MCP flexion. It was found that, with the handle in the second position, the grasp values were the strongest, and values obtained in each of the five handle positions would represent a bell-shaped curve (Palmer & Epler, 1998).

2. The client is permitted to rest the forearm on a table if desired but not any part of the dynamometer. The elbow should be flexed to 90 degrees.

3. Three trials with each hand are performed with the mean value recorded. An alternate method is to perform two separate trials where the person exerts maximum pressure (two times with hand), and the higher score is recorded. In each method, both hands are tested alternately with care taken not to fatigue the client.

The Martin vigorometer is an air-filled bulb connected to a pressure gauge (Durward et al., 1999) with three different-sized bulbs for testing grip and pinch. Measurements are expressed in kilopascals and, unlike the Jamar dynamometer, involve an isotonic contraction due to the movement necessary to compress the bulb (Desrosiers et al., 1995). The vigorometer measures grasp with high test-retest reliability with coefficients of 0.96 for the mean of three measures on the dominant hand and 0.98 for the nondominant hand. Norms are available for age and gender (Fike & Rouseau, 1982).

The third instrument, a sphygmomanometer or modified sphygmomanometer (blood pressure cuff), is often used for clients with fragile hands. The sphygmomanometer ratings had a strong and linear relationship to Jamar, with r = 0.83 for right hand and r = 0.84 for left (Flinn et al., 2008). The sphygmomanometer measurement is in millimeters of mercury (mm Hg). The following is the procedure for the use of the sphygmomanometer in measuring hand strength:

1. Roll cuff into cylinder shape and inflate to 100 mm Hg.

2. Deflate to 30 mmHg to establish the baseline.

3. Record as maximum versus baseline average (x mm Hg/30 mmHg).

A value of 300 mmHg/30 mmHg is considered within normal limits, and conversion tables have been developed that compare the sphygmomanometer results with Jamar results.

Table 4-7

STUDIES OF GRIP STRENGTH

AUTHOR	INSTRUMENT	POPULATION	SEX	AGE	HEIGHT/ WEIGHT	HAND DOMINANCE
An et al. (1980)	Strain gauge					
Balogun et al. (1991)	Harpenden dynamometer	f=26; m=35 healthy; 16-18 years	M>F	Positive correlation	Positive correlation	Dominant only tested
Balogun et al. (1991)	Harpenden dynamometer	f=480; m=480 healthy, 7-84 years	M>F	Positive correlation up to 3rd decade; thereafter a negative correlation	Positive correlation	R>L
Balogun et al. (1991)	Harpenden dynamometer MS	f=6; m=31 healthy; 16-28 years				R only tested
Desrosiers et al. (1995)	Jamar dynamometer Martin vigorometer	f=179; m=181 mean age 73.9 healthy	M>F	Negative correlation with age		
Downie et al. (1978)	VAS	7 healthy				
Fransson and Winkel (1991)	Strain gauged pliers	f=8; m=8 healthy 18-60 years	M>F			R only tested
Fraser and Benten (1983)	Vigorometer	f=60; m=60 healthy; 20-80 years	M>F	Increase to 30 years then deterioration with age especially after 50 years		No significant differences between R and L
Gilbert and Knowlton (1984)	Load cell	f=16; m=20 healthy; 20-26 years	M>F			Dominant only
Goldman et al. (1991)	Jamar	f=16; m=10 healthy; 23-29 years; f=11, m=10 23-84 years				Both hands tested Injured less than noninjured hand patients
Jarit (1991)	Jamar	44 controls; 44 baseball players				Nondominant 89.7% of dominant in baseball players; 86.5% in control group

(continued)

Table 4-7 (continued)

STUDIES OF GRIP STRENGTH

AUTHOR	INSTRUMENT	POPULATION	SEX	AGE	HEIGHT/ WEIGHT	HAND DOMINANCE
Kellor et al. (1971)	Jamar	f=126; m=126 healthy; 18-84 years	M>F	Negative correlation		R>L
Lunde et al. (1971)	Kny-sheer dynamometer	f=57; healthy			Positively correlated with height and weight	Dominant 13% stronger than non-dominant hand
Lusardi and Bohannon (1991)	Jamar MS	f=34; healthy 19-64 years				
Mathiowetz et al. (1985)	Jamar	f=328; m=310 healthy; 20-94 years	M>F	Peaked 25-39 years, gradual decline		R>L
Masaki et al. (1999)	Smedle dynamometer	3,218 men; 45-48 years; initially healthy				
Mawdsley et al. (2000)	Jamar dynamometer	f=6; m=12; healthy; mean age 20.4 years				Dominant > than nondominant
Mathiowetz et al. (1985)	Jamar	f=29; healthy 20-34 years				R>L
Mathiowetz et al. (1990)	Jamar	49 controls 49 rehabilitation patients				
Marion and Niebuhr (1992)	Jamar	f=30; healthy 21-45 years				R only tested
Newman et al. (1984)	Strain gauged system	1417 healthy children; 5-18 years	M>F	Boys-linear increase; girls increase to 13 then plateau	Positive correlation with height and weight	Dominant hand > strength
Shifman (1992)	Jamar	f=20; m=20 healthy; 24-87 years		Negative correlation with age		R>L

(continued)

Table 4-7 (continued)

STUDIES OF GRIP STRENGTH

AUTHOR	INSTRUMENT	POPULATION	SEX	AGE	HEIGHT/ WEIGHT	HAND DOMINANCE
Solgaard et al. (1984)	Vigorometer my-gripper steel spring dynamometer	f=5; m=45; healthy; 20-87 years		Negative correlation with age and weight	Positive correlation with height	Dominant > nondominant
Somedeepti et al. (1990)	Jamar	f=30; m=30 unilaterally injured; mean age 28 years	M>F			
Speigal et al. (1987)	Sphygomoman-ometer	92 rheumatoid patients with mean age 58 years				R and L measured; results not specified
Teraoka (1979)	Smedley dynamometer	9543 healthy 15-58 year olds	M>F	Negative correlation with age		R>L
Unsworth et al. (1990)	Inflatable cuffs					

Reprinted with permission from Durward B., Back, G., & Rowe, P. *Functional Human Movement*. Oxford: Butterworth Heinemann.

The Baltimore Therapeutic Equipment (BTE) power grip attachment has also been used to assess grasp and pinch. The BTE values had a test-retest value of >0.98 on the right dominant hand in clients aged 20 to 45 with a day-to-day variability of 5% for right and 3% for left. The BTE is seen as a valid measure of grip strength (ICC = 0.978) (Beaton, O'Driscoll, & Richards, 1995).

The difficulty in comparing grasp values and determining reliability and validity is that there is such an array of methods and protocols used. Different instruments use different units of measurement, different hand configurations, and different force transmission. For example, the modified sphygmomanometer measures pressure produced within the sphygmomanometer cuff, and the Jamar dynamometer measures force exerted on the handle. These tools use different joint and muscle mechanics. Even when the same tools are used, the use of different testing procedures and positions and the lack of standardization in terms of description and testing protocols make comparison of results difficult (Durward et al., 1999).

Pinch Assessment

Assessment of pinch generally involves testing of tip, lateral, and palmar pinches, although there is much variety in the description of pinch and in terminology used. The B & L Pinchmeter is often used for testing pinch. Using these tools, the client squeezes the pinch gauge using the various pinches. One to three trials are used to test tip, lateral, and palmar pinches, and usually three trials are taken with the mean calculated. Pinchmeter values are accurate to within +1%, inter-rater reliability is >0.979, and test-retest reliability is >0.81 (Lindstrom-Hazel et al., 2009; Mathiowetz, Vizenor, & Melander, 2000).

Client results are compared to norms that are based on age and gender (Table 4-9). The pinch norms can be used with most of the different pinchmeters because of demonstrated reliability (MacDermid, Evenhuis, & Louzon, 2001). Other ways of assessing pinch and grasp would be observation of hand use during functional activities and by using standardized coordination tests and timed tests of hand function.

Table 4-8

DYNAMOMETER NORMATIVE VALUES

NORMS AT AGE (YEARS)

		20	25	30	35	40	45	50	55	60	65	70	75+
Male	Right	121	121	122	120	117	110	114	101	90	91	75	66
	SD	21	23	22	24	21	23	18	27	20	21	21	21
	Left	104	110	110	113	113	101	102	83	77	77	65	55
	SD	22	16	22	22	19	23	17	23	20	20	18	17
Female	Right	70	74	79	74	70	62	66	57	55	50	50	43
	SD	14	14	19	11	13	15	12	12	10	10	12	11
	Left	61	63	68	66	62	56	57	47	46	41	41	38
	SD	13	12	18	12	14	13	11	12	10	8	10	9

n = 628; age range 20 to 94 years

Reprinted with permission from Trombly C. *Occupational Therapy for Physical Dysfunction.* 4th ed. Philadelphia: Lippincott Williams & Wilkins; 1995; pp. 150.

SUMMARY

- Muscle strength is the amount of tension that a muscle can produce. Power is the product of force and velocity. Endurance is the ability to maintain a force over time.

- Many factors influence muscle strength. Skeletal muscles have the properties of contractility, elasticity, irritability, and distensibility.

- The sliding filament theory is used to explain the contraction of single muscle cells in an all-or-none fashion where the myosin and actin filaments increase or decrease the amount of overlap.

- Total tension a muscle can develop depends on the amount of tension developed by each fiber, the number of fibers contracting at any time, the number of fibers in each motor unit, and the number of active motor units.

- Tension in a muscle depends on the initial length of the muscle, the innervation ratio of muscle fibers to motor units, the fiber arrangement and cross-section, the type of contraction, the location of the muscle in relation to the joint axis, the type of fiber, the number of joints crossed by the muscle, and the level of fatigue. Finer control of muscle tension is possible in muscles with small motor unit size.

- The greatest tension can be produced in fast-twitch gly-colic fibers and by shorter muscle fibers that are able to generate higher levels of passive tension and higher peak tension. Muscles with greater cross-sectional areas are able to produce greater tension. Muscles stretched prior to testing are stronger because passive elastic energy in the noncontractile elements is converted to kinetic energy when stretched.

- Tension in muscles can also be increased by increasing the frequency of the firing of motor units and by increasing the number and size of motor units participating.

- Subject-related factors (age, gender, activity level) and psychological factors also influence strength and muscle testing.

- Strength testing can be done by manually applying resistance to a part (manual muscle testing) or by using instruments to measure peak strength. Screening for muscle weakness may precede strength testing where muscle weakness is not the primary limiting factor in performance.

- Manual muscle testing has face and content validity but variable reliability. Reliability is enhanced when using dynamometers or instruments.

- Pinch and grasp strength are affected by subject factors, hand dominance, position of the arm and body, devices used for measurement, and variances in testing protocols.

Table 4-9

PINCHMETER NORMATIVE VALUES

NORMS AT AGE (YEARS)

			20	30	40	50	60	70	75+
Tip									
	Male	Right	18	18	18	18	16	14	14
		Left	17	18	18	18	15	13	14
	Female	Right	11	13	11	12	10	10	10
		Left	10	12	11	11	10	10	9
		(average SD: males = 4.0; females = 2.5)							
Lateral									
	Male	Right	26	26	26	27	23	19	20
		Left	25	26	25	26	22	19	19
	Female	Right	18	19	17	17	15	14	13
		Left	16	18	16	16	14	14	11
		(average SD: males = 4.6; females = 3.0)							
Palmar									
	Male	Right	27	25	24	24	22	18	19
		Left	26	25	25	24	21	19	18
	Female	Right	17	19	17	17	15	14	12
		Left	16	18	17	16	14	14	12
		(average SD: males = 5.1; females = 3.7)							

n = 628; age range = 20-94 years

Reprinted with permission from Trombly C. *Occupational therapy for physical dysfunction*. 4th ed. Philadelphia, PA: Lippincott Williams & Wilkins; 1995.

REFERENCES

Ali, N. A., O'Brien, J. M., Hoffmann, S. P., Phillips, G., Garland, A., Finely, J. C. W., et al. (2008). Acquired weakness, handgrip strength, and mortality in critically ill patients. *American Journal of Respiratory Critical Care Medicine, 178,* 261-268.

Anderson, J. L., Erzis, G., & Kryger, A. (1999). Increase in the degree of coexpression of myosin heavy chain isoforms in skeletal muscle fibers of the very old. *Muscle and Nerve, 22,* 440-454.

Balogun, J. A., Akomolafe, C. T., & Amusa, L. A. (1991). Grip strength: effects of testing posture and elbow position. *Archives of Physical Medicine and Rehabilitation, 72,* 280-283.

Bassey, E. J., & Harries, U. J. (1993). Normal values for handgrip strength in 920 men and women aged over 65 years, and longitudinal changes over 7 years in 620 survivors. *Clinical Science, 84,* 331-337.

Bear-Lehman, J., & Abreu, B. C. (1989). Evaluating the hand: Issues in reliability and validity. *Physical Therapy, 69*(12), 1025-1033.

Beasley, W. C. (1961). Quantitative muscle testing: Principles and application to research and clinical services. *Archives of Physical Medicine and Rehabilitation, 42,* 398-425.

Beaton, D. E., O'Driscoll, S. W., & Richards, R. R. (1995). Grip strength testing using the BTE Work Simulator and the Jamar Dynamometer: A comparative study. *Journal of Hand Surgery, 20A,* 293-298.

Bellace, J. V., Healy, D., Bryon, T., & Hohman, L. (2000). Validity of the Dexter evaluation system's Jamar dynamometer attachment for assessment of handgrip strength in a normal population. *Journal of Hand Therapy, 13,* 46-51.

Bello-Haas, V. D. (2009). Neuromusculoskeletal and movement functions. In B. R. Bonder & V. D. Bello-Haas (Eds.), *Functional performance in older adults* (3rd ed.). Philadelphia, PA: F. A. Davis Co.

Biel, A. (2001). *Trail guide to the body* (2nd ed.). Boulder, CO: Books of Discovery.

Bohannan, R. W. (1998). Hand-grip dynamometry provides a valid indication of upper extremity strength impairment in home care patients. *Physical Therapy, 11,* 258-260.

Bonder, B. R., & Bello-Haas, V. D. (2009). *Functional performance in older adults* (3rd ed.). Philadelphia, PA: F. A. Davis Company.

Brandsma, J. W., Schreuders, T. R., Birke, J. A., Piefer, A., & Oostendorp, P. (1985). Manual muscle strength testing: Intraobserver and interobserver reliabilities for the intrinsic muscles of the hand. *Journal of Hand Therapy, 8,* 185-190.

Cetinus, E., Buyukbese, M. A., Uzel, M., Ekerbicer, H., & Karaoguz, A. (2005). Hand grip strength in patients with type 2 diabetes mellitus. *Diabetes Research and Clinical Practice, 70,* 278-286.

Chau, N., Petry, D., Bourgkard, E., Huguenin, P., Remy, E., & Andre, J. M. (1997). Comparison between estimates of hand volume and hand strengths with sex and age with and without anthropometric data in healthy working people. *European Journal of Epidemiology, 12*, 309-316.

Cooper, C. (Ed.). (2007). *Fundamentals of hand therapy: Clinical reasoning and treatment guidelines for common diagnoses of the upper extremity.* St. Louis, MO: Mosby Elsevier.

Crosby, C. A., Wehbe, M. A., & Mawr, B. (1994). Hand strength: normative values. *Journal of Hand Surgery, 19A*, 665-670.

Cyriax, J. (1982). *Textbook of orthopaedic medicine.* London: Bailler Tindal.

Dalbo, V. J., Roberts, M. D., Lockwood, C. M., Tucker, P. S., Kreider, R. B., & Kerksick, C. M. (2009). The effects of age on skeletal muscle and the phosphocreatine energy system: can creatine supplementation help older adults? *Dynamic Medicine, 8*, 6.

Daniels, K., & Worthingham, C. (1986). *Muscle testing techniques of manual examination* (5th ed.). Philadelphia, PA: W. B. Saunders.

Davis, J. W., Ross, P. D., Preston, S. D., Nevitt, M. C., & Wasnich, R. D. (1998). Strength, physical activity, and body mass index: relationship to performance-based measures and activities of daily living among older Japanese women living in Hawaii. *Journal of American Geriatric Society, 42*, 274-279.

Desrosiers, J., Bravo, G., Hebert, R., & Dutil, E. (1995). Normative data for grip strength of elderly men and women. *Amer J Occup Ther, 49*(7), 637-643.

Doherty, T. J. (2001). The influence of aging and sex on skeletal muscle mass and strength. *Current Opinion in Clinical Nutrition and Metabolic Care, 4*(6), 503-508.

Donatelli, R. A., & Wooden, M. J. (2010). *Orthopaedic physical therapy* (4th ed.). St. Louis, MO: Churchill Livingstone Elsevier.

Durward, B. R., Baer, G. D., & Rowe, P. J. (1999). *Functional human movement: Measurement and analysis.* Oxford: Butterworth-Heinemann.

Fess, E. E. (1992). Grip strength. In J. S. Casanova (Ed.), *Clinical assessment recommendations* (2nd ed., pp. 28-32). Chicago, IL: American Society of Hand Therapists.

Fike, M. L., & Rouseau, E. (1982). Measurement of adult hand strength: A comparison of two instruments. *Occupational Therapy Journal of Research, 2*, 43-49.

Flinn, N. A., Latham, C. A. T., & Podolski, C. R. (2008). Assessing abilities and capacities: range of motion, strength, and endurance. In M. V. Radomski & C. A. T. Latham (Eds.), *Occupational therapy for physical dysfunction* (6th ed., pp. 91-185). Philadelphia, PA: Wolters Kluwer/Lippincott, Williams and Wilkins.

Florence, J. M., Pandya, S., King, W. M., et al. (1992). Interrater reliability of manual muscle test (Medical Research Council Scale) grades in Duchene's muscular dystrophy. *Physical Therapy, 72*(2), 115-122.

Frese, E., Brown, M., & Norton, B. (1987). Clinical reliability of manual muscle testing: Middle trapezius and gluteus medius muscles. *Journal of Physical Therapy, 67*(7), 1072-1076.

Frontera, W. R., Hughes, V. A., Fielding, R. A., Fiatarone, M. A., Evans, W. J., & Roubenoff, R. (2000). Aging of skeletal muscle: a 12-yr longitudinal study. *Journal of Applied Physiology, 88*(4), 1321-1326.

Gaur, D. K., Shenoy, S., & Sandhu, J. S. (2007). Effect of aging on activation of shoulder muscles during dynamic activities: An electromyographic analysis. *International Journal of Shoulder Surgery, 1*(2), 51-57.

Gench, B. E., Hinson, M. M., & Harvey, P. T. (1995). *Anatomical kinesiology.* Dubuque, IA: Eddie Bowers Publishing, Inc.

Giampaoli, S., Ferrucci, L., Cechi, F., LoNoce, C., Poce, A., & Dima, F. (1999). Hand-grip strength predicts incident disability in non-disabled older men. *Age and Ageing, 28*, 283-288.

Gutierrez, Z., & Shechtman, O. (2003). Effectiveness of the five-handle position grip strength test in detecting sincerity of effort in men and women. *Physical Medicine and Rehabilitation, 82*(11), 847-855.

Gutman, S. A., & Schonfeld, A. B. (2003). *Screening adult neurological populations.* Bethesda, MD: American Occupational Therapy Press.

Guyton, A. C. (1991). *Medical physiology.* Philadelphia, PA: W. B. Saunders.

Hall, C. M., & Brady, L. T. (2005). *Therapeutic exercise: moving toward function* (2nd ed.). Philadelphia, PA: Lippincott, Williams & Wilkins.

Hall, S., Lewith, G., Brien, S., & Little, P. (2008). A review of the literature in applied and specialized kinesiology. *Forsch Komplementmed, 15*(6), 40–46.

Hamill, J., & Knutzen, K. M. (1995). *Biomechanical basis of human movement.* Baltimore, MD: Williams and Wilkins.

Hamill, J., & Knutzen, K. M. (2003). *Biomechanical basis of human movement.* Philadelphia, PA: Lippincott, Williams & Wilkins.

Harries, A. D. (1985). A comparison of hand grip dynamometry and arm muscles size amongst Africans in North East Nigeria. *Human Nutrition and Clinical Nutrition, 39*, 309-313.

Herbison, G. J., Isaac, Z., Cohne, M. E., & Ditunno, J. F. (1996). Strength post-spinal cord injury: myometer versus manual muscle test. *Spinal Cord, 34*, 543-548.

Hughes, V. A., Frontera, W. R., Wood, M., Evans, W. J., Dallal, G. E., Roubenoff, R., et al. (2001). Longitudinal muscle strength changes in older adults: Influence of muscle mass, physical activity, and health. *The Journals of Gerontology, Series A, 56*(5), B209-B217.

Iddings, D. M., Smith, L. K., & Spencer, W. A. (1961). Muscle testing. Part 2: Reliability in clinical use. *Physical Therapy Review, 41*, 249.

Irion, G. (2000). *Physiology: The basis of clinical practice.* Thorofare, NJ: SLACK Incorporated.

Jenkins, D. B. (1998). *Hollingshead's functional anatomy of the limbs and back* (7th ed.). Philadelphia: W. B. Saunders Co.

Johannson, C. A., Kent, B. E., & Shepard, K. F. (1983). Relationship between verbal command volume and magnitude of muscle contraction. *Physical Therapy, 63*, 1260-1265.

Josty, I. C., Tyler, M. P. H., Shewell, P. C., & Roberts, A. H. N. (1997). Grip and pinch strength variation in different types of workers. *Journal of Hand Surgery, 22B*, 266-269.

Kendall, F. P., McCreary, E. K., Provance, P. G., Rogers, M. M., & Romani, W. A. (2005). *Muscles: testing and function with posture and pain* (5th ed.). Baltimore, MD: Lippincott, Williams & Wilkins.

Kendall, F. P., & McCreary, E. K. (1983). *Muscles: testing and function* (3rd ed.). Baltimore, MD: Williams and Wilkins.

Kisner, C., & Colby, L. A. (2007). *Therapeutic exercise: Foundations and techniques* (5th ed.). Philadelphia, PA: F. A. Davis Company.

Kolber, M. J., & Cleland, J. A. (2005). Strength testing using hand-held dynamometry. *Physical Therapy Review, 10*, 99-112.

Konin, J. G. (1999). *Practical kinesiology for the physical therapist assistant.* Thorofare, NJ: SLACK Incorporated.

Lamb, R. (1985). Manual muscle testing. In J. M. Rothstein (Ed.), *Measurement in physical therapy.* New York, NY: Churchill Livingstone.

Li, R., Jasiewicz, J. M., Middleton, J., Condie, P., Barriskill, A., Hebnes, H., et al. (2006). The development, validity, and reliability of a manual muscle testing device with integrated limb position sensors. *Archives of Physical Medicine and Rehabilitation, 87*(3), 411-417.

Lieber, R. L. (2002). *Skeletal muscle structure, function and plasticity.* Philadelphia, PA: Lippincott, Williams & Wilkins.

Lindle, R. S., Metter, E. J., Lynch, N. A., Fleg, J. L., Fozard, J. L., Tobin, J., et al. (1997). Age and gender comparisons of muscle strength in 654 women and men aged 20-93 years. *Journal of Applied Physiology, 83*(5), 1581-1587.

Lindstrom-Hazel, D., Kratt, A., & Bix, L. (2009). Interrater reliability of students using hand and pinch dynamometers. *American Journal of Occupational Therapy, 63*(2), 193-197.

Lu, T.-W., Hsu, H.-C., Chang, L.-Y., & Chen, H.-L. (2007). Enhancing the examiner's resisting force improves the reliability of manual muscle strength measurements: Comparison of a new device with hand-held dynamometry. *Journal of Rehabilitation Medicine, 39*(9), 679-684.

Lynch, N. A., Metter, E. J., Lindle, R. S., Fozard, J. L., Tobin, J. D., Roy, T. A., et al. (1999). Muscle quality. I. Age-associated differences between arm and leg muscle groups. *Journal of Applied Physiology, 86*(1), 188-194.

MacDermid, J. C., Evenhuis, W., & Louzon, M. (2001). Inter-instrument reliability of pinch strength scores. *Journal of Hand Therapy, 14*, 36-42.

Magee, D. J. (2008). *Orthopedic physical assessment.* Philadelphia, PA: Saunders.

Marieb, E. N. (1998). *Human anatomy and physiology.* Glenview, IL: Addison Wesley Longman Publishers.

Marx, R. G., Bombadier, C., & Wright, J. G. (1999). What do we know about the reliability and validity of physical examination tests used to examine the upper extremity. *Journal of Hand Therapy, 24A*, 185-193.

Mathiowetz, V., Weber, K., Volland, G., & Kashman, N. (1984). Reliability and validity of grip and pinch strength evaluations. *Journal of Hand Surgery, 9A*: 222-226.

Mathiowetz, V. (2002). Comparison of Rolyan and Jamar Dynamometers for measuring grip strength. *American Journal of Occupational Therapy, 9*, 201-209.

Mathiowetz, V. N. K., Volland, G., Weber, K., Dowe, M., & Rogers, S. (1985). Grip and pinch strength: Normative data for adults. *Archives of Physical Medicine and Rehabilitation, 66*, 69-74.

Mathiowetz, V., Vizenor, L., & Melander, D. (2000). Comparison of base-line instruments to Jamar dynamometer and the B & L Engineering pinch gauge. *Occupational Therapy Journal of Research, 20*, 147-162.

Mathiowetz, V., Wiemer, D. M., & Federman, S. M. (1986). Grip and pinch strength: norms for 6 to 19 year olds. *American Journal of Occupational Therapy, 40*, 705-711.

McGinnis, P. M. (1999). *Biomechanics of sport and exercise.* Champaign, IL: Human Kinetics.

McPhee, S. D. (1987). Functional hand evaluations: A review. *American Journal of Occupational Therapy, 41*(3), 158-163.

Milne, J. S., & Maule, M. M. (1984). A longitudinal study of handgrip and dementia in older people. *Age & Aging, 12*, 42-48.

Muscolino, J. E. (2006). *Kinesiology: The skeletal system and muscle function.* St. Louis, MO: Mosby.

Nalebuff, E., & Philips, C. A. (1990). The rhematoid thumb. In J. Hunter, J. Schneider, E. Mackin, & A. Callahan (Eds.), *Rehabilitation of the hand: Surgery and therapy.* St. Louis, MO: Mosby.

Neumann, D. A. (2002). *Kinesiology of the musculoskeletal system: Foundations for Physical Rehabilitation.* St. Louis, MO: Mosby.

Nicholas, J. A., Sapega, A., Kraus, H., & Webb, J. N. (1978). Factors influencing manual muscle tests in physical therapy. *Journal of Bone and Joint Surgery, 60-A*(2), 186-190.

Nordin, M., & Frankel, V. H. (2001). *Basic biomechanics of the musculoskeletal system.* Philadelphia, PA: Lippincott, Williams & Wilkins.

Nordin, M., & Frankel, V. H. (1989). *Basic biomechanics of the musculoskeletal system* (2nd ed.). Philadelphia, PA: Lea and Febiger.

Norkin, C. C., & Levangie, P. K. (1992). *Joint structure and function: A comprehensive analysis* (2nd ed.). Philadelphia, PA: F. A. Davis Co.

Nyland, J. (2006). *Clinical decisions in therapeutic exercise: Planning and implementation.* Upper Saddle River, NJ: Pearson Prentice Hall.

O'Driscoll, S. W., Horii, E., Ness, R., Callahan, T. D., Richards, R. R., & Kai-Nan, A. (1992). The relationship between wrist position, grasp size and grip strength. *Journal of Hand Surgery, 17A*, 169-177.

Oatis, C. A. (2004). *Kinesiology: the mechanics and pathomechanics of human movement.* Philadelphia, PA: Lippincott, Williams & Wilkins.

Palmer, M. L., & Epler, M. E. (1998). *Fundamentals of musculoskeletal assessment techniques* (2nd ed.). Philadelphia, PA: Lippincott.

Park, S. W., Goodpaster, B. H., Strotmeyer, E. S., deRekeneire, N., Harris, T. B., & Swartz, A. V. (2006). Decreased muscle strength and quality in older adults with type 2 diabetes: The Health, Aging and Body Composition Study. *Diabetes, 55*, 1813-1818.

Payton, O. (1979). *Research: The validation of clinical practice.* Philadelphia, PA: F. A. Davis.

Peolsson, A., Hedulnd, R., & Obert, B. (2001). Intra- and inter-tester reliability and reference values for hand strength. *Journal of Rehabilitation Medicine, 33*, 36-41.

Petersen, P., Petrick, M., Connor, H., & Conklin, H. (1989). Grip strength and hand dominance: challenging the 10% rule. *American Journal of Occupational Therapy, 43*, 444-447.

Philips, C. A. (1990). The management of patients with rheumatoid arthritis. In J. Hunter, J. Schneider, & E. Mackin (Eds.), *Rehabilitation of the hand: Surgery and therapy* (pp. 903-907). St. Louis, MO: Mosby.

Pollard, H., Lakay, B., Tucker, F., Watson, B., & Bablis, P. (2005). Interexaminer reliability of the deltoid and psoas muscle test. *Journal of Manipulative Physiologic Therapy, 28*(1), 52-56.

Pousson, M., Lepers, R., & Hoecke, J. V. (2001). Changes in isokinetic torque and muscular activity of elbow flexors muscles with age. *Experimental Gerontology, 36*(10), 1687-1698.

Pryce, J. C. (1980). The wrist position between neutral and ulnar deviation that facilitates the maximum power grip strength. *Journal of Biomechanics, 13*, 505-511.

Radomski, M. V., & Latham, C. A. T. (2008). *Occupational therapy for physical dysfunction.* Philadelphia, PA: Wolters Kluwer/Lippincott, Williams & Wilkins.

Rantanen, T., Guralnik, J. M., Foley, D., Masaki, K., Leveille, S., & Curb, J. D. (1999). Midlife hand grip strength as a predictor of old age disability. *Journal of the American Geriatric Society, 49*(1), 21-27.

Rantanen, T., Guralnik, J. M., Sakar-Rantala, R., Leveille, S., Simonsick, E. M., & Ling, S. (1999). Disability, physical activity, and muscle strength in older women: The women's health and aging study. *Archives of Physical Medicine and Rehabilitation, 80*, 130-135.

Redmond, C. L., Bain, G. I., Laslett, L. L., & McNeil, J. D. (2009). Hand syndromes associated with diabetes: Impairments and obesity predict disability. *Journal of Rheumatology, 36*(12), 2766-2771.

Reese, N. B. (2005). *Muscle and sensory testing.* St. Louis, MO: Elsevier Saunders.

Rice, C. L. (2000). Muscle function at the motor unit level: consequences of aging. *Topics in Geriatrics and Rehabilitation, 15*, 70-82.

Rice, M. S., Leonard, C., & Carter, M. (1998). Grip strengths and required forces in accessing everyday containers in a normal population. *American Journal of Occupational Therapy, 52*(8), 621-626.

Richards, L. G., Olson, B., & Palmiter-Thomas, P. (1996). How forearm position affects grip strength. *American Journal of Occupational Therapy, 50*(2), 133-138.

Rothstein, J. M., Roy, S. H., Wolf, S. L., & Scalzitti, D. A. (2005). *The Rehabilitation Specialist's Handbook* (3rd ed.). Philadelphia, PA: F. A. Davis Company.

Runnels, E. D., Bemben, D. A., Anderson, M. A., & Bemben, M. G. (2005). Influence of age on isometric, isotonic, and isokinetic force production characteristics in men. *Journal of Geriatric Physical Therapy, 28*(3), 5.

Savic, G., Bergstrom, E. M. K., Frankel, H. L., Jamous, M. A., & Jones, P. W. (2007). Inter-rater reliability of motor and sensory examinations performed according to American Spinal Injury Association standards. *Spinal Cord, 45*, 444-451.

Sayer, A. A., Dennison, E. M., Syddall, H. E., Gilbody, H. J., Philips, D. I., & Cooper, C. (2005). Type 2 diabetes, muscle strength, and impaired physical function: the tip of the iceberg? *Diabetes Care, 29*, 2541-2542.

Schechtman, O., Davenport, R., Malcolm, M., & Nabavi, D. (2003). Reliability and validity of the BTE-Primus grip tool. *Journal of Hand Therapy, 16*, 36-42.

Schechtman, O., Mann, W. C., Justis, M., & Tomita, M. (2004). Grip strength in the frail elderly. *Physical Medicine and Rehabilitation, 83*(11), 819-826.

Sisto, S. A., & Dyson-Hudson, T. (2007). Dynamometry testing in spinal cord injury. *Journal of Rehabilitation Research and Development, 44*(1), 123-136.

Smith, L. K., Weiss, E. L., & Lehmkuhl, L. D. (1996). *Brunnstrom's clinical kinesiology.* (5th ed.). Philadelphia, PA: F. A. Davis Co.

Snyder, D. C., Conner, L. A., & Lorenz, G. (2007). *Kinesiology foundations for OTAs.* Clifton Park, NY: Thomson Delmar Learning.

Stuberg, W. A., & Metcalf, W. K. (1988). Reliability of quantitative muscle testing in healthy children and in children with Duchenne Muscular Dystrophy using a hand-held dynamometer. *Physical Therapy, 68*, 977-982.

Su, C. Y., Lin, J. H., Chien, T. H., Cheng, K. F., & Sung, Y. T. (1994). Grip strength in different positions of elbow and shoulder. *Archives of Physical Medicine and Rehabilitation, 75*, 812-815.

Sunderland, A., Tinson, D., Bradley, K., & Hewer, R. L. (1989). Arm function after stroke: an evaluation of grip strength as a measure of recovery and a prognostic indicator. *Journal of Neurology, Neurosurgery and Psychiatry, 52*, 1267-1272.

Swartz, S., Cohen, M. E., Herbison, G. J., & Shah, A. (1992). Relationship between two measures of upper extremity strength: manual muscle testing compared to hand held myometry. *Archives of Physical Medicine and Rehabilitation, 73*, 1063-1068.

Tan, J. C. (1998). *Practical manual of physical medicine and rehabilitation.* St. Louis, MO: Mosby.

Toji, H., & Kaneko, M. (2007). Effects of aging on force, velocity, and power in the elbow flexors of males. *Journal of Physiological Anthropology, 26*(6): 587-592.

Trew, M., & Everett, T. (2005). *Human movement: An introductory text.* London: Elsevier Churchill Livingstone.

Trombly, C. A. (Ed.). (1995). *Occupational therapy for physical dysfunction.* (4th ed.). Baltimore, MD: Williams & Wilkins.

Van Deusen, J., & Brunt, D. (1997). *Assessment in occupational therapy and physical therapy*. Philadelphia, PA: W. B. Saunders Co.

Wallström, Å., & Nordenskiöld, U. (2001). Assessing hand grip endurance with repetitive maximal isometric contractions. *Journal of Hand Therapy, 14*(4), 279-285.

Williams, M. (1956). Manual muscle testing: Development and current use. *Physical Therapy Review, 36*, 797-805.

Zimmerman, M. E. (1969). The functional motion test as an evaluation tool for patients. *American Journal of Occupational Therapy, 23*(1), 49-56.

ADDITIONAL RESOURCES

Asher, I. E. (1996). *Occupational therapy assessment tools: An annotated index* (2nd ed.). Bethesda, MD: American Occupational Therapy Association.

Baxter, R. (1998). *Pocket guide to musculoskeletal assessment*. Philadelphia, PA: W. B. Saunders Co.

Berne, R. M., & Levy, M. N. (1998). *Physiology* (4th ed.). St. Louis, MO: C. V. Mosby.

Brand, P. W., & Hollister, A. M. (1999). *Clinical mechanics of the hand* (3rd ed.). St. Louis, MO: C. V. Mosby.

Christiansen, C., & Baum, C. (1997). *Occupational therapy: Enabling function and well being* (2nd ed.). Thorofare, NJ: SLACK Incorporated.

Demeter, S. L., Andersson, G., & Smith, G. M. (1996). *Disability evaluation*. St. Louis, MO: C. V. Mosby.

Epler, M. (2000). *Manual muscle testing: An interactive tutorial*. Thorofare, NJ: SLACK Incorporated.

Esch, D. L. (1989). *Musculoskeletal function: An anatomy and kinesiology laboratory manual*. Minneapolis, MN: University of Minnesota.

Galley, P. M., & Forster, A. L. (1987). *Human movement: An introductory text for physiotherapy students* (2nd ed.). New York, NY: Churchill-Livingstone.

Goss, C. M. (Ed.). (1976). *Gray's anatomy of the human body* (29th ed.). Philadelphia, PA: Lea and Febiger.

Greene, D. P., & Roberts, S. L. (1999). *Kinesiology: Movement in the context of activity*. St. Louis, MO: C. V. Mosby.

Hinkle, C. Z. (1997). *Fundamentals of anatomy and movement: A workbook and guide*. St. Louis, MO: C. V. Mosby.

Hinojosa, J., & Kramer, P. (1998). *Evaluation: Obtaining and interpreting data*. Bethesda, MD: American Occupational Therapy Association.

Hislop, H., & Montgomery, J. (1995). *Daniels and Worthingham's muscle testing: Techniques of manual examination* (6th ed.). Philadelphia, PA: W. B. Saunders Co.

Klein, R. M., & Bell, B. (1982). Self-care skills: Behavioral measurement with Klein-Bell ADL Scale. *Archives of Physical Medicine and Rehabilitation, 63*, 335-338

Mahoney, F. I., & Barthel, D. W. (1965). Functional evaluation: The Barthel Index. *Maryland State Medical Journal, 14*, 61-65.

Neumann, D. A. (2002). *Kinesiology of the musculoskeletal system: Foundations for physical rehabilitation*. St. Louis, MO: Mosby, Inc.

Pedretti, L. W. (1996). *Occupational therapy: Practice skills for physical dysfunction* (4th ed.). St. Louis, MO: C. V. Mosby.

Smith, R. O., & Benge, M. W. (1985). Pinch and grasp strength: Standardization of terminology and protocol. *American Journal of Occupational Therapy, 39*(8), 531-535.

Tennant, A., Geddes, J. M. L., et al. (1996). The Barthel Index: An ordinal or interval measure? *Clinical Rehabilitation, 10*, 301-308.

SECTION II

Normal Joint Movement

5

THE SHOULDER

The shoulder is a complex joint made up of many articulations capable of a wide variety of motions. It is widely accepted that the shoulder joint complex consists of four major joints, and other sources cite at least one and as many as three additional articulations that allow the wide variances of movement at the shoulder (Cailliet, 1980; Norkin & Levangie, 1992; Smith, Weiss, & Lehmkuhl, 1996; Perry, Rohe, & Garcia, 1996). The connections of the thorax with the humerus and scapula enable the positioning of the arm and hand in functional positions (Hartley, 1995).

The bones and articulations of the shoulder form a kinematic chain. The chain starts with the trunk (sternum), where forces are transmitted to the sternoclavicular joint. From there, force is transmitted from the clavicle to the acromioclavicular joint. The scapula is next in the chain, incorporating the glenohumeral joint and finally the humerus. Each part of the chain is essential for normal shoulder function. Figure 5-1 shows the shoulder in action. This kinematic chain of the upper extremity, trunk, and lower extremity works in wondrous synchrony to accomplish the activity of raising the arm over the head with the rope. The weight is shifted to the right to free up the left arm to move freely. The shoulder acts as a stable base to enable elbow extension to reach closer to the target while the hands hold onto the rope. This same action is repeated every day as we reach into cupboards, comb the back of our hair, and bend down to tie a shoelace.

Functional activities also require coordination between the visual field and use of the arms and hands. Because of the scapula's position on the anterior aspect of the thorax, this facilitates the use of the hands working in front of the body where we can see what we are doing. The position of the scapula enables the shoulder muscles to work in their stronger, middle ranges. The slight medial rotation of the humerus at the glenohumeral joint helps bring the hand to the mouth rather than to the anterior aspect of the shoulder (Trew & Everett, 2005).

The human shoulder, with the laterally directed glenoid cavity and longer, laterally twisted clavicle, allows much mobility and enables overhead action, which may have had a role in evolution by enabling vertical climbing. Because humans can carry objects, this may have been an incentive for bipedal locomotion (Veeger & vanderHelm, 2007).

BONES OF THE SHOULDER AND PALPABLE STRUCTURES

The shoulder complex is made up of the clavicle, the humerus, the sternum, the scapula, the ribs, and the vertebral column (Figure 5-2). Many skeletal landmarks and bony characteristics can be palpated. The humeral head is positioned by a closed chain formed by the thorax, scapula, and clavicle (Veeger & vanderHelm, 2007).

Clavicle

The clavicle provides the connection between the sternum and the scapula. It lies horizontally across the upper chest and

Rybski MF.
Kinesiology for Occupational Therapy, Second Edition (pp 97-174)
© 2012 SLACK Incorporated

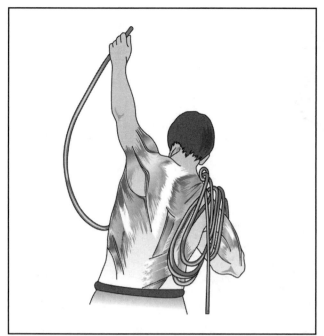

Figure 5-1. The shoulder in action. (Adapted from Biel, A. (2001). *Trail guide to the body*. Boulder, CO: Books of Discovery.)

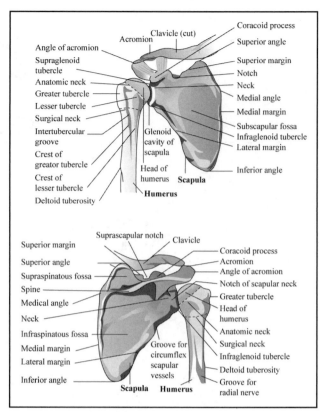

Figure 5-2. Bones of the shoulder.

has an "s" shape (Biel, 2001). The "s"-shaped bone helps to hold the scapula in the proper position for abduction by means of suspensor ligaments that help to produce maximum range of motion at the glenohumeral joint. The clavicle increases the mobility of the glenohumeral joint to permit reaching and climbing activities. The clavicle may rotate as much as 50 degrees with the shoulder in abduction and may elevate as much as 40 degrees, primarily at the acromioclavicular joint.

Lateral/Acromial End

The lateral end of the clavicle, which articulates with the acromion process of the scapula and projects above it, is easily palpable. It is relatively flat and rises slightly above the acromion (Biel, 2001).

Medial/Sternal End

Palpate the rounded projection above the superior aspect of the manubrium sterni. The line of the sternoclavicular joint can be identified (Esch, 1989). The sternal end curves inferiorly.

Shaft

It is possible to palpate the anterior and superior surfaces from medial to lateral. Note that the anterior surface is convex medially and concave laterally (Esch, 1989). Have your subject elevate/depress and protract/retract the scapula and notice how the clavicle moves.

Scapula

The scapula acts as a platform on which movements of the humerus are based (Gench, Hinson, & Harvey, 1995). It is a flat surface that allows for smooth gliding of the scapula on the

thoracic wall and provides a large surface for muscle attachments (Kibler, 1998). Motions of scapula are constrained by the medial border of scapula, which is pressed against thorax by the serratus anterior and rhomboid muscles and by external loads of the arm. Scapular motions are also constrained by clavicle, which allows the acromion to move on a sphere around the sternoclavicular joint (Veeger & vanderHelm, 2007).

Medial/Vertebral Border

This border is easily palpated about 1.5 inches lateral and parallel to the vertebral column.

Inferior Angle

The inferior angle is superficial and is located on the scapula at the medial border's lower end (Biel, 2001). Glide your fingers inferiorly along the medial border and palpate the lowest portion of the scapula (i.e., the junction of the medial and lateral borders of the scapula). If your subject consciously relaxes the shoulder girdle musculature, the angle will be more easily palpated (Esch, 1989).

Acromion Process

This structure is located at the top of the shoulder and the lateral aspect of the spine of the scapula. Palpate this flat process at the lateral point of the shoulder where it forms a shelf over the glenohumeral joint. This is the origin of the middle fibers of the deltoid muscle and the insertion of the trapezius muscle.

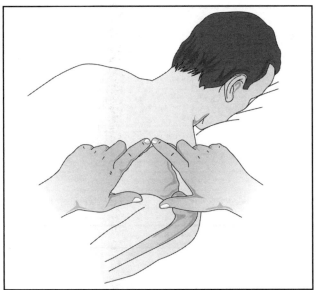

Figure 5-3. Palpation of the infraspinatus fossa. (Adapted from Biel, A. (2001). *Trail guide to the body.* Boulder, CO: Books of Discovery.)

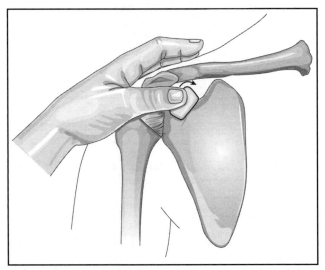

Figure 5-5. Palpation of the coracoid process. (Adapted from Biel, A. (2001). *Trail guide to the body.* Boulder, CO: Books of Discovery.)

Spine of the Scapula

This is the superficial ridge that ends at the top of the shoulder and runs at an oblique angle to the medial border of the scapula (Biel, 2001). Palpate from the acromion process to its base on the vertebral border.

Infraspinatous Fossa

Located inferior to the spine of scapula (Figure 5-3), this triangular depression can be felt above the inferior angle and between the medial and lateral borders (Biel, 2001).

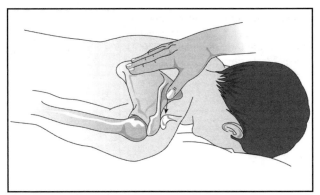

Figure 5-4. Palpation of the supraspinous fossa. (Adapted from Biel, A. (2001). *Trail guide to the body.* Boulder, CO: Books of Discovery.)

Supraspinous Fossa

This small, yet deep depression is located superior to the spine of scapula. It is difficult to access directly because it is covered by trapezius and supraspinatus muscles (Figure 5-4).

Subscapular Fossa

Because of its location on the scapula's anterior (underside) next to ribcage, it is difficult to palpate.

Coracoid Process

This pointed projection is found inferior to the shaft of clavicle and often is tender when palpated, so use care. Palpate with deep pressure through the medial border of the anterior deltoid muscle, just inferior to the clavicular concavity. It may be palpated approximately 2.5 cm below the junction of the lateral one third and the medial two thirds of the clavicle. If you have difficulty, ask your subject to protract the shoulder slightly. The coracoid process provides the origin for the coracobrachialis muscle and insertion for pectoralis minor muscle. It is the only hard, bony surface in this area (Figure 5-5).

Sternum

The sternum is made up of the manubrium, the body, and the xiphisternum. The manubrium and body are palpable as is the suprasternal notch, which is located on the superior aspect of the manubrium. The xiphisternum is less easily palpated (Lumley, 1990).

Humerus

The humerus "represents the first link in the chain of bony levers of the upper limbs; this is the only bone of the upper arm" (Esch, 1989).

Greater Tubercle

The greater tubercle of the humerus is located inferior and lateral to the acromion. With the arm in internal rotation, palpate just distal to the anterior portion of the acromion process.

Figure 5-6. Palpation of greater and lesser tubercles and the intertubercular groove. (Adapted from Biel, A. (2001). *Trail guide to the body.* Boulder, CO: Books of Discovery.)

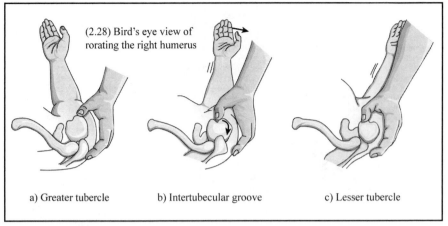

a) Greater tubercle   b) Intertubecular groove   c) Lesser tubercle

Figure 5-7. Articulations of the shoulder.

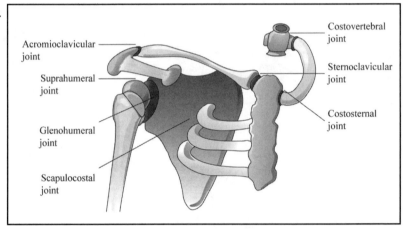

As your subject internally rotates the arm, you will feel it move under your fingers (Esch, 1989). The significance of the greater tubercle is that this is where supraspinatus, infraspinatus, and teres minor insert. This is often a site of impingement.

Lesser Tubercle

This small mound is the attachment site for the fourth rotator cuff muscle, subscapularis. With the humerus in external rotation, palpate anterior to the greater tubercle.

Intertubercular Groove

You feel this indentation between the greater and lesser tubercles. The long head of the biceps lies in this groove and can be tender with palpation (Figure 5-6).

ARTICULATIONS OF THE SHOULDER

The shoulder is capable of a wide variety of movements at as many as seven articulations as shown in Figure 5-7. It is generally accepted that there are three synovial joints (glenohumeral, acromioclavicular, and sternoclavicular) and two functional joints (scapulothoracic and scapulohumeral). The additional movement that occurs during shoulder motions

occurs between the ribs and the sternum (costosternal) and between the ribs and vertebral column (costovertebral).

Glenohumeral Joint

The glenohumeral joint is the major joint of the shoulder complex. The joint includes the glenoid fossa of the scapula articulating with the head of the humerus. The glenoid fossa is retroverted 7 degrees (slight medial rotation) and superiorly tilted 5 degrees relative to the plane of the scapula (Magee, 2002; Nordin & Frankel, 2001).

The glenohumeral joint is considered an incongruous joint because the articulating surfaces are not in direct contact. The greatest amount of articular contact is in mid-elevation between 60 and 120 degrees of motion (Wilk, Arrigo, & Andrews, 1997a). As a way of visualizing the incongruency, the glenohumeral joint has been compared to a ball on a plate. In fact, two thirds of the humeral head is not covered by the glenoid fossa of the scapula, which creates a marked discrepancy between the curvature of the glenoid fossa and the convex surface of the humeral head (Cailliet, 1980) as shown in Figure 5-8. This lack of congruence can be worsened by reduced humeral retroversion, by decreased curvature of glenoid fossa, or by an anteriorly tilted glenoid fossa, which was found in 80% of unstable shoulders as compared with 27% of normal shoulders (Nordin & Frankel, 2001).

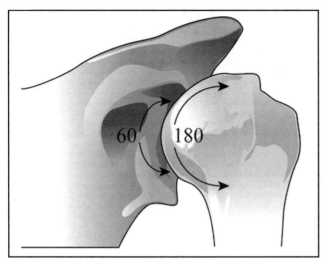

Figure 5-8. Incongruence of glenohumeral bony articulation.

The location of the head of the humerus in the glenoid provides a wide range of motion and shock-absorbing capability for the joint (Matsen, Harryman, & Sidles, 1991). The position of the humerus in the glenoid also helps to resist inferior subluxation or dislocation. However, the humeral head is particularly incongruent when the shoulder is 1) adducted, flexed, and internally rotated; 2) abducted and elevated; or 3) adducted at the side with the scapula rotated downward (Saidoff & McDonough, 1997, p. 196).

Matsen, Harryman, and Sidles noted that "it is amazing that this seemingly unstable joint can center itself to resist the gravitational pull on the arm hanging at the side for long periods, to permit lifting of large loads, to permit throwing a baseball at speeds approaching 100 miles an hour, and to hold together during the application of an almost infinite variety of forces of differing magnitude, direction, duration, and abruptness" (1991, p. 783).

The glenoid fossa is deepened somewhat by the glenoid labrum (Figure 5-9), which is comprised of dense fibrous tissue with few elastic fibers. The labrum joins with the glenohumeral joint capsule, glenohumeral ligaments, the long head of the biceps, and rotator cuff muscles, but the glenoid is still shallow, allowing only a small surface area of bone-to-bone contact.

The superior attachment of the labrum is loose while the inferior attachment is firm and unmoving (Wilk, Arrigo, & Andrews, 1997a). The role of the labrum as a passive stabilizer of the glenohumeral joint is debated (Hess, 2000; Veeger & vanderHelm, 2007). It appears to aid in controlling glenohumeral translations, acts as a load-bearing structure, protects the edges of the bones, assists in joint lubrication, provides an attachment for glenohumeral ligaments, and increases the contact area between articular surfaces (Wilk, Arrigo, & Andrews, 1997a; Hess, 2000). The intact labrum resists tangential forces of approximately 60% of compressive loads placed on the shoulder (Nordin & Frankel, 2001). Damage to the superior labrum may occur with anteroposterior extension (i.e., a "slap" lesion), repetitive overhead activities, a sudden pull on the arm, and compression (such as a fall on an outstretched arm) and may result in pain and shoulder instability (Nordin & Frankel, 2001).

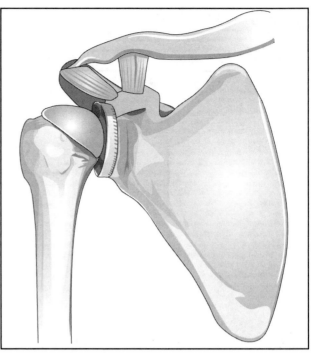

Figure 5-9. Labrum of the glenohumeral joint.

Osteokinematics

The glenohumeral joint is a ball and socket, freely movable, synovial joint with three degrees of freedom or motion in all planes. The glenohumeral joint permits rotation around all three axes, all of which pass through the head of the humerus. Glenohumeral flexion and extension move the humerus in a sagittal plane around a coronal axis, glenohumeral abduction and adduction move the humerus in a coronal plane around a sagittal axis, and internal and external rotation of the humerus moves the humerus in a horizontal plane around a vertical axis. Accessory motions (including rolling, spinning, and gliding and combinations of these movements) help to produce the diverse mobility seen at this joint.

Because this ball-and-socket joint is a synovial joint with a joint capsule and synovial fluid, friction is decreased. There are a number of bursae in the capsule that aid joint mobility. These include the subdeltoid, subcoracoid, coracobrachial, subacromial, and subscapular bursae. The bursae are formed by the synovial membrane of the joint capsule and function to decrease friction between two bony surfaces at points where muscles, ligaments, and tendons glide over bones. The subacromial bursa, which comprises the subacromial and subdeltoid bursae, allow small motions between the rotator cuff muscles and the acromion and the acromioclavicular joints. The subscapular bursae protect the tendon of the subscapularis muscle and go under the coracoid and the neck of the humerus.

The close-packed position of the joint in which the bones have the greatest congruency is in maximum abduction and external rotation. Accessory movements of the glenohumeral joint are possible in the loose-packed position of 55 degrees of abduction and 30 degrees of horizontal adduction. Table 5-1 summarizes the osteokinematics of the glenohumeral joint.

Table 5-1

SUMMARY OF THE OSTEOKINEMATICS OF THE GLENOHUMERAL JOINT

Functional joint	Diarthrotic multiaxial	
Structural joint	Synovial ball and socket	
Close-packed position	Horizontal abduction with external rotation (Hertling & Kessler, 1996) Flexion and internal rotation (Culham & Peat, 1993)	
Resting position	55° abduction, 30° horizontal adduction (scapular plane)	
Capsular pattern	External rotation, abduction, IR	
Primary muscles	External rotators	Supraspinatus Infraspinatus Teres minor
	Internal rotators	Teres major Subscapularis
	Elevation	Upper trapezius Levator scapulae Rhomboids
	Depression	Lower trapezius Latissimus dorsi Pectoralis minor
	Retractors	Rhomboids Middle trapezius
	Protractors	Pectoralis major Serratus anterior
	Upward rotation	Trapezius (upper and lower) Serratus anterior
	Downward rotation	Rhomboids Levator scapulae Pectoralis minor Latissimus dorsi

(continued)

Table 5-1 (continued)

SUMMARY OF THE OSTEOKINEMATICS OF THE GLENOHUMERAL JOINT

Primary muscles	Flexion	Biceps Deltoid (anterior)
	Extension	Triceps Deltoid (posterior)
	Abduction	Deltoid (middle)
	Adduction	Pectoralis major, triceps

Arthrokinematics

Passive motion of the glenohumeral joint produces rolling of the convex humeral head and downward gliding on the scapula's concave glenoid fossa as seen in Figure 5-10. This means that the motions of the distal and proximal humerus are reciprocally opposite during glenohumeral movement. For example, during shoulder abduction, when the humerus moves up, the head of the humerus slides inferiorly. Likewise, during adduction, as the arm comes to rest at the side of the body, the humeral head slides upward or superiorly (Shafer, 1997). However, when the humerus is stabilized and the scapula moves, the concave glenoid fossa slides in the same direction as the scapula (Kisner & Colby, 2007). Because the convex head of the humerus is not parallel to the concave glenoid fossa, rotation of the joint cannot take place as pure spin but requires that motions of the humerus be accompanied by combined rolling and gliding of the head of the humerus on the glenoid fossa in a direction opposite to the movement of the shaft of the humerus (Levangie & Norkin, 2005). This prevents impaction of the humeral head on either the acromion or the coracoacromial ligament in the normal glenohumeral joint (Norkin & Levangie, 1992).

To elevate the humerus, either in flexion or abduction, the muscles must accommodate the spin, roll, and glide of the head of the humerus. The forces that guide the arthrokinematics are the rotator cuff muscles and the glenohumeral ligaments (Table 5-2).

The concave-convex rule states that the humeral head slides inferiorly during abduction, anteriorly during external rotation, and posteriorly during internal rotation. However, other research has shown that, during the initial 30 degrees to 60 degrees of elevation in the scapular plane, the humeral head moves superiorly 3 mm then stays centered within 1 mm. During horizontal plane movement, the humeral head stays centered until maximal extension and external rotation occurs (such as the cocking phase of pitching) when 4 mm of posterior translation occurs (Howell, Galinat, Renzi, & Marone,

1988). These studies suggest that movement of the humeral head is related to tightness in the joint capsule, supporting the importance of joint mobility testing. Joint mobilization techniques need to consider the direction of force rather than just the convex-concave rule (Kisner & Colby, 2007). Kirby, Showalter, and Cook (2007) add that, because support for the convex-concave rule for the glenohumeral joint is poor, joint mobilization techniques based solely on this rule may not yield outcomes any better than other glenohumeral movement patterns.

Flexion and Extension

Flexion and extension of the glenohumeral joint, defined as the rotation of the humerus in the sagittal plane around a medial-lateral (frontal or coronal) axis of rotation (Neumann, 2002, p. 112), is primarily a spinning motion that occurs around a fixed point on the glenoid.

The anterior deltoid, pectoralis major, and coracobrachialis muscles flex the humerus, and as this occurs, the humeral head slides posteriorly and rolls anteriorly. In a normal glenohumeral joint, passive flexion produces about 4 mm of anterior translation of the humeral head on the glenoid while extension produces approximately 4 mm of posterior translation (Uhl, Kibler, Gecewich, & Tripp, 2009).

Abduction and Adduction

For abduction and adduction, defined as the rotation of the humerus in the frontal plane around an anterior-posterior axis (sagittal) (Neumann, 2002, p. 110), the middle deltoid muscle and supraspinatus are active, and the physiologic motions of the humeral head include sliding inferiorly and rolling superiorly. The convex humeral head rolls upward and slides downward on the scapula's concave glenoid fossa. Without sufficient inferior slide during abduction, the rolling action of the humeral head would impinge on the supraspinatus muscle on the coracoacromial arch, which is painful and limits abduction (Neumann, 2002). The humeral flexors and abductors do not act without the rotator cuff muscles and the long head of the

Table 5-2

SUMMARY OF THE ARTHROKINEMATICS OF THE GLENOHUMERAL JOINT

Physiologic movement of **convex** head rolls in the same direction and slides in the opposite direction of the **concave** glenoid fossa	Roll	Slide	Resultant Movement
Flexion	Anterior	Posterior	Opposite
Horizontal adduction	Anterior	Posterior	Opposite
Internal rotation	Anterior	Posterior	Opposite
Extension	Posterior	Anterior	Opposite
Horizontal abduction	Posterior	Anterior	Opposite
External rotation	Posterior	Anterior	Opposite
Abduction	Superior	Inferior	Opposite

Adapted from Kisner, C., & Colby, L. A. (2007). *Therapeutic exercise: Foundations and techniques* (5th ed.). Philadelphia, PA: F. A. Davis Company.

Figure 5-10. Glenohumeral arthrokinematics for abduction. (Adapted from Neumann, D. A. (2002). *Kinesiology of the musculoskeletal system: Foundations for physical rehabilitation.* St. Louis, MO: Mosby.)

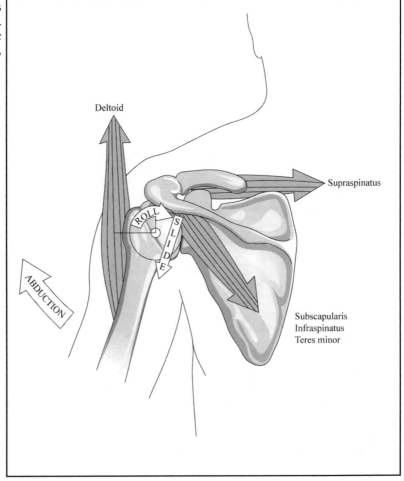

biceps, because this would cause compression of the subacromial space and there would be very little abduction or flexion. This information is invaluable to the therapist who is trying to provide mobilization to the shoulder by means of passive range of motion or soft tissue stretching (Uhl et al., 2009).

Elevation in the Plane of the Scapula

When asked to raise the arm up, most people do not elevate their arms precisely in the sagittal plane and frontal axis as occurs in forward flexion nor do they purely abduct the arm in a frontal plane and sagittal axis. Instead, the movement is more likely to be elevation in the plane of the scapula. This movement is approximately 30 degrees to 45 degrees anterior to the coronal or frontal plane and is sometimes called "scaption." The forward elevation of the plane of the scapula is considered a more functional movement because the inferior portion of capsule is not twisted and the musculature of the shoulder is optimally aligned for elevation of the arm (Nordin & Frankel, 2001). The center of the humeral head remains centered in the glenoid cavity throughout the arc of motion except during initiation of elevation (Howell et al., 1988). Movement in this plane enables upward rotation of the scapula, posterior tilt, and external rotation with clavicular elevation and retraction (McClure, Michener, Sennett, & Karduna, 2001).

The plane of the scapula is clinically significant because the length-tension relationship between the shoulder abductors, rotators, and posterior rotator cuff muscles is at an optimum length as compared with function in the coronal plane. With the shoulder in this plane, bony impingement of the greater tuberosity against acromion does not occur because of the alignment of the tuberosity and the acromion. There is optimal bony congruence in this position, which decreases anterior capsular stress (Ellenbecker & Ballie, 2010).

The amount of abduction that occurs in the frontal plane depends on rotation of the humerus. If the humerus is in internal rotation, there is 60 degrees to 90 degrees of abduction, while if in external rotation, 90 degrees to 120 degrees abduction is possible (not accounting for scapulothoracic contributions). Abduction in the scapular plane is not dependent upon humeral rotation, and there is less restriction of motion. Average maximal range of motion for abduction in the scapular plane is 107 degrees to 112 degrees (McClure et al., 2001; Zatsiorsky, 1998).

External and Internal Rotation

External rotation occurs when the humeral head simultaneously rolls posteriorly and slides anteriorly on the glenoid fossa (Neumann, 2002). Internal rotation occurs when the humeral head simultaneously rolls anteriorly and slides posteriorly on the glenoid fossa. If external rotation occurs by posterior roll without anterior slide, this amount of translation of joint surfaces can disarticulate the joint (Neumann, 2002). Internal rotation with the arm at the side produces only about 2 mm of anterior translation, with external rotation producing the same amount in a posterior direction (Uhl, 2009).

Supporting Structures

Unlike the hip joint, the other ball-and-socket joint in the body, the stability of the glenohumeral joint is not accomplished due to the articulation of bony segments but is achieved instead by capsular, ligamentous, and particularly muscular structures. Zuckerman and Matsen (1984) identified five factors that are important for the stability of the glenohumeral joint:

1. Adequate size of the glenoid fossa.
2. Posterior tilt of the glenoid fossa.
3. Humeral head retroversion.
4. Intact capsule and glenoid labrum.
5. Function of muscles that control the anteroposterior position of the humeral head.

The first four factors are considered passive supporting structures while the function of muscles is a dynamic support for the glenohumeral joint.

Passive Structures

The passive structures that aid in glenohumeral stability include the bony geometry, the glenoid labrum, and the glenohumeral joint capsuloligamentous structures.

Bone Support

Kibler (1998) and Kibler and McMullen (2003) identified five roles that the scapula assumes in the function of the glenohumeral joint. Several of these contribute to glenohumeral stability. First, the scapula is considered the stable portion of the joint. The humerus and scapula move so that the center of rotation of the joint is constrained within a physiologic pattern throughout the full range of motion (Kibler, 1998, p. 325). This allows compression of the humeral head into the glenoid fossa. In addition, the weight of the upper extremity creates a downward and forward tipping of the scapula (Kisner & Colby, 2007), creating a cohesive force of the subscapular bursa. Second, the scapula permits protraction and retraction along the thoracic wall without which there would not be full humeral elevation. The third role of the scapula is to enable elevation of the acromion. The scapula must rotate and be tilted to avoid impingement of the rotator cuff muscles. The scapula also provides a base for muscle attachments as the fourth important role. The scapular muscles attach to the medial, superior, and inferior borders and are intrinsically aligned so they are most efficient between 70 degrees and 100 degrees of abduction, acting as a compressor cuff (Kibler, 1998). The final role of the scapula is as a link in the proximal-to-distal sequencing of velocity, energy, and force transmission (Kibler, 1998). Kibler and McMullen add that the "scapula is thus pivotal in transferring large forces and high energy from the legs, back and trunk to the delivery point, the arm and hand, allowing more force to be generated in activities such as throwing than could be done by the arm musculature alone. The scapula, serving as a link, also stabilizes the arm to more effectively absorb loads that may be generated through the long lever of the extended or elevated arm" (2003, p. 143).

The glenoid concavity is formed by a combination of the shape of the bone and overlying cartilage and labrum. The effective glenoid arc, or the deformable rim under load, may be compromised due to congenital deficiencies (glenoid hypoplasia), excessive compliance, traumatic lesion, or wear.

Figure 5-11. Joint capsule of the glenohumeral joint.

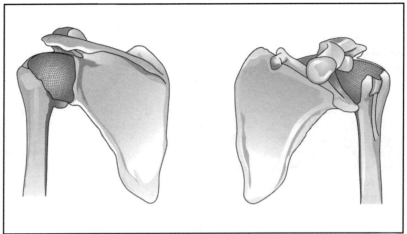

Joint Capsule

The glenohumeral joint capsule (shown in Figure 5-11) is a large, loose structure enabling much motion at the glenohumeral joint. It is a lax structure, and anterior, superior, and posterior aspects of the capsule are reinforced by the tendons of the rotator cuff and by coracohumeral and superior glenohumeral ligaments, but there is no reinforcement inferiorly, so this is an area of weakness (Wilk et al., 1997a).

The capsule has multilayered collagen fiber bundles, and the anteroinferior portion is the thickest and strongest, with densely organized fibers. Fiber arrangement varies and serves different stabilizing functions. With the radially oriented collagen fibers, rotational forces produce tension within these fibers, which leads to compression of joint surfaces and a centering of the joint. Circular fiber bundles appear to contribute to absorption of stress and tension, and spiral-shaped cross-linked collagen fibers assist with joint stability (Wilk et al., 1997b, p. 369).

With the arm in adduction, the capsule is taut superiorly and slack inferiorly (Figure 5-12). The inferior portion lies in folds with the arm adducted so it can adhere to itself, possibly leading to adhesive capsulitis when disease or trauma is present. With increasing abduction, the capsule becomes tight inferiorly and lax superiorly. This provides passive joint stability, because the capsule holds the humeral head to the glenoid. Abduction is accompanied by external rotation in the normal shoulder joint. When abducting the humerus, a twist occurs on the joint capsule, and tension develops in the joint capsule. This tension increases with abduction, pulling the humerus into external rotation and allowing greater range of motion into abduction because the greater tubercle is able to clear the coracoacromial arch (Hertling & Kessler, 1996).

There is a similar taut/loose pattern with internal and external rotation and the joint capsule. When the arm is in external rotation, the anterior capsule tightens. Conversely, in internal rotation, the capsule tenses posteriorly, resisting anterior translation of the humeral head. The passive restraints act not only to restrict movement but also to reverse the humeral head movement (Wilk et al., 1997b, p. 366).

As a rule, the superior capsular structures have a role in joint stability when the arm is adducted, and inferior capsular structures assist with joint stability from 90 degrees

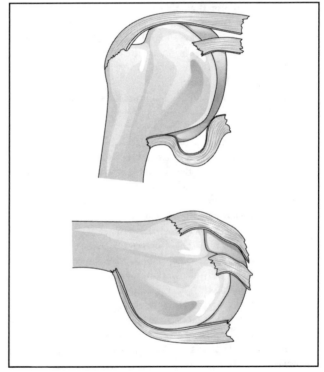

Figure 5-12. Capsular support at the glenohumeral joint.

abduction or forward humeral flexion. The posterior capsule is seen as crucial to maintaining glenohumeral stability as a secondary restraint to anterior dislocation and is seen as a primary posterior stabilizing structure (Nordin & Frankel, 2001). With pathologic shortening of the glenohumeral structures, the humerus moves to the position of least restriction. For example, if there was shortening of the posterior capsule, the humerus would move to an anterior position. Capsular adhesions and patterns for the glenohumeral joint are in abduction and in internal and external rotation. Abnormal tightness of the joint capsule can greatly impair the normal biomechanical motion of the shoulder. Patterns of capsular mechanics contributing to glenohumeral stability are summarized in Table 5-3.

Table 5-3

CAPSULAR MECHANICS FOR THE GLENOHUMERAL JOINT

CAPSULE PART	TAUT
Superior	Adduction or resting position
Inferior	90 degrees abduction
Anterior	External rotation
Posterior	Internal rotation
Posteroinferior	Restraint to posterior dislocation

An additional function of the glenohumeral joint capsule is that, because the capsule is sealed tight, this creates a relative vacuum that resists large glenohumeral joint translations. Small translations are possible and can be balanced by fluid flow in the opposite direction. Negative joint pressure pulls the capsule inward toward the joint space, creating a suction effect of the glenoid labrum with the humeral head and an adhesion-cohesion relationship of the synovial surfaces (Kisner & Colby, 2007; Matsen, Harryman, & Sidles, 1991). Even though the magnitude of the pressure is small, when the joint capsule is punctured, the humeral head tends to sublux, regardless of where puncture occurs, resulting in loss of glenohumeral stability (Wilk et al., 1997a). The joint volume effect can be compromised in clients with capsular defects, joint effusion, and in excessively compliant joint capsules (Matsen et al., 1991; Wilk et al., 1997a).

Gravity

Gravity also acts in static stability by pulling the humeral head downward in a direction parallel to the humeral shaft, moving the humerus into adduction. Gravity is offset by the superior joint capsule, superior and middle portions of the glenohumeral ligaments, and the coracohumeral ligaments, which are tight when the arm is adducted. This provides stabilization as well as limits external rotation in the lower ranges of abduction. If there are additional loads to the limb in addition to the force of gravity, dynamic stabilizers provide the additional stabilization. Supraspinatus contracts to aid with stabilization and may be assisted by the posterior deltoid, which also helps to prevent downward displacement of the humerus (Wilk et al., 1997b; Matsen et al., 1991; Kisner & Colby, 2007).

Ligaments

Ligaments resist tension in one direction and serve to connect bone-to-bone, which adds to the stabilization of the joint. Ligaments are as strong as tendons with many collagen fibers but also with fibroelastic tissues providing some elasticity to these structures. The primary ligaments adding support to the glenohumeral joint are the coracohumeral ligament and the glenohumeral ligaments.

Glenohumeral Ligament

The glenohumeral ligament is part of the glenoid labrum, and it has three parts: superior, inferior, and middle. There is much variability in size and attachment of the glenohumeral ligaments, and the clinical significance of these structures is yet to be fully explained (Muscolino, 2006). The glenohumeral ligaments are the thickenings of the anterior and inferior joint capsules, which help to prevent dislocation of the humeral head anteriorly and inferiorly. Essentially, as a group, the glenohumeral ligaments serve to limit extremes in glenohumeral motions. The ligament is lax enough to permit motion, so the ligament cannot prevent glenohumeral translation when the joint is moving through most of its range of motion. The ligament exerts an effect only when it is under tension, usually at the extremes of range (Matsen et al., 1991).

The superior glenohumeral ligament (Figure 5-13) is the smallest of the glenohumeral ligaments, with fibers running from an inferior-medial to superior-lateral orientation. It originates from the upper part of glenoid cavity and the base of coracoids and attaches to the middle glenohumeral ligament, the biceps tendon, and the labrum, inserting just superior to the lesser tuberosity in the region of the bicipital groove. The primary role of the superior glenohumeral ligament is to limit inferior translation of the humeral head in adduction and act as a restraint to anterior translation of the humeral head when the humerus is in adduction (Magee, 2002; Matsen et al., 1991). The superior glenohumeral ligament, together with the coracohumeral ligament and supraspinatus muscle, aids in prevention of downward displacement of the humeral head and limits external rotation between 0 degrees and 60 degrees (Wilk et al., 1997b).

The middle glenohumeral ligament (see Figure 5-13) is dense but variable in size and thickness. It is poorly defined or absent in 30% of normal shoulders (Matsen et al., 1991). The middle glenohumeral ligament attaches to the anterior aspect of the anatomic neck of the humerus, just medial to the lesser tuberosity of the humerus, and it arises from glenoid via the labrum. The middle glenohumeral ligament acts as a restraint to inferior translation with the arm adducted and in external rotation. In addition, the middle glenohumeral ligament can

Figure 5-13. Glenohumeral ligament.

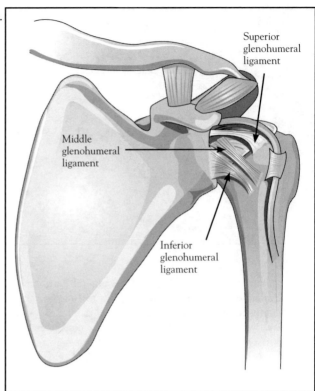

Figure 5-14. Inferior glenohumeral ligaments, anterior and posterior bands.

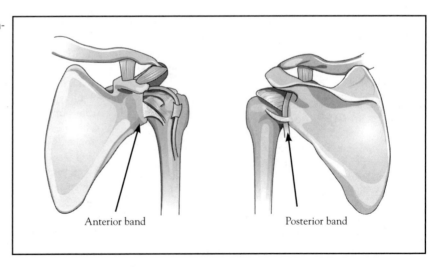

act to restrain anterior movement with the maximal effectiveness between 45 degrees and 90 degrees of abduction (Magee, 2002; Wilk et al., 1997a).

The inferior glenohumeral ligament (Figure 5-14) is the largest and most important of the glenohumeral ligaments. It consists of three parts: anterior, posterior, and the axillary pouch. The anterior and posterior portions contribute to the anterior and posterior labrum, and the axillary pouch is located between the anterior and posterior bands and attaches to the inferior two thirds of the glenoid via labrum. The axillary pouch acts like a sling: supporting the humeral head above 90

degrees of abduction, limiting inferior translation; tightening anterior band on external rotation; and tightening the posterior band on internal rotation (Magee, 2002).

The inferior glenohumeral ligament (and posteroinferior capsule) stabilizes against posterior instability, subluxation, and inferior translation with the arm in 90 degrees of abduction (Wilk et al., 1997b). If the arm is in less than 90 degrees of humeral abduction, the posterior joint capsule and the anterior portion of the inferior glenohumeral ligament are the primary restraints to anterior translation of the humeral head (Magee, 2002). In addition, tension in the anterior ligament creates a

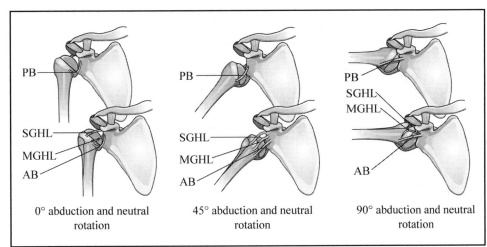

PB
SGHL
MGHL
AB

0° abduction and neutral rotation

PB
SGHL
MGHL
AB

45° abduction and neutral rotation

PB
SGHL
MGHL
AB

90° abduction and neutral rotation

Figure 5-15. Ligamentous constraints to inferior-superior translation of the humerus. SGHL = superior glenohumeral ligament; MGHL = middle glenohumeral ligament; AB = anterior band of inferior glenohumeral ligament; PS = posterior band of inferior glenohumeral ligament. (Adapted from Bowen, M. K., & Warren, R. F. (1991). Ligamentous control of shoulder stability based on selective cutting and static translation experiments. *Clinical Sports Medicine. 10*(4), 757-782.)

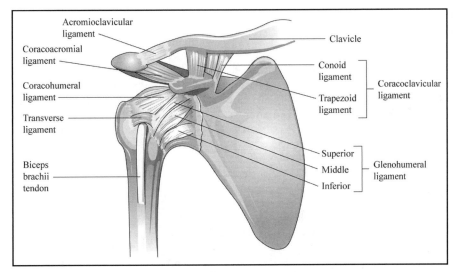

Acromioclavicular ligament

Coracoacromial ligament

Coracohumeral ligament

Transverse ligament

Biceps brachii tendon

Clavicle

Conoid ligament — Coracoclavicular ligament

Trapezoid ligament

Superior — Glenohumeral ligament
Middle
Inferior

Figure 5-16. Anterior shoulder ligaments.

posterior-directed force on the head of the humerus, and it elongates with and limits glenohumeral extension (Matsen et al., 1991). In positions of abduction and external rotation, the anterior band moves forward and stabilizes the joint anteriorly, and the fibers of the posterior band are pulled under the head to stabilize it inferiorly. The posterior band elongates with and limits glenohumeral flexion (Palmer & Epler, 1998). With internal rotation, these bands shift in the opposite direction as the anterior fibers move inferiorly and the posterior band shifts posteriorly (Figure 5-15).

The inferior glenohumeral ligament is lax in adduction. Because the inferior glenohumeral ligament is taut when the arm is abducted to 90 degrees or more, it provides anterior and posterior stabilization to the joint. It reinforces the capsular area between the subscapularis muscle and the origin of the long head of the triceps.

Coracohumeral Ligaments

The coracohumeral ligament (shown in Figure 5-16) and the acromion of the scapula form an arch that prevents excessive upward dislocation of the humeral head. The coracohumeral ligament is located between the coracoid process of the scapula and the greater tubercle of the humerus, spanning the bicipital groove. It represents the folded thickening of the gle-

nohumeral capsule in the area of the rotator interval between subscapularis and supraspinatus muscles. The coracohumeral ligament blends with the rotator cuff muscles, which provides stability superiorly and joins with the joint capsule (Cooper, O'Brien, & Warren, 1993).

When the humerus is externally rotated, the ligament is elongated, thereby limiting external rotation below 60 degrees (Magee, 2002). When the arm is externally rotated, flexed, or extended, the coracohumeral ligament gets taut, and this helps to resist inferior subluxation and dislocation of the humeral head anteriorly. Because this ligament helps to limit extremes in flexion, extension, and external rotation by virtue of its strength and strategic position, it is considered one of the most important ligamental structures in maintaining glenohumeral integrity and stabilization (Gench et al., 1995; Edelson, Taitz, & Grhishkan, 1991).

Coracoacromial Ligaments

The coracoacromial ligament (Figure 5-17) is a triangular band that originates from the lateral aspect of the coracoid process and attaches to anterior, medial, and inferior surfaces of acromion. The coracoacromial arch is formed by the coracoid process, acromion, clavicle, acromioclavicular joint, and

Figure 5-17. Rotator cuff muscles.

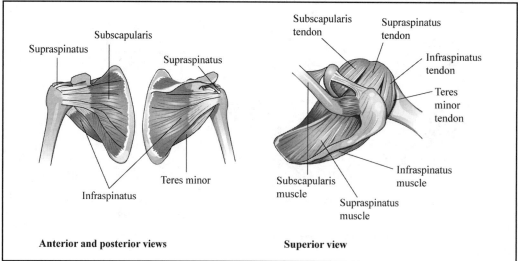

Anterior and posterior views

Superior view

the coracoacromial ligament. The arch functions to protect superior structures of the glenohumeral joint, stabilizes the humeral head, and prevents upward translation. If not for the coracoacromial arch, carrying a heavy bag on the shoulder might damage the superior structures of shoulder (Muscolino, 2006).

Biceps Tendon

The long and short head of biceps serve as anterior stabilizers of the glenohumeral joint in the movements of abduction and external rotation.

Overall, the anterior and posterior capsule and capsular ligaments limit translation and rotation of the humerus. When these capsule and ligamentous constraint mechanisms are excessive, this can do joint damage. If there is a tight posterior capsule, this imposes an anterior-superior translation with flexion, causing impingement against the acromion. If the anterior capsule is too tight, posterior subluxation of the humeral head may result with the potential of developing degenerative joint disease (Matsen et al., 1991). A summary of the passive structures of the glenohumeral joint is provided in Table 5-4.

Dynamic Structures

Dynamic stability is the combined efforts of muscles, intact proprioception, and the musculoligamentous relationship of the glenohumeral structures that stabilize the humeral head in glenoid through compression, enabling activity and movement to occur (Kibler & McMullen; 2003; Kibler, Uhl, Maddux, Brooks, Zeller, & McMullen, 2002; Neumann, 2002; Uhl et al., 2009; Wilk et al., 1997a; Wilk et al., 1997b).

Muscles

Primary dynamic muscles providing support to the glenohumeral joint are the rotator cuff muscles (supraspinatus, infraspinatus, teres minor, and subscapularis [SITS] or intrinsic muscles), the deltoid muscle, the long head of the biceps and triceps (considered extrinsic muscles), trapezius, rhomboids, levator scapula, and serratus anterior (considered periscapular muscles), which all act to stabilize and allow rotation of the scapula.

The importance of rotator cuff muscles lies in their strength, dynamic efficiency, and endurance in providing stability and unloading stress on capsular ligaments (Wilk et al., 1997a). Because the rotator cuff muscles blend into the shoulder joint capsule, these muscles create active and passive barriers to humeral head translation as well as act to absorb and dissipate repetitive microtraumatic stresses (Wilk et al., 1997a). The rotator cuff muscles provide a balanced force to the humeral head, keeping it secure in the glenoid cavity while also protecting the labrum, joint capsule, and ligaments from damage during activities (Donatelli & Wooden, 2010). Not only do the individual rotator cuff muscles contribute to dynamic glenohumeral stability, but they also provide passive stability by muscular bulk.

The rotator cuff stabilizes the glenohumeral joint through force couples in both the coronal and transverse planes. Figure 5-18 illustrates the location of the rotator cuff muscles on the scapula. Subscapularis is seen as essential for joint stability and acts both as a passive stabilizer (depressor of the humeral head that aids in prevention of subacromial and posterosuperior glenoid impingement) and an internal rotator, preventing further anterior and superior translation of the humeral head as the arm is moved.

Infraspinatus muscle, a weak muscle, is a primary external rotator of the humerus and acts as a depressor during elevation. It provides a stabilizing effect by preventing posterior subluxation of the humeral head in internal rotation and creates an anterior force by tightening posterior structures. Infraspinatus also has an additional role of preventing anterior translation during external rotation and abduction.

Supraspinatus, due to its superior location, is most frequently involved in rotator cuff muscle tears. It helps to stabilize the humeral head and initiates abduction and some external rotation. Supraspinatus is active during any elevation. Supraspinatus, infraspinatus, and teres minor are the major dynamic structures limiting internal rotation during the first half of abduction. Once abduction or flexion occurs, the

Table 5-4

SUMMARY OF PASSIVE STRUCTURES OF THE GLENOHUMERAL JOINT

STRUCTURE	*TYPE OF STABILITY PROVIDED*
Bone support (scapula)	• Compression of humeral head into glenoid due to centering of glenoid • Weight of arm produces downward and forward tipping of scapula creating a cohesive force • Size, tilt, and amount of deformation of glenoid fossa
Joint capsule	• Compression of joint surfaces and centering of glenoid • Spiral-shaped collagen fiber aid in stability • Taut superior capsule in adduction • Taut inferior capsule in abduction • Taut anterior capsule in external rotation • Taut posterior capsule in internal rotation • Posteroinferior capsule is a restraint to posterior dislocation • Vacuum function of sealed capsule creates suction effect
Gravity	• Pulls humeral head downward parallel to humeral shaft into adduction
Glenohumeral ligament	
Superior	• Limits inferior translation of humeral head in adduction • Restraint to anterior translation of humeral head • Prevents downward displacement of humeral head • Limits external rotation between 0 to 60 degrees
Middle	• Restraint to inferior translation with arm adducted and in external rotation • Restraint to anterior movement at 45 to 60 degrees of abduction

Inferior	• Stabilizes against posterior instability, subluxation
	• Stabilizes against inferior translation with arm in 90 degrees of abduction
	     o Anterior band: restricts abduction and external rotation
	     o Posterior band: restricts abduction and internal rotation

Coracohumeral ligament	• Strengthens the superior portion of the joint capsule • Limits external rotation below 60 degrees • Resists inferior subluxation and dislocation
Coracoacromial ligament	• Protects superior glenohumeral structures • Stabilizes the humeral head • Prevents upward translation of the humeral head

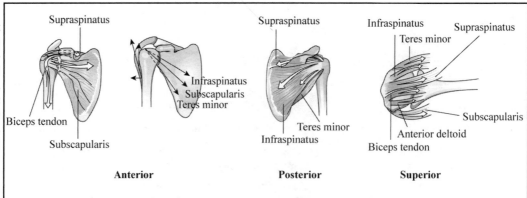

Figure 5-18. Co-contraction of dynamic stabilizers producing humeral head compression. (Adapted from Muscolino, J. E. (2006). *Kinesiology: The skeletal system and muscle function.* St. Louis, MO: Mosby.)

passive support of the superior joint capsule and supraspinatus muscle is eliminated, and stabilization is due to muscles. Given the location of the muscles and fiber architecture, one way the rotator cuff muscles help to stabilize the glenohumeral joint is by passive bulk.

Rotator cuff muscles also help to stabilize the glenohumeral joint by developing muscle tension to compress joint surfaces together. When rotator cuff muscles contract simultaneously, the humeral head is pressed into the glenoid socket. The combination of the deltoid muscle and the supraspinatus form a coronal force couple so that when the arm is abducted, the resultant joint reaction force is directed toward the glenoid fossa, which compresses the humeral head and improves stability when the arm is abducted and overhead (Parsons et al., 2002).

Rotator cuff muscles help to stabilize the glenohumeral joint by selective contraction of muscles that resist displacing forces. By selective contraction of these muscles so as to resist displacing forces (as when the lateral deltoid muscle initiates shoulder abduction), the supraspinatus muscle and the long biceps tendon actively resist displacement of the humeral head relative to the fossa. Another example is when the pectoralis major and anterior deltoid elevate and flex the shoulder; they tend to push the humeral head posteriorly out the back of the fossa. Subscapularis, infraspinatus, and teres minor muscles resist this displacement.

It is possible to identify the rotator cuff muscles by placing the hand over the shoulder as in the superior view of Figure 5-19. Place your thumb anteriorly, and the biceps tendon will be between the thumb and just anterior to the index finger. The thumb will be over the subscapularis muscle, the index finger will be over the supraspinatus muscle, the middle finger will be located over the infraspinatus muscle, and the ring finger will be over teres minor (Magee, 2002).

The tendon of the biceps brachii muscle arches over the head of the humerus and under the joint capsule. When there is a strong contraction of the biceps muscle, as in flexion with a load in the hand, there is depression of the head of the humerus, which prevents elevation of the humeral head (Nordin & Frankel, 2001).

Secondary dynamic muscles providing glenohumeral stability are teres major, latissimus dorsi, and pectoralis major muscles (Wilk et al., 1997b). While not seen as primary or

secondary stabilizers, the scapulothoracic muscles (serratus anterior muscle, rhomboid muscle, trapezius, levator scapula, pectoralis minor, subclavius muscles) do provide a significant role in shoulder stability. These muscles provide a stable base of support for glenohumeral muscles on which to fixate and function from, and they help to maintain sufficient length-tension relationships (Alexander, Miley, Stynes, & Harrison, 2007; Wilk et al., 1997b). Weakness in these muscles can contribute to loss of stability of the scapula and potentially to glenohumeral instability (Wilk et al., 1997b).

These muscles, acting together, are force couples, defined as "two forces whose points of application occur on opposite sides of an axis and in opposite directions to produce rotation of the body" (Nordin & Frankel, 1989, p. 239). Glenohumeral force couples combine to move the clavicle, scapula, and humerus to permit both stability and mobility of the arm. The muscles responsible for glenoid positioning are the trapezius, levator scapulae, serratus anterior, and rhomboids. The force couples responsible for scapular upward rotation are the upper and lower trapezius and the serratus anterior. Stability in rotation is accounted for by the rhomboids, trapezius, and the pectoralis minor (Wilk et al., 1997b).

Force couples are important in dynamic stabilization by establishing a dynamic equilibrium (Wilk et al., 1997b). Subscapularis action is counterbalanced by infraspinatus and teres minor in external rotation, and similarly, the deltoid actions are counterbalanced by the inferior rotator cuff muscles (infraspinatus, teres minor) in internal rotation of the humerus (see Figure 5-19). Deltoid is counterbalanced by supraspinatus in abduction, and scapular stabilization is accomplished by the combined efforts of the upper and lower trapezius and rhomboids countering the action of serratus anterior. Seen in this way, the force couples in the glenohumeral joint act as synergists and prime movers, producing the motion while the rotator cuff muscles act as a stable fulcrum for the motion (Wilk et al., 1997b). Wilk, Andrews, and Arrigo (1997b) suggest that a better term for this relationship would be "balance of forces" rather than "force couple" (p. 373). Force couples that occur in the shoulder are listed in Table 5-5.

The trapezius and serratus anterior muscles form a force couple to produce elevation of the arm due to combining forces to create lateral, superior, and rotational movements of the scapula, producing abduction and upward rotation. The deltoid and supraspinatus muscles contract together to

Table 5-5

FORCE COUPLES OF THE SHOULDER

MOVEMENT	AGONIST/STABILIZER	ANTAGONIST/STABILIZER
Protraction (scapula)	Serratus anterior*, pectoralis major**, pectoralis minor**	Trapezius Rhomboids
Retraction (scapula)	Trapezius Rhomboids	Serratus anterior*, pectoralis major**, pectoralis minor**
Elevation (scapula)	Upper trapezius**, levator scapula**	Serratus anterior*, lower trapezius*
Depression (scapula)	Serratus anterior*, lower trapezius*	Upper trapezius*, levator scapula*
Lateral rotation (upward rotation of inferior angle of scapula)	Trapezius (upper**, lower*), serratus anterior*	Levator scapula**, rhomboids
Medial rotation (downward rotation of inferior angle of scapula)	Levator scapulae**, rhomboids, pectoralis minor**	Trapezius (upper**, lower*), serratus anterior*
Scapular stabilization	Upper trapezius**, lower trapezius*, rhomboid	Serratus anterior*
Abduction of humerus	Deltoid	Supraspinatus
IR humerus	Subscapularis**, pectoralis major**, latissimus dorsi, anterior deltoid	Infraspinatus*, teres minor, posterior deltoid
External rotation humerus	Infraspinatus, teres minor, posterior deltoid	Subscapularis**, pectoralis major**, latissimus dorsi, anterior deltoid

*muscle prone to weakness; **muscle prone to tightness

Reprinted with permission from Magee, D. J. (2002). *Orthopedic physical assessment*. Philadelphia, PA: WB Saunders.

produce abduction or flexion at the glenohumeral joint (Smith et al., 1996), "...but their tendency to move the least massive segment rotates the scapula downward; isometric contractions of upward scapular rotators produce scapular stabilization preventing downward rotation while teres minor and deltoid work to depress humeral head and stabilize it" (Greene & Roberts, 1999, p. 83). This illustrates that the rotator cuff muscles are unique in that not only do these muscles produce (or con-

tribute) to specific joint motions, but they are also considered dynamic stabilizers of the glenohumeral joint because they combine to stabilize and resist displacement of the humeral head (Greene & Roberts, 1999, p. 240). The combinations of movements are achieved by collaboration of many muscles, which favorably position the separate joint articulations for greater movement. Some muscles act simultaneously, and others follow in sequence.

Table 5-6

SUMMARY OF THE DYNAMIC STABILIZERS OF THE SCAPULA AND GLENOHUMERAL JOINT

JOINT	DYNAMIC STABILIZERS
Scapula	• Upper trapezius and serratus anterior • Middle trapezius and rhomboids

Glenohumeral joint	• Rotator cuff
	  o Supraspinatus compresses head of humerus into glenoid fossa
	  o Subscapularis, infraspinatus, teres minor: inferior directed translation force on humeral head
	  o Infraspinatus and teres minor: rotates humeral head externally
	• Deltoid
	• Long head of biceps brachii

Humeral elevation	• Rotator cuff and deltoid
Upward rotation of the scapula	• Long head of biceps stabilizes against humeral elevation • Long head of triceps stabilizes against inferior translation

Adapted from Kisner, C., & Colby, L. A. (2007). *Therapeutic exercise: Foundations and techniques* (5th ed.). Philadelphia, PA: F. A. Davis Company.

In early flexion or abduction, teres minor and deltoid work together to depress the humeral head and stabilize it. Because the muscle force of teres minor is equal and opposite to deltoid, this is a force couple. Subscapularis and infraspinatus muscles join later in flexion or abduction to assist with humeral head stabilization; latissimus dorsi contracts eccentrically to assist with stabilization, and this muscle increases in activity as the angle of motion increases; deltoid and rotator cuff work together in flexion and abduction in that the rotator cuff acts to depress the humeral head and the deltoid elevates the arm (see Figure 5-18). Serratus anterior and trapezius now form a force couple to create lateral, superior, and rotational movements of scapula after deltoid and teres minor have initiated elevation. Figure 5-19 illustrates how the humeral head is compressed into the glenoid fossa due to co-contraction of the dynamic stabilizers of the shoulder.

A functional example of the force couples that operate in the shoulder would be when one places the hand behind the head when combing the hair. There involves elbow flexion, sternoclavicular elevation with upward rotation, scapular elevation with upward rotation and abduction, and glenohumeral abduction and external rotation. The biceps muscle is flexing the elbow, while the trapezius and serratus anterior are acting as a force couple at the scapula. The deltoid and supraspinatus form a force couple at the glenohumeral joint, as do infraspinatus and teres minor muscles. When the arm is overhead, there would also be a contraction of the triceps muscle. These muscles all cooperate to enable successful performance of daily activities, while providing stability of the glenohumeral joint surfaces.

Failure of rotator cuff muscles to maintain humeral congruency may lead to glenohumeral joint instability, rotator cuff pathology, and labral injury (Donatelli & Wooden, 2010).

Active and Passive Barriers

The blending of the rotator cuff muscles into the joint capsule is another method of achieving active glenohumeral stabilization. This creates both active and passive resistance to humeral head translation and acts to absorb and dissipate microtraumatic forces (Wilk et al., 1997b). Anteriorly, the glenohumeral ligaments blend with the attachment of the subscapularis muscle, and posteriorly, the tendons of infraspinatus and teres minor are combined (Wilk et al., 1997b).

Neuromuscular Control

The mechanical restraint interaction of the passive and dynamic structures of the shoulder is mediated by the sensorimotor system. Not only do the structures provide mechanical restraint of the humeral head but also provide neural feedback to the central nervous system influencing the efferent output to dynamic shoulder stability structures (Myers, Wassinger, & Lephart, 2006). Based on a study by Lephart and Jari (2002), Myers, Wassinger, and Lephart state that "neuromuscular control is the subconscious activation of the dynamic restraints about the shoulder in preparation and in response to joint motion and loading for the purpose of maintaining joint stability" (Myers et al., 2006, p. 198). It is speculated that capsular or ligamentous injuries result from loss of proprioception, but this is an equivocal conclusion in the literature (Nyland, 2006). A summary of the dynamic stabilizers of the scapula and glenohumeral joint is provided in Table 5-6.

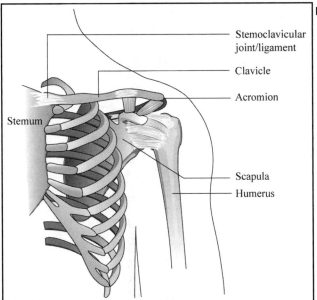

Figure 5-19. Sternoclavicular joint.

Stemoclavicular joint/ligament

Clavicle

Acromion

Stemum

Scapula

Humerus

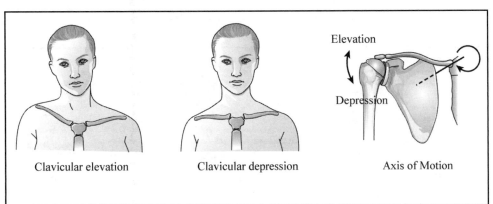

Clavicular elevation

Clavicular depression

Elevation

Depression

Axis of Motion

Figure 5-20. Clavicular elevation and depression at the sternoclavicular joint. (Adapted from Muscolino, J. E. (2006). *Kinesiology: The skeletal system and muscle function.* St. Louis, MO: Mosby.)

Sternoclavicular Joint

The sternoclavicular joint is the articulation of clavicle and manubrium of sternum and first rib cartilage as shown in Figure 5-19. This is the only true bony articulation of the shoulder girdle with the trunk and is considered the "base of operation" (Nordin & Frankel, 2001). There are intra-articular disks between the manubrium, first costal cartilage, and clavicle. These articular disks are comprised of fibrocartilage, which acts as a shock absorber of forces transmitted along clavicle from upper limb.

These disks divide the sternoclavicular joint into two functional units for gliding. The first unit is at the sternal end of the clavicle. This is considered an incongruous joint because not all of the articulating surfaces are in contact. In fact, the superior portion of the medial clavicle serves only as an attachment for the disk and the interclavicular ligament and does not contact the sternum at all (Nordin & Frankel, 1989).

The second unit is formed by the manubrium of the sternum and the first costal cartilage. These two saddle-shaped surfaces permit the movement of the clavicle on the disk and of the disk on the sternum. Because this articulation is consid-

ered incongruous, stabilization depends upon ligaments. The structures providing stability are the articular capsule, sternoclavicular ligaments, interclavicular ligament, costoclavicular ligament, and the articular disk (Figure 5-20).

Osteokinematics

While the sternoclavicular joint is an incongruent saddle-shaped (sellar) joint, it acts like a double gliding joint or, as some authors suggest, as a modified ball-and-socket joint (Greene & Roberts, 1999) or as a functional triaxial joint (Donatelli & Wooden, 2010; Kisner & Colby, 2007). As such, this joint can move in all axes with three degrees of freedom. The clavicle moves in the sternoclavicular joint. The movements of elevation and depression and protraction and retraction are described by the movement of the distal segment of the lever, i.e., the movements are visualized as movements of the lateral end of the clavicle (Nordin & Frankel, 2001). In addition, rotation of the clavicle around its own longitudinal axis occurs, but as an accessory motion when the humerus is elevated above 90 degrees and the scapula is upwardly rotated (Kisner & Colby, 1990). Table 5-7 provides a summary of the osteokinematics of the sternoclavicular joint.

Table 5-7

SUMMARY OF OSTEOKINEMATICS OF THE STERNOCLAVICULAR JOINT

Functional joint	Diarthrotic triaxial
Structural joint	Synovial
Close-packed position	Maximum shoulder elevation (full rotation of clavicle)
Resting position	Arm at side in normal physiologic position
Capsular pattern	Pain at extreme ROM, especially horizontal adduction and full elevation

Figure 5-21. Clavicular protraction and retraction at the sternoclavicular joint. (Adapted from Muscolino, J. E. (2006). *Kinesiology: The skeletal system and muscle function*. St. Louis, MO: Mosby.)

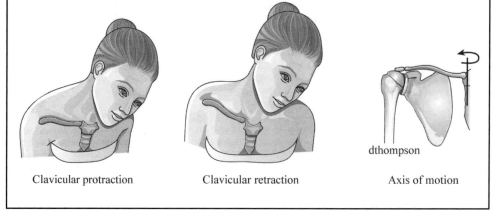

Clavicular protraction   Clavicular retraction   Axis of motion

dthompson

The sternoclavicular joint is capable of considerable mobility. Motions of the clavicle occur as a result of the scapular movements of elevation, depression (Figure 5-21), protraction, and retraction (Figure 5-21). The rotation of the clavicle happens as an accessory motion when the humerus is elevated above 90 degrees and there is upward rotation of the scapula. Clavicular elevation occurs between 15 degrees and 45 degrees, depression between 5 degrees and 15 degrees, protraction at 30 degrees, retraction between 15 degrees and 29 degrees during arm elevation and up to 40 degrees of axial rotation (Inman, Saunders, & Abbot, 1944; Veeger & vanderHelm, 2007).

Elevation and depression occur in the frontal plane around a sagittal axis. The clavicle moves on the disk as a hinge, and there is superior-inferior gliding between clavicle and meniscus or disk. The axis is oblique through the sternal end of clavicle and takes a backward and downward course. Due to this orientation, elevation is actually an upward-backward movement, and depression is a movement in a forward-downward direction. The motion of the clavicle is stopped by the first rib. Excessive clavicular elevation is often found in people with shoulder pain (Ludewig & Reynolds, 2009).

Protraction and retraction occur between the articular cartilage, disk, and sternum in a horizontal plane around a nearly vertical axis, which produces an anterior-posterior gliding as the disk moves with clavicle on the manubrium. The vertical axis lies at the costoclavicular ligament. The ROM for protraction is 0 degrees to 30 degrees, with further movement limited by posterior sternoclavicular ligament and costoclavicular ligament. Retraction has a range of 15 degrees to 29 degrees, with the anterior sternoclavicular ligament restraining further movement (Inman, Saunders, & Abbot, 1944; Smith et al., 1996; Veeger & vanderHelm, 2007).

Approximately 30 degrees to 40 degrees of transverse rotation of scapula on the clavicle occurs in a sagittal plane around the long frontal axis of the clavicle following 90 degrees of shoulder flexion or abduction. This motion occurs in a frontal axis as the disk and clavicle roll on the sternum. This rotational element is essential for full flexion or abduction and,

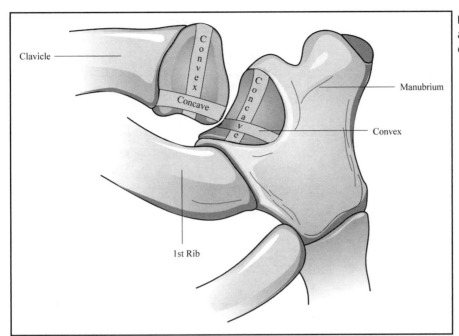

Figure 5-22. Anterior-lateral view of articular surfaces of the right sterno-clavicular joint.

should the rotation not occur, only 110 degrees of flexion or abduction would be possible (Smith et al., 1996). Upward rotation occurs due to the tightening of the acromioclavicular ligament (trapezoid and conoid). As the conoid ligament tightens, this becomes the axis for upward rotation of the sternoclavicular joint. Given the "s" shape of the clavicle, the acromial end is higher and therefore able to further elevate and upwardly rotate the scapula. Posterior rotation is produced by the pull of the coracoclavicular and acromioclavicular ligament by muscles that move the scapula on the thorax. There are no muscle groups that cross the sternoclavicular joint that can produce active posterior axial rotation, so the clavicle moves as an intercalated segment (Ludewig, Phadke, Braman, Hassett, Cieminski, & LaPrade, 2009).

The close-packed position for the sternoclavicular joint is with the shoulder in maximum elevation (either abduction or forward flexion), which puts the clavicle in full rotation. The resting position is with the humerus at the side in adduction. Capsular restrictions would be suspected when there is pain at the extreme ends of range of motion, especially in horizontal adduction and full humeral elevation.

Arthrokinematics

Given that the medial end of the clavicle is convex top to bottom (superiorly to inferiorly) and concave front to back (anterior to posterior) (Figure 5-22) and that the manubrium and first costal cartilages are concave top to bottom and convex front to back, the physiologic motions of the clavicle depend upon the direction of slide of the clavicle on the manubrium.

Because there is a convex superior-inferior clavicular surface and concave surface formed by manubrium and first costal cartilage in a frontal plane around a sagittal, anterior-posterior axis and with inferior-superior motion of the clavicle, arthrokinematically, the convex surface of the clavicle must slide on the concave manubrium and first costal cartilage in the direction opposite to movement of the lateral head of the clavicle. In elevation and depression, the medial clavicle glides in a superior-to-inferior direction on the upper attachment of the disk. For example, elevation of the clavicle results in a downward sliding of the medial clavicular surface on manubrium and first costal cartilage (Kisner & Colby, 2007; Nordin & Frankel, 2001).

Conversely, the medial end of anteroposterior clavicle is concave, and the manubrial side is convex; arthrokinematically, the clavicular surface will now slide on manubrium and first costal cartilage in the same direction as the lateral end of the clavicle. The movement of these surfaces allows protraction/retraction or horizontal forward/backward motion in a horizontal plane around a vertical axis. In protraction and retraction, the clavicle and disk glide anteroposteriorly on the manubrium as a unit, pivoting around the inferior attachment of the disk. It can be considered that the disk functions as part of the manubrium during elevation and depression and acts as a part of the clavicle during protraction and retraction. For example, protraction of the clavicle is accompanied by anterior sliding of the medial clavicle on the manubrium and first costal cartilage (Kisner & Colby, 2007; Nordin & Frankel, 2001).

Rotation of the scapula around its own axis results in a spin motion. This rotation occurs in one direction only from posterior placement to neutral position, which brings the anterior surface of the clavicle facing toward the front or anteriorly (Nordin & Frankel, 2001). A summary of the arthrokinematics of the sternoclavicular joint is presented in Table 5-8.

Supporting Structures

Because of the changing function of the disk, acting as part of the manubrium during elevation and depression and as part of the clavicle during protraction and retraction, mobility

Table 5-8

SUMMARY OF THE ARTHROKINEMATICS OF THE STERNOCLAVICULAR JOINT

Physiologic movement of clavicle	Roll	Slide	Resultant Movement
Protraction	Anterior	Anterior	Same
Retraction	Posterior	Posterior	Same
Elevation	Superior	Inferior	Opposite
Depression	Inferior	Superior	Opposite

Adapted from Kisner, C., & Colby, L. A. (2007). *Therapeutic exercise: Foundations and techniques* (5th ed.). Philadelphia, PA: F. A. Davis Company.

Table 5-9

SUMMARY OF SUPPORTING STRUCTURES OF THE STERNOCLAVICULAR JOINT

Articular disk	• Increases contact surfaces • Shock absorber • Prevents medial displacement of the clavicle
Sternoclavicular ligaments (anterior and posterior)	• Limits rotation of sternoclavicular joint during depression of clavicle • Resist anterior and posterior translation and superior displacement of the joint • Posterior sternoclavicular ligament prevents upward and lateral displacement of clavicle
Costoclavicular ligament	• Limits clavicular elevation and superior glide of the clavicle • Principle stabilizing structure • Restricts downward rotation of medial clavicle
Interclavicular ligament	• Limits shoulder depression or downward glide with the articular disk • Helps to protect the brachial plexus and subclavian artery

of the joint is maintained, and stability is enhanced (Norkin & Levangie, 1992). The sternoclavicular joint is a bit more protected than the acromioclavicular joint due to its more medial location. The bony surfaces are incongruent, adding little intrinsic joint stability. There is also little dynamic stability, because there are no muscles crossing the joint (Kisner & Colby, 2007). Tissues that stabilize the sternoclavicular joint are the anterior and posterior sternoclavicular ligaments, the interclavicular ligament, costoclavicular ligament articular disk, and sternocleidomastoid, sternothyroid, and sternohyoid muscles (Neumann, 2002) (Table 5-9).

The fibrocartilaginous articular disk (or meniscus) increases contact between incongruous joint surfaces. The disk also limits shoulder depression as well as serves as a hinge for motion and as a shock absorber for forces transmitted along the clavicle from the lateral end (Nordin & Frankel, 2001; Norkin & Levangie, 1992; Muscolino, 2006). Due to the attachments of the disk above, below, and to the sternoclavicular and interclavicular ligaments, the disk also adds strength to the joint and helps to prevent medial displacement of the clavicle (Nordin & Frankel, 2001). The articular capsule varies in thickness and strength and forms the anterior and posterior

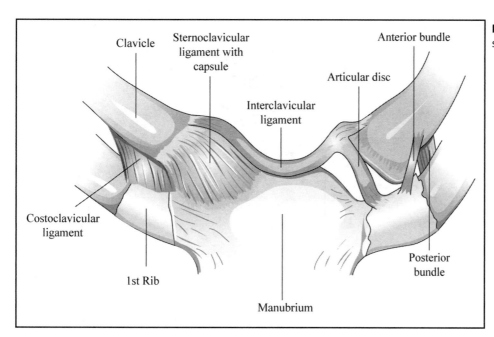

Figure 5-23. Ligaments of the sternoclavicular joint.

sternoclavicular ligaments. The ligaments and capsule are reinforced by attachment of the sternocleidomastoid muscle.

The anterior and posterior sternoclavicular ligaments attach the clavicle to sternum and reinforce the joint capsule (Figure 5-23). Both ligaments resist anterior and posterior translation and superior displacement of the joint (Nordin & Frankel, 2001). The anterior ligament covers the anterior surface of the articulation, while the posterior ligament covers the posterior portion of the joint. This posterior ligament tends to limit rotation of the sternoclavicular joint during depression of clavicle and is the strongest supporting structure, preventing upward and lateral displacement of clavicle.

The interclavicular ligament is a curved ligament that goes from the superior portion of the sternal end of one clavicle to that of the other. Due to the attachment to the superior margin of sternum, the interclavicular ligament serves to limit shoulder depression or downward glide along with the articular disk, which helps to protect the brachial plexus and subclavian artery (Levangie & Norkin, 2005). The posterior portion of the interclavicular ligament assists with anterior restraint of the joint, because the ligament tightens when the arm is lowered and becomes lax when the arm is elevated.

The costoclavicular ligament provides an axis for the movements of elevation and depression and for protraction and retraction. This ligament is attached inferiorly to the first rib and costal cartilage and superiorly to the inferior surface of the medial end of the clavicle. This ligament serves to limit clavicular elevation and superior glide of the clavicle (Levangie & Norkin, 2005) and is considered the "principal stabilizing structure" (Nordin & Frankel, 2001) of the joint. The costoclavicular ligament assists in restricting upward displacement and downward rotation of the medial clavicle by its attachment to the superomedial part of first rib carriage and to costal tuberosity of the inferior surface of the clavicle. While the costoclavicular and interclavicular ligaments restrain upward and downward movements of the clavicle, they have little effect on anterior or posterior translation (Spencer, Kuhn, Carpenter, & Huges, 2002).

Acromioclavicular Joint

The acromioclavicular (AC) is a small joint formed by the articulation of the acromial (lateral) end of the clavicle with the acromion of the scapula. Articular disks may or may not be present at this articulation. There is very little motion between the clavicle and the acromion, so fusion of the AC would produce little loss of shoulder function (Nordin & Frankel, 2001). At the AC, ligaments suspend the scapula from the clavicle. Like the glenohumeral joint, the articular capsule is weak and encloses the joint. Due to the size and shapes of articulating bones, this joint is considered incongruent. The AC capsule is more lax than the sternoclavicular joint capsule, so there is greater incidence of dislocation of the AC (Donatelli & Wooden, 2010).

Movements at the AC are movements of the scapula relative to the clavicle in early arm elevation and to allow rotation on the thorax in later stages of elevation (Levangie & Norkin, 2005). The scapula has five degrees of movement, which include two degrees of translator motion and three degrees of rotary movement. Figure 5-24 illustrates these five movements.

The primary movement of the AC is scapular rotation, which occurs in the frontal plane and anteroposterior (sagittal) axis (Figure 5-25). Rotating the scapula allows the glenoid to tilt up (upward rotation of scapula) or down (downward rotation). This motion is synonymous with and identical to the rotational movements that occur at the scapulothoracic joint (Norkin & Levangie, 1992). The scapula can move an average of 30 degrees in upward rotation and 10 degrees to 30 degrees of lateral and medial and upward/downward tilt at the AC. Without motion at the AC, the scapula and clavicle would always move as one unit (Muscolino, 2006).

Figure 5-24. Movements of the scapula.

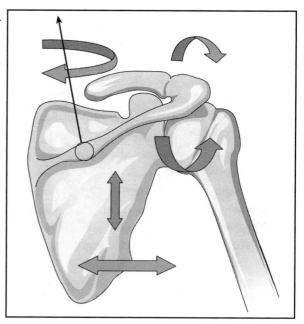

Figure 5-25. Axis of motion at the acromioclavicular joint.

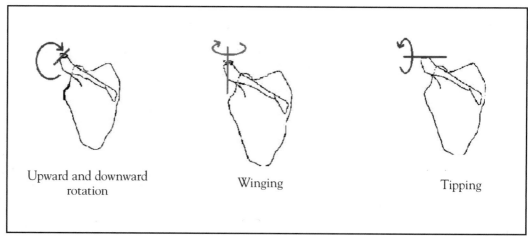

Upward and downward rotation

Winging

Tipping

In upward rotation, the scapula moves so that the glenoid cavity faces upward while the inferior angle moves laterally. Depending on the point of reference, the movement can be considered either lateral rotation of the scapula or upward rotation. This motion at the AC occurs during humeral forward flexion and abduction. Similarly, downward rotation moves the scapula so that the glenoid faces inferiorly (downward rotation) and the inferior angle moves medially (medial rotation).

Downward rotation increases the movement of humerus when it moves on the scapula during shoulder extension and adduction. The trapezoid ligament (part of the coracoclavicular ligament) acts as a hinge for this scapular motion, and the movement occurs due to the tightening of this ligament.

Scapular rotation is described based on the movement of the inferior angle (medial rotation and lateral rotation) or based on movement of the glenoid fossa (upward and down-

ward rotation). The point of reference is essential to accurately describe the movement. Medial rotation is when the inferior angle moves toward the midline, and lateral rotation is when the inferior angle moves away from the midline. Upward and downward rotation allows the glenoid fossa to tilt upward or downward. Upward rotation involves rotating the glenoid cavity while moving the inferior angle laterally, and, similarly, downward rotation involves rotating the glenoid cavity downward while the inferior angle moves medially.

Winging of the scapula is the movement of the medial border of the scapula away from the chest wall. It occurs in a horizontal plane around a vertical axis. Winging is the "normal posterior movement of the vertebral border of the scapula (or anterior movement of glenoid fossa) that must occur to maintain contact of the scapula with the horizontal curvature of the thorax as the scapula slides around the thorax in adduction and abduction" (Norkin & Levangie, 1992, p. 215).

Table 5-10

SUMMARY OF THE OSTEOKINEMATICS OF THE ACROMIOCLAVICULAR JOINT

Functional joint	Diarthrotic, triaxial
Structural joint	Synovial, plane
Close-packed position	Humerus abducted to 90 degrees
Resting position	Humerus resting by side in normal physiologic position
Capsular pattern	Pain at extreme ROM, especially horizontal adduction and full elevation

Winging occurs when there is anterior movement of the lateral end of the clavicle (protraction) and occurs naturally when the humerus is horizontally adducted. The medial border lifts off the thoracic wall. The motion of abduction (protraction) and adduction (retraction) is accomplished by the conoid ligament, which acts as a longitudinal (vertical) axis for scapular rotation (Nordin & Frankel, 2001). Often, the term *winging* refers to a pathological posterior displacement of the vertebral border of the scapula often attributed to weakness in the serratus anterior muscle and damage to the long thoracic nerve, but this motion occurs in nonpathological joints (Donatelli, 1997; Donatelli & Wooden, 2010; Ellenbecker & Ballie, 2010; Magee, 2002). Patients with excessive winging often demonstrate excessive scapular internal rotation relative to the clavicle because the slight external rotation on the thorax during elevation is not offset by the normal internal rotation at the AC (Ludewig & Reynolds, 2009).

When there is posterior displacement of the inferior angle of the scapula, this is *tipping*. This motion occurs around a coronal axis, which passes through the AC and results in movement of the superior border of the scapula moving anteriorly. The top of the scapula moves posteriorly while the bottom moves towards the ribs. Tipping of the scapula occurs in conjunction with internal rotation and hyperextension of the humerus when reaching the hand behind the back (Kisner & Colby, 2007). Both winging and tipping of the scapula function to position the glenoid fossa toward a more anterior or inferior position as well as changing the position of the humeral head.

Osteokinematics

The AC is a gliding joint with three degrees of freedom in three axes of motion. Movements of this articulation are seen as two different types: 1) a gliding motion of the clavicle and

the acromion and 2) rotation of the scapula on the clavicle (Goss, 1976). Because the articulating surfaces of this articulation are small and there are wide individual variations, there are inconsistencies in identifying the movements and axes of motion for this joint (Levangie & Norkin, 2005). Rotation of the scapula is the primary motion of the AC, with contribution to tipping and winging movements of the scapula. The joint position of greatest stability is the close-packed position, which is 30 degrees to 90 degrees of glenohumeral abduction (Hartley, 1995). A summary of the osteokinematics of the AC is shown in Table 5-10.

Arthrokinematics

Movements at the AC involve the convex portion on the lateral end of the clavicle and a concave facet on the acromion (Table 5-11). The movements of tipping, rotation, and winging of the scapula and the clavicle are in the same direction. For example, if the scapula rotates downward, then the clavicle also rotates in a downward direction.

Supporting Structures

The AC is primarily stabilized by the coracoacromial, coracoclavicular, and acromioclavicular ligaments (Table 5-12). The weak joint capsule is reinforced by the superior and inferior acromioclavicular ligaments that restrict anteroposterior horizontal movements of the joint. The acromioclavicular ligament is supported by the strong coracoclavicular ligament (Kisner & Colby, 2007). While there are no muscles that directly cross this joint, upper trapezius and deltoid muscles add to the stability of the superior portion of the joint (Neumann, 2002).

The coracoclavicular ligament binds the clavicle to coronoid process and serves as the suspensory ligament of the upper extremity (Fukuda, Craig, An, Cofield, & Chao, 1986).

Table 5-11

SUMMARY OF THE ARTHROKINEMATICS OF THE ACROMIOCLAVICULAR JOINT

Physiologic movement of concave acromion of scapula on convex clavicle	Movement of acromion	Movement of clavicle	Resultant movement
Upward rotation	Upward	Upward	Same
Downward rotation	Downward	Downward	Same
Winging	Posterior movement of vertebral border	Posterior	Same
Tipping	Anterior movement of superior border	Anterior	Same

Table 5-12

SUMMARY OF THE STABILIZING STRUCTURES OF THE ACROMIOCLAVICULAR JOINT

STRUCTURE	METHOD OF SUPPORT
Joint capsule	Weak
Coracoclavicular ligament	The primary stabilizer of the acromioclavicular joint as it connects the coracoid process and the clavicle Controls vertical stability (restrains superior and anterior displacement
Conoid	Prevents anterior and superior rotation and displacement
Trapezoid	Restricts medial displacement of scapula on clavicle Resists joint compression
Acromioclavicular ligament	Restricts anteroposterior horizontal movements of the joint Often first injured when joint stressed

It is a vertically directed ligament (Figure 5-26) that is strong but not stiff (Veeger & vanderHelm, 2007). It serves as a link between scapula and clavicle and connects the coracoid process to the inferior surface of the clavicle. It is the primary stabilizer of the AC. The coracoclavicular ligament has two parts: the conoid and trapezoid ligaments.

The conoid ligament is the most important structure preventing significant injuries and anterior and superior rotation and displacement of clavicle from the scapula. A triangular-shaped ligament, it runs between the posterior surface of coracoid and attaches on conoid tubercle on the posterior clavicle and base of the coracoid. It aids in producing the motions of protraction and retraction by producing posterior rotation of the clavicle.

The trapezoid ligament, which is located anterolateral to the conoid ligament, is broad, thin, and quadrilateral in shape. It is located on the inferior surface of the clavicle and tends to restrict medial displacement of scapula on the clavicle and

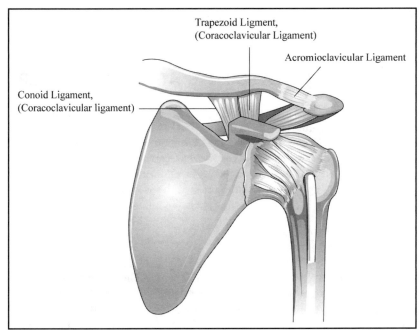

Figure 5-26. Ligaments of the acromioclavicular joint.

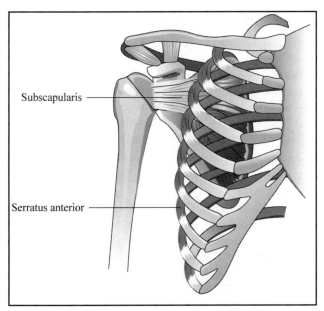

Figure 5-27. Scapulothoracic joint. (Adapted from Neumann, D. A. (2002). *Kinesiology of the musculoskeletal system: Foundations for Physical Rehabilitation.* St. Louis, MO: Mosby.)

resists joint compression (Fukuda et al., 1986). Of the two parts of the coracoclavicular ligament, this is the larger and stronger of the two. Nordin and Frankel (2001) describe the action of this ligament as "a hinge for scapular motion about a transverse (horizontal) axis in the frontal plane" (p. 231).

The AC is very susceptible to injury (i.e., fall on outstretched hand) and degeneration (Muscolino, 2006). When the AC is dislocated, it is often due to a torn coracoclavicular ligament. The ligaments together contribute to the horizontal stability of the joint and are critical to preventing superior dis-

location of the clavicle on the acromion (Levangie & Norkin, 2005). Both the conoid and trapezoid ligaments limit scapular rotation, and these ligaments assist in transmission of compression forces from the scapula to the clavicle.

The AC is subject to high loads from the chest musculature to the upper extremity (Nordin & Frankel, 2001). Acromioclavicular joint separations are not uncommon, accounting for about 12% of all injuries to the shoulder and often occurring secondary to a fall when the adducted shoulder hits a firm object (Dvir, 2000).

In addition to the coracoclavicular ligament and joint capsule, the joint is reinforced superiorly by the acromioclavicular ligament, which acts to restrain axial rotation and posterior translation of the clavicle (Nordin & Frankel, 2001). The acromioclavicular ligament, directed horizontally, is instrumental in providing horizontal stability. It is palpable as a shallow depression between the end of the clavicle and the acromion. The superior acromioclavicular ligament is a very important ligament in stabilizing the AC for normal activities (Fukuda et al., 1986).

Scapulothoracic Joint

The scapulothoracic (also known as scapulocostal) joint is a physiologic or functional joint where the scapula glides on the thorax to enable greater motion of the scapula. This joint provides a movable base for the humerus and permits wide ranges of movement with the scapula and thorax, as well as providing stability in the glenohumeral joint for overhead activities. Rather than being a bone-to-bone articulation, this is a bone-muscle-bone articulation between the scapula and thoracic wall. The serratus anterior and subscapularis muscles glide on each other (Nordin & Frankel, 2001) (Figure 5-27). Because of this articulation, deltoid tension can be maintained regardless of the arm position (Nordin & Frankel, 2001).

The scapula moves relative to the clavicle at the acromioclavicular joint but it must also move relative to the ribs at the scapulothoracic joint. Scapulothoracic motion includes forward elevation in which the scapula rotates, increasing the stability of the glenohumeral joint and decreasing the tendency for impingement of the rotator cuff muscles beneath the acromion (Nordin & Frankel, 2001). This functional articulation also acts as shock absorber for forces applied to outstretched hands and permits elevation of the body when crutch walking and during depression transfers.

The scapulothoracic joint is not a true joint and, as such, has no capsular patterns or closed-packed position. The resting position is the same as that of the acromioclavicular joint or with the humerus resting by the side in normal physiologic position (Magee, 2002).

The scapulothoracic joint allows two translatory motions (protraction/retraction, elevation/depression). Other scapular movements of rotation, winging and tipping, are secondary and occur in combination with the glenohumeral, sternoclavicular, and acromioclavicular joints (Muscolino, 2006).

Elevation occurs when the superior border of the scapula and acromion move in an upward direction as when you shrug your shoulders. Depression occurs when the superior border of scapula and acromion move in a downward direction to lower the scapula. Protraction (also considered abduction) is the movement of the scapula away from the midline, whereas retraction (or adduction) is movement of the scapula toward the midline.

Sternocostal and Vertebrocostal Articulations

The ability to achieve full shoulder flexion and abduction is accomplished by the ribs gliding on both the sternum and on the vertebrae. This occurs at the sternocostal (costosternal) articulation, where the ribs glide on the sternum, and the vertebrocostal (costovertebral) articulation, where the ribs glide on the vertebrae (Flowers, 1994; Gench et al., 1995).

The sternocostal joint is a series of gliding joints with the exception of the first rib, which is fused, making this a synchondrosis joint. The vertebrocostal, also a gliding joint, allows some rotation during depression and elevation and some rotation around its own axis as contributory movements for the total shoulder motion.

Suprahumeral Articulation

The suprahumeral (also referred to as subdeltoid or subacromial) articulation is a functional joint (as opposed to an anatomic joint) serving in a protective capacity. This is the articulation between the acromion and the coracoacromial ligament and arch. The head of the humerus slides beneath the acromion, and the tendon of the long head of the biceps muscle slides in the bicipital groove. Tendons of the rotator cuff muscles (supraspinatus, infraspinatus, teres minor, and subscapularis), the long head of the biceps, the joint capsule, capsular ligaments, subdeltoid, and subacromial bursae lie in this area and may be susceptible to impingement or compression syndromes. This articulation prevents trauma from above,

prevents upward dislocation of the humerus, and mechanically limits abduction of the humerus.

MOVEMENTS AT THE SHOULDER

The shoulder girdle is capable of much mobility due to the integrated, coordinated, and synchronous action of all of the joints acting together. A summary of the movements possible at each joint of the shoulder complex is shown in Table 5-13.

Motion of scapula relative to thorax only occurs because of the simultaneous motion at the acromioclavicular and sternoclavicular (Ludewig et al., 2009). Full range of motion of the humerus is also reliant on movement at the scapulothoracic joint, sternocostal, vertebrocostal, and suprahumeral articulations (Hess, 2000).

The contributions of each joint at any time are dependent upon the plane and axis of motion, the amount of elevation of the humerus, the total load applied to the extremity, and individual anatomic variations (Hamill & Knutzen, 1995; Kisner & Colby, 2007). The contributions of the acromioclavicular and sternoclavicular joints are that movement at these joints permits movement of the scapula, which places the glenoid fossa facing downward, upward, or forward while the costal surfaces remain close to the thorax for stabilization.

Movement at the scapula occurs because a muscle contracts. Whether the origin or insertion moves depends upon which moves easier (Greene & Roberts, 1999). The acceleration of movement is related to its mass according to Newton's Second Law and "since the scapula is less massive than the entire upper extremity when the glenohumeral muscles contract, they move the scapula unless other factors intervene" (Greene & Roberts, 1999, p. 83).

The contributions from these joints allow the synchronous gliding of the scapula, which permits the humerus to move freely in a large arc of motion at the glenohumeral joint. Four joint mechanics events must occur simultaneously to produce smooth motion during full arm elevation (Uhl, 2009):

1. Posterior rotation of the clavicle at both the acromioclavicular and sternoclavicular joints.
2. Depression of the proximal clavicle at the sternoclavicular joint.
3. The acromion must glide superiorly to the clavicle.
4. The humeral head rolls superiorly and translates.

The coordinated action of the muscles at the various articulations occurs in a smooth way with varying contributions of different muscles at different joints for different joint motions. This coordinated action occurs in a sequence with scapulothoracic rhythm.

Scapulohumeral (Scapulothoracic) Rhythm

The scapulothoracic joint, with the acromioclavicular and sternoclavicular joints, contributes to the motions of flexion and abduction of the humerus, which elevates the arm by upwardly rotating the glenoid fossa for a total of 60 degrees

Table 5-13

MOVEMENTS OF THE SHOULDER COMPLEX

ARTICULATION	MOVEMENTS POSSIBLE
Glenohumeral	Flexion/extension Abduction/adduction Internal/external rotation
Sternoclavicular	Elevation/depression Protraction/retraction Rotation of clavicle
Acromioclavicular	Rotation of the scapula (acromion) Protraction/abduction and retraction/adduction Upward/downward rotation
Scapulothoracic (bone-muscle-bone)	Elevation/depression Protraction/retraction Upward/downward (medial/lateral) rotation Winging Tipping
Suprahumeral (functional joint)	Prevents superior dislocation
Sternocostal (costosternal)	Slight gliding
Vertebrocostal (costovertebral)	Slight gliding

from resting position. Scapulothoracic rhythm is the movement of the scapula across the thoracic cage in relation to the humerus. The glenohumeral joint then moves an additional 120 degrees to give a total range of motion for flexion and abduction of 180 degrees (Nordin & Frankel, 2001). This relationship of glenohumeral contribution to shoulder motion with scapulothoracic, acromioclavicular, and sternoclavicular motions is called scapulothoracic or scapulohumeral rhythm.

For motion in the sagittal plane at the glenohumeral joint, flexion couples with protraction and upward rotation of the scapula at the scapulothoracic joint while extension is accomplished by glenohumeral extension with retraction and downward rotation of the scapula at the scapulothoracic joint. Hyperextension of the arm is accompanied by upward tilt of the scapula at the scapulothoracic joint. Abduction of the glenohumeral joint is made possible by upward movement of the

arm in a frontal plane and upward rotation of the scapula at the scapulothoracic joint. Internal and external rotation of the glenohumeral joint is possible in part due to protraction and retraction at the scapulothoracic joint (Howell et al., 1988).

Norkin and Levangie (1992) identified three purposes of scapulothoracic rhythm. First, distributing motion between two joints permits a large range of motion with less compromise of stability than would occur if the same range of motion occurred in one joint. Second, maintaining the glenoid fossa in optimal position to receive the head of humerus increases joint congruency while decreasing shear forces. Finally, permitting muscles that act on the humerus to maintain good length-tension relationships minimizes or prevents active insufficiency of glenohumeral muscles.

Overall, the ratio of glenohumeral contribution to scapulothoracic is 2:1 (i.e., if there are 15 degrees of motion, 10 degrees

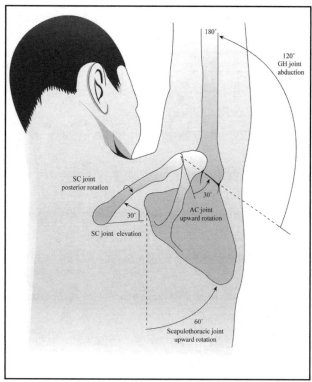

Figure 5-28. Scapulothoracic rhythm.

are due to the glenohumeral joint, and 5 degrees are due to the scapulothoracic articulations) (Ludewig et al., 2009). This relationship is shown in Figure 5-28, where 30 degrees of joint elevation occurs at the sternoclavicular joint and 30 degrees of upward rotation occurs at the acromioclavicular joint for a total of 60 degrees of combined scapulothoracic joint movement. There are also 120 degrees of humeral abduction for a total range of motion for normal shoulder abduction of 180 degrees. The 60 degrees scapulothoracic and 120 degrees glenohumeral motions demonstrate the 2:1 ratio. Scapulothoracic rhythm can be compromised by anything that changes the position of the scapula that might include muscle imbalance and pain.

This ratio has been debated, with some authors stating the ratio is 3:2, while others state the ratio is closer to 5:4 (Nordin & Frankel, 2001; Norkin & Levangie, 1992; Smith et al., 1996; Poppen & Walker, 1976). Some of the differences are due to the measurement of abduction (measured in the plane of the scapula versus the frontal plane, which yields different values for range of motion) and are due to the fact that each joint contributes differently to the motion depending upon when in the arc of motion the measurement is taken. Ludewig and colleagues (2009) found that, while the overall scapulothoracic rhythm for abduction was 2:1:1, for flexion, it was 2:4:1, and for scapular plane abduction, it was 2:2:1.

Not only do the glenohumeral and scapulothoracic joints contribute differently, so do the sternoclavicular and acromioclavicular joints. Hamill and Knutzen (1995) indicate that for flexion and abduction of the humerus, of the 60 degrees contributed by the sternoclavicular and acromioclavicular joints, 65% of the motion was due to the sternoclavicular joint, and 35% was due to the acromioclavicular joint, although the

amount of contribution of these joints is also disputed (Hamill & Knutzen, 1995; Smith et al., 1996). It is important to realize, too, that these synchronous and coordinated movements occur concomitantly and not sequentially, which produces smooth movements.

Combined movements of the sternoclavicular and acromioclavicular joints result in specific movements at each of the joints. At the acromioclavicular joint, which relates to motion of the scapula relative to the clavicle, as the arm elevates overhead, the scapula internally and upwardly rotates with a posterior tilt relative to the clavicle. The scapula is more internally rotated relative to the clavicle in flexion than in either abduction or scapular plane abduction (Ludewig et al., 2009). For scapulothoracic joint movement, defined as motion of scapula relative to thorax, during elevation of the arm, there is decreased internal rotation of the scapula and increased posterior scapular tilt relative to the thorax (Ludewig et al., 2009). Motion at the sternoclavicular joint is reciprocal with the acromioclavicular joint for protraction and retraction and elevation/depression; however, this reciprocity is not true for rotational movement. For example, if the clavicle moves up at the medial end, it moves down at the lateral end.

The muscles of scapula act in synchrony so that the scapulohumeral muscles can maintain effective length-tension relationships as they function to stabilize and move the humerus (Kisner & Colby, 2007). During abduction or flexion, the upper and lower trapezius and serratus anterior muscles upwardly rotate the scapula to aid in movement. Serratus anterior protracts the scapula on the thorax to align the scapula during flexion or activities that require a pushing action. During extension or pulling, the rhomboids downwardly rotate and retract with the latissimus dorsi, teres major, and rotator cuff muscles to control acceleration motions of scapula (Kisner & Colby, 2007).

Tables 5-14 and 5-15 summarize motions produced at each joint, the plane and axis of motion, normal limiting factors to joint movement, normal end feel, and primary muscles producing movement at each joint for the scapula and glenohumeral joints.

Phases of Elevation

The coordinated actions can be visualized by separating humeral flexion and abduction into different phases (Cailliet, 1980; Hamill & Knutzen, 2003; Inman et al., 1944; Levangie & Norkin, 2005; Ludewig et al., 2009; Magee, 2002; Muscolino, 2006). Table 5-16 summarizes the four phases of abduction.

Phase 1

The first phase is considered the resting phase. There is no glenohumeral elevation, no scapular rotation, and no clavicular elevation or rotation. The arm is held at the side in full adduction. The superior glenohumeral ligament and anterior capsule are limiting external rotation while abducting; the coracohumeral ligament, superior glenohumeral ligament, and posterior and anterior portions of the joint capsule limit movement from neutral during abduction; and the posterior capsule is limiting movement in internal rotation during abduction (Magee, 2002).

Table 5-14

Summary of Scapular Movements

MOTION	JOINT	PLANE/AXIS	NORMAL LIMITING FACTOR	END FEEL	PRIMARY MUSCLES
Elevation	Acromioclavicular Sternoclavicular Scapulothoracic	Frontal/sagittal	Tension in: • Costoclavicular ligament • Inferior sternoclavicular joint capsule • Trapezius • Pectoralis major • Subclavius	Firm	• Upper trapezius • Levator scapulae • Rhomboids
Depression	Acromioclavicular Sternoclavicular Scapulothoracic	Frontal/sagittal	Tension in: • Interclavicular ligament • Sternoclavicular ligament • Articular disk • Trapezius • Levator scapulae Bony contact between first rib and clavicle	Firm/hard	• Pectoralis minor • Lower trapezius • Pectoralis major • Latissimus dorsi (These act on the humerus)
Protraction	Acromioclavicular Sternoclavicular Scapulothoracic	Horizontal/vertical	Tension in: • Trapezoid ligament • Posterior sternoclavicular ligament • Posterior costoclavicular ligament • Trapezius • Rhomboids	Firm	• Serratus anterior • Protraction • Serratus anterior • Pectoralis minor • Pectoralis major

(continued)

Table 5-14 (continued)

SUMMARY OF SCAPULAR MOVEMENTS

MOTION	JOINT	PLANE/AXIS	NORMAL LIMITING	END FEEL FACTOR	PRIMARY MUSCLES
Retraction	Acromioclavicular Sternoclavicular Scapulothoracic	Horizontal/vertical	Tension in: • Conoid ligament • Anterior costoclavicular ligament • Anterior sternoclavicular ligament • Pectoralis minor • Serratus anterior	Firm	• Trapezius • Rhomboids
Upward rotation (lateral rotation)	Acromioclavicular Sternoclavicular Scapulothoracic	Frontal/sagittal	Tension in: • Trapezoid ligament • Rhomboid • Levator scapulae	Firm	• Middle Trapezius • Serratus Anterior
Downward Rotation (medial rotation)	Acromioclavicular Sternoclavicular Scapulothoracic	Frontal/sagittal	Tension in: • Conoid ligament • Serratus anterior	Firm	• Levator scapula • Rhomboid

(continued)

Table 5-15

Summary of Glenohumeral Movements

MOTION	JOINT	PLANE/AXIS	NORMAL LIMITING FACTOR	END FEEL	PRIMARY MUSCLES	NORMAL ROM/ FUNCTIONAL ROM
Abduction	Glenohumeral Acromioclavicular Sternoclavicular Scapulothoracic Suprahumeral	Frontal/ sagittal	Tension in: • Middle and inferior bands of glenohumeral ligament • Inferior joint capsule • Shoulder adductors Greater tuberosity of humerus contacting upper portion of glenoid and glenoid labrum or lateral surface of the acromion Scapular movement limited by tension in: • Rhomboids • Levator scapulae • Trapezoid ligament	Firm Hard	• Middle fibers of deltoid • Supraspinatus • Infraspinatus • Subscapularis • Teres minor • Biceps (long head)	0-180°/0-120°
Adduction	Glenohumeral	Frontal/sagittal	Tension in: • Conoid ligament • Anterior costoclavicular ligament • Posterior sternoclavicular ligament • Pectoralis major • Serratus anterior	Firm	• Trapezius • Rhomboid	180-0/ 120°-0

(continued)

Table 5-15 (continued)

Summary of Glenohumeral Movements

MOTION	JOINT	PLANE/AXIS	NORMAL LIMITING FACTOR	END FEEL	PRIMARY MUSCLES	NORMAL ROM/FUNCTIONAL ROM
External	Glenohumeral	Horizontal/vertical	Tension in: • Glenohumeral ligament • Coracohumeral ligament • Anterior joint capsule • Subscapularis • Pectoralis major • Teres major • Latissimus dorsi	Firm	• Infraspinatus • Teres minor • Deltoid (Posterior)	0-90°/0-60°
Internal (medial) rotation	Glenohumeral	Horizontal/vertical	Tension in: • posterior joint capsule • infraspinatus • teres minor	Firm	• Subscapularis • Latissimus dorsi • Teres major • Deltoid (anterior)	0-70°/0-70°
Flexion	Glenohumeral Acromioclavicular Sternoclavicular Scapulothoracic	Sagittal/frontal	Tension in posterior band of coracocohumeral ligament, posterior joint capsule, shoulder extensors Scapular movement limited by tension in rhomboids, levator scapulae, trapezoid ligament	Firm	• Deltoid (anterior) • Pectoralis major (clavicular part) • Biceps brachii • Coracobrachialis	0-180°/0-125°
Extension	Glenohumeral	Sagittal/frontal	Tension in anterior band of coracohumeral ligament, anterior joint capsule, pectoralis major (clavicular fibers)	Firm	• Deltoid (posterior) • Latissimus dorsi • Teres major • Triceps (long head)	0-60°/60°-0

(continued)

Table 5-15 (continued)

Summary of Genohumeral Movements

MOTION	JOINT	PLANE/AXIS	NORMAL LIMITING FACTOR	END FEEL	PRIMARY MUSCLES	NORMAL ROM/ FUNCTIONAL ROM
Horizontal abduction	Glenohumeral	Horizontal/ vertical	Tension in the anterior joint capsule, glenohumeral ligament, pectoralis major	Firm	• Deltoid (posterior) • Teres major • Teres minor • Infraspinatus	0-90°/0-90°
Horizontal adduction	Glenohumeral	Horizontal/ vertical	Tension in posterior joint capsule, soft tissue apposition	Firm/soft	• Pectoralis major • Deltoid (anterior)	0-45°/0-45°

Adapted from Gutman, S. A., & Schonfeld, A. B. (2003). Screening adult neurological populations. Bethesda, MD: American Occupational Therapy Press; Magee, D. J. (2002). Orthopedic physical assessment. Philadelphia, PA: Saunders; and Roy, J.-S., MacDermid, J. C., Boyd, K. U., Faber, K. J., Drosdowech, D., & Athwal, G. S. (2009). Rotational strength, range of motion, and function in people with unaffected shoulders from various stages of life. Sports Medicine, Arthroscopy, Rehabilitation, Therapy & Technology, 1(4), 1-7.

Table 5-16

Phases of Abduction of the Shoulder

PHASE	GLENOHUMERAL JOINT (GH)	SCAPULOTHORACIC JOINT (ScT)	CLAVICULAR MOVEMENT	STERNOCLAVICULAR JOINT	ACROMIOCLAVICULAR JOINT	ANTERIOR AND POSTERIOR VIEWS OF STRUCTURES DURING ABDUCTION
1	No abduction	0 degree scapular rotation	No elevation	0 degree movement No elevation of the clavicle	0 degree spinoclavicular angle	
2	0 to 30 degrees abduction	Winging and tipping of scapula 30 degrees upward rotation	Outer end elevation elevated 12 to 5 degrees; no rotation	Movement produces elevation	Angle increases 0 to 10 degrees	
3	30 to 90° abduction (60° GH.; 30° ScT)	0 to 30 degrees scapular rotation rotation of the clavicle	15 to 30° degrees elevation; 35 degrees posterior	5 degrees elevation	No change in spino-clavicular angle or movement	
4	90 - 180 degrees abduction (120° G.H. 60° ScT)	Coracoid depresses	30 to 50 degrees posteriorclavicular rotation and winging of 40 degrees	No further elevation	Spinoclavicular angle increased to 20 degrees Maximal tipping of 20 degrees; 25 degrees upward rotation	

Phase 2

In phase 2, considered the setting phase, the initial elevation occurs within the first 30 degrees of humeral abduction or the first 45 degrees to 60 degrees of forward flexion (Hamill & Knutzen, 1995). In early elevation, the pull of deltoid produces upward force on the humeral head. Subscapularis, infraspinatus, and teres minor counteract this by providing a depressive force on the humeral head in glenoid fossa. Subscapularis provides an anteriorly directed force, while infraspinatus and teres minor provide posterior forces. Because these muscles have approximately the same cross-sectional area, the forces are balanced and can resist both anterior and posterior translation of the humeral head.

During this phase, the scapula moves toward or away from the vertebral column in order to find a position of stability on the thorax (Hamill & Knutzen, 1995; Norkin & Levangie, 1992). During this early phase of elevation and upward rotation, the scapula and clavicle move together around an oblique axis near the spine of the scapula and through the sternoclavicular joint. A force couple is formed by the upper and lower trapezius muscles and the serratus anterior muscle, which produces movement at the sternoclavicular and acromioclavicular joints (Norkin & Levangie, 1992). The upper and lower trapezius and the upper and lower serratus anterior apply a rotary force on the scapula at the sternoclavicular joint, but further movement is prevented by the conoid and trapezoid (coracoclavicular) ligaments.

The movements that occur at the acromioclavicular joint are 0 degrees to 10 degrees of an increase in the spinoclavicular angle (from the spine of the scapula to the clavicle) but minimal movement due to the influence of the coracoclavicular ligament. Movement at the scapulothoracic joint includes about 10 degrees of winging and tipping of the scapula, which maintains the scapula against the ribs, and upward rotation. The outer end of the clavicle is elevated 5 degrees to 10 degrees, indicative of superior movement of the scapula and likely posterior tilt due to the taut coracoclavicular ligament, but there is little clavicular rotation.

Structures limiting abduction movement from 0 degrees to 45 include the coracohumeral ligament, superior glenohumeral ligament, and anterior joint capsule for external rotation; middle glenohumeral ligament, posterior capsule, subscapularis, infraspinatus, and teres minor for neutral position; and the posterior capsule for internal rotation.

Phase 3

In phase 3, the humerus moves an additional 60 degrees, so total abduction is 90 degrees. Of this 90 degrees of the total humeral abduction, 60 degrees are due to movement at the glenohumeral joint, and 30 degrees are due to scapulothoracic contributions. At 90 degrees of humeral elevation, external rotation produces scapular movement. There is an abrupt posterior tilt and upward rotation. The scapula will move laterally, anteriorly, or superiorly during movements of upward rotation, protraction (abduction), and elevation. The upward rotary force created by serratus anterior and trapezius muscles continues, and this produces movement at the sternoclavicular joint, which forces the clavicle to elevate 5 degrees and posteriorly rotate 35 degrees. Because the clavicle is attached to

the scapula, elevation of the clavicle produces elevation of the scapula as it is carried through 30 degrees of upward rotation. Further elevation of the clavicle is prevented when the coracoclavicular ligament becomes taut (Norkin & Levangie, 1992). No further elevation at the sternoclavicular joint and upward rotation at the acromioclavicular joint is now possible. There is no increase in the spinoclavicular angle.

Maximal range of motion for abduction requires external rotation of the humerus in order for the greater tubercle of the humerus to clear coracoacromial arch once the arm is elevated above the horizontal. If there is weakness or inadequate external rotation, there will be impingement of soft tissues in suprahumeral space, causing pain, inflammation, and loss of function (Kisner & Colby, 1990). Once the humerus is externally rotated, an additional 30 degrees of abduction is possible.

Similarly, internal rotation of the humerus is required to achieve full elevation through flexion, and this rotation occurs at 50 degrees of passive shoulder. Because most of the shoulder flexor muscles are also medial rotators, as the arm elevates above horizontal in the sagittal plane, the anterior capsule and ligaments get taut, causing the humerus to medially rotate (Kisner & Colby, 1990).

Structures limiting movement between 45 degrees and 60 degrees of abduction are middle glenohumeral ligament, coracohumeral ligament, inferior glenohumeral (anterior band), and anterior joint capsule. Additional structures limiting movement at this range are middle glenohumeral ligament, inferior glenohumeral ligament (anterior portion), subscapularis, infraspinatus, and teres minor for neutral rotation during abduction; inferior glenohumeral ligament (posterior band) and posterior joint capsule limit internal rotation during abduction at this range. At 60 degrees to 90 degrees of abduction, the following structures limit abduction: the inferior glenohumeral ligament and anterior capsule limit external rotation; the inferior glenohumeral ligament (posterior portion) and middle glenohumeral ligament limit neutral rotation; and the inferior glenohumeral ligament (posterior band) and posterior joint capsule limit internal rotation (Magee, 2002).

Phase 4

Phase 4 completes the abduction range to 180 degrees, with the glenohumeral joint providing 120 degrees of the total and the scapulothoracic joints contributing 60 degrees. The coracoid process of scapula is pulled down, tugging on the coracoclavicular ligament.

Thirty degrees to 50 degrees of clavicular rotation occurs around the longitudinal axis, which elevates the lateral end of the clavicle without additional elevation of the sternoclavicular joint. Because the lateral end of the clavicle is elevated, the scapula will be carried through an additional 20 degrees of upward rotation around an anteroposterior axis through the acromioclavicular joint, where the scapula will reach the maximum of 20 degrees of tipping and 40 degrees of winging. Given that 180 degrees is considered the maximum range of motion for humeral abduction and flexion (some sources say 170 degrees with the additional 10 degrees due to trunk movement), horizontally, 60 degrees of glenohumeral and 30 degrees of scapulothoracic motion occurred, with scapular movement due to clavicular elevation at the sternoclavicular joint. Horizontally to vertically, an additional 60 degrees of glenohumeral motion is produced (plus medial rotation in the sagittal plane for flexion and lateral rotation for abduction in frontal plane), and 30 degrees of scapular movement is produced due to clavicular rotation and acromioclavicular motion (Norkin & Levangie, 1992).

The final elevation from 90 degrees to 180 degrees is due to infraspinatus and teres minor with subscapularis preventing impingement during external rotation (Hess, 2000). With continued elevation, by 90 degrees of elevation, the compression forces are maximal so the joint is stable (Hess, 2000). The axis of motion is now at the acromioclavicular joint, because tension in the costoclavicular ligament prevents further elevation of the clavicle at the sternoclavicular joint, and serratus anterior and lower trapezius continue to produce upward rotation moments. The inferior glenohumeral ligament (anterior band) and anterior joint capsule limit external rotation during abduction at this range. The inferior glenohumeral ligament limits movement in both neutral and internal rotation during abduction at this range, and the posterior capsule also prevents further internal rotation (Magee, 2002) in this 90 degrees to 180 degrees range of abduction.

INTERNAL KINETICS

During hand use, there are large forces that occur in the shoulder because the resistance arm of the lever that is formed can be as long as 2 feet when using a tool, but the force arms of the muscles are measured in inches (Smith et al., 1996). Shoulder muscle is typified as being relatively small with large moment arms (Veeger & vanderHelm, 2007). The physiologic cross-section of serratus anterior is much smaller than the deltoid, but the moment arm around the sternoclavicular joint is much larger than moment arms crossing the knee or ankle (Veeger & vanderHelm, 2007). The scapula helps to provide the large moment arms for the scapulothoracic muscles: Serratus anterior for elevation; trapezius for elevating the clavicle to allow rotation of the scapula; adduction via the pectoralis and latissimus dorsi muscles to direct forces to the thorax; gravitational forces that will lift the scapula from the thorax; and serratus anterior and rhomboids that press the scapula on the thorax to provide a stable base (Veeger & vanderHelm, 2007).

Calculating the reaction forces of the shoulder is challenging because of the large number of muscles, and the force contribution of each muscle varies with differing loads, planes of shoulder elevation, and degrees of elevation (Nordin & Frankel, 2001). Smith and colleagues (1996) indicate that the greatest strength of shoulder muscles occurs when muscles contract in an elongated position and torque decreases as the muscles shorten. They add that favorable length-tension relationships over such a large range of motion are accomplished by movement of the base of support of the humerus by the scapula and by changes in muscle lever arms. For example, lever arm lengths for the deltoid increased with the motion of abduction: the middle deltoid almost doubled its leverage, and the anterior deltoid increased leverage by eight times (Smith et al., 1996). Forces in the shoulder joint at 90 degrees of abduction have been shown to be close to 90% of body weight,

whereas the forces are diminished to half of that if the forearm flexes to 90 degrees at the elbow (Hamill & Knutzen, 1995).

Hamill and Knutzen (1995) indicate that the greatest strength output in the shoulder muscles occurs in adduction due to latissimus dorsi, teres major, and pectoralis muscles acting as major contributors. The strength of adduction is twice that of abductor strength even though abduction is used more frequently in activities of daily living or sports.

Extension is the next strongest action because muscles of extension involve the same muscle as adduction. Extension is seen as slightly stronger than flexion. The weakest muscles of the shoulder are the rotators. The muscles of external rotation are weaker than the muscles of internal rotation. The motions of internal and external rotation are most influenced by arm position, with the greatest internal rotation occurring in neutral and the greatest external rotation in 90 degrees of flexion (Hamill & Knutzen, 1995; Kibler & McMullen, 2003). Rotator cuff muscles as a group can generate forces of 9.6 times the weight of the limb. Because each arm weighs 7% of body weight, the rotator cuff muscles can generate forces in the shoulder joint of approximately 70% of body weight.

The location of the muscle in relation to the joint axis will also determine the direction of the force. For example, anterior deltoid, pectoralis major, coracobrachialis, and biceps all cross the glenohumeral joint anteriorly and function to flexion the joint. Similarly, posterior deltoid, latissimus dorsi, teres major, and the long head of the triceps cross the glenohumeral joint posteriorly and aid in extension. Muscles that abduct (deltoid, supraspinatus) cross superiorly to the joint axis, and muscles that adduct (pectoralis major, latissimus dorsi, and teres major) cross below the center of the joint (Muscolino, 2006).

While it is generally true that one can infer the action of a muscle given knowledge of origin and insertion, this is not always the case in shoulder muscles. "For example, when the arm is at the side, contraction of the fibers of the middle portion of the deltoid lifts the humerus along its axis but does not produce the motion of elevation because the line of action of the middle deltoid fibers is essentially parallel to the long axis of the humerus and if 'coupled' with other muscles, elevation can occur" (Nordin & Frankel, 1989, p. 239). This is due to the unusual characteristics of shoulder muscles. Nordin and Frankel (1989) indicate that muscle actions at the shoulder have three unusual aspects:

1. Because the glenohumeral joint lacks rigid stability, muscles exerting an effect on the humerus must act in concert with other muscles to avoid producing a dislocating force on the joint.

2. The existence of multiple linkages in the shoulder (clavicle, humerus, and scapula) gives rise to an interesting situation in which a single muscle may span several joints, exerting an effect on each.

3. The extensive range of shoulder motion causes muscle function to vary depending on the position of the arm in space.

The large numbers of muscles at the shoulder joint produce large moment arms that develop force dependent upon the load applied to the muscle, the plane of motion, and the amount of elevation. The position of the arm in space determines the force produced and which muscles generate the

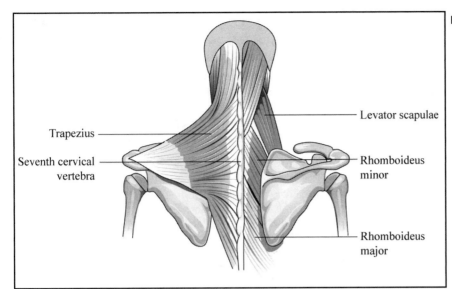

Figure 5-29. Elevators of the scapula.

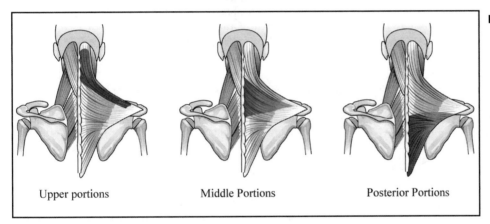

Figure 5-30. Trapezius muscle.

force. An additional unique feature of the shoulder muscles is the large number of force couples that act to provide both stabilizing forces as well as movement.

The integrated muscular activity of the shoulder ensures that the force generated at one muscle requires antagonistic (and usually eccentric contraction) activation so that dislocation does not occur or the neutralizing force of a force couple is produced (Nordin & Frankel, 2001). The actions of many muscles working alone or in combination produce movement and stability of the shoulder. By knowing the angle of muscle pull, the length and cross-section, type of muscle fiber, and location of the muscle in relation to the joint axis, one can tell much about the action and strength of that muscle.

The following discussion about muscles acting on the shoulder girdle is organized according to the muscle action at the joint. Directions for muscle palpation are included the first time the muscle is mentioned.

Elevators of the Scapula

Elevators of the scapula attach from the scapula to a superior structure, and these are considered scapulothoracic muscles and include trapezius, levator scapula, and rhomboid muscles (Figure 5-29).

Upper Trapezius

Because the upper trapezius attaches on the base of the skull and clavicle, these fibers will pull upward on the clavicle when it contracts. The resolution of the line of pull force on the clavicle is transferred to the scapula via the acromioclavicular joint to elevate the scapula (Gench et al., 1995). The upper trapezius is seen as solely responsible for elevating the lateral angle of the scapula (Jenkins, 1998). Isolated weakness is unusual for the upper trapezius muscle but could result in a position of increased scapular depression (Magee, 2002). Muscle tightness can occur in an elevated shoulder or asymmetrical head position, with restricted head and neck range of motion, and is likely to be accompanied by upward rotation of scapula (Oatis, 2004) (Figure 5-30).

Palpation of Upper Trapezius

The trapezius is a superficial muscle of the neck and upper back, often called the "shawl" muscle. The upper fibers lie along the posterior neck, and by grasping the superficial tissue on the top of the shoulder, you can feel the thin upper trapezius muscle fibers. Follow the fibers to the lateral clavicle as you ask your subject to extend the neck.

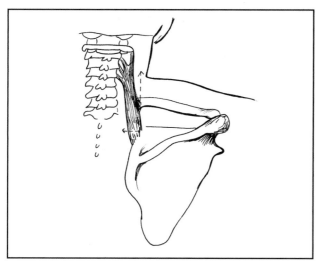

Figure 5-31. Levator scapula force resolution.

Upper/Lower Trapezius and Serratus Anterior

The upper portions of these muscles form a force couple that moves the scapula during elevation. These muscles, with levator scapula, support the shoulder girdle against the pull of gravity and act as stabilizing synergists for deltoid action working on the glenohumeral joint. This force couple is antagonistic to scapular movement and synergistic to glenohumeral moment, where trapezius and serratus anterior produce scapular upward rotation while preventing undesired motion of the deltoid as it elevates the glenohumeral joint (Nordin & Frankel, 2001; Smith et al., 1996). An imbalance of muscle forces between the upper and lower trapezius may result in poor posture with the upper trapezius being more dominant.

Levator Scapula

When the diagonal line of pull of the muscle is resolved into mobilizing and stabilizing forces, there is a long vertical component and a relatively short horizontal component. The long vertical component enables levator scapula to be an elevator of the scapula (Gench et al., 1995) (Figure 5-31).

Palpation

Levator scapula is covered by the upper trapezius and sternocleidomastoid muscles; in elevation of the shoulder girdle, the upper trapezius and levator contract together. It is difficult to isolate and palpate this muscle. By placing the forearm behind you in the small of your back and then shrugging the shoulder, you will feel this muscle when you palpate in the neck region anterior to trapezius but posterior to the sternocleidomastoid (Esch, 1989; Lumley, 1990; Magee, 2002).

An alternative method for palpation is to have the subject prone, supine, or in sidelying. Locate the upper fibers of trapezius, move anteriorly off of trapezius, and strum your fingers anteriorly and posteriorly across the levator fibers. Follow these fibers toward the lateral neck and inferiorly under the trapezius (Biel, 2001).

With weakness of levator scapulae, rhomboid major, and rhomboid minor, pulling actions would be affected, and a posture of rounded shoulders may result (Oatis, 2004). Muscle tightness does produce rounded shoulder posture with eleva-

tion, adduction, and downward rotation, causing the scapula to tilt anteriorly.

Rhomboids (Both Minor and Major)

These two muscles actually function as one muscle. There is a diagonal pull of the muscle that yields both elevation and adduction actions, but the muscle acts only on the medial border of the scapula (Jenkins, 1998). The rhomboids' downward rotation of the scapula offsets the undesired force of teres major and contributes to depression of the shoulder. Rhomboids major and minor connect the scapula with the vertebral column, and they lie under the trapezius. The upper portion of the muscle is rhomboid minor, while the lower portion is rhomboid major (Gench et al., 1995; Hinkle, 1997; Jenkins, 1998; Smith et al., 1996).

Palpation

These muscles can best be palpated when the trapezius is relaxed. Have a subject place his or her hand at the small of the back. The therapist then places fingers under the medial border of the scapula. If the subject raises his or her hand just off the small of the back, the rhomboid major contracts as downward rotator and pushes the fingers out from the medial border (Esch, 1989; Lumley, 1990; Magee, 2002).

Depressors of the Scapula

Muscles that depress the scapula attach from the scapula to a structure located more inferiorly. The muscles responsible for depression of the scapula include trapezius, pectoralis minor, subclavius, pectoralis major (sterna portion), and latissimus dorsi. These muscles are shown in Figure 5-32.

Trapezius, Lower Fibers

The lower fibers of the trapezius produce a diagonal force vector. The origin of this portion of the muscle is lower than insertion, so there are two component forces: one directed downward for depression and the other toward the spine for adduction (Gench et al., 1995) (see Figure 5-32).

Palpation of Lower Trapezius

Locate the spinous process of T12. Palpate the superficial lower fibers while the subject holds his or her arm out in front in the plane of the scapula (Biel, 2001).

Pectoralis Minor

These fibers are directed downward from the attachment on coracoid, and their function as a depressor can be easily seen. Pectoralis minor is entirely covered by pectoralis major (Gench et al., 1995; Hinkle, 1997; Jenkins, 1998; Smith et al., 1996). A person with a weak pectoralis minor muscle will have difficulty controlling the shoulder girdle, particularly during weight-bearing activities (Oatis, 2004) and may have difficulty with downward rotation of the scapulothoracic joint. Muscular tightness will result in a posture characterized by downward rotation and forward tilt of the scapula, which may impinge on the brachial plexus or axillary blood vessels leading to a form of thoracic outlet syndrome (Oatis, 2004).

Palpation

Have the subject place the forearm at the small of the back, which relaxes pectoralis major. Place a finger just below the

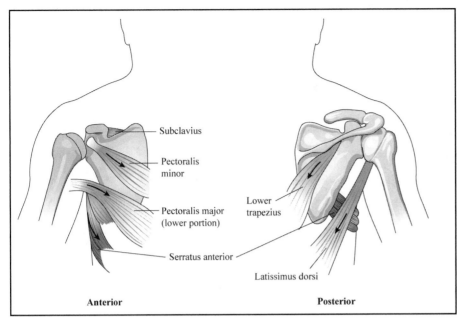

Figure 5-32. Depressors of the scapula.

Subclavius

Pectoralis minor

Lower trapezius

Pectoralis major (lower portion)

Serratus anterior

Latissimus dorsi

Anterior

Posterior

coracoid process. When the subject lifts the arm off of the back, the tendon of pectoralis minor becomes taut.

Subclavius

Although small, this muscle does produce a downward force when the line of pull is resolved, making the subclavius a weak depressor (Gench et al., 1995). This muscle has not been studied in much depth, but weakness in the subclavius muscle might limit elevation at the sternoclavicular joint, thereby causing shoulder girdle dysfunction. Its fibers run parallel to the clavicle and can be difficult to palpate.

Pectoralis Major

Jenkins (1998) indicates that the inferior fibers of pectoralis major protract the scapula as it assists in depressing it. While the pectoralis major assists the deltoid with flexion of the humerus, the sternal and abdominal portions serve as depressors of the shoulder complex (Norkin & Levangie, 1992). Both portions depress the shoulder complex in weight bearing on the hands, while anterior and posterior movement of the humerus and abduction and adduction of the scapula are neutralized (Gench et al., 1995; Hinkle, 1997; Jenkins, 1998; Smith et al., 1996).

Palpation

This muscle is easily observed and palpated because it is superficial in location and bulk. The upper portion can be seen when the arm is brought obliquely upward toward the head against resistance. The lower portion contracts separately when the arm is adducted in a lower position. You can feel the antagonistic movements of the upper and lower fibers by having your subject lie supine and asking the subject to abduct the humerus to 90 degrees and attempt to horizontally adduct the arm against resistance. The upper fibers will contract while the lower fibers are relaxed. Next, try to see if the subject can attempt to move from flexion to extension against resistance, and the lower fibers will contract while the upper fibers will be lax (Biel, 2001).

Latissimus Dorsi

Latissimus dorsi is the broadest muscle of the back. It is a thin, sheet-like muscle, and it forms the posterior fold of the axilla. Latissimus dorsi and teres major contract when adduction or extension is resisted, as when the subject presses down on the examiner's shoulder. Latissimus dorsi is seen to retract the scapula as it depresses it (Jenkins, 1998) and serves to adduct and medially rotate as well as extend the humerus and adduct and depress the scapula (Smith et al., 1996). Some studies say latissimus dorsi is active in abduction and flexion of the arm and may contribute to joint stability because compression of the glenohumeral joint occurs when the arm is overhead (Smith et al., 1996). Latissimus dorsi attaches to the crest of the ileum so when the arms are stabilized, the distal attachment can aid in lifting the pelvis, as occurs when a client places his or her hands on the armrests of a wheelchair. This is helpful in that the client can do a sitting pushup for pressure relief. This is also useful for clients with injuries to the spinal cord C8 and below because latissimus dorsi is innervated by the thoracodorsal nerve (C6, C7, and C8) (Gench et al., 1995; Hinkle, 1997; Jenkins, 1998; Smith et al., 1996).

Palpation

With the subject prone, locate the thick muscle lateral to the lateral border of the scapula. Ask the subject to internally rotate the shoulder against resistance, and follow the fibers superiorly to the axilla and inferiorly on the ribs (Biel, 2001).

Protraction (Abduction) of the Scapula

Muscles that abduct or protract the scapula run from the scapula to anterior surfaces and include pectoralis minor and serratus anterior (Figure 5-33).

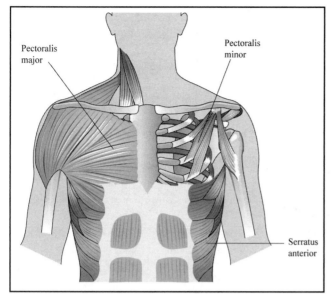

Figure 5-33. Abductors of the scapula.

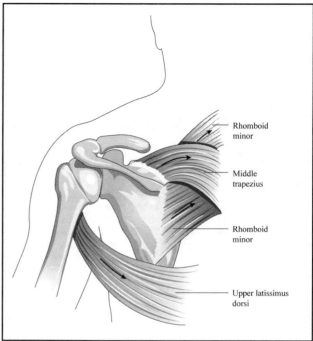

Figure 5-34. Adductors of the scapula.

Pectoralis Minor

The fibers of the pectoralis major pull downward, inward, and forward. Parts of this muscle pull the coracoid medially, which enables the scapula to slide laterally along the ribs while the acromion glides forward against the distal end of the clavicle (Gench et al., 1995). Both the clavicular and the sternocostal portions draw the arm medially, which abducts the scapula (Gench et al., 1995; Hinkle, 1997; Jenkins, 1998; Smith et al., 1996).

Pectoralis major was cited as a protractor by two authors (Jenkins, 1998; Smith et al., 1996).

Serratus Anterior

Referred to as the "saw" muscle because of the multiple digitations, serratus anterior allows one to raise the arm overhead. Because the fibers are nearly horizontal, this is an effective scapular abductor (Gench et al., 1995). When serratus anterior is injured and this motion is attempted, pathological winging of the scapula occurs because the scapula fails to slide forward on the rib and doesn't stay anchored on the thorax (Gench et al., 1995; Hinkle, 1997; Jenkins, 1998; Smith et al., 1996).

Palpation

The lower digitations of the muscle can be seen and palpated near their proximal attachment on the ribs when the arm is overhead. The middle and upper portions can be palpated in the axilla close to the ribs and posterior to pectoralis major if the arm is elevated to horizontal in the plane of the scapula (between flexion and abduction) while reaching forward.

Retractors (Adduction) of the Scapula

Muscles that act to pull the scapulae toward the vertebral column (adduct or retract) arise from the scapula and insert in the posterior midline (Figure 5-34). These include the trapezius and rhomboids.

Rhomboids

The diagonal line of pull of this muscle produces two components, which explains the role in both elevation and adduction actions.

Trapezius

The middle fibers of trapezius adduct and stabilize the scapula (Gench et al., 1995; Hinkle, 1997; Jenkins, 1998; Smith et al., 1996). Fibers 2, 3, and 4 may be seen to function in retraction in the following ways:

- Fiber 2 (upper): Fibers are in a diagonal direction, which yields two component forces, one of which is directed toward the spine for adduction.

- Fiber 3 (middle): Fibers are nearly horizontal so this muscle functions only in adduction.

- Fiber 4 (lower): Diagonal forces with two forces directed down (depression) and toward the spine (adduction) (Esch, 1989; Lumley, 1990; Magee, 2002)

Isolated weakness of the middle trapezius is unusual and would result in a significant decrease in scapular adduction strength. In addition, weakness in the middle trapezius may result in difficulty contracting the scapulohumeral muscles, such as the infraspinatus and posterior deltoid muscles (Oatis, 2004).

Palpation

Have the subject abduct the shoulder and retract the scapula. If the subject is prone or if the trunk is inclined

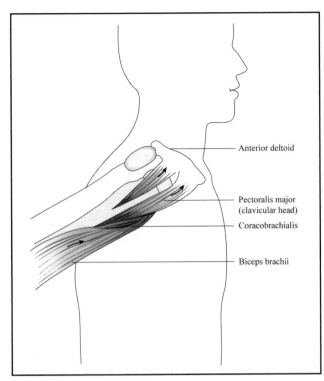

Figure 5-35. Flexors of the humerus.

forward, the muscle works against gravity so the intensity of the contraction increases and the muscle will be more easily identified (Smith et al., 1996). If you move your hand medially from the spine of the scapula, you can feel the superficial and thin middle trapezius muscle fibers (Biel, 2001).

Levator Scapula

This muscle, with a diagonal pull, has a short horizontal component that weakly adducts and a long vertical component that elevates (Gench et al., 1995).

Upward Rotators of the Scapula

Muscles that work to produce upward rotation of the scapula include the upper and lower portions of the trapezius muscle and serratus anterior.

Trapezius, Upper and Lower

The upper portion of the trapezius (part 2) crosses the acromioclavicular joint above the sagittal axis of that joint and pulls the acromion toward the neck by pivoting the acromion around its articulation with the clavicle (Gench et al., 1995). The lower portion attaches medially to the acromioclavicular joint and is below the joint so the muscle upwardly rotates the scapula by pulling downward on the base of the spine of the scapula (Gench et al., 1995).

Serratus Anterior

These horizontal fibers have a lateral pull on the inferior angle, pulling this portion of the medial border laterally and forward to depress the scapula and upwardly rotate the scapula (Jenkins, 1998). The lower fibers are effective as upward rota-

tors because they exert a lateral pull on the inferior angle (Gench et al., 1995).

Downward Rotators of the Scapula

Downward rotators of the scapula include the rhomboids, levator scapulae, and pectoralis minor.

Rhomboids

Both rhomboid major and rhomboid minor contribute to downward rotation of the scapula because their lower fibers, which pull medially and upward on the inferior angle of the scapula, raise the medial border, resulting in downward rotation of the scapula.

Levator Scapula

This muscle pulls upward on the vertebral border of the scapula to lower the inferior angle and raise the medial border.

Pectoralis Minor

With a downward pull on the end of the coracoid and the lateral angle, the scapula returns to the anatomical position (Gench et al., 1995). Pectoralis major and latissimus dorsi were also identified as downward rotators (Jenkins, 1998) because these muscles pull down on the lateral angle of the scapula.

Flexors of the Humerus

Muscles that flex the humerus lie anterior to the glenohumeral joint axis, and those fibers further from the axis will be the most effective. The flexors of the humerus include the anterior deltoid, pectoralis major, coracobrachialis, and biceps (Figure 5-35).

Anterior Deltoid

The deltoid muscle resembles the Greek letter D (δ) in shape and comprises 40% of the mass of the scapulohumeral muscles. The deltoid is responsible for the roundness of shoulder. Its multipennate structure and considerable cross-section compensate for small mechanical advantage and less-than-optimal length tensions. This muscle is seen by some as the most important flexor of the humerus (Jenkins, 1998). Weakness of the anterior deltoid muscle would result in weaker glenohumeral flexion, medial rotation, glenohumeral abduction, and horizontal adduction. If the anterior deltoid becomes tight, there will be diminished shoulder extension and lateral rotation (Oatis, 2004).

Palpation (Anterior)

Move the humerus into a horizontally abducted position, and the anterior portion contracts vigorously when horizontal adduction is resisted.

Pectoralis Major–Clavicular Head

Because this muscle crosses the shoulder in front of the frontal axis and attaches to the clavicle, it acts as a powerful flexor. Peak muscle strength is at 75 degrees and again at 115 degrees of flexion. The clavicular head of pectoralis major

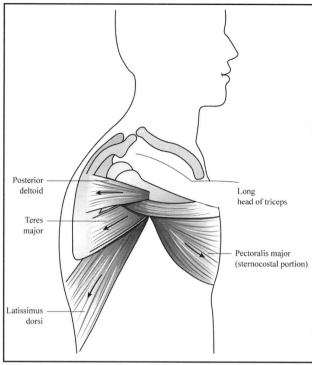

Posterior
deltoid

Teres
major

Latissimus
dorsi

Long
head of triceps

Pectoralis major
(sternocostal portion)

Figure 5-36. Extensors of the humerus.

as a shoulder flexor. This capacity is enhanced by maintaining the elbow in extension, thereby putting some stretch on the muscle (Gench et al., 1995). It is important to consider, however, that complete flexion of the shoulder is impossible when the elbow is extended unless accompanied by internal rotation to diminish the pull of the biceps against the front of the humerus (Jenkins, 1998).

Biceps can also aid in preventing subluxation of the glenohumeral joint because when the muscle contracts, tension occurs to produce downward and inward force on the head of the humerus, compressing it against the glenoid cavity (Smith et al., 1996).

Palpation

Ask your partner to flex the elbow against resistance. The biceps will be felt as a hard, round structure on the anterior surface of the arm.

Extensors of the Humerus

Posterior deltoid, latissimus dorsi, teres major, and the long head of the triceps cross the glenohumeral joint posteriorly, so are extensors of the humerus. These muscles are shown in Figure 5-36. Some sources also include pectoralis major (sternocostal fibers), coracobrachialis, and subscapularis as assisting with extension of the humerus.

Posterior Deltoid

The posterior fibers lie posterior to the frontal axis, and they function to extend the humerus. Fibers further from the joint axis will be more effective in this action, and posterior deltoid is capable of extending the humerus back further than other extensors (Gench et al., 1995; Jenkins, 1998). Weakness of this muscle would result in decreased extensor strength while restricted shoulder flexion and medial rotation range of motion might be expected when this muscle is tight (Oatis, 2004).

Palpation (Posterior)

This portion of the deltoid is most easily seen when the arm is hyperextended against resistance or resistance given to horizontal abduction (Esch, 1989; Lumley, 1990; Magee, 2002).

Latissimus Dorsi

This muscle is an excellent extensor because it is located inferior and anterior to the frontal axis, and its origin is lower than the insertion.

Teres Major

This muscle is located along the axillary border of the scapula and contributes to humeral extension only when resistance is applied to the arm. Teres major also adducts and medially rotates the humerus and is active only in static positions of humerus. The proximal attachment of teres major must be stabilized, because unopposed motion would upwardly rotate the scapula (Gench et al., 1995; Hinkle, 1997; Jenkins, 1998; Smith et al., 1996).

works synchronously with the anterior deltoid in forward flexion.

Coracobrachialis

Because of the size and location of the two attachments, this muscle has limited effectiveness, but it does cross anterior to the frontal axis so it functions in humeral flexion. Studies have not confirmed the effects of weakness or tightness of the coracobrachialis muscle, but it has been suggested that weakness would lead to diminished strength in flexion and adduction of the shoulder and tightness would lead to decreased range of motion in abduction and extension of the shoulder (Oatis, 2004).

Palpation

Coracobrachialis is covered by deltoid and pectoralis major. It is possible to palpate this muscle in the distal portion of the axillary region if the arm is externally rotated and abducted to 45 degrees. Find the pectoralis major muscle, which forms the axilla's anterior wall. Palpate the coracobrachialis by having your subject horizontally adduct the arm against resistance, and find the muscle posterior to the pectoralis fibers (Biel, 2001).

Another method is to find the short head of the biceps. Coracobrachialis lies medially and parallel to the tendon of the short head of the biceps, which is seen by supination of the forearm, and follows the short head proximally (Esch, 1989; Lumley, 1990; Magee, 2002).

Biceps Brachii

The line of pull of the biceps fibers is similar to coracobrachialis and, like coracobrachialis, has limited effectiveness

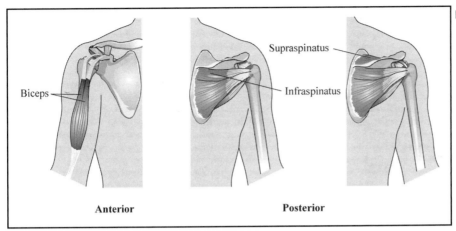

Figure 5-37. Abductors of the humerus.

Biceps

Supraspinatus

Infraspinatus

Anterior   **Posterior**

Palpation

Palpate along the inferior aspect of the axillary border of the scapula when prone, with the arm hanging over the side or in a forward inclined trunk position. If there is internal rotation of the glenohumeral joint, then teres major rises. If you palpate higher on the axillary border and externally rotate, teres minor can be felt as teres major relaxes. Teres major acts in pulling motions when the shoulder is extended or adducted against resistance (Smith et al., 1996).

Triceps—Long Head

Triceps is ineffective as a mover of the humerus because it passes directly over the axis. Because the line of pull passes slightly posteriorly to the frontal axis, triceps contributes to humeral extension. It is possible to increase the effectiveness of triceps in extension by flexing the elbow and thereby putting a stretch on the triceps muscle (Gench et al., 1995).

Palpation

Palpate the posterior aspect of the arm and locate the olecranon process. Ask the subject to extend the elbow against resistance, and you can palpate the medial and lateral heads of the triceps. The long head of the triceps is the only band of muscle on the posterior arm that runs superiorly along the proximal and medial part of the arm (Biel, 2001).

Pectoralis Major—Sternocostal Fibers

Because the fibers are located anterior to the frontal axis with the origin lower than insertion, these fibers pull downward on the humerus to extend it. With weakness, there would be decreased strength in medial rotation, adduction, and horizontal adduction of shoulder. This muscle can become tight as is seen following thoracic or breast surgery, which limits the range of motion of lateral rotation, horizontal abduction, and possibly shoulder flexion (Oatis, 2004).

Coracobrachialis

Because of the size and location of the two attachments, the line of pull of this muscle is quite close to respective axes, and so acts with limited effectiveness to assist other muscles. Not all sources indicate that coracobrachialis is an extensor of the humerus (Gench et al., 1995; Hinkle, 1997; Jenkins, 1998; Smith et al., 1996).

Abductors of the Humerus

Muscles that abduct the humerus (deltoid, supraspinatus, biceps, infraspinatus) cross superiorly to the joint axis and are shown in Figure 5-37.

Deltoid—Middle and Lateral Fibers

As a whole, the deltoid acts as the chief abductor of the glenohumeral joint. The middle portion is the most important as an abductor, because most of the anterior and posterior fibers are close to the axis, which diminishes their effectiveness. In shoulder abduction and flexion, the deltoid contributes half of the muscle force for elevation, and the contribution increases as abduction increases, with the muscle most active between 90 and 180 degrees. However, the deltoid is most resilient to fatigue at 45 to 90 degrees, so this is the more popular position for arm-raising exercises (Hamill & Knutzen, 1995) and is relevant in developing physical abilities in areas of occupation such as combing one's hair or upper extremity dressing.

Weakness in the middle deltoid weakens abduction and contributes to decreased shoulder flexion. Tightness in this muscle does not restrict adduction but may cause pain or disruption to the deltoid tendon or bursae (Oatis, 2004).

Palpation (Middle)

The middle deltoid can be seen when the humerus is maintained in an abducted or adducted position.

Supraspinatus

Located above the spine of the scapula and hidden by trapezius and deltoid, supraspinatus is an abductor of the glenohumeral joint (Smith et al., 1996). While the supraspinatus fibers are well superior to the sagittal axis, there is a pull on the head of the humerus into glenoid fossa, which acts in a complementary way with the deltoid muscle as it initiates abduction (Gench et al., 1995). Supraspinatus was considered the initiator of abduction, but this has not been shown to be true (Inman et al., 1944). Supraspinatus works with deltoid progressively through the entire range.

When the deltoid is paralyzed, supraspinatus alone can bring the humerus through much of the range, but it is weak. Supraspinatus also compresses the glenohumeral joint and acts as the vertical steerer for humeral head (Smith et al., 1996).

Figure 5-38. Adductors of the humerus.

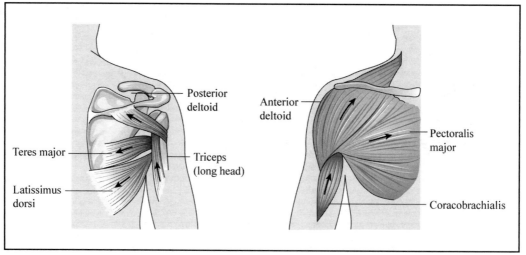

Weakness of the supraspinatus is common, resulting from entrapment of the supraspinatus nerve or from mechanical disruption of the tendon or insertion into the glenohumeral joint capsule. This can cause pain due to tendonitis, degeneration of tendons, and a significant decrease in the strength of shoulder abduction. Tightness of the supraspinatus can occur after surgical repair of the rotator cuff muscles, so it is important to avoid medial rotations and horizontal adduction (Oatis, 2004). Tears of rotator cuff muscles typically begin in the supraspinatus tendon (Clark & Harryman, 1992), which may be related to avascularity of the tendon, age-related changes in collagen, and mechanical trauma. The supraspinatus tendon often degenerates with age, especially in people whose livelihood has depended largely on the forceful use of their upper limb (Clark & Harryman, 1992; Oatis, 2004).

Palpation

The deepest portion is too deep to palpate, but the more superficial fibers can be felt through the trapezius muscle. Find the spine of the scapula, and place your fingers above the spine. Have the subject perform a quick abduction movement in a short range, and you will feel momentary contraction of the supraspinatus. You can also test for this by having the subject lie prone with the arm over the edge of a table, which causes upward rotation of scapula where one can feel supraspinatus without trapezius.

Biceps Brachii—Long Head

Biceps assists with abduction if the humerus is externally rotated. It is important to note that lateral rotation of the humerus is always accompanied by complete abduction of the arm to allow the greater tubercle to slide under, and not hit against, the acromion (Jenkins, 1998).

Infraspinatus

Infraspinatus is a flat muscle located in the infraspinous fossa. The muscle belly is superficial with a medial portion deep to trapezius and a lateral portion under deltoid (Biel, 2001). Infraspinatus is seen as a significant contributor to abduction in the scapular plane (Reinold, Escamilla, & Wilk, 2009).

Palpation

Locate the spine, medial border, and lateral border of the scapula, and form a triangle around these structures. Ask your partner to protract the scapula, and you will feel the infraspinatus contract (Biel, 2001).

Adductors of the Humerus

Muscles that adduct the humerus (pectoralis major, latissimus dorsi, and teres major) cross below the center of the joint (Figure 5-38). Other muscles that assist with adduction include triceps, teres minor, and infraspinatus.

Pectoralis Major—Sternal Portions

Because the sternal portions cross inferior to the sagittal axis, pectoralis major functions as an adductor muscle.

Latissimus Dorsi

Due to its location well inferior to the sagittal axis, latissimus dorsi is a powerful adductor. Weakness in the latissimus dorsi would result in decreased strength not only in the adductors, but also in shoulder extension, medial rotation, and scapular depression. A tight latissimus dorsi muscle would limit shoulder range of motion in flexion, lateral rotation, and abduction (Oatis, 2004).

Teres Major

"The relationship of the teres major to the three axes of the shoulder is the same as that of the latissimus dorsi and is frequently called the latissimus dorsi's 'little helper.' This conception has been largely in error, however, because the teres major contracts only when resistance has been applied to the arm and only when positions are reached and held in the ranges of movement in adduction, internal rotation and extension" (Gench et al., 1995, p. 75).

Coracobrachialis

This has limited effectiveness and assists with adduction.

Triceps—Long Head

As previously discussed, the triceps is ineffective at the glenohumeral joint and generally is seen to assist with adduction.

Deltoid—Posterior

Deltoid assists with adduction.

Lateral (External) Rotators of the Humerus

Lateral rotators of the humerus include deltoid, supraspinatus, infraspinatus, and teres minor. Infraspinatus and teres minor are closely related in location and action to externally rotate and adduct the glenohumeral joint as they both pass posterior to long axis of the shoulder joint regardless of the position of the humerus to become external rotators and horizontal abductors. That they are not also extensors, adductors, or abductors is explained by the fact that they pass directly over both the sagittal and frontal axes of the shoulder (Gench et al., 1995; Smith et al., 1996).

It is unusual to have isolated weakness in the infraspinatus muscle, but, should this occur, there would be a significant reduction in the strength of lateral rotation. Weakness in teres minor would only result in a slight decrease of lateral rotation. A tight infraspinatus muscle results in decreased range of motion of shoulder medial rotation and horizontal adduction, while tightness in the teres minor muscle is unlikely and would probably accompany infraspinatus tightness, limiting medial rotation (Oatis, 2004).

Palpation of Infraspinatus and Teres Minor

Some parts of these muscles are covered by trapezius and posterior deltoid, but most parts are superficial and can be palpated. Have the subject lie prone or stand with the trunk inclined forward and the arm hanging vertically. Find the margin of posterior deltoid, and place your fingers below the deltoid on the scapula near the lateral margin. Have the subject externally rotate, and these two muscles will rise under the fingers with teres minor next to infraspinatus but farther from the spine.

Deltoid—Posterior Fibers

Along with extension and horizontal abduction, the posterior fibers of the deltoid muscle externally rotate the humerus.

Palpation

It is possible to feel the antagonistic characteristics of the anterior and posterior fibers. Have the subject keep his or her arm in humeral adduction, and have the person medially and laterally rotate against resistance. The anterior fibers will contract upon medial rotation, and the posterior fibers will be taut during lateral rotation (Biel, 2001).

Supraspinatus

Due to the distal attachment of the posterior portion on the greater tubercle, the line of pull of this muscle is slightly posterior to the long axis of the shoulder joint, which makes this muscle an external rotator.

Medial (Internal) Rotators of the Humerus

Subscapularis

Due to the location of this muscle (medial to the long axis and wrapping around the anterior aspect of the upper humerus), subscapularis has a medial rotation action. There would be a significant decrease in shoulder medial rotation due to weakness in subscapularis, and this would contribute to anterior instability of the glenohumeral joint. Weak subscapularis and teres major with tight infraspinatus and teres minor would result in anterior humeral translation (Magee, 2002). When the subscapularis is tight, there is decreased external rotation of the shoulder (Oatis, 2004). Subscapularis has been identified as being the muscle providing the greatest resistance to posterior subluxation of the humerus (Blasier, Soslowsky, & Malicky et al., 1997).

Pectoralis Major

Because the fibers are anterior relative to the long axis of the glenohumeral joint, the pectoralis major is effective as an internal rotator. A short pectoralis major and/or latissimus dorsi muscle would result in decreased lateral rotation (Magee, 2002).

Latissimus Dorsi

The line of pull of this muscle is medial to the long axis when in the anatomical position, so the latissimus dorsi medially rotates as it adducts, flexes, or extends.

Deltoid—Clavicular Fibers

The clavicular fibers rotate the humerus as it flexes.

Teres Major

The teres major muscle medially rotates with resistance against the arm but is weak for pure medial rotation. It has been hypothesized by Oatis (2004) that weakness in the teres major muscle would result in limited shoulder medial rotators, extensor, and hyperextension. Tightness would be expected to result in restricted range of motion in shoulder lateral rotation, flexion, and abduction, which might contribute to a posture of rounded shoulders.

Biceps (short head) and coracobrachialis muscles have also been identified as weak medial rotators. A summary of the muscles producing movement of the scapula and humerus are shown in Table 5-17.

EVALUATION OF THE SHOULDER

Evaluation of the shoulder is complicated by several factors. First, many articulations at the shoulder are required to produce the wide variety of movements. Evaluation of the acromioclavicular, scapulothoracic, sternoclavicular, and gle-

Table 5-17

SUMMARY OF MUSCLES OF THE SCAPULA AND HUMERUS

JOINT MOTION	MUSCLES
SCAPULA	Trapezius, upper
Elevation	Levator scapulae
	Rhomboids, major and minor
Depression	Trapezius, lower
	Latissimus dorsi
	Pectoralis major
	Pectoralis minor
Protraction	Serratus anterior
	Pectoralis major and minor
Retraction	Trapezius, upper and lower
	Rhomboids, major and minor
Upward rotation	Serratus anterior
	Trapezius, upper and lower
Downward rotation	Levator scapulae
	Rhomboids, major and minor
	Pectoralis minor
	Latissimus dorsi
SHOULDER	Deltoid, anterior
Flexion	Coracobrachialis
	Pectoralis major
	Biceps
Extension	Deltoid, posterior
	Teres major
	Latissimus dorsi
	Pectoralis major
	Coracobrachialis
	Infraspinatus
	Triceps
Abduction	Deltoid
	Supraspinatus
	Biceps
	Infraspinatus
Adduction	Pectoralis major
	Teres major
	Latissimus dorsi
	Latissimus dorsi

(continued)

Table 5-17 (continued)

SUMMARY OF MUSCLES OF THE SCAPULA AND HUMERUS

JOINT MOTION	MUSCLES
External rotation	Infraspinatus
	Teres minor
	Deltoid
	Supraspinatus
Internal rotation	Subscapularis
	Pectoralis major
	Latissimus dorsi
	Teres major
	Deltoid
	Biceps

nohumeral, as well as movement along ribs and thorax, are an essential part of this evaluation. Posture and core stability of the client needs to be part of the assessment, as does an evaluation of muscle patterning, structural laxity, and proprioception.

Second, many muscles add not only mobility to the joint, but also add to the stability. The function of these muscles varies depending upon the amount of elevation of the humerus, the load in the distal extremity, the axis of movement, and the position of the scapula. Any interruption in the coordinated synchrony of these muscles and their actions will interfere with shoulder function.

Evaluation of the shoulder should start with an understanding of the client and client priorities. This can be done through an interview with the client and family and an understanding of the client's occupational profile. Observation and palpation of the shoulder region would then be followed by an assessment of joint movement, muscle strength, joint stability, and pain.

Occupational Profile

As with any occupational therapy assessment, the evaluation should start with chart review, pertinent history, and discussions with the client and family. Although a person's diagnosis may be specific to the shoulder, it is important that the therapist not be so focused on the shoulder joint that other areas of involvement may be missed. As an occupational therapist, the value of the evaluation is not only on the specific limitations of shoulder girdle musculature and dynamics, but also on how these limitations prevent the client from engaging in valued occupations.

Specific information that should be collected from the client answers the following open-ended questions:

- Who is the client?
- Why is the client seeking services?
- What is the client's concern about engagement in occupations?
- What areas of occupation are successful and which are not?
- What contexts support engagement in occupation?
- What is the client's occupational history?
- What are the client's priorities? (American Occupational Therapy Association, 2008)

The answers to these questions, in addition to an assessment of movement and shoulder structures, will enable a client-centered intervention plan with meaningful outcomes for the client. The outcome of occupational therapy is to return the client to participation in meaningful activities, so assessment of the shoulder is needed but also a careful analysis of the tasks required by the client in daily activities. Treatment would then be focused toward maximizing shoulder range of motion, strength, and endurance and minimizing pain by using activities related to the client's chosen occupations.

Observation and Palpation

Much information can be gathered by looking at the symmetry and alignment of shoulder joint structures. Note the characteristics of the joints, and visualize the anatomy involved in joint movements. Pay attention to the rhythm, smoothness, and patterns of motion, watching for any nonverbal indications of pain (wince, grimace, etc.). Note any skin discoloration, swelling, abrasions, or masses, and note the relationship of the neck, shoulder, and thorax. Observe the shoulder girdle from all sides and viewpoints.

Palpation should be done to discriminate differences in muscle tension to assess if there is effusion, edema, or muscle spasm as well as muscle tone (spasticity, rigidity, flaccidity). Palpation can help distinguish differences in tissue texture, thickness, direction, and shapes of structures and tissue types. By using the back of the hand or fingers, variations in temperature can be ascertained. Tissue integrity should also be assessed by looking at the condition of the tissues, amount of moisture, ecchymosis (bruising), and any abnormal sensations (Magee, 2002).

By having a client engage in activities relative to areas of meaningful occupation, observation of active, coordinated patterns of movement, compensatory and substitution movements, and indications of the amount of pain the motion elicits can be obtained. A sense of the client's cooperation is gained and a sense of the amount of functional loss can be suggested by watching the client move actively.

Posture

General posture and symmetry can influence shoulder motion and is an important part of the shoulder assessment. Poor joint position sense and balancing mechanisms are predisposing factors to muscle patterning disorders (Lewis, Kitamura, & Bayley, 2004). The position of the neck and head determine how well the arm, wrist, and hand will work. The neck and shoulder are dynamic aspects of daily movement, but are often required to act as a static base, while the hands perform skilled tasks. Spinal alignment influences scapular position, and both spinal alignment and scapular position influences shoulder function (Kebaetse, McClure, & Pratt, 1999). This is due to the numerous muscular connections between the spine, scapula, clavicle, and humerus and is due to the need for coordinated scapulothoracic rhythm for full shoulder movement.

Often, tasks (such as working on a computer) require a forward position of the head, which causes a shift of weight that makes the neck and upper back work harder with decreased control and circulation to the arms. Positions of sustained elevation of the arms may cause supraspinatus tendonitis due to compression of the humeral head against the coracoacromial arch or bicipital tendonitis due to repeated friction between the synovial sheath of the long head and lesser tuberosity.

In the position of rounded shoulders and a forward head position ("slouch"), the scapula is more elevated during 0 degrees to 90 degrees of abduction, and between 90 degrees and 180 degrees, there was less scapular posterior tilt. This resulted in less active shoulder abduction and a decrease in horizontal muscle force. By assuming this position, the normal scapulohumeral relationship was altered (Kebaetse et al., 1999).

Observable deformities, such as squaring of the shoulder (suggestive of an anterior dislocation) or extreme scapular "winging" (associated with shoulder instability and serratus anterior or trapezius dysfunction or scapulothoracic dysfunction) or rounded shoulder (possibly due to short pectoralis minor muscle or lengthened post rotator cuff muscles), can be seen. Look for both shoulders to be at the same height and for symmetrical muscle development. Elevated shoulders may

be due to a shortened levator scapula, upper trapezius and rhomboids resulting in scapular elevation, scapulothoracic dysfunction when flexing or with upward rotation. A lengthened upper trapezius, resulting in scapulothoracic dysfunction with upward rotation will be seen in a depressed scapula. If the scapula is downwardly rotated, this can result in scapular dysfunction with upward rotation due to shortened levator scapulae, shortened rhomboids or lengthened trapezius or serratus anterior (Kendall, McCreary, Provance, Rodgers, & Romani, 2005).

Scapula

Assessment of the scapula would include "any observable alterations in the position of the scapula and the patterns of scapular motion in relation to the thoracic cage" (Kibler & McMullen, 2003, p. 143). By observation, one may note leg-length discrepancies or hip rotational abnormalities and the degree of lumbar lordosis, which would result in lack of symmetry between the two scapulae. Cervical lordosis affects scapular retraction and protraction. Observe for excessive winging, tipping, elevation, or rotation. In chronic scapular dyskinesis, winging may be seen in the resting position (Kibler, 1998).

Scapula dyskinesis (differences in position and/or motion of the two scapulae) interferes with the closed kinetic chain coupling of scapula motion with humeral motion (Kibler et al., 2002). Most frequent dyskinesis is due to changes in muscle coordination between upper and lower trapezius and between rhomboids and serratus anterior (Kibler et al., 2002). Kibler and colleagues (2002) developed a classification system based on observations of the area of the scapula that is visually prominent during evaluation. Type I is characterized by the prominence of the inferior angle of the scapula with the arm at rest. There is a posterior tilting of the inferior angle around a horizontal axis through the scapula (Kibler et al., 2002). Type II is characterized by a prominent medial border at rest. Upon moving, the medial scapular border tilts off the thorax. In type III, the superior border of the scapula is elevated, and the scapula may be anteriorly displaced. In type IV, both scapulae are relatively symmetrical. The scapulae rotate symmetrically upward, and the medial border of the scapula remains against the thoracic wall (Kibler et al., 2002). Use of this classification system can make observational descriptions of scapular function more consistent.

Scapulothoracic Motion

Observe scapulothoracic motion during humeral elevation. This would include the movements of upward rotation of the scapula, external rotation of the humerus, and posterior tilting. In the standing position, compare the outlines of each scapula for bilateral symmetry regarding scapular position and prominence (Ellenbecker & Ballie, 2010). Standing behind the client, have the client place his or her hands on his or her hips while you observe the movement of the scapula through concentric elevation and eccentric lowering (Ellenbecker & Ballie, 2010). Shoulder pain is often aggravated by humeral motion, particularly abduction.

An assessment of muscular forces involved in scapulothoracic motion would include discerning if the upper trapezius is most active with abduction less than 60 degrees; that lower trapezius is active when abduction is greater than 90 degrees;

that middle trapezius is most active during abduction at 90 degrees; that serratus anterior is most active in forward flexion; and that rhomboids are most active during flexion and abduction end range (Uhl et al., 2009).

Humerus

When observing the humerus, a sharp change in the deltoid contour suggests dislocation (Shafer, 1997; Woodward & Best, 2000b). This might result in the ability to passively or actively externally rotate the affected arm.

A loose glenohumeral joint may result from repeated subluxations without dislocation. There may be an audible "click" as the head of the humerus glides back into the glenoid fossa. Pain or clunking sounds with overhead motion may also be indicative of labral disorders (Woodward & Best, 2000b). Young clients with glenoid-rim fractures or labrum tears often retain residual capsule weakness and excessively wide ranges of motion, further encouraging future dislocations.

Acromioclavicular Joint

Forces that are directed toward separating the clavicle from the scapula will cause severe sprain to the acromioclavicular, coronoid, and trapezoid ligaments unless the clavicle fractures first. Because the acromioclavicular ligament is a part of the acromioclavicular joint's capsule, a sprain would also involve a capsular tear to some degree. Repeated trauma to this joint would result in contusion, sprain, separation, and post-traumatic arthritis (Shafer, 1997). Minor sprains are characterized by minimal local swelling and tenderness, moderate pain when moving, but no limitations in joint mobility. A major sprain consists of stretching and tearing of the coracoclavicular ligaments, and there will be acute tenderness and swelling near this ligament and below the clavicle. There will be abnormal mobility of the clavicle relative to the acromion, and a subcutaneous discoloration will appear after a week or more post-injury (Shafer, 1997).

Neither the acromioclavicular nor sternoclavicular can be moved by voluntary action, so dynamic assessment includes evaluation of the small but essential joint play at both the lateral and medial ends of the clavicle. The client needs to be fully relaxed, and joint play is elicited as a slight inferior and superior glide.

Clavicle

A freely moveable clavicle is necessary for full shoulder function. The clavicle needs to pivot and rotate, and limitations at either the acromioclavicular or sternoclavicular joints will limit glenohumeral motion.

A fracture or dislocation at either end of the clavicle would result in changes in the contour of the shoulder. Look for asymmetrical roundness or fullness of the anterior deltoid and an exaggerated, rounded shoulder (Shafer, 1997). When the tip of the clavicle fractures, bony fragments can be felt under the skin, and often the client is unable to raise the involved arm above 90 degrees. The most common site of clavicle fracture is near the midpoint or outer third of the clavicle that may also result in inferior, anterior, and medial displacement of the lateral clavicle. If this injury is the result of a fall on an outstretched hand, all structures within the kinematic chain need to be evaluated because the force of impact is transferred from the palm to the carpal bones, to the radius and ulna, to the elbow, the humerus, acromioclavicular joint, and sternoclavicular joint (Shafer, 1997).

Clavicle dislocations, which can occur because of injuries in football, soccer, horse racing, bicycling, gymnastics, wrestling, and unusual accidents at work or in the home, are not as common as fractures. If the injury is to the lateral clavicle, the bone will be elevated with an increase in the distance between the clavicle and the coracoid process. This is the palpable and visible "step" deformity. The scapula will be separated from the clavicle, the acromion will lie below and anterior to the clavicle, and fracture of the coracoid process often is associated with clavicle dislocations (Shafer, 1997).

Subluxation of the clavicle will be felt as a characteristic down step. The clavicle tends to displace superiorly and anteriorly as a common result from falls, blows, and contact injuries. Often, the clavicle subluxation is accompanied by joint ligament separations.

Localized swelling at the tip of the shoulder near the lateral aspect of the clavicle is a common site of painful and tender contusions to the trapezius. You might observe shoulder girdle depression in an attempt to alleviate the pain associated with the contusion. These symptoms may be similar to acromioclavicular separation, so careful evaluation is warranted.

Range of Motion

Because the complex series of articulations of the shoulder allows a wide range of motion, the affected extremity should be compared with the unaffected side to determine the patient's normal range. Active and passive ranges should be assessed. For example, a patient with loss of active motion alone is more likely to have weakness of the affected muscles than joint disease. By observing how the person moves both actively and passively, an understanding of synchronous action of the integrated parts of the shoulder girdle can be determined. Compensatory movements can be observed as clients substitute these motions for impaired function. Information can be obtained as to the stability and symmetry of joint structures.

Active Range of Motion

Evaluation of active range of motion (AROM) is an assessment of voluntary and physiologic movement. When assessing AROM, look at both the quality and quantity of movement. It would involve elevation of the arm (forward flexion and abduction), horizontal abduction and adduction, and internal and external rotation (Wilk, Arrigo, & Andrews, 1997b). Also, it is necessary to differentiate between the scapular and glenohumeral motions because movements of the scapula may compensate for restrictions in glenohumeral motion (Magee, 2002). Because tension produces joint restrictions, client relaxation is necessary for an accurate assessment. If active motion is normal, assessment of passive motion is usually not necessary. As with palpation, visualizing the underlying anatomy is important to assessing joint motion.

If there is uncomplicated muscle weakness, passive range of motion may be normal but active may be limited. Spasm, contractures, fracture, and dislocation are the common causes of motion restriction and muscle weakness. Active and passive restriction is likely from a bone or soft tissue block, and any atrophy will likely be from disuse (Magee, 2002).

Assessment of Active Range of Motion

Elevation, depression, abduction, adduction, extension, flexion, internal rotation, and external rotation are the basic movements of the shoulder girdle. Other movements normally tested are scapular retraction (military position of attention) and shoulder protraction (reaching). The patient may be in either the standing or sitting position during testing.

Elevation and Depression

Elevation and depression are checked by having the patient hunch the shoulders and return to the normal position.

Abduction/Adduction

Full active bilateral abduction is tested by having the patient abduct the arms horizontally to 90 degrees while keeping the elbows straight and the palms turned upward then continuing abduction in an arc until the hands meet in the middle over the head. Shoulder abduction involves the glenohumeral joint and the scapulothoracic articulation. Glenohumeral motion can be isolated by holding the patient's scapula with one hand while the patient abducts the arm. The first 20 degrees to 30 degrees of abduction should not require scapulothoracic motion.

If impairments or pain occur in the first phase of shoulder elevation, this may indicate severe restrictions in the glenohumeral joint. It may also indicate restriction of the sternoclavicular joint in rare cases (Donatelli, 1997). If the client reports acute pain above 110 degrees, this may be indicative of acromioclavicular arthritis. If the client is able to hold a position of horizontal abduction above 90 degrees but not below, this may indicate a rotator cuff tear, and if there is pain throughout the full range of horizontal abduction, this may suggest bicipital tendonitis or bursitis (Shafer, 1997).

Pain in a particular part of the arc of movement can occur for many reasons (acromion neoplasm, capsule laxity, cervical disk lesion, cervical subluxation syndrome, infraspinatus tendonitis, intracapsular bicipital tendon, subdeltoid bursitis, subscapular tendonitis, or supraspinatus tendonitis) (Shafer, 1997). A painful arc may also be observed if the client is able to abduct the arm 45 degrees to 60 degrees with little difficulty, but between 60 degrees and 120 degrees of abduction, subacromial bursa and rotator cuff tendon insertions (supraspinatus in particular) may become impinged. Once past 120 degrees, the tissues are not compressed beneath the acromion process. A second painful arc occurs in between 160 degrees and 180 degrees, caused by pathology in the acromioclavicular joint (lesion, impingement syndromes) (Magee, 2002).

Flexion/Extension

Flexion and extension are assessed by seeing if the patient can bring the humerus up or down in the sagittal plane and frontal axis.

Internal/External Rotation

Rotation movements are assessed by seeing how the client can move in the horizontal plane and vertical axis. Internal rotation is the motion one uses to fasten a bra or tuck a shirt into pants (Gutman & Schonfeld, 2003). External rotation is the motion needed when one is braiding the hair (Gutman & Schonfeld, 2003). If there is acute pain that is increased by humeral rotation, synovitis may be the cause (Shafer, 1997).

Protraction/Retraction

Protraction and retraction are movements of the scapula toward or away from the vertebral column. When one brings the arms to the center of the body in a horizontal plane and vertical axis, this produces protraction. When you bring your arms back behind you in a horizontal plane, this is retraction.

Combined Movements

External rotation and abduction can be tested bilaterally at the same time by having the patient place both hands behind the neck with interlocking fingers, and then the elbows, which are initially pointing forward, are moved laterally and posteriorly in an arc (Magee, 2002). Active external rotation and abduction are easily tested by having the patient reach up and over the shoulder and attempt to touch the spinal border of the opposite scapula (Magee, 2002). Abduction, flexion, and external rotation or adduction combined with extension and internal rotation are needed to comb hair, zip a back zipper, or reach for a wallet in the back pocket. With the arm internally rotated (palm down), abduction continues to 120 degrees. Beyond 120 degrees, full abduction is possible only when the humerus is externally rotated (palm up).

Restrictions in the second phase of elevation, the most common phase for shoulder dysfunction, may be due to limitations in acromioclavicular or sternoclavicular joints limiting scapular rotation due to restricted clavicular elevation and rotation. Limited scapulothoracic rotation may be due to weakness in the levator scapulae, serratus anterior, and/or upper and lower trapezius muscles. Pathological winging of the scapula would occur in this phase. In the last phase of shoulder elevation, much flexibility is required of the muscles (teres major, subscapularis, pectoralis major, latissimus dorsi, teres minor, and infraspinatus) involved in the primarily glenohumeral movements of the final elevation.

Internal rotation and adduction can be tested as a combined motion by having the patient reach across his or her chest (internal rotation) and, keeping the elbow as close to the chest as possible (adduction), touch the opposite shoulder. Another method is to have the patient reach behind the back and attempt to touch the bottom angle of the opposite shoulder blade.

The Apley scratch test (Figure 5-39) is a useful maneuver to assess abduction, external rotation, internal rotation, and adduction shoulder range of motion. In this test, the client is asked to reach behind the back and touch the bottom of the opposite scapula (internal rotation and adduction) and then to reach behind the head and touch the top of the opposite scapula (abduction and external rotation). The inability to perform these maneuvers indicates loss of AROM and possible rotator cuff muscle limitations.

Table 5-18 describes a screening test for AROM of the shoulder. You can also observe the client during functional activities to ascertain visually the amount of AROM. For example, grooming tasks such as hair care and facial hygiene

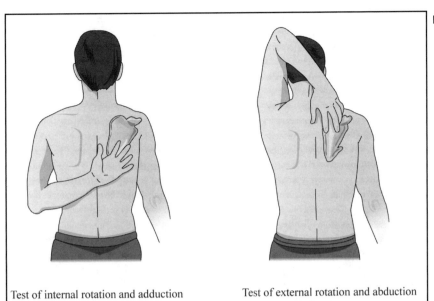

Figure 5-39. Apley scratch test.

Test of internal rotation and adduction   Test of external rotation and abduction

Table 5-18

ACTIVE RANGE OF MOTION SCREEN FOR THE SHOULDER

Ask the client to shrug his or her shoulders up toward the ear.	Scapular elevation
Ask the client to squeeze his or her shoulder blades together.	Scapular retraction
Ask the client to raise his or her arm in front of his or her body to shoulder height. Then, ask them to reach forward.	Scapular protraction
Ask the client to raise his or her arm straight out in front of his or her body and raise the arm toward the ceiling.	Humeral flexion
Ask the client to raise his or her arm out to the side and up to touch his or her ear.	Humeral abduction
Ask the client to place his or her hands behind his or her head or to the back of the neck with elbows pointed out.	Humeral external rotation
Ask the client to place his or her hands behind the back.	Humeral internal rotation
Ask the client to bring his or her arm across the chest to the opposite shoulder.	Horizontal adduction

require shoulder abduction and flexion. Adjusting a blouse or tucking in a shirt into pants in the back requires internal rotation, and adjusting a collar on a shirt in the back may involve shoulder adduction and external rotation (Constant & Murley, 1987; Magermans, Chadwick, Veeger, & vanderHelm, 2005).

Active range of motion within normal limits is not often

required for many activities of daily living. Humeral abduction values range from 30 degrees to 45 degrees for perineum care to 105 degrees to 120 degrees for combing hair (Veeger, Magermans, Nagels, Chadwick, & vanderHelm, 2006); 110 degrees to 125 degrees of shoulder flexion is needed to reach the hand behind the head, but only 60 degrees to 90 degrees is needed to reach the opposite shoulder. Eating requires approximately 70 degrees to 100 degrees of horizontal adduction and 45 degrees to 60 degrees of abduction (Matsen, Lippitt, Sidels, & Harryman, 1994). Full range of motion is needed for the rotational movements of internal and external rotation in many daily activities such as combing hair or tucking in a shirt (Harryman, Clark, McQuade, Gibb, & Matsen, 1990; Magee, 2002).

Dynamic flexibility is the resistance of tissue to movement. How muscles respond to movement is related to age, gender, levels and types of activities, tissue temperature, heredity, injury, and pain. Flexibility screens can be done to test how individual muscles respond to movement. For example, with your client in supine, by passively abducting and externally rotating the arm to 140 degrees, you can test the flexibility of the pectoralis major muscle (sternal portions). In the same position, moving the client passively to 90 degrees of horizontal abduction and external rotation, you could test pectoralis major clavicular fibers. In each case, if the client cannot reach the frontal plane, this would indicate a flexibility deficit in the pectoralis major muscle. Other specific tests have been used for pectoralis minor, subscapularis, internal and external shoulder rotators, latissimus dorsi, upper trapezius, and levator scapulae muscles.

More commonly, shoulder flexibility tests may include holding a stick (or dowel or golf club) out in front of you with both hands on either end of the stick. Bring the stick up and overhead, and then repeat by moving the stick behind you and up. Another flexibility test is to raise your right arm overhead. Bend your right elbow, and let your palm rest on the back of your neck. Then, slide it between your scapulae. Then, reach behind you so that the back of your hand rests on the middle of your back. Reach back with both hands to the middle of your back, and touch the fingers of both hands. Excellent flexibility is when fingers overlap, good is when fingers touch, average is when fingers are less than 2 inches apart, and poor is when there is more than 2 inches between fingers.

Validity and Reliability of Active Range of Motion of the Shoulder

Reliability of shoulder range of motion measurement is affected by many factors, including the complexity of the movement, variations in joints, active versus passive motion, which joint is tested, variations among clients, and testing by the same or different examiners (Gajdosik & Bohannon, 1987; Hayes, Walton, Szomor, & Murrell, 2001). In addition, different tools have been used to measure shoulder range of motion, including goniometry, visual estimation, high-speed cinematography, still photography, and functional motion tests (Hayes et al., 2001). Of the different tools used, there is fair to good reliability for visual estimation, goniometry, still photography, and the stand and reach test for shoulder motions of flexion, abduction, external rotation, and overhead reach (Hayes et al., 2001).

Both intra-rater and inter-rater reliability needs to be considered. Intra-rater reliability is higher than inter-rater (Gajdosik & Bohannon, 1987; Sabari, Maltzev, Lubarsky, Liszkay, & Homel, 1998). Intra-rater reliability interclass correlation co-efficients (ICC) for the measurement of shoulder motion ranged from 0.87 to 0.99, indicating good intra-rater reliability. Inter-rater reliability was good for passive range of motion for shoulder flexion abduction and external rotation (ICC values ranged from 0.84 to 0.90), but less for shoulder horizontal abduction and adduction, extension, and internal rotation (ICC values ranged from 0.25 to 0.55) (MacDermid, Chesworth, Patterson, & Roth, 1999; Riddle, Rothstein, & Lamb, 1987). Overall, inter-rater reliability is better for the upper extremity than for the lower (Hellebrandt, Duvall, & Moore, 1949).

Passive range of motion is seen to be more difficult to measure reliably because of the stretch of soft tissues and the control of force required, which is difficult to measure and must be carefully controlled (Gajdosik & Bohannon, 1987). Also, complex passive movement is more difficult than simple passive movement, and this affects reliability. Consider the difference between the measurement of passive range of motion of the elbow and that of the shoulder in terms of complexity of movement, and issues of reliable measurement become more apparent.

The validity of shoulder range of motion measurements may be affected by edema, pain, strength deficits, and muscle hypertrophy and patient characteristics (Gajdosik & Bohannon, 1987). The validity of the measurement is based on the clinician's knowledge of anatomical structures, experience in visually inspecting and palpating the skeletal landmarks, and accurately aligning and reading the goniometer (Gajdosik & Bohannon, 1987). Of critical consideration in both reliability and validity is having the same examiner measure the range of motion of the same client, using the same positions and standardized techniques in an optimal testing environment.

Age and gender were found to affect range of motion values for the shoulder internal and external rotation. Women were more flexible and had more range of motion in a seated position during external rotation. The reason posited for the difference is that muscle bulk may limit flexibility in stronger individuals (Roy, MacDermid, Boyd, Faber, Drosdowech, & Athwal, 2009). In a study by Walker and colleagues (1984), women had greater range of motion in all shoulder motions as compared to men. Age was also found to be a factor in a study by Roy and colleagues (2009) when measuring external rotation and internal rotation. Decreased internal rotation on the dominant side was found in people older than 60 years as compared to those 18 to 39 years of age, and there was a negative correlation between age and external rotation in a supine testing position (Roy et al., 2009).

Passive Range of Motion

An assessment of passive range of motion (PROM) can provide different information about the shoulder joint than AROM. When moved passively, there is a pulling of the joint capsule and antagonist muscles and ligaments. An evaluation of PROM looks at joint play and accessory movements possible at each joint. The joint is positioned in the resting position (or the loose-packed position) so that there is minimal congruency of articular surfaces and the joint capsule to enable slide, spin, and rolling motions. PROM should not be performed if your client is at risk for fractures (e.g., osteoporosis), dislocation, tears, or advanced bone pathology (Magee, 2002).

The application of overpressure is helpful in determining if the end feel of the joint is pathological or physiologically normal. With passive movement, bone blocks will feel like an inflexible obstacle that abruptly halts movement, while extra-articular soft-tissue blocks will be less abrupt and slightly flexible upon additional pressure (Magee, 2002).

With PROM, end-feel sensations in the shoulder can be helpful in suggesting joint dysfunction. In a normally firm end-feel joint, instability may be considered if the end feel exhibited is that of a spasm or empty end feel. If there is excessive external rotation with soft end feel, this may suggest hypermobility and hyperelasticity, which can lead to anterior glenohumeral displacement (Wilk et al., 1997a).

A screening test for PROM would be the same as described for AROM with the exception that the client is not producing the movement and the therapist is moving the arm. All shoulder motions should be assessed. If there is mild to moderate pain when you are assessing the shoulder in all directions, adhesive capsulitis might be suspected. If crepitus is noted with acutely painful abduction above 110 degrees, acromioclavicular arthritis may be the reason. If the joint elicits acute pain at all levels of abduction, there may be joint synovitis (Shafer, 1997). Limited passive motion without a capsular pattern may be due to a variety of deficits, including acromioclavicular sprain, capsule adhesions, costocoracoid contracture, first rib fracture, pulmonary neoplasm, subacromial bursitis, or subdeltoid bursitis (Donatelli, 1997; Ellenbecker & Ballie, 2010; Shafer, 1997).

Joint play movements, an important part of normal shoulder motion, are also evaluated passively, and there is a comparison of the available motion and end feel of both sides of the body. Commonly, the following five joint-play movements are tested: Lateral glide of the humerus relative to the glenoid cavity; medial glide of the humerus relative to the glenoid cavity; anterior glide of the humerus relative to the glenoid cavity; posterior glide of the humerus relative to the glenoid cavity; and downward separation from the glenoid cavity (Shafer, 1997).

In assessing posterior glide of the humerus, the client is supine. With one of the examiner's hands over the anterior humeral head and the other around the humerus above the elbow, a backward force is applied (Magee, 2002). A posterior glide assessment can occur during external rotation and abduction of the humerus (Shafer, 1997). Anterior glide would entail a forward force against the client's arm while controlling for rotation. Lateral, upward, and downward glides can also be used to assess additional glenohumeral joint play movement. Acromioclavicular and sternoclavicular joint play can be assessed by gently grasping the clavicle as close to the joint as possible and moving it in and out or up and down, which may be uncomfortable for the client, so prior warning to the client and careful observation of nonverbal client cues is warranted (Magee, 2002). In addition, passive movement of the scapula should also be assessed as it moves posteriorly, medially, laterally, caudally, cranially, and away from thorax (Magee, 2002). Passive movement of the scapula can be tested by having the

therapist stand behind the client, stabilize the scapula with one hand, and move the client's arm through shoulder movements.

If both active and passive range of motion are restricted, and passive mobility is limited in a capsular pattern of external rotation, abduction, and internal rotation, adhesive capsulitis may be suspected (Donatelli, 1997). The client may also describe a vague, dull pain over the deltoid, which increases with motion and disturbs sleep. This is also known as frozen shoulder, periarthritis, stiff and painful shoulder, periarticular adhesions, Duplay's disease, scapulohumeral periarthritis, tendonitis of the short rotators, adherent subacromial bursitis, painful stiff shoulder, bicipital tenosynovitis, shoulder portion of shoulder-hand syndrome, bursitis calcarea, supraspinatus tendonitis, and periarthrosis humeroscapularis. By observation, the client may have a stooped posture with rounded shoulders and tends to hold the involved extremity in adduction and internal rotation. Donatelli suggests that evaluation should include active and passive elevation of the shoulder, passive scapulohumeral abduction, passive lateral and medial rotation, resisted abduction and adduction, resisted internal and external rotation, and resisted elbow extension and elbow flexion (Donatelli, 1997).

Capsular pattern problems occur in characteristic proportions and have effects throughout the entire upper extremity. In the shoulder, the greatest limitations that occur due to capsular impairments are seen in limitations in external rotation, then in abduction, followed by impairments in internal rotation and flexion. If the capsular pattern does not occur in this order, a noncapsular problem may be suspected. Magee (2002) states that the end feel of capsular tightness is different from the tissue stretch end feel of muscle tightness in that capsular tightness has a more hard, elastic feel. Using end feel as an indication of tissue function is useful clinically, but, due to individual variation, the reliability may be questionable (Donatelli, 1997). Limited passive motion that occurs in a capsular pattern may be due to arthritis, bone blocks, hemarthrosis, hemiplegia, neoplasm, neuropathic arthropathy, complex regional pain syndrome, or systemic lupus erythematosus (Donatelli, 1997; Ellenbecker & Ballie, 2010; Shafer, 1997).

The location of capsular tightness and the effects on movement are shown in Table 5-19. For example, if there is tightness in the posterior capsule, this results in excessive anterior and superior movement in the glenohumeral joint. Treatment would focus on inferior and posterior mobilization techniques at the glenohumeral joint (Uhl, 2009).

Stability Assessment

The shoulder joint is not an inherently stable joint. There are two reasons for instability: structural causes (capsule and labral damage due to injury or repetitive microtrauma) or unbalanced muscle recruitment around the shoulder (as opposed to muscle weakness) so the humeral head is displaced on the glenoid (Jobe, Kvitne, & Giangarra, 1989; Kvitne & Jobe, 1993; Matsen et al., 1991). "Trick" movements, or instability that is voluntarily controlled and deliberate, are due to unbalanced muscle action, but there is also involuntary positional instability in which the joint subluxes or dislocates during normal movement, which can also be caused by unbal-

Table 5-19

PATTERNS OF CAPSULAR TIGHTNESS AND THE EFFECT ON MOVEMENT

LOCATION OF TIGHTNESS	EFFECT ON MOVEMENT
Posterior	Horizontal adduction decreased
	Medial rotation decreased
	End range of flexion decreased
	Decreased posterior glide
	Weak ER
	Weak scapular stabilizers
Posteroinferior	Elevation interiorly
	Medial rotation of elevated arm decreased
	Horizontal adduction decreased
Posterosuperior	Medial rotation limited
Anterosuperior	Flexion end range limited
	Extension end range limited
	External rotation limited
	Abduction end range decreased
	Decreased posteroinferior glide
	Increased night pain
	Weak rotator cuff
Anteroinferior	Abduction decreased
	Extension decreased
	External rotation decreased
	Increased posterior glide

Adapted from Magee, D. J. (2002). *Orthopedic physical assessment*. Philadelphia, PA: Saunders.

anced muscle action or incongruency of joint surfaces (Lewis et al., 2004).

In trying to classify and describe instability, a distinction is often made between laxity and instability with the understanding that these terms are not synonymous. Laxity is considered a degree of translation in the glenohumeral joint that falls within a physiologic range and is asymptomatic. This is the ability of the humeral head to be passively translated on the glenoid fossa. Instability is characterized by abnormal motions, which result in pain, subluxation, or dislocation. Instability is a clinical condition in which there is unwanted translation of the humeral head on the glenoid, which compromises the comfort and function of the shoulder (Matsen et al., 1991). Further definitions of terms include subluxation, which is a symptomatic separation of the joint surfaces without disloca-

tion (which is a complete separation of joint surfaces) (Lewis et al., 2004).

Instability can be defined based on the degree of instability, onset (traumatic/atraumatic, overuse), severity, and direction. Multidirectional instability is characterized by a loose feeling in the joint with tingling, paresthesia, weakness of arm, and diffuse, poorly localized pain. In contrast, in clients with unidirectional instability, the clients report more localized pain and discomfort that occurs only in specific arm positions (Wilk et al., 1997a).

There are several systems of classification of joint instability with misunderstandings of many of the terms, different combinations of pathologies, and changes in classification through time (Lewis et al., 2004). One system classifies instability based on the direction of the instability. This includes Bankart lesions, which are unidirectional and multidirectional. The Rockwood classification is based on whether the instability is due to trauma with and without previous dislocation or is atraumatic with voluntary or involuntary subluxation.

The Thomas and Matsen classification system classifies instability as either TUBS or AMBRII. TUBS refers to a traumatic event that gives rise to a unidirectional anterior instability with a Bankart lesion, which requires surgery. The glenohumeral joint has lost the stabilizing effect of the inferior glenohumeral ligament complex and deepening of glenoid due to anterior glenoid labrum damage with the arm in abduction, extension, and external rotation. In the TUBS classification, the tissue is traumatic, "torn loose" with a Bankart lesion the pathology, resulting in unidirectional instability requiring surgery. AMBRII syndrome describes someone who was "born loose": atraumatic onset of multidirectional instability that is accompanied by bilateral laxity, rehabilitation helps, and if an operation is necessary, then the goal is to tighten the inferior capsule and rotator cuff interval (Matsen et al., 1994). AMBRII lesions are atraumatic or minor trauma resulting in multidirectional and bilateral instability. In AMBRII lesions, rehabilitation is the treatment of choice or inferior capsular surgery if rehabilitation is unsuccessful. Instability may also result from non-structural causes with no structural damage to the articular surfaces, but is the result of capsular dysfunction, abnormal muscle patterning, and, like AMBRII lesions, is often bilateral.

A final system is advocated by Schneeberger and Gerber in which the degree of joint laxity (no laxity, generalized laxity), degree of trauma (multiple events, single trauma event, minor trauma event), and direction of instability (multidirectional, unidirectional) are classification factors (Lewis et al., 2004).

Lewis and colleagues (2004) indicate that these classification systems are inadequate because they ignore that instability is a dynamic process with mixed pathologies that can change over time. They recommend the Stanmore classification, which uses classification characteristics from several of the described systems. For example, Type I is considered a true TUBS; Type II is a true AMBRII; and Type III includes muscle patterning disorders that are habitual and non-structural. This system considers the changing nature of pathology and includes traumatic/nontraumatic causes and muscle patterning (Lewis et al., 2004).

A systematic assessment of instability would first include an appraisal of the amount of passive translation between the humeral head and glenoid fossa. Second, an assessment of end feel would be necessary, as directional stresses are applied to the joint. Finally, reproduction of symptoms, subluxation, or apprehension would be done (Wilk et al., 1997a). Individual client factors also need to be considered in the assessment of instability, which would include information about the person's age, structural laxity, sport activity, and areas of occupations.

It is necessary to determine if the amount of passive translation between the humeral head and the glenoid is indicative of laxity or instability. A few tests of laxity specific to the shoulder are anterior drawer test of the shoulder, posterior drawer test of the shoulder, load and shift test (a modification of the anterior and posterior drawer tests), and the sulcus sign (test for inferior shoulder instability). Table 5-20 lists additional tests for shoulder instability.

The anterior drawer test assesses the laxity of the anterior capsule, while the posterior drawer (Gerber-Ganz Posterior Drawer) assesses the integrity of the posterior capsular structures and posterior portion of the glenoid labrum. Both tests are performed with the client in supine, in 80 degrees to 120 degrees of abduction, and in 0 degrees to 20 degrees of forward flexion. The examiner holds the client's scapula with the index and middle fingers on the spine of the scapula with the thumb applying counterpressure on the coracoid. Force is applied anteriorly for the anterior drawer test or by applying a slightly rotational force to the upper arm medially and flexing the shoulder to about 60 degrees to 90 degrees and applying a posterior force for the posterior drawer test. Both tests should be pain-free.

The load and shift test is a modification of the anterior and posterior drawer tests. Directional forces are applied anteriorly and posteriorly to relocate the humeral head in the glenoid. The test is then repeated in the supine position. Assessment of the translations in other positions such as external rotation and internal rotation is also done. In the first 60 degrees of abduction, the structures tested include the superior glenohumeral ligament, coracohumeral ligament, and the rotator interval. Between 60 degrees and 90 degrees, the middle glenohumeral ligament is tested, and at 90 degrees and greater, the inferior glenohumeral ligament is tested.

Posterior and anterior instability can also be tested by apprehension tests. The posterior apprehension test positions the client's arm in 90 degrees of abduction and 90 degrees of elbow flexion, and the therapist pushes posteriorly on the humeral head. For the anterior apprehension test, the therapist applies slight anterior pressure to the humerus and externally rotates the arm. Pain and apprehension about impending subluxation or dislocation indicates instability (Gibson, 2010).

The sulcus test (Figure 5-40) is done with downward force applied to the humerus while observing the shoulder area for a depression lateral or inferior to the acromion, which would indicate inferior translation of the humerus and suggests inferior glenohumeral instability (Gibson, 2010; Wilk et al., 1997a). The shoulder is in neutral position with stress applied above the elbow to eliminate the effects of biceps and triceps. Palpation might reveal a widening of the subacromial space between the acromion and humeral head. While the sulcus test is seen as essential to the diagnosis of multidirectional instability, this could also be a sign of a rotator interval lesion and/or an injury of the superior ligament complex.

Other symptoms that may be indicative of instability may include pain, inflammation, tendonitis, or other pathological conditions. For example, if there is multidirectional instability, the client may present with localized biceps tendon inflammation and diffuse rotator cuff muscle tendonitis. Clients with traumatic injuries with unidirectional instability may report localized pain and discomfort when the joint is placed in specific positions. The pain reported in instable shoulders may not be indicative of where the instability is occurring, such as when there is traumatic anterior instability and the client describes posterior shoulder pain secondary to impingement (Wilk et al., 1997b).

Instability may result in dislocation and subluxation of the shoulder. Careful observation, movement, and palpation of the glenohumeral head will provide valuable information about

Table 5-20

Stability Tests for the Shoulder

DIRECTION OR LOCATION	NAME OF STABILITY TEST
Scapular stability tests	Lateral scapular-slide tests
	Wall/floor pushup
	Scapular retraction test
	Scapular isometric pinch or squeeze test
	Scapular assistance test
Inferior stability tests	Sulcus test (at 0°; at 90°)
	Gagey's test
	Inferior apprehension test
Posterior stability	Posterior drawer
	Fukuda test
	Posterior fulcrum
	Load and shift test
	Posterior apprehension test
	Norwood stress test for posterior instability
	Push-pull test
	Jerk test
	Circumduction test
Anterior	Anterior drawer test of the shoulder
	Anterior apprehension test
	Load and shift test
	Crank (apprehension) and relocation test
	Fowler sign or test
	Jobe relocation test (Fulcrum test)
	Anterior release test
	Dynamic rotatory stability test
	Dynamic anterior jerk test
	Anterior instability test (Leffert test)
	Rockwood test for anterior instability
	Throwing test
	Rowe test
	Prone anterior instability test
	Andrews' anterior instability test
	Dugas test
	Protzman test for anterior instability
	Surprise/release test
Multidimensional	Rowe test for multidirectional instability
Core instability	Kibler's corkscrew test

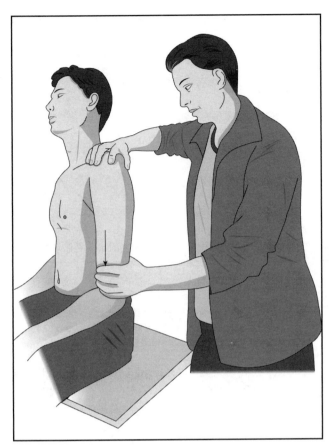

Figure 5-40. Sulcus test for instability.

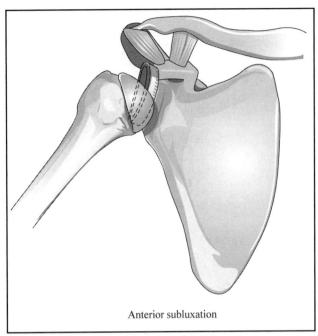

Anterior subluxation

Figure 5-41. Anterior subluxation of the glenohumeral joint.

the integrity of the joint. Dislocations and subluxations occur when the articulating surfaces of the bones that comprise a joint lose contact with each other, usually due to an external force.

Anterior humeral head displacement might be seen in clients who are unable to raise their arms overhead and is the most common type of glenohumeral dislocation (Figure 5-41). There may be a fullness noted on the upper anterior arm that is tender when palpated, the deltoid may feel tight, and the coracoid process may be sensitive to palpation and be higher up on the head of the humerus than is normally seen. There may or may not be pain, and there will be loss of upper extremity function and possibly loss of sensation. Often, anterior displacement is due to falls on outstretched arms, a fall or blow to the lateral shoulder from behind, or forced abduction with the humerus in either internal rotation with abduction or flexion with external rotation. With this type of displacement (subcoracoid is the most common but may also include intracoracoid or subclavicular), there are frequently tears in the labrum and capsule, fracture of the greater tuberosity, and tears of the rotator cuff. Any anterior subluxation can do great harm to the brachial artery, vein, or nerves. Circulation should always be checked (Shafer, 1997).

Posterior humeral head displacement (which is rare) may occur when there is direct pressure applied laterally and posteriorly to the joint or when a force is exerted in the same direction along a flexed, adducted, and internally rotated humerus. In some cases, the posterior area of the glenohumeral joint

may feel fuller, and an unusually prominent coracoid process may be felt, but often physical signs are often not noticeable. The patient's arm is abducted and rotated internally, and the elbow is directed slightly forward. The head of the humerus lies on the outer edge of the glenoid fossa. In severe cases, the lateral side of the capsule is usually torn, and there may be an associated rotator cuff tear or an avulsion fracture of the greater tuberosity resulting in persistent pain. The internal and external scapular muscles are often torn and may contain fragments of the avulsed tuberosities (Shafer, 1997).

Inferior humeral head displacement is characterized by severe pain and disability (Figure 5-42). The head of the humerus lies below the glenoid fossa, typically after a forced abduction movement followed by rotation (such as an injury to an abducted arm in a football tackle). The deltoid will often feel firm, flattened, and spastic, but other signs are often vague. Superior humeral head displacement is more likely the result of contractures within the upper humeral area that prevent the greater tuberosity from gliding smoothly under the coracoacromial ligament during abduction. The humerus cannot dislocate much superiorly due to bony arch, and a supraglenoid displacement is rare except in sports or severe accidents (Shafer, 1997).

Subluxation is commonly seen in the shoulders of people with hemiplegia, because the muscles that control the shoulder are weak or paralyzed, which allows the humeral head to be displaced inferiorly by gravity. Weakness/paresis of the rotator cuff, the prolonged gravitational pull that stretches the capsule and ligaments, also contributes to shoulder dysfunction after a stroke. Positioning is often a choice for treatment (slings, lapboards); physical agent modalities (electrical stimulation) and muscle facilitation techniques are often used in the treatment of the hemiplegic shoulder (Donatelli, 1997; Gillen & Burkhardt, 2004).

Figure 5-42. Inferior and superior subluxation of the glenohumeral joint.

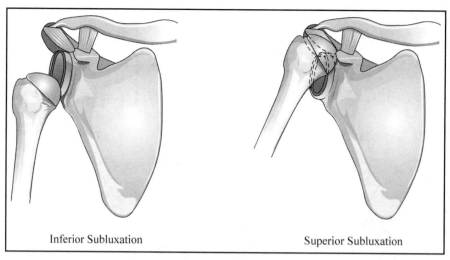

Inferior Subluxation   Superior Subluxation

Several tests are used to assess displacement, subluxation, and dislocation. The Callaway test is measurement of the circumference of the affected shoulder, measured over the acromion and through the axilla. If this measurement is greater than that on the opposite, unaffected side, this indicates dislocation. In the Dugas test, the client is asked to place his or her hand on the opposite shoulder and touch the elbow to the chest. If there is pain or an inability to perform the test, this is indicative of dislocation. If there is a lowering of the axillary fold, this is known as Byrant's Sign, also indicating dislocation on the lowered side.

Impingement Assessment

Pain, weakness, and loss of motion are the most common symptoms of impingement of the structures of the shoulder (Neer, 1972). An early classification system developed by Neer (Neer, 1972) categorized impingement as stage I if there was edema and/or hemorrhage in people younger than 25 years. This was seen as a reversible condition, and there was no permanent damage to structures or tissue tears. Stage II occurred with those 25 to 40 years of age, and there is evidence of irreversible tendon changes and fibrosis resulting in permanent scarring but no tissue tears. Stage III was seen in people over 50 years old and was a result of tendon rupture or tears often due to long-standing fibrosis and tendonosis.

Since the development of this classification, impingement classification is focused on external or internal impingement. External impingement (also known as outlet impingement or extrinsic) is the most common type of impingement. This type of impingement involves compression of the rotator cuff muscles (primarily supraspinatus and then infraspinatus and biceps tendon) by the acromion. The rotator cuff muscles and the subacromial bursa become impinged between the greater tuberosity and the anterior one third of the acromion. The compression occurs anteriorly with forward flexion. Other causes may be greater tuberosity fractures and humeral neck fractures (Woodward & Best, 2000b). There may also be coracoid impingement in which the pain is felt lower and more anteriorly than in the superior impingement. In this case,

there would be decreased horizontal adduction and pain at the tip of the coracoid and not at the acromioclavicular joint. External forces are those originating outside of the rotator cuff muscles and are largely mechanical.

Primary extrinsic causes are related to acromion and coracoacromial arch. There may be bone spurs on the acromion, or the shape of the acromion may be more prone to impingement (a beaked shape due to genetic determination or degenerative changes has a higher incidence of impingement as compared to a curved or flat acromion). Similarly, there may be spurs or bursal scarring on the coracoacromial arch because of aging, disease (such as degenerative joint disease), or as a result of poor posture (leading to rotator cuff atrophy, scapular weakness, increased thoracic kyphosis) (Woodward & Best, 2000b).

Secondary causes of external impingement are related to poor scapular stability, which changes the position of the scapula. The problem in secondary impingement is keeping the humeral head centered in the glenoid fossa. This is prevented by weakness in the rotator cuff muscles resulting in functional instability and laxity in the glenohumeral capsule and ligaments. Secondary impingement occurs in the coracoacromial space secondary to anterior translation of the humeral head (Woodward & Best, 2000b).

Internal impingement (non-outlet or intrinsic impingement) occurs in the posterosuperior portion of the joint. It is more common in younger people and those involved in overhead sports. It is also known as an athletic impingement because of adaptive shortening and scarring of posterior rotator cuff and scapular muscles. SLAP lesions, or injury to the superior labrum in the anterior-posterior portion of the joint, occur in overhead athletes. There is a loss of internal rotation and upward translation of the humeral head, and there may be excessive external rotation and/or recurrent anterior instability (Bach & Goldberg, 2006; Woodward & Best, 2000b). This is due to impingement on the posterior labrum and glenoid and irritation of the rotator cuff and biceps tendons. The client will describe pain in the back of the joint due to irritation of the posterior fibers of the supraspinatus and irritation of the anterior fibers of the infraspinatus muscle. In advanced cases, the pain may shift to the front of the joint due to biceps

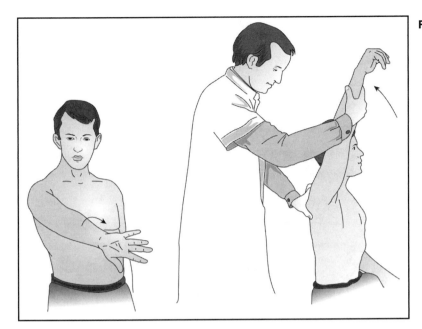

Figure 5-43. Neer impingement test.

tendon and labral involvement. Internal impingement results in a mechanical pain, activated upon specific movements; otherwise, it is asymptomatic (Woodward & Best, 2000b).

Jobe's classification of impingement combines the factors of age, impingement, and instability. Grade I in the Jobe system is joint impingement without instability, usually in people older than 40 with degenerative changes in the rotator cuff, coracoid, and anterior tissues. There may be intrinsic changes due to rotator degeneration, or there may be extrinsic factors (such as the shape of acromion or degeneration of the coracoacromial ligament). Grade II is characterized by secondary impingement and instability caused by chronic capsule and labral trauma. Grade III is the result of secondary impingement and instability due to generalized hypermobility or laxity. Grade IV is primary instability with no impingement (Jobe et al., 1989; Kvitne & Jobe, 1993).

Deficient shoulder biomechanics, a cause of impingement, results when there is weakness in or tearing of the rotator cuff muscles (especially supraspinatus), glenohumeral instability, an anatomically prominent acromion process, inflammation of any of the subacromial structures (often the subacromial bursa), or stiffness of the posterior joint capsule (Ellenbecker & Ballie, 2010; Magee, 2002; Shafer, 1997; Wilk et al., 1997a). The compression of subacromial space may be due to inadequate rotation of humerus during flexion and abduction. This results in movement of the greater tubercle closer to the acromion or excessive superior glide of the humeral head. The result may be abnormal scapulothoracic joint motion and rhythm and pain in the superior aspect of the glenohumeral joint during the mid-range of shoulder elevation. This in turn can lead to an inflammatory response and impingement syndromes (Oatis, 2004).

Because the cause of impingement is often faulty biomechanics, improving the joint biomechanics is an important part of treatment. Correcting joint capsule tightness and muscle imbalance (between deltoid and rotator cuff muscles and strengthening scapular stabilizers) while using anti-inflammatory medications can help correct the structural limitations. Joint protection techniques and modification of activities are important ways to restrict motion through the painful ranges and ensure inferior joint capsule laxity. Postural evaluation is also an important part of ensuring full normal glenohumeral movement.

Many tests have been developed to assess shoulder impingement. Tests for shoulder impingement are listed in Table 5-21. The Neer impingement test is done by having the client fully pronate the arm (Figure 5-43). The examiner then stabilizes the scapula to prevent scapulothoracic motion and then forcefully elevates the arm in flexion, causing impingement of the rotator cuff tendons under the coracoacromial arch. When the arm is put in this position under the coracoacromial arch, supraspinatus, infraspinatus, and the long head of the biceps are at risk for impingement, degenerative tendonitis, and tendon ruptures (Gibson, 2010).

In the Hawkins-Kennedy impingement test, pain occurs as the greater tuberosity and the supraspinatus tendon impinge under the anterior surface of the coracoacromial ligament and coracoacromial arch. In this test, the arm is moved into 90 degrees of forward flexion and then forcibly internally rotated.

Strength/Muscle Assessment

Mobility in the shoulder is achieved by sufficient movement at joint surfaces and appropriate tension and movement in soft tissues. Testing the strength of shoulder muscles can be performed via manual muscle tests (muscle groups or isolated muscles) or using functional tests. In these tests, evidence of a contraction, the effect of gravity, and external resistance are used to assess the strength of the muscles. Observing the client during daily activities will also give information about how the client moves, if there is pain with movement, and if the movement is coordinated. It is also crucial that clients are asked about the areas in which they are limited and about their status in personally relevant outcomes. Impairment measures alone are not sufficient to base treatment (Roddey, Cook, O'Malley, & Gartsman, 2005).

Table 5-21

IMPINGEMENT TESTS FOR THE SHOULDER

Neer impingement test

Hawkin's-Kennedy impingement test

Reverse impingement sign (impingement relief test)

Clancy impingement test

Copeland impingement test

Horizontal impingement test

Posterior impingement test

Coracoid impingement test

Subcoracoid impingement

Glenohumeral Internal Rotation Deficit (GIRD)

Posterior internal impingement test

Muscle strength screening requires that the client assume a particular position and is instructed to hold that position as the therapist applies resistance. Muscle screens are a quick way to get an overall appraisal of the strength of muscle groups involved in particular shoulder motions (Table 5-22). If limitations are noted or more specific information is needed about the degree to which a specific muscle is limited, a manual muscle test would then be done. Many activities of daily living require only a fair muscle grade, such as placing a garment into a closet, washing the opposite side of the body, or reaching back to put an arm into a sleeve gut (Gutman & Schonfeld, 2003).

Resistive tests may also be done for evaluation of strength and provocation of pain. These are isometric tests, advocated by Cyriax (1982), to assess shoulder dysfunction. Six tests are recommended and include shoulder adduction, abduction, external rotation, internal rotation, elbow flexion, and elbow extension. If the client responds to the resisted movement with a strong, painless contraction, that is considered normal. If the result is a strong but painful contraction, this may indicate a minor muscle or tendon lesion. For example, painful resisted abduction and external rotation may indicate rotator cuff disease. Weak and painful responses may indicate a gross traumatic lesion, such as a fracture or the rupture of muscle or tendon. A weak and painless result may indicate neurologic dysfunction or muscle and tendon rupture (Donatelli, 1997). A description of various shoulder disorders by characteristic motion patterns with resistance is shown in Table 5-23.

Another aspect of muscle evaluation is the evaluation of the synchronized muscle contractions and relaxation during movement, which is known as muscle patterning. If the contraction/rest patterns of the shoulder muscle are altered, then shoulder instability can result. While muscle pattern instability may begin in younger clients who voluntarily dislocate or

sublux their shoulder, joint mechanics may deteriorate so that the shoulder dislocates repeatedly and involuntarily. If this pattern continues, the client may not perceive the involuntary dislocation to be abnormal and may easily dislocate the shoulder by coughing or sneezing. The focus of intervention for positional or patterning instability is to regain normal neuromuscular control to allow full return to daily occupations (Funk et al., 2005).

Specific tests for muscle tendon pathology are listed in Table 5-24. Manual resisted muscle testing is used to help isolate the etiology of the impingement and in making differential diagnoses of muscle pathology. An example of this is Speeds test, which is used to assess the proximal tendon of the long head of the biceps. In this test, the examiner resists forward flexion of the shoulder distally at the wrist while the client is in supination and the elbow is extended. Pain would be elicited in the bicipital groove or cubital fossa for a positive sign (Gibson, 2010; Shafer, 1997).

The Yergason test (Figure 5-44) is a test of biceps tendon instability or tendonitis. The client's elbow is flexed to 90 degrees and stabilized against the thorax with the forearm pronated. The therapist resists forearm supination while the client externally rotates the arm. A positive sign is when pain is elicited in the bicipital groove, suggesting pathology in the long head of the biceps in its sheath (Gibson, 2010). Inflammation of the biceps tendon frequently occurs when there is rotator cuff tendonitis.

The lift-off test (Gerber's test) is a test of subscapularis rupture or dysfunction. The client places his or her hand behind his or her back with the dorsum of the hand resting in the mid-lumbar region. The client then raises the hand off the back, maintaining internal rotation of the humerus. Inability to lift the hand off the back would indicate subscapularis dysfunction. A modified version of the lift-off test is useful for clients who cannot place the hand behind the back. Instead, the client places the hand on the abdomen and resists attempts by the therapist to externally rotate the arm.

There are several tests for supraspinatus. One is the empty can test (also known as Jobe test or supraspinatus test). The client attempts to elevate the arm against resistance with the elbows extended and the arms abducted and in external rotation. Other tests include Apley's Scratch Test, Dawburn's sign, Sherry Party sign, Codman's Sign (Drop Arm Sign), Rent Test, Zero Degree Abduction Test, and the Scapular Retraction Test.

Pain

There is a high incidence of shoulder pain due to a variety of reasons. Table 5-25 provides a few examples of extrinsic, intrinsic, and psychological causes of pain. Pain can be due to intrinsic or extrinsic causes, can be traumatic or atraumatic, can be chronic or acute, or even can be referred from other locations and structures. The assessment of shoulder pain necessitates an accurate history and interview with the client about when, where, and how the pain is elicited.

Intrinsic causes of shoulder pain may be due to joint and muscular impairments that are within the shoulder joint. Causes of intrinsic shoulder pain may include shoulder instability, nerve and muscle impingement, fractures, tendonitis,

Table 5-22

MUSCLE SCREEN FOR STRENGTH

INSTRUCTIONS TO CLIENT	THERAPIST'S ACTION
Ask the client to shrug his or her shoulder up toward the ear; tell him or her, "don't let me push your shoulder down"	Apply resistance with palms toward the floor
Ask the client to raise his or her shoulder to 90° in front of the body; tell him or her, "don't let me push your arm down"	Apply resistance to mid-humerus toward extension
Ask the client to raise his or her shoulder to 90° in front of the body; tell him or her, "don't let me push your arm up"	Apply resistance to mid-humerus toward flexion
Ask the client to raise his or her shoulder to 90° with the elbow extended in front of him or her; tell him or her, "don't let me push your arm"	Apply resistance to mid-humerus toward horizontal adduction
Ask the client to bring his or her hands together in front of the body; tell him or her, "don't let me separate your arms"	Attempt to separate the arms into horizontal abduction
Ask the client to raise his or her shoulder out to the side to 90° with the elbow; tell him or her, "don't let me push your arm down"	Apply resistance to the mid-humerus into adduction
Ask the client to raise his or her shoulder out to the side to 90° with the elbow; tell him or her, "don't let me push your arm up"	Apply resistance to the mid-humerus into abduction
Ask the client to bring his or her arm out to the side to 90° abduction and bend the elbow to 90° flexion and up in the air; tell him or her, "don't let me push your arm down"	Apply resistance toward internal rotation
Ask the client to bring his or her arm out to the side to 90° abduction and bend the elbow to 90° flexion and down toward the floor; tell him or her, "don't let me push your arm up"	Apply resistance toward external rotation

myositis ossificans, adhesive changes (adhesive capsulitis), or neurovascular injuries. Pain can be referred from a shoulder muscle to another part of the shoulder or arm. For example, referred pain from the rhomboids may go to the medial border of the scapula. The infraspinatus may refer pain to the antero-lateral shoulder and medial border of the scapula or down the lateral aspect of the arm. If pain is felt near the deltoid inser-tion, up the shoulder, and down the lateral arm to the elbow, this may be referred pain from teres minor (Magee, 2002). The pain the client experiences is also influenced by the emotional status of the individual, especially for those clients who are experiencing depression, anxiety disorders, adjust-ment disorders, or psychosis. The assessment and intervention is directed toward therapeutic change in the whole person, which includes awareness of the mind-body connections.

Pain is often referred to the shoulder from other areas, and the pain experienced in the shoulder region may well be due to lesions and trauma from other structures and organs of the body. For these reasons, the following extrinsic problems first must be ruled out prior to attributing the cause of shoulder pain to shoulder girdle joints or muscles:

- Heart problems: Myocardial infarction, angina, pericardi-tis, which may radiate pain to the left shoulder.

- Lung or diaphragm problems; spontaneous pneumo-thorax; pulmonary tuberculosis; lung cancer; abscesses; Pancoast tumor, which can relay pain along the same nerve roots of C4 and C5.

Table 5-23

DIFFERENTIATION OF MOTION PATTERNS AND COMMON CAUSES OF DYSFUNCTION

CHARACTERISTIC MOTION PATTERN	COMMON CAUSES
Full passive motion with painful resisted abduction	Deltoid strain
	Supraspinatus strain
Full passive motion with painful resisted adduction	Bicipital strain (long head)
	Latissimus dorsi strain
	Pectoralis major strain
	Teres (major, minor) strain
Full passive motion with painful resisted internal rotation	Latissimus dorsi strain
	Pectoralis major strain
	Subscapularis strain
	Teres major strain
Full passive motion with painful resisted external rotation	Infraspinatus strain
Full passive motion with painful resisted forward movement	Bicipital strain
	Coracobrachialis strain
Full passive motion with painful resisted elbow flexion and supination	Bicipital strain (long head)
Full passive motion with painful resisted elbow extension	Triceps strain
Full passive motion, painless weak deltoid	Axillary nerve lesion
Full passive motion, painless weak deltoid, biceps, and spinatus group	C5 lesion
	Myeloma
	Traction palsy
Full passive motion, painless weak spinatus group alone	Suprascapular neuropathy
Full passive motion, solely painless weak supraspinatus alone	Supraspinatus tendon rupture
Full passive motion, solely painless weak infraspinatus alone	Infraspinatus tendon rupture
Full passive motion, painless weak serratus anterior	Long thoracic neuropathy
Full passive motion, painless weak trapezius	Spinal accessory neuropathy
Full passive motion, painless weak biceps and forearm muscles	C5 lesion
Full passive motion, painless weak triceps and forearm muscles	C6 lesion

Adapted from Shafer, R. C. (1997). Shoulder girdle trauma: Monograph 16. Retrieved from http://www.chiro.org/ACAPress/.

- Chest problems: Aortic aneurysm, nodes in the axilla or mediastinum, breast disease, hiatal hernia, which can refer pain to the local area and shoulder.
- Cervical problems: Intervertebral disk herniation or degeneration, facet joint effusion, nerve root irritation, which may also radiate pain to shoulder and scapula.
- Pain radiating below the elbow with decreased cervical range of motion may indicate cervical disk disease.
- Spinal fracture: Cervical or thoracic fracture can radiate pain to the shoulder as well as produce local pain.
- Elbow problems: Humeroulnar, humeroradial, or radioulnar joint dysfunction or pathology can also refer pain into the shoulder and humerus.
- Temporalmandibular joint problems due to degeneration or effusion can refer pain down the neck into the shoulder.
- Rib problems; costovertebral joint or costosternal joint dysfunction.

Table 5-24

TESTS OF MUSCLE TENDON PATHOLOGY

MUSCLE	*NAME OF TEST*
Biceps	Speed's test
	Speed's maneuver
	Biceps
	Abbot-Saunders
	Transverse humeral ligament test
	Snap test
	Hueter sign
	Duga sign
	Beru sign
	Traction test
	Compression test
	Yergasons test
Infraspinatus	Infraspinatus test
	Infraspinatus/teres minor (lateral rotation)
	External rotation lag sign
Subscapularis	Lift-off test (Gerber's test)
	Internal rotation lag sign test
	Lag or "spring back" tests (subscapularis [medial rotation])
	Bear-hug test
Supraspinatus	Empty can test (Jobe's test)
	Apley's scratch test
	Dawburn's sign
	Sherry party sign
	Codman's sign (drop arm sign)
	Rent test
	Zero degree abduction test
	Scapular retraction test
Additional tests	The dropping sign (Walch)
	French horn shoulder test (internal and external rotation)
	Internal rotation resistance strength test (IRRST)
	Ludington's test
	Gilchrest's sign
	Lippman's test
	Hueter's sign
	Abrasion sign

(continued)

Table 5-24 (continued)

TESTS OF MUSCLE TENDON PATHOLOGY

MUSCLE	NAME OF TEST
Additional tests	Codman (drop-arm) test
	Abdominal compression (belly-press) test
	Lateral scapular slide test/scapular load test
	Wall pushup test
	Scapular retraction test
	Hornblower's (Signe de Clairon) sign
	Teres minor test
	Pectoralis major contracture test
	Belly off sign
	Belly press/Napoleon sign
Ligament pathology	Crank test
	Posterior inferior glenohumeral ligament test
	Coracoclavicular ligament test

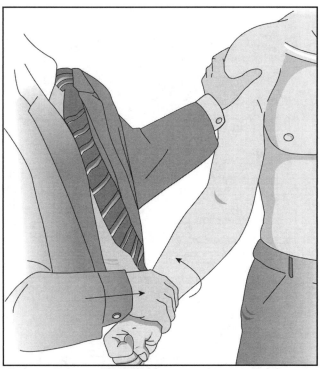

Figure 5-44. Yergason test for biceps tendonitis.

- Thoracic spine problems.
- Abdominal problems: Ruptured spleen and pancreas pathology can refer pain to the left shoulder; liver pathol-

ogy and gallbladder disease can refer pain to the right shoulder (Hartley, 1995).

Specific extrinsic causes of shoulder pain might be a herniated cervical disk (common at C5-6) where pain may radiate from the neck into the whole upper extremity accompanied by paresthesias and sensory loss in corresponding dermatomal regions. Pain is often referred from viscerosomatic reflexes from the liver, gallbladder, and right diaphragm to the right shoulder and from the stomach, left diaphragm, and heart to the left shoulder. Pain that is not easily reproduced suggests a visceral origin (Shafer, 1997).

Pain assessments start with asking the client about the pain he or she is experiencing. Often, clients are asked to show where the pain is or to indicate where the pain is on a body map or diagram. Pain is less likely to be referred if the therapist can reproduce the pain in the structure indicated by the client with gentle palpation.

Have the client describe how the pain feels. If there is a sharp, localized pain, this might suggest an acute inflammatory process, such as a muscle strain. Sharp, burning, and radiating pain may indicate a neurological disorder. A dull ache that is poorly localized may suggest referred pain from another part of the body. Differentiate stiffness from pain, and elicit information from the client about the sensations he or she may be experiencing, such as grating, clicking, or snapping sensations. For example, if pain and a clunking sound are heard with overhead movement, this may indicate a labral disorder. Table 5-26 lists some possible structures that may be responsible for pain as described by clients.

A thorough exploration of the history of the shoulder pain is essential to the assessment. Determining if this is the first

Table 5-25

CAUSES OF SHOULDER PAIN

EXTRINSIC CAUSES	*INTRINSIC CAUSES*	*PSYCHOLOGICAL CAUSES*
Brachial plexis injury	Instability	Depression
Cervical spondylosis	Impingement	Anxiety disorders
Cervical disk herniation	Fractures	Adjustment disorders
Syringomyelia	Arthritis	Somatoform disorders
Thoracic outlet syndrome	Myositis ossificans	Factitious disorders
Diaphragmatic irritation	Tumors	Malingering
Lung tumor/tumors	Neurovascular injury	Psychosis
Brachial neuropathy	Bursitis	
Myocardial ischemia	Dislocation	
Contusions/lacerations	Adhesive capsulitis	
Neuropathy	Lax shoulder stability	
Referred pain	Nerve contusion	
Vascular disorders	Complex regional pain syndrome	
	Rotator cuff lesions	
	Sprains/strains	
	Subluxation/fixation	
	Tendonitis/tendonosis	

time the client has had pain, if there was a precipitating event, or if there were any unusual activities involved are important to understanding the mechanism and underlying problems related to the client's pain. Shoulder pain may have an inflammatory, neurologic, psychologic, vascular, metabolic, neoplastic, degenerative, congenital, autoimmune, or a toxic origin in addition to pain because of direct trauma (Shafer, 1997).

A sudden onset due to acute trauma may result in muscle strain, and the client can often recount the circumstances surrounding when and how the injury occurred. A gradual onset suggests an injury that is progressively worsening or other non-musculoskeletal causes. Be sure to ask the client about possible repetitive causes of muscle strain, such as poor posture and static positioning during activities that can affect shoulder function and pain. Recurrent pain may be linked to stress, so discussion with the client about stressors may be helpful. Understanding what causes the part to feel better or worse may lead to an understanding of the type of activities or movements that aggravate the condition and the timing of the pain (morning or night, constant or intermittent). Patient-based measures of pain and function have been shown to reliably assess outcomes after surgery to the shoulder (Dawson, Hill, Fitzpatrick, & Carr, 2001).

Pathology of the Shoulder

Subluxations and dislocations of the shoulder occur most often in the anterior direction and are usually due to an external force that disrupts articulating joint surfaces. Sprains occur most frequently to the acromioclavicular joint and are usually the result of a fall or direct blow to the joint or a fall on the elbow. There is pain with joint motion, tenderness, and varying degrees of deformity. The sternoclavicular joint ligaments can be sprained by direct blows to the sternum or indirectly from a fall with the arm extended. Again, there will be localized pain over the joint, limited motion, and varying degrees of deformity.

Muscle strains frequently occur due to overuse, overfatigue, or weakness in specific muscles. Supraspinatus strain or tendonitis can be tested via the empty can test and the drop arm test. Subscapularis strain and function are assessed with the lift-off test (Gerber test), and Speed's maneuver (biceps or straight-arm test) is used to test the biceps tendon instability or tendonitis. Tendonitis and strain usually precede bursitis and may be present simultaneously, and these are sometimes difficult to differentiate on physical exam. Pain that is elicited with both passive and resistance tests is likely due to bursitis and not tendonitis. Bursitis at the shoulder usually involves

Table 5-26

DESCRIPTIONS OF PAIN AND POSSIBLE AFFECTED STRUCTURES

DESCRIPTION OF PAIN	POSSIBLE STRUCTURES THAT MAY BE AFFECTED
Sharp	Superficial muscles or tendon
	Acute bursitis
	Injury to the periosteum
	Acute inflammatory response
Dull	Tendon sheath
	Deep muscle (serratus anterior or subscapularis)
	Bone
	Referred pain
Ache	Deep muscle (subscapularis, teres minor)
	Deep ligament (glenohumeral, coracoclavicular)
	Tendon sheath or fibrous capsule
	Chronic bursitis
Radiating ache	Angina pectoris
	Bursitis
	Capsule adhesions
	Fracture
	Pancoast's tumor
	Periarthritis
	Rib subluxation
	Complex regional pain syndrome
	Subluxation (cervical or shoulder)
	Neurological process
Pins and needles	Peripheral nerve injury
	Dorsal root of a cervical nerve
Tingling	Circulatory or neural structure (thoracic outlet, brachial plexus, brachial artery, etc.)
Numbness	Cervical nerve root
	Peripheral cutaneous nerves
Twinges	Subluxation
	Muscle strain
	Ligamentous sprain
Stiffness	Capsular swelling
	Arthritic changes
	Muscle spasms
Timing of pain	Night pain may suggest inflamed bursa, vascular disease, metastatic disease
	Sleeping postures can affect shoulder pain

(continued)

Table 5-26 (continued)

DESCRIPTIONS OF PAIN AND POSSIBLE AFFECTED STRUCTURES

DESCRIPTION OF PAIN	POSSIBLE STRUCTURES THAT MAY BE AFFECTED
Timing of pain	Pain with specific movements suggests musculoskeletal problem, possible capsular or impingement problem
Swelling and tenderness (neck or shoulder)	Arthritis
	Cellulitis
	Contusion
	Dislocation
	Fracture
	Sprain
	Strain
	Subluxation (acute)

the subacromial bursa preceded by tendonitis/tenosynovitis of the rotator cuff. This occurs when weak rotator cuff muscles allow the humerus to move up into the acromion, pressing on the bursa and causing inflammation. Treatment for muscle and tendon strains and bursitis initially consists of ice, modalities, stretching, and friction massage followed by progressive eccentric strengthening.

If fibrosis of the joint capsule occurs, adhesive capsulitis may result, usually because of muscle strain, tendonitis, bursitis, or impingement. Onset is usually at 40 to 60 years of age, and there are different types of adhesive capsulitis (Table 5-27). Treatment often involves stretching the capsule, use of modalities, joint mobilization techniques, and functional activities. Tests for labral lesions, neurological involvement, thoracic outlet syndrome, and ligamental pathology and assessment for the acromioclavicular and sternoclavicular joints are listed in Table 5-28.

The biomechanical treatment approach is commonly used to treat the orthopedic joint and muscle dysfunctions of the shoulder. Please see chapter 11 for theoretic principles and treatment suggestions using this treatment approach.

Summary

- The shoulder is a complex joint with multiple articulations that enable movement in all planes and axes.
- The sternocostal and vertebrocostal joints enable full shoulder flexion and abduction due to gliding of ribs on sternum and on vertebra.
- The suprahumeral articulation is a functional joint serving in a protective capacity, preventing trauma from above, upward dislocation of the humerus, and limiting abduction of the humerus.
- The scapulothoracic joint is also a functional joint that provides a movable base for the humerus, permitting wide

ranges of movement of scapula and thorax. This joint also provides stability in the glenohumeral joint during overhead activities.

- The glenohumeral articulation is an incongruous joint because the bony surfaces are not in direct contact for most movements. Stability is achieved by the joint capsule and muscles of the joint as well as ligaments. Surface motion is mainly rotational but also includes some rolling and gliding. Shoulder motion, especially elevation, is governed by force couples, with the interaction of the deltoid muscle and oblique rotator cuff muscles a good example. Because this is a ball-and-socket joint, all motions are possible.
- The acromioclavicular joint is also considered an incongruent joint, and the weak joint capsule is reinforced by ligaments and muscles adding to the stability of the joint.
- The sternoclavicular joint is the only true articulation of the shoulder girdle with the trunk. Motions possible at this joint include elevation and depression and protraction and retraction. These motions demonstrate a reciprocal movement pattern between the sternoclavicular and the acromioclavicular joints.
- All of the joints work together smoothly and synchronously with scapulohumeral rhythm. Generally, a ratio of 2:1 is accepted as the degree of contribution of the glenohumeral articulation and the contributions of the scapulothoracic articulations.
- Shoulder flexion/extension, abduction/adduction occur in phases with differing levels of contribution of different joints during different phases.
- Many muscles contribute to the movement and stability of the shoulder. How much contribution a specific muscle makes to a particular movement is dependent upon the

Table 5-27

SUMMARY OF SHOULDER PATHOLOGY

SENSATIONS	POSSIBLE CAUSES OF DYSFUNCTION
Clicking	Glenoid labrum tear
	Glenohumeral subluxation
Snapping	Biceps tendon
	Thickened bursa under the acromion during abduction
Grating	Osteoarthritic changes
	Calcium in the joint
	Thickened bursa
Locking or catching	Calcium in joint
	Piece of articular cartilage fractured off of the humerus or labrum
Warmth	Inflammation
	Infection

CONTUSION

Acromioclavicular joint most frequently affected

Frequent deltoid contusions if shoulder pad not worn in sports activities

Contusions on upper arm over biceps and triceps common because these are below shoulder pads

Deltoid mid-belly contusions in hockey and lacrosse where sticks are used

Coracoid process can be contused when marksmen gun recoils

FRACTURE

Clavicle: Greenstick fracture of shaft frequent in children and preadolescents

Distal end of clavicle when acromion hits downward

Most common site of clavicular fracture is the junction of the middle and outer third of the clavicle

Acromion can fracture when there is a fracture of the distal end of the clavicle or due to a direct blow down over acromion

Scapula: rare

Direct blows, with the exception of the glenoid rim fractures, which are usually in association with dislocations

Humerus: due to a fall on outstretched arm but can be direct blow

Concerns due to the relation of the axillary nerve proximally and the radial nerve in the middle third as it passes through the spiral groove

Humeral neck fracture or avulsion of greater tuberosity

Hill-Sach's lesion, which is a compression fracture or defect in posterolateral humeral head due to repeated trauma

Brachial plexus or axillary nerve damage may occur

Falls on shoulder or outstretched arm as in contact sports, skiing, wrestling, cycling

Stress fractures are rare

(continued)

Table 5-27 (continued)

SUMMARY OF SHOULDER PATHOLOGY

SPRAINS

POSSIBLE CAUSES OF DYSFUNCTION
Degenerative and traumatic injuries, especially sports

Fall on point of shoulder with arm adducted

Fall on outstretched arm

Fall on olecranon of elbow

Blow from behind with ipsilateral arm fixed on ground

Traction of humerus pulling acromion away from clavicle

Direct blow over acromion

STRAINS

Frequently seen with the varying grades of impingement

Often with an etiology of repetitive intrinsic loading of the muscle/overuse, which may be associated with over-fatigue or weakening of specific muscles Almost all tears of the rotator cuff, except those due to direct trauma, occur at or near the insertion of the muscle and are due to overuse: the cuff wears before it tears.

Supraspinatus tendonitis/strain:

Common sites are just before the insertion on the greater tuberosity and at the musculotendinous junction.

Bicipital tendonitis/strain:

Often associated with tendonitis of the rotator cuff; cuff tendonitis can be misdiagnosed as biceps tendinitis.

True tenosynovitis of the long head of the biceps is found under and just distal to the transverse humeral ligament.

It is usually the result of overuse, a direct blow, or laxity of the transverse humeral ligament resulting in subluxation of the tendon.

Overstretch injuries:

Often involve nerve damage or injuries and is due to overly forceful muscle contractions

Most common strains are to:

internal rotators: subscapularis, latissimus dorsi, teres major, pectoralis major

external rotators: infraspinatus, supraspinatus teres minor, long head of biceps at superior tip of labrum

OVERUSE INJURIES

Impingement syndromes as seen in overhand tennis stoke, front crawl, and butterfly swimming strokes and side arm and overhand throwing actions

overuse → microtrauma and inflammation → instability → subluxation → impingement (repetitive trauma without repair) → rotator cuff tear

(continued)

Table 5-27 (continued)

SUMMARY OF SHOULDER PATHOLOGY

TENDONITIS   *POSSIBLE CAUSES OF DYSFUNCTION*

Bankart lesion

Hill Sach's lesion

"SLAP" lesion

Painful arc with range of motion and with isometric resistance and when muscle is stretched

Muscle contraction strong unless there is a tear

Tenderness over tendon when palpated

BURSITIS

Due to trauma or overuse so that the bursae become inflamed

Similar to acute tendonitis

ACUTE CONDITIONS

Rheumatoid arthritis (acute exacerbation)

Osteoarthritis

Trauma

Diabetes mellitus

Mircrotrauma from poor posture

Immobilization

Ischemic heart disease or stroke

SUBACUTE AND CHRONIC CONDITIONS

Osteoarthritis

Metastasis in acromion

Rheumatoid arthritis

FROZEN SHOULDER:

CAPSULAR PATTERN/   Age 40-60, unknown cause

ADHESIVE CAPSULITIS

Freezing = intense pain even at rest and limitation of motion 2 to 3 weeks following onset.

Frozen = pain only with movement; substitute patterns to achieve motions of the scapula. Atrophy of the deltoid, rotator cuff, biceps, and triceps muscles occurs.

Thawing = no pain but significant capsular restrictions.

Spontaneous recovery occurs in an average of 2 years after onset.

Inappropriately aggressive treatment will prolong this condition, and treatment at the wrong time may prolong symptoms.

(continued)

Table 5-27 (continued)

SUMMARY OF SHOULDER PATHOLOGY

FROZEN SHOULDER CAPSULAR PATTERN/ ADHESIVE CAPSULITIS	*POSSIBLE CAUSES OF DYSFUNCTION*

POSSIBLE CAUSES OF DYSFUNCTION

If there is no external rotation, abduction is limited to 75°, and flexion to 100° = severe

If there is 30° external rotation, abduction to 100°, flexion to 120° = moderate impairment.

Types:

Primary: Idiopathic, spontaneous onset of painful shoulder; insidious for no reason

Progressive: Women more than men; symptom is pain, nontraumatic, and the condition runs its course; steroids help a little but don't change rate of recovery

Secondary/traumatic adhesive capsulitis: Other injury leads to immobilization; shoulder becomes stiff and joint tissues shortened

SHOULDER IMPINGEMENT

Biceps tendon and rotator cuff often affected as these pass through the acromial space

May be due to an insufficiently stabilized humeral head (due to weakened rotator cuff or biceps, which allows more movement; weak supraspinatus or infraspinatus)

May be inflammation, tendon damage, bony malformations, poor posture

Impingement classification:

Grade I: Reversible inflammation and edema, frequently of the long head of the biceps or supraspinatus tendon.

Pain is in a specific arc of movement.

Most common in people between the ages of 16 and 20.

Grade II: Inflammation with slight tearing/fraying of the rotator cuff and occasionally a subacromial bursitis.

May be degenerative changes, such as osteophyte formation.

Usually involves people between 30 and 45 years of age.

Grade III: Severe fraying or a complete tear of the tendon of the supraspinatus or, less frequently, the long head of the biceps.

Bone spurs present.

More often seen in people older than 45 years.

NEUROLOGICAL

Thoracic Outlet Syndrome:

Compression, irritation, or direct injury of major structures within the thoracic outlet such as the subclavian vein, subclavian artery, and the brachial plexus.

(continued)

Table 5-27 (continued)

SUMMARY OF SHOULDER PATHOLOGY

NEUROLOGICAL   *POSSIBLE CAUSES OF DYSFUNCTION*

Thoracic Outlet Syndrome:

May be neurogenic, venous (effort induced or Paget Schroetter's Syndrome), or arterial.

Symptoms often develop due to poor posture and muscle tone.

Burning, numbness in shoulder and ulnar side of forearm and hand, and decreased radial pulse.

Anterior Scalene Syndrome

Costoclavicular Syndrome

Pectoralis Minor

Cervical Rib Syndrome

axis and plane of motion, the position of the humerus, what other muscles are involved, the resolution of angle of pull of the muscle fibers, and the size of the muscle.

- Evaluation of the shoulder would include an assessment of pain; observation for changes in soft tissue, stability, symmetry, and alignment; active and passive range of motion; and the client's participation in areas of occupation.

REFERENCES

Alexander, C., Miley, R., Stynes, S., & Harrison P. J. (2007). Differential control of the scapulothoracic muscles in humans. *Journal of Physiology 580*, 777-786.

American Occupational Therapy Association. (2008). Occupational therapy practice framework: Domain and process, second edition. *American Journal of Occupational Therapy, 62*, 625-683.

American Occupational Therapy Association (2002). Occupational therapy practice framework. *American Journal of Occupational Therapy, 56*(6), 609-639.

Bach, H. G., & Goldberg, B. A. (2006). Posterior capsular contracture of the shoulder. *Journal of the American Academy of Orthopaedic Surgery, 14*(5), 265-277.

Biel, A. (2001). *Trail guide to the body* (2nd ed.). Boulder, CO: Books of Discovery.

Blasier, R. B., Soslowsky, L. J., Malicky, D. M., & Palmer, M. L. (1997). Posterior glenohumeral subluxation acitve and passive stabilization in a biomechanical model. *Journal of Bone and Joint Surgery, 79A*, 433-440.

Bowen, M. K., & Warren, R. F. (1991). Ligamentous control of shoulder stability based on selective cutting and static translation experiments. *Clinical Sports Medicine, 10*(4), 757-782.

Cailliet, R. (1980). *Soft tissue pain and disability.* Philadelphia, PA: F. A. Davis.

Clark, J. M., & Harryman, M. D. (1992). Tendons, ligaments and capsule of the rotator cuff. *The Journal of Bone and Joint Surgery, 74-A*(5), 713-725.

Constant, C. R., & Murley, A. H. (1987). A clinical method of functional assessment of the shoulder. *Clinical Orthopaedics, 214*, 160-164.

Cooper, D. E., O'Brien, S. J., & Warren, R. F. (1993). Supporting layers of the glenohumeral joint: An anatomic study. *Clinical Orthopaedics and Related Research, 289*, 144-155.

Cyriax, J. (1982). Diagnosis of soft tissue lesions. In: *Textbook of orthopaedic medicine* (8th ed.). London: Bailliere Tindall.

Dawson, J., Hill, G., Fitzpatrick, R., & Carr, A. (2001). The benefits of using patient-based methods of assessment. Medium-term results of an observational study of shoulder surgery. *Journal of Bone and Joint Surgery, 83*(6), 877-882.

Donatelli, R. A. (1997). *Physical therapy of the shoulder.* Philadelphia, PA: Churchill Livingstone.

Donatelli, R. A., & Wooden, M. J. (2010). *Orthopaedic physical therapy* (4th ed.). St. Louis, MO: Churchill Livingstone Elsevier.

Dvir, Z. (2000). *Clinical biomechanics.* Philadelphia, PA: Churchill Livingstone.

Edelson, J. G., Taitz, C., & Grishkan, A. (1991). The coracohumeral ligament: Anatomy of a substantial but neglected structure. *Journal of Bone and Joint Surgery, 73*(1), 150-153.

Ellenbecker, T. S., & Ballie, D. S. (2010). The shoulder. In R. A. Donatelli & M. J. Wooden (Eds.), *Orthopaedic physical therapy* (4th ed., pp. 197-236). St. Louis, MO: Churchill Livingstone Elsevier.

Esch, D. L. (1989). *Musculoskeletal function: An anatomy and kinesiology laboratory manual.* Mineapolis, MN: University of Minnesota.

Flowers, K. R. & LaStayo, P. (1994). Effect of total end range time on improving passive range of motion. *Journal of Hand Therapy 7*(3), 150-157.

Fukuda, K., Craig, E. V., An, K. N., Cofield, R. H., & Chao, E. Y. (1986). Biomechanical study of the ligamentous system of the acromioclavicular joint. *Journal of Bone and Joint Surgery, 68*(3), 434-440.

Funk, L., & K. N. (2005). Atraumatic Shoulder Instability. Accessed April 14, 2010 from www.shoulderdoc.co.uk.

Gajdosik, R. L., & Bohannon, R. L. (1987). Clinical measurement of range of motion: Review of goniometry emphasizing reliability and validity. *Physical Therapy, 67*(12), 1867-1872.

Gench, B. E., Hinson, M. M., & Harvey, P. T. (1995). *Anatomical kinesiology.* Dubuque, IA: Eddie Bowers Publishing, Inc.

Gibson, J. (2010). Shoulder examination tests. Shoulderdoc. Accessed April 14, 2010, from www.shoulderdoc.co.uk.

Gillen, G., & Burkhardt, A. (2004). *Stroke rehabiliation: A function-based approach.* St. Louis, MO: Mosby.

Goss, C. M. (1976). *Gray's anatomy of the human body,* 29th ed. Philadelphia, PA: Lea and Febiger.

Greene, D. P., & Roberts, S. L. (1999). *Kinesiology: Movement in the context of activity.* St. Louis, MO: Mosby.

Table 5-28

TESTS OF SHOULDER PATHOLOGY AND TESTS FOR SPECIFIC JOINTS

Neurological involvement	Upper limb tension test (brachial plexus tension)
	Tinel's sign (at the shoulder)
Thoracic outlet syndrome	Roos tests (EAST)
	Wright test or maneuver
	Costoclavicular syndrome (military brace) test
	Provocative elevation test
	Shoulder girdle passive elevation
	Adson maneuver
	Halstead maneuver
Fracture	Comolli sign
Bursitis	Dawburn's Test
Ligament pathology	Crank test
	Posterior inferior glenohumeral ligament test
	Coracoclavicular ligament test
Labral lesions	Clunk test
	Anterior slide test
	The SLAP-prehension test
	Biceps tension test
	Posterior labral tear
	Push-pull test
	Jahnke jerk test
	Biceps load test 1
	Biceps load test 2
	Pain provocation test
Acromioclavicular joint	Acromioclavicular shear test
	Anterior/posterior AC shear test
	Acromioclavicular crossover, crossbody or horizontal adduction test
	Ellman's compression rotation test
	Alternate test for the AC joint/shoulder depression test (for sprains)
	Cross-chest adduction (scarf/forced adduction test)
	Cross-arm test
	Forced adduction test on hanging arm
	Dugas test
	AC distraction (bad cop) test
	Paxinos test
	O'Briens test
Sternoclavicular joint	Test for the sternoclavicular joint/SC squeeze test (sprains)
	SC joint stress test
	Scapula pinch/retraction test (for scapula stability)
	Thompson and Kopell horizontal flexion test
Cervical spine	Spurling's test

Gutman, S. A., & Schonfeld, A. B. (2003). *Screening adult neurological populations.* Bethesda, MD: American Occupational Therapy Press.

Hamill, J. & Knutzen, K. M. (2003). *Biomechanical Basis of Human Movement,* 2nd ed. Philadelphia, PA: Lippincott, Williams & Wilkins.

Hamill, J. & Knutzen, K. M. (1995). *Biomechanical Basis of Human Movement.* Philadelphia, PA: Lippincott, Williams & Wilkins.

Harryman, D. T., Clark, J. M., McQuade, K. J., Gibb, T. D., & Matsen, F. A. (1990). Translation of the humeral head on the glenoid with passive glenohumeral motion. *Journal of Bone and Joint Surgery, 72,* 1334-1343.

Hartley, A. (1995). *Practical joint assessment: Upper quadrant: A sports medicine manual* (2nd ed.). Philadelphia, PA: C. V. Mosby Co.

Hayes, K., Walton, J. R., Szomor, Z. L., & Murrell, G. A. (2001). Reliability of five methods for assessing shoulder range of motion. *Australian Journal of Physiotherapy, 47,* 289-294.

Hellebrandt, R. A., Duvall, E. N., & Moore, M. L. (1949). The measurement of joint motion: Part 3: Reliability of goniometry. *Physical Therapy Review, 29,* 302-307.

Hertling, D., & Kessler, R. M. (1996). *Management of common musculoskeletal disorders: Physical therapy principles and methods.* New York, NY: Churchill Livingstone.

Hess, S. A. (2000). Functional stability of the glenohumeral joint. *Manual Therapy, 5*(2), 63-71.

Hinkle, C. Z. (1997). *Fundamentals of anatomy and movement: A workbook and guide.* St. Louis, MO: C. V. Mosby Co.

Howell, S. M., Galinat, B. J., Renzi, A. J., & Marone, P. J. (1988). Normal and abnormal mechanics of the glenohumeral joint in the horizontal plane. *Journal of Bone and Joint Surgery, 70A*(2), 227-232.

Inman, V. T., Saunders, J. B., & Abbot, L. C. (1944). Observations on the function of the shoulder joint. *Journal of Bone and Joint Surgery, 26,* 1-30.

Jenkins, D. B. (1998). *Hollingshead's functional anatomy of the limbs and back* (7th ed.). Philadelphia, PA: W. B. Saunders Co.

Jobe, F. W., Kvitne, R. S., & Giangarra, C. E. (1989). Shoulder pain in the overhand or throwing athlete: the relationship of anterior instability and rotator cuff impingement. *Orthopaedic Review, 18,* 963-975.

Kebaetse, M., McClure, P., & Pratt, N. A. (1999). Thoracic position effect on shoulder range of motion, strength, and three-dimensional scapular kinematics. *Archives of Physical Medicine and Rehabilitation, 80,* 945-950.

Kendall, F. P., McCreary, E. K., Provance, P. G., Rodgers, M. M., & Romani, W. A. (Eds.) (2005). *Muscles: Testing and function with posture and pain* (5th ed.). Baltimore, MD: Lippincott Williams & Wilkins.

Kibler, W. B. (1998). The role of the scapula in athletic shoulder function. *The American Journal of Sports Medicine, 26*(2), 325-337.

Kibler, W. B., & McMullen, J. (2003). Scapular dyskinesis and its relation to shoulder pain. *Journal of the American Academy of Orthopaedic Surgeons, 11,* 142-151.

Kibler, W. B., Uhl, T. L., Maddux, J. W. Q., Brooks, P. V., Zeller, B., & McMullen, J. (2002). Qualitative clinical evaluation of scapular dysfunction: a reliability study. *Journal of Shoulder and Elbow Surgery, 11*(6), 550-556.

Kirby, K., & Showalter, C. (2007). Assessment of the importance of glenohumeral peripheral mechanics by practicing physiotherapists. *Physiotherapy Research International 12*(3), 136-146.

Kisner, C., & Colby, L. A. (2007). *Therapeutic exercise: Foundations and techniques* (5th ed.). Philadelphia, PA: F. A. Davis Company.

Kisner, C., & Colby, L. A. (1990). *Therapeutic exercise: Foundations and techniques* Philadelphia, PA: F. A. Davis Company.

Kvitne, R. S., & Jobe, F. W. (1993). The diagnosis and treatment of anterior instability in the throwing athlete. *Clinical Orthopaedics, 291,* 107-123.

Lephart, S. M. & Jari, R. (2002). The role of proprioception in shoulder stability. *Operative Techniques in Sports Medicine, 10*(1), pp 2-4.

Levangie, P. K., & Norkin, C. C. (2005). *Joint structure and function: A comprehensive analysis* (4th ed.). Philadelphia, PA: F. A. Davis Company.

Lewis, A., Kitamura, T., & Bayley, J. I. L. (2004). The classification of shoulder instability: new light through old windows! *Current Orthopaedics, 18,* 97-108.

Ludewig, P. M., Phadke, V., Braman, J. P., Hassett, D. R., Cieminski, C. J., & LaPrade, R. F. (2009). Motion of shoulder complex during multiplanar humeral elevation. *Journal of Bone and Joint Surgery, 91,* 378-389.

Ludewig, P. M., & Reynolds, J. F. (2009). The association of scapular kinematics and glenohumeral joint pathologies. *Journal of Orthopedic and Sports Physical Therapy, 39*(2), 90-104.

Lumley, J. S. P. (1990). *Surface anatomy: The anatomical basis of clinical examination.* New York, NY: Churchill Livingstone.

MacDermid, J. C., Chesworth, B. M., Patterson, S., & Roth, J. H. (1999). Intratester and intertester reliability of goniometric measurement of passive lateral shoulder rotation. *Journal of Hand Therapy, 12,* 187-192.

Magee, D. J. (2002). *Orthopedic physical assessment.* Philadelphia, PA: Saunders.

Magermans, D. J., Chadwick, E. K. J., Veeger, H. E. J., & vanderHelm, F. C. T. (2005). Requirements for upper extremity motions during activities of daily living. *Clinical Biomechanics, 20,* 591-599.

Matsen, F. A., Harryman, D. T., & Sidles, J. A. (1991). Mechanics of glenohumeral instability. *Clinics in Sports Medicine, 10*(4), 783-788.

Matsen, F. A., Lippitt, S. B., Sidles, J. A., & Harryman, D. T. (1994). *Practical evaluation and management of the shoulder.* Philadelphia, PA: W. B. Saunders Co.

McClure, P. W., Michener, L. A., Sennett, B. J., & Karduna, A. R. (2001). Direct 3-dimensional measurement of scapular kinematics during dynamic movements in vivo. *Journal of Shoulder and Elbow Surgery, 10*(3), 269-277.

Muscolino, J. E. (2006). *Kinesiology: The skeletal system and muscle function.* St. Louis, MO: Mosby.

Myers, J. B., Wassinger, C. A., & Lephart, S. M. (2006). Sensorimotor contribution to shoulder stability: effect of injury and rehabilitation. *Manual Therapy, 11,* 197-201.

Neer, C. S. (1972). Anterior acromioplasty for the chronic impingement syndrome in the shoulder: a preliminary report. *Journal of Bone and Joint Surgery, 54,* 41-50.

Neumann, D. A. (2002). *Kinesiology of the musculoskeletal system: Foundations for physical rehabilitation.* St. Louis, MO: Mosby.

Nordin, M., & Frankel, V. H. (2001). *Basic biomechanics of the musculoskeletal system.* Philadelphia, PA: Lippincott, Williams & Wilkins.

Nordin, M., & Frankel, V. H. (1989). *Basic biomechanics of the musculoskeletal system* (2nd ed.). Philadelphia, PA: Lea and Febiger.

Norkin, C. C. & Levangie, P. K. (1992). *Joint structure and function: A comprehensive analysis.* Philadelphia, PA: F. A. Davis Co.

Nyland, J. (2006). *Clinical decisions in therapeutic exercise: Planning and implementation.* Upper Saddle River, NJ: Pearson Prentice Hall.

Oatis, C. A. (2004). *Kinesiology: The mechanics and pathomechanics of human movement.* Philadelphia, PA: Lippincott, Williams & Wilkins.

Palmer, M. L., & Epler, M. E. (1998). *Fundamentals of musculoskeletal assessment techniques* (2nd ed.). Philadelphia, PA: Lippincott.

Parsons, I. M., Apreleva, M., Fu, F. H., & Woo, S. L. (2002). The effect of rotator cuff tears on reaction forces at the glenohumeral joint. *Journal of Orthopaedic Research, 20,* 439-446.

Perry, J. F., Rohe, D. A., & Garcia, A. O. (1996). *Kinesiology workbook.* Philadelphia, PA: F. A. Davis Co.

Reinold, M. M., Escamilla, R., & Wilk, K. E. (2009). Current concepts in the scientific and clinical rationale behind exercises for glenohumeral and scapulothoracic musculature. *Journal of Orthopedic and Sports Physical Therapy, 39*(2), 105-117.

Riddle, D. L., Rothstein, J. M., & Lamb, R. L. (1987). Goniometric reliability of shoulder measurements in a clinical setting. *Physical Therapy, 6*(7), 668-673.

Roddey, T. S., Cook, K. F., O'Malley, K. J., & Gartsman, G. M. (2005). The relationship among strength and mobility measures and self-report outcome scores in persons after rotator cuff repair surgery: Impairment measures are not enough. *Journal of Shoulder and Elbow Surgery, 14*(1 Suppl S), 95S-98S.

Roy, J.-S., MacDermid, J. C., Boyd, K. U., Faber, K. J., Drosdowech, D., & Athwal, G. S. (2009). Rotational strength, range of motion, and function in people with unaffected shoulders from various stages of life. *Sports Medicine, Arthroscopy, Rehabilitation, Therapy & Technology, 1*(4), 1-7.

Sabari, J. S., Maltzev, I., Lubarsky, D., Liszkay, E., & Homel, P. (1998). Goniometric assessessment of shoulder range of motion: Comparison of testing in supine and sitting positions. *Archives of Physical Medicine and Rehabilitation, 79,* 647-651.

Saidoff, D. C., & McDonough, A. L. (1997). *Critical pathways in therapeutic intervention: Upper extremity.* St. Louis, MO: C. V. Mosby.

Shafer, R. C. (1997). Shoulder girdle trauma: Monograph 16. Retrieved from http://www.chiro.org/ACAPress; Accessed June 2, 2009.

Smith, L. K., Weiss, E. L., & Lehmkuhl, L. D. (1996). *Brunnstrom's clinical kinesiology* (5th ed.). Philadelphia, PA: F. A. Davis Co.

Spencer, E. E., Kuhn, J. E., Carpenter, J. E., & Huges, R. E. (2002). Ligamentous restraints to anterior and posterior translation of the sternoclavicular joint. *Journal of Shoulder and Elbow Surgery, 11*(1), 43-47.

Trew, M., & Everett, T. (Eds.). (2005). *Human movement: An introductory text.* London: Elsevier Churchill Livingstone.

Uhl, T. L., Kibler, W. B., Gecewich, B., & Tripp, B. L. (2009). Evaluation of clinical assessment methods for scapular dyskinesis. *Arthroscopy: The Journal of Arthroscopic and Related Surgery, 25*(11), 1240-1348.

Veeger, H. E. J., Magermans, D. J., Nagels, J., Chadwick, E. K. J., & vanderHelm, F. C. T. (2006). A kinematic analysis of the shoulder after arthroplasty during a hair combing task. *Clinical Biomechanics, 21*, S39-S44.

Veeger, H. E. J., & vanderHelm, F. C. T. (2007). Shoulder function: The perfect compromise between mobility and stability. *Journal of Biomechanics, 40*, 2119-2129.

Walker, J. M., Sue, D., Miles-Elkousy, N., Ford, G., & Trevelyan, H. (1984). Active mobility of the extremities in older subjects. *Physical Therapy, 64*(6), 919-923.

Wilk, K. E., Arrigo, C. A., & Andrews, J. R. (1997a). The physical examination of the glenohumeral joint: Emphasis on the stabilizing structures. *Journal of Orthopedic and Sports Physical Therapy, 25*(6), 380-389.

Wilk, K. E., Arrigo, C. A., & Andrews, J. R. (1997b). Current concepts: The stabilizing structures of the glenohumeral joint. *Journal of Orthopedic and Sports Physical Therapy, 25*(6), 364-379.

Woodward, T. W., & Best, T. M. (2000a). The painful shoulder: Part I. Clinical evaluation. *American Family Physician, 61*(10), 3079-3088.

Woodward, T. W., & Best, T. M. (2000b). The painful shoulder: Part II: Acute and chronic disorders. *American Family Physician, 61*, 3079-3088.

Zatsiorsky, V. M. (1998). *Kinematics of human motion.* Champaign, IL: Human Kinetics.

Zuckerman, J. E., & Matsen, F. A. (1984). Complications about the glenohumeral joint related to the use of screws and staples. *Journal of Bone and Joint Surgery, 66A*, 175-180.

ADDITIONAL RESOURCES

Andrews, R. A., & Harrelson, G. L. (1991). *Physical rehabilitation of the injured athlete.* Philadelphia, PA: W.B. Saunders.

Aydin, T., Yildiz, Y., Yanmis, I., Yildiz, C., & Kalyon, T. (2001). Shoulder proprioception: a comparison between the shoulder joint in healthy and surgically repaired shoulders. *Journal Archives of Orthopaedic and Trauma Surgery, 121*(7), 422-425.

Bak, K., & Fauno, P. (1997). Clinical findings in competitive swimmers with shoulder pain. *Am J Sports Med, 25*, 254-260.

Barham, J. N., Wooten, E. P. (1973). *Structural kinesiology.* New York, NY: Macmillan.

Basmajian, J. V. (1969). Recent advances in the functional anatomy of the upper limb. *American Journal of Physical Medicine, 48*, 165-177.

Basmajian, J. V., & DeLuca, C. J. (1985). *Muscles alive* (5th ed.). Baltimore, MD: Williams and Wilkins.

Baxter, R. (1998). *Pocket guide to musculoskeletal assessment.* Philadelphia, PA: W. B. Saunders Co.

Bergfeld, J. A., Andrish, J. T., & Clancy, W. G. (1978). Evaluation of the acromioclavicular joint following first- and second-degree sprains. *American Journal of Sports Medicine, 6*, 153-159.

Birnbaum, J. S. (1986). *The musculoskeletal manual* (2nd ed.). Orlando, FL: Grune & Stratton.

Bjerneld, H., Hovelius, L., & Thorling, J. (1983). Acromio-clavicular separations treated conservatively: A five-year follow-up study. *Acta Orthopaedica Scandinavica, 54*, 743-745.

Black, J., & Dumbleton, J. H. (Eds.) (1981). *Clinical biomechanics: Case history approach.* New York, NY: Churchill Livingstone.

Boissonnault, W. C., & Janos, S. C. (1989). Dysfunction, evaluation, and treatment of the shoulder. In: Donatelli, R., & Wooden, M. (Eds.), *Orthopaedic physical therapy.* New York, NY: Churchill Livingston.

Bowerman, J. W. (1977). *Radiology and injury in sport.* New York, NY: Appleton-Century-Crofts.

Branch, W. T. Jr. (1987). *Office practice of medicine.* Philadelphia, PA: W. B. Saunders.

Burns, L., Chandler, L. C., & Rice, R. W. (1948). *Pathogenesis of visceral disease following vertebral lesions.* Chicago, IL: American Osteopathic Association.

Burstein, A. H., & Wright, T. M. (1994). *Fundamentals of orthopaedic biomechanics.* Baltimore, MD: Williams and Wilkins.

Cailliet, R. (1981). *Neck and arm pain.* Philadelphia, PA: F. A. Davis.

Cailliet, R. (1981). *Shoulder pain* (2nd ed.). Philadelphia, PA: F. A. Davis.

Chaitow, L. (2004). *Maintaining body balance, flexibility and stability.* St. Louis, MO: Churchill Livingstone.

Clarkson, H. M. (2000). *Musculoskeletal assessment, joint range of motion and manual muscle strength.* Philadelphia, PA: Lippincott Williams & Wilkins.

Conolly, W. B., & Prosser, R. (Eds.) (2003). *Rehabilitation of the hand and upper limb.* St. Louis, MO: Butterworth Heinemann.

Corso, G. (1995). Impingement relief test: An adjunctive procedure to traditional assessment of shoulder impingement syndrome. *Journal of Orthopedic and Sports Physical Therapy, 22*(5), 183-192.

Craig, A. S. (1964). Elements of kinesiology for the clinician. *Physical Therapy, 44*, 470-473.

Crawford, J. P., Hwang, B. Y., Asselbergs, P. J., & Hickson, G. S. (1984). Vascular ischemia of the cervical spine: A review of relationship to therapeutic manipulation. *Journal of Manipulative and Physiological Therapeutics, 7*(3), 149-154.

Culham, E., & Peat, M. (1993). Functional anatomy of the shoulder complex. *Journal of Orthopedic and Sports Physical Therapy, 18*, 342-350.

Dalinka, M. K. (1978). Ankle fractures and fractures and dislocation about the shoulder. In: Feldman, F. (Ed.), *Radiology, pathology, and immunology of bones and joints.* New York, NY: Appleton-Century-Crofts.

Dempster, W. T. (1965). Mechanisms of shoulder movement. *Archives of Physical Medicine and Rehabilitation, 46A*, 49.

Deutsch, A., Altchek, D. W., Veltri, D. M., Potter, H. G., & Warren, R. F. (1997). Traumatic tears of the subscapularis tendon: clinical diagnosis, magnetic resonance imaging findings, and operative treatment. *Am J Sports Med, 25*, 13-22.

Ellenbecker, T. (2004). *Clinical examination of the shoulder.* St. Louis, MO: W. B. Saunders.

Emery, R. H., & Mullaji, A. B. (1991). Glenohumeral joint instability in normal adolescents. *J Bone Joint Surg [Br], 73B*, 406-408.

Garrick, J. G., & Webb, D. R. (1990). *Sports injuries: Diagnosis and management.* Philadelphia, PA: W. B. Saunders, 55-98.

Gerber, C., & Krushell, R. J. (1991). Isolated rupture of the tendon of the subscapularis muscle: clinical features in 16 cases. *J Bone Joint Surg [Br], 73B*, 389-394.

Gillet, H., & Liekens, M. (1981). *Belgian chiropractic research notes.* Huntington Beach, CA: Motion Palpation Institute.

Glasgow, E. F., Twomey, L. T., Scull, E. R., Kleynhans, A. M., & Idczak, R. M. (1985). *Aspects of manipulative therapy* (2nd ed.). New York, NY: Churchill Livingstone.

Glasgow, S. G., Bruce, R. A., Yacobucci, G. N., & Torg, J. S. (1992). Arthroscopic resection of glenoid labral tears in the athlete: a report of 29 cases. *Arthroscopy, 8*, 48-54.

Goodheart, G. J. (1969). *Collected published articles and reprints.* Montpelier, OH: Williams County Publishing.

Hadler, N. M. (1984). *Medical management of the regional musculoskeletal diseases.* New York, NY: Grune & Stratton.

Harrelson, G. L. (1991). Shoulder rehabilitation. In: Andrews, R. A., & Harrelson, G. L., *Physical rehabilitation of the injured athlete* (pp. 367-394, 409-418). Philadelphia, PA: W. B. Saunders.

Harrington, I. J. (1982). Biomechanics of joint injuries. In: Gonzna, E. R., & Harrington, I. J., *Biomechanics of musculoskeletal injury* (pp. 67-79). Baltimore, MD: Williams & Wilkins.

Harryman, D. T., Sidles, J. A., Harris, S. L., et al. (1992). The role of the rotator interval capsule in passive motion and stability of the shoulder. *J Bone Joint Surg Am, 74*, 53-66.

Harryman, D. T., Sidles, J. A., Harris, S. L., & Matsen, F. A. (1992). Laxity at the normal glenohumeral joint: a quantitative in-vivo assessment. *J Shoulder Elbow Surg, 1*, 66-76.

Hawkins, R., & Kennedy, J. (1980). Impingement syndrome in athletes. *American Journal of Sports Medicine, 8(3)*, 151-158.

Hirata, I. Jr. (1974). *The doctor and the athlete* (2nd ed.). Philadelphia, PA: J. B. Lippincott.

Hole, J. D. Jr. (1978). *Human anatomy and physiology*. Dubuque, IA: William C. Brown Co. Publishers.

Inman, V. T., Saunders, M., & Abbot, LC. (1944). Observations of the function of the shoulder joint. *Journal of Bone and Joint Surgery, 26A*, 1-30.

Iverson, L. D., & Clawson, D. K. (1977). Manual of acute orthopaedic therapeutics. Boston, MA: Little, Brown.

Jacobs, R. C. (1979). Painful shoulder. In: Friedman, H. H. (Ed.), *Problem-oriented medical diagnosis* (2nd ed.). Boston, MA: Little, Brown.

Jaquet, P. (1978). *Clinical chiropractic: A study of cases*. Geneva, Switzerland: Grounauer.

Kendall, F. P., McCreary, E. K., & Provance, P. G. (1993). *Muscle testing and function*. Baltimore, MD: Williams & Wilkins.

Kessler, R. M., & Hertling, D. (Eds.) (1983). *Management of common musculoskeletal disorders*. Philadelphia, PA: Harper & Row.

Konin, J. G. (1999). *Practical kinesiology for the physical therapist assistant*. Thorofare, NJ: SLACK Incorporated.

Kozin, F. (1983). Two unique shoulder disorders. *Postgraduate Medicine, 73(5)*, 211-215.

Krupp, M. A., & Chatlon, M. J. (1974). *Current medical diagnosis and treatment*. Los Altos, CA: Lange Medical Publications.

Leek, J. C., Gershwin, M. E., & Fowler, W. M. Jr. (1986). *Principles of physical medicine and rehabilitation in the musculoskeletal diseases*. Orlando, FL: Grune & Stratton.

Lehrman, R. L. (1998). *Physics the easy way* (3rd ed.). Hauppauge, NY: Barron's Educational Series, Inc.

LeVeau, B. F. (1992). *Williams and Lissner's biomechanics of human motion* (3rd ed.). Philadelphia, PA: W. B. Saunders Co.

Loth, T., & Wadsworth, C. T. (1998). *Orthopedic review for physical therapists*. St. Louis, MO: C. V. Mosby Co.

Machner, A., Merk, H., Becker, R., Rohkohl, K., Wissel, H., & Pap, G. (2003). Kinesthetic sense of the shoulder in patients with impingement syndrome. *Acta Orthopedic Scand, 74(1)*, 85-88.

MacKenna, B. R., & Callender, R. (1990). *Illustrated physiology* (5th ed.). New York, NY: Churchill Livingstone.

Mennell, J. M. (1964). *Joint pain*. Boston, MA: Little, Brown.

Mercier, L. R. (1980). *Practical orthopedics*. Chicago, IL: Year Book Medical.

Miniaci, A., & Salonen, D. (1997). Rotator cuff evaluation imaging and diagnosis. *Orthop Clin North Am, 28*, 43-58.

Morehouse, L. E., & Cooper, J. M. (1950). *Kinesiology*. St. Louis, MO: C. V. Mosby.

Morrey, B. F., Itoi, E., & An, K. (1998). Biomechanics of the shoulder. In: Rockwood, C. A., & Matsen, F. A. (Eds.), *The shoulder*. Philadelphia, PA: W. B. Saunders.

Nicholas, J. A., & Hershman, E. B. (1990). *The upper extremity in sports medicine*. St. Louis, MO: C. V. Mosby Co.

O'Donoghue, D. H. (1984). *Treatment of injuries to athletes* (4th ed.). Philadelphia, PA: W. B. Saunders.

O'Driscoll, S. W. (1991). A reliable and simple test for posterior instability of the shoulder. *J Bone Joint Surg [Br], 73B*(suppl 1), 50.

Palastanga, N., Field, D., & Soames, R. (1989). *Anatomy and human movement: Structure and function*. Oxford: Heinemann Medical Books.

Palmer, M. L., & Blakely, R. L. (1986). Documentation of medial rotation accompanying shoulder flexion. A case report. *Physical Therapy, 66*, 55-58.

Pollock, M. L., & Wilmore, J. H. (1990). *Exercise in health and disease* (2nd ed.). Philadelphia, PA: W. B. Saunders.

Resnick, D. (1983). Shoulder pain. *Orthopaedic Clinics of North America, 14*, 81-97.

Riggins, R. S. (1977). The shoulder. In: D'Ambrosia, R. D., *Musculoskeletal disorders: Regional examination and differential diagnosis*. Philadelphia, PA: J. B. Lippincott.

Rizk, T. E., & Pinals, R. S. (1982). Frozen shoulder. *Arthritis and Rheumatism, 11*, 440-452.

Rocks, J. A. (1979). Intrinsic shoulder pain syndrome. *Physical Therapy, 59*, 153-159.

Rosse, C. (1980). The shoulder region and the brachial plexus. In: Rosse, C., & Clawson, D. K. (Eds.), *The musculoskeletal system in health and disease*. Hagerstown, PA: Harper & Row.

Rowe, C. R., & Zarins, B. (1981). Recurrent transient subluxation of the shoulder. *Journal of Bone & Joint Surgery, 63A*, 863- 871.

Sarrafian, S. K. (1983). Gross and functional anatomy of the shoulder. *Clinical Orthopaedics & Related Research, 173*, 11-19.

Schafer, R. C. (1982). Hot shots and brachial plexus traction. *Journal of the American Chiropractic Association*, September.

Schafer, R. C. (1989). *Chiropractic management of extraspinal articular disorders*. Arlington, VA: American Chiropractic Association.

Schafer, R. C. (1986). *Chiropractic management of sports and recreational injuries* (2nd ed.). Baltimore, MD: Williams & Wilkins.

Schafer, R. C. (1987). *Clinical biomechanics: Musculoskeletal actions and reactions* (2nd ed.) Baltimore, MD: Williams & Wilkins.

Schafer, R. C. (1991). *Clinical chiropractic: The management of pain and disability—Upper body complaints*. Huntington Beach, CA: The Motion Palpation Institute.

Schneider, R. C., Kennedy, J. C., & Plant, M. L. (Eds.) (1985). *Sports injuries: Mechanisms, prevention, and treatment*. Baltimore, MD: Williams & Wilkins.

Schultz, A. L. (1969). *The shoulder, arm, and hand syndrome*. Stickney, SD: Argus.

Scott, W. N., Nisonson, B., & Nicholas, J. A. (Eds.) (1984). *Principles of sports medicine*. Baltimore, MD: Williams & Wilkins.

Shands, C. (1975). *Chiropractic rehabilitation* (2nd ed.). Danville, CA: Life at Its Peak.

Simon, W. H. Soft tissue disorders of the shoulder. *Orthopaedic Clinics of North America, 6*, 521-539.

Slocum, D. B. (1959). The mechanics of some common injuries to the shoulder in sports. *American Journal of Surgery, 98*, 394-400.

Tullos, H. S., & Bennett, J. B. (1984). The shoulder in sports. In: Scott, W. N., et al (eds.), *Principles of sports medicine*. Baltimore, MD: Williams & Wilkins.

Turek, S. L. (1977). *Orthopaedics: Principles and their application* (3rd ed.). Philadelphia, PA: J. B. Lippincott.

Warner, J. J., Lephart, S., & Fu, F. (1996). Role of proprioception: Pathoetiology of shoulder instability. *Clinical Orthopaedics & Related Research, 330*, 35-39.

Warren, R. F. (1983). Subluxation of the shoulder in athletes. *Clinical Sports Medicine, 2*, 339-354.

Watkins, J. (1999). *Structure and function of the musculoskeletal system*. New York, NY: Human Kinetics.

Weed, N. D. (1983). When shoulder pain isn't bursitis. *Postgraduate Medicine, 73(3)*, 97-104.

Williams, J. G. P., & Sperry, P. N. (Eds.) (1976). *Sports medicine* (2nd ed.). Baltimore, MD: Williams & Wilkins.

Yergason, R. M. (1931). Supination sign. *J Bone Joint Surg [Am], 13*, 160.

Yoneda, B., Welsh, R. P., & McIntosh, D. L. (1982). Conservative treatment of shoulder dislocation in young males. *Journal of Bone and Joint Surgery, 64B*, 254-255.

6

THE ELBOW

The elbow joint and motion of the forearm are important in enabling proper positioning of the hands and fingers in space and permitting height and length adjustments to be made during activities. Rotation of the forearm helps to place the hand closer to the face and to position the arm to enable the most beneficial length-tension relationships in muscles. The elbow joint assists the shoulder with force distribution and in stabilizing the upper extremity for both power and fine coordination.

The elbow joint is not one joint, but actually three: the humerus articulates with the radius (humeroradial or radio-humeral) and the ulna (humeroulnar or ulnohumeral), and the ulna and radius articulate with each other (the superior/proximal radioulnar articulation) (Figure 6-1).

BONES OF THE ELBOW AND PALPABLE STRUCTURES

The bones of the elbow include the humerus, ulna, and radius (Figure 6-2). Many of the bony structures are easily palpated.

Humerus

Epicondyles

These are the distal enlargements of the humerus. When the humerus is externally rotated, the medial epicondyle lies close to the body and is found easily as you move medially from the olecranon. It is large, superficial, and rounded. The medial epicondyle is known as the flexor epicondyle, because many of the wrist and fingers attach here.

The lateral epicondyle points to the back when the humerus is externally rotated and can be found by moving laterally from the olecranon. The lateral epicondyle is known as the extensor epicondyle and is smaller than the medial epicondyle (Biel, 2001; Smith, Weiss, & Lehmkuhl, 1996) (Figure 6-3).

The two epicondyles and the tip of the olecranon form an equilateral triangle when the elbow is flexed to 90 degrees and a straight line when the elbow is in extension (Figure 6-4).

Ulna

Olecranon Process

This is the "point" of the elbow or the upper and posterior aspect of elbow, which is easily felt when the forearm is flexed to 90 degrees. The ulnar nerve runs in this area, which, when compressed, produces the tingling sensation and is referred to as the "funny bone." This is the attachment site for the triceps brachii muscle. The olecranon fossa can also be felt around the top of the olecranon process as a small "crescent-shaped ditch" as you press through the triceps tendon (Biel, 2001) (Figure 6-5).

Body of the Ulna

On the dorsal surface of the forearm, you can palpate the shaft of the ulna from the olecranon process to the distal end or head.

Rybski MF.
Kinesiology for Occupational Therapy, Second Edition (pp 175-208)
© 2012 SLACK Incorporated

Figure 6-1. Articulations of the elbow.

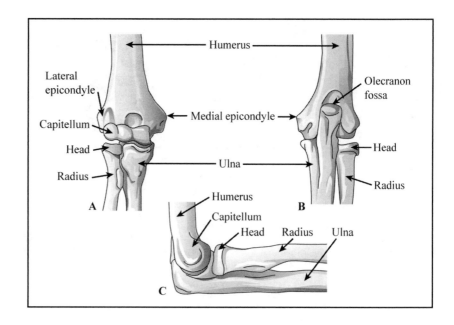

Figure 6-2. Bones of the elbow.

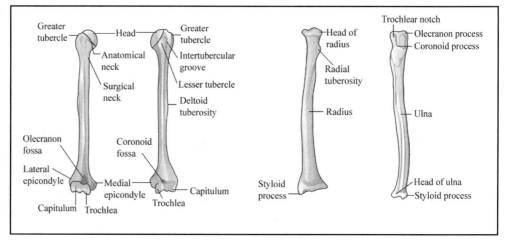

Figure 6-3. Palpation of epicondyles of the humerus. (Adapted from Biel, A. (2001). *Trail guide to the body.* Boulder, CO: Books of Discovery.)

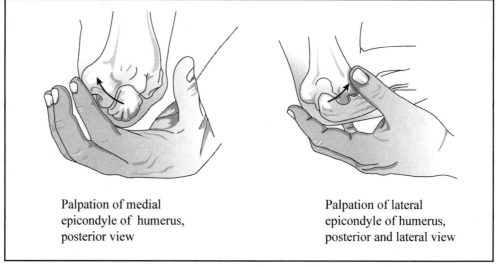

Palpation of medial epicondyle of humerus, posterior view

Palpation of lateral epicondyle of humerus, posterior and lateral view

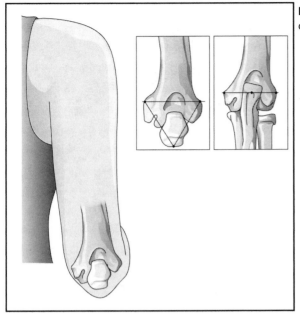

Figure 6-4. Equilateral triangle formed by epicondyles and olecranon process.

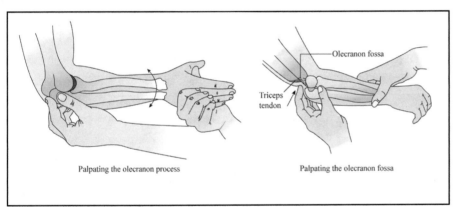

Figure 6-5. Palpation of the olecranon process and fossa. (Adapted from Biel, A. (2001). *Trail guide to the body.* Boulder, CO: Books of Discovery.)

Olecranon fossa

Triceps tendon

Palpating the olecranon process

Palpating the olecranon fossa

Head of the Ulna

This is the rounded projection on the dorsal surface of the forearm just proximal to the wrist. It is visible along the posterior and medial side of the forearm when in pronation (Figure 6-6).

Styloid Process of Ulna

This is a small projection on the posterior and medial aspect of the head of the ulna. This is a pronounced projection pointing distally off of the head of the ulna (Biel, 2001) (Figure 6-7).

Radius

Head

The head of the radius is distal to the lateral epicondyle of the humerus. With the elbow in extension, palpate the dorsal surface just distal to the lateral epicondyle of humerus. Supinating and pronating the forearm will help in identifying

this structure (Smith et al., 1996) (Figure 6-8). Even minor effusions or mild synovitis can be palpated in the triangle formed by the lateral epicondyle, the radial head, and the olecranon (Figure 6-9).

Styloid Process of the Radius

Both the radius and ulna have styloid processes at their distal ends, but the radial styloid process extends further distally and is wider than the styloid process of the ulna.

The styloid process of the radius is located on the lateral aspect of the wrist proximal to the first metacarpal. It can be felt by following the distal radial shaft along the lateral side of the forearm and is surrounded by extensor tendons and is the attachment site for the brachioradialis muscle (Biel, 2001) (Figure 6-10).

Shaft of the Radius

You can feel the radial shaft on the lateral side of the forearm in the lower half of the forearm. Most of the shaft of the radius is located beneath muscle tissue and needs to be palpated at the distal portion of the shaft.

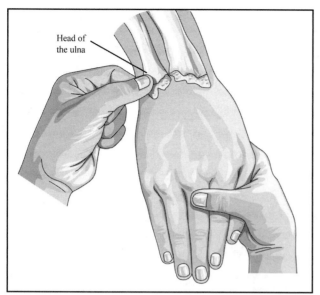

Figure 6-6. Palpation of the head of the ulna. (Adapted from Biel, A. (2001). *Trail guide to the body.* Boulder, CO: Books of Discovery.)

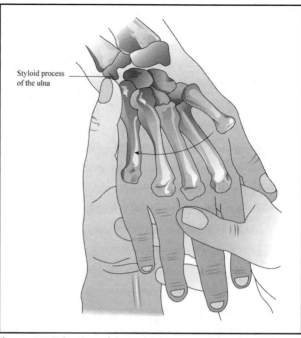

Figure 6-7. Palpation of the styloid process of the ulna. (Adapted from Biel, A. (2001). *Trail guide to the body.* Boulder, CO: Books of Discovery.)

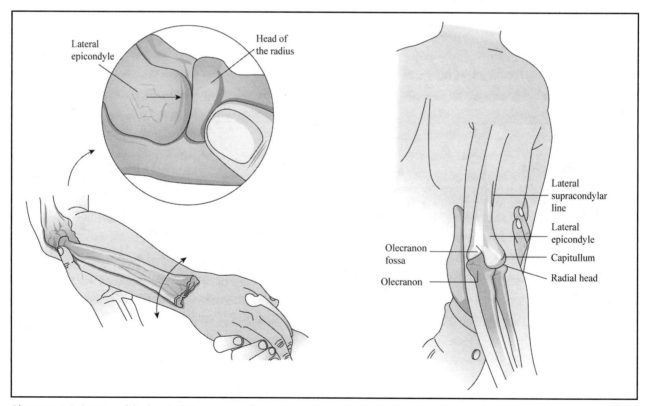

Figure 6-8. Palpation of the head of the radius. (Adapted from Biel, A. (2001). *Trail guide to the body.* Boulder, CO: Books of Discovery.)

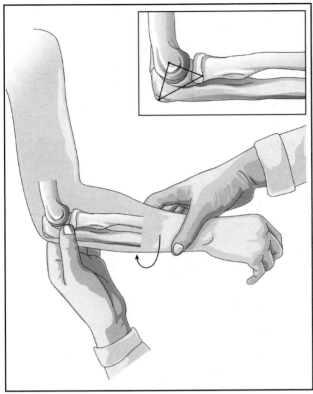

Figure 6-9. Palpation of the head of the radius, lateral epicondyle, and olecranon.

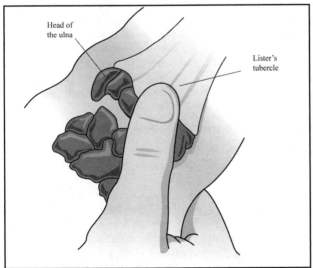

Figure 6-11. Palpation of Lister's tubercle. (Adapted from Biel, A. (2001). *Trail guide to the body.* Boulder, CO: Books of Discovery.)

Lister's Tubercle

The tubercle of Lister is located on the dorsum of the radius, about 1 inch laterally from the head of the ulna. The tendon of extensor pollicis longus lies on the ulnar side of this prominence. It can be found by palpating on the dorsal surface of the radius moving toward the head of the ulna, and you will feel a superficial ridge. This tubercle can serve as a landmark

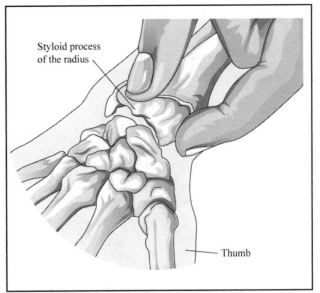

Figure 6-10. Palpation of the styloid process of the radius. (Adapted from Biel, A. (2001). *Trail guide to the body.* Boulder, CO: Books of Discovery.)

to help in identifying the lunate and capitate carpal bones (Biel, 2001) (Figure 6-11).

Nonpalpable Structures

The following are structures that are either too deep in the forearm to palpate or are difficult to differential due to overlying muscles and tendons.

- Neck of radius
- Radial tuberosity
- Trochlear notch
- Capitulum of humerus
- Coronoid process of ulna (Smith et al., 1996)

ARTICULATIONS

The elbow joint has been classified as a compound uniaxial hinge joint (Gench, Hinson, & Harvey, 1995; Nordin & Frankel, 2001), as a multiarticulating biaxial joint (Hamill & Knutzen, 2003), and as a "composite trochleoginglymoid joint" (Nordin & Frankel, 2001). The articulations of the humerus with the ulna and radius are uniaxial hinge/ginglymus joints, whereas the articulation of the ulna with the radius is a uniaxial pivot/trochoid joint (Figure 6-12). The articulations together provide two degrees of freedom at the elbow with the movements of flexion and extension and of pronation and supination.

Of the three articulations of the elbow, it is the humeroulnar and the humeroradial joints that produce flexion and extension of the elbow. While the humeroulnar and humeroradial joints are generally seen to be uniplanar hinge

Table 6-1

SUMMARY OF THE OSTEOKINEMATICS OF THE HUMERORADIAL JOINT

Functional joint	Diarthrotic uniaxial
Structural joint	Synovial hinge
Close-packed position	Elbow flexion to 90º, supinated 5º
Resting position	Full elbow extension, full supination
Capsular pattern	Flexion more limited than extension
Primary muscles	Biceps brachii, brachialis, and pronator teres

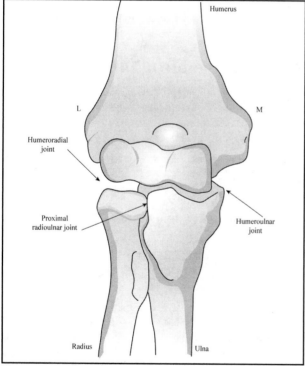

Figure 6-12. Separate joints of the elbow complex.

joints, there are slight axial and side-to-side rotational movements that occur during flexion and extension (Magee, 2002; Neumann, 2002). Prosthetists need to consider these accessory motions in fabrication of elbow joint prostheses to avoid premature loosening of the orthotic device (Neuman, 2002).

Humeroradial Joint (Radiocapitular)

This joint is composed of the articulation of the capitulum and capitulotrochlear groove of distal humerus with the head of radius. In full extension, there is no contact between the capitulum and the radial head. In full flexion, the rim of the radial head slides into the capitulotrochlear groove and contacts the radial fossa. This joint is important is transmitting axial loads imposed on the elbow and accounts for about 60% of the loads imposed on the elbow (40% through the humeroulnar joint) (Dumontier, 2010).

Osteokinematics of the Humeroradial Joint

The humeroradial joint is a uniaxial ginglymus (hinge) joint that allows for flexion and extension of the elbow in a sagittal plane around a frontal axis. The joint is structurally a synovial joint, and functionally it is a diarthrotic uniaxial joint. This joint has one degree of freedom, and the resting position is with the elbow extended and the forearm fully supinated. The position of greatest stability, the close-packed position, is with the elbow flexed to 90 degrees and supinated to 5 degrees. The muscles primarily producing movement at this joint are the biceps, brachialis, and pronator teres. When there is synovitis or capsulitis of the elbow joint, the capsular pattern is that flexion will be more limited and painful than extension. A summary of the osteokinematics of the humeroradial joint is provided in Table 6-1.

Arthrokinematics of the Humeroradial Joint

Flexion and extension of the humeroradial joint occur as the concave head of the radius rotates with the convex capitulum of the humerus (Kisner & Colby, 2007). As the elbow

Table 6-2

SUMMARY OF ARTHROKINEMATICS OF THE HUMERORADIAL JOINT

PHYSIOLOGIC MOVEMENT OF CONCAVE RADIUS ON CONVEX HUMERUS	DIRECTION OF SLIDE OF RADIUS ON CAPITULUM	RESULTANT MOVEMENT
Flexion	Anterior roll, anterior slide	Movement in same direction
Extension	Posterior roll, posterior slide	Movement in same direction

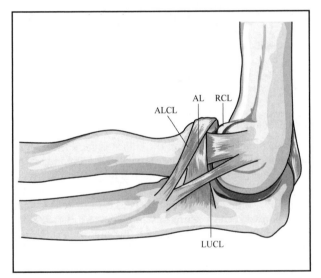

Figure 6-13. Supporting structures of the humeroradial joint.

flexes and extends, the concave radial head slides on the convex capitulum in the same direction (Kisner & Colby, 2007). The surface of the moving bone is concave, so sliding occurs in the same direction as the movement of the bone (convex-concave rule #2 from Chapter 2). A summary of the arthrokinematics of the humeroradial joint is provided in Table 6-2.

Supporting Structures

The primary supporting structures of the humeroradial joint are the lateral (radial) collateral ligament complex, the articular capsule, and the anconeus muscle (Figure 6-13). The anconeus muscle, which covers the joint capsule and collateral ligaments on the lateral side of the joint, acts as an active joint stabilizer against varus stress (Levangie & Norkin, 2005). It is often torn when the lateral collateral ligament complex is ruptured, but the role of this muscle is controversial (Dumontier, 2010).

Lateral (Radial) Collateral Ligament Complex

On the lateral side of the forearm, there are no discrete collateral ligaments as are found on the medial side. Instead, the lateral collateral ligament complex is composed of four ligaments: the radial collateral ligament, the lateral ulnar collateral ligament, the annular ligament, and the accessory lateral collateral ligament. These components vary widely among individuals (Carrino, Morrison, Zou, Steffen, Snearly, & Murray, 2001). The complex stabilizes the joint against varus forces and combined varus and supination forces and provides some resistance to longitudinal distraction. In addition, it reinforces the humeroradial joint and stabilizes the radial head for rotation (Levangie & Norkin, 2005).

Lateral (Radial) Collateral Ligament

The lateral (radial) collateral ligament (LCL) is a short, narrow fibrous band that attaches to the lateral epicondyle of the humerus and to the annular ligament below and inserts into the lateral margin of the ulna. The origin of the radial collateral ligament (RCL) lies over the center of rotation for the elbow so the ligament is tight throughout flexion and extension (Carrino et al., 2001; Gray & Lewis, 2000). This ligament provides reinforcement for the humeroradial joint and some protection against varus stress and longitudinal distraction of the joint (Levangie & Norkin, 2005). The LCL can be palpated as it extends from the lateral epicondyle of the humerus to the annular ligament and lateral surface of the ulna.

Lateral Ulnar Collateral Ligament

The lateral ulnar collateral ligament (LUCL) is a thick triangular band consisting of two portions: an anterior and posterior united by a thinner intermediate portion. It attaches anteriorly to the lateral epicondyle of the humerus above. Anterior fibers also attach to the medial margin of the coronoid below, while posterior fibers attach to the medial margin of the olecranon below (Gray & Lewis, 2000; Muscolino, 2006). The LUCL reinforces the humeroradial joint and is seen as the primary lateral stabilizer (Ellenbecker, Shafer,

Table 6-3

SUMMARY OF OSTEOKINEMATICS OF THE HUMEROULNAR JOINT

Functional joint	Diarthrotic uniaxial
Structural joint	Synovial hinge
Close-packed position	Extension with supination
Resting position	Elbow flexed to 70 degrees, supination to 10 degrees
Capsular pattern	Flexion, then extension
Primary muscles	Brachialis, triceps, and anconeus

& Jobe, 2010), although this is debated (Nordin & Frankel, 2001).

Annular Ligament

The annular ligament (AL) is a thick, strong band of fibers that attaches to the anterior and posterior radial notch, and a few of the fibers continue around the head of the radius to form a complete fibrous ring. It makes up four fifths of the fibro-osseous ring that encircles the head of the radius (Gray & Lewis, 2000). The AL provides stability to the radius against distal dislocation and is the main support of the radial head in the radial notch of the ulna (Nordin & Frankel, 2001). It can be palpated distal to the lateral epicondyle and can be palpated around the radial head. Pronating and supinating the forearm will facilitate finding these structures.

Accessory Lateral Collateral Ligament

The accessory lateral collateral ligament (ALCL) arises from the distal fibers of the annular ligament and attaches distally on the lateral ulnar collateral ligament. The role of the ALCL is poorly understood, but because it is taut only with varus stress, it may act as a stabilizer of the annular ligament when a varus stress is imposed upon the joint (Dumontier, 2010).

Humeroulnar Joint (Ulnotrochlear)

The humeroulnar joint is the articulation of the trochlea and trochlear groove of the distal humerus with the trochlear (semilunar) notch and trochlear ridge of the ulna. It is seen as the primary joint in the elbow complex for flexion and extension (Kisner & Colby, 2007). During extension, the olecranon process of the ulna fits into the olecranon fossa of the humerus. During flexion, the coranoid process of the ulna articulates with the coranoid fossa of the humerus. Some people are able to hyperextend the elbow, which may be due partially to either a smaller olecranon process or a larger olecranon fossa.

Osteokinematics

Like the humeroradial joint, the humeroulnar joint is a uniaxial hinge joint permitting the motions of flexion and extension in the sagittal plane around a frontal axis. The resting position of this joint is with the elbow flexed to 70 degrees and the forearm supinated to 10 degrees. The close-packed position is extension with the forearm in supination, and the capsular pattern of limitation and pain is seen first in flexion and then in extension. The primary muscles acting at the humeroulnar joint are the brachialis, triceps, and anconeus (Table 6-3).

Humeroulnar Joint Axis

The axis of the humeroulnar joint is not horizontal but is directed in a downward and medial direction due to the protrusion of the trochlea. This valgus (outward) position of the elbow and forearm in full extension and supination is known as cubitus valgus or carrying angle. This outward deviation is not apparent when the forearm is pronated in extension or in full flexion. With the elbow extended and the forearm supinated, the carrying angle in males is 5 degrees to 10 degrees, with a slightly larger lateral deviation angle of 10 degrees to 15 degrees in females (Figure 6-14). While it has been suggested that the purpose of the carrying angle is to prevent objects that are held in the hand from coming into contact with the body and that the wider angle in women is to accommodate the female pelvis, a definitive function of this angle has not been identified (Hamill & Knutzen, 1995). Roy and colleagues cite other examples of the functional significance of this outward deviation of the extended elbow, which includes pulling a wagon, skipping rope, increasing the reach of the arms, and providing a role in the muscular lever arms (Roy, Baeyens, Fauvart, Lanssiers, & Calrijs, 2005).

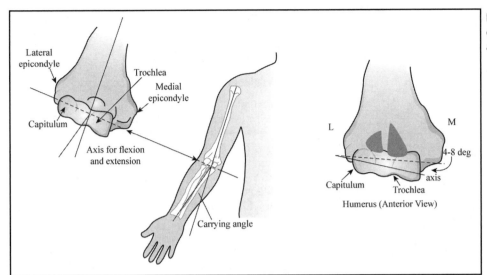

Figure 6-14. Axis of the humeroulnar joint and the carrying angle.

Table 6-4

SUMMARY OF ARTHROKINEMATICS OF THE HUMEROULNAR JOINT

PHYSIOLOGIC MOVEMENT OF CONCAVE ULNA TROCHLEAR FOSSA/NOTCH ON CONVEX TROCHLEA OF THE HUMERUS	DIRECTION OF SLIDE OF RADIUS ON CAPITULUM	RESULTANT MOVEMENT
Flexion	Anterior roll, anterior/distal slide	Movement in same direction
Extension	Posterior roll, posterior/proximal slide	Movement in same direction

A deformity of the elbow where the forearm deviates towards rather than away from the midline when the elbow is extended is called cubitus varus or gun stock deformity. This may be due to condylar fracture (Figure 6-15).

Arthrokinematics

Because the trochlea at the distal end of the humerus is convex and it articulates with the concave trochlear notch on the proximal ulna, the concave notch slides in the same direction in which the ulna moves. There is also a slight medial and lateral sliding of the ulna (see Table 6-4 for a summary of the humeroulnar joint arthrokinematics).

Supporting Structures

The primary supporting structures of the humeroulnar joint are the medial ulnar collateral ligament and the articular capsule.

The Medial (Ulnar) Collateral Ligament

The medial (ulnar) collateral ligament (MCL) attaches the medial epicondyle of humerus to the coranoid and olecranon process of ulna. It is a triangular shape and runs anteriorly, posteriorly, and obliquely (Figure 6-16). The humeral origin of MCL lies posterior to the axis of elbow flexion, resulting in stress to the anterior fibers in extension, and posterior fibers are stressed in flexion. This ligament serves as the primary constraint of the elbow joint to valgus stress with the radial head as a secondary constraint (Ellenbecker et al., 2010; Morrey, Tanaka, & An, 1991). In addition, the MCL functions to prevent abduction of forearm at the elbow and lateral movement of proximal ulna in all positions and supports the medial side of the joint. It is seen as the strongest ligament in the elbow complex. MCL laxity most often results from repetitive valgus loading, such as in throwing activities.

The MCL can be palpated as it traverses from the medial epicondyle to the medial margin of the coronoid process anteriorly and to the olecranon process posteriorly (Magee, 2002).

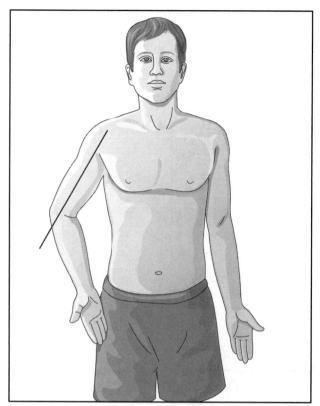

Figure 6-15. Gun stock deformity or cubitus varus.

The anterior medial collateral ligament (AMCL) is taut in extension and is considered the primary stabilizer to valgus stress in elbow flexion from 20 degrees to 120 degrees (Nordin & Frankel, 2001; Regan, Korinek, Morrey, & An, 1991; Schwab, Bennett, Woods, & Tullos, 1980). The AMCL is considered to be the strongest and stiffest of the collateral ligaments (Hotchkiss & Weiland, 1987; Regan et al., 1991).

The posterior medial collateral ligament (PMCL) is a weak, fan-shaped thickening of the medial portion of the joint capsule. It arises at the posterior aspect of the medial epicondyle and inserts over the olecranon, forming the floor of the cubital tunnel. It is taut when the elbow is in a flexed position and is less significant in valgus (outward) stability than the anterior medial collateral ligament.

The oblique (transverse) portions of the medial collateral ligament run between the olecranon and the ulnar coronoid process and assist in valgus stability and in keeping the joints aligned (Levangie & Norkin, 2005).

Proximal (Superior) Radioulnar Joint

This is the articulation of the radial notch of the ulna with the radial head and rim. The joint is encircled by the annular ligament, which takes up to approximately four fifths of the joint, and the radial head rotates around the annular ligament (Magee, 2002). In supination, the radius and ulna lie parallel to each other, but in pronation, the radius crosses over the ulna diagonally at this joint. When the distal extremity is fixed and not free to move (a closed kinetic chain movement such as when you put weight on your hand and rotate the forearm), a reverse action can occur in which the ulna moves instead of the radius at the superior radioulnar joint (Muscolino, 2006).

Osteokinematics

The proximal radioulnar joint provides the second of the two degrees of freedom possible at the elbow joint by enabling pronation and supination at this uniaxial pivot/trochoid joint. This joint is considered a uniaxial joint, meaning that the proximal and distal radioulnar joints function as one joint. This joint plays no part in producing the movements of flexion or extension at the elbow or in providing additional stability to the joint. The radial head rotates within the annular ligament and spins on the capitulum to produce pronation and supination.

The resting position of the proximal radioulnar joint is 35 degrees of supination and 70 degrees of elbow flexion. The position of greatest stability (close-packed position) is in 5 degrees of supination. The primary muscles involved at the radioulnar joint are the supinator and the pronator quadratus. The interosseous membrane between the radius and ulna helps to transmit some of the forces from the radius to the ulna. This prevents forceful contact of the radial head with the capitulum of the humerus (Ellenbecker et al., 2010) (Table 6-5).

Arthrokinematics

With pronation and supination of the forearm, the direction of slide of the proximal radius on the ulna is opposite to the motion (Figure 6-16). This is because the convex rim of the radial head articulates with the concave radial notch on the ulna. With rotation of radius, the convex rim moves opposite to the bone motion (Kisner & Colby, 2007).

However, Baeyens and colleagues found that there was anterior spinning with anterior gliding during pronation, which is in contrast to the convex-concave rule. While their study included only three people, they reported consistent findings of anterior translation of the radial head about the ulna, which "gives strong evidence to revise concave-convex rule for pronation-supination of the forearm" (Baeyens, Glabbeck, Goossens, Gielen, Roy, & Clarys, 2006). Further research is needed to validate this. Table 6-6 provides a summary of the arthrokinematics of the proximal radioulnar joint.

Supporting Structures

The supporting structures for the proximal radioulnar joint include the annular ligament, fibrous capsule, oblique cord, and proximal portion of the interosseous membrane. The interosseous membrane not only connects the ulna and radius but also prevents proximal displacement of radius on ulna (Ellenbecker et al., 2010).

Middle (Intermediate) Radioulnar Joint

The middle radioulnar joint is not a true joint, but is a fibrous, ligamentous synarthrodial articulation of the shaft

Table 6-5

SUMMARY OF OSTEOKINEMATICS OF THE PROXIMAL RADIOULNAR JOINT

Functional joint	Diarthrotic, uniaxial
Structural joint	Synovial, pivot
Close-packed position	5° supination
Resting position	35° supination, 70° elbow flexion
Capsular pattern	Equal limitation in pronation and supination
Primary muscles	Supinator, pronator quadratus

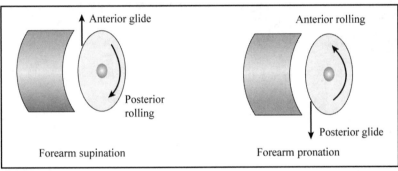

Figure 6-16. Medial collateral ligament. Diagrammatic representation of rotation at the proximal radioulnar joint.

of the radius with the shaft of the ulna and the interosseous membrane and oblique cord. There is no movement at this joint, but the interosseous membrane is tense in neutral position, helping to prevent proximal displacement of the radius on the ulna (Magee, 2002). The interosseous membrane serves to stabilize the elbow and radioulnar joints, to transmit forces from the hand to the humerus, and as a surface for attachment of deeper muscles (Magee, 2002; Muscolino, 2006; Premkumar, 2003).

Distal (Inferior) Radioulnar Joint

This joint is at the distal end of the forearm near the wrist where the head of the ulna articulates with the ulnar notch and articular disk of radius. This is an extremely stable joint due to the articular disk, the triangular fibrocartilage, the interosseous between the radius and ulna, and the pronator quadratus muscle. This is an important joint in the transmission of load via the triangular fibrocartilage and is an intricate part of wrist function (Ozer & Scheker, 2006; Palmer & Werner, 1984; Shaaban, Giakas, Bolton, Williams, Scheker, &

Lees, 2004). Further information about the distal radioulnar joint will also be covered in Chapter 7.

Osteokinematics

Like the proximal radioulnar joint, the distal radioulnar joint is a uniaxial pivot joint. In order to produce pronation and supination, the distal end of the radius must be free to move about the ulna at its distal end as well as at the proximal portions (Jenkins, 1998) (Table 6-7). Rotation of the lower end of the radius around the head of the ulna occurs at the distal radioulnar joint. The resting position is in 10 degrees of supination, and the close-packed position is in 5 degrees of supination.

Arthrokinematics

The articulating surface of the radius slides in the same direction as the bone motion because the concave ulnar notch on the distal radius articulates with the convex portion of the head of the ulna (Kisner & Colby, 2007) (Figure 6-17). The ulnar head, in a rolling, sliding motion, moves from the dorsal to the volar rim of the sigmoid notch as the joint moves from pronation to supination.

Table 6-6

SUMMARY OF ARTHROKINEMATICS OF THE PROXIMAL RADIOULNAR JOINT

PHYSIOLOGIC MOVEMENT OF OF CONVEX RIM OF RADIAL HEAD ON CONCAVE NOTCH OF ULNA	DIRECTION OF SLIDE OF PROXIMAL RADIUS ON ULNA	RESULTANT MOVEMENT
Pronation	Anterior roll, posterior/dorsal slide	Movement in opposite direction
Supination	Posterior roll, anterior/volar slide	Movement in opposite direction

Table 6-7

SUMMARY OF OSTEOKINEMATICS OF THE DISTAL RADIOULNAR JOINT

Functional joint	Diarthrotic, uniaxial
Structural joint	Synovial, pivot
Close-packed position	5 degrees supination
Resting position	10 degrees supination
Capsular pattern	Pain at extremes of rotation
Primary muscles	Pronator quadratus

There is slight movement of the ulna at the distal radioulnar joint. The ulnar head moves laterally in a direction opposite to the movement of the distal radius of up to 8 degrees. The ulna moves slightly dorsally in pronation and slightly toward the palm in pronation (Palmer & Werner, 1984; Shaaban et al., 2004). A summary of the arthrokinematics of the distal radioulnar joint is provided in Table 6-8.

Supporting Structures

The distal radioulnar joint is supported by the anterior/posterior radioulnar joint capsule and the interosseous membrane. The articular disk, which attaches to the ulnar notch and styloid process of the ulna, helps to hold the distal ends of the radius and ulnar together as well as separate the ulna from the carpal bones.

The articular disk at the distal radioulnar joint is also known as the triangular fibrocartilage. The lateral side of the disk attaches along the rim of the ulnar notch of the radius, and the main body of the disk spreads into a triangular shape. The anterior and posterior edges of the disk join with the palmar and dorsal radioulnar joint capsule ligaments, which hold the head of the ulna in the ulnar notch of the radius. The disk plus several wrist ligaments form the triangular fibrocartilage complex (TFCC) seen as a primary stabilizer of the distal radioulnar joint. Further discussion of the TFCC is in Chapter 7.

Table 6-8

SUMMARY OF ARTHROKINEMATICS OF THE DISTAL RADIOULNAR JOINT

PHYSIOLOGIC MOVEMENT OF OF CONCAVE NOTCH OF DISTAL RADIUS ON CONVEX HEAD OF ULNA	DIRECTION OF SLIDE OF DISTAL RADIUS ON ULNA	RESULTANT MOVEMENT
Pronation	Anterior roll, anterior slide	Movement in same direction
Supination	Posterior roll, posterior slide	Movement in same direction

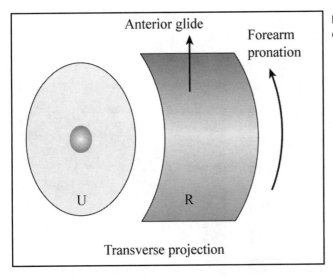

Figure 6-17. Diagrammatic representation of rotation at the distal radioulnar joint.

ELBOW STABILITY

The elbow joint is an inherently stable joint due to the bony configuration as well as ligamental support. The role of each of these structures varies with the degree of flexion or extension of the elbow (Dumontier, 2010). Overall, the three primary static constraints in the elbow are the humeroulnar joint, the medial collateral ligament, and the lateral collateral ligament complex. Other structures that provide secondary restraints include the radial head and joint capsule, flexor-pronator muscles, extensor-supinator muscles, and anconeus, triceps, and brachialis muscles (Table 6-9) (Gray & Lewis, 2000; Yallapragada & Patko, 2009).

Bone Support

The bony support is attributed to the articulation of the trochlea of the humerus with the trochlear fossa of the ulna and the head of the radius with the capitulum of the humerus. The amount of contact of the bony surfaces of the elbow increases from full extension to full flexion, which puts the radius in more contact with the capitulum, providing greater stability for the joint, especially in flexion (Figure 6-18).

The elbow is more stable when in flexion than in extension because there is more contact of the bony surfaces. In flexion, the coronoid process locks into the coronoid fossa, while the medial rim of the radial head engages in the trochleocapitellar groove. The lateral part of the olecranon process of the ulna is not in contact with the trochlea of the humerus during full flexion. In full extension, the medial part of the olecranon process is not in contact with the trochlea, but the apex of the olecranon is held in the olecranon fossa of the humerus. In full extension, there is no contact between the capitulum and the radial head. Posterior displacement of the elbow is prevented by the coronoid process of the ulna and by the humeroradial articulation.

Table 6-9

PRIMARY AND SECONDARY STABILIZERS OF THE ELBOW

Primary stabilizers	Humeroulnar joint	Provides bony stability
	Medial collateral ligament	Stabilizes against valgus stress
	Lateral collateral ligament	Stabilizes against varus stress
Secondary stabilizers	Radial head and joint capsule	Prevent hyperextension of the elbow
	Flexor-pronator muscles	Stabilizes against valgus stress
	Extensor-supinator muscles	Stabilizes against varus stress
	Anconeus, triceps, and brachialis muscles	Dynamic stabilizers that cross the joint provide compressive forces to the joint

Figure 6-18. Contact of bones in flexion and extension of the elbow.

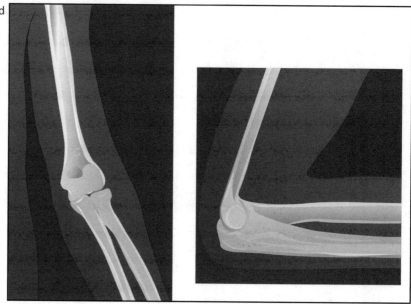

Elbow stability is improved by the congruency between the radial head and the radial notch of the ulna, which accounts for approximately 50% of the mediolateral stability of the elbow (Dumontier, 2010; Yallapragada & Patko, 2009).

Bony stabilization to valgus stress is provided by the proximal portion of the trochlear notch of the ulna, and varus stability is primarily a function of the distal part of the trochlear notch. Bony stability is also provided by the coronoid process during extension (Dumontier, 2010).

Ligamental and Soft Tissue Supports

The joint capsule is continuous for all three articulations, and the capsule is fairly large, loose, and weak, which allows for free movements. Anteriorly and posteriorly, the capsule is protected by muscles, but medially and laterally, it is reinforced by ligaments (Hartley, 1995; Nordin & Frankel, 2001). The joint capsule is important for joint support but also because it is highly innervated and is seen as the "neurologic link between shoulder and hand" (Ellenbecker et al., 2010).

Both of the collateral ligaments (medial collateral and lateral collateral) are strong fan-shaped thickenings of the fibrous joint capsule. The medial and lateral collateral ligaments provide a stabilizing force to medial and lateral stresses of the joint.

Valgus stability of the elbow is provided to a large extent by the medial collateral ligament. The bony surfaces of the humeroradial joint have only an accessory function in resisting valgus elbow stresses and only in less than 20 degrees and greater than 120 degrees of flexion (Morrey & Sanchez-Sotelo, 2009).

The roles of the different ligamentous structures in resisting varus stress are not as clear as for valgus stability. The lateral collateral ligamental complex as a whole stabilizes to varus and extension loads. The lateral collateral ligament prevents excessive supination in addition to its role in varus stability. It was initially thought that the annular ligament was responsible for resistance to varus stress between 40 degrees and 60 degrees of elbow flexion, but this has been contested, and it appears as though the primary role of the annular ligament is for stabilization of the radioulnar joint (Dumontier, 2010; Morrey & Sanchez-Sotelo, 2009; Nordin & Frankel, 2001).

The stability of the proximal and distal radioulnar joints is accomplished by the interosseous membrane, annular ligament, and triangular fibrocartilage complex (TFCC).

Dynamic stabilization of the elbow is achieved by the synergistic actions of the biceps, brachialis, brachioradialis, triceps, and anconeus muscles (Ellenbecker et al., 2010).

MOVEMENTS AT THE ELBOW JOINT

Flexion and Extension

The flexion and extension movements that occur at the humeroradial and humeroulnar joints are primarily gliding motions until the last 5 degrees to 10 degrees, when the joint surface motion changes to rolling. During flexion, there is distal (inferior) glide of the ulna in the trochlea. The trochlear ridge glides inferiorly on the trochlear groove until the coronoid process comes into contact with the coronoid fossa of the humerus. The lateral part of the olecranon is not in contact with the trochlea. The lack of full contact of the olecranon with the humerus during flexion and extension permits the small amount of joint play necessary for full pronation and supination (Magee, 2002). The rolling in flexion occurs when the coronoid process of the ulna comes into contact with the floor of the coronoid fossa of the humerus. Supination and adduction of the ulna in the trochlea and pronation of the radius on the humerus also occurs at the same time (Hamill & Knutzen, 1995). The capsule limits flexion more than extension, and total range of elbow flexion ranges from 135 degrees to 146 degrees.

Normal limitations for flexion are due to the apposition of anterior forearm and upper arm soft tissue, the posterior joint capsule, extensor muscles, and ultimately the coronoid process contacting coronoid fossa. This produces the normal soft end feel for elbow flexion, which is due to compression of forearm against upper arm. If there is little muscle bulk, the end feel may be hard due to contact of radius and radial fossa of humerus and contact of the coronoid process with coronoid.

In extension, the gliding motion occurs until the olecranon process of the ulna goes into the olecranon fossa of the humerus. The rolling that occurs in extension takes place when the olecranon of the ulna is received by the floor of the olecranon

fossa of the humerus (Esch, 1989; Levangie & Norkin, 2005). The movements occurring during elbow extension are proximal glide of the ulna on the humerus, pronation and abduction of the ulna on the humerus, and pronation of the radius on the humerus (Hamill & Knutzen, 1995). The medial part of the olecranon is not in contact with the trochlea in full extension. In addition, 10 degrees to 15 degrees of hyperextension of the elbow may occur and may be due to a shortened olecranon process, enlarged fossa, or to lax ligaments.

The normal limiting factor of elbow extension is the olecranon process contacting the olecranon fossa that results in a hard end feel. Occasionally, the end feel may be firm due to the anterior joint capsule, anterior band of the middle collateral ligament, and tension in the biceps and brachialis muscles.

The joint axis for flexion and extension goes through the middle of the trochlea and the capitulum, which bisects the longitudinal angle of the forearm. It is possible to feel this axis by grasping the elbow from side to side distal to the lateral and medial epicondyle. If muscles are located posterior to this axis, the muscles are extensors; if the muscles are anterior to the axis, they are flexor muscles.

Normative values for elbow flexion and extension range of motion are from 9 to 150 degrees. The useful arc of motion (or functional range) is between 30 and 130 degrees of elbow flexion (Morrey, Askew, & Chao, 1981; Morrey & Sanchez-Sotelo, 2009; Nordin & Frankel, 2001). Most activities of daily living require an arc of only 60 degrees to 120 degrees (Dumontier, 2010; Vasen, Lacey, Keith, & Shaffer, 1995). If elbow motion is restricted, compensatory motion is provided by increased shoulder motion. This can lead to shoulder pain and overuse, so it is important to consider both the shoulder and elbow when assessing shoulder complaints (Nordin & Frankel, 2001). Elbow flexors are brachialis, biceps (when supinated), and brachioradialis (when rapid flexion and when weight is lifted during slow flexion). Extensor carpi radialis and pronator teres also assist with flexion movements. Primary extensor muscles are triceps and anconeus (especially in initiating and maintaining extension) with some assistance from the extensor carpi ulnaris muscle (Levangie & Norkin, 2005). Generally, flexor muscles are 30% stronger than extensor (Nordin & Frankel, 2001) (Table 6-10).

Pronation and Supination

During pronation and supination, the distal and proximal radioulnar joints work together to enable the head of the radius to spin, roll, and glide in the radial notch. The axis line for pronation and supination goes through the center of the head of the radius proximally and through the center of the head of the ulna distally (Smith et al., 1996). It does not run parallel to the longitudinal axis of the forearm (Figure 6-19). Therefore, the axis is oblique to the longitudinal axis center of the radius and ulna (Levangie & Norkin, 2005). The muscles that cross anterior to this axis are pronators; those crossing posterior to the joint axis are supinators (Gench et al., 1995) (Figure 6-20).

It is important to realize that, although the radius rotates over the ulna, the ulna moves, too. The ulna moves laterally during pronation and medially during supination (Palmer & Werner, 1984; Shaaban et al., 2004).

Table 6-10

SUMMARY OF FLEXION-EXTENSION MOVEMENTS

MOTION	PLANE/AXIS	JOINT	NORMAL LIMITING FACTORS TENSION IN	END FEEL	PRIMARY MUSCLES	NORMATIVE ROM/ FUNCTIONAL ROM
Flexion	Humeroulnar Humeroradial	Sagittal/ frontal	• Soft tissue • Posterior joint capsule • Extensor muscles • Coronoid process contacting coronoid fossa	Soft	• Biceps brachii • Brachialis • Brachioradialis	0º to 150º/30º to 130º
Extension	Humeroulnar Humeroradial	Sagittal/ frontal	• Olecranon process contacting olecranon fossa	Hard	• Triceps • Anconeus	150º to 0º/130º to 30º

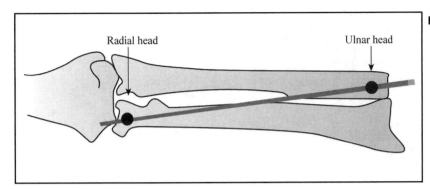

Figure 6-19. Axis for forearm rotation.

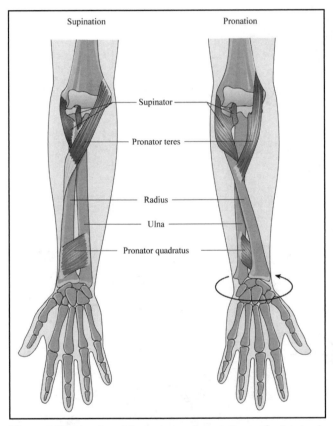

Figure 6-20. Muscles of forearm rotation relative to the joint axis.

Tension in the pronator muscles, the palmar radial ligament of the distal radioulnar joint, the oblique cord, and the interosseous membrane restrict further motion in supination as normal limiting factors. This results in a firm end feel. Primary muscles producing supination include the supinator muscle and biceps (during fast and resisted motion) (Levangie & Norkin, 2005).

Due to contact of the radius on the ulna and tension in the dorsal radioulnar ligament of the distal radioulnar joint, the interosseous membrane, and the biceps muscle, further pronation is prevented. The dorsal and volar radioulnar ligaments at the perimeter of the triangular fibrocartilage complex provide primary constraints to the distal radioulnar joint (Linscheid,

1992). The dorsal and palmar radioulnar ligaments are the primary stabilizers of the joint and aid in wrist stabilization. Each of these ligaments also consists of deep and superficial portions. The deep portions originate from the radius and conjoin to insert into the fovea. The superficial portions also originate from the radius, but separately insert into the base of the styloid. In supination, deep portions of the palmar radioulnar ligament and a superficial portion of the distal radioulnar ligament become taut. In pronation, deep portions of the distal radioulnar ligaments and a superficial portion of the palmar radioulnar ligaments are under tension. This produces a firm end feel due to tension in the ligamentous structures. Primary muscles involved in pronation are pronator quadratus and pronator teres (during fast or resisted movements) with secondary contributions from flexor carpi radialis and anconeus muscles.

Range of motion values for pronation and supination range from 0 degrees to 71 degrees to 0 degrees to 90 degrees for pronation and from 0 degrees to 81 degrees to 0 degrees to 90 degrees for supination. Functional range of motion for both supination and pronation for most activities is 0 degrees to 50 degrees (Vasen et al., 1995). A summary of pronation and supination is provided in Table 6-11.

INTERNAL KINETICS: MUSCLES

According to Levangie and Norkin (2005), there are six factors that influence motion at the elbow:
1. Location of muscles
2. Position of elbow and adjacent joints
3. Position of forearm
4. Magnitude of applied force
5. Types of muscle contraction
6. Speed of motion

The location of the muscles has a direct relationship with the motion possible at the joint. For example, pronators cross the joint axis anterior to the axis, whereas supinators cross posteriorly to the joint axis of rotation. Location of muscles as well as the position of the forearm and other joints helps to determine the line of pull of the muscles, which influences the power of the force generated by the muscle due to length-tension relationships. In addition, some of the forearm muscles are active only when the movement is unresisted or when there is

Table 6-11

Summary of Pronation and Supination Movements

MOTION	PLANE/AXIS	JOINT	NORMAL LIMITING FACTORS TENSION IN	END FEEL	PRIMARY MUSCLES	NORMATIVE ROM/ FUNCTIONAL ROM
Pronation	Radioulnar (proximal, middle, distal)	Horizontal/ longitudinal	• Dorsal radioulnar ligament • Interosseous membrane • Biceps muscle • Radius on ulna	Firm	• Pronator quadratus • Pronator teres	0 to 90/0 to 50 degrees
Supination	Radioulnar (proximal, middle, distal)	Horizontal/ longitudinal	• Pronator muscles • Palmar radial ligament • Oblique cord • Interosseous membrane	Firm	• Supinator • Biceps	0 to 90/0 to 50 degrees

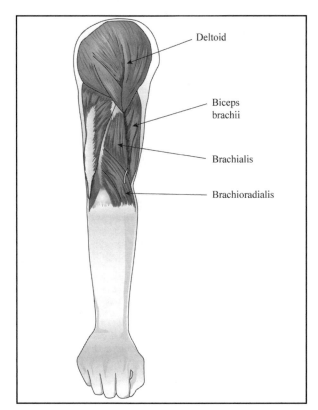

Figure 6-21. Flexors of the elbow.

external load and only in certain positions. For example, pronator quadratus will pronate the forearm regardless of forearm position or amount of flexion of the elbow, whereas pronator teres contributes to pronation where there is an external load or when rapid pronation is required.

The nervous system uses efficient economy of energy when determining which muscles will produce specific movements. Unskilled movements waste energy because muscles that are not needed also contract. As skill increases, the selection of muscles improves, and the gradation of contraction becomes more refined, which results in smoother movements (Smith et al., 1996). The number of muscles is determined by the effort needed, and the nervous system prefers to have only one muscle perform the task if possible. For example, if flexion of the elbow and supination of the forearm is the desired motion, biceps would be a good choice because the biceps muscle both flexes the elbow and supinates the forearm. The two actions could also be done by brachialis and supinator, but at the expense of energy required to produce contractions of two muscles rather than just one (Smith et al., 1996). Likewise, if only flexion without supination or pronation was the required motion, the biceps would be wasteful because the supination function of biceps would need to be neutralized by pronator muscles.

The stresses applied to the elbow vary based on the load applied, the resultant force vector, and the length of the lever arm (Dumontier, 2010). Small loads applied to the hand dramatically increase the elbow joint reaction force. In studies by Nicol, it was shown that a common activity, such as support-

ing oneself when pushing up out of a chair, generates loads of more than twice the body weight, which may challenge the view that the elbow is not a load-bearing joint (Norkin & Levangie, 1992). Compressive loads are observed during simple activities of daily living, such as dressing or feeding, and the use of crutches transfers 40% to 50% of the body weight onto the upper extremity (Dumontier, 2010).

At the elbow joint, there are two muscles that are multijoint muscles (i.e., muscles that cross two or more joints). The biceps is a multi-joint muscle that can develop active insufficiency when the elbow is in full flexion while the shoulder is in full humeral flexion and the forearm is supinated. This puts the biceps muscle in a shortened position, resulting in marked loss of force and some leverage loss. Passive tension in triceps may also limit elbow flexion (Levangie & Norkin, 2005). Triceps, also a multijoint muscle, is actively insufficient when the elbow is in full extension with the shoulder in extension (shortened position). Passive tension in the long head of the biceps brachii by passive shoulder hyperextension may limit full elbow extension (Levangie & Norkin, 2005).

The significance of this intervention is to provide activities or adaptations that will not cause muscle shortening over the multiple joints, resulting in decreased strength and force production.

Muscles must act in synergistic ways to produce motion. For biceps to supinate at the radioulnar joint without flexing the elbow, it works in synergy with the triceps. For biceps to flex the elbow without supinating, it works in synergy with an elbow pronator. For pronation, the humeral head of pronator teres works in synergy with an elbow extensor to counteract the flexion function of the pronator teres muscle. Overshortening and loss of force production in the biceps is prevented by the synergistic action of the triceps muscle, and similarly, the synergy of triceps with the biceps muscle prevents overshortening and loss of force production in triceps brachii (Thompson, 2001).

The location of muscles and position of the forearm, elbow, and adjacent joints, as well as the angle of the pull of muscle fibers all contribute to the actions of the muscles of the elbow joint.

Flexors of the Forearm

Primary flexors of the elbow are the brachialis, biceps (when supinated), and brachioradialis (when rapid flexion and a load is lifted during slow flexion) (Nordin & Frankel, 2001) (Figure 6-21). The greatest force of elbow flexion is produced when there is 90 degrees of flexion accompanied by supination of the forearm. The dominant side produces higher flexion torque, work, and power, but there is no difference between right and left for extension, pronation, or supination. Flexor muscles are stronger than extensor muscles, and flexion force is 20% to 25% higher in a supinated position due to the increased flexor moment arm of the biceps and brachioradialis (Magee, 2002; Vasen et al., 1995).

Brachialis

The brachialis is considered the primary flexor of the elbow and, as such, is always a flexor regardless of whether there is an

Figure 6-22. Palpation of brachialis muscle.

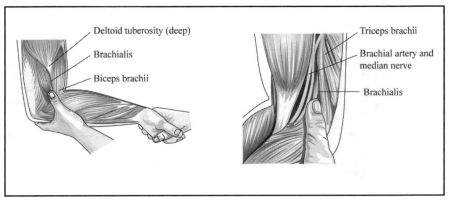

external load and whether or not the rate of movement is fast or slow. An appropriate nickname is "workhorse of elbow," as this muscle has the greatest work capacity of the elbow flexors. Because this fusiform, one-joint, spurt (mobility) muscle crosses the elbow closer to the flexion and extension axis than brachioradialis and biceps, it is less favorably located to produce force. The moment arm is greatest at about 100 degrees of elbow flexion, so the greatest torque is produced in this position (Smith et al., 1996). However, the distal attachment is on the ulna rather than the radius, which means that arm position is inconsequential to flexion and extension (Gench et al., 1995).

Brachialis also contributes to pronation and supination, but the contribution is weak. Brachialis works well eccentrically but better concentrically (Gench et al., 1995; Hinkle, 1997; Jenkins, 1998; Smith et al., 1996).

Palpation: Brachialis can be palpated just lateral to the biceps when resistance is applied to the wrist (Gench et al., 1995) (Figure 6-22). If the examiner places the palpating fingers laterally and medially to the biceps and flexes the elbow with minimal effort, contraction of brachialis may be felt. Quick contraction in a small range will result in a stronger contraction by this muscle (Esch, 1989; Lumley, 1990; Magee, 2002; Smith et al., 1996). You may also feel this muscle by sliding about ½-inch from the distal biceps in a relaxed arm. The edge of the brachialis can be felt as you roll your fingers across the surface (Biel, 2001).

Biceps Brachii

This fusiform, spurt (mobility) muscle is a multijoint muscle whose contraction produces shoulder flexion, elbow flexion, and forearm supination. The force components of the pull of the biceps muscle can vary in length so that, at full extension, biceps is accompanied by a strong stabilizing component, while at full flexion there is a strong dislocating component. It is for this reason that the biceps is most efficient as an elbow flexor when in 90 degrees of flexion, because the only component is an angular one (Gench et al., 1995) and the moment arm is greatest between 90 degrees and 100 degrees of elbow flexion.

The biceps creates an isolated, unopposed contraction when there is simultaneous flexion of shoulder, elbow, and supination. The biceps does not function in slow flexion with the arm pronated, but functions in slow flexion with the forearm supinated as well as in fast flexion with and without an

external load. As the speed of the movement increases and as the load increases, the biceps may act even in pronation. When the elbow is flexed with the forearm in pronation, the biceps is nearly inactive (Greene & Roberts, 1999). An example of how to use the actions of the biceps in a therapeutic situation is provided by Greene and Roberts (1999):

When Mary Smith arrived at Maple Grove Skilled Care Facility, she held her right arm (elbow) flexed tightly against her body. She refused to let any nurses or aides move her arm through passive range of motion, so the nursing supervisor contacted the occupational therapy department. The occupational therapy assistant who went to see Mary explained that putting her arm through range of motion probably would help her not to feel so stiff and sore all of the time.

The occupational therapy assistant promised to try some new techniques that would make range of motion less painful than it had been in the past. First, the OT assistant gave Mary's right arm a gentle rubdown to gain Mary's trust and then slowly pronated her forearm to inhibit the biceps. The OT assistant explained to Mary that in the past, medical practitioners may have attempted to extend her elbow with her palm up in supination, a position associated with increased activity in the biceps. By pronating the forearm first, the OT assistant inactivated the biceps and made elbow extension easier and more comfortable. The OT assistant continued to explain that elbow extension in pronation stretches the tight biceps tendon even further because both of the biceps' antagonistic movements—extension and pronation—occur simultaneously (Greene & Roberts, 1999).

The biceps also often works eccentrically in a protective capacity to slow down the rate of elbow extension. An example of the biceps working eccentrically would be when a person who is on a ladder slips and the fall necessitates a quick, forced extension of the elbow. The biceps is contracting while lengthening to slow the fall (Konin, 1999).

Palpation

With the forearm in supination and the elbow flexed, the tendon of insertion can be felt anterior to the elbow joint (Jenkins, 1998). When one "makes a muscle" by bending the elbow, it is the biceps muscle that is seen (Esch, 1989; Lumley, 1990; Magee, 1992). Alternate pronation and supination while palpating the biceps muscle belly against resistance to feel the biceps contraction upon supination (Biel, 2001).

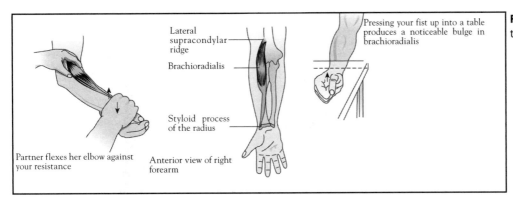

Lateral supracondylar ridge

Brachioradialis

Styloid process of the radius

Partner flexes her elbow against your resistance

Anterior view of right forearm

Pressing your fist up into a table produces a noticeable bulge in brachioradialis

Figure 6-23. Palpation of the brachioradialis muscle.

Brachioradialis

This fusiform, shunt (stability) muscle has its distal attachment farther from the elbow than other elbow flexors. This means that the force arm and moment arm are increased. Because the proximal attachment is closer to the axis and the muscle lies close to the joint axis, there is a small angular component at the distal attachment when compared to a larger stabilizing component.

Brachioradialis tends to pronate as it flexes, so the angular component can be increased by placing the forearm in neutral (midposition between pronation and supination), which will lead to a stronger contraction (Gench et al., 1995). Brachioradialis is active during rapid flexion or when an external load is lifted during slow flexion, adding speed and power to elbow flexion.

Palpation

Brachioradialis lies between the triceps and brachialis muscles. At and just below the elbow, brachioradialis forms the lateral border of the antecubital fossa (Gench et al., 1995; Smith et al., 1996) (Figure 6-23). Brachioradialis is seen when the elbow is flexed to 90 degrees and the forearm is placed in neutral position. When resistance is applied to the wrist, brachioradialis contracts and may be visible as a bulge on the lateral side of the elbow. Another way to see the muscle is to press your fist up into a table, and the brachioradialis will be visible.

Pronator Teres

This muscle lies quite close to the axis throughout the range of motion, making it inefficient as an elbow flexor, and it may actually serve to neutralize the supination action of biceps rather than actually flex the elbow. Pronator teres may not be able to flex the extremity alone against gravity (Jenkins, 1998).

Palpation

Pronator teres forms the medial margin of the antecubital fossa, and the fibers can be identified when the forearm is pronated while the elbow is flexed (Figure 6-24). When resistance is applied to the forearm in this position, pronator teres can easily be identified. Place one finger on the humeral medial epicondyle and one on the midpoint of the radius, and the

ridge of this muscle will be seen running obliquely between the two fingers (Gench et al., 1995; Smith et al., 1996). More distally, the pronator crosses to the radial side and is covered by brachioradialis. If the forearm is pronated with little effort, pronator teres will contract (Esch, 1989; Lumley, 1990; Magee, 1992).

Extensor Carpi Radialis Longus

Due to the anterior location and origin on the humeral epicondyle, both extensor carpi radialis longus and pronator teres may assist with elbow flexion. However, because extensor carpi radialis longus arises too low on the humerus, attaches too close to the flexion axis, and produces a short moment arm, this muscle fails to be too important in flexion of the forearm (Greene & Roberts, 1999; Hinkle, 1997).

Extensor Carpi Radialis Brevis

Due to this muscle crossing the elbow anterior to the axis, this muscle may serve to assist in elbow flexion (Gench et al., 1995); however, because the muscle's proximal attachment is on the epicondyle and therefore very close to the elbow's axis, it is questionable what effect this muscle has on elbow flexion.

Flexor Carpi Radialis, Palmaris Longus, Flexor Carpi Ulnaris

These muscles were also identified as having assistive actions with elbow flexion (Gench et al., 1995; Lumley, 1990; Smith et al., 1996).

Extensors of the Forearm

Primary extensors of the elbow are triceps and anconeus (Figure 6-25).

Triceps

The triceps muscle has three heads, with the medial head being the primary extensor of the forearm. The size of the muscle as well as the angle of attachment on olecranon give the triceps power in spite of poor leverage (Gench et al., 1995).

Figure 6-24. Palpation of pronator teres muscle.

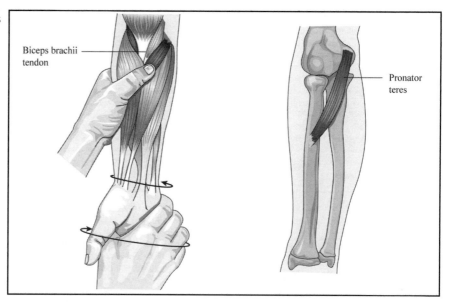

The long head of the triceps is affected by positions of the shoulder because this portion crosses both the shoulder and elbow joints. This long head can become actively insufficient when the elbow and shoulder are both extended because this shortens the muscle over both joints. The middle and lateral heads of the triceps muscle are not affected by shoulder position. All three heads of the triceps muscle extend the elbow when there is heavy resistance or when quick extension is required.

Frequently, the triceps muscle works eccentrically, as when one lowers the body to the floor with the elbows flexed and the triceps lengthen (Gench et al., 1995; Hinkle, 1997; Konin, 1999; Loth & Wadsworth, 1998; Smith et al., 1996).

Palpation

The lateral head can be palpated between the posterior deltoid and the lateral epicondyle. It appears separated from the deltoid by a groove.

The long head can be palpated between the axilla and the olecranon process. This is the contour at the lower portion of the arm. The long head of the triceps is the only band of muscle on the posterior surface that runs superiorly along the proximal and medial aspect of the arm (Biel, 2001).

The medial head is located beneath the long head and is difficult to palpate. It is best palpated in its distal portion near the medial epicondyle. By placing the dorsum of the wrist on the edge of a table and applying resistance downward, the medial head may be felt contracting when extension is resisted (Esch, 1989; Lumley, 1990; Magee, 1992; Smith et al., 1996). It can also be felt by placing the subject prone and having the person extend the elbow as you apply resistance toward flexion. You can palpate the medial and lateral heads on either side of the broad distal triceps tendon (Biel, 2001).

Anconeus

This small, triangular muscle is active in initiating and maintaining extension and in joint stabilization. Because it is

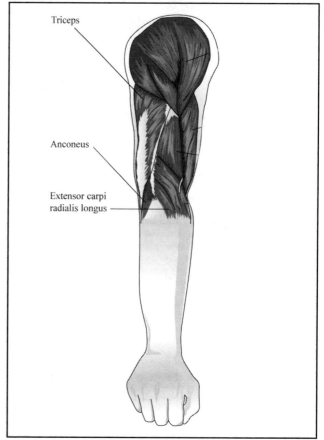

Figure 6-25. Extensors of the elbow.

very close to the axis and is small, it is weak in these actions (Jenkins, 1998). Thompson and Floyd (1994) stated that the chief function of anconeus was to pull the synovial membrane out of the way of the advancing olecranon process during extension of the elbow.

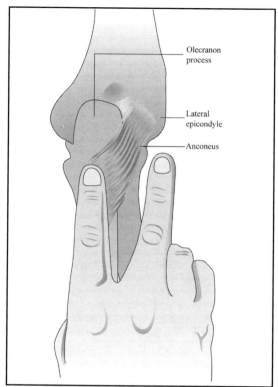

Figure 6-26. Palpation of anconeus muscle (posterior view, right arm).

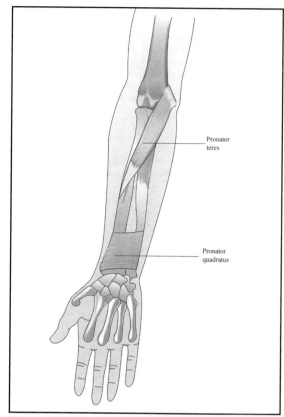

Figure 6-27. Pronators of the forearm.

Palpation

By placing the thumb and index finger on the olecranon process and lateral epicondyle, one can feel the base of this equilateral triangular muscle (Gench et al., 1995) (Figure 6-26). Extensor carpi ulnaris lies close to anconeus, and while the two muscles may appear as one, identification can be made by keeping in mind that the direction of the line of muscle pull varies and that anconeus lies more proximally and is very short (Esch, 1989; Lumley, 1990; Magee, 1992; Smith et al., 1996).

Pronators of the Forearm

The primary pronators of the forearm are pronator quadratus (primary) and pronator teres (when there is rapid or resisted pronation) (Figure 6-27). Other muscles that may contribute to pronation include brachioradialis, flexor carpi radialis, flexor carpi radialis palmaris longus, and extensor carpi radialis longus.

Pronator Quadratus

A rhomboidal or quadrangular muscle, pronator quadratus muscle is considered to be the primary pronator. Due to the size of the muscle and the pull of the fibers, this muscle is a pronator of the radioulnar joint without help and in movements that are slow and unresisted. With the triceps muscle, it extends and pronates the elbow. The action of pronator quadratus can be seen in using a screwdriver to remove a screw and in throwing a screwball in baseball (Gench et al., 1995;

Hamill & Knutzen, 1995; Hinkle, 1997; Jenkins, 1998; Smith et al., 1996).

Palpation

Because it is too deeply located, palpation is not possible (Gench et al., 1995).

Pronator Teres

This muscle lies anterior to the axis at the proximal radio-ulnar joint and has a long angular component. Due to its strength and efficiency, pronator teres does not participate in pronation in slow or unresisted motion but instead contracts during rapid pronation or in pronation against resistance. It is considered to be a secondary pronator (Norkin & Levangie, 1992). If pronator teres were to act alone, it would bring the back of the hand to the face as it contracts (Smith et al., 1996).

Palpation

See the palpation description in the earlier section on elbow flexors.

Brachioradialis

While primarily a flexor, the brachioradialis muscle is also a weak pronator. The brachioradialis directs a force volar to the joint axis, pulling the radius toward the ulna in pronation (Greene & Roberts, 1999). In other words, brachioradialis acts in pronation from a position of supination with the ability to pronate decreasing as the forearm rotates to neutral.

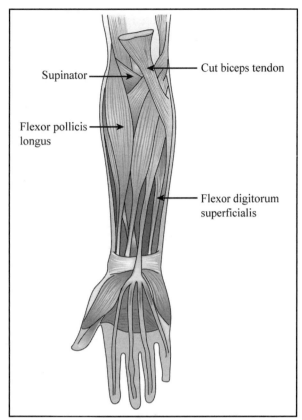

Supinator

Flexor pollicis longus

Cut biceps tendon

Flexor digitorum superficialis

Figure 6-28. Supinators of the forearm.

Flexor Carpi Radialis

Flexor carpi radialis assists with pronation because the tendon passes obliquely across the wrist (Jenkins, 1998).

Palmaris Longus, Extensor Carpi Radialis Longus

These muscles may assist with pronation (Gench et al., 1995; Hinkle, 1997; Jenkins, 1998; Smith et al., 1996).

Supinators of the Forearm

The primary supinator of the forearm is the biceps with secondary contributions from the supinator muscle and brachioradialis (Figure 6-28). The force of the muscles that produce supination is 25% stronger than elbow flexors, which helps to explain why we use a clockwise motion when using a screwdriver to tighten a screw (Nordin & Frankel, 2001).

Supinator

The supinator muscle (see Figure 6-29) is in the best position to supinate the radioulnar joint when the elbow is extended, and this position puts the muscle on more stretch than other positions (Gench et al., 1995). Two common examples of how this muscle may be seen are turning a screwdriver or throwing a curve ball in baseball. The primary function of this muscle is supination of the forearm, and it contracts when supination occurs slowly and without resistance. The supinator requires

assistance from the biceps with resisted motion (Gench et al., 1995; Hinkle, 1997; Jenkins, 1998; Smith et al., 1996).

Palpation

The supinator is located deeply and is difficult to palpate (Gench et al., 1995).

Biceps

Biceps is the strongest and most efficient supinator, especially when the elbow is flexed to 90 degrees. Biceps is four times more efficient than the supinator muscle in this flexed position and only twice as effective as the supinator muscle with the elbow extended and supinated (Gench et al., 1995). Triceps is needed to counteract the flexor action of biceps when the muscle is recruited as a supinator. Biceps assists with rapid, unresisted supination with the elbow flexed or with any supination against resistance (Gench et al., 1995; Hinkle, 1997; Jenkins, 1998; Smith et al., 1996).

Brachioradialis

In addition to flexing the elbow, brachioradialis can function as both a pronator and a supinator. As the forearm pronates, the brachioradialis directs a force on the dorsal side of the axis, and supination of the forearm occurs (Greene & Roberts, 1999). Brachioradialis can supinate the forearm from a position of pronation, with the ability to supinate decreasing as the joint moves toward neutral (Thompson & Floyd, 1994).

Other Assistive Supinators

Abductor pollicis longus, extensor pollicis brevis, extensor indicus proprius, and flexor carpi ulnaris were also identified by some sources as assistive supinators (Jenkins, 1998; Smith et al., 1996).

ASSESSMENT OF THE ELBOW

Assessment of the elbow includes an understanding of the client as a person so that treatment goals match the occupational performance needs of the client. Observation and palpation of the structures of the elbow can provide information about pathologic conditions, such as inflammation and edema. Both active and passive range of motion provides information about joint movement and both contractile and noncontractile elements of the joint. Range of motion and strength testing can be quickly assessed with screens unless more formal testing is warranted. An appraisal of the client's pain is also a vital part of the evaluation as pain can invalidate tests, cause the client discomfort, and result in functional limitations in the use of the elbow.

Occupational Profile

The essential first step in evaluation is to understand the client's occupational history, experiences, patterns of living, interests, values, and needs (American Occupational Therapy Association, 2008). It is crucial to establish what the client's priorities are for treatment so that these are integrated into the treatment plan, because it is the client's perspective about his or her priorities that drive any intervention. How limitations in elbow function affect occupational performance is the goal of assessment of the elbow. Does the client now use the nondominant extremity for activities? If so, this reversal of natural dominance may indicate that function has been severely impaired. It is also important to establish the duration of the complaint and the time since onset of elbow-related symptoms (Dumontier, 2010).

How well the person can move the elbow joint and the quality of movement is vital to understanding how the limitations will interfere with functional activity. Observe how the person uses the elbow and arm in activities. Does the client seem willing to move the joint, and does the client use the arm automatically in activities and without discomfort? Does the client support one arm or use one arm more than the other? Look at how well the wrist, elbow, and shoulder all work together, and observe for lack of symmetry or muscle imbalance.

Observation and Palpation

Many of the disorders of the elbow can be detected by observation and palpation. Observe the client's position in extension and supination. The cubitus valgus (carrying angle) should not be more than 15 degrees to 20 degrees, and cubitus varus should not exceed 5 degrees. The cubitis valgus angle has been found to be greater in athletes in unilaterally dominant upper extremity sports (baseball, tennis) and may also have more osseous changes, such as bone spurs and osteophytes, which would limit motion and present with abnormal end feel (Dumontier, 2010; Ellenbecker et al., 2010). Changes in the carrying angle may also be indicative of elbow instability or of malunion (Dumontier, 2010).

Many of the landmarks will be more easily seen by standing or sitting behind the client. The prominence of the olecranon is a sign of posterior subluxation of the elbow, which is not uncommon in clients with rheumatoid arthritis (Dumontier, 2010; Magee, 2002). Subcutaneous nodules are also common and can be found on the posterior aspect (extensor surface) of the elbow. Note the location, number, size, and other characteristics of these nodules (i.e., freely moveable, hard, tender, etc.). Bursitis may also be seen and is sometimes described as a "goose egg" formation at the back of the elbow. Other conditions that may cause nodular swelling of the elbow include rheumatoid arthritis, gout and rarely systemic lupus erythematosus, rheumatic fever, and sarcoidosis.

Typical structures that are palpated during a physical examination of the elbow are included in Table 6-12.

When palpating the anatomic landmarks, look for points of tenderness, and compare these with the opposite extremity. Palpate the radial and ulnar margins, noting any synovial thickening that may be present or that may appear as a bulge on either side of the olecranon process (Magee, 2002; Vasen et al., 1995). Lateral epicondylitis (tennis elbow) will be suggested if there is pain when the wrist is extended against resistance with the elbow in extension. Medial epicondylitis (golfer's elbow) is tested by resisted flexion of the wrist with the elbow in extension. Bicipital tendonitis is found by flexing the supinated forearm against resistance while palpating the tendon insertion.

Look for normal bony and soft tissue contours anteriorly and posteriorly. Swelling in the elbow is not uncommon, especially because all three joints share the same joint capsule. Swelling may be most evident in the triangular space between the radial head, tip of the olecranon, and lateral epicondyle. Because the elbow joint is so superficial, swelling and redness of the olecranon bursa are easy to observe. It is important to ascertain when the swelling started and make note of the location of the swelling. Additional information that can be gleaned from the physical assessment of the elbow is if any locking occurs during movement, the temperature of the skin, client descriptions of paresthesias, and any abnormal sound (such as grating or grinding) (Table 6-13).

Distal radioulnar pathologies are also visible. The "piano key" sign is seen when the client places the palm flat on the table and is asked to press to the table using shoulder depression with the forearm in pronation. If the head of the ulna is elevated, this is considered a positive sign and suggests destabilizing dysfunction of the triangular fibrocartilage complex (Ellenbecker et al., 2010). Distal radioulnar subluxation or dislocation may only be noticeable by the swelling. There would also be functional changes, such as painful wrist motions, restricted pronation if there is volar dislocation, and supination is affected if there is dorsal dislocation (Ozer & Scheker, 2006).

Range of Motion

While there are normative data available for range of motion for elbow flexion/extension, pronation, and supination, variance from these norms should be considered for intervention only when the limitations prevent the client from successful engagement in work, play, or self-care.

Most activities of daily living require 30 degrees to 130 degrees of elbow extension and flexion and only 0 degrees to 50 degrees of pronation or supination. Flexion values range from 57 degrees to open a door to 136 degrees to use a telephone. Pronation values can range from 10 degrees needed to drink from a glass to 49 degrees to read a paper. Supination is not needed at all for cutting meat or rising from a chair, but approximately 52 degrees is needed to put a fork into the mouth (Magee, 2002; Vasen et al., 1995).

Active range of motion is assessed first, because a loss of elbow extension is a sensitive indicator of intra-articular pathology (Magee, 2002). Loss of active elbow flexion is seen as more disabling due to the amount of flexion required for many daily activities, such as eating, grooming, and hygiene.

Passive movement assesses the capsule, ligaments, bursa, cartilage, and nerves. Passive movements can be done to assess end feel, sequences of pain and resistance, and capsular patterns (Palmer & Epler, 1998). End feels that are not appropriate for that particular joint or that occur sooner in the range than should normally occur are indicative of joint pathology.

Hypermobility of joints can be assessed by observation or more formally by using standardized elbow assessments or the Brighton Hypermobility Criteria. One point is added for moving into certain positions, such as hyperextension of the elbow (one point each for right and left), thumb touching the forearm, little finger bending backward past 90 degrees, hyperextension of the knee, and placing flat hand on the floor with straight legs (Grahame, Bird, & Child, 2000). The Brighton

Table 6-12

PALPATED STRUCTURES DURING ASSESSMENT

LOCATION	STRUCTURE
Anterior	Biceps tendon
	Median nerve
	Anterior capsule
Posterior	Triceps tendon
	Olecranon fossa
Medial	Medial epicondyle
	Forearm flexor and pronator tendons
	Medial collateral ligament
	Ulnar nerve
Lateral	Lateral epicondyle
	Radiocapitellar joint
	Radial head
	Radial nerve

Table 6-13

EVIDENCE OF EDEMA AND ABNORMAL SENSATIONS OF THE ELBOW

SWELLING

Type
- Local: may be due to bursae, muscle strains, or contusions to tendons, muscle bellies; or may be due to intracapsular effusion.
- Diffuse: may be due to severe hematoma, dislocations, or fractures.

Time
- Immediate to within first 2 hours: damage to a structure with rich blood supply.
- 6 to 24 hours: suggests synovial irritation as may occur with bone chips, capsular sprains, ligament sprain, or joint subluxation.
- After activity: chronic bursitis or repeated trauma to bursae.

SENSATIONS

Locking
- May occur if loose body is in the joint

Warmth
- May suggest inflammation or infection

Tingling, numbness
- May suggest C5-8 nerve root compression, thoracic outlet syndrome, injury to peripheral nerves, or neuritis or neuropathy

Grating
- Osteoarthritic changes or damage to articular surface (chondromalacia, osteochondritis, osteoarthritis).

Table 6-14

RANGE OF MOTION SCREEN FOR THE ELBOW AND FOREARM

Bend and straighten the elbows	Elbow flexion and extension
Turn your arm so your palm is up, facing the ceiling	Forearm supination
Turn your arm so your palm is down, facing the floor	Forearm pronation

Scale is seen as a simple system to quantify joint laxity and hypermobility, with higher scores representing greater laxity. Scores for young adults range from 4 to 6. Hypermobility syndrome may cause other conditions as a result of their unstable joints, which might include joint instability causing frequent sprains, tendonitis, or bursitis; early-onset osteoarthritis; frequent subluxations or dislocations; pain; crepitus in the joints; and increased nerve compression disorders.

Hypomobility of joints (those with decreased range of motion) may be more susceptible to muscular strains, overuse tendonitis, chronic tendonosis, and nerve entrapment syndromes (Palmer & Epler, 1998).

Joint-play movements can be assessed. Deviations of the ulna and radius on the humerus are examined by having the examiner stabilize the elbow, placing the other hand above the wrist, and abducting/adducting the forearm. The end feel should be hard. Distraction of the olecranon from the humerus can be done by having the subject flex the elbow to 90 degrees and applying a distractive force (but with no rotation). Additional joint-play movements of anteroposterior glide can also be tested.

Functional range-of-motion tests for the elbow are listed in Table 6-14.

Stability Testing

Instability is defined as an abnormal path of articular contact occurring during or at the end of the range of motion (Ozer & Scheker, 2006). Instability may be due to changes in joint surface congruence, in ligamental pathology, or both.

Instability testing of the elbow requires rotationally stabilizing the humerus. From full extension to about 20 degrees of elbow flexion, the collateral ligaments cannot be isolated due to bony contributions to stability. Mediolateral varus stability is tested by having the patient move into slight elbow flexion (20 degrees-30 degrees), which disengages the olecranon from the fossa and allows testing of the collateral ligaments. The humerus is in maximum internal rotation, and the forearm is also pronated. Stress is applied, and if there is instability, the client will report pain on the medial side.

Mediolateral valgus stability is tested in full external rotation (Dumontier, 2010) and with the forearm pronated to test the medial collateral ligament. Continue testing the valgus stability by slightly flexing the elbow, with the humerus in

external rotation and with the forearm supinated to test the lateral collateral complex (Andersson, Nordin, & Pope, 2006; Dumontier, 2010). Apply radial compression and if there is valgus instability, there will be tenderness 4 to 5 cm distal to the lateral epicondyle. The tenderness experienced due to valgus instability is distal to the epicondyle, while lateral epicondylitis tenderness is adjacent to the lateral epicondyle (Andersson, Nordin, & Pope, 2006). For assessing anteroposterior stability, the elbow is flexed to 90 degrees and is held by the examiner while applying an anteroposterior stress to the joint.

Instability can lead to dislocations. Dislocations can be either acute (such as an isolated injury, fracture of radial head, distal radius, ulna, or radius) or chronic (such as an isolated injury, distal radius malunion, incongruent distal radioulnar articulations, or unstable radiocarpal or ulnocarpal ligaments).

If the distal radioulnar joint subluxes or is dislocated as an isolated injury, only swelling may become apparent with painful wrist motions. The rotation of the forearm is always restricted. Complex distal radioulnar dislocations usually involve the ruptured triangular fibrocartilage ligaments, the extensor digiti minimi tendon, the extensor carpi ulnaris tendon, and the extensors of the ring and little fingers and are frequently associated with ulnar styloid fracture. Chronic instability of the distal radioulnar joint can also occur as an isolated injury or may be associated with the distal radius malunion. Chronic instability of the distal radioulnar joint can progress toward osteoarthritis of the joint (Andersson et al., 2006; Ozer & Scheker, 2006).

Strength

Assessment of strength can occur by observing the client perform daily tasks or by resisted manual muscle testing. Testing can be done either to muscle groups or individual muscles. In general, resisted testing of muscle groups includes medial (epicondyle) flexors and pronators and lateral (epicondyle) extensors and supinators. Daily tasks such as drinking from a glass or self-feeding require fair+ elbow flexor strength, and pulling up knee-high socks requires poor+ strength for elbow extensors. Fair supinator strength is required for brushing teeth, and fair pronator strength is necessary for locking a door with a key (Gutman & Schonfeld, 2003). A strength screen for the elbow can be found in Table 6-15.

Table 6-15

STRENGTH SCREEN FOR THE ELBOW AND FOREARM

Ask the client to bend the elbow (to 90°), and tell the client "don't let me straighten your arm" as you apply resistance toward elbow extension	Elbow flexion
Ask the client to bend the elbow (to 90°), and tell the client "don't let me push your arm" as you apply resistance toward elbow flexion	Elbow extension
Ask your client to bend the elbow to 90° with the forearm in neutral position. Tell your client, "don't let me turn your palm down" as you apply resistance in the direction of pronation	Forearm supination
Ask your client to bend the elbow to 90° with the forearm in neutral position. Tell your client, "don't let me turn your palm up" as you apply resistance in the direction of supination	Forearm pronation

In addition, Table 6-16 shows special tests that can be done to assess reflexes, ligamental stability, epicondylitis, and neurological status of the elbow.

Pain

An assessment of the elbow should include gathering information about the client's level of pain. Determining the location of the pain can help to identify the cause of the pain. For example, if the pain is localized, it may be indicative of bursitis or epicondylitis; if the pain is more diffuse, it may signify a subluxation or fracture (Table 6-17). Ask the client when the pain was first noticed, when it occurs, and if there was an immediate onset of pain or if the onset was gradual. Differentiating the type of pain the client is experiencing as well as eliciting descriptions about sensations being felt is also helpful. Getting an idea of the experience of the pain for the client is often done by asking the client to rate his or her pain on a scale from 1 to 10 with 1 indicating no pain and 10 being the worst pain the person has ever experienced. Visual analogs may also be used to represent the client's perception of his or her pain.

The elbow has to be considered as part of the entire kinetic chain of the upper extremity. Pain can be referred to the elbow from the neck, shoulder, or the wrist. For example, a client may complain of elbow pain due to cervical nerve root irritation or even due to metastatic cancer. Elbow pain can come from a variety of conditions and pathologies that include ligamental and bursa inflammation, sprains, arthritis, acute rheumatic fever, tuberculosis, gonorrhea, dislocations and subluxations, fractures, osteosarcoma, and hemarthrosis. A partial list of elbow pathology is included in Table 6-18.

OUTCOME MEASURES

The American Shoulder and Elbow Surgeons (ASES) developed a standardized assessment for the elbow that includes both a physical assessment portion as well as a patient self-evaluation section. It includes visual analog scales for pain, a list of activities in which the client rates his or her performance, and physical assessments of strength, stability, and signs and symptoms (King, Richards, Zuckerman, Blasier, Dillman, Friedman et al., 1999). Other outcome measures for the elbow include the Mayo Elbow Performance Index (Morrey & Sanchez-Sotelo, 2009); Ewald Scoring System (Ewald, 1975); Hospital for Special Surgery (HSS) Scoring System (Figgie, Inglis, Mow, & Figgie, 1989); Flynn Criteria (Flynn, Matthews, & Benoit, 1974); the Pritchard Score (Morrey & Sanchez-Sotelo, 2009); the Brunfeld Score (Morrey & Sanchez-Sotelo, 2009); the Nevaiser Criteria (Nevaiser & Wickstrom, 1977); the Jupiter Score (Morrey & Sanchez-Sotelo, 2009); and the Khalfayan score (Khalfayan, Culp, & Alexander, 1992). More global assessments of outcome, such as the Disability of the Arm, Shoulder, and Hand (DASH) (Hudak, Amadio, & Bombardier, 1996) and SF36 (Ware & Sherbourne, 1992), are also useful as outcome measures for the elbow.

SUMMARY

- The elbow joint consists of the articulations between the ulna, radius, and humerus.

Table 6-16

SPECIAL TESTS FOR THE ELBOW

Reflex testing	Biceps brachii reflex	Tests integrity of reflex innervated by C5 nerve root
	Brachioradialis reflex	Tests integrity of reflex innervated by C6 nerve root
	Triceps reflex	Tests integrity of reflex innervated by C7 nerve root
Ligamental instability	Valgus stress test	Rupture of MCL complex
	Varus stress test	Rupture of LCL
		• Associated with radial head dislocation and annular ligament disruption
	Valgus extension overload test	Aid in identifying an osteophyte or loose body causing pain
Neurological tests	Tinel's sign	Assess ulnar nerve in ulnar groove between olecranon process and medial epicondyle
	Pinch grip test	Anterior interosseous nerve
	Elbow flexion test	Cubital tunnel syndrome/ulnar nerve entrapment
	Test for pronator teres syndrome	Median nerve entrapment
	Wartenberg's sign	Ulnar nerve neuropathy
Epicondylitis tests	Lateral epicondylitis tests	Inflammation
	Medial epicondylitis tests	Inflammation

- The elbow joint is actually made up of three articulations: humeroulnar, humeroradius, and superior radioulnar joints.
- The superior radioulnar joint does not contribute to the uniaxial hinge motions of flexion and extension at the elbow joint.
- The superior and inferior radioulnar joints produce pronation (crossing of radius and ulna) and supination (ulna and radius parallel).
- Stability of the elbow is provided by the ligaments surrounding the joint and the congruence of the distal humerus, proximal radius, and proximal ulna.
- Flexors of the elbow (biceps, brachialis, brachioradialis) are, as a group, most efficient in supination and when the elbow is flexed (also the position of greatest stability). Extensors of the elbow (triceps, anconeus) are most efficient when the elbow is flexed to 20 degrees to 30 degrees.
- Contact areas of the radius and ulna on the humerus change as one moves in flexion and extension.
- Posterior displacement of the elbow is prevented by the coronoid process of the ulna and the humeroradial articulation.

- Assessment of the elbow should include how the client moves, the goals of the client, the amount of edema, and the client's pain. Observation of how the client uses the elbow in activities is an important aspect of evaluation.

REFERENCES

American Occupational Therapy Association. (2008). Occupational therapy practice framework: Domain and process, 2nd edition. *American Journal of Occupational Therapy, 62,* 625-683.

Andersson, G. B. J., Nordin, M., & Pope, M. H. (2006). *Disorders in the workplace: Principles and practices* (2nd ed.). St. Louis, MO: Mosby.

Baeyens, J.-P., Glabbeck, F. V., Goossens, M., Gielen, J., Roy, P. V., & Clarys, J.-P. (2006). In vivo 3D arthrokinematics of the proximal and distal radioulnar joints during active pronation and supination. *Clinical Biomechanics, 21,* S9-S12.

Biel, A. (2001). *Trail guide to the body* (2nd ed.). Boulder, CO: Books of Discovery.

Carrino, J. A., Morrison, W. B., Zou, K. H., Steffen, R. T., Snearly, W. N., & Murray, P. M. (2001). Lateral ulnar collateral ligament of the elbow: Optimization of evaluation with two-dimensional MR imaging. *Radiology, 218*(January), 118-125.

Dumontier, C. (2010). Clinical examination of the elbow Matrise Orthopedique. Retrieved from http://www.google.com/imgres?imgurl=http://www.maitrise-orthop.com/corpusmaitri/orthopaedic/mo77_dumontier/images/fig5_6_7.

Table 6-17

DESCRIPTIONS OF ELBOW PAIN

Diffuse versus localized

If easily localized, more likely to be a superficial structure; deeper structures elicit pain that is more difficult to localize or radiates.

Localized pain may be due to:
olecranon bursitis
lateral epicondylitis
medial epicondylitis
muscle strains (biceps, triceps, wrist flexors, or extensors)
ulnar or radial collateral sprains

Diffuse pain may be due to referred pain from dermatomes or cutaneous nerves, joint subluxations, severe hematoma, fractures.

Diffuse pain may be due to
nerve injuries
joint subluxations or dislocations
severe hematomas
fractures

Location

Anterior

Rupture of distal biceps tendon
Brachialis muscle tear
Anterior capsule tear
Disruption of the annular ligament and radial head dislocation
Acute thrombophlebitis

Posterior

Triceps tendonitis
Triceps rupture
Valgus extension overload syndrome
Stress fracture of the olecranon
Olecranon bursitis
Medial epicondylitis
Rupture of flexor forearm
Rupture or injury to MCL

Medial

Medial epicondylar fracture
Acute ulnar neuropathy
Epitrochlear lymphadenitis

(continued)

Table 6-17 (continued)

DESCRIPTIONS OF ELBOW PAIN

Lateral	Lateral epicondylitis
	Rupture of the lateral extensor muscle group
	Disruption of the annular ligament and anterior radial head dislocation
	Entrapment of the musculocutaneous or lateral antebrachial cutaneous nerve
	Osteochondral fracture of the radial head
Onset	Immediate pain suggests acute injuries, such as hemarthroses, fractures, subluxations, or severe sprains or tears
	Gradual onset suggests overuse injuries or epicondylitis; ulnar neuritis pain 6 to 24 hours after an activity may suggest a more chronic condition
Type of pain	May suggest
Sharp	Injury to the skin, fascia, or superficial muscle or ligaments, or inflammation of bursa or periosteum
Dull ache	Subchondral bone, fibrous capsule, or chronic olecranon bursitis
Tingling	Peripheral nerve damage, irritation of nerve roots of C5-8, or a circulatory problem
Numbness	Peripheral nerves or dorsal nerve roots affecting C6-T1 dermatomes
Time of pain	Morning: rest does not alleviate the pain and may suggest that the injury is still acute, that there is an infection or is systemic in nature, or that rheumatoid arthritis may be present.
	Evening: suggests that activities aggravate the pain
	At night: may indicate bone neoplasm, local or systemic disorders

Adapted from Tan, J. C. (1998). *Practical manual of physical medicine and rehabilitation.* St. Louis, MO: Mosby.

Table 6-18

Summary of Elbow Pathology

Diffuse versus localized	If easily localized, more likely to be a superficial structure; deeper structures elicit pain that is more difficult to localize or radiates.

Localized pain may be due to
olecranon bursitis
lateral epicondylitis
medial epicondylitis
muscle strains (biceps, triceps, wrist flexors or extensors)
ulnar or radial collateral sprains

Diffuse pain may be due to referred pain from dermatomes or cutaneous nerves, joint subluxations, severe hematoma, fractures.

Diffuse pain may be due to
nerve injuries
joint subluxations or dislocations
severe hematomas
fractures

Location
  Anterior   Rupture of distal biceps tendon
                 Brachialis muscle tear
                 Anterior capsule tear
                 Disruption of the annular ligament and radial head dislocation
                 Acute thrombophlebitis
  Posterior   Triceps tendonitis
                 Triceps rupture
                 Valgus extension overload syndrome
                 Stress fracture of the olecranon
                 Olecranon bursitis
  Medial   Medial epicondylitis
                 Rupture of flexor forearm
                 Rupture or injury to MCL
                 Medial epicondylar fracture
                 Acute ulnar neuropathy
                 Epitrochlear lymphadenitis
  Lateral   Lateral epicondylitis
                 Rupture of the lateral extensor muscle group

(continued)

Table 6-18 (continued)

SUMMARY OF ELBOW PATHOLOGY

Lateral	Disruption of the annular ligament and anterior radial head dislocation
	Entrapment of the musculocutaneous or lateral antebrachial cutaneous nerve
	Osteochondral fracture of the radial head
Onset	Immediate pain suggests acute injuries such as hemarthroses, fractures, subluxations, or severe sprains or tears
	Gradual onset suggests overuse injuries or epicondylitis, ulnar neuritis
	pain 6 to 24 hours after an activity may suggest a more chronic condition
Type of pain	May suggest
Sharp	Injury to skin, fascia, or superficial muscle or ligaments, or inflammation of bursa or periosteum
Dull ache	Subchondral bone, fibrous capsule, or chronic olecranon bursitis
Tingling	Peripheral nerve damage, irritation of nerve roots of C5-8, or a circulatory problem
Numbness	Peripheral nerves or dorsal nerve roots affecting C6-T1 dermatomes
Time of pain	Morning: rest does not alleviate the pain and may suggest that the injury is still acute, that there is an infection or is systemic in nature, or that rheumatoid arthritis may be present.
	Evening: suggests that activities aggravate the pain.
	At night: may indicate bone neoplasm, local or systemic disorders.

jpeg&imgrefurl=http://www.maitrise-orthop.com/corpusmaitri/orthopaedic/mo77_dumontier/index_us.shtml&h=615&w=304&sz=34&tbnid=RIdqQk7QwQCgFM:&tbnh=320&tbnw=158&prev=/images%3Fq%3Dlateral%2Buln ar%2Bcollateral%2Bligament&usg=__KHHWBeP1WnpEGgj6NwC5KgqG3o Y=&ei=5WbsS6XGGIO0lQeXw8y1CA&sa=X&oi=image_result&resnum=1 &ct=image&ved=0CBgQ9QEwAA.

Ellenbecker, T. S., Shafer, B., & Jobe, F. W. (2010). Dysfunction, evaluation and treatment of the elbow. In: R. A. Donatelli & M. J. Wooden (Eds.), *Orthopaedic physical therapy* (4th ed., pp. 237-265). St. Louis, MO: Churchill Livingston Elsevier.

Esch, D. L. (1989). *Musculoskeletal function: An anatomy and kinesiology laboratory manual.* Minneapolis, MN: University of Minnesota.

Ewald, F. C. (1975). Total elbow replacement. *Orthopaedic Clinical Journal of North America, 3,* 685-696.

Figgie, M., Inglis, A. E., Mow, C. S., & Figgie, H. E. (1989). Total elbow arthroplasty for complete ankylosis of the elbow. *Journal of Bone and Joint Surgery in America, 71,* 513-520.

Flynn, J. C., Matthews, J. G., & Benoit, R. L. (1974). Blind pinning of displaced supracondylar fractures of the humerus in children. *Journal of Bone and Joint Surgery, 56,* 263-272.

Gench, B. E., Hinson, M. M., Harvey, P. T. (1995). *Anatomical kinesiology.* Dubuque, IA: Eddie Bowers Publishing, Inc.

Grahame, R., Bird, H. A., & Child, A. (2000). Brighton Diagnostic Criteria for the Benign Joint Hypermobility Syndrome (BJHS). *Journal of Rheumatology, 27,* 1777-1779.

Gray, H., & Lewis, W. H. (2000). Anatomy of the human body. In I. Bartleby.com (Eds.). Available from http://education.yahoo.com/reference/gray/.

Greene, D. P., & Roberts, S. L. (1999). *Kinesiology: Movement in the context of activity.* St. Louis, MO: Mosby.

Gutman, S. A., & Schonfeld, A. B. (2003). *Screening adult neurological populations.* Bethesda, MD: American Occupational Therapy Press.

Hamill, J. & Knutzen, K. M. (2003). *Biomechanical basis of human movement* 2nd ed. Philadelphia, PA: Lippincott, Williams & Wilkins.

Hamill, J. & Knutzen, K. M. (1995). *Biomechanical basis of human movement.* Philadelphia, PA: Lippincott, Williams & Wilkins.

Hartley, A. (1995). *Practical joint assessment: Upper quadrant: A sports medicine manual* (2nd ed.). Philadelphia, PA: C. V. Mosby.

Hinkle, C. Z. (1997). *Fundamentals of anatomy and movement: a workbook and guide.* St. Louis, MO: C.V. Mosby.

Hotchkiss, R. N., & Weiland, A. J. (1987). Valgus stability of the elbow. *Journal of Orthopaedic Research, 5*(3), 372-377.

Hudak, P., Amadio, P. C., & Bombardier, C. P. (1996). Development of an upper extremity outcome measure: The DASH (Disabilities of the Arm, Shoulder, and Hand). *American Journal of Industrial Medicine, 29*, 602-608.

Jenkins, D. B. (1998). *Hollingshead's functional anatomy of the limbs and back* (7th ed.). Philadelphia, PA: W. B. Saunders Co.

Khalfayan, E. E., Culp, R. W., & Alexander, H. (1992). Mason Type II radial head fractures: operative versus nonoperative treatment. *Journal of Orthopaedic Trauma, 6*, 283-289.

King, G. J. W., Richards, R. R., Zuckerman, J. D., Blasier, R., Dillman, C., Friedman, R. J., et al. (1999). A standardized method for assessment of elbow function. *Journal of Shoulder and Elbow Surgery, 8*(4), 351-354.

Kisner, C., & Colby, L. A. (2007). *Therapeutic exercise: Foundations and techniques* (5th ed.). Philadelphia, PA: F. A. Davis Company.

Konin, J. G. (1999). *Practical kinesiology for the physical therapist assistant.* Thorofare, NJ: SLACK Incorporated.

Levangie, P. K., & Norkin, C. C. (2005). *Joint structure and function: A comprehensive analysis.* Philadelphia, PA: F. A. Davis.

Linscheid, R. L. (1992). Biomechanics of the distal radioulnar joint. *Clinical Orthopaedics and Related Research, 275*, 46-55.

Loth, T., & Wadsworth, C. T. (1998). *Orthopedic review for physical therapists.* St. Louis, MO: C. V. Mosby.

Lumley, J. S. P. (1990). *Surface anatomy: The anatomical basis of clinical examination.* Edinburgh: Churchill Livingstone.

Magee, D. J. (1992). *Orthopedic physical assessment.* Philadelphia, PA: W. B. Saunders Co.

Magee, D. J. (2002). *Orthopedic physical assessment.* Philadelphia, PA: Saunders.

Morrey, B. F., Askew, L. J., & Chao, E. Y. (1981). A biomechanical study of normal functional elbow motion. *Journal of Bone and Joint Surgery, 63*, 872-877.

Morrey, B. F., & Sanchez-Sotelo, J. (2009). *The elbow and its disorders* (4th ed.). Philadelphia, PA: Saunders Elsevier.

Morrey, B. F., Tanaka, S., & An, K. N. (1991). Valgus stability of the elbow. A definition of primary and secondary constraints. *Clinical Orthopaedics & Related Research, 265*(April), 187-195.

Muscolino, J. E. (2006). *Kinesiology: The skeletal system and muscle function.* St. Louis, MO: Mosby.

Neumann, D. A. (2002). *Kinesiology of the musculoskeletal system: Foundations for physical rehabilitation.* St. Louis, MO: Mosby.

Nevaiser, J. S., & Wickstrom, J. K. (1977). Dislocation of elbow: a retrospective study of 115 patients. *South Medical Journal, 70*, 172-173.

Nordin, M., & Frankel, V. H. (2001). *Basic biomechanics of the musculoskeletal system.* Philadelphia, PA: Lippincott, Williams & Wilkins.

Norkin, C. C., & Levangie, P. K. (1992). *Joint structure and function: A comprehensive analysis* (2nd ed.). Philadelphia, PA: F. A. Davis Co.

Ozer, K., & Scheker, L. R. (2006). Distal radioulnar joint problems and treatment options. *Orthopedics, 29*(1), 38-49.

Palmer, A. K., & Werner, F. W. (1984). Biomechanics of the distal radioulnar joint. *Clinical Orthopaedics and Related Research, 187*(Jul-Aug), 26-35.

Palmer, M. L. & Epler, M. E. (1998). *Fundamentals of musculoskeletal assessment techniques.* Philadelphia, PA: Lippincott.

Premkumar, K. (2003). *The massage connection: Anatomy and physiology* (2nd ed.). Philadelphia, PA: Lippincott Williams and Wilkins.

Regan, W. D., Korinek, S. L., Morrey, B. F., & An, K. N. (1991). Biomechanical study of ligaments around the elbow joint. *Orthopaedics and Biomechanics Laboratory, Mayo Clinic/Mayo Foundation, 271*(Oct), 170-179.

Roy, P. V., Baeyens, J. P., Fauvart, D., Lanssiers, R., & Calrijs, J. P. (2005). Arthrokinematics of the elbow: study of the carrying angle. *Ergonomics, 48*(11-14), 1645-1=656.

Schwab, G. H., Bennett, J. B., Woods, G. W., & Tullos, H. S. (1980). Biomechanics of elbow instability: the role of the medial collateral ligament. *Clinical Orthopaedics & Related Research, 146*(Jan-Feb), 42-52.

Shaaban, H., Giakas, G., Bolton, M., Williams, R., Scheker, L. R., & Lees, V. C. (2004). The distal radioulnar joint as a load-bearing mechanism—a biomechanical study. *Journal of Hand Surgery, 29*(1), 85-95.

Smith, L. K., Weiss, E. L., & Lehmkuhl, L. D. (1996). *Brunnstrom's clinical kinesiology.* Philadelphia, PA: F.A. Davis Co.

Tan, J. C. (1998). *Practical manual of physical medicine and rehabilitation.* St. Louis, MO: Mosby.

Thompson, C. W. & Floyd, R. T. (1994). *Manual of structural kinesiology.* St. Louis, MO: C.V. Mosby.

Thompson, D. (2001). Kinesiology, 2010, from http://moon.ouhsc.edu/dthompso/.

Vasen, A. P., Lacey, S. H., Keith, M. W., & Shaffer, J. W. (1995). Functional range of motion of the elbow. *The Journal of Hand Surgery, 20A*(2), 288-292.

Ware, J., & Sherbourne, C. D. (1992). The MOS 36-Item Short-Form Health Survey (SF-36). *Medical Care, 30*(6).

Yallapragada, R. K., & Patko, J. T. (2009). Elbow collateral ligaments. eMedicine Specialties: Orthopedic Surgery: Elbow. Retrieved from http://emedicine.medscape.com/article/1230902-overview.

ADDITIONAL RESOURCES

Basmajian, J. V., & DeLuca, C. J. (1985). *Muscles alive* (5th ed.). Baltimore, MD: Williams & Wilkins.

Baxter, R. (1998). *Pocket guide to musculoskeletal assessment.* Philadelphia, PA: W. B. Saunders Co.

Burstein, A. H., & Wright, T. M. (1994). *Fundamentals of orthopaedic biomechanics.* Baltimore, MD: Williams & Wilkins.

Calliet, R. (1996). *Soft tissue pain and disability* (3rd ed.). Philadelphia, PA: F. A. Davis Co.

Goss, C. M. (Ed.). (1976). *Gray's anatomy of the human body* (29th ed.). Philadelphia, PA: Lea and Febiger.

Greene, D. P., & Roberts, S. L. (1999). *Kinesiology: Movement in the context of activity.* St. Louis, MO: C. V. Mosby.

Hinojosa, J., & Kramer, P. (1998). *Evaluation: Obtaining and interpreting data.* Bethesda, MD: American Occupational Therapy Association.

Kisner, C., & Colby, L. A. (1990). *Therapeutic exercise: Foundations and techniques* (2nd ed.). Philadelphia, PA: F. A. Davis.

MacKenna, B. R., & Callender, R. (1990). *Illustrated physiology* (5th ed.). New York, NY: Churchill Livingstone.

Nicholas, J. A., & Hershman, E. B. (1990). *The upper extremity in sports medicine.* St. Louis, MO: C. V. Mosby.

Nordin, M., & Frankel, V. H. (1989). *Basic biomechanics of the musculoskeletal system* (2nd ed.). Philadelphia, PA: Lea and Febiger.

Palastanga, N., Field, D., & Soames, R. (1989). *Anatomy and human movement: Structure and function.* Oxford: Heinemann Medical Books.

Perry, J. F., Rohe, D. A., & Garcia, A. O. (1996). *Kinesiology workbook.* Philadelphia, PA: F. A. Davis Co.

Saidoff, D. C., & McDonough, A. L. (1997). *Critical pathways in therapeutic intervention: Upper extremity.* St. Louis, MO: C. V. Mosby.

Watkins, J. (1999). *Structure and function of the musculoskeletal system.* New York, NY: Human Kinetics.

7

THE WRIST

The wrist allows for incremental changes in location and orientation of the hand relative to the forearm, allows placing the hand in space, and permits fine gradations of prehension as well as powerful grasp. The wrist makes a dynamic contribution to a skill or movement or to stabilization (Hamill & Knutzen, 1995) and contributes to expression and nonverbal communication. The wrist serves a kinetic function in transmitting loads and forces from the hand to the forearm and from the forearm to the hand, as well as providing stability for the hand. Linscheid (1986) considered the wrist to be mechanically the most complex joint in the body. These positional adjustments are essential for augmenting fine motor control of fingers and in allowing optimal length-tension of long finger muscles so maximal finger movement can be attained (Nordin & Frankel, 2001) (Figure 7-1).

The wrist is a multiarticulating, complex biaxial joint made up of two compound joints. Because this is a complex joint with many articulations, the joints do not act in isolation. When the hand is in a resting position, it assumes a slight ulnar and palmar posture (Caillet, 1996) because the dorsal surface is much longer than the palmar.

The position of one joint affects the action of others. For example, if the wrist is flexed, the interphalangeal (IP) joints cannot flex fully due to the insufficiency of the finger flexors as they cross both joints. The interplay of structures has implications not only for joint and muscle actions, but also is an influence in pathology of the hand and wrist.

The wrist joint is capable of much mobility and has great structural stability (Smith, Weiss, & Lehmkuhl, 1996). This highly complex area consists of 15 bones (eight carpal bones, ulna, radius, and five metacarpal bones), 17 joints, and an extensive ligamental system. The joint is formed by articulations of the hand and wrist and has four articular surfaces:

1. Distal surfaces of the radius and the articular disk.

2. Proximal surfaces of scaphoid, lunate, and triquetrum bones.

3. An S-shaped surface formed by the inferior surfaces of the scaphoid, lunate, and trapezium carpal bones.

4. The reciprocal surface of upper surfaces of the carpal bones of the second row (Goss, 1976) (Figure 7-2).

These four surfaces form the two joints of the wrist: the proximal portion is what is commonly thought of as the wrist joint and is called the radiocarpal joint; and the distal portion forms the midcarpal joint.

BONES OF THE WRIST AND PALPABLE STRUCTURES

The bones of the wrist include the ulna, radius, and eight carpal bones. The ulnar and radial styloid processes are the medial and lateral borders of the wrist with the carpal bones easily palpated between the two styloids. The two wrist creases on the volar surface of the wrist indicate the radiocarpal (proximal line) and midcarpal (distal line) joints and can be seen in Figure 7-3.

Rybski MF.
Kinesiology for Occupational Therapy, Second Edition (pp 209-234)
© 2012 SLACK Incorporated

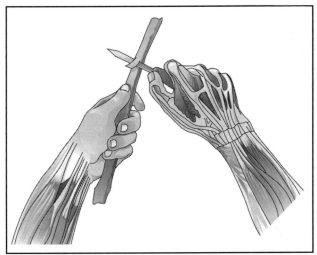

Figure 7-1. The wrist in action. (Adapted from Biel A. (2001). *Trail guide to the body* (2nd ed.). Boulder, CO: Books of Discovery.)

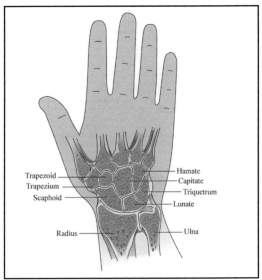

Figure 7-2. Articulating surfaces of the wrist.

Figure 7-3. Flexor wrist creases.

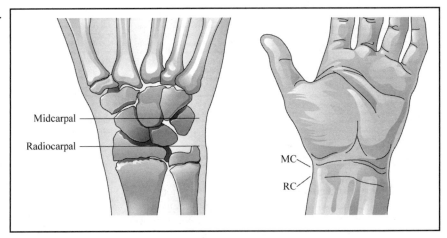

Ulna

Head of the Ulna

This can be grasped from side to side at its narrowest portion. This is the rounded projection proximal to the index finger and on the dorsum of the wrist. In a pronated position, this eminence can be seen beneath the skin. If the fingertip is placed on the highest part and the forearm is supinated, the head of the ulna can no longer be palpated because, during supination, the distal portion of the radius rotates around the head of the ulna (Smith et al., 1996).

Styloid Process of the Ulna

This is a small projection on the medial aspect of the head of the ulna, and it feels smaller and sharper than the head of the ulna. It may be palpated in pronation or supination. If the extensor carpi ulnaris tendon interferes with palpation, slide the index finger over this tendon in a palmar direction, and the styloid will be more accessible (Smith et al., 1996).

Radius

Styloid Process of the Radius

This is located at the lateral aspect of the wrist, proximal to the first metacarpal, and it extends more distally than the styloid process of the ulna. Note that the radial styloid extends further distally than ulnarly.

Lister's Tubercle (Tubercle of the Radius)

This tubercle is located on the dorsum of the radius about 1 inch laterally from the head of the ulna. The extensor pollicis longus tendon lies on the ulnar side of this prominence, and the tubercle serves as a landmark for locating many tendons in this region (extensor carpi radialis brevis [ECRB], extensor indicus proprius [EIP], and the extensor digitorum [ED] to the index finger). Locate the radial styloid process, and palpate the dorsal surface of the process directly across from the head of the ulna (Biel, 2001).

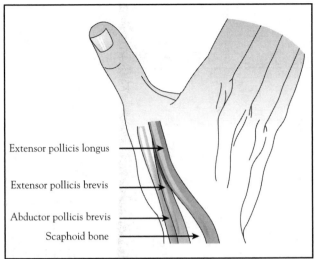

Figure 7-4. Structures in the anatomical snuffbox.

Carpal Bones

There is considerable passive accessory motion possible in the wrist with the forearm and wrist relaxed. If the distal radius and ulna are stabilized with one hand, and the other hand is placed around the proximal carpal row, the carpals can move easily in dorsal, volar, medial, and lateral translatory glides and can be distracted several millimeters (Smith et al., 1996). Due to overlying tendons and compact arrangement of the carpal bones, isolating all of the carpal bones is difficult (Biel, 2001). As a group, you can feel the shifting of the carpal bones, particularly on the dorsum of the hand just above the radius and ulna and between the proximal and distal flexor creases of the wrist.

Capitate

This bone is in the central position in the wrist in line with the middle finger. It is best approached from dorsum and is seen as a slight depression. The capitate serves as the axis for deviation motions of the wrist (Smith et al., 1996).

Scaphoid (Navicular)

The scaphoid is an important bone for wrist and hand function as it bridges and moves with the two carpal rows on the radial side. Because of this, the scaphoid helps to support the weight of the arm, transmits forces received from the hand to the forearm, and is important in wrist movements (Magee, 2002). It is also the most frequently fractured carpal bone.

This can be palpated distally to the styloid process of the radius. When in ulnar deviation, the bone becomes prominent and can be palpated, but while in radial deviation, the bone recedes. The scaphoid and trapezium make up the floor of the anatomic snuff box (fovea radialis).

Trapezium (Greater Multangular)

This bone can be palpated proximal to the first carpometacarpal of the thumb and distal to the scaphoid.

Lunate

The lunate is distal to Lister's tubercle and proximal to the capitate. It is prominent as the wrist is passively flexed and recedes as the wrist is passively extended. This is the most frequently dislocated carpal bone.

Pisiform

This is a pea-shaped bone on the palmar side of the wrist near the ulnar border. It is located on the ulnar and palmar surface of the wrist just distal to the flexor crease. It can be grasped and moved from side to side when the wrist is flexed, but is immobile when the wrist is extended due to tension of the flexor carpi ulnaris muscle, which attaches on this bone (Biel, 2001).

Triquetrum (Triangular)

The triquetrum is located on the dorsal surface of the pisiform, distal to the styloid process of the ulna. This bone is just below the ulnar styloid and is accessible for palpation when the wrist is abducted (Biel, 2001; Esch, 1989; Hamill & Knutzen, 2003; Magee, 2002; Smith et al., 1996).

Hamate

The hamate has a palpable hook on the palmar surface on which the flexor retinaculum attaches. It is located distally and laterally to the pisiform, which is often tender when palpated. The flat surface of the hamate is accessible on the dorsal surface of the bases of the fourth and fifth metacarpals and will feel like a subtle mound at the base of first finger about ½-inch from the pisiform (Biel, 2001).

Anatomical Snuff Box

This indentation forms on the dorsum of the hand when the thumb is actively extended. The tendons between extensor pollicis longus (EPL) and extensor pollicis brevis (EPB) form the medial and lateral borders of the snuffbox with the scaphoid bone inside as the floor of the snuffbox (Figure 7-4). The radial styloid is on the lateral aspect when in the anatomic position, and moving medially over the radius is Lister's tubercle. The ulnar styloid is on the medial aspect (Hinkle, 1997).

ARTICULATIONS OF THE WRIST

Radiocarpal Joint

This joint is commonly referred to as the wrist. It is made up of the distal end of the radius and the distal surface of the radioulnar articulating disk, connecting with the proximal row of carpal bones (scaphoid, lunate, and triquetrum bones). The radius articulates with scaphoid and lunate, while the lunate and trapezium articulate with the disk, not the ulna (MacKenna & Callender, 1990). The distal radius and the disk form an elliptical, biconcave surface, and the superior articulat-

Table 7-1

SUMMARY OF THE OSTEOKINEMATICS OF THE RADIOCARPAL JOINT

Functional joint	Diarthrotic, biaxial
Structural joint	Synovial, condyloid
Close-packed position	Extension
Resting position	Neutral with slight ulnar deviation
Capsular pattern	Flexion and extension

ing surfaces of scaphoid, lunate, and trapezium form a smooth biconvex surface (Goss, 1976; MacKenna & Callender, 1990). The disk binds the radius and the ulna together at their distal ends and separates the distal ulna from the radiocarpal joint.

Osteokinetics of the Radiocarpal Joint

The radiocarpal joint is a biaxial, condyloid synovial joint with two degrees of freedom capable of producing flexion and extension, abduction and adduction, and circumduction of the wrist. Circumduction is produced by the consecutive sequential combination of movements of adduction, extension, abduction, and flexion. This is a true condyloid joint with all motions except rotation. No active rotation is possible, but the effect of rotation is achieved by pronation and supination of the forearm at the radioulnar joints. Rotation is blocked by the bony fit of the radiocarpal joint and direction of fibers of the radiocarpal ligaments (Neumann, 2002). The motion at this joint is primarily gliding produced by movement of the proximal row of carpal bones on the radius and the radioulnar disk.

The close-packed position of extension with slight ulnar deviation occurs due to the elongation of the palmar radiocarpal ligament and palmar capsule and wrist and finger flexor muscles. This stabilization allows greater stability when weight bearing on hands, such as when crawling on the hands and knees or doing a transfer from a wheelchair (Magee, 2002; Neumann, 2002). A summary of the osteokinematics of the radiocarpal joint in presented in Table 7-1.

Arthrokinematics of the Radiocarpal Joint

In open chain movement, the convex surfaces of the scaphoid, lunate, and trapezium move on the concave surfaces of the radius and ulna. Because the convex surface of carpals moves on the concave distal radius, the glide of carpals is in a direction opposite to the movement of the hand (Kisner & Colby, 1990). In extension, the convex surface of the lunate rolls dorsally on the radius and simultaneously glides in a palmar direction and moves similarly but in the reverse direction for flexion

(Figure 7-5). There is a similar convex-on-concave movement of the carpals at the radiocarpal joint for ulnar and radial deviation (Neumann, 2002) (Table 7-2). Computer modeling and cadaver studies have shown that radiocarpal extension is accompanied by increased contact dorsally rather than in a volar direction, which may contradict the current understanding of the convex-concave rule and our understanding of the complexity of the wrist (Levangie & Norkin, 2005).

Stability of the Radiocarpal Joint

The radiocarpal joint is formed by the radius and radioulnar disk proximally and by the scaphoid, lunate, and triquetrum distally. No muscle forces are applied directly to the bones of the proximal row, and this row serves as a link between the radial and distal carpals; it is considered an intercalated segment. When compressive forces are applied to the wrist, the middle segment tends to collapse and move in the wrong direction. This would happen at the wrist except for the intercarpal bridge formed by scaphoid and ligaments (Norkin & Levangie, 1992).

The structures involved in the radiocarpal joint also include structures related to the distal radioulnar joint. The distal radioulnar and radiocarpal joints share the same joint fibrous, joint capsule, and radioulnar disk (triangular fibrocartilage), so these structures participate in movements at both the distal radioulnar joint and at the radiocarpal joint (Donatelli & Wooden, 2010).

The radioulnar disk (triangular fibrocartilage) runs from the distal radius to the distal ulna and blends into capsular ligamentous structure of the distal radioulnar joint. The triangular fibrocartilage covers the distal end of the ulna, creating a smooth appearance of the wrist as the ulna articulates with the proximal carpal row via the disk. The disk accounts for just more than 11% of the articulating surfaces of the radiocarpal joint (Linscheid, 1986). The disk and connective tissue continue distally to attach to triquetrum, hamate, and base of fifth metacarpals and ulnar collateral ligament, radioulnar ligament, and sheath of extensor carpi ulnaris.

Table 7-2

SUMMARY OF THE ARTHROKINEMATICS OF THE RADIOCARPAL JOINT

PHYSIOLOGIC MOVEMENT OF CONVEX CARPALS ON CONCAVE RADIUS	DIRECTION OF SLIDE OF CARPAL ON RADIUS	RESULTANT MOVEMENT
Flexion	Anterior (volar) slide, posterior roll	Movement in opposite direction
Extension	Posterior slide, anterior roll	Movement in opposite direction
Radial deviation	Ulnar slide, radial roll	Movement in opposite direction
Ulnar deviation	Radial slide, ulnar roll	Movement in opposite direction

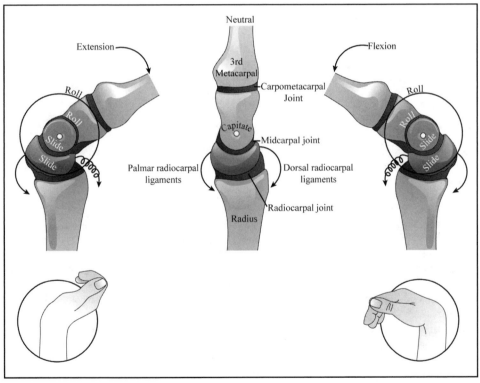

Figure 7-5. Arthrokinematics of wrist flexion and extension. (Adapted from Neumann, D. A. (2002). *Kinesiology of the musculoskeletal system: Foundations for physical rehabilitation.* St. Louis: Mosby.)

The articular disk adds stability to the joint by maintaining a mechanical relationship that binds the distal ends of the radius and ulna as well as the carpal bones and ulna. With the disk intact, the radius takes on 60% of the axial load, and the ulna takes on 40%. If the disk is injured, the radius takes on 95% of the axial load (Ishii, Palmer, Werner, Short, & Fortino, 1998; Magee, 2002; Tang, Ryu, & Kish, 1998; Wheeless, 2010). This view was challenged by Markolf who found that the distal ulna only bore 3% of the forces and that the TFCC was too compliant to bear significant loads (Markolf, Lamey, Yang, Meals, & Hotchkiss, 1998).

There is a wedge of connective tissue (meniscus homolog) that continues from the fibrocartilage that thickens medially and encloses the ulnar styloid, attaching to the articular capsule (Levangie & Norkin, 2005). The dorsal and palmar sections of the joint capsule thicken to form the dorsal radioulnar and palmar radioulnar ligaments.

Figure 7-6. Triangular fibro-cartilage complex (TFCC). (Adapted from Neumann, D. A. (2002). *Kinesiology of the musculoskeletal system: Foundations for physical rehabilitation.* St. Louis: Mosby.)

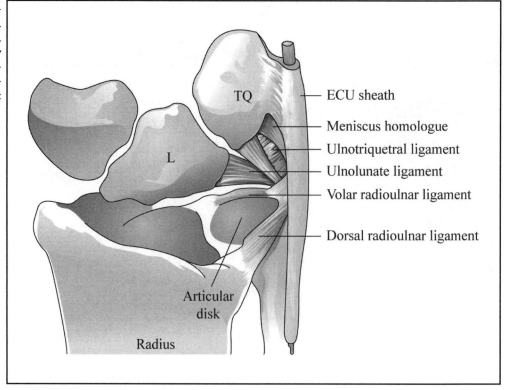

Triangular Fibrocartilage Complex (TFCC) or Ulnocarpal Articulation

Konin (1999) asserted that these structures at the wrist would comprise an additional functional articulation of the wrist commonly called the triangular fibrocartilage complex (or ulnocarpal, ulnar-meniscal-triquetrum, or ulnomenisco-carpal joint) (Figure 7-6). This articulation is a meniscal type of structure located between the distal ulna and the proximal part of the triquetrum bone that separates the radiocarpal joint from the distal radioulnar joint. The TFCC consists of radioulnar articular disk, meniscus homologue (lunocarpal), ulnocarpal ligament, dorsal and volar radioulnar ligaments, and the extensor carpi ulnaris sheath.

This complex acts as a shock absorber and binds the distal radioulnar joint. It has similar mechanics to the glenoid labrum of the shoulder and menisci of the knee. Notice, too, that the ulna has no contact with the carpal bones because they are separated by this fibrocartilaginous disk; this allows the ulna to glide on the disk for pronation and supination without influence or interference of the carpal bones. The distal end of the ulna can be surgically removed without any impairment of wrist motion because the ulna doesn't influence wrist motion at this articulation (Magee, 2002).

The TFCC is seen as a major stabilizer of the distal radioulnar joint and contributes to ulnocarpal stability. The volar portions of the TFCC prevent dorsal displacement of the ulna and are taut in pronation while the dorsal portions prevent volar displacement of the ulna and are tight in supination (Wheeless, 2010).

Midcarpal Joints

The midcarpal joints are formed as a compound articulation between the proximal and distal rows of carpal bones (with the exception of pisiform bone). On the radial (lateral) side, the scaphoid articulates with the trapezoid and trapezium bones. The scaphoid is essentially a convex surface on which the concave surfaces of the trapezoid and trapezium slide. The central portion consists of the head of capitate and superior part of hamate articulating with the deep cavity formed by scaphoid and lunate (Goss, 1976). This forms a type of ball-and-socket joint where some axial rotation may be possible (Nordin & Frankel, 1989).

The ulnar (medial) side of the joint is the articulation of the hamate with the triquetrum bone. The articulating surface of the hamate is convex as it slides on the concave articulating surface of the triquetrum.

Flexion and extension motions are possible at this joint, as is some rotation. The trapezium and trapezoid on the radial side and the hamate on the ulnar side glide forward and backward on the scaphoid and trapezium, while the head of capitate and the superior surface of hamate rotate in the cavity of the scaphoid and lunate (Goss, 1976). This joint is inherently stable because it is bound by the dorsal and palmar and ulnar and radial ligaments.

The joint capsule of the midcarpal joint is separate from the radiocarpal joint capsule and is usually divided into medial and lateral compartments. The medial compartment is larger with the concave proximal surface and convex distal surface. The smaller lateral compartment is oriented opposite to the

Table 7-3

SUMMARY OF THE OSTEOKINETICS OF THE MIDCARPAL JOINTS

Flexion	Diarthrotic, biaxial
Extension	Synovial, condyloid
Radial deviation	Extension with ulnar deviation
Ulnar deviation	Neutral or slight flexion with ulnar deviation
Flexion	Equal limitation of flexion and extension

medial surface with a proximal convex surface and concave distal surface (Muscolino, 2006).

Osteokinetics

This joint is predominantly described as a condyloid joint, although it was also cited as a sellar (saddle-shaped) joint (Gench et al., 1995). The excursion of the articulating surfaces favor the range of extension over flexion and radial deviation over ulnar, which is opposite to what occurs at the radiocarpal joint (Nordin & Frankel, 2001). There are interosseous membranes between scaphoid and trapezoid and trapezium on the lateral side and the scaphoid, lunate, triquetrum with capitates, and hamate on the medial side, but none between the proximal and distal rows of bones so there is greater movement at the midcarpal joint than between individual bones of two rows of intercarpal joint (Viegas, Patterson, Todd, & McCarty, 1993). The midcarpal joint is a biaxial synovial joint, and the bony surfaces are in most contact in extension with slight ulnar deviation (Table 7-3).

Arthrokinematics

The concave-convex relationship rules apply in the case of the midcarpal joints, but the relationships between these bones are actually more complex than is readily apparent. For example, in wrist flexion at the midcarpal joint, the capitate and hamate slide dorsally, and the trapezium and trapezoid slide in a volar direction while the distal row of carpals rolls in a volar direction. The relationship of the motion with the direction of movement of the distal carpal bones in respect to proximal carpals is shown in Table 7-4. The distal row of carpal bones moves in the same direction as the hand but in an opposite direction of the proximal carpal bones. Because of the movement at the carpal bones, there is significant range of motion produced by only moderate movements of either the radiocarpal or midcarpal joints (Neumann, 2002) (Table 7-4).

Intercarpal Joints

The intercarpal joints are those articulations between the individual bones in the proximal and distal rows of carpal bones. It is the articulation of the scaphoid, lunate, and triangular (trapezium) bones that form gliding joints and are connected by dorsal, palmar, and interosseous ligaments as well as the articulation of the bones of the distal row (trapezium, trapezoid, capitate, hamate) that also form gliding joints and are connected by dorsal, palmar, and interosseous ligaments. The intercarpal joints are bound by intercarpal ligaments and are capable of slight movements (Table 7-5).

Pisotriquetral Joint

This is a separate joint, and it does not take part in other intercarpal movements. The sesamoid bone forms an articulation is the pisiform with the triquetrum bone. This is capable of slight movement due to ligamental connections and serves to increase the contraction force of the flexor carpi ulnaris muscle and serves as an attachment for the transverse carpal ligament (Muscolino, 2006).

STABILITY

The wrist joint is very stable due to the proximal radiocarpal and distal midcarpal joints, which act together as a double hinge (Nordin & Frankel, 2001). In addition to the bony support and geometry of the sigmoid notch of the radius, wrist stability is maintained by an extensive system of ligaments. Not only do the wrist ligaments provide additional support at the wrist, but they also contribute to passive motion.

Primary stability of the joint comes from the radial and ulnar collateral ligaments, dorsal and volar radioulnar

Table 7-4

SUMMARY OF THE ARTHROKINEMATICS OF THE MIDCARPAL JOINT

PHYSIOLOGIC MOVEMENT OF CONVEX CARPALS ON CONCAVE RADIUS	DIRECTION OF SLIDE OF PROXIMAL CARPALS ON DISTAL CARPALS	RESULTANT MOVEMENT
Flexion	Posterior slide (dorsal), anterior roll	Movement in opposite direction
Extension	Anterior (volar) slide, posterior roll	Movement in opposite direction
Radial deviation	Ulnar slide, radial roll	Movement in opposite direction
Ulnar deviation	Radial slide, ulnar roll	Movement in opposite direction

Table 7-5

SUMMARY OF THE OSTEOKINEMATICS OF THE INTERCARPAL JOINTS

Functional joint	Multiaxial
Structural joint	Gliding
Close-packed position	Extension
Resting position	Neutral or slight flexion
Capsular pattern	None

ligaments, and the triangular fibrocartilage complex. The joint is enclosed by a strong but loose joint capsule and is reinforced by the dorsal radiocarpal, palmar radiocarpal, palmar ulnocarpal, and intercarpal ligaments and flexor and extensor retinaculum (Nordin & Frankel, 2001).

Ligaments in the wrist have been classified as extrinsic (connecting the distal radius and ulna to the carpal bones) or intrinsic (with origins and insertions within the carpal bones) (Kijima & Viegas, 2009).

Extrinsic Ligaments

The extrinsic ligaments tend to limit motion primarily at the radiocarpal joint. The radial and ulnar collateral ligaments provide significant passive control of radiocarpal motion in the frontal plane. The radial collateral ligament, which runs from the styloid process of the radius to the scaphoid and trapezium, limits ulnar deviation while the ulnar collateral ligament, which runs from the styloid process of the ulna to the triquetrum, limits radial deviation (Muscolino, 2006). The radial collateral ligament, which runs from the lateral epicondyle of the humerus to the annular ligament and olecranon process, can be felt as a depression between the lateral epicondyle and the head of the radius. With the elbow flexed, the fibers of the ligament run parallel to the forearm (Biel, 2001). The triangular ulnar collateral ligament is deep to the common flexor tendon but superficial to the ulnar nerve. It can be found between the medial epicondyle and the medial aspect of the olecranon process (Biel, 2001).

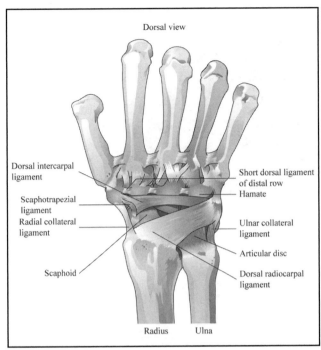

Figure 7-7. Dorsal ligaments of the wrist. Adapted from Neumann, D. A. (2002). *Kinesiology of the musculoskeletal system: Foundations for physical rehabilitation.* St. Louis, MO: Mosby.

The dorsal radiocarpal ligament, which exists as a thickening of the capsule, attaches the radius to the capitate, lunate, and scaphoid. This ligament limits full flexion and allows the hand to follow the radius during pronation of the forearm (Levangie & Norkin, 2005). Figure 7-7 shows the dorsal ligaments of the right wrist.

The palmar radiocarpal ligaments are thick and strong and are the most important ligaments for motion and stability (Norkin & Levangie, 1992). The palmar radiocarpal ligament has three distinct portions that are named for their attachments. The first is the radiolunate (radiotriquetral or long radiolunate) ligament, which is the strongest and most distinct. This ligament allows the scaphoid to rotate as well as provides stabilization for the scaphoid at the extremes of movement (Magee, 2002). The second ligament is the radiocapitate (radioscaphocapitate ligament), and the third is the radioscaphoid (short radiolunate). The radioscaphoid assists with joint motion and stability by checking movement of the joint surfaces and maintaining joint integrity (Norkin & Levangie, 1992). The palmar radiolunate and radiocapitate ligaments are the primary constraints to carpal translation (Linscheid, 1986). The radiocarpal ligaments ensure that the carpals follow the radius throughout forearm rotation. Supination tightens the palmar ligaments, and the palmar radiocarpal ligament carries the wrist with the radius.

The ulnocarpal complex includes the ulnar collateral ligament, the radioulnar articular disk, and the palmar ulnocarpal ligament. These structures are part of the TFCC, which consists of radioulnar articular disk, meniscus homologue (lunocarpal), ulnocarpal ligament, dorsal and volar radioulnar ligaments, and the extensor carpi ulnaris sheath. The ulnocar-

pal ligament is composed of the ulnolunate and ulnotriquetral ligaments of the meniscus homologue and is really portions of the medial palmar radiocarpal ligament. The ulnocarpal ligament prevents separation of the carpals to which they attach and prevents dorsal migration of the distal ulna to add to joint stability (Ishii et al., 1998; Wheeless, 2010) (Figure 7-8).

Intrinsic Ligaments

Intrinsic ligaments stabilize and limit movement between carpals. There are short intrinsic ligaments that connect the distal row of carpal bones and help to maintain the bones of the distal carpal row as an immovable unit. The intermediate intrinsic ligaments (lunotriquetral, scapholunate, and scaphotrapezial ligaments) connect the proximal row of carpal bones. The long intrinsic ligaments connect the scaphoid, triquetrum, and capitate and include the dorsal intercarpal and palmar intercarpal ligaments.

The dorsal and palmar intercarpal ligaments strengthen the hand for any impact imposed on knuckles (such as striking with a closed fist). Of these two ligaments, the palmar is seen to be more important and is called the deltoid or V ligament, which stabilizes the capitate. The dorsal intercarpal ligament and dorsal radiocarpal ligament work together to allow normal kinematics and stabilization of the proximal carpal row, and the scaphoid and these two ligaments together are also known as the dorsal radioscaphoid ligament (Kijima & Viegas, 2009; Viegas, 2001; Viegas, Yamaguchi, Boyd, & Patterson, 1999).

Dynamic Stability

Dynamic stability occurs only when the muscles are actively contracting. Pronator quadratus connects the ulna to the radius and depresses the ulna during pronation. Extensor carpi ulnaris has a moderate effect in depressing the ulna relative to the triquetrum during pronation, but the mechanical advantage of the muscle is small. Flexor carpi ulnaris also acts to depress the ulnar head but only when the hand is fixed against resistance (Linscheid, 1992).

MOVEMENTS OF THE WRIST

Movements at the wrist include flexion, extension, abduction, adduction, and circumduction. The amount of movement varies considerably from person to person, varies due to the position of the forearm, and can even vary from one hand to the other of the same person (Jenkins, 1998). Movement of the carpal row upon the radius and triangular fibrocartilage complex is a gliding translatory movement produced concomitantly with wrist movements (Hamill & Knutzen, 2003). As the hand flexes in the palmar direction, the carpal row glides dorsally; in radial abduction, the proximal carpal row glides in the ulnar direction, but this must be accompanied by an elastic capsule and lax ligaments. Wrist motion is actively driven by muscles, but is controlled by the passive tension of ligaments (Neumann, 2002). Table 7-4 provides a summary of motions of the wrist.

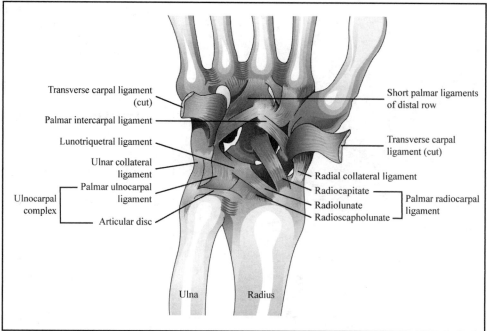

The function that occurs at the wrist depends on the muscles relative to the axes of motion. The wrist does not move in a direct plane but along a plane between radial extension (deviation) and ulnar flexion (deviation) and an opposite plane of ulnar extension and radial flexion due to direction of muscles and their tendons acting on the wrist (Caillet, 1996).

The amount of motion depends also on the shape of the articulating surfaces. The surface of the radioulnar articulation is less concave across the bony surfaces than anteroposteriorly, and the proximal row of carpal bones is more mobile than the distal. Plus, the curvature of the proximal carpal row is greater than the opposing curve of the radioulnar surface. The curvature of the proximal carpal row is greater than the opposing curve of the radioulnar surface. The distal end of the radius is concave and usually reaches further distally on the radial side than the ulnar side (although 60% of people have equal length). These discrepancies permit greater excursion of flexion and extension than of radioulnar motion. There is a greater degree of flexion as compared to extension and of ulnar abduction as compared to radial that is due to the angulation of the distal articular surface of the radius and due to the fact that dorsal wrist ligaments are more slack than is the palmar ligament (Caillet, 1996).

Grip strength is influenced by the position of the wrist. Hamill & Knutzen (1995) cited a study by Nordin and Frankel that found that when the wrist is in 40º of hyperextension (extension), grip was more than three times greater than if the wrist was in 40 degrees of wrist flexion. In addition, wrist flexion has more than twice the work capacity of extension, and radial deviation slightly exceeds ulnar deviation in work capacity (Baxter, 1998).

Flexion and Extension

The wrist has been conceptualized as being comprised of three columns: rotational (triquetrohamate), flexional and extensional (capitolunate), and radial (scaphotrapezial)

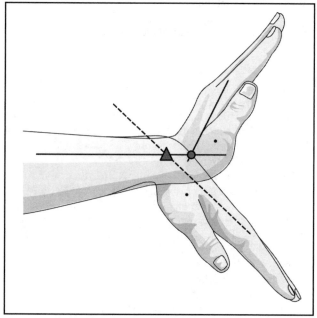

Figure 7-9. Axis of wrist flexion (triangle) and extension (circle).

(Taleisnik, 1985). The kinematic of flexion and extension is sagittal plane movement within the central column of the wrist (articulation of the radius, lunate, capitate, and third metacarpal). The axis for wrist flexion and extension of the wrist occurs in the sagittal plane and frontal or side-to-side axis through the wrist just distal to the styloid process of the radius and ulna through the capitate carpal bone (Patterson, Nicodemus, Viegas, Elder, & Rosenblatt, 1998). Smith and colleagues (1996) indicate that the axis migrates distally when moving from full flexion to extension, which is caused by translatory and rotational movements of the lunate and scaphoid (Figure 7-9). These movements change the height

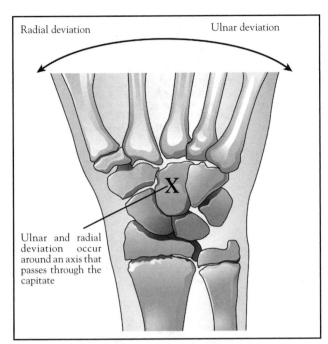

Radial deviation Ulnar deviation

X

Ulnar and radial deviation occur around an axis that passes through the capitate

Figure 7-10. Axis of radial and ulnar deviation.

of the bones of the wrist, which is necessary to maintain tension on the ligaments. Muscles that lie anterior to the wrist's flexion/extension axis are wrist flexors, whereas muscles lying posterior to the axis are wrist extensors.

Flexion of the wrist is produced when the carpals slide dorsally on the radius and the disk. It has been estimated that 60% of the flexion motion occurs at the midcarpal joint and 40% at the radiocarpal joint (Nordin & Frankel, 2001; Sarrafian, Melamed, & Goshgarian, 1977), although the exact percentage of motion at each joint has been debated (Levangie & Norkin, 2005; Neumann, 2002). The movement of flexion is often accompanied by slight ulnar deviation and supination, and the primary wrist flexors are flexor carpi radialis, flexor carpi ulnaris, and palmaris longus.

Flexor carpi radialis enables the second and third metacarpals to compress the capitate and trapezoid, minimizing their movement. The capitate flexes on the scaphoid and lunate, resulting in metacarpal joint flexion, and the scaphoid flexes and pronates on the radius while the lunate flexes and supinates on the radius. Flexor carpi ulnaris enables the proximal portion of the fifth metacarpals to glide dorsally on the hamate, and the hamate will then rotate anteriorly on the triquetrum (metacarpal joint flexion). Secondary wrist flexor muscles include flexor digitorum profundus, flexor digitorum superficialis, and flexor pollicis longus. The proximal portion of the triquetrum will flexion on the TFCC, resulting in radiocarpal joint flexion (Prosser & Conolly, 2003). End feel for wrist flexion is firm secondary to tautness of the dorsal radiocarpal ligament and dorsal joint capsule.

Wrist extension also has a firm end feel secondary to the tautness of the palmar radiocarpal ligament and palmar joint capsule. It has been estimated that 67% of the movement of extension occurs at the radiocarpal joint (Nordin & Frankel, 2001; Sarrafian et al., 1977).The motion of extension is initi-

ated by the distal carpals, with the capitate at the center as the axis, which glide on the relatively fixed proximal row of carpal bones. Extensor force, via extensor carpi ulnaris and extensor carpi radialis, moves the distal carpal row and scaphoid on relatively fixed lunate and triquetrum. At 45 degrees of wrist extension, ligaments draw the capitulum and scaphoid together, which increases the extensor force and unites the carpals, which now will act as a single unit. Midcarpal motion, estimated at 33% of the joint motion (Nordin & Frankel, 2001; Sarrafian et al., 1977), occurs with the distal carpals gliding on proximal carpals (lunate and triquetrum), which are relatively fixed.

Full wrist extension requires a slight spreading of distal radius and ulna, and if these two bones were grasped together, complete wrist extension would not be possible (Smith et al., 1996). Wrist extension is usually accompanied by slight radial deviation and forearm pronation. Pure extension (without deviation) is dependent upon the ulnar and radial extensors working together for balance.

The most powerful extensors are the extensor carpi radialis longus, extensor carpi radialis brevis, and extensor carpi ulnaris. These muscles are active in activities requiring wrist extension or stabilization against resistance, especially if pronated, as occurs when doing the backhand stroke in racquet sports (Hamill & Knutzen, 1995). The extensor carpi ulnaris enables the proximal portion of the fifth metacarpal to glide on the hamate. The hamate then will rotate posteriorly on the triquetrum (metacarpal joint extension), and the proximal portion of the triquetrum will glide on the TFCC (radiocarpal joint extension). The extensor carpi radialis produces forces that enable the second and third metacarpals to compress the capitate and trapezoid, limiting movement. The capitate extends on the scaphoid and lunate, resulting in metacarpal joint extension. This leads to scaphoid extension and supination on the radius while the lunate extends and pronates on the radius (Prosser & Conolly, 2003). Secondary wrist extensors include extensor digitorum communis, extensor indicis, extensor digiti minimi, and extensor pollicis longus.

Ulnar and Radial Deviation

Ulnar deviation and radial deviation are terms that are synonymous with ulnar flexors and extensors and wrist adduction and abduction. This motion occurs in the frontal/coronal plane at the anteroposterior/sagittal axis, with the axis of motion through the capitate at a right angle to the palm (Nicholas & Hershman, 1990) (Figure 7-10). Motion that occurs lateral to this front-to-back axis is radial deviation (abduction), and motion medial to this axis is ulnar deviation (adduction).

There is more movement in ulnar deviation than radial because the radial styloid process encounters the scaphoid in radial deviation, which prevents further motion and causes the normal hard end feel. Ulnar and radial deviation is greatest if the wrist is in neutral regarding wrist flexion or extension. If the wrist extends, very little deviation occurs, because the carpals are drawn into a locked, close-packed position. In wrist flexion, further movement is not possible because the bones are already splayed and the joint is in its loose-packed position (Levangie & Norkin, 2005).

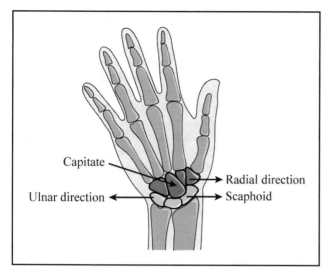

Figure 7-11. Radial deviation.

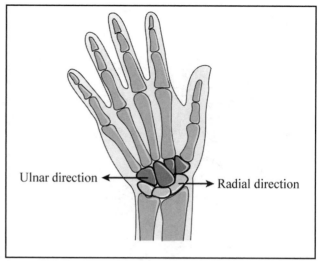

Figure 7-12. Ulnar deviation.

In radial deviation, the proximal carpal row moves ulnarly on the radius and the radioulnar disk, while the distal row of carpal bones is displaced radially. Most of the motion occurring in radial deviation occurs at the midcarpal joints (Figure 7-11). If movement occurred in a single plane (frontal), the distal row would swing radially during radial deviation to push the scaphoid into the radial styloid. Instead, what happens is that there is a shift in the position of the scaphoid. The scaphoid flexes approximately 15 degrees (palmward rotation) as the trapezium approaches the radius via the scaphoid lunate ligament (Flinn & DeMott, 2010). As the distal portion of scaphoid rotates toward the palm and the proximal row exhibits some flexion due to the scapholunate ligament. The capitate glides ulnarly toward the proximal row, causing close-packed congruity between surfaces. The lunate moves in an ulnar direction and rests on the TFCC. Radial deviation produces tension in the medial palmar intercarpal ligament and the palmar radiocarpal ligaments. Primary radial deviators are extensor carpi radialis longus, extensor carpi radialis brevis, and flexor carpi radialis.

Ulnar deviation occurs similarly to radial deviation, with the triquetrum moving much like the scaphoid in radial deviation. Most of the motion of ulnar deviation occurs at the radiocarpal joint. The proximal carpal row flexes from the neutral position about 20 degrees and extends about 20 degrees in ulnar deviation (Wheeless, 2010). The triquetrum glides distally by the proximal migration of the hamate, which brings the lunate into an extended position and rotates toward the palm (Linscheid, 1986). The capitate rotates a small amount by rotating around the vertical axis while slightly gliding around the medial and lateral portions of the joint. This produces movement toward the radial side and functionally disengages the capitate from the proximal row. The scaphoid also moves into some extension, and the distal carpal bones move ulnarly. Most of the motion of ulnar deviation occurs at the radiocarpal joint, with very little contribution of the movements of the bones of the midcarpal articulations.

Compression of the hamate against the triquetrum pushes the proximal carpal row radially against the radial styloid

process, which helps to stabilize the wrist in activities requiring large gripping forces (such as cutting meat or hammering) (Neumann, 2002).

During ulnar deviation, tension develops in the lateral part of the palmar intercarpal ligament and palmar ulnocarpal ligament. Ulnar deviation or adduction is produced primarily by the flexor carpi ulnaris and extensor carpi ulnaris working together synergistically to produce a movement neither could produce alone. Ulnar abduction has a firm end feel due to tension on the radial collateral ligament and the radial portion of joint capsule. In ulnar deviation, radiocarpal movement predominates (Figure 7-12).

INTERNAL KINETICS: MUSCLES

No wrist muscle crosses the wrist directly anterior-posterior or medial-lateral to the joint axis at the base of the capitate. Because of this, all muscles have moment arms of varying lengths and can produce force in both the sagittal and frontal planes. The wrist muscles work together to cancel one action of one muscle to produce a unique motion with another. This can be seen with the radial and ulnar wrist extensors, which work together to produce deviation. Flexor carpi radialis and extensor carpi radialis longus work together to cancel the flexion and extension actions of each other and to produce radial deviation. Similarly, flexor carpi ulnaris and extensor carpi ulnaris function antagonistically in flexion and extension but synergistically in ulnar deviation. Other synergistic actions occur when flexor carpi radialis and flexor carpi ulnaris work together to hold the wrist in neutral or flexion that is needed when one performs delicate work or when extensor carpi radialis longus and brevis and extensor carpi ulnaris work together to produce a powerful grip and maintain wrist extension even when the fingers tightly grasp an object. These are but a few of the muscular combinations that occur to provide sufficient wrist positioning for optimal and diverse hand function and the transmission of forces from the hand to the forearm.

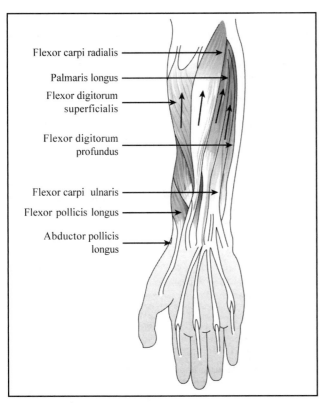

Flexor carpi radialis

Palmaris longus

Flexor digitorum superficialis

Flexor digitorum profundus

Flexor carpi ulnaris

Flexor pollicis longus

Abductor pollicis longus

Figure 7-13. Wrist flexor muscles.

Flexors at the Wrist

The following muscles flex the wrist because they lie anterior to the side-to-side axis: flexor carpi radialis (FCR), flexor carpi ulnaris (FCU), and palmaris longus (PL) are the primary flexors, with FCU considered the strongest of the three (Neumann, 2002). All three of these muscles are fusiform and are most powerful with the wrist in flexion or in stabilization of the wrist against resistance (Thompson & Floyd, 1994). Secondary wrist flexors include flexor digitorum profundus (FDP), flexor digitorum superficialis (FDS), and flexor pollicis longus (Figure 7-13). The role of the wrist flexors is to maintain appropriate length-tension in finger extensors, so they can forcefully open the hand.

FDP, FDS, and PL are multijoint muscles, and their capacity to produce effective forces at the wrist is dependent upon a synergistic stabilizer to prevent full excursion of more distal joints. If these muscles attempt to act over both the wrist and more distal joints, they will become actively insufficient (Gench et al., 1995; Hinkle, 1997; Jenkins, 1998; Norkin & Levangie, 1992; Smith et al., 1996).

The flexor group of muscles is located on the forearm's anteromedial surface between the brachioradialis muscle and the ulnar shaft (Biel, 2001). The flexor tendons are generally thicker and more pliable than the extensor tendons, but isolating specific flexor muscle bellies can be difficult. The most superficial layer of the flexor muscles includes FCR, PL, and FCU. FCR is medial to pronator teres and brachioradialis; FCU lies close to the ulnar shaft; and PL (absent in about 10% of people) runs between FCR and FCU. The middle and deep layers include FDS and FDP, and these are difficult to access directly.

Flexor Carpi Ulnaris

FCU is a superficial muscle on the palmar aspect of the ulna. It crosses the wrist anterior to the flexion/extension axis and to the ulnar side of the deviation axis, so this muscle flexes and ulnarly deviates. This muscle is active in activities requiring a stronger, sustained power grip, such as using a hammer or in the stroke of an axe (Gench et al., 1995). This muscle is considered the strongest wrist flexor, especially due to the tendon encasing the pisiform bone, which adds mechanical advantage and decreases tendon tension (Gench et al., 1995; Hinkle, 1997; Jenkins, 1998; Norkin & Levangie, 1992; Smith et al., 1996).

If FCU is impaired, there will be weakness in wrist flexion with ulnar deviation. For clients with central nervous system disorders who have spasticity and in clients with wrist instability (such as rheumatoid arthritis), tightness in the FCU muscle will result in the wrist being pulled into and held in a position of flexion and ulnar deviation, resulting in significant functional impairment (Oatis, 2004).

Palpation

This muscle lies close to the ulnar border of the forearm, and the tendon may be palpated between the ulnar styloid process and proximal to the pisiform bone with wrist flexion and adduction resisted (Gench et al., 1995; Smith et al., 1996). Pinch fingertips together or flex the wrist, and palpate proximal to the pisiform (Gench et al., 1995).

Flexor Carpi Radialis

While a primary flexor of the wrist, FCR is only 60% as strong as FCU. This muscle crosses the wrist anterior to the flexion/extension axis and to the radial side and assists with radial deviation as well as flexing the wrist (Gench et al., 1995; Hinkle, 1997; Jenkins, 1998; Norkin & Levangie, 1992; Smith et al., 1996).

Weakness in the FCR muscle only is uncommon, but would result in weakness in wrist flexion and ulnar deviation. Tendonitis of FCR can occur and results in pain with muscle contraction (Oatis, 2004).

Palpation

When resistance is applied with the wrist in flexion and radial deviation, the FCR muscle is located on the radial side, in a superficial position in the lower part of the forearm. PL is in the central location, while FCR is located radially to PL. The muscle cannot be followed to its distal attachment (Gench et al., 1995; Smith et al., 1996). Figure 7-14 illustrates the location of the flexor muscles on the distal forearm.

Palmaris Longus

PL is a small fusiform muscle that varies widely in structure and is actually absent in 10% to 15% of people (Gench et al., 1995; Thompson & Floyd, 1994). This muscle is generally seen as a contributor to wrist flexion, but because PL crosses the wrist farther from the flexion/extension axis than FCU or FCR and is small, it produces weak flexion even with its long force arm. Due to its insertion in the flexor retinaculum, PL contributes to the cupping motion of the hand.

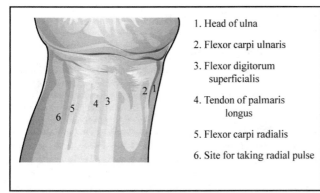

1. Head of ulna
2. Flexor carpi ulnaris
3. Flexor digitorum superficialis
4. Tendon of palmaris longus
5. Flexor carpi radialis
6. Site for taking radial pulse

Figure 7-14. Surface anatomy of flexor tendons.

Palpation

While applying resistance in wrist flexion, the centrally located tendon is palmaris longus (Smith et al., 1996). It can also be seen and palpated when the thumb is opposed to the small finger with the wrist flexed. Palmaris longus, if present, will be between the flexor carpi radialis and flexor carpi ulnaris (Gench et al., 1995).

Flexor Digitorum Superficialis (Flexor Digitorum Sublimis)

This muscle has four tendons on the palmar aspect of the hand that insert into each of the four fingers. Flexor digitorum superficialis crosses the wrist, metacarpophalangeal, and proximal interphalangeal joints anterior to the flexion/extension axis. The action of flexor digitorum superficialis is vital in gripping activities, because it acts with the flexor digitorum profundus to flex the digits (Gench et al., 1995; Hinkle, 1997; Jenkins, 1998; Norkin & Levangie, 1992; Smith et al., 1996).

Weakness in the FDS muscle would result in weak PIP flexion with the DIP joint relaxed, which would have an impact on the wrist and metacarpal joint movements and reduced grasping ability. If there is tightness in the FDS and FDP or a loss of balance between the intrinsic and extrinsic muscles of the hand, claw hand deformity may result (Oatis, 2004).

Palpation

With the fist closed tightly and wrist flexion simultaneously resisted, one or more of these tendons may become apparent between palmaris longus and flexor carpi ulnaris. The tendons may be seen to move within their sheaths as the fingers are flexed to make a fist (Gench et al., 1995). Flexor digitorum superficialis is located underneath flexor carpi radialis and palmaris longus and runs from the medial epicondyle to the center of the palmar side of the wrist. It is not possible to palpate the proximal attachment because the muscle is widespread, but one can observe it in the distal forearm and wrist (Smith et al., 1996).

Flexor Digitorum Profundus

This deep muscle, covered by flexor digitorum superficialis, flexor carpi ulnaris, palmaris longus, flexor carpi radialis, and pronator teres, assists with wrist flexion when the digits are kept extended. Flexor digitorum profundus flexes the MCP, PIP, and DIP joints, but is dependent upon wrist position

for length-tension relationships in these actions. It is the only muscle that flexes the DIP joints. The length-tension relationship is favorable for flexor digitorum profundus when the wrist is extended and the fist can be closed. As the wrist moves toward greater flexion, flexor digitorum superficialis is recruited to aid in closure of the fist. Forceful fist closure elicits high activity levels of flexor digitorum superficialis, flexor digitorum profundus, and the interossei (Smith et al., 1996). The line of pull is to the palmar side of the flexion/extension axis of each joint, and the only action is flexion, which progresses sequentially from the most distal joint (DIP joint) to the most proximal (wrist joint). The strength of the contraction lessens progressively so that flexor digitorum superficialis contributes only weakly to wrist flexion (Gench et al., 1995).

Palpation

The contracting muscle belly may be palpated, provided tension is minimal in the more superficial muscles. To achieve relaxation of the overlying muscles, the subject is seated with the forearm supinated or resting in the lap while the wrist is extended by the weight of the hand. Then, ask the subject to close the fist fully with moderate effort, and profundus may be felt under the fingers in the region between pronator teres and flexor carpi ulnaris about 2 inches below the medial epicondyle of humerus (Smith et al., 1996). Stabilize the PIP joint in extension, and flex the DIP joint (Gench et al., 1995).

Flexor Pollicis Longus

This penniform muscle lies deep on the radial side of the forearm, where it assists with wrist flexion due to the palmar location in the wrist and plays an important role in grip activities (Thompson & Floyd, 1994). Some authors disagree with the role of this muscle in wrist flexion because it lies too close to both wrist axes and the power of the muscle is already spent as it flexes the thumb (Gench et al., 1995; Goss, 1976).

If there is damage to the FPL muscle, there will be weakness in flexion of the thumb interphalangeal joint, which may occur due to anterior interosseous nerve impingement. Loss of balance of forces between extensor pollicis longus and FPL may result in the ape thumb deformity (Oatis, 2004).

Palpation

The tendon can be palpated on the palmar surface of the thumb between the metacarpal and interphalangeal joints as the IP joint is flexed (Gench et al., 1995).

Abductor Pollicis Longus

This muscle crosses the wrist directly above the axis of flexion/extension, but "bowstrings" during wrist flexion, allowing it to contribute to wrist flexion.

Palpation

The superficial tendon can be palpated as it crosses the wrist on the palmar side of extensor pollicis brevis when the thumb is forcefully abducted (Gench et al., 1995).

Extensors at the Wrist

Extensor carpi radialis longus (ECRL), extensor carpi radialis brevis (ECRB), and extensor carpi ulnaris (ECU) are the most powerful wrist extensors and are active in activities

Secondary wrist extensors include extensor digitorum communis, extensor indicis, extensor digiti minimi, and extensor pollicis longus. Extensor digitorum communis can generate significant wrist extension torque but is primarily involved in metacarpal extension (Neumann, 2002).

Extensor Carpi Ulnaris

ECU is active in wrist extension and frequently in wrist flexion, too, and adds stability to the less stable position of wrist flexion (Gench et al., 1995; Hinkle, 1997; Jenkins, 1998; Norkin & Levangie, 1992; Smith et al., 1996). When the forearm pronates, the crossing of the radius over the ulna leads to a smaller moment arm of ECU, so this muscle is less effective as a wrist extensor and is more effective as an extensor when the forearm is supinated.

Palpation of the Tendon

The tendon can be palpated between the head of the ulna and the prominent tubercle on the base of the fifth metacarpal. The tendon becomes prominent if the wrist is extended with the fist closed and is even more prominent if the wrist is simultaneously abducted ulnarward. The tendon is also easily palpable when the thumb is extended and abducted.

Palpation of the Muscle

The muscle can be palpated 2 inches (5 cm) below the lateral epicondyle of the humerus as the wrist is forcefully extended where it lies between anconeus and extensor digitorum, and it can be followed distally along the dorsoulnar aspect in the direction toward the head of the ulna (Gench et al., 1995; Smith et al., 1996).

Extensor Carpi Radialis Longus

Extensor carpi radialis longus (ECRL) lies posterior to the side/side axis so it extends the wrist and is most effective as a wrist extensor when the elbow is also extended. ECRL has increased activation when in either radial deviation or in a position supported against ulnar deviation or when forceful finger flexion is performed (Norkin & Levangie, 1992).

If there is weakness in ECRL and ECRB, there will be a loss of wrist extension and radial deviation, resulting in loss of forceful grasp and pinch (Oatis, 2004).

Palpation of the Muscle

This muscle may be palpated just superior to the elbow as the wrist is forcefully extended with the fist closed.

Palpation of the Tendon

The tendon is prominent during wrist extension on the dorsoradial aspect of the wrist located on the radial side of the capitate bone but on the ulnar side of the tubercle of radius. If the subject places a lightly closed fist on a table or in the lap with the forearm pronated and the subject alternately closes and relaxes the fist, the rise and fall of the tendon may be felt, and the muscle belly may be identified in the forearm close to brachioradialis. It may improve the accuracy of identification of the ECRL muscle by identifying the extensor pollicis longus muscle first, which is seen when the thumb is in extension (Gench et al., 1995; Smith et al., 1996).

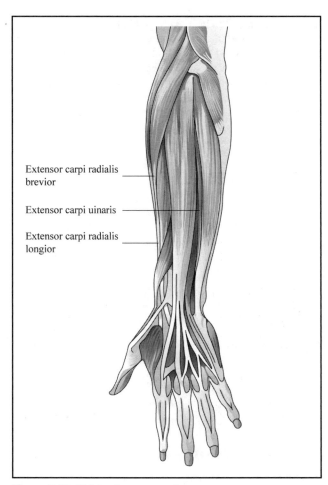

Extensor carpi radialis brevior

Extensor carpi uinaris

Extensor carpi radialis longior

Figure 7-15. Wrist extensor muscles.

requiring wrist extension or stabilization against resistance, especially if pronated (as occurs in the backhand stroke in racquet sports) (Figure 7-15). These three muscles act on the elbow, so joint position is important. ECRL and ECRB cause flexion at the elbow and so can be enhanced as wrist extensors with extension of the elbow, whereas ECU is an elbow extensor and is enhanced as a wrist extensor with the elbow in flexion (Hamill & Knutzen, 2003). The primary wrist extensor muscles cross the dorsal and dorsal-radial side of the wrist and are covered by the extensor retinaculum, which prevents "bowstringing" of the tendons during extension. The primary role of the wrist extensors is to maintain appropriate length-tension forces in the finger flexors to produce a strong grip.

The primary wrist extensors are located between the brachioradialis muscle and the shaft of the ulna on the posterolateral surface of the forearm. ECRL and ECRB are lateral and posterior to the brachioradialis muscle, while extensor carpi ulnaris lies beside the ulnar shaft. Extensor digitorum lies between these two muscles with long, superficial tendons stretching along the dorsal surface of the hand (Biel, 2001). Brachioradialis, ECRL, and ECRB are sometimes referred to as the "wad of three" and can be palpated lateral to the inner elbow, and they form a long mass of muscle distal to the supracondylar ridge of the humerus (Biel, 2001).

Extensor Carpi Radialis Brevis

This muscle crosses well posterior to the flexion/extension axis so it extends the wrist. While extensor carpi radialis brevis (ECRB) is in a central location on the forearm, it is covered by the ECRL muscle. Its tendon is crossed by the tendons of the abductor pollicis longus and extensor pollicis longus, which makes identification of the ECRB tendon difficult.

Palpation

While the tendon protrudes less than the extensor carpi radialis longus, it can be felt on the dorsum of the wrist in line with the third metacarpal during fisted wrist extension (Gench et al., 1995; Smith et al., 1996).

Extensor Digitorum (Extensor Digitorum Communis)

Extensor digitorum (ED), a fusiform muscle, crosses the wrist, MCP, PIP, and DIP joints. If all of these joints are extended simultaneously, then the ED will be contracted to its shortest length and can be only weakly responsive at the wrist as an extensor (Gench et al., 1995). The role of the ED muscle is to open the fingers and extend the wrist.

Palpation

With the wrist extended and the fist closed, the prominent tendon of extensor carpi radialis longus can be palpated at the base of the second metacarpal. If you maintain the wrist in this position and extend the fingers, the four tendons of the extensor digitorum can be seen and palpated on the dorsum of the hand (Smith et al., 1996).

Extensor Digiti Minimi and Extensor Indicis (Extensor Indicis Proprius)

Extensor digiti minimi (EM) and extensor indicis (EI) are capable of wrist extension after continued contraction, but wrist extension action is credited more to the extensor digitorum muscle. If the MCP and IP joints are held in flexion, then extensor indicis is an effective wrist extensor.

Palpation of Extensor Digiti Minimi

Located on the ulnar side of extensor digitorum, the tendon can be palpated and observed on the dorsum of the hand when the small finger is extended against resistance, especially at the fifth MCP joint (Gench et al., 1995).

Palpation of Extensor Indicis

The tendon is superficial and can be observed on the ulnar side of and parallel to the tendon of extensor digitorum as it approaches the base of the index finger (Gench et al., 1995).

Extensor Pollicis Brevis and Extensor Pollicis Longus

While specified as contributors to wrist extension due to the location dorsal to the joint axis by some sources (Hamill & Knutzen, 2003), these muscles do not contribute much to wrist function (Thompson & Floyd, 1994).

Radial Deviation/Abductors at the Wrist

The primary radial deviators of the wrist are the FCR, ECRL, and ECRB. Additional muscles that work to abduct the wrist include EPL, EPB, APL, and FPL. In neutral position, the ECRL has the largest cross-sectional area and moment arm for radial deviation followed by ECRB and APL. Of all the muscles that abduct the wrist, APL has the greatest moment arm but is not a strong radial deviator due to the small cross-sectional area. APL and EPB are important in stabilization of the radial side of the wrist together with the radial collateral ligament (Neumann, 2002). The radial deviation muscles are stronger than the ulnar (Delp, Grierson, & Buchanan, 1996) (Figure 7-16).

Flexor Carpi Radialis

Even though this muscle is located to the radial side of the deviation axis, it is not significantly effective as a radial deviator in isolated contractions due to the longer moment arm for flexion than for deviation. However with palmaris longus, flexor carpi radialis does contribute to radial deviation (Figure 7-8).

Extensor Carpi Radialis Longus

Because this muscle is located on the radial side of the deviation axis, it serves to abduct the wrist. Because this muscle has a longer moment arm for deviation than for extension, it is more effective as a deviator than as an extensor.

Extensor Carpi Radialis Brevis

The extensor carpi radialis brevis muscle is located close to the deviation axis, so it provides scant contribution to deviation (Gench et al., 1995).

Abductor Pollicis Longus

This muscle lies on the radial side of the deviation axis and assists with wrist abduction.

Extensor Pollicis Brevis

This muscle is located so that there is a favorable line of muscle action for radial deviation regardless of the position of the thumb (Gench et al., 1995). Other sources feel that, because the muscle crosses directly over the axis, this muscle is not involved in adduction or abduction (Burstein & Wright, 1994).

Extensor Pollicis Longus

This muscle lies to the radial side of the deviation axis and serves to abduct the wrist, in addition to its contribution to wrist extension.

Ulnar Deviation/Adductors at the Wrist

The two muscles that adduct the wrist are the FCU and ECU muscles (see Figure 7-16). Extensor indicis is also considered a weak wrist adductor due to location in relation to the joint axis.

Flexor Carpi Ulnaris

The FCU crosses farther away from the axis than flexor carpi radialis, so it is more effective in ulnar deviation than flexor carpi radialis is in radial deviation.

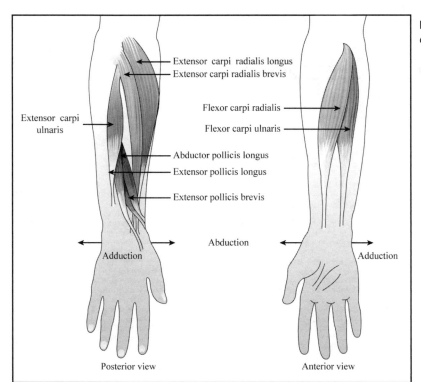

Figure 7-16. Radial and ulnar deviation muscles.

Extensor Carpi Ulnaris

Because this muscle crosses the axis posteriorly and ulnarly, it extends and ulnarly deviates the wrist. Due to the muscle's flare at the base of the fifth metacarpal, it increases the angle of the muscle's attachment and helps to stabilize the head of the ulna during wrist movements. It is a stronger ulnar deviator when the forearm is pronated (Gench et al., 1995).

Extensor Indicis

This is considered a weak adductor because the muscle crosses on the ulnar side of the adduction/abduction axis.

WRIST ASSESSMENT

Occupational Profile

Assessment of the wrist involves consideration of many bones, ligaments, muscles, and joints that are influenced by action at the elbow and shoulder and that control the actions of the fingers in functional prehensile patterns. But as with any occupational therapy assessment, it is not only the impairment of the body part that is considered but the use of the part in occupation. It is particularly important to understand how the client uses the hands in daily activities. Ask the client about particular tasks involved in his or her work activities and occupation. Those using a keyboard are more prone to repetitive strain injuries and postural changes, while those clients involved in construction work may have more trauma-related injuries.

Awkward wrist postures can lead to deformity and loss of hand function. Computers are an integral part of daily life. Keyboard users were found to exhibit 60 degrees of pronation and 20 degrees of extension, and some had 20 degrees of ulnar deviation (Baker & Cidboy, 2006).

Awkward postures can lead to elevated pressure inside the carpal tunnel, creating strain on muscle structures. Increased pressure can occur when wrist extension or pronation is past 45 degrees and when ulnar deviation is 10 degrees or more. The force exerted by the carpal bones and carpal ligament against flexor tendons increases, which can lead to inflammation and increased pressure on nerves, blood vessels, and tendons. The result may be ischemia and tendon synovial edema, pressure on the median nerve, and ultimately carpal tunnel syndrome (Marklin, Simoneau, & Monroe, 1999).

Understand the client's occupational history, experiences, patterns of living, interests, values, and needs (American Occupational Therapy Association, 2008), and use this information in collaboration with the client in developing treatment goals.

Observation and Palpation

Assessment of the wrist includes looking at the skin for blisters or lacerations and observing for any abnormal formations or nodules. Edema may be seen in local swollen ganglions (synovial hernia in tendonous sheath or joint capsule), usually on the dorsum and radial side of the hand, or there could be inflammation of the tendon or its synovial sheath. Edema may occur due to fractures, tenosynovitis, or direct trauma. Immediate swelling after injury may indicate more severe injury, while swelling that is more gradual suggests ligamental

Table 7-6

SENSATIONS DETERMINED BY PALPATION AND OBSERVATION

SENSATIONS	POSSIBLE CAUSES
Warmth	Inflammation
	Infection
Numbness	Carpal tunnel syndrome at the elbow or wrist
	Radial nerve palsy or injury
	Cervical nerve root problems
	Thoracic outlet syndrome
	Local cutaneous nerve injury
	Cubital tunnel syndrome at the elbow
Clicking	Lesion on the disk between the lunate and radius and triquetrum
	Carpal bone subluxation (lunate or capitate)
Popping	Ligament or muscle tears
	Carpal subluxation
	Joint dislocation
Grating	Osteoarthritic changes
	Cartilage deterioration
Crepitis	Tenosynovitis of tendons in tendon sheaths as the fingers move

or capsular sprains or subluxation. Diffuse swelling usually is seen on the dorsal surface of the hand because there is more room for expansion (Hartley, 1995). While swelling is difficult to observe in the wrist, edema will limit joint movement. Edema of the wrist is dangerous because it can congest the carpal tunnels and restrict extensor tendon compartments (Hartley, 1995).

Compare the right and left wrists, and observe the contours of the forearm and wrist for any deviations or changes between sides. Look for the flexor creases that distinguish the radiocarpal from the midcarpal joints. Observe the client's ability and willingness to use his or her wrist and hand. In the resting (functional) position, is there 20 degrees to 35 degrees of wrist extension with 10 degrees to 15 degrees of ulnar deviation. Dense fibrous bands across the palm of the hand may indicate Dupuytren's contracture of the palmar fascia and may result in loss of extension mobility of the fourth and fifth fingers and observable deformities of the metacarpophalangeal and interphalangeal joints (Palmer & Epler, 1998).

With the client's hand in a static, resting position, you can observe postural resting imbalances. Resting imbalances

are due to intrinsic and extrinsic forces acting on the joint. These may include factors related to articular, neurologic, and vascular structures and muscles and tendons. Intrinsic imbalances could be due to forces acting on the joint capsule, while unopposed antagonistic muscle action might be an example of an extrinsic imbalance (Flinn & DeMott, 2010).

Be attentive to the color and temperature of the skin. Is one wrist or hand warmer than the other? Is there a difference in temperature at different parts of the wrist? Does the client respond to palpation and gentle movement with a pain response? Table 7-6 summarizes sensations that can be felt when palpating wrist structures. These are important indicators for sensory and neurological deficits that would suggest further, in-depth assessment.

Range of Motion

While the wrist is important in positioning the hand in space and for prehension, very little pure wrist motion is necessary for everyday activities (Table 7-7). Even with a significant loss of motion in the wrist, adjacent joints can be used

Table 7-7

Summary of Wrist Motions

MOTION	JOINT	PLANE/AXIS	NORMAL LIMITING FACTORS	END FEEL	PRIMARY MUSCLES	NORMATIVE ROM/ FUNCTIONAL ROM
Flexion	Midcarpal Radiocarpal	Sagittal/frontal	• Posterior radiocarpal ligament • Posterior joint capsule	Firm	• Flexor carpi radialis • Flexor carpi ulnaris • Palmaris longus	0°-65°-80°/0°-10°*
Extension	Radiocarpal Midcarpal	Sagittal/frontal	• Anterior radiocarpal ligament • Anterior joint capsule • Contact between radius and carpal bones	Firm	• Extensor carpi radialis longus • Extensor carpi radialis brevis • Extensor carpi ulnaris	0°-55°-70°/0°-35°*
Radial deviation	Midcarpal Radiocarpal	Frontal/sagittal	• Ulnar collateral ligament • Ulnar portion of joint capsule • Contact between styloid process and scaphoid	Hard	• Extensor carpi radialis longus • Extensor carpi radialis brevis • Extensor pollicis longus • Extensor pollicis brevis • Flexor carpi radialis • Abductor pollicis longus • Flexor pollicis longus	0°-20°/0°-20°
Ulnar deviation	Radiocarpal	Frontal/sagittal	• Radial collateral ligament • Radial portion of joint capsule	Hard	• Extensor carpi ulnaris • Flexor carpi ulnaris	0°-30°/0°-20°

Reprinted with permission from Brumfield, R. H., & Champoux, J. A. (1984). A biomechanical study of normal functional wrist motion. *Clinical Orthopaedics, 187*(23).

Table 7-8

ACTIVE RANGE OF MOTION SCREEN FOR THE WRIST

Ask client to bend the wrist up with fingers pointing toward ceiling or make the "stop" gesture with the hand	Wrist extension
Ask the client to bend the wrist down with fingers pointing to the floor	Wrist flexion
Ask the client to move his or her hand so that the fingers move in and the palm is pointed toward the floor	Radial deviation
Ask the client to move his or her hand so that the fingers move out and the palm is pointed toward the floor	Ulnar deviation

to compensate for diminished movement of the wrist. Most activities of daily living require 10 degrees to 54 degrees of flexion, 15 degrees of extension, 40 degrees of ulnar deviation, and 17 degrees of radial deviation as minimal requirements for ROM. Most activities can be accomplished with 40 degrees of flexion and extension and 40 degrees of combined ulnar-radial deviation (Neumann, 2002; Nicholas, Sapega, Kraus, & Webb, 1978; Ryu, Cooney, Askey, An, & Chao, 1991).

Active Range of Motion

Watching the client use the wrist and hand can help to identify any dynamic postural imbalances related to the movement. Consideration of the entire kinematic chain is necessary to identify the primary mechanisms interfering with normal movement patterns. Continued postural imbalances can result in permanent soft tissue changes, degeneration of joint structures, pain, and loss of function (Flinn & DeMott, 2010).

Have the client actively make a fist and open the hand wide. Observe the movement, and note restrictions, deviations, or pain (Magee, 2002). Observe the client pronate and supinate. Approximately 75 degrees of supination or pronation is due to forearm rotation, but the remaining 15 degrees is due to wrist movement. If the client experiences pain on supination, you can differentiate between distal radioulnar joint and radiocarpal deficits by passively supinating ulna on radius with no stress to the radiocarpal joint. If this provokes the pain response, the problem is at the distal radioulnar joint, not the radiocarpal joint (Magee, 2002).

Screening tests for active range of motion are shown in Table 7-8, which is a quick way to observe wrist dysfunction.

Passive

Moving the client passively can give the therapist information about capsular patterns, end feel, and tendon extensibility. When there are differences in values for active and passive range of motion, this may indicate muscle weakness, poor

tendon excursion, swelling, nerve palsy, pain, or fear of movement (Flinn & DeMott, 2010).

End feel is assessed by moving the client's wrist in all motions and applying overpressure at the end of the range. Joint play at the radiocarpal joint is assessed by distraction forces applied to the proximal row or carpal bones via ventral glide, dorsal glide, radial glide, and ulnar glide procedures. If the client experiences pain or is unable to flexion the wrist, the lesion is more likely to be at the midcarpal joint. If there is more restriction and pain on wrist extension, the problem is more likely in the radiocarpal joint, because, with extension, more movement occurs at the radiocarpal than the midcarpal joint. If there is pain during pronation or supination, there may be lesions at the ulnomeniscocarpal or inferior radioulnar joints (Magee, 2002; Palmer & Epler, 1998).

Strength Assessment

Resisted isometric movements of the wrist are influenced by muscles acting on the forearm, wrist, and hand, so injuries to any of these structures will influence the strength of the wrist. Table 7-9 provides examples of strength screening tests for the wrist.

Assessment of strength can occur by observing the client perform daily tasks. Watching a client clean a countertop (which requires at least a Fair– muscle grade), pulling up on a car door handle to open it (requiring 30 degrees flexion), or pushing up from a chair to stand (requiring 25 degrees wrist extension) can provide valuable information directly related to functional tasks (Gutman & Schonfeld, 2003).

Stability Assessment

If there are alterations in joint surface contact, orientation, or ligamental support, an abnormal path of articular contact

Table 7-9

STRENGTH SCREEN FOR THE WRIST

DIRECTIONS TO CLIENT	THERAPIST ACTION	MUSCLE ACTION TESTED
Ask the client to bring his or her hand up and hold that position. Tell him or her, "Don't let me push your hand down."	Push in the direction of wrist flexion	Wrist extension
Ask the client to bring his or her hand down and hold that position. Tell him or her, "Don't let me push your hand up."	Push in the direction of wrist extension	Wrist flexion

Table 7-10

INSTABILITY TESTS FOR THE WRIST

TEST	STRUCTURE
Thumb ulnar collateral ligament laxity or instability test	Test of ulnar collateral ligament and accessory collateral ligament
Lunotriquetral ballottement (Reagan's) test	Lunotriquetral ligament
Murphy's sign	Lunate dislocation
Watson (scaphoid shift) test	Scaphoid subluxation
Scaphoid stress test	Scaphoid subluxation
Dorsal capitate displacement apprehension test	Capitate stability test
Supination lift test	TFCC pathology
Axial load test	Fracture of metacarpals or adjacent carpal bones
Pivot shift test of the midcarpal joint	Anterior capsule and interosseous ligaments
Sitting hands test	Wrist synovitis or wrist pathology

during or at the end of the range of motion may occur. This is instability (Ozer & Scheker, 2006). All of the ligaments providing support at the wrist except for the dorsal radioulnar ligament have equivalent tensions during joint movements indicative of their role in stabilization of the joint (DiTano, Trumble, & Tencer, 2003). Instability may result in subluxations or dislocations of the many bony articulations in the wrist complex.

Acute dislocations may be due to an isolated injury or may be the result of fractures of the radial head, the distal radius, and both bones of the forearm. In isolated injuries, ulna dorsal dislocations occur because of hyperpronation injury, whereas ulna volar dislocations occur due to hypersupination (Andersson, Nordin, & Pope, 2006; Ozer & Scheker, 2006). Chronic instability can progress to osteoarthritis in the joint, further limiting range of motion and limiting functional use of the elbow and wrist.

Tests for ligament, capsule, and joint instability are listed in Table 7-10.

Pain Assessment

A complete understanding of the pain symptoms from the client's perspective is an important part of the assessment

Table 7-11

Pain Sensations at the Wrist

Localized	Indicative of more superficial structures such as muscles, ligaments (radial and ulnar collateral ligaments), or periosteum (pisiform, styloid process, or metacarpal heads)
Deep pain	Referred pain
	Deeply located muscles, ligaments, bursae, or bones (scaphoid or radius)
Sharp	Skin
	Fascia
	Tendon (deQuervain's disease)
	Ligaments
	Muscles
	Bursitis
Dull	Neurological problem
	Bony injury
	Chronic capsular problem
	Deep muscle injury
	Tendon sheath problem
Ache	Tendon sheath
	Deep ligament
	Fibrous capsules
	Deep muscles
	Rheumatoid arthritis
	Complex regional pain syndrome (CRPS)
Pins and needles	Peripheral or dorsal nerve root damage
	Systemic condition
	Vascular occlusion (Raynaud's disease)
Tingling	Neural involvement (C7-8 dermatome or peripheral nerve involvement)
	Circulatory involvement

process. Pain can invalidate strength tests and limit functional movement of the wrist joint. Understanding and describing the exact location of the pain may help to identify structures involved. Having the client describe the type of pain that is being experienced (burning, tingling), the frequency and duration of the pain, and what movements or activities aggravate or bring relief from the pain is essential information to gather from the client. Immediate onset of pain usually indicates a more severe injury than pain occurring within 6 to 12 hours after the activity. Gradual onset of pain may also indicate overuse syndromes, neural lesions, or arthritic changes. Table 7-11 summarizes pain and sensory indications for the wrist.

If the client can localize the pain in the anatomic snuffbox and reports a fall on an outstretched hand (FOOSH injury), a scaphoid fracture or Preiser's disease (osteonecrosis or avascular necrosis of the scaphoid) might be suspected. There

would be a loss of wrist motion, decreased grasp, and pain with movement. If the client locates the site of pain over the lunate with localized pain and history of a FOOSH injury, this may be due to Keinböck's disease (osteonecrosis or avascular necrosis of the).

There may be referred pain in the deeper structures, such as in the more deeply located muscles and ligaments, in the bursae, or in bones (especially the scaphoid or radius) (Hartley, 1995) or the cervical spine or upper thoracic spine, shoulder, and elbow. The muscles acting on the wrist have specific pain referral patterns when injured, such as pain referring to the medial side of the dorsum of the wrist when the extensor carpi ulnaris is injured or a pain referral pattern to the anteromedial wrist and lateral palm for the flexor carpi ulnaris (Magee, 2002).

Pain may be due to compression of the median nerve in the carpal tunnel leading to carpal tunnel syndrome. Compression may be due to trauma (Colles' fracture or lunate dislocation), flexor tendon paratendonitis, a ganglion, arthritis, or collagen disease (Magee, 2002). Symptoms are usually worse at night and include burning, tingling, pins and needles, and numbness, which may be referred to the forearm. Specific tests for carpal tunnel syndrome include Phalen's test and the Tinel's sign at the wrist. The ulnar nerve can become compressed as it passes through the pisohamate or Guyon's canal (Guyon's Canal Syndrome), but only the fingers show altered sensation. There may be motor loss in the hypothenar, adductor pollicis, interossei, medial two lumbricals, and palmaris brevis (Magee, 2002). Special tests for the ulnar nerve would be the Froment's sign and Wartenberg's sign.

Wrist Pathology

Common injuries to the wrist include FOOSH, which can result in fracture or dislocation of the scaphoid or lunate bones, and anterior ligamental strain, which can produce synovial effusion, joint pain, and limited movement. FOOSH injuries affect not only the wrist but, due to the interrelatedness of the structures in the upper extremity kinematic chain, forces are transmitted from the scaphoid to the distal radius, across the interosseous membrane to the ulna, from the ulnar to the humerus. From the glenoid, the force then moves to the coracoclavicular ligament and clavicle to the sternum. Because of these links, evaluation of more than just the wrist is important. FOOSH injuries can result in radial head fracture if the arm is supinated with anterolateral pain and tenderness at the elbow. Distal radius fracture (Colles fracture) can occur in FOOSH injuries with forceful wrist extension. Distal radius fractures are accompanied by wrist swelling and pain upon wrist extension.

Carpal tunnel syndrome is a compression of the median nerve at the carpal tunnel of the wrist. The carpal tunnel contains (from radial to ulnar side) the flexor carpi radialis tendon, flexor pollicis longus tendon, the median nerve, tendons of the flexor digitorum superficialis and profundus, and vascular structures. Overuse of the wrist via repetitive motion can produce tenosynovitis in the tendon sheaths of the long flexor muscles. This can increase hydrostatic pressure in the tunnel, causing compression damage to the median nerve. Another overuse syndrome of the wrist and thumb is the extensor intersection syndrome, where there is pain and inflammation in the upper forearm.

The triangular fibrocartilage complex can acquire an isolated lesion but may also incur injury in association with a distal radius fracture, distal radioulnar fracture, ulnar styloid fracture, or radial shaft fracture. Injury to the TFCC is controversial and can result in prolonged disability (Tang et al., 1998). The central portion of the articular disk is avascular, and this hinders healing. Classification of TFCC lesions is based on traumatic injuries (specifying location of injury such as central, ulnar, radial) or degenerative injury, which specifies which structures are worn or perforated (Palmer & Werner, 1981). Initial treatment options are activity modification, splinting, and anti-inflammatory medications. If surgery is required, the goal is to decrease the force loading on the ulnar

side of the wrist. Additional wrist pathology is summarized in Table 7-12.

OUTCOME MEASURES FOR THE WRIST AND HAND

Outcomes measures in occupational therapy are necessary to improve decisions about specific treatment procedures with specific clients in order to provide the most cost-effective evidence-based interventions (Law & McColl, 2010). Outcomes in occupational therapy include changes in occupational performance, enhanced adaptation, improved health and wellness, increased participation, prevention of disability, improved quality of life, development of role competence, self-advocacy, and enriched occupational justice (American Occupational Therapy Association, 2008). The Levine Symptom Severity Scale (Levine et al., 1993) and related Brigham and Women's Hospital Carpal Tunnel Questionnaire (CTQ), the Michigan Hand Outcomes Questionnaire (MHQ) (Chung, Hamill, Walters, & Hayward, 1999; Chung, Pillsbury, Walters, & Hayward, 1998), and the Disabilities of the Arm, Shoulder, and Hand (DASH) (Hudak, Amadio, & Bombardier, 1996) measure limitations in activities of daily living and instrumental activities of daily living. The Patient-Rated Wrist Evaluation score (PRWE) (MacDermid, Turgeon, Richards, Beadle, & Roth, 1998) is a patient-rated assessment of pain and disability in patients with distal radius fractures. The Gartland and Werley score (Gartland & Werley, 1951) involves an objective evaluation of wrist function, and while this tool is used extensively by orthopedic surgeons as an objective assessment of outcome, there have been no validity and reliability studies completed on this tool (Changulani, Okonkwo, Keswani, & Kalairajah, 2008).

SUMMARY

- The radiocarpal joint is made up of the radius and carpal bones.
- The wrist includes three joints: radiocarpal, midcarpal, and intercarpal joints.
- Movements at the radiocarpal and midcarpal joints occur at the same time and are involved in the motions of flexion and extension.
- Adduction/abduction of the wrist (also known as ulnar and radial deviation) is accomplished when the capitate and scaphoid carpal bones move in radial deviation; triquetrum/triangular in ulnar deviation.
- Circumduction is a combined sequence of movement of adduction, extension, abduction, and flexion.
- Many muscles work together at the wrist to produce movement that a single muscle alone would not be able to produce.

Table 7-12

Summary of Wrist Pathology

Fracture	Scaphoid may be impinged between capitate and radius
	• Particularly high incidence in young athletes in contact sports
	• Point tenderness in the anatomical snuff box and history of hyperextension
	• Fracture often misdiagnosed
	• Bone heals poorly due to poor blood supply lunate fracture
Rheumatoid arthritis	Commonly affects the wrists bilaterally
	In advanced stages, there may be subluxation or deformities that may include the following:
	• Volar subluxation of the triquetrum with relation to the ulna
	• Extensor carpi ulnaris tendon displacing volarly
	• Subluxation of the carpals, which would result in radial deviation
Strains	FCR and FCU most commonly strained
Dislocation	Distal ulna dislocation can occur with ulnar styloid fracture
	Radiocarpal or midcarpal dislocations are rare
	With hyperextension, lunate dislocates anteriorly then remains stationary while the rest of the carpals dislocate anteriorly
Sprains or tears	Forced radial deviation
	• Can sprain or tear the medial ligament of the radiocarpal joint at the ulnar styloid process, the anterior band into the pisiform, or the posterior band into the triquetrum
	• May fracture the scaphoid or distal end of the radius or avulse the ulnar styloid process
	• Forced wrist ulnar deviation
	• Can sprain or tear the lateral ligament of the radiocarpal joint at the radial styloid process, the anterior band into the scaphoid, or the posterior band into the scaphoid tubercle
	• Can strain the ECRL or APL tendons or avulse the radial styloid process
Hyperpronation and hypersupination	Hyperpronation can cause dorsal subluxations or dislocations of the distal radioulnar joint
	Hypersupination is less common but can result in volar radioulnar subluxations or dislocations

(continued)

Table 7-12 (continued)

SUMMARY OF WRIST PATHOLOGY

Overuse

Carpal tunnel syndrome

- can occur in baseball, rowing, weight lifting, wheelchair athletes
- carpal tunnel can become constricted due to many conditions
- numbness or tingling in the hand and fingers that are supplied by the median nerve
- Motor weakness can develop from prolonged or severe constriction
- Extensor intersection syndrome
- Overuse of thumb or wrist causing inflammation of APL and EPL in upper forearm where they cross each other
- Common in weight lifters and paddlers
- Ulnar nerve entrapment as it passes around the hook of hamate or can be damaged with scaphoid or pisiform fractures
- Tingling and paresthesia of the hand and fingers in ulnar distribution
- Can be seen from trauma from the handle of a baseball bat or hockey stick, karate blows, or prolonged wrist extension that occurs with long distance cycling.

- Assessment of the wrist includes consideration of the movement produced, pain, edema, and functional use of the wrist in activities.

REFERENCES

American Occupational Therapy Association. (2008). Occupational therapy practice framework: Domain and process, 2nd edition. *American Journal of Occupational Therapy, 62*, 625-683.

Andersson, G. B. J., Nordin, M., & Pope, M. H. (2006). *Disorders in the workplace: Principles and practices* (2nd ed.). St. Louis, MO: Mosby.

Baker, N. A., & Cidboy, E. L. (2006). The effect of three alternative keyboard designs on forearm pronation, wrist extension, and ulnar deviation: A meta-analysis. *American Journal of Occupational Therapy, 60*(1), 40-49.

Baxter, R. (1998). *Pocket guide to musculoskeletal assessment.* Philadelphia, PA: W. B. Saunders Co.

Biel, A. (2001). *Trail guide to the body* (2nd ed.). Boulder, CO: Books of Discovery.

Burstein, A., & Wright, T. M. (1994). *Fundamentals of orthopaedic biomechanics.* Baltimore, MD: Williams & Wilkins.

Cailliet, R. (1996). *Soft tissue pain and disability.* Philadelphia, PA: F.A. Davis Co.

Changulani, M., Okonkwo, U., Keswani, T., & Kalairajah, Y. (2008). Outcome evaluation measures for wrist and hand—which one to choose? *International Orthopaedics, 32*(1), 1-6.

Chung, K. C., Hamill, J. B., Walters, M. R., & Hayward, R. A. (1999). The Michigan Hand Outcomes Questionnaire (MHQ): assessment of responsiveness to clinical change. *The Annals of Plastic Surgery, 42*(6), 619-622.

Chung, K. C., Pillsbury, M. S., Walters, M. R., & Hayward, R. A. (1998). Reliability and validity testing of the Michigan Hand Outcomes Questionnaire. *Journal of Hand Surgery, 23*(4), 575-587.

Delp, S. L., Grierson, A. E., & Buchanan, T. S. (1996). Maximum isometric moments generated by the wrist muscles in flexion-extension and radial-ulnar deviation. *Journal of Biomechanics, 29*, 1371-1375.

DiTano, O., Trumble, T. E., & Tencer, A. F. (2003). Biomechanical function of the distal radioulnar and ulnocarpal wrist ligaments. *Journal of Hand Surgery, 28*(4), 622-627.

Donatelli, R. A., & Wooden, M. J. (2010). *Orthopaedic physical therapy* (4th ed.). St. Louis, MO: Churchill Livingstone Elsevier.

Esch, D. L. (1989). *Musculoskeletal function: An anatomy and kinesiology laboratory manual.* Minneapolis, MN: University of Minnesota.

Flinn, S., & DeMott, L. (2010). Dysfunction, evaluation, and treatment of the wrist and hand. In R. A. Donatelli & M. J. Wooden (Eds.), *Orthopaedic physical therapy* (4th ed., pp. 267-312). St. Louis, MO: Churchill Livingstone Elsevier.

Gartland, J. J., & Werley, C. W. (1951). Evaluation of healed Colles fracture. *Journal of Bone and Joint Surgery, 33*, 895-907.

Gench, B. E., Hinson, M. M., Harvey, P. T. (1995). *Anatomical kinesiology.* Dubuque, IA: Eddie Bowers Publishing, Inc.

Goss, C. M. (Ed.) (1976). *Gray's anatomy of the human body* (29th ed.). Philadelphia, PA: Lea and Febiger.

Gutman, S. A., & Schonfeld, A. B. (2003). *Screening adult neurological populations.* Bethesda, MD: American Occupational Therapy Press.

Hamill, J., & Knutzen, K. M. (1995). *Biomechanical basis of human movement.* Philadelphia, PA: Lippincott, Williams & Wilkins.

Hamill, J., & Knutzen, K. M. (2003). *Biomechanical basis of human movement.* Philadelphia, PA: Lippincott, Williams & Wilkins.

Hartley, A. (1995). *Practical joint assessment: Upper quadrant: A sports medicine manual* (2nd ed.). Philadelphia, PA: C. V. Mosby.

Hinkle, C. Z. (1997). *Fundamentals of anatomy and movement: A workbook and guide.* St. Louis, MO: C. V. Mosby Co.

Hudak, P., Amadio, P. C., & Bombardier, C. (1996). Development of an upper extremity outcome measure: The DASH (Disabilities of the Arm, Shoulder, and Hand). *American Journal of Industrial Medicine, 29*, 602-608.

Ishii, S., Palmer, A. K., Werner, F. W., Short, W. H., & Fortino, M. D. (1998). An anatomic study of the ligamentous structure of the triangular fibrocartilage complex. *Journal of Hand Surgery, 23A*(6), 977-985.

Jenkins, D. B. (1998). *Hollingshead's functional anatomy of the limbs and back* (7th ed.). Philadelphia, PA: W. B. Saunders Co.

Kijima, Y., & Viegas, S. F. (2009). Wrist anatomy and biomechanics. *Journal of Hand Surgery, 34A*, 1555-1563.

Konin, J. G. (1999). *Practical kinesiology for the physical therapist assistant.* Thorofare, NJ: SLACK Incorporated.

Law, M., & McColl, M. A. (2010). *Interventions, effects and outcomes in occupational therapy.* Thorofare, NJ: Slack Incorporated.

Levangie, P. K., & Norkin, C. C. (2005). *Joint structure and function: A comprehensive analysis* (4th ed.). Philadelphia, PA: F. A. Davis Company.

Levine, D., Simmons, B., Koris, M., Daltroy, L., Hohl, G., Fossel, A., Katz, J. N. (1993). A self-administered questionnaire for the assessment of severity of symptoms and functional status in carpal tunnel syndrome. *Journal of Bone and Joint Surgery, 75*(11), 1585-1592.

Linscheid, R. L. (1986). Kinematic considerations of the wrist. Clinical Orthopaedics and Related Research, 202, 27-39.

Linscheid, R. L. (1992). Biomechanics of the distal radioulnar joint. *Clinical Orthopaedics and Related Research, 275*, 46-55.

MacDermid, J. C., Turgeon, T., Richards, R. S., Beadle, M., & Roth, J. H. (1998). Patient rating of wrist pain and disability: a reliable and valid measurement tool. *Journal of Orthopaedic Trauma, 12*, 77-86.

MacKenna, B. R., & Callender, R. (1990). *Illustrated physiology* (5th ed.). New York, NY: Churchill Livingstone.

Magee, D. J. (2002). *Orthopedic physical assessment.* Philadelphia, PA: Saunders.

Marklin, R. W., Simoneau, G. G., & Monroe, J. F. (1999). Wrist and forearm posture from typing on a split and vertically inclined computer keyboards. *Human Factors, 41*(4), 559-569.

Markolf, K. L., Lamey, D., Yang, S., Meals, R., & Hotchkiss, R. (1998). Radioulnar load-sharing in the forearm. A study in cadaver. *Journal of Bone and Joint Surgery, 80*, 879-888.

Muscolino, J. E. (2006). *Kinesiology: The skeletal system and muscle function.* St. Louis, MO: Mosby.

Neumann, D. A. (2002). *Kinesiology of the musculoskeletal system: Foundations for physical rehabilitation.* St. Louis, MO: Mosby.

Nicholas, J. A., & Hershman, E. B. (1990). *The upper extremity in sports medicine.* St. Louis, MO: C. V. Mosby Co.

Nicholas, J. A., Sapega, A., Kraus, H., & Webb, J. N. (1978). Factors influencing manual muscle tests in physical therapy. *Journal of Bone and Joint Surgery, 60-A*(2), 186-190.

Nordin, M., & Frankel, V. H. (2001). *Basic biomechanics of the musculoskeletal system*, 3rd ed. Philadelphia, PA: Lippincott, Williams & Wilkins.

Nordin, M., & Frankel, V. H. (1989). *Basic biomechanics of the musculoskeletal system.* Philadelphia, PA: Lippincott, Williams & Wilkins.

Norkin, C. C., & Levangie, P. K. (1992). *Joint structure and function: A comprehensive analysis* (2nd ed.). Philadelphia, PA: F. A. Davis Co.

Oatis, C. A. (2004). *Kinesiology: the mechanics and pathomechanics of human movement.* Philadelphia, PA: Lippincott, Williams & Wilkins.

Ozer, K., & Scheker, L. R. (2006). Distal radioulnar joint problems and treatment options. *Orthopedics, 29*(1), 38-49.

Palmer, A. K., & Werner, F. W. (1981). The triangular fibrocartilage complex of the wrist—anatomy and function. *Journal of Hand Surgery, 6A*, 151-162.

Palmer, M. L., & Epler, M. E. (1998). *Fundamentals of musculoskeletal assessment techniques* (2nd ed.). Philadelphia, PA: Lippincott.

Patterson, R. M., Nicodemus, C. L., Viegas, S. F., Elder, K. W., & Rosenblatt, J. (1998). High-speed, three-dimensional kinematic analysis of the normal wrist. *Journal of Hand Surgery, 23*(3), 446-453.

Prosser, R., & Conolly, W. B. (2003). *Rehabilitation of the hand and upper limb.* London: Butterworth Heinemann.

Ryu, J., Cooney, W. P., Askey, L. J., An, K.-N., & Chao, E. Y. S. (1991). Functional ranges of motion of the wrist joint. *Journal of Hand Surgery, 16*(3), 409-419.

Sarrafian, S. K., Melamed, J. L., & Goshgarian, G. M. (1977). Study of wrist motion in flexion and extension. *Clinical Orthopaedics, 126*, 153-157.

Smith, L. K., Weiss, E. L., & Lehmkuhl, L. D. (1996). *Brunnstrom's clinical kinesiology* (5th ed.). Philadelphia, PA: F. A. Davis Co.

Taleisnik, J. (1985). *The wrist.* New York, NY: Churchill Livingstone.

Tang, J. B., Ryu, J., & Kish, V. (1998). The triangular fibrocartilage complex: An important component of the pulley for ulnar wrist extensor. *Journal of Hand Surgery, 23A*(6), 986-991.

Thompson, C. W. & Floyd, R. T. (1994). *Manual of structural kinesiology.* St. Louis, MO: C.V. Mosby.

Viegas, S. F. (2001). The dorsal ligaments of the wrist. *Hand Clinics, 17*(1), 65-75.

Viegas, S. F., Patterson, R. M., Todd, P. D., & McCarty, P. (1993). Load mechanics of the midcarpal joint. *Journal of Hand Surgery, 18*(1), 14-18.

Viegas, S. F., Yamaguchi, S., Boyd, N. L., & Patterson, R. M. (1999). The dorsal ligaments of the wrist: Anatomy, mechanical properties, and function. *Journal of Hand Surgery, 24*(3), 456-468.

Wheeless, C. R. (2010). *Wheeless' textbook of orthopaedics.* Durham, NC: Duke University, Data Trace Internet Publishing, LLC.

ADDITIONAL RESOURCES

Basmajian, J. V., & DeLuca, C. J. (1985). *Muscles alive* (5th ed.). Baltimore, MD: Williams & Wilkins.

Greene, D. P., & Roberts, S. L. (1999). *Kinesiology: Movement in the context of activity.* St. Louis, MO: C. V. Mosby.

Loth, T., & Wadsworth, C. T. (1998). *Orthopedic review for physical therapists.* St. Louis, MO: C. V. Mosby Co.

MacKenna, B. R., & Callender, R. (1990). *Illustrated physiology* (5th ed.). New York, NY: Churchill Livingstone.

Magee, D. J. (1992). *Orthopedic physical assessment.* Philadelphia, PA: W. B. Saunders Co.

Melvin, J. L. (1982). *Rheumatic disease: Occupational therapy and rehabilitation* (2nd ed.). Philadelphia, PA: F. A. Davis Co.

Palastanga, N., Field, D., & Soames, R. (1989). *Anatomy and human movement: Structure and function.* Oxford: Heinemann Medical Books.

Palmer, M. L., & Epler, M. E. (1998). *Fundamentals of musculoskeletal assessment techniques* (2nd ed.). Philadelphia, PA: J. B. Lippincott.

Saidoff, D. C., & McDonough, A. L. (1997). *Critical pathways in therapeutic intervention: Upper extremity.* St. Louis, MO: C. V. Mosby.

Trombly, C. A. (Ed.) (1995). *Occupational therapy for physical dysfunction* (4th ed.). Baltimore, MD: Williams & Wilkins.

Watkins, J. (1999). *Structure and function of the musculoskeletal system.* New York, NY: Human Kinetics.

8

THE HAND

The hand is amazingly complex, intricately constructed, and vital to daily life activities. Capable of complex movements, the hand is a vital connection and mechanism for interaction with the environment. Not only does the hand provide mobility for the proximal joints to move the hand in large areas of space and to reach nearly all parts of the body, but the hand participates in fine motor grasp and pinch, tactile exploration of the environment, and in nonverbal communication with others.

The hand has three functional components. The first component is the thumb, which is positioned at right angles to the fingers to produce effective grip forces. The second functional component includes the index and middle fingers. These two fingers work with the thumb for precision grip. These fingers have thick, long, and strong bones with strong ligaments to add rigidity to the hand. The final component is made up of the ring and little fingers with smaller, shorter bones. This component provides flexible grip to mould around objects of different shapes. The differences in bone length of the metacarpals result in flexed fingers moving diagonally toward the thenar eminence to facilitate the grasping of objects of a variety of shapes (Trew & Everett, 2005).

Mobility within the hand is variable. The distal row of carpals and their articulation with the metacarpals of the index and long fingers are relatively fixed due to the intercarpal ligaments and distal carpal bones. The thumb is the most mobile digit, followed (in order from most to least) by the phalanges of the index, long, ring, and small fingers and then the fourth and fifth metacarpals.

BONES OF THE HAND AND PALPABLE STRUCTURES

The hand consists of five digits, or four fingers and one thumb. Each of the digits has a carpometacarpal (CMC) articulation and a metacarpophalangeal (MCP) joint. The fingers have two interphalangeal (IP) joints while the thumb has only one. There are 19 bones distal to the carpal bones of the wrist that make up the hand: five metacarpal bones, five proximal phalanges, five middle phalanges, and four distal phalanges.

Appearance

The hand has a concave appearance even when the palm is fully opened. This is due to the arches in the hand produced by the carpal bones and ligaments. There are two transverse arches and one longitudinal arch in the hand as seen in Figure 8-1.

The proximal transverse arch (or carpal transverse arch) is curved due to the shape of the carpal bones, with the flexor retinaculum (transverse carpal ligament) as the "roof" of the arch (Magee, 2008). This becomes the carpal tunnel. There is much variability in the shape of this arch due to the mobility of the bones that make up the arch.

It is situated across the hand at the level of the CMC joint with the capitate centrally located and is a fairly rigid arch (Muscolino, 2006), producing the concave appearance of the palm.

Rybski MF.
Kinesiology for Occupational Therapy, Second Edition (pp 235-276)
© 2012 SLACK Incorporated

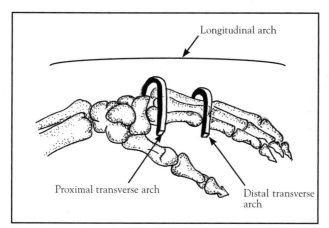

Figure 8-1. Arches of the hand.

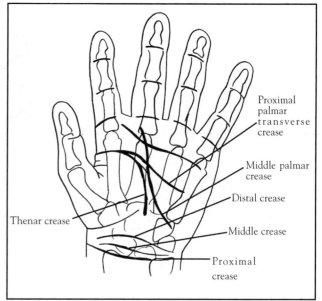

Figure 8-2. Palmar creases of the hand.

The distal transverse arch is found at the level of the metacarpophalangeal joint. The first, fourth, and fifth metacarpals rotate around the stable second and third metacarpals to cup or flatten the palm by changing the distal transverse arch (Bugbee & Botte, 1993).

The longitudinal arch runs from the wrist to the fingertips, with the apex of the arch along the row of metacarpal heads. The carpal bones, metacarpal bones, and phalanges make up this arch, with the metacarpal bones providing the stability. The length of the arch is greatest at the second metacarpal and shortest at the fifth metacarpal. Greene and Roberts (1999) caution that if the shorter length of the longitudinal arch on the ulnar side is not considered when making splints, the splint may extend too far distally, blocking flexion of the fourth and fifth MCP joints (Greene & Roberts, 1999).

Weakness or atrophy of intrinsic hand muscles will lead to loss of arches, which are most noticeable with medial and ulnar nerve damage resulting in the development of the ape hand deformity. The palmar arches provide the function of palmar cupping, which enables the hand to conform to the object being held. Maximum surface contact is made, increased stability is produced, and additional sensory input is provided in this cupped hand position.

The two transverse arches are connected by the longitudinal arch. In finger flexion, the longitudinal arch curls in a pattern called an equiangular spiral or logarithmic spiral. This spiral pattern comprises a series of isosceles triangles with angles of 36 degrees and is a biologically natural pattern, as seen in flowers and in the shell of the nautilus. It is again important to recognize the importance of maintaining these arches when splinting to enable normal hand function.

Visually, the hand and wrist have many creases that are easily seen in Figure 8-2. In the palmar view, the proximal transverse skin crease (linea carpi palmaris proximalis), located at the wrist, is the upper limit of the synovial sheaths of flexor tendons and indicates the radiocarpal joint. The middle skin crease is an indication of the location of the radiocarpal joint, and the distal skin crease (linea carpi palmaris distalis) is the location of the upper margin of flexor retinaculum and the midcarpal joint motion. The radial longitudinal skin crease (thenar crease) encircles the thenar eminence and is some-

times referred to as the life line. The proximal transverse line of the palm runs across shafts of metacarpal bones (sometimes called the head line), and the distal transverse line of the palm lies over the head of the second to fourth metacarpals (commonly called the love line). The proximal skin crease of fingers is 2 cm distal to the MCP joints, the middle skin crease of fingers indicates the location of the proximal IP joints, and the distal skin crease of fingers lies over the distal interphalangeal (DIP) joints (Greene & Roberts, 1999; Magee, 2008).

Palpation

Metacarpals

On the dorsum of the hand, palpate the superficial metacarpal shafts and the space between the metacarpals for the interosseous muscles. If your subject flexes the wrist, on the dorsum of the hand is a ridge of bony protuberances that are the bases of the metacarpals as they articulate with the carpals to form the CMC joints (Biel, 2001).

Phalanges

Distal to the metacarpals are the distal and proximal interphalangeal joints. Follow the metacarpals up to the metacarpophalangeal joint. Gently distract the joint surfaces at the metacarpophalangeal joint to differentiate the heads of the metacarpals from the phalanges.

ARTICULATIONS OF THE HAND

The articulations in the hand include the CMC joints, the MCP joints, the proximal interphalangeal (PIP) joints, and the distal interphalangeal (DIP) joints of the fingers and thumb.

Table 8-1

OSTEOKINEMATICS OF THE CARPOMETACARPAL JOINTS OF THE FINGERS

Functional joint	Synarthrotic, nonaxial
Structural joint	Synovial, plane
Close-packed position	Full flexion
Resting position	Midway between flexion and extension
Capsular pattern	Equal limitations in all directions

Carpometacarpal Joint of the Fingers

The CMC joint is the articulation of digits two through five with the distal row of carpal bones, as well as the articulation of each metacarpal bone with the base of the adjacent metacarpal bone. Specifically, the second metacarpal bone articulates primarily with the trapezoid and partially with the trapezium and the capitate and with the base of the third metacarpal. The third metacarpal articulates with the capitate and the second and fourth metacarpals. The second and third CMC joints are the primary parts of the transverse arch of the hand and are relatively rigid and are primary stabilizers of the hand. The fourth metacarpal articulates with the capitate and the hamate bones and with the side of the third and fifth metacarpal bone. The fifth metacarpal articulates with the hamate, and the ulnar side articulates with the fourth metacarpal bone. The fourth and fifth CMC joints are mobile to permit cupping of the hand. All of the CMC joints are united by the dorsal, palmar, and interosseous ligaments.

The CMC joint provides the most movement for the thumb and the least amount of movement for the fingers (Hamill & Knutzen, 1995). The most movement of the CMC joint is the metacarpal bone of the little finger, followed by the metacarpal bone of the ring finger. The metacarpal bones of the index and middle fingers are almost immovable, so the range of motion increases from the radial side to the ulnar side, which enables the hand to curve anteriorly or to cup. The relative immobility of the second and third metacarpals enables us to hold tools more firmly and enhances the function of the radial wrist flexors and extensors (FCR, ECRL, ECRB), serving as a fixed axis about which the first, fourth, and fifth metacarpals can move. This provides an increased lever arm without the loss of tension resulting from excessive range of motion (Levangie & Norkin, 2005).

Osteokinematics of the Carpometacarpal Joints of the Fingers

Movements at the CMC joints of the fingers are of the nonaxial gliding plane synovial type, although some authors suggest the fourth and fifth are a modified saddle biaxial joint (Gench, Hinson, & Harvey, 1995; Muscolino, 2006), and others classify the second and third digits as complex saddle joints (Neumann, 2002). Generally, the CMC joints of the fingers are seen to be plane joints with only gliding movements possible.

Movements produced are flexion and extension, which can be seen when the hand is cupped or flattened. There are approximately 25 to 30 degrees in the fifth carpometacarpal, 10 to 15 degrees in the fourth carpometacarpal, and minimal movements in the second and third metacarpals. A summary of the osteokinematics of the CMC joints of the fingers is presented in Table 8-1.

Arthrokinematics of the Carpometacarpal Joints of the Fingers

When the metacarpals flex (as when cupping the hand), the metacarpals slide on the carpals in a volar (palmar or anterior) direction, creating an increased transverse arch, whereas when the hand is flattened in extension, the metacarpals slide on the carpals in a dorsal or posterior direction (Kisner & Colby, 2007), resulting in a decreased transverse arch (Table 8-2).

Muscles contributing to the movement at the CMC joints of the fingers are radial wrist muscles (flexor carpi radialis, extensor carpi radialis longus, extensor carpi radialis brevis, extensor carpi ulnaris) and flexor digitorum superficialis and flexor digitorum profundus.

Carpometacarpal (Fingers) Stability

In addition to the stabilizing effect of the relatively immobile second and third carpometacarpal bony surfaces, the

Table 8-2

ARTHROKINEMATICS OF THE CARPOMETACARPAL JOINTS OF THE FINGERS

PHYSIOLOGIC MOVEMENT OF CONVEX METACARPALS ON CONCAVE CARPALS	DIRECTION OF SLIDE OF METACARPALS ON CARPALS	RESULTANT MOVEMENT
Flexion	Volar (Anterior) slide, volar roll	Movement in same direction
Extension	Dorsal (Posterior slide), dorsal roll	Movement in same direction

Figure 8-3. Ligamental support of the carpometacarpal joints. (Adapted from Muscolino, J. E. (2006). *Kinesiology: The skeletal system and muscle function.* St. Louis, MO: Mosby.)

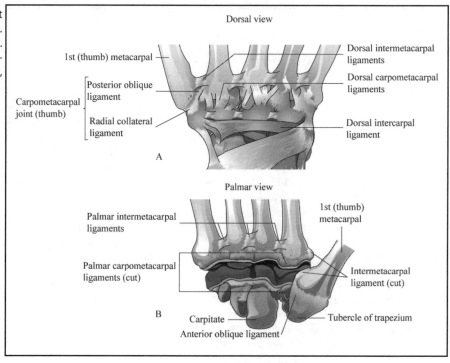

CMC joints are surrounded by fibrous articular capsules and dorsal, palmar, and interosseous carpometacarpal ligaments as seen in Figure 8-3.

The ligaments are the primary control of the amount and range of motion of the CMC joints. The dorsal carpometacarpal ligaments are particularly well-developed around the middle CMC joints. They connect the carpals and metacarpal bones on the dorsal surface (see Figure 8-3).

The palmar ligaments perform a similar function on the palmar surface of the hand. The interosseous ligament is short and thick in appearance and is limited to one part of the carpometacarpal articulation—that of connecting the inferior angles of the capitate and hamate with the third and fourth metacarpal bones (Goss, 1976). The transverse meta-

carpal (intermetacarpal) ligament is a narrow, fibrous band connecting the palmar surfaces of the heads of the second through fifth metacarpal bones. This ligament functions to prevent abduction and adduction at the CMC joint (Nordin & Frankel, 2001).

Carpometacarpal Joint of the Thumb

This articulation is formed at the juncture of the trapezium and base of the first metacarpal. The CMC joint of the thumb is capable of many movements.

Because this joint is at the base of the thumb, there are large functional demands placed on the joint, often result-

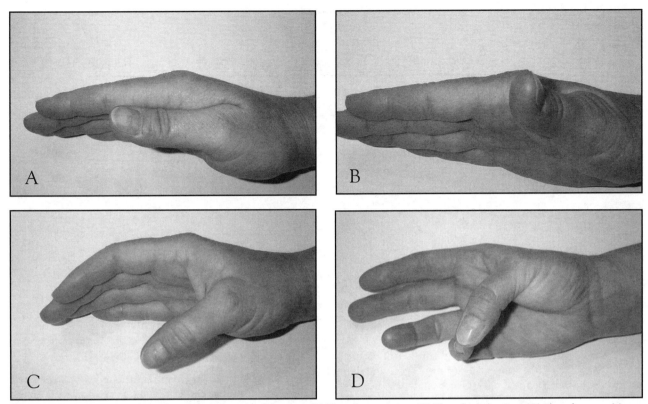

Figure 8-4. Movements of the thumb. (A) Thumb adduction. (B) Thumb extension. (C) Thumb abduction. (D) Thumb opposition.

ing in the painful condition of basilar (referring to the base) arthritis, often in women in the fifth to sixth decade of life. Without movement of the CMC joint of the thumb, firm grasp is limited.

Terminology used to describe motions of the thumb is specialized due to the rotation of the thumb in relation to the fingers. The planes of motion are in relation to the palm so that flexion-extension movements are parallel to the palm and abduction-adduction movements are perpendicular. Flexion is also known as ulnar adduction or palmar flexion and is the movement of the palmar surface of the thumb in the frontal plane across the palm. Extension (radial abduction or radial extension) returns the thumb to the anatomic position. This is also referred to as the hitchhiking position. Abduction of the CMC joint (palmar abduction) is the forward movement of the thumb away from the palm in the sagittal plane, and adduction returns the thumb to the palm (Neumann, 2002).

Circumduction (axial rotation or opposition in the CMC joint, thumb rotation, thumb pronation) is a combined movement of abduction, flexion, adduction, and extension that involves medial rotation of the thumb's metacarpal. The first metacarpal rotates on the trapezium to permit placement of the pad of the thumb on the pads of one or more fingers in a cone-shaped path. Reposition (supination of the thumb) is the combination of adduction, extension, lateral rotation of the thumb's metacarpal, abduction, and flexion. Movements of the thumb are shown in Figure 8-4.

Osteokinematics of the Carpometacarpal Joint of the Thumb

The CMC joint of the thumb is classified as a biaxial joint but actually is capable of movement in three planes and three axes. Flexion and extension occur in a frontal plane around the anteroposterior axis; abduction and adduction occur in a sagittal plane around a mediolateral axis; and medial and lateral rotation occurs in a transverse plane around the vertical axis. However, the rotational movements in the transverse plane cannot be actively isolated, which explains why the joint is considered a biaxial joint. Medial rotation of the first metacarpal must accompany flexion of the first metacarpal, and lateral rotation accompanies extension (Muscolino, 2006). The CMC joint of the thumb is capable of 60 degrees of abduction, 10 degrees of adduction, 40 degrees of flexion, 10 degrees of extension, and 45 degrees of medial rotation.

The orientation of the axis of motion in the thumb differs from the axes at other joints. The two axes of motion are offset from the cardinal planes of motion and are not perpendicular to the bones or to each other (Gench et al., 1995). Flexion and extension occur in the plane of the palm, with the axis oriented to the front and back in relation to the palmar and dorsal hand surfaces; that is, the motion is parallel to the palm (Goss, 1976; Greene & Roberts, 1999). This axis goes through the trapezium but at a right angle to the palm (Gench et al., 1995). Adduction and abduction occur in the plane at right angles to the palm, with the axis in a side-to-side orientation

Table 8-3

OSTEOKINEMATICS OF THE CARPOMETACARPAL JOINT OF THE THUMB

Functional joint	Diarthrotic, biaxial
Structural joint	Synovial, saddle
Close-packed position	Full opposition
Resting position	Midway between flexion and extension and midway between abduction and adduction
Capsular pattern	Abduction then extension

rather than front to back (Goss, 1976; Greene & Roberts, 1999). This axis is at the base of the first metacarpal and slants toward the base of the ring finger (Gench et al., 1995). A summary of the osteokinematics of the CMC joint of the thumb is provided in Table 8-3.

Arthrokinematics of the Carpometacarpal Joint of the Thumb

Because there are both convex and concave bone surfaces, this is a biaxial sellar (saddle-shaped) joint.

In flexion and extension, the convex trapezium articulates on the concave metacarpal base so the surface slides in the same direction as the angulating bone.

In flexion, the concave metacarpal rolls and slides in an ulnar (medial) direction. There is slight medial rotation of the metacarpal as it moves toward the third digit with elongation of the radial collateral ligament. In extension, the concave metacarpal rolls and slides in a lateral (radial) direction (i.e., away from the third digit) with slight lateral rotation of the metacarpal of the thumb, which requires elongation of the anterior oblique ligament. Flexion and extension of the CMC joint of the thumb can be observed by watching the change of orientation of the thumbnail during flexion and extension (Neumann, 2002).

Adduction and abduction movements occur in the sagittal and frontal axis, and, because the trapezium is concave and the metacarpal convex, the surface slides in an opposite direction of the articulating bone (Kisner & Colby, 2007) as summarized in Table 8-4. In abduction, the convex metacarpal rolls in a palmar direction with a dorsal slide on the concave surface of the trapezium. Full abduction stretches the anterior oblique ligament, the intermetacarpal ligament, and the adductor pollicis muscle with the abductor pollicis longus muscle responsible for the roll. Adduction results in a dorsal roll and palmar slide of the convex surface of the metacarpal on the concave trapezium. A summary of the arthrokinematics of the CMC joint of the thumb is provided in Table 8-4.

Carpometacarpal (Thumb) Stability

The CMC joint of the thumb is surrounded by a thick but loose fibrous joint capsule that is reinforced by radial, ulnar, volar, and dorsal ligaments, as well as an intermetacarpal ligament that holds the bases of the first and second metacarpals together. This prevents extreme radial and dorsal displacements of the base of the first metacarpal (Nordin & Frankel, 1989). The laxity of the joint capsule allows 15 degrees to 20 degrees of rotation, and the metacarpal can be distracted up to 3 mm from the trapezium.

Five major ligaments provide stability to the CMC joint of the thumb as shown in Figure 8-5. Three oblique capsular ligaments (anterior oblique, posterior oblique, and radial collateral ligament) attach the first metacarpal to the trapezium. Located from the tubercle of the trapezium to the base of the metacarpal of the thumb is the anterior oblique ligament, which is taut in abduction, extension, and opposition and is considered the strongest of the three. Laxity in these three ligaments allows for circumduction and rotation of the CMC joint. The radial collateral ligament (also called the dorsal-radial ligament), located from the radial surface of the trapezium to the base of the thumb metacarpal, is taut in all thumb movements except extension. The posterior oblique ligament and the first intermetacarpal ligaments are taut in abduction and opposition. The ulnar collateral ligament (palmar oblique ligament), located from the transverse carpal ligament to the base of the thumb metacarpal, is taut in abduction, extension, and opposition. In addition, the interosseous ligaments also provide support.

Metacarpophalangeal Joint of the Fingers

The metacarpal bones articulate into shallow cavities of the proximal ends of the first phalanges (with the exception of the thumb) at the MCP joints (Goss, 1976). The first metacar-

Table 8-4

ARTHROKINEMATICS OF THE CARPOMETACARPAL JOINT OF THE THUMB

PHYSIOLOGIC MOVEMENT OF METACARPALS ON CARPALS	DIRECTION OF SLIDE OF METACARPALS ON CARPALS	RESULTANT MOVEMENT
Flexion	Ulnar (medial) slide, ulnar roll	Movement in same direction
Extension	Radial (lateral) slide, radial roll	Movement in same direction
Abduction	Dorsal slide, volar roll	Movement in opposite direction
Adduction	Volar slide, dorsal roll	Movement in opposite direction

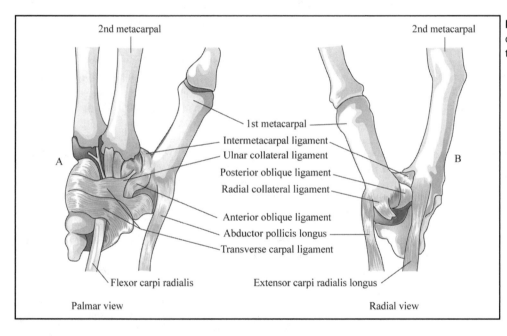

Figure 8-5. Ligaments of the carpometacarpal joint of the thumb.

pophalangeal joint is located between the first metacarpal and the proximal phalanx of the thumb, and subsequent metacarpophalangeal joints are located between the metacarpal and the proximal phalanx of the adjacent finger.

When the metacarpophalangeal joints are flexed, it is not possible to abduct and adduct the fingers due to the shapes of the bones and tension in ligaments. There is, however, substantial accessory movement possible when the fingers are extended and relaxed. There is a great deal of anteroposterior, side-to-side, and distraction gliding motions. Axial rotation of the metacarpals occurs with as much as 30 to 40 degrees possible in the ring and little fingers. These accessory motions enable the hand to conform to the shape of objects held in the hand with greater control.

Osteokinematics of the Metacarpophalangeal Joints

The MCP joints of the fingers are unicondylar diarthrodial biaxial joints with three planes of movement (Nordin & Frankel, 1989), with flexion, extension, abduction, adduction, and slight rotational movements possible. The axis for flexion and extension is in the sagittal plane around a mediolateral axis. The location of the axis permits as much as 95 to 110 degrees of flexion in the little finger, approximately 70 to 90 degrees in the index finger, and 90 to 115 degrees at the fifth finger. The mobility of the second and fifth digits is the greatest because adjacent fingers do not limit the motion (Neumann, 2002). Passive hyperextension of the metacarpophalangeal finger joints of 30 to 45 degrees is possible and is often used as a measure of generalized body flexibility. Being considered "double-jointed" is due to laxity of the ligaments

Table 8-5

Osteokinematics of the Metacarpophalangeal Finger Joints

Functional joint	Diarthrotic, biaxial
Structural joint	Synovial, condyloid
Close-packed position	Full flexion
Resting position	Slight flexion
Capsular pattern	Flexion, then extension

of the joint, and hyperextension is normally limited by the palmar plate.

The point of reference for abduction and adduction is the middle finger. Lateral movement of the middle finger is called radial abduction, and medial movement is ulnar abduction. Brand and Hollister (1999) describe the adduction/abduction axis as a cone that runs forward from the metacarpal head with an inclination distally of approximately 30 degrees above a right angle. Abduction and adduction can occur in this anteroposterior axis that passes from the palmar to dorsal surfaces when the fingers are extended because the tip of the finger is a long distance from the joint axis, thereby permitting this motion. When the fingers are flexed, the tips of the fingers are on the axis of rotation, so no further lateral movement can occur. For example, when the hand holds a hammer in a power grip, the thumb aligns with the longitudinal axis, providing maximal strength and stabilization against lateral forces. In full flexion, the phalange is parallel to the longitudinal axis, putting the joint into the most stable, close-packed position.

The close-packed position of the metacarpophalangeal joints is in full flexion, as this is when the cord portion of the collateral ligaments are taut, providing substantial stability to the base of the fingers (Neumann, 2002) (Table 8-5).

Arthrokinematics of the Metacarpophalangeal Joints

Although each metacarpal has a slightly different shape, generally, the concave articulating surface of the phalanges moves on a convex metacarpal head in the metacarpophalangeal joints of the fingers. In extension, the proximal phalanx rolls and slides dorsally via the extensor digitorum communis muscle. At 0 degrees of extension, the collateral ligaments slacken, and the palmar plate unfolds, which enables total contact with the head of the metacarpal. Flexion motion of rolling and sliding occurs in the palmar direction.

Figure 8-6 illustrates flexion at the metacarpophalangeal, proximal interphalangeal, and distal interphalangeal joints. Abduction produces movement as the proximal phalanx rolls and slides in a radial direction with the first dorsal interosseous

muscle directing both the roll and slide. Table 8-6 summarizes the arthrokinematics of the metacarpophalangeal joints of the fingers.

Stability of the Metacarpophalangeal Joints

Mechanical stability of the metacarpophalangeal joints is crucial to the hand, as the metacarpals are the support of the arches of the hand (Neumann, 2002). Each joint has a fibrous joint capsule that is lax in extension and taut in flexion. The metacarpophalangeal joints are often affected by rheumatoid arthritis because the joint capsules are not stable enough to resist ulnarly directed forces, such as when things are held between the thumb and fingers (Muscolino, 2006).

Imbedded in each capsule of each metacarpophalangeal joint are radial and ulnar ligaments and one palmar ligament or plate. The collateral ligaments are strong, and they run along the sides of the joints. The dorsal surface is covered by the extensor tendons with tissue that connects the deep surfaces of the tendons to bones. The collateral ligaments are slack in extension, although some authors disagree and say that parts provide stability throughout the range (Norkin & Levangie, 1992). The laxity in extension allows some passive axial rotation of the proximal phalanx. Because the ligament is taut with flexion, this prevents abduction and adduction, because there is a longer distance between the points of attachment when these joints are flexed than extended. Abduction and adduction can occur only when the joints are extended. This enables mechanical stabilization of the MCP joint (Smith, Weiss, & Lehmkuhl, 1996).

Each collateral ligament has two parts: the cord part, which is thick and strong and attaches to the palmar aspect of the distal end of the phalanx, and the accessory part, which is fan-shaped and attaches distally to the palmar plate.

The palmar ligaments (or volar plate) are thick, dense, multilayered, and fibrocartilaginous. These are located between the collateral ligaments to which they are connected. These ligaments are loosely connected to the metacarpal bones but are firmly attached to the base of the first phalanges, helping

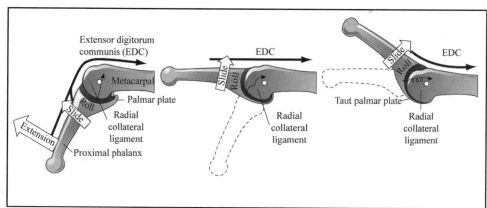

Figure 8-6. Arthrokinematics of flexion at the MCP, PIP, and DIP joints. (Adapted from Neumann, D. A. (2002). *Kinesiology of the musculoskeletal system: Foundations for physical rehabilitation.* St. Louis, MO: Mosby.)

Table 8-6

ARTHROKINEMATICS OF THE METACARPOPHALANGEAL JOINTS

PHYSIOLOGIC MOVEMENT OF METACARPALS ON CARPALS	DIRECTION OF SLIDE OF METACARPALS ON CARPALS	RESULTANT MOVEMENT
Flexion	Volar slide, volar roll	Movement in same direction
Extension	Dorsal slide, dorsal roll	Movement in same direction
Abduction	Radial roll and slide	Movement in same direction
Adduction	Ulnar roll and slide	Movement in same direction

to reinforce the joint capsule as well as provide a surface for contact in extension and in prevention of excessive hyperextension (Levangie & Norkin, 2005). Fibrous digital sheaths form tunnels or pulleys for the extrinsic finger flexors, which are anterior to the palmar plates. The purpose of the palmar plates is to strengthen the metacarpophalangeal joints and to resist hyperextension. Between the palmar plates are three deep transverse metacarpal ligaments that interconnect the second through fifth metacarpals, as seen in Figure 8-7.

Metacarpophalangeal Joint of the Thumb

The metacarpophalangeal joint of the thumb is located between the convex head of the first metacarpal and the concave proximal phalanx of the thumb. While the arthrokinematics are very similar to the metacarpophalangeal joints of the fingers, there is much less active and passive movement in the metacarpophalangeal joints of the thumb.

Osteokinematics of the Metacarpophalangeal Joint of the Thumb

The articulation of the first metacarpal bone and the proximal phalange produces a condyloid joint that acts like a ginglymus (hinge) joint. Goss (1976) states this is a "ginglymoid" joint, and Hamill and Knutzen (1995) say this is a hinge joint, while Norkin and Levangie (1992) and Kisner and Colby (1990) indicate this is a condyloid joint. Movements include flexion/extension and abduction/adduction (considered an accessory motion), with an insignificant amount of passive motion possible. Close-packed position is in full opposition, and resting position is with the joint in slight flexion. The joint is more limited in flexion followed by extension in capsular restrictions (Table 8-7).

Figure 8-7. Ligaments of the metacarpophalangeal and intermetacarpal joints. (Adapted from Muscolino, J. E. (2006). *Kinesiology: The skeletal system and muscle function.* St. Louis, MO: Mosby.)

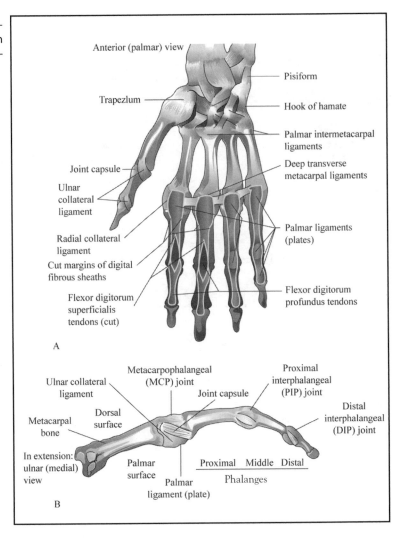

Arthrokinematics of the Metacarpophalangeal Joint of the Thumb

As with the other metacarpals, the metacarpal head is convex and fits into the concave base of the first phalange, so, arthrokinematically, the MCP joint of the thumb slides and rolls as do the phalanges as previously discussed regarding the MCP joints of the fingers.

Stability of the Metacarpophalangeal Joint of the Thumb

Motion at the MCP joint of the thumb is more restricted than at the fingers. The collateral ligaments restrict abduction and adduction. Adduction and abduction forces are transferred to the CMC joint (Neumann, 2002). The joint capsule and ligaments of the thumb at this joint are similar to the other MCP joints with the addition of two sesamoid bones that act as additional reinforcement for thumb stability.

Interphalangeal Joints

The articulation of the interphalangeal joints with each other and proximally with the phalanges produces two inter-phalangeal joints for digits two through five and one interphalangeal joint for the thumb. There are nine interphalangeal joints in the hand (one interphalangeal joint, four proximal interphalangeal joints, and four distal interphalangeal joints). The thumb interphalangeal joint is structurally and functionally identical to the DIP joints of the fingers (Levangie & Norkin, 2005; Muscolino, 2006; Neumann, 2002).

Osteokinematics of the Interphalangeal Joints

Flexion and extension movements in the sagittal plane around a coronal axis are more extensive between the proximal and middle phalanges than between the middle and distal joints. In addition, flexion is considerable, while extension is limited by the ligamental tautness. Because there is a larger proximal articular surface than distal, there is very little hyperextension and practically no passive hyperextension at the proximal interphalangeal (PIP) joint. The joints are closely congruent during movement, which adds stability to these joints. Full extension results in the close-packed position, and slight flexion is the resting position of these joints (Table 8-8).

There is greater range of motion going from the radial to ulnar joint, permitting more opposition in the ulnar fingers

Table 8-7

OSTEOKINEMATICS OF THE METACARPOPHALANGEAL JOINT OF THE THUMB

Functional joint	Diarthrotic, biaxial
Structural joint	Synovial, condyloid
Close-packed position	Full opposition
Resting position	Slight flexion
Capsular pattern	Flexion then extension

Table 8-8

OSTEOKINEMATICS OF THE INTERPHALANGEAL JOINTS

Functional joint	Diarthrotic, uniaxial
Structural joint	Synovial, hinge
Close-packed position	Full extension
Resting position	Slight flexion
Capsular pattern	Flexion, extension

by angling these fingers toward the thumb so one can get a tighter grip on objects from the ulnar side. Average ranges of motion are 0 to 100 degrees to 120 degrees for proximal interphalangeal flexion, 0 to 70 degrees to 90 degrees for distal interphalangeal flexion, and 0 degrees to 80 degrees for thumb flexion. Flexion and extension of the interphalangeal joints of the ring and little fingers are accompanied by slight axial rotation of the phalanx, which allows greater contact with the thumb during opposition (Neumann, 2002). The rotation of the joints during flexion is called the cascade sign, and failure of the joints to converge toward the scaphoid tubercle may indicate trauma or possibly a fracture.

Because of the greater movement in the ulnar fingers, this reinforces the idea that digits one and two are primarily functional in prehension and precision movements, while digits three through five are used for power. Notice how many tools have handles that are narrower at the ring and little fingers and wider at the radial side along the base of the long and index fingers. These also are concepts to be remembered in making adaptations to tools and utensils.

Arthrokinematics of the Interphalangeal Joints

With flexion of the phalanx, the base of the phalanx slides in a volar direction, while, during extension, the base of the phalanx slides dorsally because the distal end of each phalanx is convex and the articulating surface at the proximal end of each phalanx is concave.

Stability of the Interphalangeal Joints

There are two collateral ligaments and a palmar (volar plate) ligament for each interphalangeal joint with a similar arrangement as those of the MCP joint. The proximal interphalangeal joints have an additional structure, the check-rein ligament, to restrict hyperextension at those joints. In athletic injuries, often, there is an injury to both the palmar plate and check-rein ligament.

The radial collateral ligaments of the interphalangeal joints are essentially identical to the radial collateral ligaments of the metacarpophalangeal joints. There is a cord part that limits abduction and adduction and an accessory portion that blends with and reinforces the palmar plate. The radial collateral ligament limits the motion of the phalanx to the ulnar side, and the ulnar collateral ligament limits motion of the phalanx to the radial side. The palmar plate stabilizes the interphalangeal joints and resists hyperextension.

Intermetacarpal Joints

Additional synovial plane joints are formed between the metacarpal bones and the hand. There are four proximal intermetacarpal joints, as the metacarpals articulate with each other at their bases, and three distal intermetacarpal joints, as metacarpals two through five articulate with each other at the metacarpal heads.

The intermetacarpal joints produce slight nonaxial gliding motions. The metacarpals at the intermetacarpal joints on either side of the hand (first and fifth) are the most mobile (Muscolino, 2006).

These joints are bound together by palmar and dorsal intermetacarpal ligaments, interosseous ligaments, and dense fibrous joint capsules. The dorsal and palmar intermetacarpal ligaments and interosseous ligaments connect the base of each of the five metacarpals to the base of the adjacent metacarpal. The deep transverse metacarpal ligament connects the heads of metacarpals two through five.

STABILITY OF THE HAND

Many ligaments operate within the hand to provide stability and to permit mobility. At the wrist, there is a concavity formed by the arched carpal bones that is spanned by the transverse carpal ligament (flexor retinaculum or anterior annular ligament creating the carpal tunnel). The flexor retinaculum covers the scaphoid, trapezium, pisiform, and hamate carpal bones. The tendons of flexor digitorum profundus, flexor digitorum superficialis, flexor carpi radialis, flexor pollicis longus, and the medial nerve run through this tunnel. The transverse carpal ligament restricts bowing of the long finger flexor tendons when the wrist is flexed, disallows abduction, protects the median nerve, and is the site of origin for thenar and hypothenar muscles (Gench et al., 1995; Muscolino, 2006).

The proximal band of the transverse carpal ligament becomes taut when flexor carpi ulnaris contracts and when the hand is held in an ulnarly flexed position. It is located at the distal skin crease of the wrist and forms the border on the palmar side of the carpal tunnel. The distal band is always taut, acts as a pulley for flexor tendons (Gench et al., 1995), and holds down the extrinsic finger flexor muscles to prevent bowstringing of the tendons. The palmar plate supports the anterior metacarpophalangeal, distal interphalangeal, and proximal interphalangeal joints. There are also five dense annular pulleys and three cruciform pulleys to allow for a smooth curve for tendon excursion.

The second connective structure in the palm is the heavy fibrous palmar aponeurosis (palmar fascia), a thick and deeply located continuation of the flexor retinaculum. It forms the ridges in the palm, which increases friction for firm grasp of objects. Palmaris longus inserts into the palmar aponeurosis. This tissue receives the insertion of the palmaris longus muscle, and it merges with the flexor retinaculum at its distal edge.

Firm collateral ligaments and a thick anterior capsule that is reinforced by the palmar fibrocartilaginous (volar) plate provide support as do the collateral ligaments at the metacarpophalangeal, proximal interphalangeal, and distal interphalangeal joints. The collateral ligaments support the sides of the fingers and restrict varus and valgus forces.

Additional support comes from the transverse intermetacarpal ligaments. The dorsal intercarpal ligaments keep the carpals together.

The extensor retinaculum (dorsal carpal ligament or posterior annular ligament) goes across the dorsal surface of the wrist and forms a roof for extensor tendons. Fibers of the retinaculum insert on the pisiform and triquetrum on the ulnar side, and on the radial side, the retinaculum blends with flexor retinaculum. The extensor retinaculum holds the extensor muscle tendons in place and acts as a pulley mechanism for the extensor tendons. Each extensor tendon is surrounded by a synovial sheath, and these form into six compartments that contain the following:

1. First: Radial nerve, tunnel for abductor pollicis longus, extensor pollicis brevis.
2. Second: Extensor carpi radialis longus, extensor carpi radialis brevis.
3. Third: Extensor pollicis longus.
4. Fourth: Extensor indicis, extensor digitorum.
5. Fifth: Extensor digiti minimi.
6. Sixth: Extensor carpi ulnaris.

A helpful way to remember the contents in each dorsal compartment is to remember the numbers 22, 12, 11, which correspond to the numbers of tendons in each compartment (Fess, Gettle, Philips, & Janson, 2005).

MOVEMENTS OF THE HAND

The amount and type of movement of the hand depends on the joint and digit. The most mobility of the CMC joints occurs in the thumb followed by the little and ring fingers. This enables cupping of the hand to cup around a stable core of the index and middle CMC joints. At the metacarpophalangeal joints, there is much less movement possible at the thumb. The second and fifth fingers have the greatest mobility, because movement is not hampered by adjacent fingers. There is much accessory motion and passive hyperextension possible at the metacarpophalangeal joints of the fingers but not at the thumb. The greater movement of the interphalangeal joints from the radial to ulnar direction permits the fingers to angle their location to permit more opposition and grasp. All of these factors enable prehension. A summary of the movements of the hand is presented in Tables 8-9 and 8-10.

Prehension

The hand is capable of many movements and variations in strength or precision. Many efforts have been made to categorize these movement patterns so that consistent use of descriptive terminology can be used. Movements can be produced where the hand as a whole pushes, pulls, or moves an object and where no actual grasp is involved. This would essentially be a nonprehensile movement pattern.

Prehension includes reaching, grasping, carrying, and releasing and not just pushing or pulling on the object. The thumb tends to be the defining factor in prehension as to stabilization, control of direction, and power. When the thumb is not functional in the grasp pattern, the prehension is nonmanipulative and essentially one of power. When the thumb and one or more fingers are involved in the action, the pattern is one of manipulative prehension and precision.

Often, the distinction is made between grasp (power grip), precision grip, and pinch to differentiate those actions with the thumb and those without. In grasp, all digits and the palm are used, but pinch typically uses the digits on the radial side of the hand. The area of contact within the hand is just one of the factors that distinguish pinch from grasp. Other factors include the number of fingers involved, the amount of finger flexion, the position of the thumb, and the position of the wrist (Oatis, 2004).

Prehension requires the coordinated interaction of muscles, joints, and sensation. Peripheral mobility is achieved by the movement of the CMC joint of the thumb and the fourth and fifth metacarpophalangeal joint movement. The rigidity of the CMC joints of digits two and three provides central stability on which to base movement. The stability of the arches of the hands assists with cupping the hand to hold objects. There is a synergistic balance between the agonists and antagonists of the extrinsic and intrinsic muscle so that optimal length-tension relationships are maintained. Finally, intact proprioception and sensation are necessary to perceive the object, its characteristics, and the relationship of the object to the hand for skillful object manipulation (Konin, 1999; Nordin & Frankel, 2001).

Gripping an object occurs in four stages. Opening the hand requires simultaneous activity of the intrinsics and long extensor muscles and is considered the first stage. Closing the hand around the object is stage two, requiring flexion of the fingers to grasp it. The third stage involves application of the correct amount of force upon the object based on the weight and size of the object and its surface characteristics, fragility, and use. In stage four, the hand opens to release the object (Magee, 1992).

Prehensile Nonmanipulative Patterns

Prehensile patterns can be non-manipulative or manipulative patterns of movement. In prehensile non-manipulative patterns, three types of power grips are used to seize and hold objects in the hand for objects requiring firm control. The digits position the object against the palm, and the thumb is essentially nonfunctional. The joint position brings the hand into line with the forearm with the fingers flexed and the wrist in ulnar deviation and extended (Magee, 1992). In the hook grip, digits two to five are used as a hook (Smith et al., 1996) to carry objects, such as a purse or briefcase, where force is sustained for long periods. The extended thumb is nonfunctional, and the heel of the hand may provide some counterpressure. The metacarpophalangeal joints are in neutral or are extended with the fingers flexed at the proximal interphalangeal and distal interphalangeal joints. Primarily, the muscles involved include flexor digitorum superficialis, flexor digitorum profundus, extensor pollicis longus and brevis, extensor digitorum, and fourth lumbrical and interossei. If a powerful grip is needed, the fingers will flex at all three joints, and the thumb will be adducted (Jenkins, 1998).

In the second and third types of prehensile nonmanipulative grip, the fingers are used to hold an object in the palm of the hand so that the actual position of the hand varies according to the size, shape, and weight of the object being held. The palm contours to the object, and the thumb provides an additional surface for the object by adducting against it. The thumb applies counterpressure to the partially flexed fingers but again is essentially not functional as a manipulative force.

Cylindrical grasp and spherical grasp are considered palmar prehension or power grips. In cylindrical grasp, the thumb is in opposition and the fingers are adducted and flexed; this can be visualized by the position of the hand in holding a beverage can. Flexor pollicis longus and adductor pollicis, flexor digitorum profundus, and the fourth lumbrical are active in this grasp pattern. Spherical grasp is seen when one holds a round object, such as a ball, where the thumb is in opposition with fingers flexed and abducted. Flexor pollicis longus and adductor pollicis, flexor digitorum profundus, and fourth lumbrical are active in this prehension pattern. Some sources state that cylindrical and spherical grasps are merely descriptions of the objects held in the hand and are not distinct grasp patterns. The third type of power grip is the fist grasp or digital palmar prehension pattern when the hand grips a narrow object such as a broom handle.

Prehensile Manipulative Patterns

Prehensile manipulative pinch patterns include lateral pinch, palmar pinch, and tip pinch. Lateral pinch (also known as key pinch, pad-to-side, subterminolateral opposition, or pulp pinch) is used when a thin object is held in the hand, such as a card or a key. It is the least precise of the manipulative prehension patterns and is the finest grasp that can be accomplished without active hand musculature via tenodesis action (Norkin & Levangie, 1992). The object is grasped between the palmar surface of the thumb and the lateral side of the index finger (Smith et al., 1996). The thumb is adducted with IP flexion with the index finger flexed and abducted, which involves the flexor pollicis longus and brevis, adductor pollicis, flexor digitorum profundus and superficialis, as well as first dorsal interossei.

Palmar pinch, also known as subterminal opposition or pad-to-pad prehension, occurs when the palmar surfaces of the distal phalanges contact the palmar surface of the thumb. The thumb is in opposition and slight flexion with the fingers in flexion at the MCP and PIP joints. This grip pattern can

Table 8-9

Summary of Hand Movements

MOTION	PLANE/AXIS	JOINT	NORMAL LIMITING FACTORS TENSION IN	END FEEL	PRIMARY MUSCLES	NORMATIVE ROM/ FUNCTIONAL ROM
Flexion	Sagittal/frontal	MCP	• Posterior joint capsule • Collateral ligaments • Contact between the proximal phalanx and the metacarpal	Firm	• Flexor digitorum superficialis • Flexor digitorum profundus • Lumbrical • Flexor digiti minimi brevis • Dorsal interossei • Palmar interossei • Opponens digiti minimi	0-90/0-61 degrees
		PIP	• Posterior joint capsule • Collateral ligaments • Contact between the middle and proximal phalanx • Soft tissue apposition of the middle and proximal phalange	Hard		0-100/0-60 degrees
		DIP	• Posterior joint capsule • Collateral ligaments • Oblique retinacular ligament	Firm		0-90/0-39 degrees
Extension	Sagittal/frontal	MCP	• Anterior joint capsule • Palmar fibrocartilaginous plate (palmar ligament)	Firm	• Extensor digitorum communis • Extensor inidicis proprius • Extensor digiti minimi	90-0/61-0 degrees

(continued)

Table 8-9 (continued)

SUMMARY OF HAND MOVEMENTS

MOTION	PLANE/AXIS	JOINT	NORMAL LIMITING FACTORS TENSION IN	END FEEL	PRIMARY MUSCLES	NORMATIVE ROM/ FUNCTIONAL ROM
		PIP	• Anterior joint capsule • Palmar ligament	Firm	• Lumbricals • Dorsal interossei • Palmar interossei	100-0/60-0 degrees
		DIP	• Anterior joint capsule • Palmar ligament			90-0/39-0 degrees
Abduction of the fingers	Frontal/sagittal	MCP	• Collateral ligaments • Fascia and skin of the web spaces • Palmar interosseous ligament	Firm	• Abductor digiti minimi • Dorsal interossei • Extensor indicis	No norms
Adduction of the fingers	Frontal/sagittal	MCP			• Palmar interossei • Opponens digiti minimi	No norms
Opposition of the little finger	Frontal and sagittal planes and axes	MCP CMC			• Opponens pollicis • Flexor digiti minimi • Palmaris longus • Palmaris brevis	No norms

Table 8-10

SUMMARY OF THUMB MOVEMENTS

MOTION	PLANE/AXIS	JOINT	NORMAL LIMITING FACTORS TENSION IN	END FEEL	PRIMARY MUSCLES	NORMATIVE ROM/ FUNCTIONAL ROM
Flexion	Frontal/sagittal	CMC	• Posterior joint capsule • Extensor pollicis brevis • Abductor pollicis brevis • Soft tissue apposition between the thenar eminence and the palm	Soft	• Adductor pollicis • Flexor pollicis longus • Flexor pollicis brevis	0-15 degrees
		MCP	• Posterior joint capsule • Collateral ligaments • Extensor pollicis brevis • Contact between the first metacarpal and the proximal phalanx	Hard	• Adductor pollicis • Flexor pollicis brevis • Abductor pollicis brevis • Flexor pollicis longus	0-50/0-21 degrees
		IP	• Collateral ligaments • Posterior joint capsule • Contact between the distal phalanx, fibrocartilaginous plate, and the proximal phalanx	Firm	• Flexor pollicis longus	0-80/0-18 degrees
Extension	Frontal/sagittal	CMC	• Anterior joint capsule • Flexor pollicis brevis • First dorsal interosseous	Firm	• Extensor pollicis brevis • Extensor pollicis longus • Abductor pollicis longus	0-20 degrees
		MCP	• Anterior joint capsule • Palmar ligaments • Flexor pollicis brevis	Firm	• Extensor pollicis longus • Extensor pollicis brevis	50-0/21-0 degrees
		IP	• Anterior joint capsule • Palmar ligaments	Firm	• Extensor pollicis longus • Abductor pollicis brevis	80-0/18-0 degrees

(continued)

Table 8-10 (continued)

SUMMARY OF THUMB MOVEMENTS

MOTION	PLANE/AXIS	JOINT	NORMAL LIMITING FACTORS TENSION IN	END FEEL	PRIMARY MUSCLES	NORMATIVE ROM/ FUNCTIONAL ROM
Abduction	Sagittal/frontal	CMC MCP	• Fascia and skin of the first web space • First dorsal interosseous • Abductor pollicis longus	Firm	• Abductor pollicis brevis • Abductor pollicis longus	No norms
Adduction	Sagittal/frontal	CMC MCP	• Soft tissue apposition between the thumb and index finger	Soft	• Adductor pollicis • Extensor pollicis longus • First dorsal interossei	No norms
Opposition	Frontal and sagittal planes and axes	CMC MCP			• Opponens pollicis • Flexor pollicis brevis • Abductor pollicis brevis • Flexor pollicis longus • Abductor pollicis longus	No norms
Reposition	Frontal and sagittal planes and axes	CMC MCP			• Extensor pollicis longus	No norms

Table 8-11

FREQUENCY OF USE OF PINCH PATTERNS

Pulp to pulp pinch (lateral)	20%
Three lateral pinch	20%
Five-finger pinch	15%
Fist grip	15%
Cylindrical grip	14%
Three-fingered pinch	10%
Spherical grip	4%
Hook grip	2%

Adapted from McPhee, S. D. (1987). Functional hand evaluations: A review. *American Journal of Occupational Therapy, 41*(3), 158-163.

be seen in picking up and holding a coin between the thumb and fingers or even with larger objects. From a radial view, this grip pattern forms an oval. The muscles involved in this pinch pattern are flexor pollicis longus, select interossei, and flexor digitorum superficialis of the involved fingers (flexor digitorum profundus in DIP flexion is also present). If the index and middle fingers meet the opposed thumb, this is called three-jaw chuck, three-point chuck, three-fingered or digital prehension, or the dynamic tripod, and can be considered a "precision grip with power" (Magee, 1992, p. 187). This type of grip pattern is seen when one holds a pencil.

Tip pinch is described as the movement of the tip of the thumb against the tip of another finger to pick up a small object, such as a pin. From the radial side, the finger and thumb form a circle. As with palmar pinch, the thumb is in opposition with slightly greater flexion, and the fingers are flexed at the MCP, DIP, and PIP joints. The same muscles are involved in tip pinch as are in palmar pinch. Tip pinch is also referred to as terminal opposition, and tip pinch is involved in activities requiring fine coordinated movement rather than power.

A study by Smith (1995) shows the frequency of use of the three types of prehension patterns for picking up and holding objects. Palmar pinch is used 50% of the time to pick up objects and 88% of the time to hold objects for use. Lateral pinch was the second most frequently used prehension pattern for picking up and holding objects with tip pinch used only 17% of the time to pick items up and 2% of the time to hold them for use. The frequency of pinch patterns used in daily activities such as buttoning, tying a shoelace, and tracing activities found by McPhee (1987) is shown in Table 8-11. From these studies, it could be stated that different types of pinch are used to pick up and hold objects and that daily activities did not significantly indicate one pinch preference over another.

Other researchers disagree with the distinction of power and prehension, contending that hand function is far more complex than these descriptions portray and that these words limit descriptions of the hand to static positions. Cassanova (1989) proposes a classification system based on the contact of the hand surfaces using anatomical terminology. The fingers would be designated by roman numerals (thumb = I, index = II, middle = III, ring = IV, and little = V), and contact areas would be noted as well (such as palmar, surface, or pad). Examples of the different pinch patterns are shown in Table 8-12.

INTERNAL KINETICS: MUSCLES OF THE HAND

Muscles of the hand are often divided into intrinsic (originating within the hand) and extrinsic (originating outside of the hand). Extrinsic muscles can be divided into those muscles that flex the digits (flexor digitorum superficialis, flexor digitorum profundus, and flexor pollicis longus); that extend the digits (extensor digitorum communis, extensor indicis, and extensor digiti minimi); and extensors of the thumb (extensor pollicis longus, extensor pollicis brevis, and abductor pollicis longus). The actions of the extrinsic muscles are influenced by the position of the wrist.

Intrinsic muscles can be organized into those muscles acting on the thenar eminence (abductor pollicis brevis, flexor pollicis brevis, and opponens pollicis); those influencing the hypothenar eminence (abductor digiti minimi, flexor digiti minimi, opponens digiti minimi, and palmaris longus); adductor pollicis; four lumbricals; and palmar and dorsal interossei.

An acronym for remembering the intrinsic muscles is "A of A of A," which stands for abductor pollicis brevis, opponens pollicis, flexor pollicis brevis (thenar muscles), adductor pollicis, opponens digiti minimi, flexor digiti minimi brevis, abductor digiti minimi (hypothenar muscles). Intrinsic muscles are

Table 8-12

EXAMPLES OF PINCH PATTERNS

LATERAL PINCH

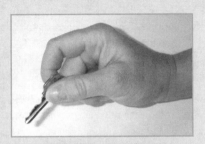 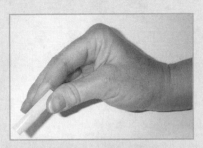

TIP PINCH

 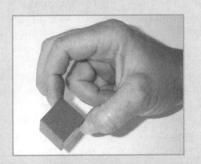

PALMAR PINCH

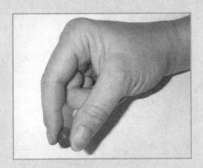 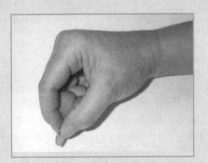

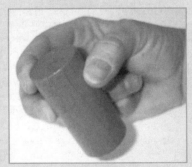 

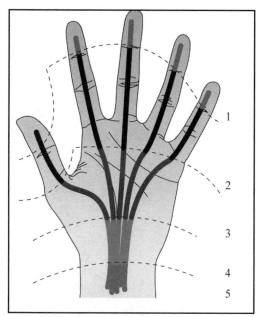

Figure 8-8. Flexor zones of the hand.

influenced by different positions of the metacarpophalangeal, proximal interphalangeal, and distal interphalangeal joints rather than changes in wrist position (Flinn & DeMott, 2010). Just remember to include the interossei and lumbricals with the acronym! For precision movements of the hand, cooperation between the intrinsic and extrinsic muscles is necessary.

Because the hand is capable of such diverse movements, there is much intricacy in the relationships of the many muscles involved. The finger muscles have effects on the wrist and actually have moment arms as long as those of the wrist (Greene & Roberts, 1999). The finger muscles would move both the wrist and joints of the fingers if they were not stabilized. For example, the finger flexors are multijoint muscles, and without stabilization at the wrist, these flexors would flex each of the joints they cross. By doing this, the fingers would be unable to fully flex due to active insufficiency of the flexors because they can't fully flex at each of the joints they cross. This inability to fully flex all of the joints crossed by the finger flexors is actually a combination of insufficient excursion and insufficient strength capability.

The insufficient strength production is due to poor length-tension relationships that limit the amount of force that can be produced. By flexing over all of the joints, the flexors are attempting to produce strength in a shortened position at the lower end of the length-tension curve.

Excursion refers to the distance that each tendon slides as the fingers move. Excursion takes place simultaneously in flexor and extensor tendons, and this is an important concept in determining muscle forces, fabricating splints, and in rehabilitation as well as in surgical procedures. Measurement of tendon excursion is in relation to angular motion. Nordin and Frankel (1989) state that "when a lever rotates around an axis of an angle, the distance moved by every point on the lever is proportional to its own distance from the axis" and that "every point on the lever moves through a distance equal to its own distance from the axis—its moment arm" (p. 285). Given this, the moment arms and the excursion are larger in muscles that

are more proximally located. For example, flexor superficialis has a longer tendon excursion than does flexor profundus. The excursion of the flexor tendons is larger than the extensors, and the excursion of the extrinsic muscles is generally larger than the intrinsic muscles.

When the finger flexors attempt to flex fully over all of the joints they cross, there is insufficient excursion of the flexor muscles. This insufficiency explains tenodesis grasp where passive tension yields finger extension and forces the fingers to release an object. Wrist extension stretches the flexor digitorum profundus, producing finger flexion. Passive tension, not contraction, produces the finger movement. The same is true for the passive insufficiency that develops when the wrist is flexed and passive tension develops to produce finger extension.

There also exists a relationship between adduction and abduction with flexion and extension. In abduction and adduction with the MCP joints extended, the movements are free because the collateral ligaments are loose. When there is flexion at the MCP joint, the fingers automatically adduct, which limits the range of abduction because the collateral ligaments become tight. There is a natural tendency to abduct the fingers when they are also extended. Another combined action that occurs is that, when the fingers flex, the hand is cupped, and the hand is flattened when the fingers extend.

Synergistic relationships exist between the muscles of the wrist and the finger muscles. For example, when the little finger is abducted by means of abductor digiti minimi, the flexor carpi ulnaris contracts to provide counter-traction on the pisiform. In order to prevent flexor carpi ulnaris from abducting the wrist in an ulnar direction, abductor pollicis longus contracts.

Synergistic actions occur in the thumb as well. When the thumb is moved in a palmar direction as in flexion, palmaris longus contributes to the movement by tensing the fascia of the palm, while extensor carpi radialis brevis contracts to prevent palmaris longus from flexing the wrist. Another example is the thumb extension. The extensor carpi ulnaris contracts to prevent radial deviation of the wrist by the abductor pollicis longus muscle.

Flexors of the Fingers

On the volar surface of the hand, there are five flexor tendon zones as shown in Figure 8-8. The first zone is distal to the insertion of the flexor digitorum superficialis tendon located halfway between the proximal interphalangeal and distal interphalangeal crease. This zone is not affected by the excursion of the flexor tendons once they enter the hand (Flinn & DeMott, 2010). The second zone is between the A1 pulley and the insertion of the flexor digitorum superficialis tendon near the distal palmar crease to the middle of the third phalanx. This zone includes the flexor pollicis longus and the median nerve. If there is pain with passive extension of the digit, this may suggest flexor tendon sheath infection (Flinn & DeMott, 2010). In zones one and two, the tendons are protected in sheaths, and the annular and cruciate pulleys provide smooth gliding surfaces. Zone three is between the flexor retinaculum and the A1 pulley. In this zone, the lumbrical muscle originates from the flexor digitorum profundus, and the lubricating sheath does not protect the flexor tendons to the fingers in this zone, making them prone to adhesions when injured (Flinn & DeMott, 2010). Within the carpal tunnel, zone four is located. Zone five is proximal to the flexor retinaculum. An area encompassing much of the length of the flexor tendons is known as "no man's land" due to poor healing of tendons, precarious blood supply, and easy rupture of joint structures (Oatis, 2004).

The anterior compartment of the forearm contains the flexor-pronator group of muscles, which arise from a common flexor attachment on the medial epicondyle of the humerus. There are eight muscles in this anterior compartment, and three of these flex the digits: flexor digitorum profundus, flexor digitorum superficialis (sublimis), and flexor pollicis longus. Flexor pollicis longus is located deep and lateral to flexor digitorum profundus and is the sole flexor at the interphalangeal joint of the thumb.

The flexor digitorum superficialis is deep to the three wrist flexors and pronator teres. The flexor digitorum superficialis tendon splits into four tendons and passes under the flexor retinaculum in the palm of the hand. At the base of the first phalange, each tendon divides into two slips to allow passage of the flexor digitorum profundus tendon. Three quarters of the superficialis fibers continue to the middle phalanx to insert, and one quarter of the fibers cross under flexor digitorum profundus to insert into the proximal phalange. While flexor digitorum superficialis primarily flexes the proximal interphalangeal joint, it can also flexion all of the joints it crosses.

The flexor digitorum profundus is deep to flexor digitorum superficialis. Once in the hand, the tendon inserts into the base of the distal phalanges of the digits after passing through the openings in the flexor digitorum superficialis tendons just opposite to the first phalange. Flexor digitorum profundus is the only flexor of the distal interphalangeal joint, but can assist in flexion of every joint it crosses. Flexor digitorum profundus of the index finger is relatively independent of the other profundus tendons. An example of the interconnectedness of the other finger flexor digitorum profundus tendons can be seen when all joints are extended and you attempt to actively flex only the distal interphalangeal joint of the ring finger. Isolated distal interphalangeal flexion is difficult due to excessive elongation of the muscle belly of the flexor digitorum profundus due to extension of the middle finger (Neumann, 2002, p. 215).

Distal to the carpal tunnel is the ulnar synovial sheath, which surrounds flexor digitorum superficialis and flexor digitorum profundus. The radial synovial sheath is in contact with the flexor pollicis longus. The flexor tendons are restrained by tendon sheaths and retinaculum that keep the tendons close to the bones and to the planes of motion, while maintaining a relatively constant moment arm rather than producing "bowstringing" of the tendons. There are five dense annular pulleys and three thinner cruciform pulleys that allow for smooth curves and no sharp bends in the tendon excursions (Neumann, 2002; Nordin & Frankel, 1989).

Flexion of the DIP joint produces flexion of the PIP joint because flexion of the DIP is produced by flexor digitorum profundus with a simultaneous flexor force produced by the muscle as it crosses the PIP. Levangie and Norkin (2005) add that when the DIP joint begins flexing, the terminal tendon

and lateral bands stretch over the dorsal aspect of DIP, which pulls the extensor hood (from which the lateral bands arise) distally. This causes the central tendon of the extensor expansion to relax, creating a flexor force at the PIP joint. Active (FDP) and passive forces (release of the central tendon) occur when the lateral bands migrate volarly. The coupling of the flexor action at the DIP and PIP joint can be overridden, as some people can actively flex the DIP while the PIP extends. To achieve full flexion of all joints, the long finger flexors must override the extension components of the lumbricals and interossei, which is easier if in some wrist extension (Hamill & Knutzen, 1995) (Figure 8-9).

Intrinsic finger flexors include flexor digiti minimi brevis, abductor digiti minimi, opponens digiti minimi, palmar and dorsal interossei, and the lumbricals.

Flexor Digitorum Superficialis

The largest of the flexor muscles of the forearm, the flexor digitorum superficialis muscle crosses the wrist, metacarpophalangeal, and proximal interphalangeal joints anterior to the flexion-extension axis, so this muscle flexes these joints and assists with extension at the wrist. Flexor digitorum superficialis is capable of flexing each finger individually at the proximal, but not distal, interphalangeal joints.

Palpation

In its location underneath flexor carpi radialis and palmaris longus, it is difficult to palpate at the more proximal location because the attachment is widespread. Movement of the tendon can be seen and palpated on the palmar surface of the wrist in the space between the flexor carpi radialis and flexor carpi ulnaris tendons as the fingers flex to make a firm fist with the wrist flexed (Gench et al., 1995; Smith et al., 1996).

Flexor Digitorum Profundus

Due to the line of pull to the palmar side of the flexion-extension axis, the muscle acts in flexion of proximal interphalangeal, distal interphalangeal, metacarpophalangeal, and finally at the wrist with the strength of the contraction lessening progressively. This is because the muscle gradually shortens and can therefore contribute only weakly to wrist flexion. There is a single muscle belly, so contraction of the muscle causes movements in all fingers simultaneously. As an example, flex the long finger; notice that the ring finger also moves as would the index and small fingers if not for the neutralizing effects of extensor indicis and extensor digiti minimi. Weakness in the flexor digitorum profundus muscle would result in decreased strength of distal interphalangeal flexion, which would affect grasp and pinch.

Palpation

Flexor digitorum profundus lies deep and is covered by flexor digitorum superficialis, flexor carpi radialis, flexor carpi ulnaris, palmaris longus, and pronator teres. A contracting muscle belly can be palpated if tension is minimal in the more superficial muscles. Have the subject supinate the forearm and rest the hand in the lap while the wrist is extended by weight of the hand. Have the subject close the fist fully, and the profundus may be felt under the fingers in the region between pronator teres and FCU about 2 inches (5 cm) below the medial epicondyle of the humerus (Gench et al., 1995; Smith et al., 1996).

Figure 8-9. Extrinsic flexors of the hand.

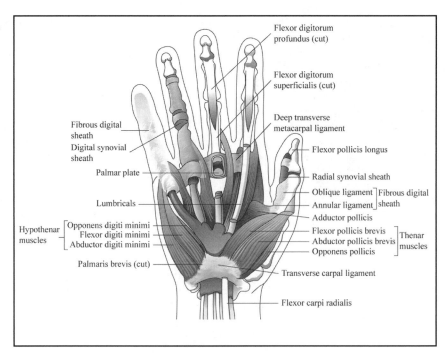

Flexor Digitorum Superficialis and Flexor Digitorum Profundus

FDP is more active than FDS in finger flexion. FDS works alone in finger flexion only when flexion of the DIP joint is not required. When simultaneous PIP and DIP flexion is required, FDS acts as reserve, joining FDP when greater force is needed or when finger flexion with wrist flexion is required (Smith et al., 1996).

FDP and FDS are dependent on wrist position for an optimal length-tension relationship. Finger flexor effectiveness would be reduced by 25% without the counterbalancing effect of extensor muscles (extensor carpi radialis brevis or extensor digitorum), because wrist flexion would occur, reducing tension at the more distal joints. The wrist extensors also serve to stabilize the wrist as well as provide optimal length-tension relationships for these long finger flexors.

Flexor Digiti Minimi Brevis

This muscle lies parallel to and on the radial side of abductor digiti minimi. While superficially located, it is easily confused with abductor digiti minimi. Crossing on the palmar side of the flexion/extension axis, flexor digiti minimi brevis (FDMB) flexes the fifth MCP joint.

Opponens Digiti Minimi/Opponens Digiti Quinti

The proximal fibers of opponens digiti minimi (ODM) flex the fifth CMC joint while the distal fibers adduct it. This combination of flexion and adduction produces opposition if both contract simultaneously, actually drawing the fifth metacarpal forward, which deepens the hollow of the hand (Gench et al., 1995; Smith et al., 1996).

Palpation

This muscle lies beneath abductor digiti minimi and flexor digiti minimi brevis in hypothenar eminence, so it is not palpable.

Palmar Interossei

Although somewhat farther from the flexion/extension axis than are the dorsal interossei, the palmar interossei are more effective as flexors. Both the lumbricals and palmar interossei are located on the palmar side of the axis for flexion and extension, and so mechanically are capable of flexion, with the lumbricals more favorably located. The role of the interossei in flexion and adduction at the metacarpophalangeal joint is thought to be from passive stretch (Smith et al., 1996).

Palpation

The three palmar interossei are located deep in the palm beneath lumbricals and dorsal interossei, so these are very difficult muscles to palpate (Gench et al., 1995; Smith et al., 1996).

Dorsal Interossei

These four muscles lie between the metacarpals but are very difficult to palpate except for the first dorsal interossei. The dorsal interossei serve to flex the metacarpophalangeal joints of the second, third, and fourth digits (Gench et al., 1995).

Lumbricals

The four lumbricals are not active in metacarpophalangeal flexion unless the interphalangeal joints are extended. The lumbricals do not participate in grip and rarely contract synchronously with the flexor digitorum profundus (Smith et al., 1996).

Palpation

While one source says that the lumbricals are too deep to palpate, others state that these muscles may be palpated on the radial side of the long finger flexors and are best visible in the claw hand position of MCP hyperextension and IP flexion (Smith et al., 1996).

Abductor Digiti Minimi

Located superficially on the ulnar border of the hypothenar eminence, this muscle is closer to the flexion/extension axis than to the abduction/adduction axis, so abductor digiti minimi (ADM) is primarily an abductor with a secondary role as a flexor of the proximal interphalangeal joint when the long extensor is relaxed.

Palpation

ADM can be palpated next to the fifth metacarpal during resisted abduction of the small finger (Gench et al., 1995).

Flexors of the Thumb

Flexors of the thumb include flexor pollicis longus, flexor pollicis brevis, abductor pollicis brevis, adductor pollicis, and dorsal interossei.

Flexor Pollicis Longus

The most radial of the tendons of the carpal tunnel, flexor pollicis longus (FPL) crosses the interphalangeal and metacarpophalangeal joints of the thumb to the palmar side of the flexion-extension axis and acts as a flexor of the interphalangeal joint. FPL also crosses to the ulnar side of the CMC axis for flexion-extension, so this muscle also flexes the CMC joint but is not credited with wrist motion because the muscle lies too close to both wrist axes (Gench et al., 1995).

Flexor Pollicis Brevis

Flexor pollicis brevis is a strap-like muscle with long parallel fibers with two heads. The deep head crosses only the metacarpophalangeal joint on the ulnar side and serves to adduct the metacarpophalangeal joint; the superficial head crosses the metacarpophalangeal and CMC joints. At the CMC joint, flexor pollicis brevis crosses on the palmar side of the flexion-extension axis and is a flexor of the metacarpophalangeal and CMC joints. There is not always a clear distinction between flexor pollicis brevis and opponens pollicis because the muscles become continuous on the ulnar side and both muscles flex the metacarpals on the carpus. Interestingly, the superficial portion is innervated by the median nerve, whereas the deep portion is innervated by the ulnar nerve.

Abductor Pollicis Brevis

Abductor pollicis brevis (APB) lies to the palmar side of the flexion-extension axis and assists with flexion and abduction of the metacarpophalangeal joint of the thumb (Gench et al., 1995). Weakness of the APB occurs with medial nerve palsy, and there can be atrophy of the muscle belly, resulting in flattening of the thenar eminence (Oatis, 2004).

Palpation

This is the most superficial muscle of the thenar eminence, so it can be palpated in the center of the eminence during resisted abduction of the CMC joint of the thumb.

Adductor Pollicis

Called the "pinching muscle" by Brand and Hollister (1999), adductor pollicis (AP) is a fan-shaped penniform muscle favorably located on the palmar side of the flexion/extension axis to flex the first CMC joint (Gench et al., 1995).

Palpation

Even though the muscle is deep in the palm, AP can be palpated between the first and second metacarpal on the palmar surface of the thumb between the metacarpophalangeal and interphalangeal joints as the thumb presses against the tip of the index finger. Or, you can ask the subject to hold a piece of paper between the thumb and radial aspect of the proximal phalanx of the index finger (Smith et al., 1996; Gench et al., 1995).

Opponens Pollicis

Triangular in shape, the upper portion flexes the CMC joint while the lower portion abducts so the muscle acts to oppose the CMC of the thumb. Opponens pollicis usually works with abductor pollicis brevis and flexor pollicis brevis. When the injury is to the opponens pollicis muscle, the hand has difficulty positioning and stabilizing the CMC joint of the thumb during grasp and pinch (Oatis, 2004).

Palpation

While opponens pollicis is located beneath abductor pollicis brevis, it is superficial along the radial border of the thenar eminence next to the first metacarpal and can be palpated when the thumb is pressed firmly against the tip of the long finger (Gench et al., 1995).

Dorsal Interossei

While normally considered a muscle acting on the index finger, the first dorsal interossei is important in stabilization of the thumb CMC joint. The first dorsal interossei pulls ulnarly and distally against the forces of the adductor-flexors that push the metacarpal base dorsally and radially (Brand & Hollister, 1999). Weakness in the dorsal interossei can lead to a claw hand deformity and loss of pinch and grasp (Oatis, 2004).

Palpation

The first dorsal interossei can be palpated in the space between the metacarpal bones of the thumb and index finger when resistance is applied to abduction of the index finger (Gench et al., 1995).

Extensors of the Fingers

The finger flexors follow a well-defined sheath toward a discrete attachment with pulleys to provide smooth movement. The finger extensor tendons do not have a defined sheath or pulley system. There are six separate compartments that describe the extensor tendons on the dorsum of the hand and five that identify the tendons of the thumb shown in Figure 8-10. The first compartment contains the abductor pollicis longus and extensor pollicis brevis, and inflammation in this compartment can result in de Quervain's syndrome. The second compartment contains extensors carpi radialis longus and brevis; the third compartment contains extensor pollicis longus; the extensor digitorum indicis and extensor digitorum communis are within the fourth compartment; extensor digiti minimi is in the fifth; and extensor carpi ulnaris is in the sixth. If there is limited excursion of the extensor carpi ulnaris tendon in the sixth compartment, this might suggest possible radioulnar joint dysfunction (Flinn & DeMott, 2010).

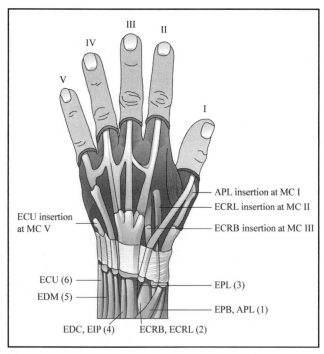

ECU insertion
at MC V

APL insertion at MC I
ECRL insertion at MC II
ECRB insertion at MC III

ECU (6)
EDM (5)

EPL (3)

EPB, APL (1)

EDC, EIP (4)   ECRB, ECRL (2)

Figure 8-10. Extensor compartments.

Tendons of extensor digitorum, extensor indicus, and extensor digiti minimi cross the wrist in synovial-lined sheaths in the extensor retinaculum. Distal to the extensor retinaculum, the extensor digitorum is often interconnected by several juncturae tendinae that stabilize the tendons. Extensor indicis and extensor digitorum usually travel in parallel.

Instead of having well-defined sheaths as the flexor tendons, the extensor tendons continue as a fibrous expansion along the length of each finger, and this is the extensor apparatus (also called the extensor hood, extensor expansion, extensor mechanism, dorsal aponeurosis, or dorsal finger apparatus). This moveable hood is in motion during flexion and extension. It begins on the dorsal, medial, and lateral sides of the proximal phalange of each finger and attaches to the dorsal side of the middle and distal phalanges. The extensor mechanism serves as a primarily distal attachment for the extensor digitorum (communis) muscle and several intrinsic hand muscles (lumbricals, palmar interossei, dorsal interossei, and abductor digiti minimi). There is a similar mechanism for the thumb, which is formed by the distal tendon of the extensor pollicis longus muscle.

The extensor mechanism is made up of the extensor digitorum, connective tissue, and expansion fibers from the dorsal interossei, volar interossei, and lumbricals (Figure 8-11). Extensor digitorum tendons pass through the first dorsal compartment with extensor indicis.

At the proximal phalanx, each tendon divides into three components. One portion, the central slip, inserts into the dorsum of the proximal end of the middle phalange. This central slip acts to extend the proximal phalange and to stabilize the proximal finger joint so that the lumbricals and interossei can extend distal interphalangeal and proximal interphalangeal joints and laterally move the digit. Two lateral bands pass on either side of the central tendon, cross the proximal

phalange, reunite as a single terminal tendon on the distal phalange, then unite with the lumbricals and interossei. Fascia extends laterally from the extensor tendon that forms a hood that encircles the intrinsics (interossei and lumbricals).

If only the proximal interphalangeal joint is flexed, the entire trifurcated extensor assembly is pulled distally following the central slip, which is taut while a distal pull occurs at the middle phalanx. The lateral bands are slack. This creates a tension force at the proximal interphalangeal joint, so flexion of this joint is unavoidable. If the distal interphalangeal is actively flexed, the entire assembly is displaced distally as the central slip relaxes, creating increased tension on the retinacular ligaments. This creates a flexion force at the proximal interphalangeal joint that is important in grasp and pinch (Nordin & Frankel, 2001). In addition, simultaneous flexion of metacarpophalangeal, proximal interphalangeal, and distal interphalangeal joints forces extensor digitorum to stretch to its fullest. Full flexion of the wrist can only occur if fingers are allowed to uncurl.

As extensor digitorum passes dorsally to the metacarpophalangeal joint, the contraction of the muscle creates tension on the extensor hood, which is pulled proximally over this joint and acts to extend the proximal phalange. The intrinsics pass volar/anterior to the metacarpophalangeal joint axis, so a flexor force is created. When all muscles contract simultaneously, the metacarpophalangeal joint will generally extend because extensor digitorum generates greater torque (Levangie & Norkin, 2005).

Extensor digitorum, the interossei, and the lumbricals produce extensor forces on the PIP joint due to their attachments on the extensor hood. Extensor digitorum alone cannot produce sufficient tension in either the central slip or lateral bands to overcome the passive forces of flexor digitorum profundus and superficialis. When the extensor digitorum contracts, the forces are distributed along all three phalanges of each finger. If extensor digitorum contracts unopposed, a claw hand position will result (metacarpophalangeal hyperextension and interphalangeal flexion) due to the passive pull of the extrinsic flexor tendons (Levangie & Norkin, 2005). The intrinsics (interossei and lumbricals) act as moderators between flexion and extension forces, and the lumbricals are said to have a counterclawing bias (Nordin & Frankel, 2001).

The influence of the extensor mechanism on distal interphalangeal and proximal interphalangeal extension is interdependent; that is, when the proximal interphalangeal joint is actively extended, the distal interphalangeal joint also extends because of the combined active and passive forces applied to the lateral bands and the terminal tendon (Norkin & Levangie, 1992). Extension of the interphalangeal joints can occur only if there is tension at the extensor digitorum because this forms a firm base for the interossei and lumbricals to execute their pull (Gench et al., 1995).

The extensors of the fingers include extensor digitorum, extensor digiti minimi, extensor indicis, lumbricals, and interossei as shown in Figure 8-12.

Extensor Digitorum/Extensor Digitorum Communis

A fusiform muscle on the dorsal surface of the forearm, extensor digitorum extends the MCP and proximal and distal IP joints of the four fingers and, if contraction continues, can

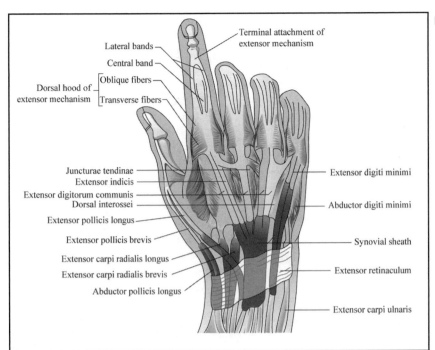

Figure 8-11. Extensor mechanism.

Labels on figure:
Lateral bands
Central band
Oblique fibers
Transverse fibers
Dorsal hood of extensor mechanism
Terminal attachment of extensor mechanism
Juncturae tendinae
Extensor indicis
Extensor digitorum communis
Dorsal interossei
Extensor pollicis longus
Extensor pollicis brevis
Extensor carpi radialis longus
Extensor carpi radialis brevis
Abductor pollicis longus
Extensor digiti minimi
Abductor digiti minimi
Synovial sheath
Extensor retinaculum
Extensor carpi ulnaris

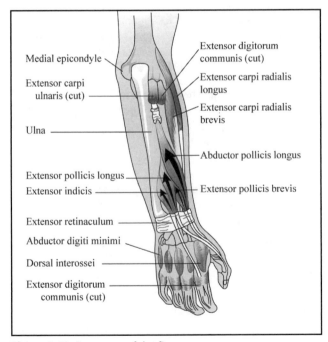

Labels on figure:
Medial epicondyle
Extensor carpi ulnaris (cut)
Ulna
Extensor pollicis longus
Extensor indicis
Extensor retinaculum
Abductor digiti minimi
Dorsal interossei
Extensor digitorum communis (cut)
Extensor digitorum communis (cut)
Extensor carpi radialis longus
Extensor carpi radialis brevis
Abductor pollicis longus
Extensor pollicis brevis

Figure 8-12. Extensors of the fingers.

extend the wrist. The extension of the IP joints occurs due to the attachment on the extensor hood. Several different sources cite how the function of the extensor digitorum can be used in self-defense maneuvers. Because it is not possible to simultaneously and fully flex the wrist and the interphalangeal and metacarpophalangeal joints, by forcing the wrist into flexion, a villain will be forced to loosen the grip on any weapon (Gench et al., 1995; Greene & Roberts, 1999; Norkin & Levangie, 1992; Smith et al., 1996).

Trauma (such as tendon lacerations) to extensor digitorum may result in weakness or loss of extension of the metacar-

pophalangeal joints of the fingers. If the extensor digitorum is tight, this would limit full flexion range of motion of the fingers, and extension of the metacarpophalangeal joint would be accompanied by flexion of the interphalangeal joint due to the pull of the extensor digitorum and antagonist pull of flexor digitorum profundus (Oatis, 2004).

Palpation

Extensor digitorum can be palpated except where it is covered by extensor carpi radialis longus. The four tendons can be easily seen and palpated as they cross the second through fifth metacarpal joints and are especially prominent when the MCP joint is fully extended (Gench et al., 1995) or when one finger is extended while the others are flexed into the palm (Smith et al., 1996).

Interossei

The dorsal and palmar interossei extend the IP joints of the index, long, and ring fingers by their attachments to the extensor hood.

Lumbricals

The lumbricals extend the IP joints of the index, long, and ring fingers by their attachments to the extensor hood. The lumbricals have better leverage for extension than do the interossei (Smith et al., 1996). Some authors indicate that the function of the lumbricals is to pull the flexor digitorum profundus tendon distally to decrease passive tension and therefore facilitate extension by the extensor digitorum. Other researchers indicate that the prevention of hyperextension of the MCP joint by extensor digitorum contractions is an additional function of these muscles (Nordin & Frankel, 1989; Norkin & Levangie, 1992; Smith et al., 1996). Of special note, the lumbricals have a high rate of variability with a low number of muscle fibers per motor unit, indicative of a skilled muscle with a high number of spindles. The lumbricals are richly innervated, and it has been hypothesized that the lumbricals may have a specialized proprioceptive function.

Smith and colleagues (1996) indicate that the line of pull of interossei and lumbricals is dorsal to the joint centers of the interphalangeal joints, so these muscles are mechanically capable of extension. While the intrinsics are called the primary extensors of interphalangeal joints, this is not always true. In unresisted extension, the long extensor and lumbricals only are active. The interossei are not active unless there is forceful or resisted extension (Smith et al., 1996).

Extensor Digiti Minimi/Extensor Digiti Quinti Proprius

A slender muscle, extensor digiti minimi is located to the ulnar side of the extensor digitorum and acts to extend and abduct the interphalangeal joints of the little finger due to the attachment in the extensor hood. Extensor digiti minimi also extends the fifth metacarpophalangeal joint and, with continued contractions, contributes to wrist extension. Weakness or injury to EDM would result in the inability to extend the little finger of the metacarpophalangeal joint independently (Oatis, 2004).

Palpation

Extensor digiti minimi is superficially located slightly proximal to the wrist and can be palpated and observed on the dorsum of the hand at the fifth metacarpal when the small finger is extended against resistance (Gench et al., 1995).

Extensor Indicus/Extensor Indicis Proprius

Extensor indicis extends the metacarpophalangeal joint of the index finger and, with the connection to the extensor hood, extends the proximal interphalangeal and distal interphalangeal joints. With continued contraction, the extensor indicis can contribute to extension of the wrist. If the metacarpophalangeal joint is held in flexion, the ability to extend the interphalangeal joint and the wrist is improved. Because extensor indicis inserts on the extensor digitorum tendons of the first and fourth fingers, this adds independence of action to these fingers rather than strength or additional actions (Levangie & Norkin, 2005).

Palpation

This muscle is superficial and runs on the ulnar side and parallel to the extensor digitorum on the dorsal aspect. It can be seen when one extends the index finger with the other fingers flexed into the palm (Gench et al., 1995; Smith et al., 1996).

Extension of the Thumb

Abductor pollicis longus, extensor pollicis longus, and extensor pollicis brevis extend the thumb.

Abductor Pollicis Longus

Because abductor pollicis longus inserts on the radial side of the metacarpophalangeal joint, it acts to pull the thumb into extension at the CMC joint in addition to its actions of abduction of the CMC and radial deviation and flexion of the wrist. The thumb cannot function without the abductor pollicis longus muscle, as is seen in de Quervain's disease (Brand & Hollister, 1999).

Extensor Pollicis Longus

Extensor pollicis longus (EPL) has its own compartment in the retinaculum on the dorsum of the wrist and is located dorsal to the flexion/extension axis, so it extends both the interphalangeal and metacarpophalangeal joints of the thumb. Weakness in the EPL muscle would result in weak extension at the interphalangeal joint of the thumb. Tightness is often seen with the flexor pollicis longus, resulting in weakness of the intrinsic muscles of the thumb (Oatis, 2004).

Palpation

While muscle belly is difficult to palpate, the tendon can be palpated when the thumb is fully extended (Gench et al., 1995).

Extensor Pollicis Brevis

This muscle extends the metacarpophalangeal and CMC joints of the thumb. At the wrist, the tendon passes posterior to the flexion/extension axis and to the radial side of the deviation axis, so extensor pollicis brevis (EPB) extends and radially deviates the wrist. At the CMC joint, it passes on the radial side of the flexion/extension axis so it extends this joint. At the metacarpophalangeal and interphalangeal joints, EPB is dorsal to flexion/extension axis, so it extends these joints. EPB is closely associated with abductor pollicis brevis with similar actions at the wrist and CMC joints, and its primary action is extension of the metacarpophalangeal joint. Weakness in the EPB would result in decreased metacarpophalangeal and CMC extension of the thumb (Oatis, 2004).

Gench and colleagues (1995) indicate that EPB is different from EPL in three ways (Burstein & Wright, 1994):

1. The tendon of EPL crosses the wrist posterior to the flexion/extension axis to become a wrist extensor; the tendon of EPB crosses directly over the axis and is ineffective as an adductor or abductor.

2. The tendon of EPL passes dorsal to the axis of adduction/abduction of the CMC joint to act as an adductor; the tendon of EPB crosses directly over and is ineffective as an abductor or adductor of the CMC joint.

3. EPB does not cross the IP joint of the thumb as does extensor pollicis longus, so EPB has no action at that joint.

Palpation

Beneath and to the radial side of EPL, the EPB tendon is superficial as it crosses the wrist and can be seen during forced extension of the thumb. EPB and EPL form the medial and lateral borders of the anatomical snuff box (Gench et al., 1995).

Abductors of the Fingers

Abductor Digiti Minimi

This muscle abducts and flexes the MCP joint of the little finger. The muscle's line of pull is more favorably placed for abduction than for flexion because muscle is further from the adduction/abduction axis than from the flexion/extension axis, so this is primarily an abductor with secondary flexion action (Gench et al., 1995).

Dorsal Interossei

The dorsal interossei have relatively long force arms so they act to abduct the second through fourth MCP joints in addition to radial and ulnar deviation of the third metacarpal joint (Gench et al., 1995).

Extensor Digiti Minimi/Extensor Digiti Quinti Proprius

This muscle extends the metacarpophalangeal and the interphalangeal joints as well as abducts the little finger.

Abductors of the Thumb

Abductor Pollicis Brevis

While small and weak in size and tension, this muscle is considered by Brand & Hollister (1999) to be important for opposition and therefore in grasp and pinch. Opponens pollicis is located underneath abductor pollicis brevis so when opponens pollicis contracts, it pushes APB farther from the axis of the CMC joint and increases the moment arm and its effectiveness (Brand & Hollister, 1999). Because this is a fan-shaped muscle, different fibers contribute differently to the muscle action; the most radial portions are abductors of the CMC joint; the most distal are adjacent to FPB and so are flexor-abductors of the MCP joint. The strength of the abductor pollicis brevis is relatively weak because the action of abduction of the thumb is not one usually done against resistance, and this demonstrates that the primary function of this muscle is to position the thumb for action rather than performing the action itself. In action, it is the EPL muscle that most directly opposes APB, not the adductors.

Abductor Pollicis Longus

A stout muscle, abductor pollicis longus abducts and extends the MCP joint and stabilizes the first metacarpal joint. Because the muscle spirals from the dorsal radius to the lateral aspect of the first metacarpals, fibers are variable in length, and moment arms vary. The name of this muscle doesn't reflect the true action. The tendon pulls the thumb laterally or radially, which is abduction in body terms but not in terms of the thumb. The muscle pulls on the back or extends the thumb so it extends the thumb and abducts it. Stronger than FPL, abductor pollicis longus works to oppose the adductor and short flexor at the CMC joint and allows them to flex the MCP joint (Brand & Hollister, 1999).

Opponens Pollicis

The lower portion of this muscle abducts the CMC joint of the thumb. With the upper portion, which flexes the CMC, this muscle is capable of opposition. There is considerable variation in fiber length, and it is capable of producing greater tension than abductor pollicis brevis. The action of opponens pollicis is swinging the thumb in an arch toward the fingers (Gench et al., 1995; Hinkle, 1997; Jenkins, 1998; Nordin & Frankel, 1989; Riley, 1998).

Adductors of the Fingers

Adduction of the fingers is achieved by contraction of the opponens digiti minimi, extensor indicis, and the palmar interossei.

Opponens Digiti Minimi/Opponens Digiti Quinti

While the proximal fibers flex the CMC joint of the fifth joint, the distal fibers adduct. By drawing the fifth metacarpal forward, this deepens the hollow of the hand (Gench et al., 1995; Hinkle, 1997; Jenkins, 1998; Riley, 1998; Nordin & Frankel, 2001).

Extensor Indicis

Extensor indicis extends and adducts the metacarpophalangeal joint of the index finger, but because it passes just to the ulnar side of the axis, it is weak in adductor action. Weakness in the EI would result in difficulty with independent movement of the index finger and some weakness in extension of the metacarpophalangeal joint. Tightness in the EI alone is unlikely, but with limitations in extensor digitorum would contribute to hyperextension of the metacarpophalangeal joint (Oatis, 2004).

Palmar Interossei

Given the medial location to the adduction/abduction axis, the palmar interossei adduct the metacarpophalangeal joints of the index, ring, and little fingers (Gench et al., 1995). If the palmar interossei become tight, there would be weakness in finger adduction, and this would contribute to weakness of metacarpophalangeal flexion with interphalangeal extension. Without these movements, there would be limited grasp and pinch capacity, and the claw hand deformity may result (Oatis, 2004).

Adductors of the Thumb

Adductors of the thumb include flexor pollicis brevis, adductor pollicis, extensor pollicis longus, and flexor pollicis longus.

Flexor Pollicis Brevis

The deep head of flexor pollicis brevis crosses only the metacarpophalangeal joint and is on the ulnar side of the abduction/adduction axis, so it adducts the metacarpophalangeal joint. Injury to the FPB muscle would result in decrease flexion at the CMC and metacarpophalangeal joint of the thumb with resultant decreased pinch (Oatis, 2004).

Adductor Pollicis

Adductor pollicis (AP) slides the thumb across the palm and bases of fingers toward the ulnar side as it adducts the CMC joint of the thumb. This muscle is most effective when the joint is fully abducted, which pulls the metacarpal closer to the palm. When the thumb is even with the palm, AP is not effective because the position aligns the muscle with the adduction/abduction axis (Gench et al., 1995). Weakness of the AP muscle would result in weakness in flexion and adduction of the CMC joint and limited metacarpophalangeal thumb flexion. If this muscle becomes tight, the result would be limited abduction and extension of the CMC joint and limited extension range of motion of the metacarpophalangeal joint, preventing movement of the thumb away from the palm (Oatis, 2004).

Extensor Pollicis Longus

Not only an extensor at the CMC and metacarpophalangeal joint, extensor pollicis longus (EPL) also acts an adductor at these joints. It is an adductor at the CMC joint because it crosses to the dorsal side of the abduction/adduction axis (Gench et al., 1995; Bellace, Healy, Bess, Bryon, & Hohman, 2000; Jenkins, 1998).

Flexor Pollicis Longus

While not the primary action, with continued contraction, flexor pollicis longus can adduct the metacarpal joint (Goss, 1976).

HAND ASSESSMENT

The hand is the terminal part of the upper limb and is essential for functional activities requiring both strength and precision. The loss of hand function accounts for 90% of the loss associated with upper extremity function (Hume, Gellman, McKellop, & Brunfield, 1990). A majority of decline in occupational performance is due to disuse (Fiatorone & Evans, 1993; Rider, 2004).

The thumb is the most important digit functionally due to the relationship with the other digits in force production and mobility. With the loss of the movement of the thumb, 40% to 50% of hand function is lost (Hume et al., 1990). The index finger coordinates precise movements with the thumb, and loss of index finger function results in decreased pinch and power grips. The middle finger provides a stable base for the hand and is vital for power and prehension. The little finger has the least functional role but, due to its peripheral position, enables the hand to mould around objects and hold items against the hypothenar eminence. The loss of function of the index, middle, ring, and little fingers has much less impact on the functional use of the hand.

Occupational Profile

Kimmerle, Mainwaring, and Borenstein (2003) state that "the role of the occupational therapist or the hand therapist who assesses hand function is to delineate functional abilities, limitations, and activities within a meaningful and purposeful context" (p. 490). A clear understanding of how the client uses the hands in everyday activities is an essential part of the assessment of the hand. Inquiries about the client's history should reflect the needs and goals of the client in the tasks and activities associated with his or her desired occupations. Because the hand is so important to everyday activities and to upper extremity function, a clear understanding of the use of the hands is crucial.

Consideration of the psychosocial adaptation as well as physical recovery from hand injuries is important in engagement of occupations, relationships, and outcomes (Chan & Spencer, 2004). Chan and Spencer conducted a qualitative study of clients with acute hand injuries, and their findings support "the value of individualized, occupation-based therapy that addresses the mind and spirit as well as physical recovery in occupational therapy practice with hand injuries" (p. 128).

Understand the meaning that the hand injury has to the individual to gain the client's perspective of the illness experience (Cooper, 2003).

In addition, specific attention should be paid to the client's age, because some conditions are more common (such as arthritis) after 40 years of age (Magee, 2008). Arthritis and cumulative trauma disorders may be more prevalent after years of trauma. There are physiologic changes in the elderly that may influence the ability to heal after a hand injury, such as decreased blood flow, decreased cellular activity, dryness, loss of skin elasticity, and changes in muscle (Rider, 2004).

Discuss how the injury or dysfunction occurred, which can lead to an understanding of underlying pathology and guide intervention. How long and under what conditions the symptoms occur and the preferred or dominant hand for activities are also important pieces of information to know and can be gathered by careful interview of the client with attention to symptoms and activity preferences.

The Functional Repertoire of the Hand Model (Kimmerle et al., 2003) can be used when assessing the hand. In this model, four key components are identified. First, identification of personal constraints includes both physical and psychological limitations to hand function. The second component is that of hand roles and includes hand preference and bimanual use of the hands. Object-related hand actions is the third component and consists of reaching, grasping, and manipulating objects. Finally, inclusion of task parameters takes into account characteristics of the objects, movement patterns, and performance demands involved in tasks using the hands. A model like this can provide a consistent language and comprehensive assessment of the hand.

Observation and Palpation

The hand is in a slightly cupped position, and the wrist is positioned to provide an optimal length for the finger flexors (Neumann, 2002). The position of function of the hand illustrates the interaction of the wrist and hand motion. The wrist extensors are synergistic to the finger flexion (and the wrist flexors are synergistic to finger extensors). Grasp is greatest when the wrist is in 20 degrees of extension, and the wrist needs to be stable with slight ulnar deviation so that loads can be transmitted through the TFCC structures (Nordin & Frankel, 2001) (Figure 8-13).

Note the position of the hand at rest. The fingers should be progressively more flexed from the radial to ulnar side. If this is not observed, it could be due to a lacerated tendon, Dupuytren's contracture, or other pathology (Magee, 2008). Postural imbalances during movement or rest can become fixed over time and less responsive to treatment, so early detection is important (Flinn & DeMott, 2010).

Comparison of both hands usually shows that the dominant hand is slightly larger than the non-dominant or non-preferred hand. Observe the normal creases and lines of the hand, and palpate the bony structures taking note of any tenderness. Look at the eminences of the hand: muscle wasting of the thenar eminence (median nerve) may be indicative of C6 nerve root problems or first dorsal interosseous muscle (C7 nerve), whereas wasting of the hypothenar eminence (ulnar nerve) may indicate C8 nerve root damage (Magee, 2008).

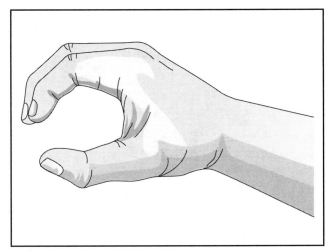

Figure 8-13. Position of the function of the hand.

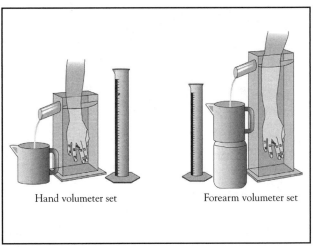

Hand volumeter set   Forearm volumeter set

Figure 8-14. Volumeter.

Notice if there are any differences between the two hands in color, temperature, appearance of the skin (shiny, sweating), loss of hair on the hand, or brittle fingernails, which may be indicative of vasomotor (blood vessels), sudomotor (sweat glands), pilomotor (hair and postganglionic sympathetic nerves that innervate them), circulatory, and trophic changes. Hot skin may indicate inflammation or rheumatoid arthritis, and cold, damp hands may indicate neurocirculatory aesthesia or Raynaud's disease (Magee, 2008). Take note of the shape of the fingers, because variations in the size and shape of the hand may indicate different pathology. For example, large blunt fingers may suggest acromegaly (enlargement of the bones) or Hurler's disease, while slender fingers may indicate hypopituitarism, tuberculosis, or osteogenesis inperfecta (Magee, 2008). The thickness of the nail, the presence and direction of ridges, spots or changes in color, or loosening of the nail bed can all be symptoms of pathologic processes.

Observe if there are any contusions on the hand. If there are repeated direct blows to the hand, this can cause vascular damage. If contusions occur to the palmar surface, there can be dorsal hand swelling. Thenar and hypothenar eminences can be bruised by means of a direct blow, which can occur in racquet sports or in catching a ball. Bruising of the metacarpophalangeal joints is common in boxing and football (Oatis, 2004).

Observation and palpation of any abnormal formations should also be noted in the documentation as to location and description. For example, a description may be "single nodule, freely movable, nontender, 0.5 centimeters in diameter located over the dorsum of the ring proximal interphalangeal joint" (Melvin, 1982, p. 212). Nodules can occur in several conditions including Dupuytren's contracture, which is the progressive fibrosis of the palmar aponeurosis and usually affects the ring and little fingers first. Fibrous nodules can develop in the flexor tendons that can catch on the annular sheath opposite the metacarpal head, causing trigger fingers or thumb. Swelling and bony enlargement of the proximal interphalangeal joint may indicate secondary synovitis from rheumatoid arthritis or may be due to osteoarthritis (Bouchard nodes) (Hartley, 1995). Heberden nodules are seen and palpated on

the dorsal surface of the distal interphalangeal joints and are associated with osteoarthritis, which weakens and destroys the articular cartilage.

Swelling of the hand can be assessed by circumferential measurement or by using a volumeter of either the hand or forearm as depicted in Figure 8-14. The hand or forearm is inserted into a rectangular container with water until in contact with the fixed bar, and the amount of water displaced is measured. Improvement in the amount of edema is objectively determined by less water displaced in subsequent measurements.

Obvious hand deformities can be seen. Swan neck deformity is a result of contracture of the intrinsic hand muscles or tearing of the volar plate (Figure 8-15). This results in flexion of the metacarpophalangeal and distal interphalangeal joints but extension of the proximal interphalangeal joint. Swan neck deformity often occurs in clients following trauma, rheumatoid arthritis, cerebral palsy, mallet finger, or congenital joint laxity (Ehlers-Danlos syndrome). Swan neck deformity of the thumb is called duck-bill deformity and is more common in osteoarthritis than rheumatoid arthritis. It results in CMC joint dysfunction. There is erosion of articular surfaces, stretching of the joint capsule, and dorsal and radial subluxation of the metacarpal joint (Wheeless, 2010).

Boutonnière deformity (also shown in Figure 8-15) occurs when the central slip of the extensor hood ruptures. The lateral bands of the extensor tendons separate to allow the joint to protrude, resulting in extension of the metacarpophalangeal and distal interphalangeal joints and flexion of the proximal interphalangeal joint. Boutonnière deformity commonly occurs after trauma or due to rheumatoid arthritis.

Another hand deformity that is easily observed is mallet finger, which usually occurs from injury to the tendons. As seen in Figure 8-15, the distal interphalangeal joint of the finger is unable to extend. Ulnar drift occurs when the extensor digitorum tendon slips off of the dorsal portion of the metacarpophalangeal joint and moves toward the ulnar side. When digits two through five are involved, this is called a wind-swept hand. Other observable hand deformities include extensor plus deformity, claw fingers (intrinsic minus hand), ape hand

Table 8-13

ACTIVE RANGE OF MOTION SCREEN FOR THE HANDS

DIRECTIONS TO CLIENT	MOTION TESTED
Ask the client to spread his or her fingers apart and bring them back together	Finger abduction and adduction
Make a tight fist and open fingers	Finger flexion and extension
Ask the client to bend his or her thumb across the palm and out to the side	Thumb flexion and extension
Ask the client to position his or her thumb in line with the index finger and then move it straight out	Thumb abduction and adduction

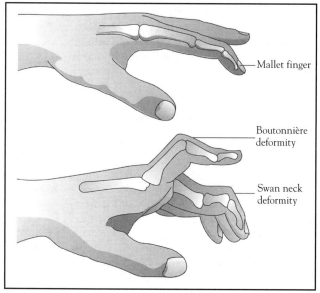

Figure 8-15. Swan neck, Boutonnière, and mallet finger deformities.

deformity, bishop's hand or benediction hand deformity, drop-wrist deformity, Dupuytren's contracture, and "Z" deformity of the thumb.

Range of Motion

As with other joints, active and passive range of motion are important assessments of the hand. Observation of the active use of the hand as well as passive movements and joint play provide initial information about the structures of the hand. Active movements will enable the assessment of physiologic movements and will provide insight into the client's willingness to use the hands.

A gross screen of hand function may involve having the client make a fist and then open the hand, noting any restrictions, deviations, or pain. Additional active range of motion screening tests are listed in Table 8-13.

Each motion at each joint is assessed either by observation or by measuring via the use of a goniometer, ruler, and/or tape measure. Most activities of daily living require 10 degrees wrist flexion, 35 degrees wrist extension, 60 degrees of MCP and PIP flexion, 40 degrees DIP flexion, and 20 degrees of thumb MCP flexion and 18 degrees of IP flexion of the thumb (Hume et al., 1990; Magee, 2008). A functional test of hand function might be to observe the client buttoning, zipping, holding a razor, or wringing out a sponge as indicative of finger flexion. Releasing objects and pushing a door open with the hand might entail the finger extensors. Opening a bag of chips would entail lateral pinch as would holding a key. Palmar pinch is involved in holding a pen and tip pinch in plugging in an appliance. Gross grasp can be observed by watching someone open a jar, hold a laundry basket, or drain water from a pot (Gutman & Schonfeld, 2003).

Overpressure applied at the end of passive range of motion will assist with assessment of end feel at each joint and the presence of capsular patterns. The test for tight retinacular ligaments is used to differentiate tight retinacular ligaments from capsular tightness. Accessory motions can be tested by assessing the long axis extension (distraction) of the fingers, anteroposterior glide, side glide, and slight rotation possible at the joints of the fingers.

Strength Assessment

Strength may be assessed by individually testing muscles (as in a manual muscle test), or a brief screening can be done

Table 8-14

STRENGTH SCREEN OF HANDS

DIRECTIONS TO CLIENT	THERAPIST ACTION	MUSCLE ACTION TESTED
Ask the client to squeeze your fingers or to make a fist and hold that position. Tell him or her, "Don't let me pull your fingers."	Attempt to pull the fingers into extension	Finger flexion
Ask the client to straighten his or her fingers and hold that position. Tell him or her, "Don't let me bend your fingers."	Attempt to push the fingers at each joint into flexion	Finger extension
Ask the client to squeeze his or her fingers together and hold that position. Tell him or her, "Don't let me pull your fingers apart."	Attempt to abduct the fingers	Finger adduction (palmar interossei)
Ask the client to open their fingers apart and hold that position. Tell him or her "Don't let me pull your fingers together."	Attempt to push the fingers into adduction	Finger abduction (dorsal interossei)
Ask the client to put his or her little finger and thumb together and hold that position. Tell the client, "Don't let me pull your thumb and little finger apart."	Attempt to reposition the thumb and finger	Opposition (opponens pollicis, opponens digiti minimi)

(as in a functional motion test or as seen in Table 8-14). Dynamometers are typically used to evaluate grasp, and pinch meters are used to measure pinch strength with norms established by age and gender.

Functional muscle strength of the hand can be seen in daily activities, such as maintaining the grip on a bar of soap, squeezing a tube of toothpaste, writing (requires Fair+ strength of finger flexors), releasing a can of pop, or releasing the arm of a chair when standing up (requires Fair– strength in finger extensors) (Gutman & Schonfeld, 2003). A quick screen for function grip strength would be observing whether a client can open a large jar, hold a gallon of milk, and open plastic containers (Gutman & Schonfeld, 2003).

An important consideration in the assessment of hand strength is that the hand is the final link in the kinematic chain. Dysfunction in other parts of the chain can influence hand strength. For example, the use of the hand is dependent upon proximal shoulder stability, and in the case of rotator cuff or humeral fractures, grip strength can be affected on the same side as the injury to the shoulder (Flinn & DeMott, 2010). Assessment of the other joints in the chain may be warranted.

Specific tests for muscles and tendons include the Finklestein test to detect the presence of de Quervain's or Hoffman's disease in the thumb. The Sweater finger sign is seen when one of the distal phalanges of the fingers doesn't flex when making a fist. When the finger being tested is placed in 90 degrees of proximal interphalangeal flexion and the client is asked to extend the proximal interphalangeal joint, if the therapist feels little pressure from the distal interphalangeal joint, then there is likely a torn central extensor hood (test for extensor hood rupture). Boyes test also tests the extensor hood, specifically the central slip. The Bunnel-Littler (Finochietto-Bunnel) test assesses the structures around the metacarpophalangeal joint (Magee, 2008).

Additional special tests of the hand are listed in Table 8-15.

Table 8-15

Neurological and Circulatory Tests for the Hand

TYPE OF TEST	NAME OF TEST	STRUCTURE ASSESSED
Neurological	Tinel's sign (at the wrist)	Thumb, index finger, middle and lateral half of the ring finger (median nerve distribution)
	Phalen's (wrist flexion) test	Thumb, index finger, middle and lateral half of the ring finger (median nerve) indicative of carpal tunnel syndrome
	Reverse Phalen's (prayer) test	Thumb, index finger, middle and lateral half of the ring finger (median nerve) indicative of carpal tunnel syndrome
	Carpal compression test	Carpal tunnel syndrome
	Froment's sign	Paralysis of adductor pollicis longus and possibly ulnar nerve damage
	Egawa's sign	Interossei involvement and possible ulnar nerve palsy
	Wrinkle (shrivel) test	Denervated fingers do not shrivel when placed in water
	Pinch test	Compromised anterior interosseous nerve
	Ninhydrin Sweat Test	Nerve lesion indicated
	Weber's (Moberg) two-point discrimination test	Hand sensation
	Dellon's moving two-point discrimination test	Mechanoreceptor system
	Warentberg's sign	Ulnar nerve neuritis/paralysis
Circulation and edema	Digit blood flow	Distal blood flow
	Hand volume test	Edema
	Circumferential measurement	Edema
	Allen's test	Occlusion of radial or ulnar artery

Reprinted with permission from Magee, D. J. (2002). *Orthopedic physical assessment.* Philadelphia, PA: Saunders.

Stability Assessment

There are specific tests that, when positive, suggest that a problem exists. These might include the Test for Tight Retinacular (collateral) ligaments, which tests the structures around the proximal interphalangeal joint. The axial load test is positive for a fracture of the metacarpal or adjacent carpal bones if there is pain or crepitation when axial compression is applied to the client's thumb. A similar test can be done with the fingers. In the grind test, the therapist applies axial compression and rotation to the metacarpophalangeal joint. If pain is elicited, then this may indicate degenerative joint disease in the metacarpophalangeal or metacarpotrapezial joint (Magee, 2008). The Linscheid test detects ligamentous instability in the second and third CMCs (Magee, 2008). Varus and valgus stress tests are used to assess the ligamental instability of digital collateral ligaments.

Pain Assessment

Synovitis, degenerative joint diseases, and complex regional pain syndrome often present with the expression of pain. Pain may be due to blisters, lacerations of the fascia, disruption of superficial ligaments, disruption of superficial muscles (e.g., extensor digitorum, palmaris longus, or opponens), or due to a disruption in the periosteum. Pain may also be the result of a neural problem, bony injury, chronic capsular problems, deep muscle injury or tendon sheath problem, deep ligaments, or deep muscles. Peripheral nerve conditions, dorsal nerve root problems, systemic conditions (e.g., diabetes), and vascular occlusion can cause painful conditions in the hand. Table 8-16 lists types of pain and possible structures involved.

It is important to differentiate pain from stiffness and articular from periarticular pain (as in subcutaneous nodules, synovial cysts, osteophytes, or muscular conditions). Local tenderness may be indicative of injury to more superficial structures,

Table 8-16

PAIN IN THE HAND

TYPE OF PAIN	POSSIBLE STRUCTURES AFFECTED
Sharp	Skin, fascia
	Tendons
	Superficial ligaments
	Superficial muscles
	Acute bursitis
Numbness in elbow to fingers	"Opera glove" anesthesia (hysteria, leprosy, diabetes)
Dull	Neural problem
	Bony injury
	Chronic capsular problems
	Deep muscle injury
	Tendon sheath problems
Painful, swollen, hot	Causalgic states
Cold	Raynaud's disease
	Neurocirculatory aesthesia
Pins and needles	Peripheral nerve
	Dorsal nerve root
	Systemic condition
	Vascular occlusion

while deep pain may be due to dysfunction of deeper muscles or ligaments. Immediate onset may be indicative of a more severe injury while gradual onset suggests overuse syndromes, neural lesions, or arthritis.

Pain can be referred to other parts from specific muscles. For example, if the extensor digitorum muscle is injured, pain may be referred to the forearm, wrist, or digit. If extensor indicis is injured, pain may be experienced on the dorsum of the index finger (Magee, 2008).

Hand Pathology

The hand is directly involved in nearly every activity we do associated with work, play, and self-care. Trauma from direct blows or overuse of the hand in repetitive patterns as well as disease can limit the use of the hand.

Tendons can shorten secondary to immobilization, soft tissue contractures, and spasticity. Tenosynovitis (inflammation of the tendon and sheath) can cause fluid to become trapped in the synovial sheath, leading to a proliferation of tissue interfering with tendon gliding. The integrity and strength of the tendon is compromised, which can lead to rupture over rough or subluxed bones or spurs. The rupture of extensor tendons is more common than flexors because the extensors go over the carpal bones.

Sensory problems can be due to injury or compression. The most commonly affected nerve is the median nerve. The median nerve innervates the flexor portion of forearm and hand (pronator teres and quadratus, flexor carpi radialis, palmaris longus, flexor digitorum superficialis and profundus, flexor pollicis longus, abductor pollicis, opponens pollicis, flexor pollicis brevis, and first and second lumbricals). Carpal tunnel syndrome is an example of compression of the median nerve affecting the motor and sensory abilities of the hand and fingers. The radial nerve (which innervates the extensor muscles of the arm and forearm including triceps, anconeus, brachioradialis, extensor carpi radialis, and part of brachialis) is the least commonly injured nerve. The radial nerve innervates the skin on the dorsum of the wrist and hand to the lateral surface of the thumb and the index and middle fingers. The muscles and skin on the ulnar side of the hand and forearm are innervated by the ulnar nerve. The muscles innervated include flexor carpi ulnaris, flexor digitorum profundus, palmaris brevis, muscles of the little finger, third and fourth lumbricals, all of the interossei, adductor pollicis, and flexor pollicis brevis. A loss of the intrinsics, coupled with the inability to abduct the fingers or adduct the thumb with hyperextension of the metacarpophalangeal and flexion at the proximal interphalangeal joint, is claw hand due to damage to the ulnar nerve as a result of trauma or nerve entrapment behind the medial epicondyle with loss of motor and sensory innervation.

While fractures do occur in the hand, fractures and subluxation of the metacarpals are more common than phalangeal fractures, often due to direct blows to the metacarpal shaft or metacarpal head. Proximal phalange fractures are more common than middle or distal fractures that cause damage to flexor or extensor tendons. Table 8-17 provides an overview of pathology of the hand.

OUTCOMES MEASURES FOR THE HAND

Outcome measures have been developed to assess change in a client over time. The changes need to impact the meaningful areas of the person's life based on collaboration between the therapist and the client throughout the treatment process. Not only is quality of service delivery measured as improvements in physical function and engagement in occupations, but client satisfaction and active involvement are important parts of the intervention outcome.

Some specific hand outcome measures have been developed. The Arthritis Hand Function Test (AHFT) consists of 11 test items designed to measure pure and applied strength and dexterity of both hands in clients with rheumatoid arthritis and osteoarthritis (Backman, Mackie, & Harris, 1991). The Jebsen Hand Function Test involves seven hand function tests designed to simulate activities of daily living, such as feeding, writing, and turning pages (Jebsen, Taylor, Trieschmann, & Howard, 1969). The Rheumatoid Hand Function Disability scale is a questionnaire assessing the client's perception of functional difficulty in activities of daily living (Duruoz, Poiraudeau, Fermanian, Menkes, Amor, Cougados et al., 1996). Disabilities of the Arm, Shoulder, and Hand (DASH) (Hudak, Amadio, & Bombardier, 1996) measure limitations in activities of daily living and instrumental activities of daily living.

Most outcome measures cover more than one component of the World Health Organization *International Classification of Functioning, Disability, and Health* but can be loosely classified according to their main purpose and component of measurement. For example, outcome measures for rheumatoid arthritis include the client's own assessment of the level of disability, patient and physician global assessments, and evaluation of specific physical characteristics (joint pain/tenderness, joint swelling) (Brooks & Hochberg, 2001). The Goal Attainment Scaling (GAS) is designed to be used collaboratively by client and therapist in establishing individualized goals and outcomes (Malec, 1999).

MacDermid, Richards, and Rother (2001) compared the results of the DASH, Patient Related Wrist Evaluation, and Short Form Health Survey (SF-36) for clients who had distal radius fractures and found that the most gains were made in wrist function and the least in role resumption and quality of life. Case-Smith (2003) conducted a study to determine changes in performance, satisfaction, and health-related quality of life in clients with upper extremity injury. In addition to the DASH and SF-36, clients were also asked to develop individualized client-centered goals by using the Canadian Occupational Performance Measure (COPM). This tool identifies occupational performance problem areas and considers the client's perception of his or her level of performance and satisfaction with everyday task completion. The tool helps in clarifying client-centered goals and evaluating the success of goal attainment over time. The use of the COPM ensures that intervention "remains true to the client's goals and priorities" (Case-Smith, 2003, p. 499), and the COPM "helped therapists focus intervention and supported development of measureable, achievable goals that were meaningful to clients" (Case-Smith, 2003, p. 504). Several tools ensure a comprehensive assessment of hand rehabilitation outcomes.

SUMMARY

The hand is amazingly complex with many articulations between carpal bones, metacarpals, and phalanges. As a result, the following joints are formed: carpometacarpal, metacarpophalangeal, distal interphalangeal, and proximal interphalangeal.

- By observation, one can identify three arches in the hand: two transverse and one longitudinal. Many creases are visible in the hand as well.

- The carpometacarpal joint provides the most movement for the thumb and the least amount for the fingers.

- The thumb is capable of flexion/extension, adduction/abduction, circumduction, and opposition/reposition.

- There is less mobility in the index and middle fingers so that these digits can provide stabilization while permitting greater mobility in the little and ring fingers.

- The interphalangeal joints of the fingers are closely congruent during movement so these are relatively stable joints.

- The flexor apparatus is made up of the flexor digitorum superficialis and flexor digitorum profundus muscles.

- The extensor apparatus or extensor hood comprises the extensor digitorum, connective tissue, fibers from the interossei muscles, and the lumbricals.

- Hand muscles, while many in number, are often distinguished by those that are intrinsic to the hand and those that are extrinsic.

Table 8-17

OVERVIEW OF HAND PATHOLOGY

NEUROLOGICAL

Radial nerve	Injuries	• Cervical cord injuries
		• Dermatome sensory and brachial plexus lesions
		• Peripheral nerve injuries
		• Dislocation of the shoulder
		• Fractured humerus and radius
		• Radial nerve palsy
	Severance	• Extensor paralysis
		• Inability to extend thumb or proximal interphalangeal joint
		• Wrist drop (inability to extend wrist)
		• Possibly loss of elbow extension
		• Loss of grip since lacks stabilizing function of wrist extension
	Compression (entrapment neuropathy)	• Radial tunnel syndrome
		• Thoracic outlet syndrome (C8T syndrome)
		• Posterior interossei syndrome
	Diseases	• Lead poisoning
		• Alcoholism
		• Polyneuritis
		• Trauma
		• Diphtheria
		• Polyarteritis
		• Neurosyphilis
		• Anterior poliomyelitis
Median nerve	Injuries	• Cervical cord and brachial plexus
		• PNI
		• Prolonged compression
		• Dislocated ulna
		• Fractured elbow or lower radius
	Severance	• Weak wrist flexors
		• Inability to flex and abduct thumb
		• Inability to flex index and middle finger
		• Tendency for thumb and index to hyperextend
		• Loss of thumb opposition

(continued)

Table 8-17 (continued)

OVERVIEW OF HAND PATHOLOGY

NEUROLOGICAL

Median nerve	Compression	• Carpal tunnel syndrome
		• Pronator teres syndrome
		• Anterior interosseus N. syndrome
		• Rheumatoid arthritis
	Diseases	• Amyloidosis
		• Gout
		• Plasmacytoma
		• Anaphylactic reaction
		• Myxedmea
Ulnar nerve	Lesions	• Cervical cord and brachial plexus
		• Peripheral nerve injuries
		• Fracture and dislocation of humeral head and elbow
		• Pressure during sleep
	Severance	• Inability to flex ring and little fingers
		• Loss of hypothenar muscles and interossei
		• Hyperextension of ring and little fingers
		• Limited radial deviation
		• Loss of thumb adductor
	Diseases	• Polyneuritis
		• Trauma
	Complex Regional Pain Syndrome (shoulder-hand syndrome, RSD)	• Myocardial infarction
		• Pancoast tumor
		• Brain tumor
		• Neoplasm
		• Spondylosis
		• Vascular occlusion
		• Hemiplegia
		• Osteoarthritis
	Polyneuritis	• Carcinoma of the lung or gastric
		• Hodgkin's disease
		• Pregnancy
		• Diabetes mellitus
		• Chemical neuritis
		• Arteriosclerosis

(continued)

Table 8-17 (continued)

OVERVIEW OF HAND PATHOLOGY

MUSCLES AND TENDONS

Tendons	Mallet finger	• Extensor tendon torn from insertion
		• Distal interphalangeal drops into flexion
	Trigger thumb	• Thumb snaps as it flexes
		• May become locked in flexion or extension
		• Due to thickening of the sheath or tendon or nodule
		• Prevents gliding of tendon
	Trigger finger (digital tenovaginitis stenosis)	• Thickening of flexor tendon sheath and tendons stick
		• Usually occurs in middle or ring fingers
		• Often due to direct, severe, or multiple traumas
	Extensor mechanism pathology	• Boutonnière deformity
		• Ulnar drift
	Flexor apparatus pathology	• Swan neck deformity
		• Wrist drop
		• Z deformity of thumb
		• Dupuytren's contracture
Sprains	Thumb	• Skier's thumb
		• Forceful hyperextension often combined with abduction of first MCP joint
		• Also seen in baseball, basketball, and volleyball, which can lead to sprain of ulnar collateral ligament or fracture or displacement of proximal phalanx or thumb
	Fingers	• Second to fifth MCP commonly injured through hyperextension usually resulting in ligamental damage
		• With PIP hypertension, the joint capsule, transverse retinacular ligaments, or volar plate can be injured
		• Hyperextension of the DIP is common in basketball and volleyball; sprained with anterior capsular damage, ligament, and sometimes volar plate damage
		• A flexed finger that is violently extended can cause the flexor digitorum profundus to rupture from insertion on distal phalange as in football or rugby
		• With forced ulnar deviation of the PIP joint, radial collateral ligaments can be injured, volar plate can be ruptured, or complete dislocation can occur
Muscular	Muscle wasting or atrophy	• Hypothenar: C8 nerve root problem
		• Thenar: C6 nerve root problem
		• Hypothenar, interossei, and medial lumbricals: median nerve

(continued)

Table 8-17 (continued)

OVERVIEW OF HAND PATHOLOGY

MUSCLES AND TENDONS

Muscular	Muscle wasting or atrophy	• Amyotrophic lateral sclerosis
		• Charcot-Marie-Tooth peroneal atrophy
		• Syringomylia
		• Neural leprosy
	Muscle loss	• Intrinsic loss: clawed position overpowered by extrinsic muscles
		• Dupuytren's contracture of palmar aponeurosis pulling fingers into flexion
		• Intrinsic minus hand position
		• Extrinsics acting alone lead to a position of MCP (hyper) extension called and DIP flexion
		• Intrinsic plus hand position
		• Intrinsics acting alone would give a position of MCP flexion and interphalangeal extension
	Ligamental	• Incomplete injuries at PIP most common but usually no dislocation of PIP joint due to capsular support
		• MCP—hyperextension injuries to radial collateral ligament
		• Thumb—collateral ligament on the ulnar side often requires surgery

OVERUSE/CUMULATIVE TRAUMA

	Carpal tunnel syndrome	• Any lesion that significantly reduces the size of the carpal tunnel
		• Swelling of tendons in synovial sheaths
		• Paresthesia, hypoesthesia, or anesthesia may occur
		• Motor loss may result
	Extensor intersection syndrome	• Overuse of thumb or wrist
		• Inflammation of abductor pollicis longus and extensor pollicis brevis
		• Often seen in paddlers or weight lifters
	de Quervain disease (constrictive tenosynovitis)	• Overuse of thumb or wrist
		• Tendonitis of abductor pollicis longus and extensor pollicis brevis as pass through first compartment
		• Seen in paddling, baseball, javelin, hockey

HAND DEFORMITIES

	Dupuytren's contracture	• Progressive shortening, thickening, and fibrosis of the palmar fascia and aponeurosis
		• Ring and little fingers into partial flexion at the MP and PIP joints

(continued)

Table 8-17, Continued

OVERVIEW OF HAND PATHOLOGY

HAND DEFORMITIES

Dupuytren's contracture
- Frequently bilateral
- Common in men older than 50 years

Claw (ulnar) hand
- Ulnar nerve injury
- Extensive motor and sensory loss to the hand
- Difficulty making a fist because they cannot flex their fourth and fifth digits at the distal interphalangeal joints.

Simian or ape hand
- Thumb movements being limited to flexion and extension in the plane of the palm
- Inability to oppose the thumb

Bishops (benediction) hand
- Wasting of hypothenar muscles, interossei muscles, and two medial lumbricals
- Due to ulnar nerve palsy

VASOMOTOR CHANGES

Redness, blanching diseases
- Suspect circulatory problem
- Raynaud's disease
- Rheumatoid disease
- Causalgia
- Acromegaly

Adapted from Magee, D. J. (2002). *Orthopedic physical assessment.* Philadelphia, PA: Saunders.

- Synergistic relationships exist between the muscles of the wrist and the finger muscles.
- Prehension is the use of the hand in precise ways. Defining how the hand moves has proved challenging. Grasp usually involves more fingers and often the thumb, whereas pinch usually only involves one or two fingers. Pinch is usually described as tip, palmar, lateral, and pulp. Cassanova has described a more detailed description of grasp and pinch patterns that demonstrates the variety of ways hands are used in everyday activities.
- Evaluation of the hand includes assessment of pain, edema, symmetry, movement, and observation and palpation. In addition, grasp and pinch measurements are taken to determine the strength of the muscles in the hand in addition to strength.

REFERENCES

Backman, C., Mackie, H., & Harris, J. (1991). Arthritis hand function test: development of a standardised assessment tool. *The Occupational Therapy Journal of Research, 11*(4), 245-256.

Bellace, J. V., Healy, D., Bess, M. P., Bryon, T., & Hohman, L. (2000). Validity of the Dexter Evaluation System's Jamar dynamometer attachment for assessment of hand grip strength in a normal population. *Journal of Hand Therapy, 13*(1), 46-51.

Biel, A. (2001). *Trail guide to the body* (2nd ed.). Boulder, CO: Books of Discovery.

Brand, P. W. & Hollister, A. M. (1999). *Clinical mechanics of the hand.* St. Louis, MO: Williams & Wilkins.

Brooks, P., & Hochberg, M. (2001). Outcome measures and classification criteria for the rheumatic diseases. A compilation of data from OMERACT (Outcome Measures for Arthritis Clinical Trials), ILAR (International League of Associations for Rheumatology), regional leagues and other groups. *Rheumatology, 40,* 896-906.

Bugbee, W. D., & Botte, M. J. (1993). Surface anatomy of the hand. The relationships between palmar skin creases and osseous anatomy. *Clinical Orthopaedics and Related Research, 296,* 122-126.

Burstein, A. H., & Wright, T. M. (1994). *Fundamentals of orthopaedic biomechanics.* Baltimore, MD: Williams & Wilkins.

Case-Smith, J. (2003). Outcomes in hand rehabilitation using occupational therapy services. *American Journal of Occupational Therapy, 57*(5), 499-506.

Cassanova, J. S., & Gernert, B. K. (1989). Adult prehension: Patterns and nomenclature for pinches. *Journal of Hand Therapy,* 231-243.

Chan, J., & Spencer, J. (2004). Adaptation to hand injury: An evolving experience. *American Journal of Occupational Therapy, 58*(2), 128-139.

Cooper, C. (2003). Hand therapy. *Occupational Therapy Practice, 9*(10), 17-20.

Duruoz, M. T., Poiraudeau, S., Fermanian, J., Menkes, C. J., Amor, B., Dougados, M., Revel, M. (1996). Development and validation of a rheumatoid hand functional disability scale that assesses functional handicap. *The Journal of Rheumatology, 23*(7), 1167-1172.

Fess, E. E., Gettle, K. S., Philips, C. A., & Janson, J. R. (2005). *Hand and upper extremity splinting: Principles and methods* (3rd ed.). St. Louis, MO: Elsevier Mosby.

Fiatorone, M. A., & Evans, W. J. (1993). The etiology and reversibility of muscle dysfunction in the aged. *Journal of Gerontology, 48*, 77-83.

Flinn, S., & DeMott, L. (2010). Dysfunction, evaluation, and treatment of the wrist and hand. In R. A. Donatelli & M. J. Wooden (Eds.), *Orthopaedic physical therapy* (4th ed., pp. 267-312). St. Louis, MO: Churchill Livingstone Elsevier.

Gench, B. E., Hinson, M. M., Harvey, P. T. (1995). *Anatomical kinesiology.* Dubuque, IA: Eddie Bowers Publishing, Inc.

Goss, C. M. (Ed.) (1976). *Gray's anatomy of the human body* (29th ed.). Philadelphia, PA: Lea and Febiger.

Greene, D. P., & Roberts, S. L. (1999). *Kinesiology: Movement in the context of activity.* St. Louis, MO: Mosby.

Gutman, S. A., & Schonfeld, A. B. (2003). *Screening adult neurological populations.* Bethesda, MD: American Occupational Therapy Press.

Hamill, J., & Knutzen, K. M. (1995). *Biomechanical basis of human movement.* Baltimore, MD: Williams & Wilkins.

Hartley, A. (1995). *Practical joint assessment: Upper quadrant: A sports medicine manual* (2nd ed.). Philadelphia, PA: C. V. Mosby.

Hinkle, C. Z. (1997). *Fundamentals of anatomy and movement: A workbook and guide.* St. Louis, MO: C. V. Mosby.

Hudak, P., Amadio, P. C., & Bombardier, C. (1996). Development of an upper extremity outcome measure: The DASH (Disabilities of the Arm, Shoulder, and Hand). *American Journal of Industrial Medicine, 29*, 602-608.

Hume, M. C., Gellman, H., McKellop, H., & Brunfield, R. H. (1990). Functional range of motion of the joints of the hand. *Journal of Hand Surgery, 15*(2), 240-243.

Jebsen, R. H., Taylor, N., Trieschmann, R. B., & Howard, L. A. (1969). An objective and standardised test of hand function. *Archives of Physical Medicine and Rehabilitation, 50*(6), 311-319.

Kimmerle, M., Mainwaring, L., & Borenstein, M. (2003). The functional repertoire of the hand and its application to assessment. *American Journal of Occupational Therapy, 57*(5), 489-498.

Kisner, C., & Colby, L. A. (1990). *Therapeutic exercise: Foundations and techniques* (2nd ed.). Philadelphia, PA: F. A. Davis.

Kisner, C., & Colby, L. A. (2007). *Therapeutic exercise: Foundations and techniques* (5th ed.). Philadelphia, PA: F. A. Davis Company.

Konin, J. G. (1999). *Practical kinesiology for the physical therapist assistant.* Thorofare, NJ: SLACK Incorporated.

Levangie, P. K., & Norkin, C. C. (2005). *Joint structure and function: A comprehensive analysis* (4th ed.). Philadelphia, PA: F. A. Davis Company.

MacDermid, J. C., Richards, R. S., et al. (2001). Distal radius fracture: A prospective outcome study of 275 patients. *Journal of Hand Therapy, 14*, 154-169.

Magee, D. J. (1992). *Orthopedic physical assessment.* Philadelphia, PA: W. B. Saunders Co.

Magee, D. J. (2008). *Orthopedic physical assessment.* Philadelphia, PA: Saunders.

Malec, J. F. (1999). Goal attainment scaling in rehabilitation. *Neuropsychological Rehabilitation, 9*, 253-275.

McPhee, S. D. (1987). Functional hand evaluations: A review. *American Journal of Occupational Therapy, 41*(3), 158-163.

Melvin, J. L. (1982). *Rheumatic disease: Occupational therapy and rehabilitation* (2nd ed.). Philadelphia, PA: F. A. Davis Co.

Muscolino, J. E. (2006). *Kinesiology: The skeletal system and muscle function.* St. Louis, MO: Mosby.

Neumann, D. A. (2002). *Kinesiology of the musculoskeletal system: Foundations for physical rehabilitation.* St. Louis, MO: Mosby.

Nordin, M., & Frankel, V. H. (1989). *Basic biomechanics of the musculoskeletal system* (2nd ed.). Philadelphia, PA: Lea and Febiger.

Nordin, M., & Frankel, V. H. (2001). *Basic biomechanics of the musculoskeletal system.* Philadelphia, PA: Lippincott, Williams & Wilkins.

Norkin, C. C., & Levangie, P. K. (1992). *Joint structure and function: A comprehensive analysis* (2nd ed.). Philadelphia, PA: F. A. Davis Co.

Oatis, C. A. (2004). *Kinesiology: the mechanics and pathomechanics of human movement.* Philadelphia, PA: Lippincott, Williams & Wilkins.

Rider, D. A. (2004). Hands-on experience: Specific hand conditions affect the aging population. *Rehab Management*, October, 38-41.

Riley, M. A. (1998). The effects of medical conditions and aging on hand function. *OT Practice, 3*(6), 24-27.

Smith, L. K, Weiss, E. L., & Lehmkuhl, L. D. (1996). *Brunnstrom's clinical kinesiology* (5th ed.). Philadelphia, PA: F. A. Davis Co.

Trew, M., & Everett, T. (Eds.) (2005). *Human movement: An introductory Text.* London: Elsevier Churchill Livingstone.

Wheeless, C. R. (2010). *Wheeless' textbook of orthopaedics.* Durham, NC: Duke University, Data Trace Internet Publishing, LLC.

ADDITIONAL RESOURCES

Aaron, D. H., & Jansen, C. W. S. (2000). Hand rehabilitation: Matching patient priorities and performance with pathology and tissue healing. *OT Practice*, 10-15.

Basmajian, J. V., & DeLuca, C. J. (1985). *Muscles alive* (5th ed.). Baltimore, MD: Williams and Wilkins.

Baxter, R. (1998). *Pocket guide to musculoskeletal assessment.* Philadelphia, PA: W. B. Saunders Co.

Caillet, R. (1971). *Hand pain and impairment.* Philadelphia, PA: F. A. Davis Co.

Calliet, R. (1996). *Soft tissue pain and disability* (3rd ed.). Philadelphia, PA: F. A. Davis Co.

Cooper, C., & Evarts, J. L. (1998). Beyond the routine. *OT Practice*, 18-22.

Esch, D. L. (1989). *Musculoskeletal function: An anatomy and kinesiology laboratory manual.* Minneapolis, MN: University of Minnesota.

Fuller, Y., & Trombly, C. (1997). Effects of object characteristics on female grasp patterns. *American Journal of Occupational Therapy, 51*(7), 481-487.

Jenkins, D. B. (1998). *Hollingshead's functional anatomy of the limbs and back* (7th ed.). Philadelphia, PA: W. B. Saunders Co.

Jones, L. A. (1989). The assessment of hand function: A critical review of techniques. *Journal of Hand Surgery, 14A*(2), 221-228.

Kellor, M., Frost, J., Silberberg, N., Iversen, I., & Cummings, R. (1971). Hand strength and dexterity. *American Journal of Occupational Therapy, 25*(2), 77-83.

Loth, T., & Wadsworth, C. T. (1998). *Orthopedic review for physical therapists.* St. Louis, MO: C. V. Mosby Co.

Lumley, J. S. P. (1990). *Surface anatomy: The anatomical basis of clinical examination.* New York, NY: Churchill Livingstone.

MacKenna, B. R., & Callender, R. (1990). *Illustrated physiology* (5th ed.). New York, NY: Churchill Livingstone.

Masaki, K., Leveille, S., Curb, J. D., & White, L. (1999). Midlife hand grip strength as a predictor of old age disability. *JAMA*, 558-560.

Nicholas, J. A., & Hershman, E. B. (1990). *The upper extremity in sports medicine.* St. Louis, MO: C. V. Mosby Co.

Norkin, C., & White, D. J. (1989). *Measurement of joint motion: A guide to goniometry.* Philadelphia, PA: F. A. Davis Co.

Oxford, K. L. (2000). Elbow positioning for maximum grip performance. *Journal of Hand Therapy, 13*(1), 33-36.

Palastanga, N., Field, D., & Soames, R. (1989). *Anatomy and human movement: Structure and function.* Philadelphia, PA: Lippincott Williams and Wilkins.

Palmer, M. L., & Epler, M. E. (1998). *Fundamentals of musculoskeletal assessment techniques* (2nd ed.). Philadelphia, PA: J. B. Lippincott.

Perry, J. F., Rohe, D. A., & Garcia, A. O. (1996). *Kinesiology workbook.* Philadelphia, PA: F. A. Davis Co.

Petersen, P., Petrick, M., Connor, H., & Conklin, D. (1989). Grip strength and hand dominance: Challenging the 10% rule. *American Journal of Occupational Therapy, 43*(7), 444-447.

Rice, M. S., Leonard, C., & Carter, M. (1998). Grip strengths and required forces in accessing everyday containers in a normal population. *American Journal of Occupational Therapy, 52*(8), 621-626.

Richards, L. G., Olson, B., & Palmiter-Thomas, P. (1996). How forearm position affects grip strength. *American Journal of Occupational Therapy, 50*(2), 133-138.

Saidoff, D. C., & McDonough, A. L. (1997). *Critical pathways in therapeutic intervention: Upper extremity.* St. Louis, MO: C. V. Mosby.

Smith, R. (1985). Pinch and grasp strength: Standardization of terminology and protocol. *American Journal of Occupational Therapy, 39*(8), 531-535.

Thompson, C. W., & Floyd, R. T. (1994). *Manual of structural kinesiology* (12th ed.). St. Louis, MO: C. V. Mosby.

Torrens, G. E., Hann J., Webley, M., & Sutherland, I. A. (2000). Hand performance assessment of ten people with rheumatoid arthritis when using a range of specified saucepans. *Disability and Rehabilitation, 22*(3), 123-134.

Toth-Fejel, G. E, Toth-Fejel, G. F., & Hendricks, C. A. (1998). Occupation-centered practice in hand rehabilitation: Using the experience sampling method. *American Journal of Occupational Therapy, 52*(5), 381-385.

Trombly, C. A. (Ed.) (1995). *Occupational therapy for physical dysfunction* (4th ed.). Baltimore, MD: Williams and Wilkins.

Watkins, J. (1999). *Structure and function of the musculoskeletal system.* New York, NY: Human Kinetics.

9

POSTURE
FUNCTIONAL INTERACTION OF
THE SPINE AND PELVIS

The function of the upper extremity is dependent upon a stable base on which to work. Poor pelvic stability and posture will result in inefficient and uncoordinated activity of hands and arms (Trew & Everett, 2005). Symmetry and alignment of the spine with the pelvic girdle provides proximal stability, which in turn supports distal control of the extremities. A symmetrical position can minimize the risk of subsequent orthopedic deformities by keeping the bones and joints in alignment. Good posture minimizes fatigue because muscles are used correctly, with less strain and overuse in potentially deforming positions. There is a reduction of stress on ligaments, minimizing the likelihood of injury. Good posture also enables full use of the upper extremities and hands and aligns vision with the salient features of the environment (Figure 9-1).

Posture and balance depend on the functional relationship between the spine and the hips as it relates to movement of both the upper and lower extremities. Balance needs to be maintained between the pelvis and the feet, the trunk and the pelvis, and the head on the trunk. Weight is distributed over each foot and between the two feet. Postural adjustments are made based on sensory input from the visual, proprioceptive, and vestibular systems.

This chapter will discuss variables related to posture. The axial skeleton consists of the cranium, vertebral column, ribs, and sternum and is important in providing stability and movement while transferring loads from the extremities. The portion of the axial skeleton that pertains to posture, the vertebral column, will be discussed. The sacroiliac joint, which is the transition between the caudal end of the axial skeleton

and lower appendicular skeleton, aids posture because this joint is designed for stability and is relatively rigid. The pelvis includes bones of the axial skeleton of the spine (sacrum and coccyx) and of the appendicular skeleton (ileum, ischium, and pubis), which are pelvic girdle bones of the lower extremity.

THE SPINE

The vertebral column and ribs serve to protect the spinal cord and internal organs, provide a means for breathing, support the head and extremities, transmit loads between the extremities, and stabilize and mobilize the body for hand function and ambulation (Smith, Weiss, & Lehmkuhl, 1996).

The spine is formed by 33 vertebral bones that are labeled according to location: seven cervical, 12 thoracic, five lumbar, five sacral, and four coccygeal (Gench, Hinson, & Harvey, 1995). Each vertebra consists of a body, disks, pedicles, spinous process, vertebral arch, and laminae. The body is the largest part of the vertebra and is essentially a cylindrical shape. In the midthoracic vertebrae, the body is the primary weight-bearing portion of the vertebral column (Neumann, 2002). The intervertebral disks of fibrocartilage attach to the superior and inferior surfaces, and this enables each vertebra to articulate with the vertebra above and below it (Gench et al., 1995). These disks serve to absorb shock and provide spaces throughout the vertebral column. Two pedicles project dorsally from the body, which serves to connect the vertebral bodies to the posterior part of the vertebral column. The pedicles merge

Rybski MF.
Kinesiology for Occupational Therapy, Second Edition (pp 277-298)
© 2012 SLACK Incorporated

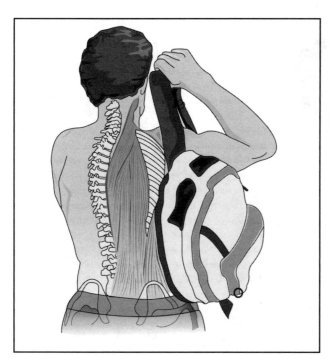

Figure 9-1. The spine, pelvis, and thorax in action.

with a pair of lamina (which protect the posterior aspect of the spinal cord), creating the vertebral arch with the vertebral foramen (enabling passage of nerve roots in and out of the vertebral canal) in the center. The vertebral arch supports seven processes: four articular, two transverse, and one spinous process. The processes help to increase the leverage for the muscles of the trunk and extremities (Smith et al., 1996).

There are 24 pairs of apophyseal (facet) joints in the vertebral column. An apophyseal joint is the articulation between the articular facets of adjacent vertebrae.

They have thin and loose articular capsules, synovial membranes, and menisci. Each vertebra has a pair of apophyseal joints, one on the left and one on the right. A superior articular facet joint faces up, and an inferior articular faces down. These joints are hinge-like and allow flexion, extension, and torsion. In addition to these movements, the facets interlock to make the spine more stable.

There are distinct differences in the vertebrae in each of the spinal segments (Figure 9-2). The differences depend on the purpose and function of each vertebral segment, how movement occurs, and the level of participation in load bearing in which the vertebra is involved. In general, the posterior portions of the vertebrae serve to protect the spinal cord and stabilize the spine, while anterior portions function as a shock absorber, to bear weight and loads, and to mobilize the trunk (Smith et al., 1996). The articular processes allow some movements but restrict others. Horizontal facet surfaces enable axial rotation, and vertical facet structures in either the frontal or sagittal plane block axial rotation. This helps to explain why axial rotation is greater in the cervical region than in the lumbar. Other things influencing motion include the variability in the size of the disks, the different shapes of the vertebrae, the different muscle actions, and the attachment of the rib and ligaments (Neumann, 2002).

The average range of motion for the entire spine is flexion 135 degrees, extension 120 degrees, lateral flexion 90 degrees, and rotation 120 degrees (Muscolino, 2006).

Cervical Vertebrae

The cervical vertebrae are the smallest of the vertebrae and are different from the thoracic and lumbar vertebrae due to a foramen in each of the transverse processes. The body is generally smaller, oval, and broader than other vertebrae.

The first two cervical vertebrae are peculiar, too. The first cervical vertebra, C1 or atlas, has no body and resembles a bony ring. This vertebra supports the head and has a shorter (and some say no) spinous process but long transverse processes (Gench et al., 1995; Goss, 1976). There are two large concavities on the superior surface that articulate with occipital condyles of skull, allowing flexion and extension around a frontal axis (Gench et al., 1995). The second cervical vertebra, C2 or the axis, forms a pivot around which the first vertebra, carrying the head, rotates. This rotational movement permits extensive range of motion around a vertical axis so that we can look side to side by turning our heads. The most conspicuous difference of C2 is the superior extension of the body called the dens where it articulates with the atlas.

The seventh cervical vertebra is also distinctive in having a long and prominent spinous process. Because this spinous process protrudes further than all other cervical vertebrae, it is easily palpated and useful as a skeletal landmark, especially when describing to clients where their injuries occurred in relation to this landmark.

The cervical facets are oriented in an oblique plane 45 degrees between transverse and frontal planes. This allows movement in both the transverse and frontal planes, including rotation and lateral flexion.

Flexion and extension occur within a sagittal plane around a mediolateral axis. In the craniocervical region, there is 45 degrees to 50 degrees of flexion and 70 degrees to 85 degrees of extension. At the atlanto-occipital joint, the convex occipital condyle rolls backwards in extension and forward in flexion on concave superior facets of the atlas. Simultaneously, the condyle slides slightly in the direction of the roll. Further rolling motion of the condyle is due to tension in the tectorial membrane, the articular capsule, and the atlanto-occipital membranes (Neumann, 2002). The atlanto-axial complex has slight flexion and extension (about 15 degrees). At the intercervical articulations of C2-7, 70 degrees of extension occurs as the inferior articulating surfaces slide inferiorly. Extension is considered the close-packed position for most of the spine. Thirty-five degrees of flexion occurs as the inferior articular facets of the superior vertebrae slide superiorly and anteriorly on the superior articular facets of the inferior vertebrae (Neumann, 2002).

Protraction and retraction of the head within a sagittal plane also occurs. Protraction flexes the lower to mid cervical spine and extends the upper craniocervical region; retraction extends the lower to mid cervical spine and flexes upper craniocervical region (Neumann, 2002).

Rotation within a transverse plane around a vertical axis occurs at the cervical region. Axial rotation is seen as an important mechanism for attaining visual and auditory cues

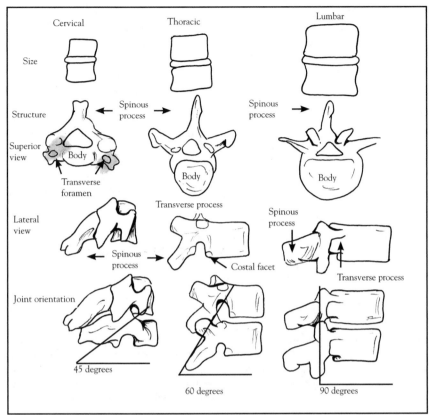

Figure 9-2. Comparison of vertebrae.

from the environment (Neumann, 2002). A total of 180 degrees of total rotation is possible, 90 degrees to each side. The atlanto-axial complex permits 40 degrees to 45 degrees of rotation in each direction as the concave inferior articular facets of the atlas slide on the superior convex facets of the axis. There is also a slight lateral flexion to the opposite side. At the intracervical articulations, the inferior facet slides posteriorly and inferiorly to the same side for about 45 degrees of rotation and anteriorly and superiorly on the opposite side to the rotation. Rotation is greater in the more cranial portions of the cervical region.

Lateral flexion permits one to touch the ear to the shoulder. In the craniocervical region, 40 degrees of lateral flexion to each side is possible.

Thoracic Vertebrae

The thoracic vertebrae are intermediate in size and have four articular facets not found on other vertebrae. These form the articulation with the 12 ribs. Another difference is that the spinous processes project downward, especially T2-T10, which limits the ability of the thoracic spine to hyperextend.

The thoracic spine is less mobile than either the cervical or lumbar portions of the spine. Because it is more stable, it is injured less often. The decreased mobility is due to the ribcage (formed by the ribs, vertebrae, and sternum), which limits lateral flexion in the frontal plane and rotation in the transverse plane (Neumann, 2002).

The thoracic facets are oriented in the frontal plane around an anteroposterior axis for 25 degrees of lateral flexion. In the

sagittal plane around a mediolateral axis, there is 35 degrees of flexion and 25 degrees of extension. At the thoracic joint, 30 degrees of axial rotation in a transverse plane around a vertical axis is possible as is gliding translational movements in all directions (Muscolino, 2006; Neumann, 2002).

The thoracolumbar region is considered the trunk. It consists of the thoracic and lumbar spine. These joints together produce 85 degrees of flexion, 40 degrees of extension, 45 degrees of lateral flexion, and 35 degrees of rotation.

Lumbar Vertebrae

The lumbar vertebrae are the largest of the movable vertebrae and are important because they support the weight of the body. However, the lumbar vertebrae are also quite mobile, moving freely in all ranges except rotation. The orientations of the facets are nearly vertical, permitting flexion and extension until the lower lumbar region, where the facet orientation changes to the frontal plane. Because of the change in orientation of the facets and differences in movement and due to joint capacity for both mobility and stability, this area of the spine is injured a lot.

In the sagittal plane around a mediolateral axis, there is a total of 30 degrees to 50 degrees of flexion and 15 degrees to 25 degrees of extension. Greater flexion and extension are possible in the upper portions of the lumbar spine. In the frontal plane around an anteroposterior axis, 25 degrees of lateral flexion are possible. Rotation in a transverse plane around a vertical axis of 5 degrees occurs at the lumbar joint, and there is gliding translational movement in all directions.

The lumbosacral joint is located between the fifth lumbar vertebra and the sacrum. While not structurally special, this is the joint at which the pelvis can move relative to the trunk (Muscolino, 2006; Neumann, 2002). The L5 vertebra articulates distally with the sacrum through the lumbosacral joint, so there is greater movement here than elsewhere in the lumbar spine (Gench et al., 1995).

Sacral Vertebrae

The fusion of the five sacral vertebrae forms a triangular bone in the dorsal part of the pelvis located like a wedge between the two coxal bones of the hip. The sacrum articulates with the last two lumbar vertebrae proximally and the immovable coccyx distally. The coccyx is formed by three to five rudimentary vertebrae, the last of which is only a nodule of bone.

Stability of the Vertebral Column

Tissues that may limit motions of the vertebral column include ligaments, muscles, and joint capsules. For flexion, the ligamentum nuchae, interspinous and supraspinous ligaments, ligamentum flava, capsule of apophyseal joints, posterior annulus fibrosus, and the posterior longitudinal ligament limit further motion. The esophagus and trachea, anterior annulus fibrosus, and anterior longitudinal ligament limit extension. Axial rotation is limited by annulus fibrosus, capsule of the apophyseal joints, and the alar ligaments. The intertransverse ligaments, contralateral annulus fibrosus, and capsule of the apophyseal joints limit lateral flexion (Neumann, 2002).

Movements of the Spine

The movements of the spine are made by the articulations of the vertebral bodies and by the articulations with the vertebral arches. While motion of the spinal column is small within each vertebra, the combined motion of the spine is much more. Flexion, extension, and hyperextension occur in the sagittal plane around multiple axes through nucleus pulposa, and lateral flexion occurs in the frontal plane around sagittal axes, also through nucleus pulposa.

Flexion and extension have a range of 110 degrees to 140 degrees, with free movement in cervical and lumbar regions but limited motion in the thoracic area. The axis of rotation moves anteriorly with flexion and posteriorly with extension. Flexion of the whole trunk occurs primarily in the lumbar area for the first 50 degrees to 60 degrees, and then more flexion is achieved by forward tilting of the pelvis and backward movement of the sacrum. Similarly, greater extension is achieved by posterior pelvic tilting and anterior sacral movement, followed by lumbar spine extension. The additional range of motion when considering both spinal and pelvic movements demonstrates the functional relationship between these two parts of the skeleton.

Lateral flexion of 75 degrees to 85 degrees occurs mainly in cervical and lumbar and is often accompanied by rotation.

Rotation of the spine is described as right or left rotation and occurs in a transverse plane around the long axis from the top of the head through the disks to the sacral region. Rotation of 90 degrees is possible with free movements in the cervical area and with lateral flexion in the thoracic and lumbar regions. Rotation is generally more limited in the lumbar region (Gench et al., 1995).

Functional active range of motion for the cervical spine was 20 degrees of flexion and extension, 9 degrees to 21 degrees for lateral flexion, and 13 degrees to 57 degrees for rotation. In a review of 15 activities of daily living including standing to sitting, backing up a car, cutting food with a knife, tying shoelaces, washing hair, walking, and picking up objects from the floor, backing up a car required the most active range of motion. Personal hygiene items entailed more active range of motion than ambulation. From this study, the authors concluded that most people only use a small percentage of cervical range of motion for most activities of daily living (Bible, Biswas, Miller, Whang, & Grauer, 2001).

While little discussion will be made about the spinal muscles, generally, the muscles are anterior or posterior in location. Anterior muscles flex the spine, whereas posterior muscles extend or hyperextend spinal segments. Lateral flexion occurs when one side of each pair of anterior and posterior muscles contracts, and rotation occurs when anterior and posterior muscles contract but only if they do not lie parallel to the long axis (Gench et al., 1995).

THE PELVIC GIRDLE

The lower extremity comprises the hip and ankle joints as well as multiple joints in the feet. The lower extremities absorb high forces and support the body weight, as well as providing a mechanism for movement. The joints and muscles of the lower extremity are important in balance and posture. Every movement made or posture assumed in the lower extremity or trunk is interrelated (Hamill & Knutzen, 1995). Forces from the lower extremity are transmitted through the hips to the pelvis and trunk, and the hips have the function of supporting the head, trunk, and upper extremities (Jenkins, 1998). When considering posture or positioning of a client, the relationship between the trunk and the hip is essential.

The pelvic girdle is made up of the sacrum, the coccyx, and two coxal (innominate) bones, which comprise the fused ilium, ischium, and pubis. These bones form seven joints: the lumbosacral, sacroiliac (two in number), sacrococcygeal, symphysis pubis, and the hip (two of them). While small amounts of movement are possible at the sacroiliac, symphysis pubis, and sacrococcygeal joints, the ability to have even these small amounts of movement is very important. The pelvic girdle attaches posteriorly to the axial skeleton at the sacroiliac joint. The two coxal bones attach anteriorly to the pubic symphysis and form a strong arch that is slightly movable and is able to transmit the weight of the body to the femur (Jenkins, 1998).

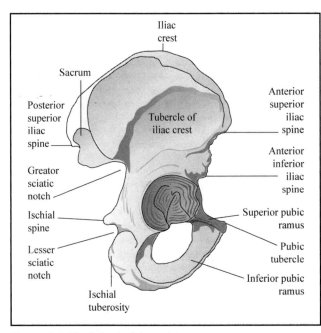

Figure 9-3. The bones of the pelvis.

Bones and Palpable Structures

Ilium

Crests of Ilia

There are two of these, one on each side, and they are easily seen and palpated by placing the thumbs laterally. These two crests should be level in a standing position, and the highest point on the crest is at the level of the spinous process of the fourth lumbar vertebra (Esch, 1989; Smith et al., 1996). These can be seen in Figure 9-3.

Anterior Superior Spines of Ilia

With the thumbs on the front of the crest, follow downward from the crest, and trace the crest to the most anterior point to the rounded anterior superior spines of ilia (ASIS) (Esch, 1989; Smith et al., 1996).

Posterior Superior Iliac Spine

If the crest is followed in a posterior direction about 1.5 inches from the midline, the broad and sturdy posterior superior iliac spine will feel rough. Below each is a depression that is a posterior landmark for the sacroiliac joint.

Sacrum

This is the flat bone at the center of the back between the PSIS (Esch, 1989).

Ischium

Tuberosities of Ischia

Also called the "sit bones," these are easy to locate when sitting on a hard chair or sidelying with hips and knees flexed. These are the large bony prominences at the midline of the buttocks just below the gluteal fold, and they can be palpated when the gluteus maximus and hamstrings are relaxed (Esch, 1989; Sine, Liss, Rousch, Holcomb, & Wilson, 2000).

Femur

Greater Trochanter of the Femur

With the thumb on the crest laterally, reach down on the thigh as far as possible with the middle finger or 4 to 5 inches inferior to the most lateral portion of the iliac crest. These large bony prominences can be felt if one stands on the opposite leg and rotates the femur in internal and external rotation. Also, this is marked by the depression that appears when the thigh is abducted (Esch, 1989).

Sacroiliac Joint

The sacroiliac joint helps to form the arch between the two pelvic bones and with symphysis pubis. The joint helps to transfer weight from the spine to the lower extremities and acts to decrease forces and provide shock absorption. The sacroiliac joint is formed by the articulation between the sacrum and the ilium, which forms a synchondrosis joint. Small amounts of translational movement are possible at this joint in the sagittal plane: 0.2 degrees to 2 degrees rotation and 1 to 2 millimeters for translation (Neumann, 2002). Nutation is anterior tilt of sacrum relative to ileum, and counternutation is relative posterior tilt of sacrum relative to ileum (Neumann, 2002). The close-packed position is nutation, and loose-packed is counternutation.

The amount of movement varies from individual to individual and between genders. Men have stronger and thicker sacroiliac ligaments so they have less mobile sacroiliac joints, and as many as three out of 10 men have fused joints. Men have a higher precedence of developing osteophytes and ankylosis in older age than do women. Women have more mobile sacroiliac joints due to greater ligamental laxity, which increases with monthly cycles of hormones and with pregnancy. The male sacroiliac joint is also more stable. In women, the center of gravity is in the same plane as the sacrum, and in men, the center of gravity is more anterior, so there is more of a load on the male sacroiliac joint than the female (Hamill & Knutzen, 1995).

The sacroiliac joint is part of the pelvic ring, which also includes the ileum, pubis, and ischium (hemipelvis). The pelvic ring transfers body weight bidirectionally between the trunk and femurs (Neumann, 2002), so is an important part of posture and balance.

In childhood, the sacroiliac joint is a synovial joint but changes from a diarthrodial to a modified amphiarthrodial joint in adulthood. The joint is reinforced by the anterior sacroiliac, interosseous, and posterior sacroiliac ligaments, and the sacrotuberous and sacrospinous ligaments provide secondary joint stability (Neumann, 2002). The long posterior sacroiliac ligaments limit anterior pelvic rotation (counternutation), and the short posterior sacroiliac ligament limits all pelvic and sacral movement (Magee, 2008).

Figure 9-4. Ligaments of the hip.

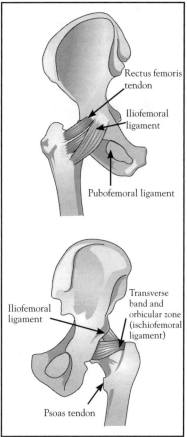

Rectus femoris tendon

Iliofemoral ligament

Pubofemoral ligament

Iliofemoral ligament

Transverse band and orbicular zone (ischiofemoral ligament)

Psoas tendon

The Hip

The hip joint, or coxofemoral joint, permits simultaneous movement between the lower extremity and the pelvis. It is a joint often compared to the glenohumeral joint. Both are ball-and-socket joints capable of a variety of movements. However, the hip acts more often as a closed-chain mechanism, which is opposite to the open kinetic chain of the shoulder. This demonstrates the different functions of these two ball-and-socket joints. The shoulder functions to enable the hands to be used to their most efficient capacity by means of significant mobility, whereas the hip provides stability, balance, and weight-bearing to provide postural accommodations and locomotion.

The hip joint is made up of the convex acetabulum of the pelvic bone and the convex head of the femur. The acetabulum is formed by the fusion of part of the ileum, ischium, and pubis and is sometimes called innominate bone or the pelvis. The acetabulum is deepened by a ring of fibrocartilage, the acetabulum labrum, which is located in the lateral aspect of the pelvis. The acetabulum faces laterally, anteriorly, and inferiorly (Kendall & McCreary, 1993). The synovial capsule encloses the entire joint, and it thickens and forms ligamentous bands (Gench et al., 1995). This strong articular capsule is reinforced by iliofemoral, pubofemoral, and ischiofemoral ligaments (Kendall & McCreary, 1993).

The iliofemoral ligament, also known as the Y ligament of Bigelow, covers the hip anteriorly and is considered the strongest ligament of the body (Figure 9-4). In standing, the iliofemoral ligament (especially the inferior band) prevents posterior motion of the pelvis on the femur (hyperextension of hip). The pubofemoral ligament is anterior and inferior to the hip (Hamill & Knutzen, 1995), and this ligament assists in checking extreme abduction and some external rotation. The ischiofemoral ligament is in a posterior and inferior location and is considered a weak ligament. This ligament limits internal rotation and helps to stabilize the hip in extension (Magee, 2008). All of these ligaments are slack when the hip is flexed, and all ligaments become taut with hyperextension. Abduction is limited by the pubofemoral and ischiofemoral ligaments, and adduction is limited by the superior or iliotrochanteric portion of iliofemoral (Y) ligament. Under low loads, the joint surfaces of the hip are incongruous, but under heavy loads, there is maximum surface contact (Magee, 2008).

The head of the femur fits deeply in the acetabulum. The femoral head is attached to the femoral neck, which projects anteriorly, medially, and superiorly at an angle of inclination of 125 degrees. This angle of inclination is between the axis of the femoral neck and the medial side of the shaft of the femur within the frontal plane. Angles greater than 125 degrees are called coxa valga; and angles less than 125 degrees are called coxa vara. Angles of inclination greater or lesser than 125 degrees can cause inequities in the hip joint, leading to deformity and pain.

Osteokinematics of the Hip

Movements that tilt the pelvis occur in a sagittal plane around a coronal axis. These movements can occur simultaneously or in a single limb around one axis. Anterior pelvic tilt is the forward movement of the pelvis and is produced when the anterior superior iliac spine moves in an anterior and inferior direction. The ASIS is then in a position closer to the anterior portion of the femur as the pelvis rotates around the transverse axis of hip joint (Kisner & Colby, 2007). The symphysis pubis is more inferior in location and because of anterior pelvic tilt, hip flexion and lumbar spine extension (hyperextension) occur. Posterior pelvic tilt occurs when the posterior superior iliac spine moves posteriorly and inferiorly, bringing the pelvis closer to the posterior aspect of the femur as the pelvis rotates backward around the axis of hip joints (Kisner & Colby, 2007). There is an upward and forward movement of the symphysis. This results in hip extension and lumbar spine flexion.

The resting position of the hip is 30 degrees flexion, 30 degrees abduction with slight external rotation. The close-packed position is in extension, internal rotation, and abduction, and this is the position of greatest bony congruency. The capsular pattern is flexion, abduction, and internal rotation, and the order may vary.

Movements of the femur in the acetabulum at the hip joint include flexion, extension, abduction, adduction, internal and external rotation, and circumduction. Because the hip is a ball-and-socket or triaxial joint, it is capable of all movements in three planes and has three degrees of freedom.

The axis for flexion and extension is around a coronal axis in a sagittal plane through the head and neck of the femur. Flexion ROM is generally considered to be 90 degrees with the knee extended and 120 degrees to 140 degrees with the knee flexed. The difference in ranges is due to the effect of the hamstrings, which restrict the motion when the knee is extended. With hip flexion, anterior pelvic tilt and increased lumbar extension will occur unless the pelvis is stabilized by the abdominal muscles (Kisner & Colby, 2007). The range for hip extension is 10 degrees to 30 degrees of motion. Posterior pelvic tilt and decreased lumbar extension will also occur with extension. Flexion motion is limited due to the inferior fibers of the ischiofemoral ligament and the inferior capsule.

The range of motion for hip extension is 20 degrees with the knee in extension. Further extension is limited by the iliofemoral ligament, anterior capsule, and parts of the pubofemoral and ischiofemoral ligaments (Neumann, 2002).

Abduction and adduction occur in a frontal plane around a sagittal (anteroposterior) axis. The femur can be abducted 30 degrees to 50 degrees, and adduction can occur in a range from 10 degrees to 30 degrees of motion. Abduction movement is limited by the pubofemoral ligament; inferior capsule; adductor and hamstring muscles; and adduction by the superior fibers of the ischiofemoral ligament, iliotibial band, and extensor fasciae latae.

Internal rotation of the hip occurs in a transverse plane around a longitudinal axis with a range of 30 degrees to 70 degrees. Further internal rotation is limited by the ischiofemoral ligament and by the piriformis muscle. External rotation occurs in a range from 45 degrees to 90 degrees (differences in normative values are due to differences in measurement procedures). External rotation is limited by the lateral fasciculus of the iliofemoral ligament, iliotibal band, gluteus minimus, and tensor fascia latae (Neumann, 2002).

Arthrokinematics of the Hip

The head of the femur is convex, while the acetabulum is concave, so most motions of the femoral head will be in a direction opposite to the motion of the distal end of the femur (Norkin & Levangie, 1992) (Table 9-1). The motions of flexion and extension are almost purely spin, with spinning occurring in a posterior direction for flexion and anteriorly for extension. Abduction and adduction are combined spin and glide, but again in a direction opposite to the motion of the distal end of the femur when the femur is the moving segment. However, when the hip is weightbearing, the femur is fixed. The motion of the hip is produced by the movement of the pelvis on the femur. While this motion is more common, the acetabulum now moves in the same direction as the opposite side of the pelvis (Norkin & Levangie, 1992).

Movements of the Hip

There is a functional relationship between the hips, pelvis, and spine so that with pelvic motion, the angle of the hip and the lumbar spine changes. Movements of the pelvis include pelvic tilt, pelvic shift, rotation, and lumbar-pelvic rhythm. Due to these pelvic movements, flexion, extension, abduction and adduction, and internal and external rotation of the hip occur.

When one hip is moved into a position that is higher than the other side, this occurs due to lateral pelvic tilt. On one side, there is hip hiking and hip adduction; on the other, there is hip drop and hip abduction. This occurs when the lumbar spine laterally flexes toward the side of the elevated pelvis. The

pelvis can tilt anteriorly 30 degrees and posteriorly 15 degrees, and the movement is in a sagittal plane around a mediolateral axis. Posterior tilt at the lumbosacral joint is equivalent to flexion of the trunk at the lumbosacral joint. This means that the muscles that flex the trunk will also tilt the pelvis because this is the same action. Also, posterior tilt of the pelvis results in extension of the thigh. Similarly, anterior tilt of the pelvis is analogous to extension of the trunk at the lumbosacral joint. Anterior tilt of the pelvis produces flexion of the thigh at the hip. Pelvic tilt is shown in Figure 9-5.

Forward translatory pelvic shifting occurs in standing when one is in a relaxed or slouched position, resulting in extension of the hip and extension of LE spinal segments. There is a compensatory posterior shifting of the thorax on the upper lumbar spine with increased flexion of these segments. This position requires little muscle action, and the position is maintained by the iliofemoral ligaments of the hip, anterior longitudinal ligaments of lower lumbar spine, and posterior ligaments of the upper lumbar and thoracic spine (Kisner & Colby, 2007).

Approximately 15 degrees of pelvic rotation occurs to the left or right at the lumbosacral joint, resulting in a pivotal movement of hips around the long axis (Gench et al., 1995). This movement generally occurs in a transverse plane around a vertical axis. When one leg is fixed on the ground, the other unsupported leg swings forward or backward. If the leg moves forward, there is forward rotation with the trunk rotating backwards on the stabilized side with the femur on the stabilized side internally rotated. With backward rotation, there is posterior rotation, and the femur externally rotates with the trunk moving in the opposite direction. The motion of rotation of the right pelvis is the same movement as internal or medial rotation of the right thigh and to left trunk rotation at the lumbosacral joint (Muscolino, 2006).

Depression and elevation in the frontal plane around the anteroposterior axis also occurs. There is 30 degrees of elevation and depression. Elevation of the right pelvis is the same as right lateral flexion at the lumbosacral joint, and depression of the right pelvis at the hip is analogous to abduction of the right thigh at the hip (Kvitne & Jobe, 1993; Muscolino, 2006).

The coordinated movement between the lumbar and pelvic segments is known as lumbar-pelvic rhythm. Due to this coordinated muscular activity, maximal forward flexion is possible, enabling us to pick up items from the floor and to touch our toes. It is the combined movements of hip flexion, pelvic tilt, and flexion of the lumbar spine seen as an analog to the scapulohumeral rhythm of the upper extremity, except there are no proportional contributions of each motion nor a set sequence of occurrence as is true in the shoulder (Levangie & Norkin, 2005). Once the head and upper trunk initiate the flexion movement, the pelvis shifts posteriorly to maintain the center of gravity over the feet. At approximately 45 degrees of forward flexion, the ligaments become taut, the vertebrae become stabilized, and the muscles relax. Anterior pelvic tilt occurs once all vertebral segments have been stabilized.

The posterior ligaments and the pelvis rotate forward. Forward movement continues until the full length of the muscles is reached and is influenced by muscle extensibility in the back and hips (Kisner & Colby, 2007). Similar combined

Table 9-1

SUMMARY OF FAULTY POSTURES

POSTURE	DESCRIPTION	MUSCLE IMBALANCE AND STRUCTURAL ISSUES	POTENTIAL PAIN	POTENTIAL CAUSES
Lordotic posture	Increase in • Lumbosacral angle lumbar lordosis • Anterior pelvic tilt hip flexion	• Stress to anterior and posterior longitudinal ligament • Stress to iliofemoral ligament • Narrowing of the intervertebral foramen • Compression of the dura and blood vessels of nerve roots	• Tight hip flexors • Tight lumbar extensors • Imbalanced or stretched abdominal muscles • Synovial irritation and joint inflammation	• Faulty posture • Pregnancy • Obesity • Weak abdominal muscles
Kyphosis: Flat low back	Decrease in • Lumbosacral angle • Lumbar lordosis • Hip extension posterior tilt	• Lack of normal physiologic lumbar curve • Stress to posterior longitudinal ligament • Increase in posterior disk space (nucleus puposus may protrude posteriorly)	• Tight trunk flexors and hip extensors • Stretched and weak lumbar extensor and hip flexor muscles • Reduced shock absorption	• Continuous slouching • Over-emphasis on flexion exercises
Kyphosis: Flat upper back	Decrease in • Thoracic curve • Depressed scapulae and clavicle	• Fatigue of muscles • Compression of neurovascular bundle in thoracic outlet between clavicle and ribs	• Tight thoracic erector spinae, intercostals, scapular protractors, and potentially restricted scapular movement	Exaggerating the upright posture
Kyphosis: Round back	Increase in • Pelvic angle • Thoracic curve • Protracted scapulae • Forward head	• Stress to posterior longitudinal ligament • Fatigue in thoracic erector spinae and rhomboid muscles • Thoracic outlet syndrome • Cervical posture syndromes	• Tight anterior thorax (intercostal) muscles • Pain in muscles of cervical spine	• Continuous relaxed lumbar posture or the flat low back posture

(continued)

Table 9-1 (continued)

SUMMARY OF FAULTY POSTURES

POSTURE	DESCRIPTION	MUSCLE IMBALANCE AND STRUCTURAL ISSUES	POTENTIAL PAIN	POTENTIAL CAUSES
				• Continuous slouching • Over-emphasis on flexion exercises
Kyphosis: Lordosis posture	• Forward head • Hyperextension of cervical spine and lumbar spine • Knees extension • Increased flexion of thoracic spine	• See both lordosis and kyphosis	• Weakness in anterior neck and upper back muscles and muscles of lower abdomen	• See both lordosis and kyphosis

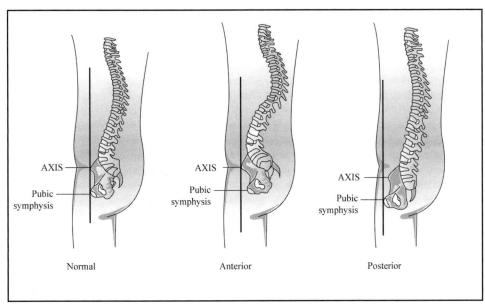

Figure 9-5. Pelvic tilt.

Normal   Anterior   Posterior

Figure 9-6. Anterior hip muscles.

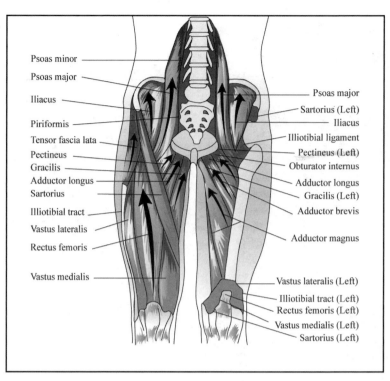

movements of hip abduction, lateral pelvic tilt, and flexion of the lumbar spine occur when one is sidelying and attempting maximal hip abduction. Full abduction would not be possible without the lateral tilting of the pelvis and the lumbar spine.

While the range of motion values reflect norms for each motion, daily activities involving the lower extremities require much less range of motion. For example, while the normal value for hip abduction is 30 to 50 degrees, tying a shoe with the foot on the floor requires only about 19 degrees of range of motion in the frontal plane. This illustrates that while a client may not have full or normal ROM, the client may be able to perform daily tasks successfully.

It is important to consider, too, that norms are culturally determined. Mulholland and Wyss (2001) found that in Asian and Middle Eastern cultures, many activities are performed while squatting, kneeling, or sitting cross-legged. Squatting requires 130 degrees to full hip flexion, and sitting cross-legged requires 90 to 100 degrees of hip flexion. Both positions require 111 to 165 degrees of knee flexion.

Internal Kinetics: Muscles

Hip flexors are those muscles on the anterior or anteromedial surface of the hip region (Figure 9-6). The rectus femoris and iliopsoas muscles both cross the hip anterior to the frontal axis and act as hip flexors. The iliopsoas muscle (sometimes referred to as two separate muscles, the iliacus and psoas muscles) is a powerful muscle that is especially useful in the initial part of hip flexion. As flexion progresses, the iliopsoas muscle becomes shorter and less effective. This mechanism has led to the development of the "crunch" to exercise the abdominal muscles as they flex the spine. By bending the knees while in a supine position, the iliopsoas is rendered ineffective so that the "crunch" involves the abdominals and not synergistic muscles (Gench et al., 1995). The rectus femoris exerts little power in flexion until other muscles have initiated the flexion action, and its action is complementary to the iliopsoas muscle. Pectineus, with a relatively long force arm and advantageous angle of insertion, is a powerful flexor as well as adductor muscle. Sartorius, the longest muscle in the body, is an effective hip flexor because its force arm is longest to the frontal axis, and tensor fascia latae also participates with this movement. Anterior fibers of gluteus medius and gluteus minimus contribute to hip flexion, as do adductor magnus, adductor longus, and adductor brevis. The gracilis muscle, depending upon the location of the femur, can assist with flexion and extension, in addition to the adductor action at the hip (Gench et al., 1995).

The muscles that extend the thigh lie on the posterior aspect of the thigh. Gluteus maximus, while a large muscle on the posterior buttocks, is a strong extensor. However, Gench and colleagues (1995) state that this muscle is "notoriously lazy during actions associated with daily living," which may account for the ease with which this muscle loses its firmness and attracts fat deposits (p. 197). The posterior portions of gluteus medius also contribute to hip extension, as do the lower portions of the adductor magnus muscle. Posterior hamstring muscles (semimembranosus, semitendinosus, and biceps femoris) lie posterior to the joint axis and can act as extensors. All three muscles have essentially the same mechanical advantage in moving the femur. Because the posterior hamstrings extend the thigh and flex the leg, they are not strong contributors to extension unless the knee is kept in extension. When the leg is kept extended and the thigh is flexed (as in touching one's toes or high-kicking activities), pain may be felt behind the knee

Figure 9-7. Posterior hip muscles.

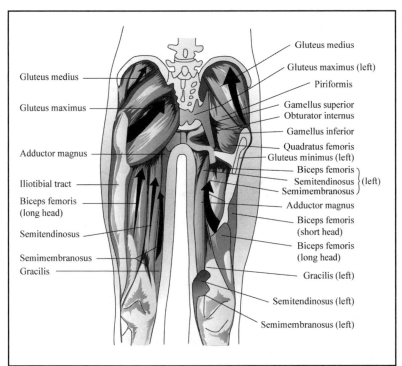

Gluteus medius

Gluteus maximus

Adductor magnus

Iliotibial tract

Biceps femoris
(long head)

Semitendinosus

Semimembranosus

Gracilis

Gluteus medius

Gluteus maximus (left)

Piriformis

Gamellus superior
Obturator internus

Gamellus inferior

Quadratus femoris
Gluteus minimus (left)

Biceps femoris
Semitendinosus }(left)
Semimembranosus

Adductor magnus

Biceps femoris
(short head)

Biceps femoris
(long head)

Gracilis (left)

Semitendinosus (left)

Semimembranosus (left)

due to the posterior hamstrings. The posterior hamstrings are active in any forward bending at the hips and act as antigravity postural muscles (Jenkins, 1998) (Figure 9-7).

The function of the muscles that abduct the thigh is to keep the pelvis in a horizontal placement when weight is put on the limb (Jenkins, 1998) (Figure 9-8). Two muscles are particularly strong in supporting the pelvis: gluteus medius and gluteus minimus. Both of these muscles pass the sagittal axis of the hip joint to the lateral side, so they are strong abductors, and both have similar angles of attachment, which give these muscles superior mechanical advantages at the hip (Gench et al., 1995). Because the gluteus minimus is a smaller muscle, it is less powerful than the medius. Tensor fascia latae also contribute to the movement of the limb in abduction, but only when the limb is weightbearing (Jenkins, 1998). Sartorius, and to a lesser degree the piriformis, obturator internus, and the upper fibers of the gluteus maximus, assists with abduction. When the thigh is flexed to 90 degrees, the gluteus maximus then becomes an abductor; in other positions, this muscle is an adductor (Jenkins, 1998).

Adductor muscles are anteriorly placed on the thigh and include pectineus, adductor longus and brevis, and the obturator portion of the adductor magnus (Figure 9-9). The adductor magnus is well-placed to adduct, but is recruited only when resistance is met (Gench et al., 1995). Adductor longus and brevis are active in all stages of adduction whether there is resistance or not (Gench et al., 1995), and pectineus has both a long force arm and advantageous angle of insertion, so it is a strong adductor. Crossing the hip joint on the medial side of the sagittal axis, gracilis is well placed for adduction action. Muscles that also contribute to adduction include gluteus maximus, quadratus femoris, obturator externus, the hamstrings, gracilis, and iliopsoas (when the thigh is flexed), although not all sources indicate that these muscles function in adduction

(Gench et al., 1995; Goss, 1976; Jacobs & Bettencourt, 1995; Sine et al., 2000).

The muscle actions of medial and lateral rotation are often the secondary result of other movements, and there is some variability in the categorization of muscles that rotate the thigh. Muscles that laterally rotate the thigh are numerous. Gluteus maximus is a powerful lateral rotator. Piriformis, obturator internus, obturator externus, quadratus femoris, and superior and inferior gemelli are small muscles that lie deep in the pelvis with the primary action of external rotation. These six muscles are often grouped together and called the outward rotators (Gench et al., 1995). These six small rotator muscles hold the femur in acetabulum just like the rotator cuff muscles of the glenohumeral joint, and they either laterally rotate the femur or balance the pelvis and trunk. Biceps femoris, posterior fibers of the gluteus medius, and sartorius also produce weak lateral rotation. Medial rotation is produced by gluteus minimus, tensor fascia latae, and anterior fibers of gluteus medius. Semimembranosus and semitendinosus are credited with weak internal rotation.

Due to the gluteals and hamstrings, the hip is very strong in extension. Perhaps extension of the hip is even more critical than hip flexion because extension is needed for an upright posture as well as sitting and standing. The hip extensors are active when gravity pulls the upper body into flexion, and they prevent this forward motion of the trunk (Hinkle, 1997). The hip extensors and flexors also maintain the lumbar curve, with iliopsoas seen as a key postural muscle as it pulls anteriorly on lumbar vertebrae and ilium to reinforce lumbar curve and anterior pelvic tilt (Hinkle, 1997), and the flexors help to maintain the anterior and posterior trunk balance.

The two-joint function of rectus femoris and hamstrings (semitendinosus, semimembranosus, and biceps femoris) requires further discussion. When the hip is flexed, the rectus

Figure 9-8. Normal curves of the spine.

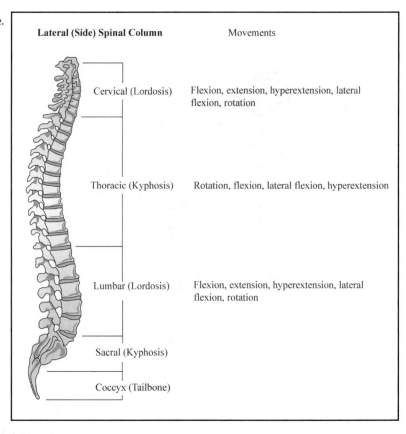

Lateral (Side) Spinal Column	Movements
Cervical (Lordosis)	Flexion, extension, hyperextension, lateral flexion, rotation
Thoracic (Kyphosis)	Rotation, flexion, lateral flexion, hyperextension
Lumbar (Lordosis)	Flexion, extension, hyperextension, lateral flexion, rotation
Sacral (Kyphosis)	
Coccyx (Tailbone)	

femoris acts as the agonist, and hamstrings work antagonistically. In hip extension, the hamstrings are acting as agonists, while the rectus femoris is the antagonist. In knee flexion, the hamstrings have an agonist function, and rectus is antagonist, and in knee extension, the muscle roles reverse. If hip and knee actions are performed simultaneously, the relationships are more complex. For example, a place kick in soccer requires hip flexion and knee extension, so the rectus femoris contracts to flex the hip and extend the knee, whereas the hamstrings must relax to allow the leg to move. The hamstrings are capable of generating more force than the rectus femoris at the hip due to a longer force arm, just as the rectus femoris is more forceful at the knee.

Simultaneous hip and knee flexion or hip and knee extension requires that both hamstrings and rectus femoris be agonists at one joint and antagonists at the other. This is contradictory to muscle action in the upper extremity, where a muscle will act as an agonist at all of the joints it crosses. For example, the extensor digitorum longus crosses five joints, causing extension at each; flexor digitorum profundus crosses four joints, causing flexion at each joint. Due to this, these upper extremity muscles can be easily overstretched or overshortened.

In walking, the abductors control the pelvis, and the hamstrings control the amount of hip flexion and some propulsion. Abduction of the femur separates the feet, providing a wider base of support. Without adequate abductor strength, the hip of a swinging leg would drop (Trendelenburg sign), and ambulation would be more cumbersome and less energy-efficient (Hinkle, 1997). Lateral rotation accompanies abduction and

extension of hip and provides stability by allowing the feet and lower leg to be positioned before striking the ground or in balancing on one leg as we walk and run. When lateral rotation is diminished, the foot points toward the midline, which adds stress to the knees (Hinkle, 1997). Muscles controlling hip abduction and adduction are also important in maintenance of dynamic sitting balance, as are medial rotators, which control diagonal balance and weight shifts.

Gench and colleagues (1995) state that "the muscles of spine and pelvic girdle are at the mercy of the demands of upright posture" (p. 184), which constantly needs to overcome the forces of gravity and sustain loads. If the muscles are unable to overcome these forces, the forces and movements tend to be transmitted to the spine rather than the pelvis, creating pain, decreased movement, and structural damage. Tight hip extensors can cause an increase in lumbar flexion when the thigh is flexed, and tight hip flexors will cause increased lumbar extension as the thigh extends. If the adductor muscles are tight, this can cause lateral pelvic tilt on the opposite side and side bending of the trunk toward the side of tightness. The opposite result occurs with tight abductors (Gench et al., 1995; Hinkle, 1997).

POSTURE

Posture is the position of the head, limbs, and trunk and their relationships to each other. Changes in posture occur when any part is moved—adjusting the position of the head and limbs in relation to trunk, adjusting the trunk in relation

Normal Spinal Curves

The spine has four naturally occurring curves as seen in Figure 9-8: the thoracic and sacrococcygeal curve in a convex direction posteriorly; cervical and lumbar bend in a convex direction anteriorly. The cervical and lumbar curves exist before birth and are called primary curves, whereas the thoracic and sacrococcygeal curves are secondary curves that develop in infancy and young childhood (Gench et al., 1995). Other curves associated with the spine occur when there is an increase in an anterior curve (lordosis), an increase in the posterior curve (kyphosis), or the existence of a lateral curve (scoliosis).

Standing

Standing upright depends on the weight distribution on each foot, between the two feet, and the balance of the pelvis over the feet, the trunk over the pelvis, and the head over the trunk. Posture and balance have been compared to stacking movable blocks, where balance is achieved between each block, and the weight is distributed between two sides of the body. Changes in position of the pelvis result in automatic realignment of the spine, especially in the lumbar region. Being in an upright stance has definite advantages for using one's hands in complex tasks, but also creates a position that is inherently unstable because the center of gravity of the body is over a relatively small base of support (Galley & Forster, 1987).

Ideal erect standing posture is when the line of gravity passes through the mastoid process of the jaw as seen in Figure 9-9. If the head is too far forward or too far back, tension, strain, and pain in neck muscles, headache, and eye strain may result. The line of gravity continues to a point just in front of the shoulder, a point just behind the center of the hip joint, a point just in front of the center of knee joints, and approximately 5 to 6 cm in front of the ankle (Cook & Hussey, 1995). The cervical, thoracic, and lumbar regions of the spine maintain the normal curve, and the scapula is flat against the upper back (no winging or tipping). The pelvis and hips are in neutral position, and the hip and knee are neither flexed nor extended. The ankle is also in neutral position relative to plantarflexion and dorsiflexion.

Anterior muscles that are involved in posture include prevertebrals (four on each side), scaleni (three on each side), sternocleidomastoid, and abdominals (rectus abdominis, external oblique, internal oblique, and psoas minor). The anterior muscles flex the spine. The abdominal muscles pull upward, while the hip flexors pull downward in a synergist action to maintain posture. The posterior trunk muscles include levator scapulae, splenus, suboccipitals, erector spinae, semispinalis, multifidus, long rotator, short rotator, intertransversii, interspinalis, quadratus lumborum, and trapezius. Posterior muscles extend or hyperextend spinal segments. A similar muscle synergy exists for the posterior muscles where the back muscles pull upward and the hip extensors pull downward to create stability. Lateral flexion occurs when one side of each pair of anterior and posterior muscle contracts. Rotation occurs when anterior and posterior muscles contract but only if they do not lie parallel to the long axis.

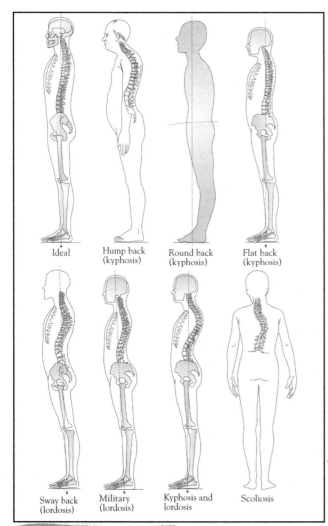

Ideal

Hump back (kyphosis)

Round back (kyphosis)

Flat back (kyphosis)

Sway back (lordosis)

Military (lordosis)

Kyphosis and lordosis

Scoliosis

Figure 9-9. Standing postures.

to the head or limbs when these are fixed, and making finely adjusted vertebral movements. Posture also communicates nonverbal body language to observers, reflecting self-esteem and attitude. If someone has a slouched position, it may suggest a negative attitude; if one stands upright with shoulders back, it may suggest attentiveness.

A balanced posture reduces work done by muscles, and not all muscles are active in standing posture. The intrinsic muscles of the feet are quiescent as are the quadriceps and hamstrings. There is slight activity in the gluteus maximus, and the abdominal muscles are quiescent except for the lower fibers of internal obliques, which are active to protect the inguinal canal. The soleus is continuously active (as is iliopsoas) in all cases because gravity tends to pull the body forward over feet. Erector spinae are active, counteracting gravity's effect to pull the body forward. Gastrocnemius and the posterior tibial muscles are less frequently active, and tibialis anterior is quiescent, but this changes if high heels are worn. Lateral postural sway is counteracted by gluteus medius and tensor fascia latae.

Faulty postures occur when one or more of the natural curves of the spine are exaggerated or diminished (see Figure 9-9). These can include kyphosis, lordosis, and scoliosis. Kyphosis is a pathological curve in which there is a forward curving of the spine that causes bowing of the back. Often, this posture is accompanied by posterior pelvic tilt. Parts of the spine lose the normal lordotic curve, which can cause slouching, breathing difficulties, and discomfort. Most descriptions of kyphosis pertain to the thoracic region (dowager's hump, hump back, or round back), resulting in abnormal rounding of the spine greater than 50 degrees, resulting in extreme curvature of the upper back. If a client has a flat scapula or winging of the scapula, this can give the appearance of a kyphotic curve (Magee, 2002, 2008). In round back, there is decreased pelvic inclination with thoracolumbar or thoracic kyphosis usually caused by tight soft tissues. In a round back, increased compression on the anterior bodies of the vertebrae may lead to degenerative osteoarthritic changes (Muscolino, 2006). Hump back usually presents with a localized, sharp posterior protuberance called a gibbus.

Kyphosis can also occur in the lumbar region, where it is known as flat back when there is decreased lumbar lordosis. In the flat back posture, there is decreased pelvic inclination. People with flat back have a tendency to lean forward when walking or standing. People who have a habitual slouching position of the spine may develop this posture. There can also be a curve in the cervical region, and this is called torticollis. In addition, there may be one or more lateral curves in the lumbar or thoracic spine (notably in the Hunchback of Notre Dame).

Kyphosis may be due to degenerative diseases (arthritis), congenital or developmental problems, osteoporosis with compression fractures, or trauma. Older women with osteoporosis often get kyphotic postures (called dowager's hump or hyperkyphosis), and it is often caused by a wedge fracture where the front of a vertebra collapses. In adults, quite a few diseases may contribute to the degeneration of the vertebra and development of kyphosis, which may include hyperparathyroidism, Cushing disease, prolonged corticosteroid treatment, Paget disease, polio, fractured vertebrae, cancer, and tuberculosis. Scheuermann kyphosis is considered a form of juvenile osteochondrosis of the spine and can affect teenagers between the ages of 12 and 16 years of age (girls more often than boys). It may be due to growth retardation, blood circulation problems during rapid growth spurts, or poor posture. In this disease, infection, inflammation, and disk degeneration put the vertebrae under stress. There can also be congenital kyphosis in which infant spines do not develop normally, are malformed, or are fused.

When there is an extreme curve in the lumbar area of the spine, this produces a faulty lordotic posture (hyperlordosis). Usually, this spinal movement is accompanied by anterior pelvic tilt. Other names attributed to this posture are hollow back, saddle back, sway back, or relaxed. In the sway-back position, the cervical spine is slightly extended, there is increased flexion in the thoracic spine, and there is increased flexion with flattening of the lower lumbar curve and extended hips and knees. In the sway-back posture, the weightbearing is shifted through the facet joints rather than the disks, which can lead to degenerative osteoarthritic changes (Muscolino,

2006). Common causes might include tight low back muscles, excessive visceral fat, or pregnancy, and conditions such as achondroplasia, discitis, obesity, osteoporosis, and spondylolisthesis may contribute to the development of this posture.

Exaggerated lumbar curvature is also called military curve. In this position, there is a slight posterior tilt of the head, normal curves in the cervical and thoracic spines, hyperextension and increased lumbosacral angle in the lumbar spine, and the lumbosacral angle is determined by drawing a line parallel to the ground and the other line along the base of the sacrum). In the military stance, the pelvis is in anterior tilt, and the hip is in flexion. The knee is slightly hyperextended, and the ankle is slightly plantarflexed. Causes of this posture may include pregnancy, obesity, or weak abdominal muscles. Lordosis and kyphosis can also be present in the same individual.

Scoliosis is the lateral curvature of the spine or a frontal plane distortion in the thoracic or lumbar spinal regions. If viewed from the side, there would be mild roundness in the cervical or thoracic regions and a degree of sway back in the lumbar area of the spine. Causes of scoliosis are unknown and generally fall into the idiopathic scoliosis category. Neuromuscular scoliosis is seen in people with cerebral palsy, spinal bifida, and muscular dystrophy. Degenerative scoliosis may be due to trauma, bone collapse, or osteoporosis. The abnormal curve tends to develop during growth spurts just before puberty. A summary of faulty postures is provided in Table 9-1.

One cannot maintain a symmetrical stance for long, and standing is not a static activity. There is a continuous slight sway, and the magnitude of the sway tends to be larger in those who are very old or very young. The alternating activity enables muscle spindles to be activated irregularly, so fatigue in any one motor unit is prevented and venous return is assisted. Sitting provides a more stable posture, because the supporting surface area of the buttocks, back of thighs, and feet is greater. The center of gravity is lowered, and this posture allows relaxation of lower extremity muscles and less energy use. Pelvic stability, though, is greater in standing than sitting due to the passive locking mechanism at the hip by ligaments when the hips are fully extended during standing. When seated, the hips are flexed, and the locking mechanism is lost.

Sitting

Effective seating promotes posture and comfort. It enables physiologic maintenance and tissue protection. In an optimal seated position, vision, breathing, swallowing, and upper extremity function is optimized. A stable base of support can also enhance appearance and possibly social acceptance. Occupational therapists are able to adapt seating around existing deformities to achieve these objectives.

Various sitting postures are possible. One can be sitting supported, in a relaxed unsupported or erect unsupported position, or in a forward sitting posture. This has been called the functional task position or the posture of readiness (Cook & Hussey, 1995), in which the person's center of gravity shifts in the direction of the activity. In unsupported sitting, the muscles of the spine are activated to overcome the backward rotation of the pelvis and lumbar kyphosis and achieve erect or lordotic sitting posture. The pelvis rotates forward, and the line of gravity is through the ischial tuberosities (Cook & Hussey, 1995).

Optimal sitting posture is described as no pelvic obliquity or rotation with a slight anterior pelvic tilt. The trunk will have slight lumbar lordosis, slight thoracic kyphosis, and slight cervical extension. There will be neutral hip rotation (no internal or external rotation) and slight hip abduction. There will be 90 degrees of flexion of the hips, knees, and ankles. The hips should be higher than the knees, and the back should be supported. In the upper extremity, the elbows should be located slightly forward of the shoulders, and the forearms should be supported with the hands in the midline. The head is also in midline with eyes facing forward (Perr, 1998).

When seated and working, the head and neck should be vertically in line with the spine. The trunk should be straight with no torsion. The elbows should be close to the body with forearms approximately parallel to the floor. Wrists should be in neutral. Feet should be flat on the floor with thighs parallel to the floor.

The correct sitting posture has the following health benefits: decreased ligamentous strain to prevent overstretching; decreased muscular strain and overstretching of the back muscles; decreased intradiskal pressure. This results in a healthy spine along the whole kinetic chain because of a reduction in stress on the thoracic and cervical spine and shoulder girdle. The muscles work more efficiently with a reduction in fatigue because the postural muscles are used to support the spine and rib cage while the extremities are used to conduct work. Greater range of motion of the upper extremities is possible, and more efficient diaphragmatic breathing is achieved. More air entering the lungs ensures more oxygenated blood to vital organs, including the brain. With proper seat depth and tilt, there is improved lower extremity circulation (Jacobs & Bettencourt, 1995).

Prolonged sitting can have deleterious effects on the lumbar spine. Unsupported sitting requires higher muscle activity in the thoracic region and more of a load on the lumbar spine because of the backward tilt and flattening of the low back and the forward shifting of the center of gravity, which adds an additional load on the disks. Jacobs and Bettencourt (1995) indicate that "research has provided a dichotomy: disk pressures are reduced when a person sits in an erect posture and maintains the three natural spinal curves but the trunk muscles exert less energy when a person sits in a slightly flexed or slouched position" (p. 140). For this reason, periodic changes in position and properly designed chairs are essential to the maintenance of good posture.

The height of a properly fitted chair should be equal to length of the leg from the back of the knee to the base of the heel with knee bent to 90 degrees and feet flat on the floor. If the seat is too long, there is pressure behind the knee. If too short, there will be pressure on the posterior surface of the thigh and more pressure on the feet. The chair back should be inclined back slightly and with a lumbar support.

An ergonomically well-designed chair would include the following:

- Seat height that is easily adjustable, as in a pneumatic chair.
- Backrest that is easily adjustable to support the lumbar spine vertically (height) and horizontally (forward and backward) and is narrow enough so that the operator's arm or torso does not strike it if rotation is required.
- The seat tilts forward and backward independently of the backrest. This feature is useful with fine detail work and office work.
- The seat edge is curved to reduce pressure under the legs.
- There is enough space between the back of the chair and the seat to accommodate the buttocks.
- The adjustable armrests (optional) are small and low enough to fit under the work surface and to support the back when the worker works close to the work surface.
- The base has five points of contact on the floor (safety).
- The worker can make adjustments easily with one hand while seated.
- The upholstery fabric is comfortable, reduces heat transfer in warm climates and static electricity in cold weather, and is stain resistant and easily cleaned.
- Training is provided to ensure that workers are familiar with the features and adjustments of an optimally fitting chair.

By having an adjustable chair, you can avoid many of the awkward seating positions. A seat that is too low may cause you to laterally flexion the trunk or to rest the body on the arm, fatiguing the neck, shoulders, and back. A seat that is too high will elevate the scapula, resulting in muscle tension and fatigue in the neck and shoulders. Armrests made of hard materials or with sharp corners can irritate the nerves and occlude blood vessels, creating pain or tingling in the fingers, hand, and arm.

Sitting in a wheelchair presents several challenges that this text will only briefly discuss. Many wheelchairs have a sling seat, which is good for transport of the chair because it can be easily folded and placed in a trunk or backseat. However, the position of the person in the wheelchair is one where the hips slide forward and rotate inward, which brings the knees together. This position encourages "sacral sitting," which is an exaggerated posterior pelvic tilt. This position in the wheelchair is not particularly comfortable for long periods of time and makes it more difficult to effectively use one's hands in activities. There are ramifications related to pressure distribution as well as to posture and comfort. With the posterior pelvic tilt, there is additional weight on the ischial tuberosities, coccyx, and possibly lower sacrum, which may precipitate scoliosis or promote pelvic obliquity. Cook and Hussey (1995) add that a "sling seat conforms to the forces generated by the individual instead of providing forces that resist and stabilize. This is an application of Newton's Third Law: Each action (force) has an equal but opposite reaction. The internal forces of the body are not balanced by external forces from the sling seat" (p. 257). These present challenges for the comfort, safety, and function of our clients in wheelchairs.

EVALUATION

Posture is influenced by many factors, including general health status, body build, gender, strength and endurance, visual and kinesthetic awareness, personal habits, and maintenance of one's center of gravity over a sufficient base of support. Anatomic factors that affect posture are body contours, laxity of ligamentous structures, musculotendinous tightness (e.g., hamstrings, tensor fascae latae, pectorals, hip flexors), muscle tone (particularly in gluteus maximus, abdominals, erector spinae), pelvic angle, joint mobility, and neurogenic flow (Magee, 2002). Poor balance coupled with decreased vision and other sensations (especially proprioception and vestibular), lower extremity muscle loss, cognitive impairment, blood pressure fluctuations, and medication side effects are often cited as intrinsic fall risks.

Postural assessment is multidimensional. Because of the functional triad of vestibular, visual, and proprioceptive sensory modalities, these are all specific areas of assessment. Spatial awareness and spatial orientation are perceptual aspects of balance. Awareness of posture and position in space is fundamental to the assessment. Kinesiologic factors, such as achieving and maintaining the center of gravity over the base of support and the ability to move easily and without restriction, are important to consider.

Strength and muscle length assessments of not only the lower extremity but also grip and the upper extremity are done. Essential strength tests would include rectus abdominis, internal oblique, external oblique, and transverse abdominis muscles. In the lower extremity, the strength of iliopsoas, gluteus maximus, the hamstrings, gluteus medius, quadriceps, and tensor fascia latae should be assessed. Serratus anterior, trapezius, infraspinatus, teres minor, and subscapularis should also be tested for strength in the upper extremity (Hall & Brady, 1999). Testing the muscle length of the hamstrings, gastrocnemius, tensor fascia latae, hip flexors, and hip rotators in the lower extremity should be done as well as testing of teres major, latissimus dorsi, rhomboid major, rhomboid minor, levator scapula, pectoralis major, pectoralis minor, and shoulder rotators in the upper extremity.

Range of motion and flexibility, particularly for knee extension and dorsiflexion/plantarflexion, are evaluated with careful attention to the amount of ankle control in ambulation. Muscle tone, peripheral innervation, and involuntary movements are also assessed. Vital signs in lying, sitting, and standing are taken to determine if there are any differences based on position because fluctuations in blood pressure can directly influence balance.

Discuss with the client any fears he or she has about moving or falling, and identify if the client engages in any risk-taking behaviors that would increase the risk of falls (such as climbing on high ladders). Anticipatory and reactive sensory processing can be seen in functional tasks, such as opening a door and lifting objects of different weights. Observation during activities of daily living can also reveal balance and postural dysfunction.

Occupational Profile

A comprehensive evaluation would include both the intrinsic and extrinsic elements related to posture and balance. A thorough history of falls, periods of dizziness, and frequency of episodes of imbalance are important pieces of information to gather from the client. Does the client have a history of depression or cognitive impairments?

Cognition is often overlooked as a variable in balance, posture, and falls. If proprioception diminishes (as occurs with advanced age), people tend to look at the feet when walking as a compensatory strategy. This requires attention because walking is no longer automatic, and there are limits to the number of tasks one can attend to simultaneously (Chronister, 2004). Establishing the client's level of alcohol consumption and the medications that he or she uses is also a part of the occupational profile. Understanding what occupations the client does and when and where they are done will facilitate any necessary environmental adaptations that may be necessary.

Extrinsic factors that can lead to postural disturbance and falls include unstable chairs, steep stairs, poor or inadequate lighting, loose rugs or cords on the floor, improper footwear, or excessive clutter. Visual aspects in the environment, such as patterned carpets or flooring, glare, and poor lighting, also contribute to falls. Diminished sensory and neurological function can impair balance as seen in slowed reaction times, failure of protective responses, and lack of awareness of body parts in space. Coupled with intrinsic factors of decreased muscle strength, diminished bone strength, and failure of shock absorbing structures, it is not difficult to see why there are fractures of the hip following a fall in the elderly that are functionally devastating. If a home visit is not possible, a home assessment completed by the client or family member can be very helpful in identifying environmental hazards.

Observation

Watch the client move and observe the client's posture in sitting and standing, from the front and back. Consider the upper and lower extremity together, because movement and positions in the upper extremity, neck, and back influence posture and balance of the trunk and lower extremity. Look to see if the right and left sides are symmetrical and at the same height at the hip and the shoulder. Observe if the knee joints are straight and aligned. Notice if each body part (or blocks stacked upon each other) is balanced over the part below. Are the normal spinal curves present, or are they excessive or decreased? Is the head protracted? These postural relationships can easily be observed.

A faulty scapular position can influence posture. If there is a forward tilt of the scapula, this results in a forward head position and increased thoracic kyphosis (Figure 9-10). This is a position often assumed when working at a desk on a computer. This creates a muscle length and strength imbalance of not only the scapular muscles but also in the humeral structures. This position is associated with decreased flexibility in pectoralis minor, levator scapula, and scalenius muscle, and weakness in serratus anterior or trapezius muscles. This posture is

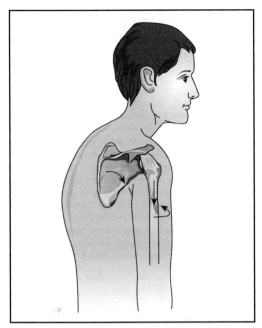

Figure 9-10. Forward head posture. (Adapted from Kisner, C., & Colby, L. A. (2007). *Therapeutic exercise: Foundations and techniques* (5th ed.). Philadelphia, PA: F. A. Davis Company.)

accompanied by cervical lorsis in the midcervical region and extension of the atlanto-occipital joint, which can lead to sub-occipital headaches if the greater occipital nerve is pinched, possible temporomandibular joint pain, and increased lower cervical mobility that may result in degenerative changes in the intervertebral disks. Tight anterior cervical and chest musculature might lead to thoracic outlet syndrome and diminished respiratory capacity. Changes in scapular position can lead to decreased shoulder range of motion (Konin, 1999) with the glenohumeral joint in a relatively abducted and internally rotated position relative to the scapula. Glenohumeral internal rotation is less flexible, and external rotation may be weaker (Kisner & Colby, 2007).

The forward head position is also a posture commonly seen in the elderly (see Figure 9-11). The altered lordotic curve (either flattened or exaggerated) and rounded shoulders are often accompanied by flexed hips and knees. Flexed or stooped posture can result in a chronically stretched neck, trunk, hip, and knee extensor muscles, and a compensatory shift in the vertical displacement of the center of gravity in a backwards direction (Bonder & Bello-Haas, 2009).

Wearing high heels is another example of how changes in one part of the body affect balance and posture in other areas. When high heels are worn, the body weight is displaced anteriorly, leading to an increased curve in the lumbar spine (hyperlordodic lumbar spine or sway back) in order to counterbalance the weight of the body. Due to the increased lumbar curve, the thoracic spine curvature also increases (round back or hyperkyphotic thoracic curve), leading to a rounded shoulder position and anterior movement of the neck. The wearing of heels raises the center of gravity but over an unnaturally narrow base of support, placing an unstable mass onto the

forefoot. There is also a narrow side-to-side base of support, with the tendency to invert or evert the ankle. The weight is shifted onto the toes, and the arch of the foot collapses into inversion. Prolonged lordosis causes a shortening of the hip flexors, which can ultimately lead to degenerative changes in the disks. The Achilles tendon will shorten as will the gastrocnemii and hamstrings. Combined with a weakened quadriceps, this can lead to low back pain, injury, and alterations in gait (Greene & Roberts, 1999).

Balance can be assessed by means of standardized assessments and in the course of activities. Static posture is posture without motion, whereas dynamic posture occurs during active movement (Jacobs & Jacobs, 2009). Simple balance screens can be done by observing the client sitting or standing, bending to reach something from the floor, having the client look over his or her shoulder and up at the ceiling, or having the client reach for something at the side. Ask the client if he or she felt unstable or dizzy in addition to observing for any loss of balance or unsteadiness.

Different scales have been developed to classify balance. The Performance Oriented Mobility Assessment (POMA) is a task-oriented scale that measures gait and balance of older adults (Tinetti, 1986). The balance portion of this test uses nine different maneuvers to test balance including sitting, standing, sit to stand, turning around 360 degrees, receiving a nudge on the sternum, turning the head, leaning back, standing on one leg, and reaching objects from the floor and from overhead. This test has good inter-rater reliability (85% agreement between raters) and good concurrent validity with the Berg Balance Scale and is predictive of fall risks.

A grading scale for balance is used as a basic guideline in many U.S. facilities although evidence to support the use of the scale has not been documented (Jacobs & Jacobs, 2009). The scale is similar to manual muscle grades (zero, trace, poor, fair, good, normal) with pluses and minus given to further refine the scale. A minus rating in the fair and good ranges indicates an inconsistent ability, and notation of assistive devices needs to be included when assessing the client's abilities. Table 9-2 provides the balance guidelines.

Tests of static balance include the Frailty and Injuries: Cooperative Studies of Intervention Techniques (FICSIT-4) (Rossiter-Fornoff, Wolf, Wolfson, & Buchner, 1995), the Clinical Test of Sensory Interaction and Balance (CTSIB) or Sensory Organization Test (Shumway-Cook & Horak, 1986), and the Romberg test. In the FICSIT, there are two balance scales comprising three or four tests of static balance. This tool has acceptable reliability, validity, and discriminate ability (Rossiter-Fornoff et al., 1995). The CTSIB test assesses the interaction of somatosensory, visual, and vestibular elements on stability. The Romberg test assesses stance as a test of proprioception when a client stands with his or her eyes open and then with the eyes closed.

There are many dynamic assessments of balance. The Berg Balance Scale (Berg, Wood-Dauphinee, Williams, & Gayton, 1995) looks at balance components common to many functional tasks, such as reaching, bending, transfer, stand, arise. It has high inter-rater reliability (0.98, ICC -0.98), high internal consistency (Cronbach's = 0.96), and good concurrent validity (r = 0.91 with Tinetti, r = 0.76 with Get Up and Go). The Four Square Step Test (FSST) is a test of dynamic

Table 9-2

BALANCE GUIDELINES

GRADE	POSTURE	MOVEMENT	ABILITY OF INDIVIDUAL
0	Sitting	Static	Needs max assistance to maintain sitting without back support
0	Sitting	Dynamic	N/A
0	Standing	Static	Needs max assistance to maintain
0	Standing	Dynamic	N/A
P	Sitting	Static	Needs moderate assistance to maintain
P	Sitting	Dynamic	N/A
P	Standing	Static	Needs moderate assistance to maintain
P	Standing	Dynamic	Needs moderate assistance during gait
P+	Sitting	Static	Needs minimum assistance to maintain
P+	Sitting	Dynamic	N/A
P+	Standing	Static	Needs minimal assistance to maintain
P+	Standing	Dynamic	Needs minimal assistance during gait
F	Sitting	Static	Maintains without assistance but unable to take any challenges
F	Sitting	Dynamic	N/A, cannot move trunk without losing balance
F	Standing	Static	Maintains without assistance but unable to take any challenges
F	Standing	Dynamic	Needs contact guard assistance during gait
F+	Sitting	Static	Able to take minimum challenges from all directions
F+	Sitting	Dynamic	Maintains balance through minimal active trunk motion

(continued)

Table 9-2 (continued)

BALANCE GUIDELINES

GRADE	POSTURE	MOVEMENT	ABILITY OF INDIVIDUAL
F+	Standing	Static	Takes minimal challenges from all directions
F+	Standing	Dynamic	Needs close supervision during gait; able to right self with minor loss of balance
G	Sitting	Static	Takes moderate challenges in all directions
G	Sitting	Dynamic	Maintains balance through moderate excursion of active trunk motion
G	Standing	Static	Takes moderate challenges in all directions
G	Standing	Dynamic	Needs supervision only during gait; is able to right self with moderate loss of balance
G+	Sitting	Static	Takes maximal challenges in all directions
G+	Sitting	Dynamic	Maintains balance through maximum excursion of active trunk motion
G+	Standing	Static	Takes maximum challenges in all directions
G+	Standing	Dynamic	Independent gait with/without devices
N	No deviation seen in posture held statically or dynamically		

stepping ability and standing balance (Dite & Temple, 2002). The Functional Reach Test measures self-initiated movement within a fixed base of support while standing (Duncan, Weiner, Chandler, & Studenski, 1990). The Functional Reach Test has 89% test-retest reliability with ICC 0.99. The Physical Performance Test (PPT) has seven out of nine items that are related to static and dynamic balance including making a 360 degrees turn, donning/doffing a jacket, picking up a heavy book, picking up a penny from the floor, stair climbing, and walking 50 feet (Ruben & Slu, 1990). The Timed Up and Go Test (Podsiadlo & Richardson, 1991) times how long it takes a person to go from sit to stand, walk 3 meters, and return. Inter-rater reliability is 0.99 with ICC=0.99 with 0.81 concurrent validity with the Berg balance scale. The complexity of the test can be modified by having the client count backwards from a randomly selected number between 20 and 100 (cognitive aspects) or by carrying a full glass of water during the test. A self-administered interview about the client's confidence in doing activities without loss of balance or unsteadiness is the activities-specific balance confidence (ABC) (Powell & Myers, 1995). Many standardized balance measures have established norms for age and gender.

In a review of balance assessments, Whitney, Poole, and Cass (1998) concluded that many of the balance assessments require minimal equipment and time to administer. Timed or ratio measurement tools, such as the Timed Up and Go Test and the Physical Performance Test, were more sensitive to change. The Berg Balance Test, the Functional Reach Test, and the Tinetti Balance Test are helpful in setting goals and seem to be good predictors of falls.

Sensory Triad: Vestibular, Visual, and Proprioceptive Assessment

The physiologic basis for many of the balance tests (particularly the Romberg test) is the combination of proprioception, vestibular, and visual sensory systems. One part of the Romberg test is done with the client's eyes open, which tests whether the sensorimotor integration (cerebellum and dorsal column-medial lemniscus tract) and motor pathways (corticospinal tract and medial and lateral vestibular tracts) are functioning. Vision is occluded in another part of the test, so a loss of proprioception or vestibular sensation is more obvious. Maintaining balance relies on these intact sensory pathways, sensorimotor integration, and motor pathways.

In the sensory triad related to balance, each sensory modality adds an important piece to the ability to remain balanced. The vestibular system detects motion and the position of the head in space in relation to gravity. This influences muscle tone and helps to maintain stable visual perception when moving. Proprioceptive input informs the body where the head is in relation to the body and monitors the intensity, velocity, and tension of body segments. Vision tells us the relation of the head to the object and aids in head and neck alignment.

Visual Assessment

Visual assessment related to balance and posture would focus primarily on assessment of oculomotor skills. These would include an evaluation of smooth visual pursuit, saccades, nystagmus, gaze stabilization, and convergence/divergence. Visual field and visual acuity testing is also important. In addition, eye-hand coordination and visual-vestibulocular reflex interaction would be additional areas to assess.

Proprioception Assessment

Tests of gross sensory evaluation often include tests of proprioception (position sense) and kinesthesia (movement sense). Tests of proprioception have the therapist move the client in one position, and then the client describes the position or assumes the position on the opposite side.

Vestibular Assessment

A comprehensive vestibular assessment includes many of the motor and sensory components already discussed. In addition, gross cerebellar tests of coordination are helpful to identify specific deficits in eye-hand function. Primary symptoms of vestibular dysfunction include vertigo or dizziness, imbalance, decreased safety, falls, visual-motor dysfunction (e.g., blurred vision, spatial disorientation), and nystagmus. Secondary symptoms may also be present, which may include nausea, muscle tension, postural rigidity, anxiety, decreased conditioning, decreased activity levels, and fear of movement (Cohen, Burkhardt, Cronin, & McGuire, 2006). The assessment of the vestibular system needs to involve moving the head to elicit these feelings of spinning or whirling and its intensity and frequency.

Clients might report avoiding tasks that require bending over or bending down (such as looking into low cupboards or bending to put on a pair of socks), bending forward (as when grooming the hair or brushing teeth), bending the head back (to swallow medication), repetitive head movements (driving), navigating their way in stores, carrying objects, or scanning grocery shelves (Cohen et al., 2006).

A vertigo assessment would start first with an evaluation of the vertebrobasilar artery to determine if it is safe for more provocative testing. The client will be seated and is asked to actively rotate and hyperextend the head to each side for 3 to 5 seconds. A positive sign is if the client experiences vertigo, nystagmus, nausea, pupil dilation, or syncope indicative of vertebral, basilar artery insufficiency. If these symptoms occur, do not proceed with additional testing. Other tests of vascular insufficiency resulting in symptoms of vertigo, nystagmus, or dizziness include George's Screening Procedure, Barre-Leiou test, Dekleyn's test, and Hallpike's test.

In the Hallpike's test, the client is in a supine position with the head extended off the end of the table. The head is then passively moved into extension, rotation, and lateral flexion to each side for 15 seconds. A Modified Hallpike test has the client in supine with the head tilted comfortably and then gently passively moved 45 degrees to one side. Other modifications include testing the client in sidelying. The Dix-Hallpike test (Nylen-Barany test) is performed with the client seated with the legs extended and the head rotated 45 degrees. The client then lies down backwards with the head in 20 degrees of extension.

Have the client change the position of the head, and assess the intensity of the symptoms and the time required to recover. Moving the head from side to side with the eyes open and closed, moving the head up and down with eyes open and closed, tilting the head to the right and left with the eyes open and closed, bending forward from the waist with eyes open and closed, and moving from sit to supine and then supine to sit are positional head movements that may elicit vestibular dysfunction symptoms (Cohen et al., 2006; Cohen & Gavia, 1998).

Vestibular intervention done by occupational therapists includes pre-positioning strategies, vertigo habituation exercises, gaze stabilization activities, balance therapy, environmental modification and assistive devices, patient education, mobility training, and task modification (Cohen, Miller, Kane-Wineland, & Hatfield, 1995; Cohen et al., 2006). Evidence supports the use of vestibular home programs to decrease symptoms of vertigo and improve ADLs (Cohen & Kimball, 2003).

Summary

- The spine is formed by 33 bones, including the cervical, thoracic, lumbar, sacral, and coccygeal vertebral bones. There are distinct differences in each of the spinal segments dependent upon the purpose and function of these segments.

- The spine has four naturally occurring curves. If there is more or less of a curve than what naturally occurs, the balance of the spine, pelvis, and body is disturbed. This can result in pain, deformity, and loss of function.

- The lower extremities absorb high forces and support the body weight as well as providing a mechanism for movement.

- The hip, a ball-and-socket joint, is capable of a wide variety of movement. The hip generally serves to provide stability, weightbearing, and balance.

- Movements of the pelvis include pelvic tilt, pelvic shift, rotation, and lumbar-pelvic rhythm. These movements enable flexion/extension, abduction/adduction, and internal/external rotation of the hip.

- The two-joint function of rectus femoris and the hamstrings (semitendinosus, semimembranosus, and biceps femoris) can act as an agonist at one joint and as an antagonist at another. This action is contradictory to the muscle action in the upper extremity, where a two-joint muscle acts as an agonist at all of the joints it crosses.

- Correct posture in sitting, standing, or lying maintains forces in the body and expends the least amount of energy. Postural control is maintained by automatic postural adjustments and is affected by body build, natural movement ability, central nervous system functioning, vision, vestibular system function, perception, muscle tone, and other factors.

- Faulty sitting or standing postures limit function and expend excess energy. Good posture is achieved when the body weight is borne evenly on various surfaces and natural spinal curves are maintained. Careful attention to posture, whether seated or standing, is essential to functional performance.

References

Berg, K. O., Wood-Dauphinee, S. L., Williams, J. I., & Gayton, D. (1995). Measuring balance in the elderly: Validation of an instrument. *Canadian Journal of Public Health, 83*, S7-S11.

Bible, J. E., Biswas, D., Miller, C. P., Whang, P. G., & Grauer, J. N. (2001). Normal functional range of motion of the cervical spine during 15 activities of daily living. *Journal of Spinal Disorders and Techniques, 23*(1), 15-21.

Bonder, B. R., & Bello-Haas, V. D. (2009). *Functional performance in older adults* (3rd ed.). Philadelphia, PA: F. A. Davis Company.

Chronister, K. (2004). Cognition: The missing link in fall-prevention programs. *Occupational Therapy Practice*, November 29.

Cohen, H., & Kimball, K. (2003). Increased independence and decreased vertigo after vestibular rehabilitation. *Otolaryngology-Head and Neck Surgery, 128*, 60-70.

Cohen, H. S., Burkhardt, A., Cronin, G. W., & McGuire, M. J. (2006). Specialized knowledge and skills in adult vestibular rehabilitation for occupational therapy practice. *American Journal of Occupational Therapy, 60*(6), 669-678.

Cohen, H. S., & Gavia, J. A. (1998). A task for assessing vertigo elicited by repetitive head movements. *American Journal of Occupational Therapy, 52*(8), 644-649.

Cohen, H., Miller, L. V., Kane-Wineland, M., & Hatfield, C. L. (1995). Vestibular rehabilitation with graded occupations. *American Journal of Occupational Therapy, 49*(4), 362-367.

Cook, A. M. & Hussey, S. M. (1995). *Assistive technologies: Principles and practice.* St. Louis, MO: C.V. Mosby.

Dite, W., & Temple, V. A. (2002). A clinical test of stepping and change of direction to identify multiple falling older adults. *Archives of Physical Medicine and Rehabilitation, 83*(11), 1366-1371.

Duncan, P., Weiner, D., Chandler, J., & Studenski, S. (1990). Functional reach: A new clinical measure of balance. *Journal of Gerontology, 45*(6), M192-M197.

Esch, D. L. (1989). *Musculoskeletal function: An anatomy and kinesiology laboratory manual.* Mineapolis, MN: University of Minnesota.

Galley, P. M., & Forster, A. L. (1987). *Human movement: an introductory text for physiotherapy students.* New York: Churchill-Livingstone.

Gench, B. E., Hinson, M. M., & Harvey, P. T. (1995). *Anatomical kinesiology.* Dubuque, IA: Eddie Bowers Publishing, Inc.

Goss, C. M. (1973). *Anatomy of the human body* (29th ed.). Philadelphia, PA: Lea & Febiger.

Greene, D. P., & Roberts, S. L. (1999). *Kinesiology: Movement in the context of activity.* St. Louis, MO: Mosby.

Hall, C. M., & Brady, L. T. (1999). *Therapeutic exercise: Moving toward function.* Philadelphia, PA: Lippincott, Williams & Wilkins.

Hamill, J. & Knutzen, K. M. (1995). *Biomechanical basis of human movement.* Philadelphia, PA: Lippincott, Williams & Wilkins.

Hinkle, C. Z. (1997). *Fundamentals of anatomy and movement: A workbook and guide.* St. Louis, MO: C.V. Mosby.

Jacobs, K. & Bettencourt, C. M. (1995). *Ergonomics for therapists.* Boston, MA: Butterworth-Heinemann.

Jacobs, K., & Jacobs, L. (Eds.) (2009). *Quick reference dictionary for occupational therapy* (5th ed.). Thorofare, NJ: SLACK Incorporated.

Jenkins, D. B. (1998). *Hollingshead's functional anatomy of the limbs and back.* (7th ed.). Philadelphia: W. B. Saunders Co.

Kendall, F. P. & McCreary, E. K. (1993). *Muscle testing and function.* Baltimore, MD: Williams & Wilkins.

Kisner, C., & Colby, L. A. (2007). *Therapeutic exercise: Foundations and techniques* (5th ed.). Philadelphia, PA: F. A. Davis Company.

Konin, J. G. (1999). *Practical kinesiology for the physical therapy assistant.* Thorofare, NJ: SLACK Incorporated.

Kvitne, R. S., & Jobe, F. W. (1993). The diagnosis and treatment of anterior instability in the throwing athlete. *Clinical Orthopaedics 291*, 107-123.

Levangie, P. K., & Norkin, C. C. (2005). *Joint structure and function: A comprehensive analysis* (4th ed.). Philadelphia, PA: F. A. Davis Company.

Magee, D. J. (2002). *Orthopedic physical assessment.* Philadelphia, PA: Saunders.

Magee, D. J. (2008). *Orthopedic physical assessment.* Philadelphia, PA: Saunders.

Mulholland, S. & Wyss, U. P. (2001). Activities of daily living in non-western cultures: Range of motion requirements for hip and knee joint implants. *International Journal of Rehabilitation Research, 23*(3), 191-198.

Muscolino, J. E. (2006). *Kinesiology: The skeletal system and muscle function.* St. Louis, MO: Mosby.

Neumann, D. A. (2002). *Kinesiology of the musculoskeletal system: Foundations for physical rehabilitation.* St. Louis, MO: Mosby.

Norkin, C. C., & Levangie, P. K. (1992). *Joint structure and function: A comprehensive analysis* (2nd ed.). Philadelphia, PA: F. A. Davis Co.

Perr, A. (1998). Elements of seating and wheeled mobility intervention. *OT Practice*, October, 16-24.

Podsiadlo, D., & Richardson, S. (1991). The timed "Up and Go": a test of basic functional mobility for frail elderly persons. *Journal of American Geriatric Society, 39*, 142-148.

Powell, L. E., & Myers, A. M. (1995). The Activities-specific Balance Confidence (ABC) Scale. *Journal of Gerontological Medical Science, 50*(1), M28-M34.

Rossiter-Fornoff, J. E., Wolf, S. L., Wolfson, L. I., & Buchner, D. M. (1995). Tests of static balance: parallel, semi-tandem, tandem and one-legged stance tests. *Journals of Gerontology Series A: Biological Sciences and Medical Sciences, 50*(6), M291-M297.

Ruben, R. B., & Slu, A. L. (1990). An objective measure of physical function of elderly outpatients. *Journal of American Geriatric Society, 38*, 1105-1112.

Shumway-Cook, A., & Horak, F. B. (1986). Assessing the influence of sensory interaction on balance. *Physical Therapy, 66*, 1548-1550.

Sine, R., Liss, S. E., Roush, R. E., Holcomb, J. D., & Wilson, G. (2000). *Basic rehabilitation techniques: A self-instructional guide.* Gaithersburg, MD: Aspen Publishers.

Smith, L. K., Weiss, E. L., & Lehmkuhl, L. D. (1996). *Brunnstrom's clinical kinesiology* (5th ed.). Philadelphia, PA: F. A. Davis Co.

Tinetti, M. E. (1986). Performance-oriented assessment of mobility problems in elderly patients. *Journal of the American Geriatric Society, 34*(2), 119-126.

Trew, M., & Everett, T. (Eds.) (2005). *Human movement: An introductory text.* London: Elsevier Churchill Livingstone.

Whitney, S. L., Poole, J. L., & Cass S. P. (1998). A review of balance instruments for older adults. *American Journal of Occupational Therapy, 52*(8), 666-671.

ADDITIONAL RESOURCES

AOTA. (2003). Fall prevention for people with disabilities and older adults [consumer tip sheet]. American Occupational Therapy Association.

Basmajian, J. V., & DeLuca, C. J. (1985). *Muscles alive* (5th ed.). Baltimore, MD: Williams & Wilkins.

Bassett, J. (2002). A strong balance: exercise can eliminate a major cause of falls. *Advance for Occupational therapy practitioners*, December 16, 2002.

Baxter, R. (1998). *Pocket guide to musculoskeletal assessment.* Philadelphia, PA: W. B. Saunders Co.

Blanche, E. I., & Reinoso, G. (2008). The use of clinical observations to evaluate proprioceptive and vestibular functions. *Occupational Therapy Practice, 13*(17), CE1-CE7.

Burstein, A. H., & Wright, T. M. (1994). *Fundamentals of orthopaedic biomechanics.* Baltimore, MD: Williams & Wilkins.

Ceranski, S., & Haertlein, C. (2002). Helping older adults prevent falls. *OT practice*, July 22, 2002.

Clemson, L., Cumming, R. G., & Heard, R. (2003). The development of an assessment to evaluate behavioral factors associated with falling. *American Journal of Occupational Therapy, 57*(4), 380-388.

Cronin, G. W. (2003). Occupational therapy in vestibular rehabilitation. *Occupational Therapy Practice*, August 4, 21-25.

Cumming, R. G., Thomas, M., Szonyi, G., Frampton, G., Salkeld, G., & Clemson, L. (2001). Adherence to occupational therapist recommendations for home modifications for falls prevention. *American Journal of Occupational Therapy, 55*(11), 641-647.

Daniels, L., & Worthingham, C. (1977). *Therapeutic exercise for body alignment and function* (2nd ed.). Philadelphia, PA: W. B. Saunders Co.

Demeter, S. L., Andersson, G. B. J., & Smith, G. M. (1996). *Disability evaluation.* St. Louis, MO: C. V. Mosby.

Diffendal, J. (2001). Before the fall: OTs should discuss fall risk factors and prevention with appropriate clients. *Advance for Occupational Therapy Practitioners*, June 25, 2001.

Durward, B. R., Baer, G. D., & Rowe, P. J. (1999). *Functional human movement: Measurement and analysis.* Oxford: Butterworth Heinemann.

Funk, K. P. (2008). Managing vestibular disorders. *Occupational Therapy Practice*, November 17, 11-15.

Gans, R. E. (2003). In a spin: using vestibular rehabilitation therapy to acclimate to conflicting input. *Advance for Occupational Therapy Practitioners*, August 11, 2003.

Greene, D. P., & Roberts, S. L. (1999). *Kinesiology: Movement in the context of activity.* St. Louis, MO: C. V. Mosby.

Hartley, A. (1995). *Practical joint assessment: Lower quadrant* (2nd ed.). St. Louis, MO: C. V. Mosby.

Kangas, K. (2002). Seating for task performance. *Rehab Management*, June/July 2002, 54-74.

Kisner, C., & Colby, L. A. (1990). *Therapeutic exercise: Foundations and techniques* (2nd ed.). Philadelphia, PA: F. A. Davis Co.

Lange, M. (2001). Positioning philosophies. *OT practice*, October 1, 15-16.

Lange, M. L. (2000). Dynamic seating. *OT practice*, July 2, 2122.

Loth, T., & Wadsworth, C. T. (1998). *Orthopedic review for physical therapists.* St. Louis, MO: C. V. Mosby.

Lumley, J. S. P. (1990). *Surface anatomy: The anatomical basis of clinical examination.* New York, NY: Churchill Livingstone.

MacKenna, B. R., & Callender, R. (1990). *Illustrated physiology* (5th ed.). New York, NY: Churchill Livingstone.

Magee, D. J. (1992). *Orthopedic physical assessment.* Philadelphia, PA: W. B. Saunders Co.

Morris, P. A. (1991). A habitual approach to treating vertigo in occupational therapy. *American Journal of Occupational Therapy, 45*(6), 556-558.

Morris, P. A. (1991). A habituation approach to treating vertigo in occupational therapy. *American Journal of Occupational Therapy, 45*(6), 556-558.

Morse, J. (1997). *Preventing patient falls.* Thousand Oaks, CA: Sage Publisher.

Nordin, M., & Frankel, V. H. (1989). *Basic biomechanics of the musculoskeletal system* (2nd ed.). Philadelphia, PA: Lea and Febiger.

Palastanga, N., Field, D., & Soames, R. (1989). *Anatomy and human movement: Structure and function.* Philadelphia, PA: Lippincott Williams & Wilkins.

Palmer, M. L., & Epler, M. E. (1998). *Fundamentals of musculoskeletal assessment techniques* (2nd ed.). Philadelphia, PA: J. B. Lippincott.

Perry, J. F., Rohe, D. A., & Garcia, A. O. (1996). *Kinesiology workbook.* Philadelphia, PA: F. A. Davis Co.

Smith, L. K., Weiss, E. L., & Lehmkuhl, L. D. (1996). *Brunnstrom's clinical kinesiology* (5th ed.). Philadelphia, PA: F. A. Davis Co.

Swedberg, L. M. (1998). Low-tech adaptations for seating and positioning. *OT Practice*, October, 26-34.

Tideiksaar, R. (2009). Interruptions in function: Falls. In B. R. Bonder & V. D. Bello-Haas (Eds.). *Functional Performance in Older Adults.* Philadelphia, PA: F.A. Davis Co.

Walker, J. E., & Howland, J. (1991). Falls and fear of falling among elderly persons living in the community: occupational therapy interventions. *American Journal of Occupational Therapy, 45*(2), 119-122.

Wilson, B., Pollock, N., Kaplan, B. J., Law, M., & Faris, P. (1992). Reliability and construct validity of the clinical observations of motor and postural skills. *American Journal of Occupational Therapy, 46*(9), 775-783.

Wu, S.-H., Huang, H.-T., Lin, C.-F., & Chen, M.-H. (1996). Effects of a program on symmetrical posture in patients with hemiplegia: A single-subject design. *American Journal of Occupational Therapy, 50*(1), 17-23.

10

THE KNEE, ANKLE, AND FOOT

The knee, ankle, and foot are a continuation of the lower extremity kinetic chain. The knee contributes to the functions of walking and running by adjusting the length of the leg to absorb shocks, conserve energy, and transmit forces. The foot is pliable enough to conform to a variety of surfaces and absorb shock as well as to impart thrust when ambulating. It is also rigid enough to withstand large propulsive and compressive forces (Magee, 2008; Neumann, 2002).

THE KNEE

The knee, with the hip and ankle, supports the body. In addition, the knee "functionally lengthens and shortens the lower extremity to raise and lower the body or move the foot in space" (Kendall & McCreary, 1983, p. 345). While occupational therapists rarely remediate the knee, ankle, or foot by either stretch or strengthening, this chapter will serve as an overview of the contributions that these structures add to functional performance and skills in occupations.

While the knee is often considered a simple hinge or ginglymus joint, it is actually a joint complex made up of three articulations. Two articulations are between each condyle of the femur and condyle of the tibia and are called the medial and lateral tibiofibular joints. These are condyloid joints, and only the superior tibiofibular articulation is contained within the knee. A partly arthrodial joint, formed by the articulation between the patella and femur, forms the third articulation or the patellofemoral joint (Goss, 1976; Nordin & Frankel, 1989). The tibia and fibula have been seen as corresponding to the

ulna and radius of the upper extremity while the anterior surface of the leg is seen to correspond to the extensor surface of the forearm (Jenkins, 1998).

Of the two bones in the leg, the tibia is larger, more medially located, and articulates with both the knee and ankle. The fibula is smaller, more lateral in location, and articulates only with the ankle but not the knee. The knee joint is actually the articulation of the lower femur and upper tibia and acts as a double condyloid joint, producing the movements of knee flexion and extension in a sagittal plane around a coronal axis and medial and lateral rotation, which occurs in the transverse plane around a vertical axis.

The joint capsule does not form a complete covering around the joint but contributes to knee stability. Stability of the knee is provided primarily by the ligaments, muscles, menisci, joint capsule, and cartilage, while the bones provide mobility (Smith, Weiss, & Lehmkuhl, 1996) (Figure 10-1).

The ligaments of the knee are important, especially because there are no bony restraints to knee movements, and these are shown in Figure 10-1. The ligaments resist or control the following:

- Excessive knee extension

- Varus and valgus stresses of the knee (attempted adduction or abduction of tibia)

- Anterior or posterior displacements of the tibia beneath the femur

- Medial or lateral rotation of the tibia beneath the femur

- Combinations of anterior-posterior displacement and rotation of tibia (Norkin & Levangie, 1992, p. 347)

Rybski MF.
Kinesiology for Occupational Therapy, Second Edition (pp 299-306)
© 2012 SLACK Incorporated

Figure 10-1. Ligaments of the knee. (Adapted from Muscolino, J. E. (2006). *Kinesiology: The skeletal system and muscle function.* St. Louis, MO: Mosby.)

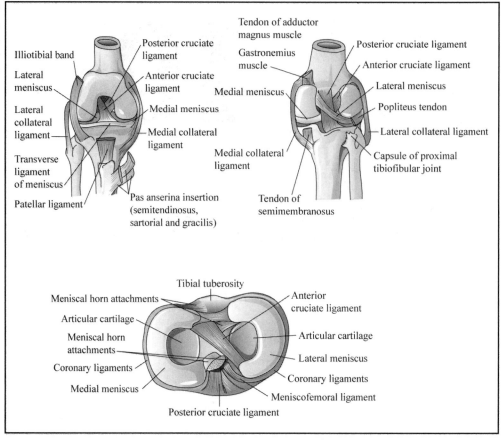

There are two collateral and two cruciate ligaments that function as passive load-carrying structures and as a back up for the muscles. The medial collateral ligament (MCL), or medial tibial ligament, attaches to the medial aspect of the medial femoral epicondyle and supports the knee against valgus forces and offers some resistance to rotational stresses. This ligament is taut with knee extension. The thinner lateral collateral ligament (LCL) provides the main resistance to varus forces and lateral rotation and is also taut in knee extension. In full extension, the collateral ligaments are assisted by the tightening of the posteromedial and posterolateral joint capsules, making extension the most stable position of the knee joint. When the knee is flexed, the ligaments are lax so that the tibia can rotate around the long axis. When the knee is extended, the tight ligaments and bony structures are more congruent, which prevents tibial rotation. Because the knee often is in a flexed position (as in dressing, bathing, etc.), effort should be made to ensure that the muscles of the knee, which are the last line of defense against injury, be maintained at peak strength to avoid injurious twisting of the tibia (Gench, Hinson, & Harvey, 1995).

The iliotibial band (ITB) is formed from the fascia from tensor fascia latae, gluteus maximus, and gluteus medius muscles. It reinforces the anterolateral aspect of the knee joint and assists with preventing posterior displacement of the femur when the tibia is fixed and the knee is in extension (Norkin & Levangie, 1992).

The two cruciate ligaments are important in controlling anterior-posterior and rotational movements. The anterior cruciate ligament (ACL) is the primary restraint for anterior movement of the tibia relative to the femur. Different parts of the ACL are tight in different positions—the anterior portions are tight in extension, the middle portions are tight in internal rotation, and the posterior portions are tight in flexion. Like the collateral ligaments, the ACL as a whole is taut in extension. The ACL has secondary functions of limiting internal and external rotation.

The posterior cruciate ligament (PCL) is the primary restraint to posterior movement of the tibia on the femur. The posterior fibers of the PCL are taut in extension, while the anterior portion is taut in midflexion and posterior in full flexion (Hamill & Knutzen, 1995). In spite of the numerous ligaments that support the knee, it is the most frequently injured joint in the body (Hamill & Knutzen, 1995). The palpable structures in the knee are listed below.

Femur

Medial and Lateral Condyles

With the knee flexed, these distal enlargements or rounded projections of femur or condyles can be easily felt anteriorly on both sides of the patella (Esch, 1989; Sine, Liss, Rousch, Holcomb, & Wilson, 2000) and are easily palpated.

Medial and Lateral Epicondyles

These are two roughened prominences proximal to the condyles.

Patella

This flat, rounded, triangular sesamoid bone is easily seen and palpated on the front of the knee. When the knee is extended, the patella is freely movable because the quadriceps are relaxed (Esch, 1989; Goss, 1976).

Tibia

Tuberosity of the Tibia

Located anteriorly on the tibia and below the tibial condyles, this roughened area is easily palpated 2 inches below the inferior border of patella with the knee flexed.

Anterior Crest (or Border) of the Tibia

This is sharp and can be followed distally to the ankle (Smith et al., 1996).

Fibula

The head is irregular and flat in shape and is located on the posterior and lateral aspect of the lateral condyle of the tibia at the level of the tibial tuberosity (Esch, 1989).

Tibiofemoral Joint

This biaxial modified hinge joint (Kisner & Colby, 1990), which some consider to be a double condyloid (Norkin & Levangie, 1992), is made up of two large condyles on the distal femur and two asymmetrical concave tibial plateaus or condyles on the proximal tibia. The medial condyle is longer than lateral and contributes to locking of the knee. The articular surfaces of the tibia and femur are not congruent, so they move different amounts. The tibia and femur approach congruency in full extension, which is the close-packed position of the joint.

There are two fibrocartilaginous joint disks called menisci that enhance joint stability by deepening the contact surface on the tibia. The menisci also aid in shock absorption, help to protect underlying articular cartilage and subchondral bone, serve to reduce the load per unit area on tibiofemoral contact sites, enhance lubrication, and limit the motion between the tibia and femur (Hamill & Knutzen, 1995). The medial menisci are subject to injury if there is a lateral blow to the knee joint.

Stability is provided by ligaments. Tibial and fibular (or medial and lateral) collateral ligaments prevent passive movements of the knee in the frontal plane (Smith et al., 1996). These ligaments prevent abduction and adduction of the tibia on the femur, which would produce genu valgum (knock knee) and genu varum (bowleg), respectively. Anterior-posterior stability is provided via the posterior and anterior cruciate ligaments. Coronary ligaments attach the menisci to the joint capsule.

Patellofemoral Joint

The patella is a sesamoid bone in the quadriceps tendon that articulates with the anterior aspect of the distal femur. This is a modified plane joint, and during flexion and extension, different parts of the patella articulate with the femoral condyles (Magee, 2008). The patella is embedded in the anterior portion of the joint capsule and is connected to the tibia by the ligamentum patellae (Kisner & Colby, 1990). Many bursae surround the patella. The patella slides caudally with flexion of the knee and slides cranially with extension. Patellofemoral surface motion simultaneously occurs in two planes, the frontal and transverse, but the greatest motion occurs in the frontal plane.

Support for this joint is by the quadriceps tendon and the patellar ligament. The angle formed between the femur and in relation to the position of the patella is known as the Q angle. The Q angle is greater in women due to wider pelvic girdles, and increases in this angle will increase the valgus stress on the knees (Hamill & Knutzen, 1995).

The patella is helpful in knee extension because it serves to extend the lever arm of the quadriceps muscle throughout the entire range of motion (ROM) and allows a wide distribution of compressive stress on the femur (Magee, 1992).

Joint action between the patella and femur is important. Vastus medialis and vastus lateralis exert opposite forces on the patella, and if there are muscular imbalances, contraction of the quadriceps will pull the patella off center. The patella can be kept in balance by exercising the knee joint through full range of extension.

Movements of the Knee

Movements that take place at the knee joint occur due to the motion at the tibiofemoral joint. These motions (primarily flexion and extension but also rotation of the tibia on the femur in nonweight-bearing positions) occur in three planes simultaneously, with the greatest amount of motion occurring in the sagittal plane. The movements of flexion and extension differ from the movements at the elbow in that: 1) the axis of motion is not a fixed axis but shifts anteriorly during extension and posteriorly during flexion and 2) the beginning of flexion and ending of extension are accompanied by rotary movements associated with fixation of the limb (Goss, 1976). This mechanical movement of extension with external rotation and flexion with internal rotation has been called the "screw-home mechanism" and provides mechanical stability that is energy efficient to withstand forces occurring in the sagittal plane.

The axis for flexion is a few centimeters above the joint line transversely through femoral condyles. Clinically, the joint axis can be located approximately through the center of the lateral and medial condyles of the femur. The shifting of the joint axis creates problems when fitting for orthoses (such as long leg braces or below-the-knee prosthetic devices) or when using goniometers or isokinetic dynamometers. When one moves from extension to flexion, the anatomic axis moves 2 cm, while the mechanical axis remains fixed. It is important to carefully align the devices to prevent discomfort and abrasions. The muscles that cross anterior to this axis are extensors, and those crossing posterior to the axis are flexors. Internal and external rotation occurs around the long axis, and muscles that insert medial to the axis will internally rotate the leg, whereas those inserting laterally will externally rotate the leg (Gench et al., 1995).

Arthrokinematically, if the motion of the tibia is in an open chain, the concave tibial plateaus slide in the same direction as the bone motion (Kisner & Colby, 1990). However, if the motions of the femur occur on a fixed tibia (or closed kinetic chain), the convex condyles slide in the direction opposite to bone motion. The large articular surface of the femur and relatively small tibial condyle create problems as the femur initiates flexion on the tibia. If there was only rolling of the fibial condyle on the tibial condyle, the femur would roll off of the tibia. For this reason, there must be simultaneous rolling and sliding during flexion and extension. The femoral condyle rolls posteriorly while simultaneously sliding anteriorly during flexion. The opposite motions occur during extension with anterior rolling of the femoral condyle while there is simultaneous sliding in a posterior direction.

Internal Kinetics: Muscles

Knee flexion occurs in a range from 120 degrees to 145 degrees. The motion is due to the three hamstring muscles (biceps femoris, semi-membranosus, and semi-tendinosus muscles) and is accompanied by internal rotation of the tibia, which is produced by sartorius, popliteus, and gracilis (as well as semi-membranosus and semi-tendinosus). The semi-tendinosus and semi-membranosus are particularly well-suited to flex and internally rotate the leg because they are located posteriorly to the frontal axis as it crosses the knee and medially to the long axis. For example, biceps femoris, while active in flexion, does not assist with internal rotation because its location lateral to the long axis makes this muscle responsible for external rotation of the tibia. It is an important muscle for knee stability because it will neutralize the internal rotation of the other knee flexors (Gench et al., 1995). Sartorius courses diagonally and medially in the front of the thigh and acts both as a flexor and internal rotator, although it should be noted that, in some individuals, the muscle crosses anterior to the knee joint rather than posteriorly, which would make this muscle an extensor of the knee (Gench et al., 1995). There are several calf muscles that extend across the knee and have a flexor action there, including gastrocnemius, plantaris, and popliteus.

The quadriceps femoris group is made up of the vastus lateralis, vastus medialis, vastus intermedius, and rectus femoris muscles, and these participate in extension of the leg. The iliotibial band, with its origins from the tensor fascia latae, gluteus maximus, and gluteus medius, has been reported to have an effect on the knee through this band (Jenkins, 1998). Some sources also cite gastrocnemius and soleus as contributors to knee extension (Jenkins, 1998).

The medial rotators are essentially the same muscles indicated as flexors. Sartorius, gracilis, semi-membranosus, and semi-tendinosus all cross the joint medial to the axis and act as internal rotators. Popliteus also crosses the knee joint distally and medially and acts as a medial rotator. Lateral rotation is achieved by the biceps femoris muscle, which passes on the lateral side of the joint axis (Jenkins, 1998).

Knee stabilization can be categorized on the basis of function, structure, or location of supporting structures. Categorization based on location refers to embryonic joint compartments (Norkin & Levangie, 1992). Functional stabi-

lizers can be static (joint capsule and ligaments) or dynamic (muscles and aponeurosis). The joint capsule and coronary, meniscopatellar, patellofemoral, middle and lateral collateral and anterior and posterior cruciate ligaments, as well as the oblique popliteal, arcuate, and transverse ligaments are all static stabilizers. Muscles providing dynamic stabilization are the quadriceps femoris, extensor retinaculum, popliteus, biceps femoris, semi-membranosus, and the pes anserinus (made up of the semi-tendinosus, sartorius, and gracilis muscles) (Norkin & Levangie, 1992).

Supporting structures can be categorized according to location. Medial compartment structures include the medial patellar retinaculum, medial collateral ligament, oblique popliteal ligament, and posterior cruciate ligament, as well as the medial head of the gastrocnemius, pes anserinus, and semimembranosus muscles. The lateral compartment structures are the iliotibial band, biceps femoris, popliteus muscles, lateral cruciate ligament, meniscofemoral ligament, arcuate ligament, and anterior cruciate ligament with the lateral patellar retinaculum (Norkin & Levangie, 1992).

Anteroposterior stabilization is achieved by the extensor retinaculum, which is comprised of fibers from quadriceps femoris and fuses with fibers from the joint capsule. This supports the anteromedial and anterolateral aspects of the knee. The medial and lateral head of gastrocnemius reinforce medial and lateral aspects of the posterior capsule, with the popliteus being an important posterolateral stabilizer with the posterior cruciate ligament. The anterior cruciate ligament and the hamstrings (especially semi-membranosus) work together to resist anterior displacement of tibia and shear forces on the femur posteriorly. The patella helps with posterior knee stability by preventing the femur from sliding forward and off the tibia (Norkin & Levangie, 1992).

Contributors to medial-lateral stabilization are the soft tissue and tibial tubercles and menisci when the knee is extended. The knee is reinforced medially and laterally by the medial and lateral collateral ligaments. Laterally, the iliotibial tract, lateral collateral ligament, popliteus tendon, and biceps tendon contribute to stability, and the posterolateral capsule is important for varus stability in extension (Norkin & Levangie, 1992).

Passive mechanisms seem to contribute to rotational stabilization. The cruciate ligaments seem particularly important, especially in the extended knee. Rotational stability is also credited to the medial and lateral collateral ligaments, the posteromedial and posterolateral capsule, and the popliteus tendon. The menisci are important in medial-lateral stability (Norkin & Levangie, 1992).

ANKLE/FOOT COMPLEX

The foot and ankle together have 26 bones, 30 synovial joints, 100 ligaments, and 30 muscles, so it a very stable complex. In fact, Norkin and Levangie (1992) state that the ankle joint is the most congruent joint in the body. Structurally, the ankle/foot complex is comparable to the wrist/hand, although the hand is more critical to daily life tasks. The ankle/foot complex needs to balance conflicting demands for stability and mobility. On one hand, the ankle and foot meet the stability demands of providing a stable base of support for the body without undue muscular or energy demands and act as a rigid lever for propulsive weight-bearing. On the other hand, mobility demands of dampening the rotations imposed by the more proximal joints of the lower extremity, having the flexibility to absorb the shock of body weight, and permitting the foot to adjust to varied terrains are also met by this section of the lower extremity (Norkin & Levangie, 1992). The flexible-rigid characteristics of the ankle/foot complex provide these functions:

- Support of superincumbent weight.
- Control and stabilization of the leg on planted foot.
- Adjustments to irregular surfaces.
- Elevation of body, as in standing on toes, climbing, or jumping.
- Shock absorption.
- Operation of machine tools.
- Substitution for hand function in people with upper extremity amputations or muscle paralysis (Smith et al., 1996).

The ankle joint is actually composed of the talocrural articulation (tibiotarsal or talotibial/tibiotalar) and the tibiofibular joints. The forces acting on the ankle can be as much as five times the body weight, and the talocrural joint transmits approximately 1/6 of the force exerted through the foot (Norkin & Levangie, 1992).

Tibiofibular Joints

There are three tibiofibular joints: proximal, middle, and distal. The proximal tibiofibular joint is located close to the knee but has a separate joint capsule and is not related functionally to the knee. All three tibiofibular joints are related to ankle movements (Muscolino, 2006).

The superior (or proximal) tibiofibular is a plane synovial joint formed by the articulation of the head to fibula with the posterior and lateral condyle of the tibia. Motions are variable and limited and have been described as superior and inferior sliding of the fibula and as fibular rotation (Norkin & Levangie, 1992).

The middle tibiofibular joint is located between the two shafts of the tibia and fibula and is a syndemosis fibrous joint created by the interosseus membrane. This joint and the interosseus membrane hold the two bones together and allow the force of muscle attachments that pull on the fibula to be transferred to the tibia, moving the leg at the knee (Muscolino, 2006).

The distal tibiofibular joint is a syndemosis or fibrous joint between the medial side of the lateral malleolus of the fibula and the fibular notch of the distal tibia. The tibia and fibula do not actually come into contact because they are separated by fibroadipose tissue.

The tibiofibular joints allow superior and inferior glide of the fibula in relation to the tibia. The tibiofibular joints are essential to dorsiflexion and plantarflexion due to the mortise-like arrangement of the bones that allows the tibia and fibula to grasp and hold onto the distal joint segments (Norkin & Levangie, 1992).

Talocrural Joint (Ankle)

The talocrural articulation is the articulation of the tibia and fibula with the talus of the foot and is designed for stability. It is supported by strong ligaments. On the medial side, major ligaments are the deltoid (medial collateral) ligament with three separate ligaments, including the tibionavicular, tibiocalcaneal, and tibiotalar ligaments, which serve to limit eversion across the talocrural, subtalar and talonavicular joints (Neumann, 2002). The tibiotalar ligament also resists translation and lateral rotation of the talus (Magee, 2008). On the lateral side, the anterior talofibular ligament resists excessive inversion of the talus, and this is the ligament that is most often injured by a lateral ankle sprain. The posterior talofibular ligament resists ankle dorsiflexion, adduction, medial rotation, and medial translation of talus, and the calcaneofibular ligament provides stability against maximum inversion at the ankle and subtalar joints (Magee, 2008).

Because this is a uniaxial modified hinge joint, there is one degree of freedom permitting flexion and extension of the foot, which is known as plantarflexion, and dorsiflexion is the motion that occurs parallel to the sagittal plane around a mediolateral axis. Approximately 50° of plantarflexion occurs and is due to the gastrocnemius, soleus, flexor digitorum longus, peroneus longus and brevis, flexor hallucis longus, and tibialis posterior (Jenkins, 1998). The large size of the gastrocnemius and the long force arm enable this muscle to be a strong plantarflexor of the ankle, but only when ankle motion is needed, not when standing at ease (Gench et al., 1995). Gastrocnemius and soleus (together known as the triceps surae) work powerfully on the calcaneus to push the foot down, and the soleus is more involved in static standing. The tendons of these two muscles form the large, easily palpated tendon on the posterior distal leg known as the Achilles tendon. Because tibialis posterior, flexor digitorum longus, and flexor hallucis longus cross the joint posterior to the joint axis, these muscles act to plantarflex the ankle in addition to other actions. These three muscles are fondly referred to as Tom for tibialis posterior, Dick for flexor digitorum longus, and Harry for flexor hallucis longus. Peroneus longus changes direction twice before insertion and is close but posterior to the axis, so it acts only as a weak plantarflexor, as does peroneus brevis. The plantarflexion action is the strongest movement of the foot and is important in propulsion force.

Dorsiflexion of the foot occurs in a range of 0 degrees to 20 degrees to 30 degrees, and the muscles producing the movement cross anterior to the joint axis. One of the most important muscles in this action is tibialis anterior; this is the muscle that gives our legs the roundness of the shank portion. Extensor digitorum longus crosses the joint anterior to the axis with a long force arm and is effective as a dorsiflexor. Peroneus tertius, which appears to be a part of the extensor digitorum longus (which, incidentally, is often missing in individuals), crosses the ankle in the same manner as the extensor digitorum longus and has similar action on the joint. Similarly, extensor hallucis longus, with primary action involving the big toe, also crosses the joint anterior to the axis and is contributory

Figure 10-2. Parts of the foot.

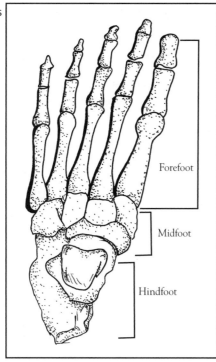

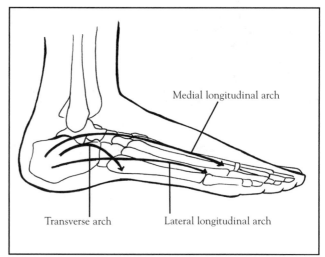

Figure 10-3. Arches of the foot.

to dorsiflexion. Generally, these muscles are weak and are not capable of generating high forces (Hamill & Knutzen, 1995).

Movements of the Ankle/Foot Complex

The ankle/foot complex is capable of moving the foot up and down as well as in and out. Much variability exists in the literature about how to define these motions. The movement of the foot and ankle toward and away from the midline in the transverse plane around a vertical axis is referred to as adduction and abduction, and there is minimal ROM in these movements of adduction and abduction at the ankle/foot complex. Frontal plane motion of the foot turning inward (inversion) or outward (eversion) occurs in the anterior-posterior axis, although some authors use the terms supination (to turn up or medial tilt or varus) and pronation (to turn down or lateral tilt or valgus) to describe this motion. Often, the terms *eversion-inversion* will be used to describe motion in open kinetic chains and *pronation-supination* to describe a closed-chain motion (Kendall & McCreary, 1983; Sine et al., 2000). Other authors use the term *pronation* to describe a combination of dorsiflexion, eversion, and abduction, while *supination* is a combination of plantarflexion, inversion, and adduction. For simplicity, the terms *inversion* and *eversion* will be used in this text to describe the motion at the ankle, although the definition based on the combined movements more aptly describes the actual motion that occurs. Inversion and eversion occur primarily around the subtalar and transverse tarsal joints with very little movement at the talocrural joint. Due to these motions, the foot can be positioned to travel uneven ground, and the ability to invert/evert one's foot is helpful in bathing and dressing of the lower extremity.

The Foot

The foot is divided into the forefoot, midfoot, and hindfoot (Figure 10-2). The anterior segment of the foot is made up of the five metatarsals and 14 phalanges, which make up the five toes. The midfoot is the middle section of the foot and is made up of five bones—the navicular, the cuboid, and the three cuneiforms. The most posterior segment, the hindfoot, is formed by the talus and calcaneus bones. Compression forces in the foot while standing are greatest at the hindfoot (60%), then the forefoot (28%), and least by the midfoot (8%) (Neumann, 2002).

As with the hand, there are arches in the foot—two longitudinal and one transverse as seen in Figure 10-3. These arches create an elastic shock-absorbing system, and the flexibility of these arches is critical in walking and running. The lateral longitudinal arch is a relatively flat arch that plays a support role in weight-bearing and has limited function in mobility. The medial longitudinal arch is more flexible and mobile and plays a greater role in shock absorption. When one takes a step, the force is absorbed by a fat pad on the inferior surface of calcaneus. This force then causes an elongation of the medial arch that continues to maximum elongation at toe contact with the ground. This portion of the foot rarely touches the ground unless the person has flat feet. The medial longitudinal arch helps to diminish the impact by transmitting vertical loads through the deflection of the arch (Hamill & Knutzen, 1995). The transverse arch supports a significant portion of weight during weight-bearing (Hamill & Knutzen, 1995).

Hindfoot

The hindfoot includes the articulation between the calcaneus and talus at the subtalar (or talocalcaneal) joint. This is a synovial gliding joint, and the talus presents with three facets that correspond to three facets on the calcaneus. With the three articulations, the convex-concave facets limit mobility of the joint. When the talus moves on the posterior facet of calcaneus, the articular surface of the talus should slide in the same direction as the bones move. However, at the middle and anterior joints, the talar surfaces should glide in a direction opposite to the movement of bone. What actually occurs is a screw-like or twisting motion around a triplanar axis (Smith et al., 1996). When inversion (supination) occurs, the posterior articulation slides in a lateral direction, and with inversion (pronation), there is a medial slide of the posterior articulating surfaces (Kisner & Colby, 1990).

Midfoot

The midfoot is made up of the transverse tarsal joint and the distal intertarsal joints. The transverse tarsal joint includes two articulations: talonavicular joint (talocalcaneonavicular) in which the head of the talus articulates with the proximal side of the navicular bone and the plantar calcaneonavicular ligament; and the calcaneocuboid joint, which is the articulation of the transverse tarsal joint with the proximal surface of the cuboid bone.

The transverse tarsal joint forms an S-shaped joint line between the talonavicular and calcaneocuboid joints. The midtarsal joint participates in gliding and rotational movements of the forefoot on the hindfoot to lower the longitudinal arch in pronation and raise it in supination. The midtarsal joint also contributes to eversion (pronation) and inversion (supination) and in absorbing forces of contact.

Inversion occurs when the muscles pass around the medial border of the foot. This includes tibialis posterior, flexor digitorum longus, flexor hallucis longus, tibialis anterior, and extensor hallucis longus. Of these, tibialis anterior is the strongest inverter of the foot, while extensor hallucis longus is the weakest. By pulling the feet inward (inversion or supination), the arches are maintained, and the body weight is distributed to the lateral sides of the feet.

Eversion of the foot is produced by peroneus longus, brevis, and tertius and by extensor digitorum longus. Extensor digitorum longus, while primarily involved in extension of the interphalangeal joints, does cross the joint anterior and lateral to the joint axis, so it has a long force arm for producing dorsiflexion and eversion of the ankle. By pulling the sole of the foot outward into eversion, body weight is distributed to the medial side of the foot.

The distal intertarsal joints include the intercuneiform and cuneocuboid joint complex, cuneonavicular joints, and the cuboideonavicular joints. All of these joints except the cuboideonavicular joint are plane synovial joints, and all of the intertarsal joints are capable of gliding and rotational movement. The talus moves with the foot during dorsiflexion and plantarflexion, but during inversion and eversion, which occurs at the intertarsal joints, the talus moves very little.

Forefoot

The forefoot includes the tarsometatarsal, intermetatarsal, metatarsophalangeal, and interphalangeal joints. Tarsometatarsal joints are where the three cuneiforms and cuboid articulate with the base of the five metatarsals. The tarsometatarsal joints function to stabilize the second metatarsal, making it a rigid forefoot bone to carry most of the load while walking. The metatarsophalangeal joints are the articulations between the metatarsals with the phalanges. Interphalangeal joints are the articulations between the phalanges, which aid in keeping the toes in contact with the ground.

During weight-bearing, pronation and supination of the foot permit the leg to rotate in all three directions relative to the calcaneus due primarily to the interaction between the subtalar, transverse tarsal joints, and the medial longitudinal arch. The distal intertarsal joints assist the transverse tarsal but have small joint movements, and the primary function of these joints is to provide stability across the midfoot (Neumann, 2002).

Essentially, the tarsals transmit the weight of the body to the heel and the ball of the foot. Tarsals are seen to correspond structurally to the carpals of the hand. The metatarsals and phalanges, then, correspond to the metacarpals of the hand, and these form the instep of the foot (Smith et al., 1996).

SUMMARY

- While not the direct focus for intervention by occupational therapists, an overview of the structures of the knee, ankle, and foot is provided to understand the contribution these structures make to function.

- The knee, with the hip and ankle, supports the body and also serves to help in foot placement in space.

- The ligaments of the knee are especially important because there are no bony restraints to knee movement.

- The ankle and foot are seen as very stable due to the large number of bones, ligaments, and muscle that comprise these joints.

- These joints balance the conflicting demands of providing a stable base of support.

REFERENCES

Esch, D. L. (1989). *Musculoskeletal function: An anatomy and kinesiology laboratory manual.* Minneapolis, MN: University of Minnesota.

Gench, B. E., Hinson, M. M., & Harvey, P. T. (1995). *Anatomical kinesiology.* Dubuque, IA: Eddie Bowers Publishing, Inc.

Goss, C. M. (Ed.) (1976). *Gray's anatomy of the human body* (29th ed.). Philadelphia, PA: Lea and Febiger.

Hamill, J., & Knutzen, K. M. (1995). *Biomechanical basis of human movement.* Baltimore, MD: Williams & Wilkins.

Jenkins, D. B. (1998). *Hollingshead's functional anatomy of the limbs and back* (7th ed.). Philadelphia, PA: W. B. Saunders Co.

Kendall, F. P., & McCreary, E. K. (1983). *Muscles: Testing and function* (3rd ed.). Baltimore, MD: Williams & Wilkins.

Kisner, C., & Colby, L. A. (1990). *Therapeutic exercise: Foundations and techniques* (2nd ed.). Philadelphia, PA: F. A. Davis.

Magee, D. J. (1992). *Orthopedic physical assessment.* Philadelphia, PA: W. B. Saunders Co.

Magee, D. J. (2008). *Orthopedic physical assessment.* Philadelphia, PA: Saunders.

Muscolino, J. E. (2006). *Kinesiology: The skeletal system and muscle function.* St. Louis, MO: Mosby.

Neumann, D. A. (2002). *Kinesiology of the musculoskeletal system: Foundations for physical rehabilitation.* St. Louis, MO: Mosby.

Nordin, M., & Frankel, V. H. (1989). *Basic biomechanics of the musculoskeletal system* (2nd ed.). Philadelphia, PA: Lea and Febiger.

Norkin, C. C., & Levangie, P. K. (1992). *Joint structure and function: A comprehensive analysis* (2nd ed.). Philadelphia, PA: F. A. Davis Co.

Sine, R., Liss, S. E., Rousch, R. E., Holcomb, J. D., & Wilson, G. (2000). *Basic rehabilitation techniques: A self instructional guide* (4th ed.). Gaithersburg, MD: Aspen Publishers.

Smith, L. K., Weiss, E. L., & Lehmkuhl, L. D. (1996). *Brunnstrom's clinical kinesiology* (5th ed.). Philadelphia, PA: F. A. Davis Co.

SECTION III

Intervention

11

BIOMECHANICAL INTERVENTION APPROACH

The outcome of occupational therapy intervention is "supporting health and participation in life through engagement in occupation" (American Occupational Therapy Association, 2008, p. 660). The types of occupational therapy interventions include occupational-based intervention, purposeful activity, preparatory methods, consultation, education, and advocacy (American Occupational Therapy Association, 2008). Intervention approaches are the strategies that direct the process of intervention, and these include create/promote, establish/restore, maintain, modify, and prevent disability.

The biomechanical approach is a remediation or restoration approach, and the intervention is designed to restore or establish client-level factors of structural stability, tissue integrity, range of motion (ROM), strength, and endurance. In particular, the focus is on performance skills, performance patterns, and client factors with the underlying belief that by establishing or restoring these factors, resumption of valued roles and successful participation in areas of occupation will be possible. In cases where full restoration is not possible or in degenerative conditions, the maintenance approach is used within the biomechanical approach to enable preservation of the client's physical performance capabilities and slow declines in impairments and task abilities. A summary of the focus, assumptions, definition of function, expected outcomes, and techniques used in the biomechanical approach are shown in Table 11-1.

CONCEPTUAL BACKGROUND

In selecting an intervention approach, remediation or restorative approaches are chosen when there is an expecta-

tion for significant reduction in the impairment that leads to prevention of further activity limitations and participation restrictions. It is assumed that resolution of physical impairments will reduce activity limitations and increase participation in areas of occupations. Intervention may involve learning new performance skills to maintain or improve the client's quality of life (McGinnis, 1999).

The biomechanical approach is used to explain function using anatomical and physiological concepts with exercise physiology, kinetics, anatomy, and kinematics as the theoretical base (Trombly, 1995). Occupational therapists use their knowledge of activity analysis and apply it to understanding movement created by muscles, joints, and soft tissues and those circumstances that prevent or permit motion to occur (Pedretti, 1996). The biomechanical approach is a study of the relationship between musculoskeletal function and how the body is designed for and used in the performance of daily occupations. The effect, purpose, and meaning of engagement in these activities influence the client's compliance, effort, fatigue, and improvement in movement capacity (Kielhofner, 1992).

Assumptions

The biomechanical approach assumes that the client has the capacity for voluntary control of the body (muscle control) and mind (motivation) (Trombly, 1995). It is anatomy and physiology that determine normal function, and humans are biomechanical beings whose range of motion (ROM), strength, and endurance have physiological and kinetic potential as well as role-relevant behaviors (Smith, Weiss, & Lehmkuhl, 1996). Humans are able to perform role-relevant behaviors most efficiently when they assume and maintain positions that are

Rybski MF.
Kinesiology for Occupational Therapy, Second Edition (pp 309-354)
© 2012 SLACK Incorporated

Table 11-1

SUMMARY OF BIOMECHANICAL APPROACH

Focus	• Bottom-up approach
	• Restore or establish client-level factors, performance skills, performance patterns
	• Teaching/training new performance skills and performance patterns
	• Musculoskeletal capacities, peripheral nerves, integumentary system, cardiopulmonary systems
	• Related fields: exercise physiology, kinetics, anatomy, and kinematics
Assumptions	• Motor activity is based on physical mobility and strength
	• Purposeful activities remediate loss of ROM, strength, and endurance
	• Purposeful activity has meaning to client
	• Activities can be graded
	• Participation in activities maintains and improves function
	• Improvement in ROM, strength, and endurance will result in improved function
	• Rest/stress principles are inherent
Function	• Capacity for movement
	• Adequate range of motion, strength, and endurance to complete activities
	• Performance requires simultaneous actions of muscles and joints
	• Functional person uses positions that are biomechanically advantageous for efficiency
	• Reliant on the principles of rest and stress
Expected outcomes	• Reduction in limitations
	• Learning new skills
	• Slowing declines
	• Maintaining or improving quality of life
Techniques	• Teaching new skills, behaviors, or habits to reduce dysfunction and/or enhance performance
	• Change the biological, physiological processes
	• Use of procedural reasoning skills to incorporate disease and prognostic information into intervention planning
	• Correlation of the physical demands of the graded activities to the subskills and role-relevant behaviors
	• Motivating and meaningful activities meeting individual needs and interests in relation to social roles
	• Provide graded activities that simulate the physical requirements of the task and demand increasing levels of ROM, strength, and endurance

biomechanically advantageous. Further, the environment can be specifically designed to facilitate the development or recovery of voluntary control of skeletal musculature (Trombly, 1995).

The biomechanical approach involves the musculoskeletal system, peripheral nerves, and integumentary and/or cardiopulmonary systems and requires an intact brain and central nervous system to produce isolated, voluntary, coordinated movements. Intervention is aimed at improving strength, range of motion, endurance, tissue integrity, structural stability, and coordination of these systems. This approach is most effective for clients with orthopedic disorders (fractures, rheumatoid arthritis) and lower motor neuron disorders resulting in weakness and flaccidity (peripheral nerve injuries and diseases, Guillain-Barré syndrome, polio, spinal cord injuries and diseases) as well as clients with hand injuries, burns, cardiopulmonary disease, and amputations.

For the intervention techniques to be effective, the client must have motor pathways available with the potential for recovery in strength, ROM, endurance, and/or coordination. Some sensory feedback must be available in order to provide information about movement to the muscles and nervous system. Because the focus of intervention is on strength, endurance, and ROM, muscles and tendons must be free to move and relatively free of pain. In addition, the client must be able to understand the directions and purpose of the intervention and be interested and motivated to perform the activities and exercises. The biomechanical approach would not be appropriate for clients with impairments in central nervous system function that may result in spasticity and lack of voluntary control of isolated motion. Clients with inflamed joints or those just out of surgery would not be appropriate for vigorous exercise regimens and activities.

The focus on musculoskeletal systems defines biomechanical intervention and includes physical fitness and health. While not the only frame of reference to accentuate the value of health promotion, many of the physical fitness goals are biomechanical in nature. In the *Guide to Occupational Therapy Practice*, health promotion and wellness often involves a "lifestyle redesign" (Moyers, 1999), underscoring the belief that occupation is an important component to staying healthy. Health involves the body, the self, and the environment and is intricately linked to sufficient cardiovascular function, age, fitness levels, hereditary factors, and level of physical health (Christiansen & Baum, 1997).

Function/Dysfunction

Function, according to the biomechanical approach, is the capacity for movement in bones, joints, muscles, tendons, peripheral nerves, heart, lungs, and skin as demonstrated by adequate range of motion, strength, and endurance needed in role-relevant behavior. The functioning person is able to assume and maintain positions that are biomechanically advantageous and promote efficiency of motion. Performance depends on simultaneous actions of muscles across joints to produce the movement and the stability that is required of the task (Trombly, 1995).

Dysfunction is characterized by the inability to demonstrate adequate ROM, strength, and endurance for physical subskills and independent life skills in role-relevant behaviors. The person is unable to assume and maintain positions that are biomechanically advantageous and promote efficiency of motion due to alterations or decreased capacity for movement in joints and bones, muscles, and tendons, resulting in impairments in strength, ROM, coordination, or endurance that interfere with occupational performance.

Assessment in this approach involves identification of ROM, strength, and endurance subskills needed to perform role-relevant behaviors (Pedretti, 1996). Activity analyses will enable an understanding of the most biomechanically advantageous and efficient positions to assume when participating in desired activities and will help ascertain discrepancies between the physical requirements of the activity and the client's baseline performance. Specific assessments usually include ROM testing and screens, strength testing (manual muscle testing and grasp/pinch via dynamometer and pinchmeter), and strength screens, as well as assessment of edema and endurance.

Strengths and Limitations of the Biomechanical Approach

Moyers (1999) identified two main techniques used in remediation approaches. The first is that of teaching new skills, behaviors, or habits where outdated or maladaptive skills, behaviors, or habits are replaced. This is done by modifying the existing performance skills and performance patterns. Client-related instruction also includes providing the client with information about the pathological process and impairments resulting in the experienced functional limitations. Clear explanations about the purpose, benefits, and complications are provided as part of the collaborative and ethical treatment planning process.

The second technique used in remediation approaches is to change the biological, physiological, psychological, or neurological processes. The focus is on decreasing the client's pain and reducing the impact of impairments on occupational performance.

While sensorimotor techniques, graded exercise and activities, physical agent modalities, and manual techniques are used as intervention strategies in this approach, Moyers adds that "...simply expecting improvements in impairments to automatically produce changes in the level of disablement without addressing performance in occupations within the intervention plan is inappropriate. The relationship between the impairment and the level of disablement is complex and is affected by many factors in addition to changes in the impairment" (Moyers, 1999, p. 276).

This typifies a cited limitation of this therapeutic approach: changes in impairments automatically lead to increased occupational performance. Intervention goals of increasing ROM, strength, and endurance are established with the targeted outcome of increased engagement in occupational performance areas. However, just because strength is increased or range of motion is within functional limits does not mean that the client will automatically or spontaneously use these to regain physical skills in work, play, or self-care. Dutton cites several studies that show low correlations of dexterity with activities of daily living (ADL) and asks the question, "How can skills such as manual dexterity and strength have such a low correlation with ADLs?" (1995, p. 31). If dexterity is not a needed client physical ability for successful performance in areas of occupation, should this be an area of intervention? This is an area needing further research to substantiate this assumption or to refute it.

Another cited limitation of this approach is that, with the focus of the intervention on the client's abilities and performance skills, the approach is reductionistic. Pedretti (1996) describes a variety of activities that are used in intervention that are, in fact, considered nonpurposeful and without an

Table 11-2

COMPARISON OF REMEDIATION INTERVENTION CONTINUA

OT FRAMEWORK	PEDRETTI	FISHER
Preparatory methods	Adjunctive methods	Exercise
Purposeful activity	Enabling methods	Contrived occupation
	Purposeful activities	Therapeutic occupation
Occupation-based activities	Occupation-based intervention	Adaptive or compensatory occupation

inherent goal of engaging both the physical and mental attributes of the individual. These intermediate activities are called enabling activities and are done in order to:

- Practice specific motor patterns
- Train in perceptual and cognitive skills
- Practice sensorimotor skills

Examples are having clients work on inclined sanding boards, cone stacking, puzzles, fastener boards, work simulators, and some computer programs as used in cognitive remediation programs. Pedretti adds that, while these activities are seen as the intermediate means to achieve long-term goals, enabling activities "should be used judiciously and are often used with adjunctive modalities and purposeful activities as part of a comprehensive treatment program" (Pedretti, 1996, p. 300). These activities may be analogous to Fisher's continua classification of "contrived occupation" (1998).

Adjunctive modalities are defined as "preliminary to the use of purposeful activities and that prepare the client for occupational performance" (Pedretti, 1996, p. 300) (Table 11-2). These modalities include exercise, orthotics, sensory stimulation, and physical agent modalities. These adjunctive modalities are necessary to provide structural stability, to position parts to prevent deformities, to provide rest for a part, and to increase function in components for use in occupations. These are roughly analogous to Fisher's (1998) "exercise" where there is little meaning to the client with the focus on remediation of impairments. The *Occupational Therapy Practice Framework: Domain and Process* would classify adjunctive modalities as preparatory methods in anticipation of future purposeful and occupation-based activities.

The use of enabling and adjunctive modalities seems inherently at odds with the core values of occupational therapy, and this would be true if the ultimate intervention goal is simply remediation of an impairment and not one of functional improvement in meaningful tasks. Occupational therapists see clients earlier in the recovery phase, see clients who are more acutely ill, and often have shorter lengths of stays that necessitate these preliminary interventions. Again, it must be reiterated that, without the ultimate focus of a functional outcome, using adjunctive or enabling activities should not be considered occupational therapy practice. Activities used

in this intervention can also be purposeful (i.e., goal directed within a designed context, or engagement in actual occupations in activities that are client directed and important to that person).

Other limitations of this approach are that it focuses only on the physical performance of occupations and does not include aspects of volition, context, role, or environment (Pedretti, 1996), and little reference is made to motivation or to the psychological, emotional, or social aspects of rehabilitation. There is no attention to the holistic values held by occupational therapists that address the need for balance in daily life and higher levels of self-esteem and self-actualization (Pedretti, 1996). This approach is seen as reductionistic, with the client being a passive recipient in a program that does not reflect the client's interests, needs, or improved occupational performance (Pedretti, 1996). Again, this would be true if exercise only is the focus of intervention at the expense of improved occupational performance (Table 11-3).

However, it must be seen clinically that the biomechanical approach is used to restore those variables that can be improved to enhance client participation in work, play, and self-care. Thoughtful clinical reasoning is used to determine what aspects of performance can be improved and which need compensatory strategies based on an in-depth understanding of the client as a person (interactive reasoning). Intervention occurs not only because there are impairments in biological, physiological, psychological, or neurological processes, but also because these impairments interfere with the successful living of this person's life. The remediation of strength or ROM is only one small part of the total intervention strategies used, and the biomechanical approach is often used with other approaches to address these other areas simultaneously. While the biomechanical approach relies heavily on procedural reasoning to incorporate disease and prognostic information into intervention planning, that is not to say that other aspects of reasoning are not also occurring concurrently as part of the total intervention process.

The biomechanical approach makes good use of media and equipment to promote physical function, and the techniques can be applied to a variety of creative and constructive activities. Knowledge of activity analysis is applied to understanding movements needed and to planning appropriate activities and

Table 11-3

STRENGTHS AND LIMITATIONS OF THE BIOMECHANICAL APPROACH

STRENGTHS	LIMITATIONS
Good use of media and equipment in a variety of creative and constructive activities	It cannot be assumed that increases in range of motion, strength, and endurance will automatically be used by the client in functional activities
Incorporates numerous knowledge bases: Activity analysis Anatomy/physiology Kinesiology—kinetics/kinematics Apparent direct cause-and-effect relationship between treatment and remediation or prevention	Requires continuous review and revision of treatment to ensure relevance to client needs
Direct cause-and-effect relationship between techniques and goals	Client-centered focus can easily be lost if not attentive to the client's needs

exercises. The utilization of anatomical, physiological, and mechanical knowledge has led to the development of specific techniques for measuring and treating movement, strength, and endurance limitations (Dutton, 1995).

A direct cause-and-effect relationship can be seen in the treatment process and prevention of deformities that can be clearly seen and demonstrated. To ensure that the activities used in intervention will achieve goals that are of value to the client, these activities must correlate the physical demands of the graded activities to subskills and role-relevant behaviors. These activities should be motivating and have meaning to the client as a means of meeting individual needs and interests in relation to social roles. Short-term goals may relate to performance of subskills in the clinic while in the hospital, but long-term goals must relate to role-relevant behaviors and performance in the community environment in which they are ultimately expected to occur. The activities should be designed to prevent or reverse dysfunction and/or to develop new skills to enhance performance in life roles. Intervention should provide graded activities that simulate or duplicate the physical requirements of the tasks and demand increasing levels of range of motion, strength, and endurance (Demeter, Andersson, & Smith, 1996; Norkin & Levangie, 1992; Smith et al., 1996; Steinbeck, 1986).

Hall and Brady cite the following benefits of physical remediation:

- Increased patient, caregiver, family, significant other knowledge about condition, prognosis, and management
- Acquisition of behaviors that foster healthy habits, wellness, and prevention

- Improved levels of performance in ADL, IADL areas of occupation
- Improved physical function, health status, and sense of well-being
- Improved safety
- Reduced disability, secondary conditions, and recurrence
- Enhanced decision-making about use of health-care resources by the patient and family
- Decreased service use and improved cost-containment (Hall & Brady, 2005)

The elements of the biomechanical approach are structural stability, tissue integrity, range of motion, strength, and endurance.

STRUCTURAL STABILITY

Occupational therapists are involved in intervention programs for the entire spectrum of clients—from wellness programs to those who are critically ill. In hospital settings, occupational therapists previously would be consulted to work with clients who were medically stable. Now, our clients are more severely disabled and acutely ill, so there is the need to balance early mobilization with rest. Clients do not always come to occupational therapy with stable cardiovascular and respiratory function. They may be referred due to shortness of breath, decreased endurance, or physical deconditioning. Shorter lengths of stay greatly influence intervention, which starts where the client currently is in the continuum of care and what further services are needed and expected.

Structural stability is necessary to regain prior to implementation of other intervention goals. Splints or devices that are worn temporarily to enforce rest to enable healing are facilitating changes in the biological, physiological, or neurological processes by temporarily immobilizing the part. Positioning devices can also serve to temporarily enforce rest to promote healing. By changing the damaged structures, this would be considered a remediation goal of the biomechanical approach.

Devices and orthoses can be used to remediate, compensate, or to prevent further disability. If the orthosis or device was used in joints too weak to resist the force of gravity to maintain functional alignment with the goal of preventing development of contractures and deformities, then this would be more a goal of prevention of disability than remediation. The use of a lapboard for a dependent hemiplegic arm, used to prevent further injury to the extremity, might also be an example of the intervention approach to prevention of disability. Neither of these devices is used to remediate the lack of function of that extremity. The use of a universal cuff with utensils inserted is a means of compensating for lost function, not to prevent disability or promote healing. However, if a foam wedge was used by a client with a hip replacement, this object will be used temporarily until the structures relative to the hip surgery are healed. The wedge does not serve to compensate for lost function (compensation) nor to prevent disability; rather, the function of the wedge is to promote healing and enforce positioning. While the distinction between the use of these devices seems academic, why one is using a particular device or positioning technique has much to do with clinical reasoning, responsible use of adaptive equipment, and client outcomes.

Postural instability and misalignment may limit upper extremity reaching movements, and discrete movement is often not well-controlled (O'Sullivan & Schmitz, 1999). Stability enables controlled mobility, and many activities require both static and dynamic mobility for skilled and effective movement patterns. To progress the client to the level of effective, controlled mobility, static then dynamic stability is achieved first, then controlled mobility follows including static-dynamic control involving the ability to shift weight onto one side and free the opposite side for nonweight-bearing dynamic activities or to fix the distal segment to enable the proximal segment to move (O'Sullivan & Schmitz, 1999).

Joint stability can be seen in a loss of control of small arthokinematic movements of roll, spin, slide, and translation (translational instability or pathological or mechanical instability); can be due to clinical or gross instability or due to pathological hypermobility (anatomic instability) resulting in excessive movement; or can result in the inability to control either arthrokinematic or osteokinematic movement in the available range of motion (functional instability) (Magee, 2008). When there is instability due to excessive flexibility due to stretched or loose ligaments, this increases the risk of sprain and subluxation.

Ligaments should be tight, and they provide the stability needed for controlled and safe movement. The fibers need to be flexible to enable movement, yet tight to protect the joints.

When ligaments are loose, too much movement at a joint can lead to ligamental, joint capsule, tendon, and muscle damage and can lead to joint instability.

Ligaments can become loose in three ways: through genetics, trauma, or by the development of distended scar tissue. Genetically, tendons can be disproportionately long as they attach to the joint surfaces, allowing greater movement but less joint stability. Trauma or injury can stretch a ligament permanently. Adhesive scar tissue can stretch over time, leaving the joint hypermobile and unstable. Weak or lax ligaments allow pathologic forces to act upon structures, altering body mechanics and creating deformities, loss of function, and painful movement.

Training for joint stability would include strengthening weak structures, limiting excessive or abnormal joint mobility, and providing noncontractile joint stability (Pendleton & Schultz-Krohn, 2006). Kinesio taping has been used to prevent joint injury and provide stability while not restricting movement. While research is limited regarding the effectiveness of taping, it has been used with glenohumeral instability to enhance joint stability with dyskinetic dynamic scapulothoracic movement. The tape is anchored at either the origin or insertion and is gently stretched and taped over/around either shortened or lengthened muscles. This method of improving joint stability is believed to affect the peripheral somatosensory receptors in superficial skin, which influences the skin and lymphatic systems and joint and muscle function as it relates to pain, proprioception, and motor control (Cooper, 2007).

Documentation

Documentation of intervention is an important part of the treatment process as a confirmation that the intervention actually occurred and clearly indicates that professional judgment and decision-making was needed by a skilled practitioner. The occupational therapist's unique skills and knowledge need to be evident in the written record with evidence that the client would not improve naturally without the intervention provided. Documentation of intervention should provide evidence that significant improvement was made in a reasonable and predictable amount of time.

The goals are outcomes of intervention, not the process of intervention, and they are concrete, measurable, testable, achievable, relevant, and predictable. Because occupational therapy is a client-centered profession, substantiation of collaboration with the client in the development of the goals should also be seen in the documentation. Goals are relevant, related to roles and the expected environment, time-limited, and related to length of stay. A short-term goal of driving would not be relevant or related to length of stay in an acute-care environment where clients stay only a few days until moving to the next level of care.

Intervention goals should have three essential parts: a statement of terminal behavior, a statement about conditions of treatment, and a criteria statement (Quinn & Gordon, 2003). A statement of terminal behavior is an indication of physical changes or changes in behavior that you expect. An example is, "The client will feel self." Further goal definition is needed, so a statement about the conditions of treatment will describe the circumstances that will facilitate achieving

Table 11-4

STRUCTURAL STABILITY

GOAL	METHOD	PRINCIPLE/RATIONALE	EXAMPLE
Client will be (independent/modified independent/require maximal/moderate or minimal assistance) in performance of ____ tasks (specify) by regaining structural stability in damaged structures by using…	Orthoses	Which are worn/used to temporarily enforce rest until structures are healed	Dorsal rubber band splint for flexor tendon repair Body jacket or cast Resting splints Wrist braces for CTS
	Positioning (static, dynamic, weight shift)	Which is used to enforce rest until damaged structures are healed	Stryker frame Wedge for hip replacement Pillow placement for CVA
	Procedures	Which gradually stress structures to new levels of adaptation	Stump toughening and shaping LE weight bearing for fracture

Adapted from Dutton, R. (1995). *Clinical reasoning in physical disabilities.* Baltimore, MD: Williams & Wilkins; and Marrelli, T. M., & Krulish, L. H. (1999). *Home care therapy: Quality, documentation and reimbursement.* Boca Grande, FL: Marrelli & Associates, Inc.

the terminal behavior. This can include how the environment will be set up to enhance performance, the use of any special devices, the degree of training, assistance, or supervision that is needed, and the types of cues that are required. An example is, "Given adaptive devices, the client will feed self." What is still not included in this goal is the acceptable level or degree of performance or competence that is required to achieve this goal. The statement of criteria answers how much, how often, how well, how accurately, or how quickly the goal is to be accomplished. An example is, "Given assistive devices, the patient will feed self independently in 30 minutes." The criteria state that the client will be independent (requiring no assistance, cues, or supervision) and will accomplish the task in a reasonable amount of time.

Borcherding (2005) uses the FEAST acronym, which covers similar essential parts of the goal. F = Function, which is the functional gain to be achieved. E = Expectation, which is a description of what the client will do as part of the goal; A = Action or what will be done; S = specific conditions, essentially the same as the statement of conditions described above; and T = Timeline as an indication of how long the goal will take to be achieved.

While goal statements seem awkward and wordy, the focus of biomechanical intervention is actually the short-term improvement of the client's physical abilities that will result in the long-term outcomes of improvements in the areas of occupation of work, play, leisure, self-care/ADLs, IADLs, and social participation. Biomechanical goals are actually short-term goals. By writing one goal that includes the short-term objectives of intervention (e.g., increasing range of motion, an aspect of motor skills, and client factors) and the reason the goal is being attempted (e.g., the functional outcome of the ability to put on a shirt or area of occupation), the focus on function is clear. This answers the question "so what"; so, what is so important about an increase of 50 degrees of additional ROM? It can mean the difference between being independent in dressing and being dependent upon others. This needs to be clear in the goal statement. Table 11-4 provides a sample goal statement, method, principle or rationale for the goal, and examples of activities or methods to achieve the goal. Each element of the biomechanical approach (structural stability, tissue integrity, range of motion, strength, and endurance) has similar tables to assist with writing functional, measurable, short-term biomechanical goals.

Tissue Integrity

Issues relative to the maintenance of soft tissue integrity involve tissue healing, remolding of scar tissue, as well as reduction in edema. Intervention can be viewed as both a biomechanical change in structures as well as prevention of further disability due to the damage that edema and scarring can do to adjacent structures and in resultant function.

Assessment of tissue integrity begins with observation of the skin. Look at the color of the skin, and make note of any areas that are blue (cyanosis), red (erythematic), or white (pallor). A tissue that is red may be due to a superficial, second-degree burn, acute fresh wound, surgical wound, or wound left open to heal by secondary intention (Cooper, 2007). Yellow skin may indicate that an exudate is present, and often pink or red granulation tissue can be seen at the wound edges. Yellow wounds may indicate a late inflammatory or early fibroplasia phase of tissue healing. Tissue that is black, brown, or grey may indicate necrosis or eschar. The wound may be in all stages of wound repair (Cooper, 2007). Wounds often have more than one color present at one time, and treatment would begin with the most serious wounds (i.e., progress from black to yellow to red) (Cooper, 2007).

Notice if the skin is unusually moist or dry, and notice the texture and temperature of the skin. If there is a wound, determine the length and width of the wound (using sterile tools to make the measurements). Cotton swabs may be used to determine the depth of the wound.

Tissue Healing

If there has been trauma or surgery to the part or in joints that have limitations in movement, it is important to be able to recognize the normal healing process of tissue. There are three phases of healing with specific characteristics in each phase. The phases overlap, and parts of the healing tissue could be in one phase while other parts are in another.

Inflammation Phase (Exudative Phase)

When a tissue is injured, inflammation begins almost immediately and lasts approximately 1 to 14 days. The inflammation phase serves to protect structures from infection and to initiate healing by clearing tissue debris from the site of injury. The inflammation phase can be acute (lasting 3 to 4 days) or chronic when there is persistent phagogenic microoganisms still present after the acute inflammation occurs (can continue for months and may have a gradual onset) (Cooper, 2007; Prosser & Conolly, 2003).

When injured, initiated within hours, and lasting about 72 hours, there is a hemorrhagic reaction due to disruption of blood flow. The wound is easily contaminated by bacteria and foreign substances. There is immediate but transient vasoconstriction, and platelet degranulation begins to initiate clotting. Activation of the platelets results in vasodilation, resulting in an increase in circulation and tissue edema. There is a migration of white blood cells to promote phagocytosis. Injured cells release chemicals that initiate the inflammatory response and eliminate bacteria and debris.

Fibroblasts are recruited, which then replace the phagocytes, and wound strength is provided as fibroblasts begin to lay down weak hydrogen bonds of collagen fibers (Cooper, 2007; Donatelli & Wooden, 2010; Hall & Brady, 2005; Konin, 1999). Due to the weak wound structure, immobilization is often advised. The joint is characterized by pain, warmth, tenderness, swelling, and loss of function. There may also be deficits in force production.

The goal of intervention at this stage is to decrease joint pain and tenderness, prevent the development of chronic inflammation, and rest the acutely injured parts while maintaining the mobility of adjacent joints (Cooper, 2007; Donatelli & Wooden, 2010; Kisner & Colby, 2007; Konin, 1999). During the first 24 to 48 hours, rest is provided (via splints, tape, casts), and cold modalities, compression, and elevation are used to decrease the inflammation (Kisner & Colby, 2007). This phase can subside and progress to phase II or can persist indefinitely, depending on the degree of bacterial contamination (Pedretti, 2001).

Fibroplasia Phase

The fibroplasia phase is when fibroblasts enter the wound and may last 2 to 4 weeks (Fess, Gettle, Philips, & Janson, 2005). In this phase, cell proliferation and granular tissue formation occurs. Fibroblasts, myofibroblasts, and endothelial cells proliferate, and temporary scaffolding called granulation tissue (type II collagen with little mechanical strength and is very fragile) is laid down. Fibers are thin and weak and have poor tensile strength. The tissue is vulnerable to breakdown, and because the vascular system is weak, the wound is subject to bleeding.

The wound begins to contract due to proliferation of epithelial cells, which have cells adjacent to the wound enlarge, flatten, and detach until cells touch each other (Prosser & Conolly, 2003). Fibroblasts begin scar formation by laying down new collagen, leading to a gradual increase in tensile strength. The collagen is primarily type III collagen, which is small, disorganized, and susceptible to disruption. Because the scar is not comprised of organized tissues, the scar is red and swollen, easily damaged, and tender to stretch and pressure (Hall & Brady, 2005). Avoid sun exposure when the scar is still immature (pink or red, thick, itchy, or sensitive) (Cooper, 2007). The fibroblasts are replaced by the scar collagen fibers, making the wound stronger. The disappearance of fibroblasts marks the end of this phase. The joint may feel warm, have edema and tenderness, and pain is felt with resistance and stretch.

Active ROM, typically used to promote balance in structures and splints, is used to protect structures (Cooper, 2007). Intervention at this point would focus on range of motion and joint mobilization with gentle resistance (Fess et al., 2005). The client is experiencing less pain so may be tempted to resume normal activities that may be detrimental to the healing process. Active movement that does not result in inflammation or decreased range of motion is encouraged to prevent new tissue adherence to surrounding structures that would limit joint mobility. As active motion and joint play improve, progress to isotonic activities and exercises to develop neuromuscular control, strength, and endurance in the affected muscles (Kisner & Colby, 2007).

Remodeling Phase

This maturation phase can start between weeks 3 and 6 and can continue for as long as 3 months to 2 years (Konin, 1999). The result is healing in all tissues in a scar (Fess et al., 2005). Factors related to scar maturation are rate of collagen turnover, increased number of stronger collagen cross-links, and linear alignment of collagen fiber (Cooper, 2007). The tissue changes as new collagen replaces the old, the collagen fibers become more organized, and there is an increase in the tensile strength of the scar.

Scar tissue reorganization is better if therapy is started as soon as the tissue can tolerate with gentle resistive activities and exercises and both dynamic and static splinting (Cooper, 2007). Scar formation occurs due to simultaneous collagen breakdown and production. If the rate of breakdown exceeds the rate of collagen production, then the scar is softer and less bulky. If the rate of production is greater than the rate of breakdown, a hypertrophic, bulky (keloid) scar may result (Fess et al., 2005). The rate and extent of scar organization varies among individuals, and the wound may remain metabolically reactive for long periods of time, possibly necessitating long-term splinting to avoid contracture development (Fess et al., 2005).

The remodeling and maturation of the collagen and scar results in more dense tissue, but the tensile strength of the scar never reaches the pre-injury state (Prosser & Conolly, 2003). By about 6 weeks after the injury, the vascularity of the scar matches adjacent skin, and there is decreased sensitivity.

Intervention is directed toward providing controlled tension on the scar by means of stretching, active movement, resistance, or electrical stimulation, in order to remodel the collagen fibers (Cooper, 2007) and make the scar more pliable. Light compression (Coban, tubigrip, pressure garments) facilitates scar maturation (Cooper, 2007). The focus of intervention is to remodel the scar tissue, increase strength and neuromuscular control, and return to engagement in occupations (Cooper, 2007). Cooper (2007) advises early mobilization within a pain-free range of motion to promote faster healing of connective tissue, stronger collagen bonds, reduced scar tissue adhesions, and improved collagen fiber orientation.

Increasing scar flexibility can be done by deep friction massage, tendon gliding exercises, and pressure garments that help to separate deep structures (tendons, nerves, blood vessels) that are stuck by the collagen bundles so that these deep structures can move separately (Demeter et al., 1996). Once the structures move separately, slow stretch forces can be applied. Rapid stretching causes scar tissue to tear, while slow stretching produces small gains and is more effective than aggressive, rapid overstretching (Demeter et al., 1996).

Scar tissue adhesions occur when scar tissue adheres to healthy tissue and limits motion. Factors that contribute to adhesion formation include location of the injury, extent of the trauma, reduced blood flow, and prolonged immobilization. Contractures due to scar tissue adhesions can occur in muscles, tendons, joint capsules, or skin. Ligamental tautness can decrease due to adhesive scar tissue, because it will stretch and distend over time, which can lead to further injury. Internal adhesive scar tissue forms within a structure (for example, within a tendon or a ligament) following injury as part of the normal healing process. When an external adhesion forms, it is between two different structures (for example, between a ligament and the bone), interfering with the normal function of smooth, friction-free gliding of cartilaginous articular surfaces and the excursion of tendons unimpeded by restrictive scars and adhesions (Cooper, 2007; Fess et al., 2005).

If the part is immobilized, the scar tissue may form in unusual places, and the pattern of collagen is irregular but when the part is allowed to move, external adhesive scar tissue is minimized. Movement can help minimize restrictive scar tissue formation, adherent tendons, shortened ligaments, and muscle contracture (Magee, 2008). Knowledge of the healing process and subsequent intervention can prevent scar tissue adhesions, and exercise is helpful once scar tissue develops.

Edema

Edema is a barrier to function because range of motion is limited, and there is decreased coordination and pain. Edema often reduces sensation, and coordination is often diminished. Untreated edema may result in permanent losses of range of motion if fluids become fibrotic. Nerve compression can cause losses in sensation. Loss of active movement combined with compromised nutrition of distal parts can ultimately lead to amputation (Dutton, 1995). Maintenance and remediation of the anatomical and physiological condition of interstitial tissue and skin are important aspects of intervention.

Causes of Edema

Edema may be the result of bone or synovial thickening or fluid accumulation in or around a joint (Magee, 2008). Edema may be localized as in a cyst, swollen bursae, or intra-articular swelling. Swelling and edema is a normal consequence of trauma, resulting from vasodilation, which leads to an increase in white blood cells. The inflammatory response is an attempt to decrease the bacteria in the injured part (Pendleton & Schultz-Krohn, 2006).

Swelling that occurs in 2 to 4 hours is usually accompanied by blood leaking into the tissues (ecchymosis) or joint. Blood swelling feels firmer, thick, and gel-like. There may be elevated skin temperature when swelling is due to blood or pus. Within 8 to 24 hours, there is inflammation and synovial swelling. Swelling that feels hard may be due to osteophytes or myositis ossificans, while soft tissue swelling may feel boggy or spongy.

Long-standing soft tissue swelling feels leathery and thick. Pitting edema is fluid that is thick and slow moving, leaving indentation when pressure is applied, often caused by circulatory stasis (Magee, 2008). Take note of the skin color, temperature, and sensory changes, as these may be signs of serious problems (for example, purple color may mean pooling of venous blood, which might mean that arterial blood flow is impaired) (Pendleton & Schultz-Krohn, 2006).

Types of Edema

Acute

Acute (transudate) edema occurs in the inflammation state of tissue healing. The edema pits deeply, rebounds quickly, and can be easily moved (Pendleton & Schultz-Krohn, 2006). The edema is comprised primarily of water and electrolytes and decreases after 2 to 5 days because intact venous capillaries can absorb the edema and the lymphatic system absorbs the large plasma proteins not removed by macrophages (Artzberger, 2005).

Subacute and Chronic Edema

The edema in this stage continues to pit but is slower to rebound with a viscous quality due to high plasma protein content. Chronic edema (edema lasting longer than 3 months) pits very little and has a hard feeling. Severe edema is characterized by no elasticity, and the skin is shiny, taut, and cannot be lifted (Pendleton & Schultz-Krohn, 2006). Chronic edema is due to the protracted presence of plasma proteins in the interstitum, causing the tissue to become fibrotic.

Complex or Combined Edema

Edema that initially begins as transudate edema may develop into an exudate edema and is often found in clients who have experienced a stroke. Overexercising a flaccid extremity can cause microscopic rupture of tissues, resulting in inflammation and increased edema. If there is no motor return in the flaccid extremity and no motor pump to remove the fluids, the edema can become fibrotic (Pendleton & Schultz-Krohn, 2006).

Other Types of Edema

When the heart is unable to pump blood effectively, fluids can accumulate in the extremities. This might be noted in the ankles along with a slight pink color. Many edema reduction techniques are contraindicated for cardiac edema because of the additional stress placed on an already impaired cardiac system (Pendleton & Schultz-Krohn, 2006).

Low protein edema is swelling due to liver disease, malnutrition, or kidney failure. In contrast to chronic edema, in low protein edema, there are too few plasma proteins in the interstitum. As with cardiac edema, many edema reduction methods are not safe for this type of edema because of the additional burden placed on the liver and kidney (Pendleton & Schultz-Krohn, 2006).

Evaluation of Edema

Evaluation of edema begins with observing and palpating the skin. Is the skin taut, shiny, and is there a loss of normal wrinkles and joint creases? Find out how the injury occurred and how long it has been since the injury happened. Describe in detail how the skin feels, looks, and responds to pressure. Is the edema displaced by pressure, leaving a pit that fills back slowly when the pressure is removed (pitting edema), or is it more gel-like and firmer (brawny)?

The edema rebound test measures the amount of time required for skin to return to its original shape after pressure is applied to an edematous structure. The more boggy the edema, the slower the skin is to rebound. A tonometer can be used to measure changes in resistance to tissue pressure, but this not widely used (Cooper, 2007).

Edema can be measured and graded by palpating and applying firm pressure for 5 seconds. The degree of edema is based on the depth of indentation in centimeters as follows:

1 cm or less = 1+ edema

2 cm = 2+ edema

3 cm = 3+ edema

4 cm = 4+ edema

5 cm or more = 5+ edema (Jacobs & Jacobs, 2009)

Volumetric measurements can be taken using a volumeter, which measures the amount of water displaced from a container when the hand is inserted. Circumferential measurements can also be taken using a tape measure. While these measurements do not provide data about the normal hand, they are very useful in documenting progress.

Treatment of Edema

The goals of treatment of edema depend upon the stage and type of edema. Often, pain reduction is noticed before ROM shows improvement and is an indication of improvement.

Acute Edema

In acute edema, the goal is to prevent excessive swelling, accumulation of blood, and additional tissue damage and to control pain. Minimizing further irritation, inflammation, and infection is important in the treatment of acute edema so be careful to disinfect any open wounds, provide protection to healing structures, and balance rest with activities.

Reduction of swelling is done by vasoconstriction, compression, and elevation. One way to do this is with bulky dressings (Artzberger, 2005). Bulky dressings can be used to reduce excessive fluid outflow and help to prevent stress on fragile tissues (Cooper, 2007). Caution needs to be taken not to wrap the dressings too tight, though, which can cause vascular and temperature changes resulting in increased edema and painful compression.

Retrograde massage is an effective method of applying compression. Teaching self-massage to the client is helpful in desensitizing painful parts and in giving the client a sense of control and responsibility in the management of the edema. Light retrograde massage with elevation facilitates diffusion of small molecules into the venous system. If the compression too tight, this will restrict fluid flow, which increases edema.

Compression can be applied by pressure garments, gloves, elastic wraps, or even string. The sequence of dressings may start with Ace wraps initially, followed by tubular/coban next, and then pressure garments. Pressure garments are worn nearly all day for as long as 6 months to 2 years. Elastomer molds may also be worn for the chest, palms of the hand, and the face. Pressure garments and elastomer molds use compression to decrease blood flow, which may slow collagen synthesis (Dutton, 1995). These garments may also produce friction and shear forces, putting the part at risk for ischemia, so wear schedules and skin should be monitored periodically. Because scar management can take such an extended period of time, patient compliance is an important issue in intervention.

When performing massage or when wrapping parts for compression, it is important that the application of the compressive force be distal to proximal. The idea is to push the fluid from the part toward the heart. For example, if the dorsum of the hand is swollen, begin at the fingertips and massage toward the wrist, from the wrist to the elbow, etc. In addition, the choice of material for application of compression needs to be considered with regard to the skin surface to which it will be applied. Coban would be contraindicated for open wounds and skin grafts, as it leaves a ribbed, uncosmetic imprint.

While pressure garments are smooth and would leave fewer marks on the skin, they also are expensive and take time to receive from vendors. Use of a fitted pressure garment may be more applicable to maintaining edema reduction than for actually reducing edema (Dutton, 1995).

Elevation is also used to reduce outflow, which reduces capillary filtration pressure and increases the gradient pressure of the lymphatic vessels (Artzberger, 2005). Optimally, the part should be elevated above the level of the heart to enhance the flow of fluid. Caution should be taken with extreme elevation in stroke clients with right-sided weakness, because this might cause the fluid to flow too quickly into the right side of the heart (Cooper, 2007).

Adaptive equipment is also used to decrease edema by means of elevation, such as an arm trough placed on the wheelchair of a client with hemiplegia. The arm trough positions the arm safely and elevates it to facilitate circulation of the fluid from the hand toward the heart. Continuous passive motion (CPM) devices are also used to prevent and decrease edema. The use of elevation is practical when the client is at rest or is not able to use the extremity. However, once the extremity can be used in activities, some adaptive equipment can actually prevent active use of the arm and should be discontinued. For example, a sling may enable a dependent extremity to be placed in an elevated position but may prevent active motion. In this case, other positioning devices would be better for the extremity that is capable of active motion, such as using a trough or pillow for optimal positioning (Dutton, 1995).

Therapeutic use of cold modalities for the treatment of pain and edema produce quick vasoconstriction and also desensitize pain receptors and decrease muscle spasms. If cold packs are used to cause vasoconstriction and reduce outflow, the temperature of the pack should not be colder than 59°F (15°C) because, below these temperatures, proteins leak into the interstitum from the lymphatic structures causing more edema. Cold packs are recommended within the first 24 to 48 hours, and a dry towel should be placed between the skin and the cold pack to prevent tissue damage.

Other forms of cryotherapy might include ice massage, ice dips, vasocoolant sprays, or use of contrast baths. Ice baths or ice dips are used by having the client immerse an edematous hand for 3 seconds in water. The client would then squeeze the hand or wiggle it while in water, and this would be repeated two to three times. Contrast baths are used if the client cannot tolerate ice or is experiencing hypersensitivity. Contrast baths alternate cold (66°F or 18.9°C) and warm (96°F or 35.6°C) water each minute for 20 minutes, which provides vasodilation/vasoconstriction acting as a pumping action.

Other modalities that are used for the treatment of edema include iontophoresis, phonophoresis, and mild pulsed alternating current ultrasound. Iontophoresis promotes healing and decreases pain and edema by using electrical currents to deliver medications.

Active motion can be used if not painful to minimize reduced mobility of joints, ligaments and tendons, tissue atrophy, and impaired gliding of structures. Active proximal trunk and shoulder motions are helpful during the acute wound stage to decongest the lymphatic vessels, remove tissue waste, and better oxygenate the healing tissues (Artzberger, 2005; Pendleton & Schultz-Krohn, 2006). Edema reduction should not begin where the edema is visible but instead should start in normal, uninvolved areas proximal to the visible edema. Diaphragmatic breathing, trunk stretching, and muscle contraction exercises and activities help to reduce edema and decongest the lymphatic system. Diaphragmatic breathing helps to decongest the lymphatic system by changing the thoracic pressure, creating a vacuum that draws the lymphatic fluid from more distal vessels (Cooper, 2007). Exercise of abdominal muscles increases the pumping of blood, which stimulates the lymph nodes, moving lymph through more rapidly (Aretzberger, 2007).

Normal muscle contraction serves to increase circulation, so active movement and use of weight-bearing positions stimulate these normal muscle actions and can be useful in decreasing edema. However, active motion that is too forceful in early stages of recovery may aggravate edema instead of reducing it, so parameters of movement need to be controlled.

Subacute and Chronic Edema

Treatment for subacute edema includes diaphragmatic breathing, trunk exercise, lymphatic massage of the uninvolved axilla, manual edema mobilization, kinesio taping, gentle myofascial release, fluidotherapy at body temperature, passive motion (via continuous passive movement [CPM] machine), fluidotherapy with the temperatures no higher than 98°F (36.7°C), and active and passive exercise and activities (Cooper, 2007).

Treatment for chronic edema is similar to subacute but also includes softening of fibrotic tissue before the tissues can be decongested. Methods to soften the fibrotic tissue include neutral warmth, elastomer, or silicone gel sheets that provide light compression and retain body heat, kinesio taping to reduce inflammation and stimulate the lymphatic system, low-stretch bandages, foam-lined splints, and gentle myofascial release (Artzberger, 2005; Cooper, 2007). For the low-stretch bandaging to work, proximal manual edema mobilization needs to be done first to decongest the lymphatic system (Cooper, 2007).

Manual edema mobilization (MEM) is used to activate only the lymphatic system in order to absorb the large plasma protein molecules and other molecules from the interstitum that are not permeable to the venous system (Artzberger, 2005). The lymphatic system does not have a continuous pump system so large molecule that cannot permeate the venous system are absorbed into the lymphatic system. It works on a negative pressure gradient so that when the lymph vessel fills, high pressure is created so fluid moves to an area of lower pressure. Excessive swelling distal to the lymph nodes does not increase their rate of filtration but causes further congestion distally (Artzberger, 2005; Cooper, 2007). Given this, proximal uninvolved structures have to be stimulated first to decongest or draw the distal edema out of the involved area, creating lower negative pressure (Artzberger, 2005). Congestion exists proximal to the visible edema. MEM decongests the more proximal edema and moves the edema proximally, thereby creating a space into which the more distal edema can flow. Absorption of the protein into the lymphatics occurs because of changes in interstitial pressure due to movement, stretching of filaments from lymphatic to connective tissue, respiration, pulsation of nearby arteries, and light massage (Artzberger, 2005).

Manual edema mobilization is considered an advanced skill requiring specialized training but includes a light, stroking massage starting proximal to distal to the edematous area.

The massage is done following the flow of lymphatic pathways. Exercise or engagement in functional activities follows the massage of each segment (Pendleton & Schultz-Krohn, 2006). Contraindications and precautions of MEM are that this method may spread infection and should not be used in areas of inflammation or in the inflammatory phase of wound healing. Do not perform MEM if there are hematomas or blood clots, as this may cause the clot to move. MEM is contraindicated in clients with cancer, renal disease, kidney disease, or lymphedema (either primary or from a mastectomy). If the client has congestive heart failure or cardiac or pulmonary problems, MEM may cause problems because these systems are already overloaded or impaired (Aretzberger, 2007; Cooper, 2007; Pendleton & Schultz-Krohn, 2006).

Lymphedema is a condition of chronic, high-protein edema that results when a permanent mechanical obstruction of the lymphatic system creates a lymphatic overload due to surgical removal of a lymph node, irradiation of nodes, or a congenital deficit. MEM is not appropriate for clients with lymphedema, because it is used with intact lymphatic systems (Cooper, 2007). Eliminating firmly entrenched external adhesive scar tissue takes a considerable force. Deep sustained friction therapy, manipulation, or injection are three techniques that are usually effective in the appropriate circumstances.

RANGE OF MOTION

Range of motion may be limited for a variety of reasons, including disease, injury, edema, pain, skin tightness, muscle spasticity, muscle and tendon shortening (tightness or contractures), and prolonged immobilization.

The effect of immobilization on muscle is that there is a loss of muscle mass due to decreased numbers of muscle fibers, which may begin as early as 48 hours after immobilization. There is a decrease in muscle force not only because of lost muscle mass but also due to metabolic, vascular, and neurologic factors. There are changes in the length and number of sarcomeres, and filaments lose the ability to slide. There is disruption of synovial fluid, membranes, and articular cartilage (Nyland, 2006). Due to these changes, muscle fatigue increases, and there is decreased functional movement. Movement of the surrounding joints and muscles helps to prevent atrophy and scar tissue that may limit motion.

Immobilized connective tissue results in reduced tensile forces applied to tendons and ligaments due to reduced structural stiffness. There is an increase in collagen synthesis and degradation, and the collagen becomes disorganized (Fess et al., 2005). Tendons atrophy, and the space between the tendon and tendon sheath is decreased because of adhesions, which restricts the gliding action of the tendon (Cooper, 2007).

Even if immobilized, isometric contractions of the immobilized muscles load the immobilized tendons, which is essential for connective tissue nutrition and repair, encourages normal collagen production, and aligns the collagen along lines of stress (Cooper, 2007). Low load and prolonged stretch are preferable to high-load, short-stretch movements for these weakened tissues. When remobilizing tissues, avoid excessive force, poor force direction, and undesirable or abnormal patterns of mobility.

Flexibility is the amount of range of motion of a joint and is determined by the elasticity of soft tissues, specific conditions within the joint, excessive body fat or muscle mass, and pain (Hamill & Knutzen, 1995). Muscle stiffness is the ability to resist stress and is a measure of the rate of increase of passive tension as a muscle is stretched (Nyland, 2006). Flexibility training is designed to decrease the stiffness of the musculotendinous unit. Flexibility can be static or dynamic. Static flexibility is measured with a goniometer, whereas dynamic flexibility is seen to be the ease of movement or amount of resistance to joint movement or the amount of resistance to passive motion of a joint or tone. Kisner and Colby define flexibility as "the ability of muscle to relax and yield to a stretch force" (Kisner & Colby, 1990, p. 110).

In interpreting ROM values made via goniometric measurements, it is important to remember that the available range varies with age, occupation, gender, specific joint, and activity levels. Earlier discussions described problems with reliability and validity, especially because normative values rarely describe how the measurement was made, from what population the values were taken, and the standard deviation for the mean values.

Limitations in ROM are the focus of intervention if the limitation prevents successful engagement in areas of occupation that are of value to the client or would present structural imbalances that might lead to deformity. Full ROM is not needed to achieve many tasks, and functional range is sufficient for many occupations. This illustrates that the clinical collaboration between the therapist and the client is vital, and use of this approach must go beyond the procedural reasoning about the knowledge of musculoskeletal function and prognosis. It is only by assessing contextual variables, knowing the patient, and using interactive reasoning that meaningful interventions take place. For example, if a client lacks 90 degrees of shoulder flexion but is still able to dress himself or herself and this is of value to the client, then increasing ROM is not indicated. If the client lacks 120 degrees of shoulder flexion and is unable to put on a shirt, but the spouse will dress the client, then increasing ROM for the shoulder may not be indicated. Intervention directed at remediation of ROM itself is not a valid long-term intervention goal because it is not directed toward a functional outcome related to work, play, or self-care.

Generally, if the limitations in movement are due to shortened tissues, the goal is to stretch the tissues to new, permanent lengths to enable function. If the limitation is due to edema, pain, or spasticity, preventing the loss of ROM is the secondary goal. If the limitation is due to permanent shortening (contracture) or chronic shortening of the tissues, the goal is to compensate or adapt to the limitations because remediation efforts would not be successful (Radomski & Latham, 2008).

Maintaining Range of Motion and Passive Range of Motion

Limitations in passive range of motion (PROM) are due to problems within the joint. This might include stiffness due to capsule, ligament, muscle, or tendon tightness, decreased joint space, or osteophytes (Cooper, 2007). If passive and active ROM are equal, the first step of treatment is to focus on

sasittal axis = Ant/post

regaining passive ROM. If active ROM is normal and active ROM is limited, the focus is on active assisted range of motion (AAROM). This progression from passive to active assistive to active to resistive is a treatment pattern for ROM and strength.

Passive ROM would be done with clients who are unable to move through full ROM actively due to weak or denervated muscles, are unconscious or in a coma, have paralysis, neurological disease, or are on enforced rest (Trew & Everett, 2005). The intention of performing passive ROM or maintaining the current length of the tissues is to prevent adhesions and contractures from developing. If no movement was done, the actin and myosin proteins would be reabsorbed, the area of cross bridge formation would be decreased, and the result would be muscle weakness and shortening. The noncontractile elements will change the collagen turnover rate so that more collagen will be produced, increasing the stiffness of the muscle and decreasing its elasticity. Passive ROM helps to maintain the integrity of soft tissue and muscle elasticity. By moving the part, even passively, venous circulation is increased, synovial fluid production is enhanced, and the joint receives needed nutrition. Movement may reduce pain and maintain functional movement patterns and can be used to increase kinesthetic awareness (Trew & Everett, 2005).

The main principle regarding ROM is that there must be movement through full ROM. There are many different ways to ensure that the movement produced is through the complete range.

Principle: Maintenance of ROM occurs when movement occurs through the full available ROM.

Exercise and activities can occur in anatomical planes of motion. For example, PROM for flexion of the shoulder would need to occur in a sagittal plane and in a frontal axis. By grasping the client's arm under the elbow and with the other hand grasping the wrist, the client's arm would be lifted straight up and parallel to the trunk. Similarly, abduction would occur in a frontal plane and a sagittal axis, and the arm would be moved in a direction perpendicular to the trunk. The therapist could perform this motion, or the client or caregiver could be instructed in the procedure through the available ROM.

Codman pendulum exercises can also be done in which the client bends forward at the waist (or while seated as in Figure 11-1) and allows the affected arm to hang down toward the floor. The body is moved so that the arm moves in a straight line in forward/backward, side-to-side, and clockwise-counterclockwise direction.

Another method of providing passive ROM would be to move the part (or have the client move) in the direction of the muscle range of elongation. Knowledge of muscle fiber composition and line of pull would be necessary because the range of motion exercise would be antagonistic to the line of pull of the muscle (Kisner & Colby, 2007).

Combined patterns of movement (as in proprioceptive neuromuscular facilitation [PNF] patterns) (as seen in Figure 11-2) are seen as a very functional way of incorporating ROM actions with movement in several planes of motion. An example might be to have the client start with the arm in shoulder flexion, adduction, and internal rotation, and then the client

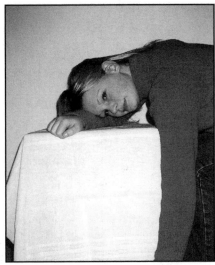

Figure 11-1. Pendulum exercises.

moves to a position of shoulder extension, abduction, and internal rotation. This D1 flexion pattern may be used when combing the hair on the opposite side, and the D1 extension pattern may be functionally used in pushing a car door open (Pendleton & Schultz-Krohn, 2006). Several motions and planes are combined in one smooth movement. The disadvantage with these patterned movements is the complexity of the movement, the planning involved, and the ability to follow multistep directions. Again, these combined patterns can be performed actively by the client as exercise or as part of an activity and can be taught to caregivers.

Functional patterns of movement associated with functional tasks can also be encouraged. This would be an appropriate screening strategy or could be considered an aspect of the top-down approach, where occupational performance areas of work, play, and activities of daily living are assessed rather than an initial focus on client factors. Observation of functional patterns of movement is actually a constant ongoing assessment that occupational therapists make and is part of what comprises conditional clinical reasoning.

A continuous passive motion (CPM) device is often used to provide constant movement through specified ranges for a predetermined duration for clients following orthopedic surgery. Muscle fatigue is not a factor because the motion is passive.

If a client experiences unilateral deficits in ROM, then self-range of motion exercises can be taught for independence in maintenance of ROM. These can be exercises that use the unaffected hand and arm to move the affected side or can use a variety of tools such as those used in wand, cane, or dowel exercises (Figure 11-3).

Other tools and devices used in ROM exercises include the following:

- Finger ladder
- Shoulder wheel
- Overhead pulleys
- Suspension (e.g., Swedish suspension sling)

Figure 11-2. PNF patterns.

	Scapula	glenohumeral	Elbow	Wrist and fingers
D₁ flexion				
	Elevation Abduction Upward rotation	Flexion Adduction External rotation	Flexion or extension	Flexion Radial deviation
D₁ extension				
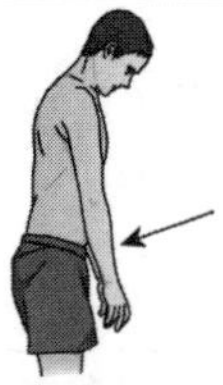	Depression Adduction Downward rotation	Extension Abduction Internal rotation	Flexion or extension	Extension Ulnar deviation
D₂ flexion				
	Elevation Adduction Upward rotation	Flexion Abduction external rotation	Flexion or extension	Flexion Radial deviation
D₂ extension				
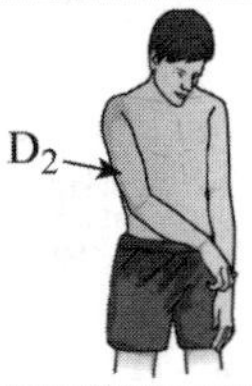	Depression Abduction Downward rotation	Extension Adduction Internal rotation	Flexion or extension	Extension Ulnar deviation

- Skate or powder board
- Tai chi
- Games
- Aquatics

The TERT (Total End Range Time) theory states that the amount of increase in ROM of a stiff joint is proportional to the amount of time the joint is held at its end range and is seen as a valid means to increase the elongation of tissue and the importance of sustained stretch (Flowers & LaStayo, 1994). TERT is based on a formula of intensity x duration x frequency. If the client could tolerate maximum stretch (intensity) for 20 minutes (duration) and frequency is three times a day, then optimal total TERT is 60 minutes per day (Davies & Ellenbecker, 1999). The plastic deformation that occurs lasts no more than 4 weeks, and passive ROM in a joint is not functional. To achieve permanent increases in ROM, the tissues in their newly elongated state need to be integrated into functional activities.

Contraindications to Passive Range of Motion

There are a few contraindications to performing passive ROM. Passive ROM can injure swollen and inflamed joints and tissues (Cooper, 2007). An inflammatory response can be triggered, causing additional scar production, pain, and stiffness and can lead to complex regional pain syndrome (CRPS) if done inappropriately or too aggressively. Other contraindications include early fractures, excessive pain, incomplete muscle or ligament tears, or areas where circulation is compromised (Cooper, 2007). Table 11-5 provides documentation suggestions and methods, principles, and examples relevant to the treatment of remediation of tissue integrity.

Increasing Range of Motion

When soft tissues around a joint are shortened and there is a loss of ROM, elongating these soft tissues will result in increased range. The shortening of soft issues may be due to prolonged immobilization, restricted mobility, connective tis-

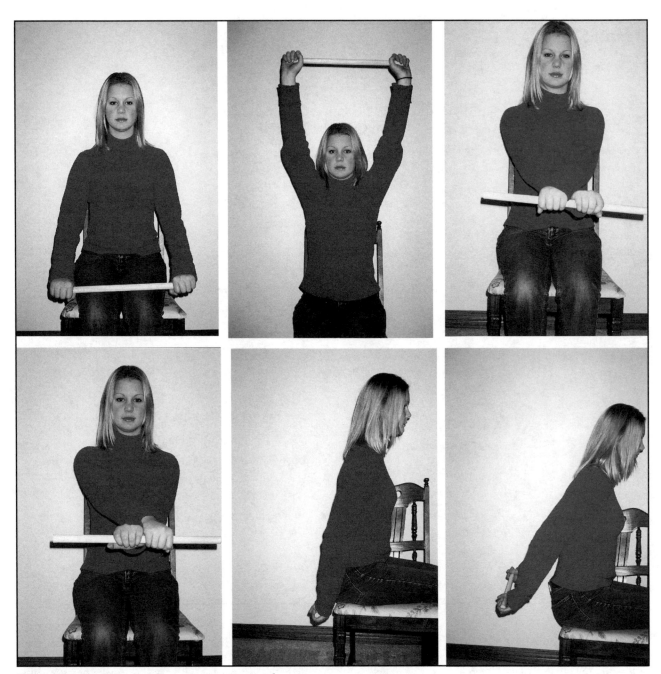

Figure 11-3. Cane exercises for maintaining range of motion.

sue or neuromuscular disease, tissue pathology secondary to trauma, or due to congenital and acquired bony deformities (Kisner & Colby, 1990).

Principle: To increase ROM, a stretch force must be applied to the tissue that is beyond the current ROM to elongate the collagen fibers in soft tissues.

Shortening of muscle or other tissues that cross a joint can produce a contracture. Note that this is not synonymous with contraction, which is the process by which tension develops in a muscle during shortening or lengthening. Contractures can be defined and classified by the soft tissues affected. A myostatic contracture has no specific tissue pathology, and yet there is loss of range that is usually mild and transient. Myostatic contractures often are resolved quickly with gentle stretching. Tightness is a term often used to describe loss of ROM at the outer limits of ROM that occurs in otherwise healthy tissues (Jacobs & Bettencourt, 1995).

Chronic inflammation and fibrotic changes in soft tissues can lead to fibrotic adhesions that are difficult to reduce. Irreversible contractures are those in which there is a permanent loss of extensibility of soft tissues (Kielhofner, 1992). Irreversible contractures often are released surgically and

Table 11-5

TISSUE INTEGRITY

GOAL	METHOD	PRINCIPLE/RATIONALE	EXAMPLE
Client will be (independent/ modified independent/ require maximal/ moderate or minimal assistance in performance of ____ tasks (specify) in reducing peripheral edema as measured by (volumeter/ circumferential measurement, etc) by using...	Elevation	Which reduces outflow, reduces capillary filtration pressure, and increases the gradient pressure of the lymphatic vessels and to reduce pain	Arm trough on wheelchair CPM machines Elevation on pillow or lapboard Sling Cold modalities
	Compression	Which permits filtration of fluids from capillaries into interstitial tissues, decreases blood flow, and controls pain	Retrograde massage Ace wraps Coban wraps Elastomer molds Pressure garments
	Temperature control	Which stimulates localized vascular responses	Thermal modalities
	Active range of motion and/or diaphragmatic breathing	Which pumps fluids out of interstitial tissues and joint structures and increases circulation	Gentle active range of motion Active assisted range of motion
	Manual edema mobilization	Which activates the lymphatic system and decreases proximal congestion	Manual edema mobilization techniques

Adapted from Dutton, R. (1995). *Clinical reasoning in physical disabilities*. Baltimore, MD: Williams & Wilkins; and Marrelli, T. M., & Krulish, L. H. (1999). *Home care therapy: Quality, documentation and reimbursement*. Boca Grande, FL: Marrelli & Associates, Inc.

occur when normal tissue is replaced by bone or fibrotic tissue. Hypertonicity can also cause a limitation of joint ROM as a result of central nervous system lesions. The muscle is in a state of high tone and constant contraction, resulting in a pseudomyostatic contracture (Kisner & Colby, 1990).

To elongate the tissues, a stretch force is applied. The force, velocity, speed, and extent of the stretch force must be controlled, and the joint movement needs to exceed the currently available range. The effects of stretching are an increase in ROM, increase in flexibility, increase in soft tissue length, relief of muscle spasm, and increase in tissue compliance (Trew & Everett, 2005). Allowing quicker recovery from workouts, reducing soreness, facilitating relaxation, maintaining balanced musculotendinous lengths between agonist/antagonist muscles, and improving motor performance are benefits of stretching (Nyland, 2006).

Some general concepts related to stretch are that static stretching is preferable to ballistic stretch techniques. Stretching will increase flexibility and ROM but the motion should not exceed the normal muscle length by more than 10%. Stretching is recommended before and after strengthening activities, although results vary as to whether stretching before exercise decreases the risk of injury (Gleim & McHugh, 1997; Shrier, 1999).

Stretching is enhanced if the tissue is warmed prior to elongation of tissues, which increases collagen extensibility and decreases muscle-tendon stiffness. The stretch should not be painful but just an overstretch of current tissue length. Stretching and warm-up are not the same, as a warm-up done prior to exercise raises the total body and muscle temperature.

Stress-Strain Relationship

When a tissue is stretched, this causes stress inside of the tissue. Stress is a force per unit area and is an internal reaction force within the material. Stress is not visible.

When stress is applied to a tissue, as when a muscle is elongated beyond the current range, the size and shape of the tissue changes or deforms. The amount of deformation depends upon

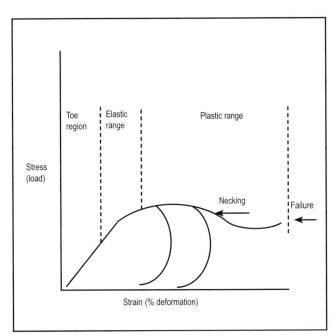

Figure 11-4. Stress-strain curve.

the amount of the load and the ability of the material to resist the load. The percentage of deformation that occurs is known as strain. These two ideas, stress and strain, are reflected in Hook's Law, which states, "deformation increases proportionally to the applied force or strain increases proportionally to the stress of resisting the applied load" (Irion, 2000, p. 228).

Both the contractile and noncontractile elements of muscle fibers influence stretching of soft tissues, and both contribute to the temporary (elasticity) and permanent (plastic) changes in tissue length. Noncontractile tissue, comprised of collagen, elastin, reticulin fibers, and ground substance (mostly gel containing water), can develop tightness and contractures and limit joint motion and function. Because collagen is the element that absorbs most of the tensile stresses, tissues with a greater proportion of collagen fibers will provide greater stability, while soft tissue with a greater proportion of elastin fibers will have greater extensibility and flexibility.

When stress is applied to a muscle, the reaction of the muscle follows a predictable pattern shown in the stress-strain curve in Figure 11-4. When a muscle is passively stretched, there is an initial lengthening of the series elastic components due to a mechanical disruption of the cross bridges as the actin and myosin proteins slide apart. This causes the sarcomere to lengthen until the stretch force is released. This is the toe region where most functional activity normally occurs (Kisner & Colby, 1990). Different connective tissues have different toe regions (e.g., ligaments have longer toe regions than tendons). Once the force is stopped, the muscle returns to its original resting length, illustrating the elastic nature of the short-term stretch and the muscle characteristic of elasticity.

If, however, the muscle is immobilized or held in a lengthened position for a prolonged period, the changes in the muscle will be more permanent, and a plastic change has occurred. This is due to the greater number of sarcomeres in series, allowing the greatest functional overlap of actin and myosin proteins (Kisner & Colby, 1990). This more permanent change is the result of sequential failure of the collagen bonds, which responds to forces by remodeling and rebonding over time in line with the application of the stress. For tissue to return to normal length after plastic changes, it must exhibit inflammation and undergo repair (Donatelli & Wooden, 2010).

The muscle may also be immobilized for prolonged periods of time in a shortened position. In this case, the muscle produces increased amounts of connective tissue to protect the muscle against stretch. There are fewer sarcomeres due to sarcomere absorption. The changes in sarcomere length are transient, and they will resume the original position if the muscle is allowed to return to normal length after immobilization.

If stress continues to be applied to a muscle in the plastic range, plastic changes can only continue up to a point, known as necking. This is the point of ultimate strength of the muscle. Past this point, there may be increased strain without an increase in stress so that even with less loads applied, the tissue is under increased strain. After this point, the tissue fails rapidly, either due to a single maximal force applied or repeated submaximal stresses. Stretching effectiveness is time dependent: use of much force for a short duration results in elastic changes; low force application for a long duration produces plastic deformation. Permanent change in the tissue length is the goal of increasing ROM.

It is important for the therapist to be aware of the way that the tissue feels when stretching because when the strain is increased but the resistance felt in the muscle decreases and when less force is needed for deformation to occur, necking may be occurring, and tissue failure may be imminent even with smaller loads (Fisher, 1998).

Creeping is the application of "low magnitude loads over prolonged periods of time [that] increases the plastic deformation of noncontractile tissue, which allows a gradual rearrangement of collagen fibers" (Kisner & Colby, 1990, p. 119), and it is the gradual increase in tissue length necessary to maintain a constant stress (Nyland, 2006). The remodeling and rebonding of the collagen fibers require time for healing. Intensive stretching is usually done every other day to allow this time for healing. This is especially true in tissues of elderly clients because a normal age-related change is a reduction in collagen, and there is decreased capillary supply that reduces the healing abilities in the elderly. In addition, there is a decrease in tensile strength and rate of adaptation to stress in the elderly (Bonder & Bello-Haas, 2009; Kisner & Colby, 2007).

Heat increases creep and therefore the elasticity of collagen fibers, making tissues more amenable to stretch (Demeter, Andersson, & Smith, 1996; Kisner & Colby, 2007) and making the stretching of tissues more comfortable for the client. Heating muscle tissue to 40°C affects the bonding between the collagen molecules, increasing the rate of creep and relaxation (Oatis, 2004). For this reason, both superficial and deep heat modalities increase the extensibility of tissue and are often used prior to stretch.

Cognitive relaxation techniques, such as disassociative visualization (pleasant memory or image), autogenic relaxation (similar to self-hypnosis), use of videotapes, audiotapes, music, or meditation, also serve to relax the client prior to stretch. Several nontraditional techniques, such as the Feldenkrais method, Alexander technique, or hatha yoga, can also be used.

Other techniques used to relax the tissues prior to stretch force application might include active inhibition techniques, local relaxation, progressive relaxation techniques (which may include controlled breathing techniques or progressive muscle relaxation), and biofeedback. Low-intensity exercise or activity can also be used to warm the tissues prior to stretch. Massage done before stretching serves to increase circulation and decrease muscle spasms and stiffness. Taylor and colleagues (1990) found that the muscle-tendon units respond viscoelastically to stretch. They also found that only a minimum number of stretches (as few as four) will lead to tissue elongation and that slow stretching (rather than an actual technique) may decrease the risk of injury.

Neurological Factors Related to Stretch

A limiting factor related to stretching a muscle is resistance secondary to reflex activity. In increasing ROM, you want to inhibit reflex activity of the muscle.

Two sensory receptors are directly involved in muscle function and to the application of stretch forces. The muscle spindle is a major sensory organ of the muscle and consists of microscopic intrafusal fibers that lie parallel to the extrafusal (skeletal) muscle fibers. These fibers are sensitive to the length and velocity of the lengthening of extrafusal muscle fibers, and the muscle spindle initiates muscle contraction. Type Ia primary afferent sensory neurons are responsive to the rate of stretch, and type II secondary afferent sensory neurons are facilitated by changes in muscle length. Once a muscle shortens, the type Ia primary sensory neurons are no longer activated because they are no longer stretched. Because the muscle spindle lies parallel to the extrafusal muscle fibers, when the muscle is lengthened, the muscle spindle is also lengthened and responds by initiating a muscle contraction (Oatis, 2004).

The muscle spindle has been called a "comparator" because the spindle compares the length of the spindle with the length of the skeletal muscle fiber. If the length of the extrafusal muscle fiber is less than the spindle, the spindle is less active. Conversely, when the spindle is stretched, more nerve impulses are sent to activate the extrafusal fibers (Pedretti, 1996). For example, when the patella is tapped by a small hammer, this illustrates the activation of the muscle spindle in a brief contraction called the stretch reflex, muscle spindle reflex (MSR), myostatic or monosynaptic stretch reflex, or deep tendon reflex (DTR).

The second sensory receptor, the golgi tendon organ (GTO) wraps around the extrafusal fibers of the muscle in neuromuscular junctions between the muscles and tendons. This sensory receptor is sensitive to the tension in muscles, either due to passive stretch or active muscle contraction. It serves a protective function to inhibit the contraction of a muscle in which it lies and has a low threshold (fires easily) after active muscle contraction. The GTO has a high threshold (fires less easily) to passive stretch. Due to the GTO, the force that can be developed in a muscle is limited.

The implications of the actions of these two receptors on the stretching of extrafusal skeletal muscles is that if a muscle is stretched too quickly (too great a velocity) or too far (too great an increase in length), the muscle spindle will be stretched. This in turn will produce increased tension in the skeletal muscle, and this is counterproductive to efforts to lengthen this muscle. Slow, static stretching is recommended because it slowly decreases the response of type Ia input, which provides interference to joint movement. When a muscle is stretched, the type II fibers respond with an inhibitory response to relax the muscle.

Stretching done at too high a velocity may actually increase the tension in a muscle that you are trying to lengthen. This relates to the viscoelastic tissue of the parallel elastic components of the epimysium, perimysium, and endomysium, which respond with greater resistance to a quick stretch (Nyland, 2006). When a muscle is stretched slowly, the GTO fires, which inhibits increased tension development in that muscle and allows lengthening to occur by permitting the sarcomeres to lengthen. The decreased reflex activity from the slow stretch results in reduced resistance to stretch, enabling gains in range of motion (Nyland, 2006).

Types of Stretch

Stretch can be applied actively or passively, statically, and dynamically. Situations for passive stretch would be in cases when the client is unable to stretch the part actively due to muscle weakness, when the client cannot learn how to prevent substitution movements, or when the goal is to increase ROM at the end of range. The evidence is not clear about the most effective method to increase flexibility due to differences in studies in experimental design, measurement instrumentation, lack of control group or inadequate control groups, and inconsistent application of stretching techniques (Cooper, 2007).

Static Stretching

Static stretching is the most common method of force application. This is a sustained controlled stretch that is done slowly. Static stretching does not require another person, there is less danger of excessive lengthening, less energy is used, and it may help to decrease muscle soreness. Static stretching is seen as more tolerable for the client and is seen as less painful or uncomfortable (Cooper, 2007; Oatis, 2004).

Dynamic Stretching

In dynamic stretching, a muscle is stretched by muscular contraction, usually as a warm-up to prepare the muscle for movements specifically required by a particular sport or activity.

Passive Stretch

Passive stretch can be applied by means of joint mobilization, which involves passive rocking and oscillatory movements aimed at increasing joint play. Joint play is not under the patient's voluntary control and is used when there is joint stiffness, pain, or reversible joint hypomobility (Trombly, 1995). Joint mobilization is believed to stimulate mechanoreceptors and inhibit nociceptive stimulation and can cause muscle relaxation. The distraction and gliding of joint surfaces improve synovial fluid movement to improve joint nutrition. Stretching the joint capsule causes permanent changes in collagen fibers to improve motion (Houglum, 2010).

Another technique, myofascial release, is the passive stretching and breaking up of adhesions of the fascia, capsule, and ligaments (Nyland, 2006). Joint manipulation is the skilled passive movement of the joint following joint arthro-kinematics (Donatelli & Wooden, 2010) and is done to break up massive adhesions, often under anesthesia. The benefits of joint mobilization are that there is better joint lubrication, impingement can be prevented, soft tissue length can be maintained, and it may reduce spasticity (Gillen & Burkhardt, 2004). Capsular tightness has its own capsular pattern, and if this is present, then joint mobilization is indicated as an intervention option. These three techniques require extensive training beyond entry-level competence and are beyond the scope of this text.

Manual Passive Stretch

Manual passive stretch is a short duration stretch (15 to 60 seconds) where the tissue is elongated beyond resting length by a therapist applying external force. The therapist controls the direction, speed, intensity, and duration of the stretch. It is important to point out that manual passive stretch is not the same as PROM. PROM moves the part through unrestricted, available range, while manual passive stretch is done to the point of maximal stretch or a few degrees past the point of discomfort (Thompson & Floyd, 1994).

Excessive overstretching occurs when the movement over-comes the natural supportive function of the joint and tissues, resulting in hypermobility, joint instability, and increased risk of strain and injury. Signs of excessive overstretching are pain, redness, and swelling that persist for several hours. Nonverbal danger signals that may be observed might include a sudden increase in sweating, visual signs of stretching on the skin, gradual loss of slack, the client looks away suddenly, or there is a sudden constriction of the pupils.

Gains in muscle elongation due to manual passive stretch are transient and are attributed to temporary sarcomere give (Kisner & Colby, 1990). The stretch force applied is dependent upon the client's tolerance and the therapist's strength and endurance. This type of stretch may be appropriate during the initial stages of intervention to determine the client's respons-es to and tolerance for stretch (Kisner & Colby, 2007).

Prolonged Mechanical Stretch

Prolonged mechanical stretch is a low-intensity external load (5 to 15 pounds) that is applied to a part for a prolonged period of time by means of positioning, traction, braces, splints (static), or serial casting. The force is applied from 20 to 30 minutes to several hours. The longer duration of low-inten-sity force application permits the sarcomeres to be added and for remodeling and rebonding of collagen fibers to occur so more plastic changes occur. Prolonged mechanical passive stretching is seen to be more effective and more comfortable for the client than manual passive stretch. This static stretch-ing technique is seen as less labor intensive because the body part or object is providing the stretch force.

Active Stretching

Active stretch involves the series elastic and parallel elastic components of the muscle as well as the contractile elements. Active stretching is done in cases where the client is afraid of passive stretch due to loss of control, fear of re-injury, or low tolerance for discomfort. When passive stretch is harmful (for

example, in newly sutured tendons or grafted skin), active stretching can be done. Because the client is providing or con-trolling the stretch force, there is less chance of overstretching. This can mean that injury and discomfort will be minimized, or it could also prevent sufficient lengthening of the soft tis-sues and for a long enough duration to enable permanent elon-gation. In active stretching, the resistance force is minimal. Active stretching can be done by self-stretching one's own muscles, by means of the shoulder wheel or finger boards, or by using PNF diagonal patterns of movement. Self-stretching is done by the client using the weight of a body part or gravity as the stretch force.

Active stretch done during occupational activities and tasks that are meaningful to the client serves to distract the patient who has pain, can be a motivational factor, and can reduce psychological dependence on the therapist (Dutton, 1995). Whether passive or active stretching is done, it is vital to incorporate tasks and activities that the client actually needs to be able to successfully perform as a major focus of interven-tion. The goal of occupational therapy is to enable clients to return to the daily activities that are important to them, and it cannot be assumed that increases in range of motion will automatically be assimilated into these meaningful tasks.

Studies to determine the optimal type of load and duration of force application have used a variety of stretch mechanisms and methods. Low load prolonged stretch (LLPS) is used to reduce contractures, often by means of an orthotic device. High load brief stretch (HLBS) and exercise in proprioceptive neuromuscular facilitation (PNF) diagonal patterns were com-pared with LLPS. In this case, LLPS was found to be superior in the results, while a study comparing the use of a Dynasplint (Severna Park, MD) with PROM and manual stretching found no difference between this regimen and LLPS (provided by the Dynasplint). Serial casting often is used as a means of provid-ing a LLPS, and this has been found to result in significant increases in passive ROM as compared with other methods (Nuismer, Ekes, & Holm, 1997).

Proprioceptive Neuromuscular Facilitation

Neuromuscular facilitation techniques include a broad range of techniques to improve neuromuscular movement by applying proprioceptive stimulation. The client's own voli-tional isometric muscle contractions are used to increase ROM by minimizing the resistance attributed to spinal reflexes. This is accomplished when the isometric contraction of the antago-nist muscle group results in reflex inhibition of the muscle being stretched (reciprocal inhibition). At the same time the agonist muscle is facilitated, the antagonist muscle is relaxed (Nyland, 2006).

Active inhibition techniques are used in collaboration with the patient to actively stretch shortened tissues. The patient must have normally innervated muscles capable of volitional control, and the stretch affects contractile elements by stretch-ing the elastic tissues. Kisner and Colby indicate that the assumption is that the sarcomere will give more easily when the muscle is relaxed, and there is less active resistance in the muscle as it is elongated (Kisner & Colby, 1990). While active inhibition is generally more comfortable for the client than other types of stretches, the gains are usually temporary.

Table 11-6

CONTRAINDICATIONS AND PRECAUTIONS FOR TISSUE ELONGATION

CONTRAINDICATIONS	PRECAUTIONS
Hypermobility	Hematomas
Joint effusion	Infectious disease
Inflammation	Neurologic dysfunction
Bony blocks	Arthritis (post-inflammatory phase)
Sharp pain	Tendon repair or tendon transfer
Arthritis (acute exacerbation)	Prolonged bed rest
Dislocation	Prolonged Immobilized tissue
Recent trauma or surgery	
Osteoporosis or bone weakness	
SCI when trying to develop teneodesis	
Unhealed fractures	

Several PNF methods are used to elongate tissue. In the contract-relax method, there is a voluntary contraction of the antagonist muscle against maximal resistance provided by the therapist. The client then relaxes the muscle as the therapist moves the limb to the end of the ROM until a stretch is felt. The client contracts against resistance for 5 to 10 seconds and is then passively moved to the new muscle length where it is held in place for 10 seconds. The contract-relax with agonist contraction is similar to the contract-relax method except that the limb is taken to the point of stretch and the therapist applies resistance to the muscle being stretched for 5 seconds. Again, the client relaxes the muscle while the agonist concentrically contracts, is pushed to the new length, and is held. The hold-relax method is also similar to the contract-relax method, but an isometric rather than dynamic contraction against resistance is applied prior to relaxation (Nyland, 2006; Smith, 2005). Whether contract-relax, hold-relax, contract-relax-contract (hold-relax-contract), or agonist-contraction methods are used, the client is asked to first relax and then elongate tight muscles. These relaxation and inhibition techniques have been used to relieve pain, muscle tension, tension headaches, high blood pressure, and respiratory distress (Kisner & Colby, 1990).

Ballistic

Ballistic stretching involves a jerking or bouncing motion and is the least desirable and most controversial method to stretch tissues. It is felt that this method of tissue elongation places tissues at risk for rupture and actually stimulates the muscle spindles, thus putting the muscle in a state of resistance to further stretch (Nyland, 2006).

Contraindications to Stretch

An obvious contraindication to stretching tissues is not to force the tissue beyond the normal limits of ROM. It is also true that vigorous stretching should be avoided in tissues that have been immobilized due to the changes in muscle, tendon, and ligamental tensile strength and joint integrity. Allowing time for the collagen to remodel requires rest after stretching, so this does not need to be done every day. If soreness or pain occurs 24 hours after stretching, too much force was used.

Avoid stretching edematous tissues because they are more susceptible to injury, and stretching may cause further pain and edema. In post-inflammatory phases of rheumatoid arthritis, gentle, prolonged stretch can help prevent the shortening of tissues into fixed deformities, but when the joints are in acute, inflammatory phases, stretching should be avoided.

In cases where shortened tissues are providing greater joint stability in lieu of normal stabilization or strength, stretching is contraindicated. This may occur in special cases, such as cervical spinal cord injuries, where intervention is aimed at the development of tenodesis action of the fingers and wrist. The development of slight finger flexion tightness enables a better passive grasp of objects when the wrist is extended. To encourage this, the fingers are flexed while the wrist is extended, or the fingers are extended while the wrist is flexed to avoid overstretching across all joints crossed. Table 11-6 summarizes the contraindications and precautions related to active tissue elongation.

Table 11-7 provides sample goal statements, methods, and principles guiding treatment that can be used in documentation and examples of activities to increase ROM.

Table 11-7

INCREASING RANGE OF MOTION

GOAL	METHOD	PRINCIPLE/RATIONALE	EXAMPLE
Joint range of motion will be increased in (specify joint; R/L/B, and motion) by using ...	Heat	Which increases the elasticity of collagen fibers in soft tissue and decreases muscle-tendon stiffness	Neutral warmth Warm water Paraffin Fluidotherapy Light activities or exercise
	Relaxation	Of soft tissues prior to stretch, which decreases tension	Active inhibition techniques Progressive muscle relaxation Controlled breathing Low-intensity activities or exercise Massage Cognitive relaxation strategies Videotape/audiotapes/music Feldenkrais method Alexander method Hatha yoga
	Passive stretch	Beyond current range, which elongates collagen fibers in soft tissues	Manual passive stretch Prolonged mechanical stretch Joint mobilization Joint manipulation Myofascial release
	Active stretch	Beyond current range, which elongates collagen fibers in soft tissues	Active activities and exercise Active inhibition techniques Finger boards Shoulder wheels PNF diagonal Self-stretch Codman exercises

(continued)

Table 11-7 (continued)

INCREASING RANGE OF MOTION

GOAL	METHOD	PRINCIPLE/RATIONALE	EXAMPLE
	Prolonged stretch via orthoses and positioning	Which maintains gains from stretching between treatment sessions *or* which provide gradually increasing amounts of prolonged static stretch	Splints Serial casting
	Role-related activities	Which will ensure gains will be generalized to daily routines	Simulated activities Engagement in occupations
Joint range of motion will be increased in (specify joint; R/L/B, and motion) by using scar remodeling procedures of...	Temperature	Which stimulates localized vascular responses and increases tissue elasticity	Neutral warmth
	Compression	Reduce inflammation and stimulate lymphatic system	Elastomer molds Silicone gel sheets Kinesio taping Low-stretch bandages Foam-lined splints Gently myofascial release Pressure garments
	Deep friction massage	Which reduces dense external adhesive scar tissue and separates deeper structures	Deep friction massage techniques

Adapted from Dutton, R. (1995). *Clinical reasoning in physical disabilities.* Baltimore, MD: Williams & Wilkins; and Marrelli, T. M., & Krulish, L. H. (1999). *Home care therapy: Quality, documentation and reimbursement.* Boca Grande, FL: Marrelli & Associates, Inc.

STRENGTHENING

Muscle strength may be limited due to many different factors, including injury, disease, disuse, immobilization, and overwork. Lower motor neuron diseases (e.g., polio, amyotrophic lateral sclerosis [ALS]), spinal cord injuries or disease, peripheral nerve damage, muscle diseases (e.g., muscular dystrophy [MD]), and stroke all can result in loss of muscle strength.

Weakness occurs in 80% to 90% of the clients with stroke (Kane & Buckley, 2004). If muscle weakness is a primary contributing factor to the inability to achieve participation in occupations of choice, then increasing strength would be a valid treatment objective. Efforts to increase strength should focus on increasing motor neuron recruitment, but the most effective means is not known (Trombly & Ma, 2002). Many clients who have had a stroke have muscle tone problems ranging from flaccidity to spasticity. The generally accepted practice is to normalize tone first and to strengthen only those muscles that are able to be controlled voluntarily by the individual. Volitional control of muscles is required for strengthening programs and activities.

Similarly, in clients with Parkinson's disease tone, specifically rigidity, is the limiting factor, not strength. Rigidity represents a problem in muscle tone that is more aptly treated using sensorimotor/neurodevelopmental approaches. Instead of focusing on increasing strength, activities and exercises should emphasize increasing mobility and rapid, rhythmical movements.

In rheumatoid arthritis, strengthening may be done if the client is in remission and not in an acute, inflammatory phase. Isometric exercises and activities are preferred because no joint movement occurs. For clients with multiple sclerosis (MS), be aware that the client with MS will more likely be fatigued in the afternoon and more energetic early in the morning and in the evening, although this may vary based on the individual. In addition, the environment is important when working with the client with MS because heat is fatiguing to these individuals. It is very important to avoid overfatiguing the recovering muscles of clients with Guillian-Barré. Clients with cardiopulmonary diseases fatigue more quickly due to decreased oxygen availability and require a longer amount of time to recover from exercise or strenuous activities.

Muscle overwork or trauma may result in injuries that disrupt the contractile unit (e.g., muscle or ligament tear), mechanically affect ability of muscle to either produce or transmit force, or disrupt the nerve supply and so do not permit recruitment, and no muscle contraction occurs (Trew & Everett, 2005). Decreased muscle force occurs because the force that each muscle fiber can produce is reduced or due to loss of motor neurons and the size of motor units.

Muscle disuse may be due to muscle imbalance, pain, denervation of motor units, spasm, habit, psychological factors, or immobilization. Muscle disuse can lead to decreased force production and strength as actin and myosin develop diminished ability to form cross bridges. The eventual re-absorption of small blood and lymph vessels leads to decreased endurance (Trew & Everett, 2005).

In the elderly, the loss of strength is multifactoral. There may be normal age-related changes, such as a decline in total muscle fibers, atrophy of muscle fibers, loss of muscle mass, changes in connective tissue and bone, decreased hydration of the joint, decreased elasticity of the joint capsule, and increased joint stiffness (Bonder & Bello-Haas, 2009). Pre-existing conditions make intervention more complex. It is important to ensure that there is postural and structural stability and that all co-morbidities are known prior to intervention implementation. Limitations in strength may be permanent or temporary, and an understanding of the cause of the deficit will influence intervention strategies. Remediation intervention will be focused on those muscles in which improvement is possible.

Maintenance of Strength

Maintenance of strength might be an intervention focus if a client has been on bed rest for 6 to 10 days or more or a specific body part has been restricted (Dutton, 1995). Active movement and participation in grooming and hygiene activities can maintain muscle grades of F+ to G. In muscle grades of F or greater, active movement is beneficial not only in maintaining strength but also in maintaining range (Dutton, 1995). Pedretti indicates that isometric activities and exercise without resistance can be used to maintain muscle strength when active motion is not possible or is contraindicated with muscle grades above T. This type of activity is especially good for those in casts, after surgery, and with arthritis or burns

(Pedretti, 1996). Because strength is maintained and not increased or changed, maintenance of strength may more aptly be viewed as a rehabilitation goal or as one to prevent further structural limitations (disability prevention).

Increasing Strength

Increasing muscle strength would be the focus of intervention when muscle weakness interferes with occupational performance in work, play, and self-care. Muscle weakness is potentially deforming due to muscle imbalances, and strengthening weak muscles will balance the forces of agonist and antagonist muscles. If muscle imbalance is present, the tight muscle must first be elongated to a normal length before strengthening programs and activities are initiated (Magee, 2008).

When a muscle is stressed, this message is sent to the central nervous system, which then stimulates ribosomes to replicate more actin and myosin. As a result, the myofibrils thicken and increase in length. The number and size of the sarcomeres increases so there is an increase of strength in the muscle. Changes in the mitochondria and increased vascularization also occur. Increasing strength occurs when more motor units are recruited due to the stress applied to a muscle to the point of fatigue. The stress to the muscle needs to be great enough to require a maximal motor unit response. A training effect occurs, which reflects the increased number and type of motor units involved in muscle contractions. The muscle fibers become larger (hypertrophy), and some believe the fibers also split (hyperplasia) as they adapt to increased stress (Figure 11-5) (Nyland, 2006).

Strengthening programs are based on Hellebrandt's (Hellebrandt, 1958; Tan, 1998) overload principle, which states that an increase in strength occurs only if the load is greater than that to which the tissue is accustomed and that the load is applied to the point of fatigue. The application of the appropriate stress will overload the system and cause adaptation. Without overload, there is no adaptation and no improvement in strength. Overload can be accomplished by increasing the resistance, increasing the number of repetitions, increasing the numbers of sets done, and decreasing the rest between sets or exercises (Nyland, 2006).

> **Principle: To increase strength, the muscle must be overloaded to the point of fatigue, which recruits more motor units and causes hypertrophy and hyperplasia of glycolic type II fast-twitch muscle fibers.**

Adaptation is defined as a fairly persistent change in the structure or function following repeated bouts of activity. Adaptation is based on Wolff's law, which states that body systems adapt over time to stresses placed on them (Kisner & Colby, 2007). Examples of adaptation include hypertrophy and increased strength, increased maximal oxygen consumption, or a change in body composition (Hamill & Knutzen, 1995). Overload is achieved by varying the parameters of intensity, duration, and frequency of the stress, with intensity being the most potent factor in yielding adaptation to stress.

Activities and exercises should be directed to increasing strength in specific muscle groups, the amount of energy used,

Figure 11-5. Effects of strengthening on muscle fibers.

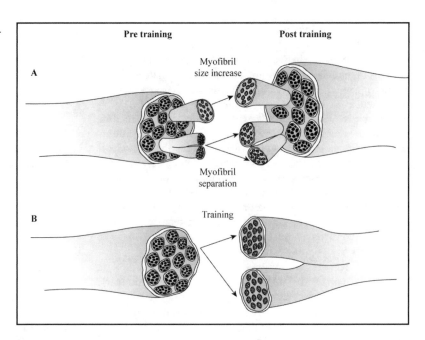

and the type of muscle contraction required for specific tasks. This reflects the SAID principle (specific adaptation to the imposed demand), which indicates that the choice of activity or exercise should be related to the activity in which you want a functional outcome. There is training specificity so an analysis of the motions required in daily tasks needs to be done in order to strengthen those muscles needed in those specific activities.

Factors Influencing Strengthening

Duration, velocity, type of muscle contraction, frequency, intensity, pain, muscle strength, and fiber type all influence a strengthening program. Designing the program is based on the client's physical needs and interests in returning to roles and occupations. At times, engagement in desired tasks is sufficient stress to muscles to increase strength and is more desirable and motivating to the client. Other times, a home exercise program can lead to gains in strength more quickly. Consideration of the client's unique needs and interests in addition to the parameters related to strengthening are all considered when establishing a strengthening program.

Duration

Duration of exercise refers to the length or total time the client participates in the activity or exercise. When the duration increases, this causes fatigue, which leads to the recruitment of additional motor units. In addition, weak muscles activate more motor units than do strong muscles. If the client is engaged in activities that are of high intensity and low duration, increases in strength are the objective. Usually, programs that are of low intensity but high duration are aimed at increasing endurance. Activities that are of interest to the client will likely be done for longer periods of time, which increases the duration as well.

Velocity

The rate or velocity is the speed of the activity or exercise. Generally, it has been found that by using slower speeds,

muscle strength can be increased more quickly and that slower, more constant rates tend to decrease the effects of momentum. Galley states that low-speed, high-load exercise produces greater increases in muscular force only at slow speed, whereas high-power (high-speed, low-load) exercise produces increases in muscular force at all speeds as well as increases in endurance (Galley & Forster, 1987). The high-speed/low-load exercises need to be done for a longer duration, because rapid movements have little strengthening capability because muscle forces are so low (Lieber, 2002).

The velocity of the contraction should be based on the client's physical capacity at the time the exercise or activity is initiated. Activity analysis is a helpful skill in determining in advance the rate of the activity for which the strengthening is directed.

Muscle Contraction

The type of muscle contraction is linked to velocity. In concentric muscle contractions, as the velocity of shortening increases, the force the muscle can generate decreases because the muscle does not have time to reach peak tension. Slow concentric contractions produce more torque than fast concentric. But, as Dutton points out, it is ironic that clients often prefer fast concentric activities and exercises because the speed recruits assistive, synergist muscles (Dutton, 1995). If using concentric contractions, the velocity should be slow.

In eccentric muscle contractions, as the velocity of lengthening increases against resistance, more tension can be generated because this serves as a protective measure when excessively loaded (Baxter, 1998; Berne & Levy, 1998; Galley & Forster, 1987). Fast eccentric contractions produce more torque than do slow eccentric muscle contractions, but ballistic movements result, as when one slams a glass on a table. While more torque is produced in the fast contraction, this is not always the most desirable motion to produce, especially for those with fragile tissues, such as clients with rheumatoid arthritis or burns (Dutton, 1995). In addition, muscle injury and soreness are associated with eccentric contraction, and muscle strengthening is greatest using eccentric contractions (Magee, 2002; Nyland, 2006). Because eccentric contractions are physiologically common, even weak clients may be able to perform the controlled lowering of a body part but not be able to hold or raise the part. For this reason, it might be advisable to start strengthening programs with eccentric muscle contractions (Hamill & Knutzen, 2003) and then move to concentric and isometric muscle contractions. Most activities and exercises use alternating concentric and eccentric contractions.

It is important to match the type of contraction used in intervention to that which is needed in the client's preferred and required occupational demands and interests. Analysis of the activities will help to determine the type of contraction needed to produce functional gains.

Fiber Type

The type of fiber that predominates in the muscle (i.e., Type I, Type II) will also be a factor. For example, larger fast-twitch muscle fibers have a greater capacity to develop tension and show a more pronounced hypertrophy in response to strength training than do smaller slow-twitch muscle fibers.

Frequency

Frequency is the number of times per day or number of days per week the activity or exercise is done. Frequency depends upon the types of fibers being strengthened, the type of activity for which the intervention is directed, the client's endurance and general health status, the muscle grades, joint mobility, diagnosis, intervention goals, position of client, desirable plane of motion, and the level of fatigue.

Intensity

Intensity is the level of exertion or energy expenditure and can be gauged by the client's heart rate, perceived exertion, and ability to speak normally. The amount of resistance needed to increase strength ranges from 50% to 67% of the maximum that the client can lift (repetition maximum [RM]). The number of repetitions varies from one to three sets of 10 repetitions with rest periods of 2 to 4 minutes between.

The intensity and number of repetitions is based on the client's failure-to-lift in which all motor units in a muscle are strengthened. Optimally, one set of 10 repetitions has been found to be adequate to increase strength if failure-to-lift occurs in the final repetitions. Signs that the client is approaching failure-to-lift include mild shaking, difficulty completing the full movement, grimacing, grunting, or sweating (Demeter et al., 1996). Watch for substitution movements because this would activate muscles other than the ones needing strengthening. Low-intensity activities affect the oxidative fibers (slow twitch), while high intensity affects the glycolic (fast twitch) muscle fibers.

Intensity can be graded by the type of exercise or activity that is performed, by the amount of resistance used, by changing the length of the lever arm, by changing the point where the load is applied, or by changing the plane of movement (Radomski & Latham, 2008). Given the variety of ways intensity can be graded, increasing strength can be accomplished by many activities as well as by exercise regimens.

Pain

Pain management is often a part of the intervention process because, without this, intervention will not yield successful outcomes. Often, it is the complaint of pain that brings the client for intervention and not the underlying deficit. Pain presents both with sensory and emotional symptoms. Lack of comfort and often sleep, loss of activities, and loss of roles may cause withdrawal, introversion and depression, and anger and aggression. Pain and lack of customary life roles may develop into learned helplessness where the client avoids strenuous activities and where secondary benefits may be gained, such as excused participation in work activities or in less desired roles.

Acute pain is experienced immediately following a physical injury and is proportional to physical findings. Chronic pain lasts months or years and can change the client's personality, cause disassociation from physical problems, and can develop into a different clinical syndrome (Smith et al., 1996).

Pain-Reduction Theories

Pain management strategies draw upon the gate control theory (Melzack & Wall, 1965). This theory indicates that impulses from large sensory nerve fibers at segmental levels of the spinal cord enable non-painful sensory input to stimulate the same transmission cells that the pain receptors do. This would inhibit the transmission of the pain input. Every time you rub your knee after banging it against something, you are demonstrating this theory; mechanoreceptors are superseding the pain messages. The activation of the mechanoreceptors competes for transmission with the pain input, so by rubbing your knee, the sensation of pressure, not pain, is felt. The faster signals of the mechanoreceptors, activated when you rubbed your knee, blocked the reception of the slower pain message. This response is believed to have evolutionary basis where movement and pressure were required for flight or fight and sensation of pain was not as necessary to survival (Muscolino, 2006). One physiologic explanation for the gate theory mechanism is that the local stimulation of nonpain-mediated sensory afferents closes the gate at the spinal cord level, thereby preventing further transmission of pain impulses.

A further elaboration on the gate theory was proposed by Melzack and Katz (2004). The neuromatrix theory indicates that the brain creates a perceptual experience in the absence of external inputs due to sensory and afferent links to large portions of the brain and thus produces the multidimensional pain experience (McCormack & Gupta, 2007). The heightened sensitivity and altered autonomic nervous system activity creates a chronic state of high alert that perpetuates the pain cycle.

Another theory about the mechanism of pain is that the body is stimulated to release endogenous opiate substances, which increases the circulation of neuropharmacologic agents (endorphins), which then decreases the pain (Bear, Conners, & Paradiso, 2001; McCaffrey, Frock, & Garguilo, 2003). It has been found that the inhibition of pain input is enhanced by the client's concentration on competing activities that have important implications for occupational therapists and the use of meaningful occupation as the means of intervention.

Pain-Reduction Treatment

Pain is a complex sensory experience influenced by the body, the mind, and culture (Chesney & Brorsen, 2000). The experience of pain is based on one's belief system, so pain can be viewed from different perspectives. Jarvis (2000) identified three different pain perspectives. The first perspective is the scientific or biomedical view. In this view, pain has a cause and effect, and the human body functions much like a machine. By observing and measuring those parts that are not functional or are painful, pain-free function can be restored. The naturalistic or holistic approach (as seen in some native Americans, Asians, Hispanics, Arabs, and Blacks) believes that human life is only one aspect of nature as a part of the larger cosmos. Pain and dysfunction are evidence of lack of harmony or imbalance. It is the individual as a whole (physical, psychological, spiritual, and social), not the particular impairment, that is significant. The third perspective, the magico-religious view, sees the world as a place where supernatural forces dominate. Jarvis includes voodoo, witchcraft, and faith healing (including Christian Scientists, Roman Catholicism, and Mormonism) as holding beliefs related to this approach. It is clear that culture and your belief about health and disease would influence your perceptions and experience of pain, and these need to be part of the client assessment.

Occupational therapists are uniquely qualified to address both the physical and psychological needs of people with acute and chronic pain. In cases where the pain cannot be completely alleviated, the occupational therapist can provide strategies for adaptation due to the holistic and multifaceted approaches used in intervention (Fisher, Emerson, Firpo, Ptak, Wonn, & Bartolacci, 2007). The unique emphasis on function and the influence of pain is important because function is how clients perceive their quality of life. Advances in disruption of the pain receptors, new medicines for the control of pain, and cognitive therapy for altering pain perception are being explored as part of the multifaceted approach to pain (Bausbaum & Julius, 2000, 2006).

Intervention for acute pain is directed toward the underlying cause of the pain. An improperly aligned joint due to lack of humeral rotation during forward flexion or improper placement of the humeral head in the glenoid fossa may require realignment to alleviate the pain. Pain disturbing sleep may be indicative of systemic pain arising from one of the body's systems other than the musculoskeletal system. Until the origin of the pain is identified, pain reduction will be minimal (this clearly reflects a scientific perspective).

The underlying assumptions to this approach are that every pain has a source and that the intervention you provide must be directed toward and influence that source. Isolate the cause of the pain, and treatment it specifically.

A variety of methods are used to treat both acute and chronic pain. Pain control modalities include ice massage, cold packs, acupressure, and self-massage.

Modalities: Deep heat for muscle spasm, sprains, strains, tendonitis to loosen soft tissue contractures; treatment of chronic arthritis, bursitis, fracture, and inflammation; iontophoresis, phonophoresis, laser therapy (Meriano & Latella, 2008; Rochman & Kennedy-Spaien, 2007). Functional body retraining might include instruction in body mechanics, energy conservation, work simplification, and proper posture (Chesney & Brorsen, 2000; Dudgeon, Tyler, Rhodes, & Jensen, 2006; Fisher et al., 2007). Cognitive/behavioral approaches might include pacing (learning to do activities without pain), assertiveness training (for the expression of needs, feelings), relaxation training, and enhancement of self-efficacy (Chesney & Brorsen, 2000; Dudgeon et al., 2006; Fisher et al., 2007). Muscle tension reduction training might involve teaching the client to recognize ineffective muscle tension and to reduce or reverse compensatory patterns (Rochman & Kennedy-Spaien, 2007). The use of virtual reality programs and games can also serve as a distraction, and users have shown decreased activity in known pain centers in the brain (Hoffman, 2004).

Intervention for chronic pain involves establishing functional goals. By working toward clear, functional goals, attention is directed away from the pain and toward an observable result of intervention. Clients may not become pain free but instead must be taught how to tolerate and manage their pain. Modalities are often used, such as localized heat (whirlpool, fluidotherapy, paraffin), deep heat (ultrasound, TENS, iontophoresis), massage, acupressure or acupuncture, and vibration. Behavioral techniques might include hypnosis, relaxation techniques, biofeedback, behavioral modification programs, and counseling. Complementary practices such as imagery, meditation, mindfulness training, spiritual practices, Qigong, reiki, yoga, and tai chi are used (MacCormac, 2010; Meriano & Latella, 2008). Mirror therapy has been used for chronic pain (in which the use of the non-injured side is reflected in mirror, giving the illusion of function in an injured hand) with some success and is another option (Grünert-Plüss, Hufschmid, Santschi, & Grünert, 2008). Lifestyle and habit changes, such as alcohol and medication reduction, smoking cessation, monitoring fluid intake and nutrition, quality and quantity of sleep, are also part of the multidimensional treatment of chronic pain (Nicosia, 2004). Intractable pain may require medication or surgery.

Muscle Fatigue and Distress

The overload principles states that muscle must be worked, but only to the point of fatigue. The question is what is muscle fatigue and how do you know when a muscle is fatigued. When there is a decline in muscle tension and the muscle has decreased shortening velocity and a slower rate of relaxation, then the muscle is fatigued. Fatigue may be caused by synaptic fatigue, depletion of glycogen, build-up of lactic acid, shunt of blood to skin to control temperature, electrolyte imbalance, increased blood viscosity due to dehydration, or blood flow constriction by prolonged muscle action.

The onset and rate of fatigue depends on the type of muscle and the duration of the contraction. A muscle that is fatigued can recover if given rest, and the rate of recovery depends on the duration and intensity of the exercise (Nyland, 2006). If the exercise was high intensity/short duration, recovery is rapid; if the exercise was low intensity/long duration, fatigue is due to the buildup of lactic acid and changes in the muscle proteins, so recovery is slower (Hall & Brady, 2005; Nyland, 2006).

Fatigue can be differentiated from overwork weakness. Fatigue is the result of a maximal motor unit response, whereas overwork is an insidious loss of strength due to an overload on individual muscle strands. Overwork weakness occurs in clients with spotty denervation, with or without sensory losses, who have impaired neuromuscular function or a systemic, metabolic, or inflammatory disease (Kisner & Colby, 2007). These clients have fewer motor units and fewer contracting muscle fibers, so each fiber must work maximally. Because there is less waste product buildup, there is less sensation of fatigue, and the client does not interpret the fatigue sensation properly. Overwork is frequently irreversible but fortunately occurs rarely.

Being aware of the level of fatigue will help prevent overwork of weak muscles and will prevent substitution movements.

Muscle fatigue levels vary from person to person. One's threshold for fatigue decreases in pathological states (Dutton, 1995), and clients may lack the sensation to be aware of their level of fatigue, while others may push themselves beyond their tolerance. Local fatigue is normal and is characterized by a diminished response of muscles to repeated stimuli (Kisner & Colby, 1990).

Signs of fatigue might include the following:

- Slowed performance
- Distraction
- Perspiration
- Increase in rate of respiration
- Decreased ROM
- Inability to complete prescribed number of repetitions
- Inability to maintain a given force
- Decreased time of contraction
- Increased time for muscle lengthening
- Tremors with contraction
- Increased heart rate and respiration with no increase in load
- General sense of tiredness
- Attention wanders
- Incoordination
- Loss of concentration
- Substitution movements (Demeter et al., 1996; Kisner & Colby, 2007; Levangie & Norkin, 2005)

As the client performs activities, observe for signs of fatigue and distress. Distress may be evident in shortness of breath (dyspnea), confusion, profuse sweating in association with cold, clammy skin (diaphoresis), and straining while holding one's breath (Valsalva maneuver), which is especially dangerous in that it raises the blood pressure. More subtle signs of distress might be seen in escalating frustration, hurrying to finish the task, less range of motion, chest pain, nausea, or lightheadedness (syncope) (Demeter et al., 1996). In the cases of chest pain or syncope, confusion, diaphoresis, or level 2 dyspnea (defined as one who needs three breaths to count to 15), it is advisable to discontinue the activity and consult with a physician for further evaluation of these symptoms (Hamill & Knutzen, 1995).

Muscle Injury, Strain, and Soreness

Injuries to muscles can occur as a result of muscle strain or microtears in the muscle fibers. Client symptoms would be pain or muscle soreness, swelling, and possible deformity and dysfunction. Muscles at greatest risk are two-joint muscles, muscles used to terminate a range of motion, and muscles used eccentrically. Two-joint muscles are especially prone to fatigue and strain because they can be stretched at two joints simultaneously. For example, to extend at the hip and flex at the knee, the rectus femoris muscle is vulnerable to injury because it is put on an extreme stretch (Hamill & Knutzen, 1995). Muscles that are used to terminate a motion are at risk because they are used to eccentrically slow a limb moving very rapidly, such as the hamstrings when slowing hip flexion or the posterior rotator cuff muscles when slowing the arm on the follow-through phase of a throw (Hamill & Knutzen, 1995).

While muscles may be the site of damage, soreness and strain are due to the connective tissues, especially at the muscle-tendon junction. Common injuries at this site are in the gastrocnemius, pectoralis major, rectus femoris, adductor longus, triceps brachii, semi-membranosus, semi-tendinosus, and biceps femoris (Hamill & Knutzen, 1995). Delayed-onset muscle soreness is associated with eccentric muscle contractions and may be because eccentric contractions are capable of greater forces than concentric muscle contractions and in a reduction in compliance of the fascia (Tweed & Barnes, 2008).

Methods of Increasing Strength

Strengthening programs can use a variety of methods to increase strength that can include both activities and exercise. The exercise or activity needs to provide sufficient stress to the muscles to require adaptation but also permit smooth, pain-free movement. Based on the client's muscle grades and on the target movement pattern required in specific activities, activities and exercises can be done actively with assistance (active assisted), actively, or against resistance. Passive movements are done with muscles graded zero to maintain muscle lengths and joint integrity. Active assisted activities and exercises are done with muscle grades of T, P-, F-. Active activities and exercises are done with muscle grades of P or F, and resistive activities and exercises are done with P+, F+, G-, G, and G+ muscle grades. Table 11-8 summarizes the relationships between muscle grades and types of activities/exercises.

Activity or Exercise

The use of therapeutic exercise is most beneficial to clients with muscle strength at the extremes of the manual muscle testing ratings (i.e., those muscles rated trace and poor or those with good muscle strength or better). Because occupational therapy intervention is based on the client's needs and interests, it is important to recognize that therapeutic exercise may be more acceptable to some clients than activities, while engaging in activities may be more meaningful to others.

Pedretti proposes eight purposes for the use of therapeutic exercise:

1. Develop awareness of normal movement patterns and improve voluntary, automatic movement responses.

2. Develop strength and endurance in patterns of movement that are acceptable and necessary and do not produce deformity.

Table 11-8

MUSCLE GRADES AND TYPES OF ACTIVITIES AND EXERCISES

	PASSIVE	ACTIVE ASSISTED	ACTIVE	RESISTIVE
Gravity eliminated	0	T	P	P+
		P-		
Against gravity		F-	F	F+
Against resistance and gravity				G-
				G
				G+

3. Improve coordination, regardless of strength.

4. Increase power of specific isolated muscles or muscle groups.

5. Aid in overcoming ROM deficits.

6. Increase strength of muscles that will power hand splints, mobile arm supports, or other devices.

7. Increase work tolerance and physical endurance through increased strength.

8. Prevent or eliminate contractures developing as a result of imbalanced muscle power by strengthening antagonistic muscles (Pedretti, 1996, pp. 300-301).

Studies support the use of activities to improve strength and endurance. Nelson and colleagues confirmed the advantages of embedding exercise within occupations over exercise alone (Nelson et al., 1996; Sietsema, Nelson, Mulder, Mervau-Scheidel, & White, 1993).

The use of object-oriented motor control training is more effective than nonobject-oriented training (Yuen, Nelson, Peterson, & Dickinson, 1994). Purposefulness in activities results in enhanced performance with greater motor skill retention and motor learning (Ferguson & Trombly, 1997; Hsieh, Nelson, Smith, & Peterson, 1996; King, 1993). Functional activities that are meaningful are more effective in increasing performance (Neistadt, 1994; Thibodeaux & Ludwig, 1988; Wennemer, Borg-Stein, Delaney, Rothmund, & Barlow, 2006), and task-oriented treatment augmented with resistive exercises results in substantial gains in ADL, IADL, manual muscle test scores, and grasp (Flinn, 1995). Hoppes (1997) combined games and play with increasing standing tolerance in geriatric clients and found that clients were able to stand up for longer periods when engaged in a game.

Active Assistive

Active assistive activities and exercises are those in which the client moves actively as far as possible and the therapist or a device completes the motion through the existing ROM. Because the patient moves actively as far as possible, there is stress to the muscle, and recruitment, hypertrophy, and hyperplasia can occur. Exercise is graded so that the device/therapist

gradually decreases the amount of support or assistance that is provided while the client provides more active (unassisted) movement.

Isotonic active assistive is appropriate for clients with T, P-, or F- muscle grades. For the trace muscle, the client contracts the muscle, and the therapist completes the motion. Active-assistive exercise in a gravity-eliminated plane is used for P-, while with F- muscles, gravity is included as a factor. Bilateral activities are useful for active assistive exercise if only one extremity is affected and the assistance can be provided by the unaffected arm.

Progressive assistive exercise (PAE) is a type of assistive exercise where equipment (such as the Swedish suspension sling) provides the minimal amount of weight required to complete a motion. A skate with weights and a pulley, dynamic orthoses, and towel and dowel exercises are other examples of exercises that can be done assistively. Few activities provide assistive exercise without resistance, although some examples cited by Trombly include adapting floor looms or polishing a smooth surface (Trombly, 1995).

Active

Active activities and exercise are those in which the client moves through the available ROM without resistance. Active activities and exercises increase muscle grades of P to F. There are very few activities or exercises that are purely active exercise, and even most ADLs are at least mildly resistive. Needlepoint completed in a gravity-eliminated plane would be an active exercise for wrist extensors or elbow extensors. A fair muscle can move the wrist against gravity, as in a mosaic tile project (Pedretti, 1996). The goal of activities and exercises at this level is to progress to a point where some resistance other than gravity can be applied.

Resistive

Resistive activities and exercise are done when an outside resistance is required to apply maximal stress to the muscle in order to promote adaptation. Benefits of resistance exercise include enhanced muscle performance, increased strength of connective tissues, greater bone mineral density, decreased stress on joints during activity, reduced risk of soft tissue injury,

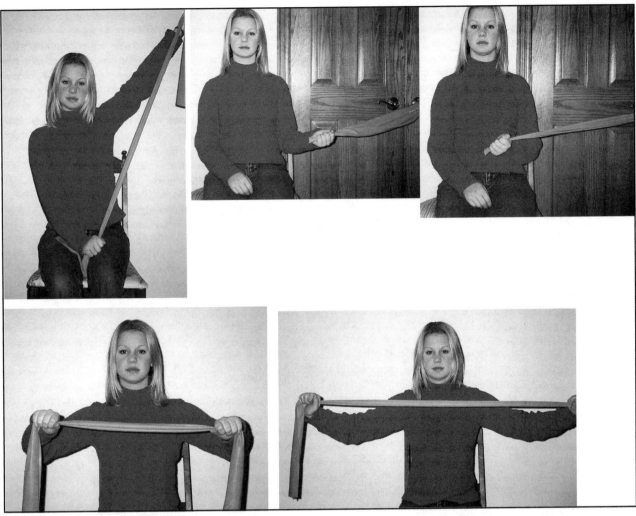

Figure 11-6. Theraband resisted exercises.

which can enhance physical performance during work, play, and self-care, feelings of well-being, and possible improvements in perceptions of quality of life (Kisner & Colby, 2007).

The resistance force may be applied manually (by the therapist) or by equipment, tools, or activity (Figure 11-6). Use of resistance is necessary to increase the strength of P+, F+, G-, G, and G+ to normal muscle strength by means of either isotonic or isometric muscle contractions.

Manual Resistive Activities and Exercise

In manual resistive activities and exercise, not only is the intensity of the resistance force controlled by the therapist, but also the site and direction of the force application. Force is usually applied at the distal end of the segment to which the muscle attaches, which provides a mechanical advantage for the therapist applying the force. The direction of force is directly opposite to the desired motion. Stabilization is important in order to avoid substitution or compensatory movements. The resistance is applied smoothly, steadily, slowly, and gradually so that the movement is pain free and through the maximal range. These criteria can be applied either to activities or to exercise regimens. For exercises, eight to 10

repetitions are completed, one to two times per session, with rests of 2 to 4 minutes minimally between sets of application of resistance.

Mechanical Resistive Activities and Exercise

Mechanical resistive activities and exercise are any form of endeavor where the resistance is applied by mechanical equipment, tools, or by the placement of the task. Isotonic resistance equipment might include free weights, elastic resistance devices (Theraband, tubing), pulley systems (using weights or springs) that may be free-standing or wall-mounted (e.g., Elgin exercise unit [Elgin Exercise Equipment, Lonbard, IL], Variable Resistance or Dynamic Variable Resistance [DVR] systems [Nautilus, Louisville, CO], Cybex Eagle [Medway, MA], Keiser Cam II [Zurich]), or cycle ergometers (stationary bicycles) (Norkin & Levangie, 1992; Rantanen et al., 2000). Many of these devices are part of circuit weight-training programs, where the client exercises in short bouts, using light to moderate workloads with frequent repetitions and short rests. A specific sequence of exercise is followed.

Adding resistance can be done by changing the effect of gravity on the activity with a gravity-eliminated plane (generally this is parallel to the horizon). Additional resistance would be achieved by using an inclined plane, additional resistance to the weight of the objects or tools used, or the duration of the activity. Activity analysis and activity adaptation are particular strengths of the occupational therapist, and these skills are especially helpful in developing intervention aimed at increasing strength by means of activities of value and interest to the client. Occupations done daily can have a strengthening benefit without being contrived and can have the added advantage of being a necessary part of one's life.

Isotonic strengthening activities and exercise require a dynamic effort through a specified ROM. The actual load varies throughout the range of motion because the resistance cannot be heavier than the muscle tension that can be developed at the weakest joint position. This means that isotonic strengthening may not be adequately overloading the muscle in midrange, where it is usually the strongest (Hamill & Knutzen, 2003). The isotonic movement may require either eccentric or concentric contractions or both.

A specific exercise regimen was developed by DeLorme and is called Progressive Resistive Exercise (PRE). DeLorme based this exercise regimen on the overload principle where several sets of repetitions are completed against a portion of the repetition maximum (RM). The modified DeLorme PRE method first determines the RM, which is the greatest amount of weight that can be lifted, pulled, or pushed 10 times through the full existing ROM. The RM is based on the muscle grades as a guide and is determined through trial and error. The client then performs 10 repetitions at 50% of the RM, 10 repetitions at 75% of the RM, then 10 repetitions at 100% of the RM with 2- to 4-minute rests in between exercise sets. These exercises should be performed once a day, four to five times per week for maximum strengthening benefit. An example might be that, if a client is able to lift 12 pounds 10 times, this would be the 10 RM. The PRE program would be 10 repetitions at 6 pounds, rest, 10 repetitions at 9 pounds, rest, 10 repetitions at 12 pounds.

Regressive resistive exercise (RRE) or Zinovieff's Oxford method is essentially the opposite of the modified DeLorme PRE program. The client completes 10 repetitions at 100%, rests, 10 repetitions at 75%, rests, then 10 repetitions at 50%. The RRE program was designed to diminish the resistance as muscle fatigue develops, but this has been disproven as a reliable form of strengthening exercise (Kisner & Colby, 2007; Nyland, 2006; Rothstein, Roy, Wolf, & Scalzitti, 2005).

Isometric activities and exercise are characterized by no visible joint movement nor appreciable change in muscle length but increased muscle tension (Figure 11-7).

The load is only applied to one joint position so strength is only enhanced at the joint angle in which the muscle is stressed, not throughout the entire range of motion.

Because the joint does not move, isometric activities and exercise are well-suited for clients with rheumatoid arthritis and those with pain or inflammation and are the only type of exercise for clients with trace muscle strength.

Isometric exercises have the advantage of being easy to perform with little setup or equipment needed, and it takes minimal time because isometric contractions can be very fatiguing (Nuismer et al., 1997). Disadvantages of isometric exercise and activity are that the gains in strength do not necessarily transfer to dynamic activities, there is no effect on improving coordination, and it does not cause hypertrophy. The most serious disadvantage is that isometric contraction causes an increase in blood pressure and so is contraindicated for patients with cardiovascular problems. Using isometric contractions in activities is easily done when one considers that holding tools requires an isometric contraction.

Different types of isometric contraction exercises include the following:

- Brief maximal isometric exercise regimen: A single isometric contraction is held against a fixed resistance for 5 to 6 seconds, once a day, five to six times per week. Longer duration contractions have been found to be more effective in increasing strength than one held for 5 to 6 seconds (Tan, 1998).

- Brief repetitive isometric exercise (BRIME) regimen: The client completes five to 10 brief but maximum isometric contractions, each held 5 to 16 seconds, and performed against resistance. This exercise regimen is done daily (Tan, 1998).

- Multiple angle isometric exercise regimen: Resistance is applied at least every 20 degrees through ROM. For example, the client completes 10 sets of 10 repetitions with each contraction lasting 10 seconds in every 10 degrees of the ROM ("rule of tens") (Tan, 1998).

- Prolonged method: Involves holding an isometric contraction as long as possible and then repeating this 10 times. The amount of time the client maintains the maximal effort for 10 repetitions is increased (Trombly, 1995).

- Weighted method: When the client holds a contraction against resistance 30 to 45 seconds with a 15-second rest between each of the 10 repetitions (Trombly, 1995).

When using exercise machines such as Cybex (Medway, MA), Biodex (Shirley, NY), Kincom (Harrison, TN), Lido, Merac, and Orthotron II, isokinetic strengthening exercise is being done. This is a type of dynamic exercise performed with constant angular joint velocity (i.e., the muscle shortens/lengthens at a constant rate) against varying resistance.

When clients are engaged in resistive, strengthening activities, it is important to caution them to avoid holding their breath while performing the task. By having them count, talk, or sing during the task, the Valsalva maneuver (holding the breath while exerting effort, which raises the blood pressure) will be avoided. This is especially important during isometric exercise and activities with heavy resistance.

Warming Up and Cooling Down

Warming up is any activity or exercise that raises total body and muscle temperature. Cardiovascular warm-up gets the heart and lungs prepared for exercise, prepares tissue for greater extensibility force, decreases viscosity, and increases tissue compliance. Increasing muscle temperature increases the uptake of oxygen, and increases the metabolic rate and blood flow (Cooper, 2007).

The benefits of warming up prior to strenuous activities include the following:

1. Cold muscles and tissues do not stretch very easily. Use of modalities will warm the tissues. Warming up the muscles

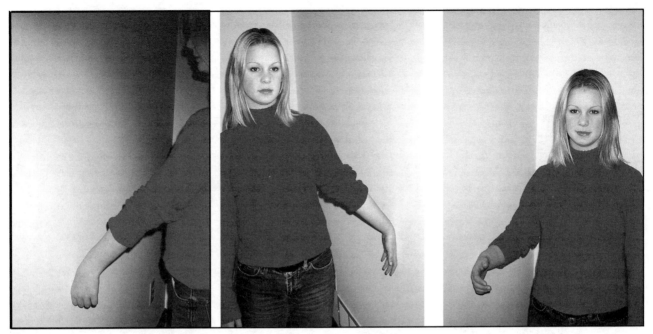

Figure 11-7. Isometric shoulder exercises.

and tissues tend to relax them, which makes them more easily stretched.

2. With aerobic activities, the warm-up period prepares the heart, cardiovascular system, and muscles for activity. It has been shown that attempting to perform strenuous activities without adequate warm-up may precipitate cardiac arrhythmia even in those without heart disease.

3. Warming up causes the blood vessels that supply muscles involved in the activity to dilate and vessels supplying less involved parts to constrict. This provides additional oxygen to the muscles, requiring the energy for the task.

4. Muscles not properly warmed up may work anaerobically without sufficient oxygen. As a result, the muscles may become prematurely fatigued, and lactic acid will accumulate.

Warming up may involve 5 to 10 minutes of gentle activities or exercises, followed by a few minutes of low-intensity aerobic activities and then some gentle stretching. Pre-stretching muscles prior to concentric muscle action results in greater force because the stretch increases the tension by release of potential elastic energy in the series elastic components of the muscle (Hamill & Knutzen, 2003). Predominantly, fast-twitch muscles benefit from high-velocity pre-stretches over a small range of motion because myosin and actin cross bridging occurs quickly, while slow-twitch muscles respond better to slower speeds and great range of motion (Hamill & Knutzen, 2003).

Cooling down is similar to warming up, but with decreasing physical demands placed on the body. Cooling down enables the body to eliminate metabolic wastes, bring in additional oxygen, and to gradually reduce the heart rate to resting level, which helps to reduce dizziness or fainting. Cooling down after exercise does not seem to have an effect on delayed-onset muscle soreness because it is not due to an accumulation of lactic acid but instead is due to damage to muscle fibers (Law & Herbert, 2007).

Muscle soreness can be minimized or prevented by ensuring a sufficient warm-up and cool-down prior to initiation of the activities, stretching prior to heavy resistance, and by gradually increasing the resistance as strength improves.

Sample goal statements, methods, principles and examples are shown in Table 11-9.

Contraindications and Precautions to Strengthening

An extensive list of precautions and contraindications are included in Table 11-10 (Hall & Brady, 2005). These are helpful guidelines for clients of all ages. Use care when developing strengthening programs for clients with osteoporosis, which may be suspected if the client has rheumatoid arthritis, if she is a postmenopausal woman, or if the client uses systemic steroids. While older adults gain less absolute strength and gains in strength are made more slowly, studies have demonstrated improvements in strength and functional activities following strengthening programs for older adults (Bamman, Hill, Adams, Haddad, Wetzstein, & Gower, 2003; Carmeli, Reznick, Coleman, & Carmeli, 2000; Charette, McEvoy, Pyka, Snow-Harter, Guido, & Wiswell, 1991; Ferri, Scaglioni, Pousson, Capodaglio, VanHoecke, & Narici, 2003; Fiatarone, O'Neill, Ryan, Clements, Solares, & Nelson, 1994; Fiatarone & Evans, 1993; Frontera, Hughes, Fielding, Fiatarone, Evans, & Roubenoff, 2000; Frontera, Meredith, O'Reilly, Knuttgen, & Evans, 1988; Khruda, Hicks, & McCartney, 2003; Meuleman, Rechue, Kubilis, & Lowenthal, 2000; Vincent & Braith, 2002).

Table 11-9

Increasing Strength

GOAL	METHOD	PRINCIPLE/RATIONALE	EXAMPLE

Strength will be increased (indicate where/which muscles, R/L/B) to _____ (what level of improvement) to perform (tasks) using...	Isometric exercise [in a trace muscle]	Against no resistance, which recruits more motor units and increases proprioceptive awareness of individual muscles as they contract.	Electric stimulation
			Biofeedback machine
			Vibrators
			Manual contacts
			Isometric exercises:
			  Brief maximal
			  Brief repetitive
			  Multiple angle
			  Prolonged method
			  Weighted methods

	Active assistive exercise with assistance to complete the motion [in a poor minus or fair minus muscle]	Which recruits more motor units, causes hypertrophy and hyperplasia of glycolytic type II fast twitch muscle fibers.	Therapeutic skate Deltoid aid Mobile arm support Manual assistance Bilateral activities: one side is affected
	Active exercise through full range of motion but against minimal resistance [in a poor or fair and greater muscle]	Which recruits more motor units and causes hypertrophy of glycolytic type II fast-twitch muscle fibers.	Shoulder wheel Combing short hair Electric typewriter Clothespin races Needlepoint Mosaic tile project Computer work

(continued)

Table 11-9 (continued)

INCREASING STRENGTH

GOAL	METHOD	PRINCIPLE/RATIONALE	EXAMPLE
[in a poor plus, fair plus, or good minus and above muscle]	Progressive or regressive resistive exercise against the maximum resistance needed to produce failure-to-lift	Which recruits more motor units, causes hypertrophy and hyperplasia of glycolytic type II fast-twitch muscle fibers.	Weight well Theraplast Theraputty Weighted weaving Overhead rug knotting ADL tasks like dressing Manual resistance via another person, activity Weighted exercises (isometric and isotonic)

Adapted from Dutton, R. (1995). *Clinical reasoning in physical disabilities.* Baltimore: Williams & Wilkins; and Marrelli, T. M., & Krulish, L. H. (1999). *Home care therapy: Quality, documentation and reimbursement.* Boca Grande, FL: Marrelli and Associates, Inc.

Table 11-10

CONTRAINDICATIONS FOR STRENGTHENING

CONTRAINDICATION	PRECAUTIONS
Severe CAD (unstable angina pectoris and acute MI)	CAD
	Marked obesity
Decompensated congestive heart failure	Congestive heart failure
Uncontrolled ventricular arrhythmias	Significant valvular heart disease
Acute myocarditis	Cardiac arrhythmias
Uncontrolled atrial arrhythmias	Hypertension
Recent pulmonary embolism of DVT	Fixed-rate permanent pacemaker
Severe valvular heart disease	Cyanotic congenital heart disease
Uncontrolled systemic hypertension (> 200/105)	Congenital anomalies of coronary arteries
Cardiomyopathy	
Pulmonary hypertension	Marfan's syndrome
Unstable angina	Peripheral vascular disease
Thrombophlebitis	Electrolyte abnormalities
Resting systolic BP over 200 mm Hg	Uncontrolled metabolic diseases (diabetes)
Resting diastolic BP over 100 mm Hg	Any serious systemic disorder (i.e., hepatitis)
Acute thyroiditis	Neuromuscular or musculoskeletal disorders
Hypokalemia	Severe obstructive or restrictive lung disease
	Anemia
	Use caution with children, older adults
	Unhealed fracture
	Avoid uncontrolled ballistic movements
	Discontinue if client experiences pain, dizziness, shortness of breath

Adapted from Hall, C. M., & Brady, L. T. (2005). *Therapeutic exercise: Moving toward function* (2nd ed.). Philadelphia, PA: Lippincott, Williams & Wilkins; Kisner, C., & Colby, L. A. (2007). *Therapeutic exercise: Foundations and techniques* (5th ed.). Philadelphia, PA: F. A. Davis Company; and Rothstein, J. M., Roy, S. H., Wolf, S. L., & Scalzitti, D. A. (2005). *The rehabilitation specialist's handbook* (3rd ed.). Philadelphia, PA: F. A. Davis Company.

COORDINATION

Coordination can be defined as the ability to produce accurate, controlled movements characterized by smoothness, rhythm, appropriate speed, appropriate muscle tension, a minimal number of muscles involved to accomplish the movement with adequate postural tone and equilibrium. A person is uncoordinated when there are errors in rate, rhythm, range, direction, and force of movement (Nuismer et al., 1997).

Notice from the definition that coordination involves the ability to move at the proper or appropriate rate (not too fast nor too slow). An example of a dysfunction in rate might be dysdiadochokinesis (the inability to perform rapidly alternating movement). The rate or timing problems may be due to the inability to generate sufficient force in muscles, decreased rate of force generation, insufficient range of motion for the movement to occur, reduced motivation, or abnormal postural control (Shumway-Cook & Woollacott, 2000).

The lack of control might be seen in a tremor or ballism (where the limb flies out suddenly) or the arm rebounds after motion (Rebound Phenomenon of Holmes). The client may be unable to reach a cup of coffee because he or she overshoots or underestimates the distance needed to reach the cup (dysmetria). Problems with coordination and the structure associated with the deficit are listed in Table 11-11.

Table 11-11

COORDINATION AND ASSOCIATED STRUCTURES

Posterior Column Dysfunction: Results in loss of proprioception, misjudgment of limb position and balance.
- Ataxia: Reeling, wide based, unsteady gait.
- Romberg: Inability to maintain balance with eyes closed.

Cerebellar Dysfunction: Seen in regulation of loss of smooth voluntary movement and maintenance of upright posture.
- Adiadochokinesis or dysdiadochokinesis: Inability or impairment in the ability to perform rapidly alternative movements.
- Dysmetria: Inability to estimate the ROM necessary to reach the target of the movement (i.e., touches cheek instead of nose); "overshooting".
- Tremor: Intention tremor during voluntary movement intensified at the termination of the movement (often seen in multiple sclerosis); tremor is in proximal parts due to lack of stability.
- Nystagmus: Involuntary movement of the eyeballs up and down and back and forth or in rotary motion.
- Hypotonia: Decreased resistance to passive stretch.
- Dysarthria: Explosive or slurred speech.
- Rebound phenomenon of Holmes: Lack of reflex to stop a strong, active motion (i.e., therapist applies pressure in the direction of elbow extension, releases pressure, and arm rebounds toward flexion).
- Asthenia: Weak and easily tired muscles.

Basal Ganglia Dysfunction: Functions in control of automatic, patterned movements of locomotion and initiation of rhythmical movements.
- Athetosis: slow, writhing motion primarily in distal parts, lack of stability in neck, trunk, and proximal joints; excessive mobility with increased speed but movement is involuntary and purposeless; not present during sleep.
- Dystonia: Form of athetosis, characterized by postures (ex: lordosis); not present during sleep; bizarre writhing and twisting movements of trunk and proximal muscles of the extremities.

Extrapyramidal Dysfunction
- Tremors: Resting tremors in pill rolling tremor in Parkinson's disease.
- Choreiform movements: Irregular, purposeless, coarse, quick, jerky, and dysrhythmic movements; may occur during sleep.
- Spasms: Involves contractions of large groups of muscles in arms, legs, and/or neck.
- Ballism: Rare; produced by abrupt contractions of axial and proximal musculature of the extremities resulting in limb flying out suddenly; hemiballism = ballism on one side caused by subthalamic lesion on opposite side.

Coordination involves smooth, rhythmical movement of multiple muscle groups, and that is a function of the cerebellum and extrapyramidal system. Also needed are intact proprioceptors and perceptual-motor systems, especially spatial orientation, vision, and body scheme (closely associated with proprioceptors). Lesions in the corticospinal tract can lead to loss of the ability to recruit muscle and to control individual joints (Shumway-Cook & Woollacott, 2000). Both the biomechanical and neuromuscular systems are needed for coordinated movement.

Coordination not only refers to the upper extremities but also to gait (ataxia), eyes (nystagmus), and facial muscles (dysarthria). Dexterity refers to coordination (smoothness, grace), especially skill and ease in using the hands.

Observation is important to identify irregular movements or sudden movements meant to compensate for incoordination.

Factors that increase incoordination may include poor balance, fear, too much resistance, pain, fatigue, strong emotions, and prolonged inactivity.

The assessment and treatment of coordination does not seem to correspond with any one frame of reference. Coordination depends on accurate somatosensory feedback for normal reciprocal innervation and co-contraction, which is inconsistent with the biomechanical approach assumption that sensation is intact for biomechanical treatment techniques. Biomechanical techniques, such as exercise, are often performed in linear patterns along anatomical planes with a constant rhythm while normal movements are often diagonal with rotary components, irregular rhythms, and large numbers of muscles working together. It is difficult to perform fine, coordinated movements when maximum resistance is applied (Demeter et al., 1996). The biomechanical approach considers structural stability, tissue integrity, range of motion, strength, and endurance. However, if you have strength and good endurance but are clumsy, you will not be able to resume satisfying life roles in the community. When coordination is an intervention goal, it is usually included after increased range and strength and may be the responsibility of the client, as an outpatient or as a home health goal, because length of stay is so reduced in inpatient facilities.

Neurodevelopmental or sensorimotor approaches (NDT, PNF, Rood) are often cited as treatment approaches to use for deficits in coordination. However, the focus of these interventions is not directed toward the improvement of coordination. These approaches assume that good coordination will naturally follow if tone is remediated (Dutton, 1995). While neurodevelopmental approaches stress the sense of normal movement, these approaches are not designed to provide intervention directly for coordination training (Dutton, 1995).

Many structural systems are involved in coordinated movement. Guidelines for evaluation suggested by Pedretti (1996) are varied and include the following:

- Assess tone and joint mobility.
- Observe client's sitting position, and locate any hypertrophied muscles.
- Observe for ataxia proximally to distally.
- Stabilize joints proximally to distally, and note differences in performance compared to performance without stabilization.
- Observe for resting or intention tremors. Are the eyes and speech involved?
- Does the client's emotional status affect the incoordination?
- How does ataxia or incoordination affect function?

There are many tests of coordination and dexterity. Many are standardized, and some are not (Table 11-12). As with any standardized test, it is important to administer the test exactly per protocols and to know with what population the test was standardized (e.g., norms for children ages 8 to 12 would not be applicable to adults). Some coordination and dexterity tests use functional tasks. For example, the Jensen Hand Function Test includes seven subtests (writing a short sentence, turning 3- x 5-inch cards over, placing small objects in a container,

stacking checkers, simulated eating, moving large cans that are light and those that are heavy). Some tools simulate ADL tasks, and others have functional tasks plus grasp and pinch evaluations. Some coordination assessments evaluate unilateral function, some only bilateral use of the hands, and some test both. Some tools evaluate gross coordination, fine coordination, or both (Norkin & Levangie, 1992; Pedretti, 1996; Trombly, 1995). Knowing the performance skills required by the client in daily activities will aid in the appropriate tool to use in assessment of coordination.

Treatment of coordination deficits may require several theoretic approaches. Lesions of the corticospinal system often use sensorimotor approaches to normalize tone and develop normal movement patterns. Normal motor learning to attain proximal stability then mobility and modulation of reflexes and synergies are also used, as are relearning of motor control mechanisms such as righting and equilibrium reactions.

Involuntary movements of cerebellar or extrapyramidal systems may be controlled pharmacologically to control tremors; weighting extremities with tremors is often suggested as a compensatory approach but is impractical in daily activities (Pedretti, 1996).

Exercises for training coordination have the general goal of achieving multimuscular motor patterns that are faster, more precise, and stronger. Repetition is used, and the parts of the activity are broken down and attempted separately at first. Initially, the tasks are simple and slow, requiring the conscious awareness of the client. Frequent rests are permitted, and when the client can perform each step precisely and independently, the steps are made more difficult by grading for speed, force, and complexity (Pedretti, 1996).

Principle: To improve coordination, repetitious activities and exercises need to produce accurate, controlled movements characterized by smoothness, rhythm, appropriate speed, appropriate muscle tension, and a minimal number of muscles involved to accomplish the movement.

Repetition and practice of functional tasks is often used, and the therapist can provide feedback to the client about performance (Shumway-Cook & Woollacott, 2000). Adaptation to tools and tasks, such as weighted utensils for tremors, is also used for some types of incoordination.

ENDURANCE

Endurance is defined as "the ability to sustain cardiac, pulmonary, and musculoskeletal exertion over time" (American Occupational Therapy Association, 1994). Endurance is the ability to maintain muscle actions for long periods of time, which requires a continuous restoration of energy resources (Cooper, 2007). It would be reasonable to expect endurance limitations with clients who have deficits in local muscle metabolism (e.g., diabetes), cardiovascular system (e.g., congestive heart failure), respiratory system (e.g., emphysema), those on total bed rest 6 or more days, and in clients with paralysis (Dutton, 1995). Endurance may be limited due to activity

Table 11-12

COORDINATION TESTS

Block and Box test
Minnesota Rate of Manipulation
Jensen-Taylor Hand Function test
Crawford Small Parts Dexterity Test
Hand Tool Dexterity Test, Bennett
O'Connor Tweezer Dexterity Test
Pennsylvania Bi-Manual Work sample
Purdue Pegboard
Stromberg Dexterity Test
Nine Hole Peg Test
Moberg's Pickup Test
Sister Kenny's Hand Function Test

COMMERCIAL DEXTERITY TESTS, WORK SAMPLES, OR WORKSTATIONS

BTE Bolt Box and Assembly Tree
Easy Street Environments
Singer Work samples
Tower Work samples
Skills Assessment Module (SAM)
Coats Work samples
Bennett Hand Tool Dexterity Test
The Work System: work simulations for sustained productivity
Valpar Component Work samples
MESA Computerized Screening Tool
JEVS Work samples
BTE Work simulator
LIDO Work set
ERGOS Work simulator
Jacobs Prevocational Skills Assessment (JPSA)
Disabilities of the Arm, Shoulder and Hand (DASH)

LIMB COORDINATION

Finger-Nose Test
Finger to Finger Test
Heel-Shin Test
Disdiadokinesia test
Rebound test

restrictions, pathology of cardiac or pulmonary systems, or muscle diseases.

Endurance is influenced by three factors: muscle function, oxygen supply, and their combined effects. With cardiorespiratory dysfunction, breathing itself may be exhaustive. Cardiovascular function is a factor of fitness. This is the ability of the body to take in, transport, and use oxygen (Christiansen & Baum, 1997). Ways to assess cardiorespiratory function include the following:

- Pulse
- Blood pressure
- Respiration
- Lung volume
- Vital capacity
- Breathing rate
- Pulmonary ventilation
- Cardiac output

Minor adds that "in the absence of pulmonary disease (e.g., chronic obstructive lung disease [COLD] or chronic obstructive pulmonary disease [COPD]), most of the limitation to endurance performance depends not on our ability to inspire and diffuse oxygen, but on the ability of the heart and circulatory system to deliver oxygen and cellular mechanisms to use the oxygen for energy production" (Christiansen & Baum, 1997, p. 261). If there are disruptions in the blood supply, there is not nutrition or waste removal, resulting in reduced respiratory capacity and decreased endurance.

Cardiorespiratory decondition is seen in increased resting heart rate, decreased heart volume, loss of blood volume, decreased stroke volume, decreased cardiac output, decreased coronary blood flow, impaired orthostatic response, and diminished aerobic capacity. Clinical manifestations of cardiorespiratory deconditioning might include the following:

- Reduced exercise tolerance demonstrated by increased heart rate and respiration at low workloads.
- Early onset of fatigue.
- Exertional dyspnea.
- Perception of doing heavy or maximal work at low to moderate loads.
- Rise in heart rate and drop in blood pressure upon standing up (orthostatic hypotension), which produces syncope and fainting.

The Borg's rate of perceived exertion (RPE) scale is often used by the client to rate his or her perception of the level of intensity of the activity or exercise as well as the level of pain. While a subjective rating by the client, this scale is commonly used to rate angina, aches in muscles, levels of pain, and in exercise tolerance testing (Tan, 1998). The client is asked to rate the intensity or pain on a scale predetermined by the clinician or the client himself or herself. Target heart rate range (THRR) is one way of monitoring aerobic activities. If the pulse rate exceeds the upper limit of the THRR, then the activity is too strenuous, and similarly, if the heart rate is below the lower limits, then the intensity should be increased (Tan, 1998).

Another way of monitoring endurance aerobic activities is to use the Karvonen (Rothstein et al., 2005) formula, which is maximal HR – resting HR (40% to 85%) + resting HR = THRR.

Formulas have been developed to account for age differences as in the age-adjusted maximal heart rate (AAMHR) where the formula is (220 – age) (65% to 85%) = THRR. This formula is commonly used but is less accurate than the Karvonen formula.

It is the interaction of the cardiovascular and respiratory systems that influence endurance. A conditioned heart produces a greater cardiac output at a lower heart rate than the untrained heart. Less work results in lowered oxygen demand. Mean and peak heart rates have been established for different types of jobs, which can be helpful in planning intervention (McKenna & Maas, 1987).

Muscular endurance is defined as the "ability to move a submaximal load repetitiously without degradation in performance" (Nyland, 2006, p. 181). Diminished muscular endurance may be due to inactivity resulting in disuse atrophy. The deficits are most notable in the muscles of locomotion following bed rest and are associated with fast-twitch/type II muscle fibers. Muscle endurance relies on ability of lungs, capillaries, and muscles to transport and uptake oxygen. The decreased maximal oxygen consumption is not only a consequence of diminished capacity of the heart to deliver oxygen but also the result of diminished capacity of muscle to use oxygen (Hamill & Knutzen, 1995). Tonic (slow-twitch) muscles primarily require oxygen from the vascular system, while phasic (fast-twitch) muscles require glucose for nutrition (Cooper, 2007). Each type of muscle will have enhanced muscular endurance when there is increased function of the cardiopulmonary system.

To increase cardiorespiratory and muscular endurance, there must be stress imposed on the systems to facilitate adaptation. The overload principle is applicable here, but the systems need far less than maximal stress applied. The amount of resistance varies according to different authors from 15% to 40% (Pedretti, 1996) of the repetition maximum to less than 50% of the repetition maximum (Trombly, 1995), or 60% to 90% of the maximum heart rate or 50% to 85% of maximum oxygen uptake.

> **Principle: To increase endurance, submaximal exertion will require adaptation to stress, resulting in hypertrophy and hyperplasia of oxidative type I slow-twitch muscle fibers and increased function in the cardiorespiratory systems.**

Clients with very low endurance often resume activities based on metabolic equivalent (MET) levels. One MET equals the basal metabolic rate, which is the amount of oxygen necessary to maintain metabolic processes, such as respiration, circulation, peristalsis, temperature regulation, and glandular functions at rest. MET levels can vary according to humidity, temperatures, and emotion (Radomski & Latham, 2008). MET levels indicate endurance and activity tolerance, and these levels are often used in cardiac rehabilitation programs. By referring to a table, one can determine the energy required to complete specific activities. Energy is measured by the amount of oxygen consumed while engaged in activities as well as the oxygen required to maintain metabolic functions (Trombly, 1995). The higher the MET level, the more vigorous the activity.

Usually, endurance activities and exercises are designed to have less load for a longer duration (low load, high duration). These activities generally target the slow-twitch/type I muscle fibers. While there is less than maximal load, there must be some resistance, or adaptation will not occur. The probability that a client will engage in activities for a longer period of time is greatly enhanced if that activity is of interest and is meaningful because the client needs to sustain effort for increasingly longer periods of time (Smith et al., 1996). For example, the client may play 2 hours of wheelchair basketball but only do wheelchair laps on a track for 30 minutes (Dutton, 1995).

To facilitate cardiorespiratory and muscular endurance, an activity should require rhythmic, dynamic contractions of large muscle groups (Christiansen & Baum, 1997). As with exercises and activities to increase strength, the intensity, duration, and frequency are carefully controlled. Gradations are based on the client's current level of function with gradual progression.

Warm-up and cool-down activities are recommended to diminish muscle cramping, soreness, and syncope. Activities and exercises involving excessive use of the arms overhead or those requiring sustained isometric contractions should be avoided because these tend to elevate the blood pressure without cardiovascular (aerobic) benefits. A general guideline of 20 full-range repetitions and sustaining the activity for at least 30 seconds done three to five times per week is suggested, but is variable based on individual client abilities. The intervention goal is to increase the client's ability to perform repeated motor tasks in daily living and to carry on sustained levels of activity.

Long-term effects of greater endurance and cardiorespiratory fitness include decreased resting and submaximal HR; submaximal effort with increased peak BP during maximal exercise; increases in cardiac output; increased coronary blood flow; improved oxygen delivery; and decreased exercise recovery time (Hall & Brady, 2005).

Low-impact aerobic activities, such as a stationary bicycle, cycling, race-walking, calisthenics, swimming, aquatic exercise, rowing, hiking, and cross country skiing, can be used for less conditioned clients. High-impact activities for more well-conditioned clients might include jogging, running, volleyball, hopping, jumping, rope skipping, aerobic dance, and downhill skiing. Tai chi has been used to increase balance control, flexibility, and cardiorespiratory fitness in older clients (Hong, Li, & Robinson, 2000). These activities can be used to increase endurance and strength and can be incorporated into discharge plans for the client's continued wellness.

The terms *tolerance* and *endurance* are often used interchangeably by therapists. For example, in documentation, it may be stated that a client has "increased sitting tolerance from 10 to 20 minutes." This ability to remain upright in the wheelchair has an influence on how well the client can feed himself or herself and work at a computer and can be linked to a functional long-term goal. Increases in the number of repetitions of an activity are indicators of increased endurance, such as increasing the number of spoonfuls per meal (Pedretti,

1996). The client is increasing aspects of endurance where the activities or exercises are being performed at less than maximal levels of intensity for increased periods of time or number of repetitions.

Clients with decreased endurance often have cardiopulmonary, cardiovascular, and musculoskeletal limitations with serious precautions related to these conditions. Clients with cardiopulmonary diseases need to avoid isometric exercise because this type of muscle contraction increases the heart rate and blood pressure. Clients with asthma, pulmonary diseases, and shortness of breath may need to be referred to a physician if they have excessive coughing, wheezing, substernal chest lightness, and elevated respiratory rate. Symptoms of headache or blurred vision, pain in the chest, substernal, or left arm, lightheadedness, irregular heart rhythms, or uncontrolled hypertension are red flags for people with cardiovascular disorders. Diabetes clients may have hypoglycemic episodes characterized by shakiness, weakness, blurred vision, anxiety, confusion, and decreased cognitive ability and should be given a carbohydrate snack or protein and not begin aerobic exercise or activities. If a client has a hypercoagulable disorder or has been immobilized by best rest or casts (calf, thigh, arms, pelvis), he or she is at risk for deep vein thrombosis characterized by pain in the calf or thigh, swelling, and tenderness. If the thrombus has traveled to the right side of the heart to occlude a pulmonary artery, the client may have symptoms of pleuritic pain, shortness of breath, fast rate of respiration, and rapid pulse rate and may be coughing up blood. This is an urgent condition and needs immediate medical care (Hall & Brady, 2005).

Table 11-13 provides sample goal statements, principles for intervention, and treatment examples to increase endurance.

CASE STUDY

The following case study is provided to demonstrate how theory influences intervention. This case study has elements that are more appropriately considered within a remediation or biomechanical intervention approach, while other aspects are better served using a compensatory/adaptation or rehabilitation construct. This same case study is used in Chapter 12 so that the elements appropriate to each theory can be seen as they relate to this case. Other theories not covered in this text may also be appropriate, and it is important to consider whether all of the approaches selected can be implemented concurrently with others or if the design of the intervention needs to be sequential in nature.

Maggie

Maggie is a 61-year-old woman referred to home health occupational therapy on October 29 with a diagnosis of right Colles' fracture. Maggie fell while shopping at her neighborhood department store on October 27. The physician has ordered occupational therapy to evaluate and treat.

On your first visit to see Maggie, you interview her and find out the following information. Maggie was in excellent health prior to this injury. Maggie's son, Fred, lives in the next town.

Table 11-13

INCREASING ENDURANCE

GOAL	METHOD	PRINCIPLE/RATIONALE	EXAMPLE
Client will initially increase endurance (indicate level) to enable performance in tasks by using...	Increased duration Increased level in a cardiac step program Increased intensity Increased repetitions	Which gradually stresses cardio-pulmonary system while ensuring rest.	Sitting tolerance in minutes Tolerance for evaluation Graduate to eating while sitting up in a chair Stand instead of sitting to shave Cardiac target heart rate Dress in 10 fewer minutes Feed self 10 additional bites of food Walk 10 additional steps
Client will maximize endurance (indicate level) to enable performance in tasks using...	Increased duration Increased level in a cardiac step program Increased intensity Increased repetitions	Which gradually stresses the cardiopulmonary system and local muscle metabolism of oxidative slow-twitch muscle fibers.	Increase the time involved in an activity (standing at table, completing a full meal) Increase levels on MET charts Perform activities at 50 to 70% maximum levels Dress 10 minutes faster than baseline Increase number of spoonfuls of food per meal Increase the number of wheelchair pushups

Adapted from Dutton, R. (1995). *Clinical reasoning in physical disabilities.* Baltimore: Williams & Wilkins; and Marrelli, T. M., & Krulish, L. H. (1999). *Home care therapy: Quality, documentation and reimbursement.* Boca Grande, FL: Marrelli and Associates, Inc.

She works part-time in her son's accounting office. A widow, she lives alone with her cat and her dog. Maggie is concerned that, because she lives alone, she needs to be able to do all of the cooking, cleaning, and care for her dog and cat. Her son is only able to come to her home every other day or every 3 days to assist with some of these activities.

Maggie is concerned about her present situation. She is right-handed and states that although she tries her best, she is not able to do very much for herself. Maggie has been very active in her local church organization. Maggie always heads up the Christmas crafts bazaar. Part of her responsibilities for the bazaar include organizing and teaching at craft nights.

Maggie also has been one of the primary cooks for the annual fish dinner held just before Easter. She states that the church especially needs her this year as the other primary cook is moving to another state. She doesn't want to let the church organization down.

In the emergency room on October 27, the emergency room physician performed a closed reduction and applied a plaster cast from mid-humeral level to the metacarpophalangeal (MCP) joints of the right hand. The elbow is casted in approximately 90 degrees of elbow flexion, and the wrist is casted in approximately 30 degrees of wrist flexion. Maggie states that the physician has instructed her to stay home for the next 2 weeks and not to drive or do any housework. This is confirmed when you speak to the physician. A summary of information about Maggie as well as analysis of additional information that is needed and ideas about the next steps are as follows:

Facts

- 61-year-old woman
- Enjoys shopping
- Right Colles fracture
- Fractures are immobilized, cause pain and edema
- Fractures may limit the use of the UE
- Lives alone
- Was independent in ADL, IADL
- Cares for dog and cat
- Active in church craft bazaar and fish dinner
- Anxious about present situation
- Immobilized in plaster cast
- Physician restricted activities for 2 weeks

Additional Information Needed

- How did Maggie fall?
- Precautions
- Medical management of Colles fracture
- Possible complications
- Prognosis to return to prior level of function
- Tasks that Maggie will need to do to resume roles

Action

- Continue dialogue with Maggie
- Review medical record
- Consult with physician
- Consult orthopedic textbooks and OT texts
- Perform OT assessments

From this information, an evaluation plan can be developed. Several assessments have been listed, but these are not the only assessments that can be used to address the concerns about Maggie.

Evaluation Plan

Concerns

1. Maggie's perception of her strengths and weaknesses and her priorities for therapy.
2. Decreased ADL performance due to right (dominant) UE in cast.
3. Unable to carry out worker and volunteer roles.
4. Unable to participate in leisure pursuits.
5. Sensorimotor concerns (sensory): possible sensory loss or altered sensation due to pressure on nerves from swelling or tight cast.
6. Sensorimotor concerns (motor): potential for edema, stiffness, and complex regional pain syndrome (CRPS) in right hand; potential for frozen right shoulder.
7. Psychological and cognitive concerns: potential for depression due to loss of role performance and feelings of helplessness.

Assessments

1. Canadian Occupational Therapy Performance Measure (COPM) or other client-centered informal interview tool.
2. Observe Maggie attempting to perform ADL tasks; Klein-Bell ADL assessment; self-report; Functional Independence Measure (FIM).
3. Interview Maggie to determine specific worker and volunteer tasks and activities.
4. Interest checklist; formal or informal interview.
5. Sensory evaluation to exposed areas; observe cast for fit, look for areas of redness where cast is pressing against the skin.
6. AROM and PROM of PIP and DIP joints of the right hand; circumferential edema measurements of the digits; pain assessment; AROM and PROM of right shoulder.
7. Formal and informal interview; observe Maggie during evaluation; speak with family members and other team members.

How Assessment Results Will Be Used

1. Set client-centered goals for therapy; establish rapport; gain better understanding of Maggie's occupational context.
2. Assess potential for teaching compensatory techniques or use of adaptive equipment; determine components of ADL tasks causing difficulty; determine baseline for therapy.
3. Determine readiness for return to work and volunteer duties; assess potential for adapting tasks and activities.

4. Determine realistic leisure interest; assess potential for adapting leisure activities.

5. Set baseline for therapy; make suggestions for alterations of cast if necessary; determine need to teach compensatory techniques or sensory re-education program if sensation is impaired or absent.

6. Set baseline for therapy; develop treatment plan that includes management of edema and home exercise program (HEP) to maintain ROM and prevent complications.

7. Determine Maggie's motivation for therapy; assess ability to follow directions and carry out HEP, set realistic goals appropriate for Maggie's cognitive and emotional status.

You evaluated Maggie and found the following:

Maggie is cognitively intact and motivated to be independent. Maggie is proud of her ability to figure out how to open the foil packet of cat food with her left hand so that she can feed her cat. She is unable to don some types of blouses and dresses over the cast and is not able to don pants or skirts with closures. She requires minimal assistance to don a housedress, underpants, and socks. She is unable to don her bra or panty hose, or tie her shoes/sneakers. Because of the inability to use her dominant right hand, Maggie is unable to manage buttons, has difficulty feeding herself with her nondominant left hand (spills food on herself frequently), requires minimal assist with personal hygiene after toileting, sponge bathing, and brushing her teeth. She requires maximal assistance in simple meal preparation. Maggie states she is able to get herself some simple food items, such as prepared foods (prepared by her son's wife) from plastic containers. She is dependent in household chores such as cleaning.

Through further interview, you learn that much of Maggie's work at her son's office involves working with Excel, a spreadsheet computer program. She has a computer at home and is an avid Internet user. She tells you that it is difficult to use the computer now because she cannot use her right hand.

Tasks associated with her responsibilities as chairperson of the Christmas craft bazaar include phoning church members to inform them of meeting times and craft supplies they will be required to bring, and delegation of other responsibilities such as making refreshments, teaching certain crafts, and clean-up duties. She is having difficulty using her rotary dial telephone to keep in touch with other church volunteers.

Maggie complains of moderate pain in her right upper extremity, mainly the wrist and fingers. She rates her pain an 8 on a scale of 1 to 10. She describes it as a burning, throbbing type of pain. Severe edema is noted in all of the digits of the right hand (approximately two times the size of the left hand). She has a great deal of difficulty flexing her fingers to make a fist (the cast restricts the MP motion). She is unable to oppose her thumb to her index or any other finger. She is unable to hold any objects in her right hand because of the cast. She states that, at times, her hand feels numb, and at other times, she has paresthesia. The fingers are discolored, a dusky reddish color.

In your conversation with the home health coordinator and the social worker, you find out that Maggie's insurance will cover home health occupational therapy for up to 3 months for the purpose of improving independence in ADLs. The following would be an example of the intervention that would address biomechanical remediation concerns in this case. This is not a complete intervention plan in that many areas that can be addressed in treatment by an occupational therapist are not included. By combining the treatment plan for the compensation/adaptation/rehabilitation treatment plan with this, a more complete intervention for Maggie can be realized.

In Table 11-14, deficits as identified previously are related to long- and short-term outcomes. Specific principles or rationales are provided so that the reason for the specific intervention is clear.

SUMMARY

- The biomechanical frame of reference is a remediation intervention approach. In this approach, there is an expectation of an improvement in a performance component that will lead to improved occupational performance.

- The biomechanical frame of reference focuses on musculoskeletal system functions that include strength, endurance, range of motion, tissue integrity, and structural stability.

- Because this frame of reference focuses on the musculoskeletal system, physical fitness and health also are parts of this approach. Strategies to improve muscle function and range of motion in those with activity and participation limitations also apply to those without restrictions as part of an overall health promotion objective.

REFERENCES

American Occupational Therapy Association. (2008). Occupational therapy practice framework: Domain and process (2nd ed.). *American Journal of Occupational Therapy, 62*, 625-683.

American Occupational Therapy Association. (1994). Uniform terminology for occupational therapy (3rd ed.). *American Journal of Occupational Therapy, 48*(11), 1047-1054.

Aretzberger, S. (Ed.). (2007). *Edema reduction techniques: A biologic rationale for selection*. St. Louis, MO: Mosby Elsevier.

Artzberger, S. M. (2005). A critical analysis of edema control techniques. *Physical Disabilities Special Interest Section Quarterly, 28*(2).

Bamman, M. M., Hill, V. J., Adams, G. R., Haddad, F., Wetzstein, C. J., & Gower, B. A. (2003). Gender differences in resistance-training-induced myofiber hypertrophy among older adults. *Journal of Gerontology: Biological Sciences, 58A*, 108-116.

Basbaum, A. I., & Julius, D. (2000). The perception of pain. In E. R. Kandel, J. Schwartz, & T. Jessel (Eds.). *Principles in neural science*. New York: McGraw-Hill.

Basbaum, A. I., & Julius, D. (2006). Toward better pain control. *Scientific American, 294*(6), 60-67.

Baxter, R. E. (2003). *Pocket guide to musculoskeletal assessment*. St. Louis, MO: W.B. Saunders.

Bear, M., Conners, B., & Paradiso, M. (2001). The somatic sensory system. In M. Bear, B. Connors, & M. Paradiso (Eds.), *Neuroscience: Exploring the brain*. Baltimore, MD: Lippincott Williams & Wilkins.

Table 11-14

TREATMENT PLAN FOR MAGGIE: BIOMECHANICAL APPROACH

DEFICITS	STAGE-SPECIFIC CAUSE	SHORT-TERM GOALS	METHODS	PRINCIPLE	SPECIFIC ACTIVITY/MODALITY
1. Potential for muscle wasting in right wrist extensors and flexors limiting functional use of right UE.	1. Immobilization in cast.	1. Client will perform HEP (home exercise program) of isometric exercises with minimal verbal cues.	1. Educate client in isometric exercises for right wrist muscles.	1. Active contraction of muscle fibers will maintain muscle bulk.	1. Client will learn techniques of performing isometric exercises for R wrist musculature using 10 repetitions of each exercise, 3 times a day.
2. Potential for active ROM and muscle strength to decrease in right shoulder, which will limit functional use of RUE.	2. Inability to use RUE for functional activities.	2. Client will perform home activity program with minimal assist to prevent decrease in AROM and muscle strength in right shoulder motions and muscle groups.	2. Client will be educated in activities/ exercises to move UE through all shoulder motions.	2. Stretch of muscles on connective tissue through full ROM daily prevents shortening of these tissues. Stress provided by weight of cast will maintain or increase muscle bulk.	2. Therapist will provide written and videotaped guide to HEP to move UE through all shoulder motions. Client will learn a HEP.
3. Client has severe edema in right digits with potential for contractures, which limit functional use of RUE.	3. Immobilization in cast.	3. Client will independently perform self retrograde massage.	3. Application of pressure distally to proximally through massage of the hand.	3. Massage will prevent filtration of fluids from capillaries into the interstitial tissues and assist with blood and lymph flow.	3. Therapist will perform retrograde massage to R hand each treatment session. Client will be educated in self retrograde massage with instructions to perform 3 times a day.

Berne, R. M. & Levy, M. (1998). *Physiology*. St. Louis, MO: C.V. Mosby Co.

Bonder, B. R., & Bello-Haas, V. D. (2009). *Functional performance in older adults* (3rd ed.). Philadelphia, PA: F. A. Davis Company.

Borcherding, S. (2005). *Documentation manual for writing SOAP notes in occupational therapy*. Thorofare, NJ: SLACK Incorporated.

Carmeli, E., Reznick, A. Z., Coleman, R., & Carmeli, V. (2000). Muscle strength and mass of lower extremities in relation to functional abilities in older adults. *Gerontology, 46*(5), 249-257.

Charette, S. L., McEvoy, L., Pyka, G., Snow-Harter, C., Guido, D., & Wiswell, R. A. (1991). Muscle hypertrophy response to resistance training in older women. *Journal of Applied Physiology, 70*, 1912-1916.

Chesney, A. B., & Brorsen, N. E. (2000). OT's role in managing chronic pain. *OT Practice Online, 10/9/00.*

Christiansen, C. & Baum, C. (1997). *Occupational therapy: Enabling function and well being* (2nd ed.). Thorofare, NJ: SLACK Incorporated.

Cooper, C. (Ed.). (2007). *Fundamentals of hand therapy: Clinical reasoning and treatment guidelines for common diagnoses of the upper extremity*. St. Louis, MO: Mosby Elsevier.

Davies, G. J., & Ellenbecker, T. S. (1999). Focused exercise aids shoulder hypomobility. *Biomechanics*, November, 77-80.

Demeter, S., & Andersson, G. (1996). *Disability evaluation* (2nd ed.). St. Louis, MO: C.V. Mosby Co.

Donatelli, R. A., & Wooden, M. J. (2010). *Orthopaedic physical therapy* (4th ed.). St. Louis, MO: Churchill Livingstone Elsevier.

Dudgeon, B. J., Tyler, E. J., Rhodes, L. A., & Jensen, M. P. (2006). Managing unusual and unexpected pain with physical disability: A qualitative analysis. *American Journal of Occupational Therapy, 60*(1), 92-103.

Dutton, R. (1995). *Clinical reasoning in physical disabilities*. Philadelphia, PA: Williams and Wilkins.

Ferguson, J. M., & Trombly, C. A. (1997). The effect of added-purpose and meaningful occupation on motor learning. *American Journal of Occupational Therapy, 51*(7), 508-515.

Ferri, A., Scaglioni, G., Pousson, M., Capodaglio, P., VanHoecke, J., & Narici, M. V. (2003). Strength and power changes of the human plantar flexors and knee extensors in response to resistance training in older age. *Acta Physiological Scandinavica, 177*, 69-78.

Fess, E. E., Gettle, K. S., Philips, C. A., & Janson, J. R. (2005). *Hand and upper extremity splinting: Principles and methods* (3rd ed.). St. Louis, MO: Elsevier Mosby.

Fiatarone, M. A., O'Neill, E. F., Ryan, N. D., Clements, K. M., Solares, G. R., & Nelson, M. E. (1994). Exercise training and nutritional supplementation for physical frailty in very elderly people. *New England Journal of Medicine, 330*, 1769-1775.

Fiatarone, M. A., & Evans, W. J. (1993). The etiology and reversibility of muscle dysfunction in the aged. *Journal of Gerontology, 48*, 77-83.

Fisher, A. (1998). Uniting practice and theory in an occupational framework: 1998 Eleanor Clark Slagle Lecture. *American Journal of Occupational Therapy, 52*(7), 509-521.

Fisher, G. S., Emerson, L., Firpo, C., Ptak, J., Wonn, J., & Bartolacci, G. (2007). Chronic pain and occupation: An exploration of the lived experience. *American Journal of Occupational Therapy, 61*(3), 290-302.

Flinn, N. (1995). A task-oriented approach to the treatment of a client with hemiplegia. *American Journal of Occupational Therapy, 49*(6), 560-569.

Flowers, K. F., & LaStayo, P. (1994). Effect of total end range time on improving passive range of motion. *Journal of Hand Therapy, 7*(3), 150-157.

Frontera, W. R., Hughes, V. A., Fielding, R. A., Fiatarone, M. A., Evans, W. J., & Roubenoff, R. (2000). Aging of skeletal muscle: a 12-yr longitudinal study. *Journal of Applied Physiology, 88*(4), 1321-1326.

Frontera, W. R., Meredith, C., O'Reilly, K. P., Knuttgen, H. G., & Evans, W. (1988). Strength conditioning in older men: skeletal muscle hypertrophy and improved function. *Journal of Applied Physiology, 64*, 1038-1044.

Galley, P. M. & Forster, A. L. (1987). *Human movement: An introductory text for physiotherapy students*. New York: Churchill-Livingstone.

Gillen, G., & Burkhardt, A. (2004). *Stroke rehabilitation: A function-based approach*. St. Louis, MO: Mosby.

Gleim, G., & McHugh, M. (1997). Flexibility and its effects on sports injury and performance. *Sports Medicine, 21*(5), 289-299.

Grünert-Plüss, N., Hufschmid, U., Santschi, L., & Grünert, J. (2008). Mirror therapy in hand rehabilitation: A review of the literature, the St. Gallen protocol for mirror therapy and evaluation of a case series of 52 patients. *British Journal of Hand Therapy, 13*(1), 4-11.

Hall, C. M., & Brady, L. T. (2005). *Therapeutic exercise: moving toward function* (2nd ed.). Philadelphia, PA: Lippincott, Williams & Wilkins.

Hamill, J., & Knutzen, K. M. (2003). *Biomechanical basis of human movement* (2nd ed.). Philadelphia, PA: Lippincott, Williams & Wilkins.

Hamill, J., & Knutzen, K. M. (1995). *Biomechanical basis of human movement*. Philadelphia, PA: Lippincott, Williams & Wilkins.

Hellebrandt, F. A. (1958). Application of the overload principle to muscle training in man. *American Journal of Physical Medicine, 37*(5), 278-283.

Hoffman, H. G. (2004). Virtual-reality therapy. *Scientific American*, 58-65.

Hong, Y., Li, J. X., & Robinson, P. D. (2000). Balance control, flexibility, and cardiorespiratory fitness among older Tai Chi practitioners. *British Journal of Sports Medicine, 34*, 29-34.

Hoppes, S. (1997). Can play increase standing tolerance? A pilot-study. *Physical and Occupational Therapy in Geriatrics 15*(1), 65-73.

Houglum, P. (2010). *Therapeutic exercise for musculoskeletal injuries* (3rd ed.). Champaign, IL: Human Kinetics.

Hsieh, C. L., Nelson, D. L., Smith, D. A., & Peterson, C. Q. (1996). A comparison of performance in added-purpose occupations and rote exercise for dynamic standing balance in persons with hemiplegia. *American Journal of Occupational Therapy, 50*(1), 10-16.

Irion, G. (2000). *Physiology: The basis of clinical practice*. Thorofare, NJ: SLACK Incorporated.

Jacobs, K. & Bettencourt, C. M. (1995). *Ergonomics for therapists*. Boston, MA: Butterworth-Heinemann.

Jacobs, K., & Jacobs, L. (Eds.). (2009). *Quick reference dictionary for occupational therapy* (5th ed.). Thorofare, NJ: Slack Incorporated.

Jarvis, C. (2000). *Physical examination and health assessment*. Philadelphia, PA: W. B. Saunders.

Khruda, K. V., Hicks, A. L., & McCartney, N. (2003). Training for muscle power in older adults: Effects on functional abilities. *Canadian Journal of Applied Physiology, 28*(2), 178-189.

Kielhofner, G. (1992). *Conceptual foundations of occupational therapy*. Philadelphia, PA: F.A. Davis.

King, T. I. (1993). Hand strengthening with a computer for purposeful activity. *American Journal of Occupational Therapy, 47*(7), 635-637.

Kisner, C., & Colby, L. A. (2007). *Therapeutic exercise: Foundations and techniques* (5th ed.). Philadelphia, PA: F. A. Davis Company.

Kisner, C., & Colby, L. A. (1990). *Therapeutic exercise: Foundations and techniques* (2nd ed.). Philadelphia, PA: F. A. Davis Company.

Konin, J. G. (1999). *Practical kinesiology for the physical therapy assistant*. Thorofare, NJ: SLACK Incorporated.

Law, R. Y. W., & Herbert, R. D. (2007). Warm-up reduces delayed-onset muscle soreness but cool-down does not: A randomised controlled trial. *Australian Journal of Physiotherapy, 53*, 91-95.

Levangie, P. K., & Norkin, C. C.(2005). *Joint structure and function: A comprehensive analysis*. Philadelphia, PA: F.A. Davis Co.

Lieber, R. L. (2002). *Skeletal muscle structure, function and plasticity*. Philadelphia, PA: Lippincott, Williams & Wilkins.

MacCormac, B. A. (2010). Reaching the unmotivated client. *Occupational Therapy Practice, 15*(4), 15-19.

Magee, D. J. (2008). *Orthopedic physical assessment*. (5th ed.). Philadelphia, PA: W. B. Saunders.

Magee, D. J. (2002). *Orthopedic physical assessment*. (4th ed.). Philadelphia, PA: W. B. Saunders.

McCaffrey, R., Frock, T., & Garguilo, H. (2003). Understanding chronic pain and the mind-body connection. *Holistic Nursing Practice*, November/December, 281-287.

McCormack, G. L., & Gupta, J. (2007). Using complementary approaches to pain management. *OT Practice, 12*(13), 9-20.

McGinnis, P. M. (1999). *Biomechanics of sport and exercise.* Champaigne, IL: Human Kinetics.

McKenna, K. T., & Maas, F. (1987). Mean and peak heart rate prediction using estimated energy costs of jobs. *Occupational Therapy Journal of Research, 7,* 323-334.

Melzack, R., & Katz, J. (2004). The gate control theory: reaching the brain. In T. Hadjistavropoulos & K. D. Craid (Eds.). *Pain: Psychological Perspectives.* Mahwah, NJ: Erlbaum.

Melzack, R., & Wall, P. (1965). Pain mechanisms: A new theory. *Science, 150,* 971-976.

Meriano, C., & Latella, D. (2008). *Occupational therapy interventions: Function and occupations.* Thorofare, NJ: SLACK Incorporated.

Meuleman, J. R., Rechue, W. F., Kubilis, P. S., & Lowenthal, D. T. (2000). Exercise training in the debilitated aged: strength and functional outcomes. *Archives of Physical Medicine and Rehabilitation, 81,* 312-318.

Moyers, P. A. (1999). The guide to occupational therapy practice. *American Journal of Occupational Therapy, 53*(5), 247-321.

Muscolino, J. E. (2006). *Kinesiology: The skeletal system and muscle function.* St. Louis, MO: Mosby.

Nielson, C., Hansen, R., Hinjosa, J., Mitchell, M., & Manoly, B. (1994). *A guide for the preparation of occupational therapy practitioners for the use of physical agent modalities.* Rockville, MD: The American Occupational Therapy Association.

Neistadt, M. E. (1994). The effect of different treatment activities on the functional fine motor coordination in adults with brain injury. *American Journal of Occupational Therapy, 48*(10), 877-882.

Nelson, D. L., & Konosky, K. (1996). The effects of an occupationally embedded exercise on bilaterally assisted supination in persons with hemiplegia. *American Journal of Occupational Therapy, 50,* 639-646.

Nicosia, M. B. (2004). The pain fighters: A team approach to car confronts all the different facets of controlling pain. *Rehab Management,* March.

Norkin, C. C. & Levangie, P. K. (1992). *Joint structure and function: A comprehensive analysis.* Philadelphia, PA: F.A. Davis Co.

Nuismer, B. A., Ekes, A. M., & Holm, M. B. (1997). The use of low-load prolonged stretch devices in rehabilitation programs in the Pacific northwest. *American Journal of Occupational Therapy, 51*(7), 538-43.

Nyland, J. (2006). *Clinical decisions in therapeutic exercise: Planning and implementation.* Upper Saddle River, NJ: Pearson Prentice Hall.

O'Sullivan, S. B., & Schmitz, T. J. (1999). *Physical rehabilitation laboratory manual: Focus on functional trainings.* Philadelphia, PA: F. A. Davis Co.

Oatis, C. A. (2004). *Kinesiology: the mechanics and pathomechanics of human movement.* Philadelphia, PA: Lippincott, Williams & Wilkins.

Pedretti, L. W. (2001). *Occupational therapy: Practice skills for physical dysfunction.* (5th ed.). St. Louis, MO: Mosby.

Pedretti, L. W. (1996). *Occupational therapy: Practice skills for physical dysfunction.* (4th ed.). St. Louis, MO: Mosby.

Pendleton, H. M., & Schultz-Krohn, W. (Eds.). (2006). *Pedretti's occupational therapy: Practice skills for physical dysfunction.* St. Louis, MO: Mosby.

Prosser, R., & Conolly, W. B. (2003). *Rehabilitation of the hand and upper limb.* London: Butterworth Heinemann.

Quinn, L. & Gordon, J. (2003). *Functional outcomes: Documentation for rehabilitation.* St. Louis, MO: W. B. Saunders.

Radomski, M. V., & Latham, C. A. T. (Eds.). (2008). *Occupational therapy for physical dysfunction.* Philadelphia, PA: Wolters Kluwer/Lippincott, Williams & Wilkins.

Rantanen, T., Guralnik, J. M., Foley, D., Masaki, K., Leveille, S., Curb, J. D., & White, L. (1999). Midlife hand grip strength as a predictor of old age disability. *Journal of the American Medical Association, 281*(6), 558-560.

Rochman, D. L., & Kennedy-Spaien, E. (2007). Chronic pain management: Approaches and tools for occupational therapy. *OT Practice, 12*(13), 9-15.

Rothstein, J. M., Roy, S. H., Wolf, S. L., & Scalzitti, D. A. (2005). *The rehabilitation specialist's handbook* (3rd ed.). Philadelphia, PA: F. A. Davis Company.

Shrier, I. (1999). Stretching before exercise does not reduce the risk of local muscle injury: a critical review of the clinical and basic science literature. *Clinical Journal of Sport Medicine, 9*(4), 221-227.

Shumway-Cook, A., & Woollacott, M. H. (2000). *Motor control: Translating research into clinical practice.* Baltimore, MD: Lippincott Williams & Wilkins.

Sietsema, J. M., Nelson, D. L., Mulder R. M., Mervau-Scheidel, D., & White, B. E. (1993). The use of game to promote arm reach in persons with traumatic brain injury. *American Journal of Occupational Therapy, 47*(1), 19-24.

Smith, J. (2005). *Structural bodywork.* St. Louis, MO: Elsevier Churchill Livingstone.

Smith, L. K., Weiss, E. L., & Lehmkuhl, L. D. (1996). *Brunnstrom's clinical kinesiology.* (5th ed.). Philadelphia: F. A. Davis Co.

Steinbeck, T. (1986). Purposeful activity and performance. *American Journal of Occupational Therapy, 40*(8), 529-34.

Tan, J. C. (1998). *Practical manual of physical medicine and rehabilitation.* St. Louis, MO: Mosby.

Taylor, D. C., & Dalton, J. D. (1990). Viscoelastic properties of muscle-tendon units: The biomechanical effects of stretching. *American Journal of Sports Medicine, 18*(3): 300-309.

Thibodeaux, C., & Ludwig, F. M. (1988). Intrinsic motivation in product-oriented and non-product oriented activities. *American Journal of Occupational Therapy, 42,* 169-175.

Thompson, C. W., & Floyd, R. T. (1994). *Manual of structural kinesiology.* St. Louis, MO: C.V. Mosby.

Trew, M., & Everett, T. (Eds.). (2005). *Human movement: An introductory text.* London: Elsevier Churchill Livingstone.

Trombly, C. A. (Ed.). (1995). *Occupational therapy for physical dysfunction* (4th ed.). Baltimore, MD: Williams and Wilkins.

Tweed, J. L., & Barnes, M. R. (2008). Is eccentric muscle contraction a significant factor in the development of chronic anterior compartment syndrome? A Review of the literature. *The Foot, 18,* 165-170.

Vincent, K. R., & Braith, R. W. (2002). Resistance exercise and bone turnover in elderly men and women. *Medical Science and Sports Exercise, 34,* 17-23.

Wennemer, H. K., Borg-Stein, J., Delaney, B., Rothmund, A., & Barlow, D. (2006). Functionally oriented rehabilitation program for patients with fibromyalgia: Preliminary results. *American Journal of Physical Medicine and Rehabilitation, 85*(8), 659-666.

Yuen, H. K., Nelson, D. L., Peterson, C., & Dickinson, A. (1994). Prosthesis training as a context for studying occupational forms and motoric adaptation. *American Journal of Occupational Therapy, 48*(1), 55-61.

ADDITIONAL RESOURCES

American Occupational Therapy Association. (1992). Use of adjunctive modalities in occupational therapy. *American Journal of Occupational Therapy, 46*(1), 1075-1081.

American Occupational Therapy Association. (1993). Position paper: Purposeful activity. *American Journal of Occupational Therapy, 47*(12), 1081-1082.

American Occupational Therapy Association. (2003). Physical agent modalities: A position paper. *American Journal of Occupational Therapy, 57*(6), 650-651.

American Occupational Therapy Association. (1995). Position paper: Occupational performance: Occupational therapy's definition of function. *American Journal of Occupational Therapy, 49*(10), 1019-1020.

American Occupational Therapy Association. (1997). Statement—fundamental concepts of occupational therapy: Occupation, purposeful activity and function. *American Journal of Occupational Therapy, 51*(10), 864-866.

Andersen, L. T. (2001). *Adult physical disabilities: Case studies for learning.* Thorofare, NJ: SLACK Incorporated.

Asher, I. E. (1996). *Occupational therapy assessment tools: An annotated index* (2nd ed.). Bethesda, MD: American Occupational Therapy Association.

Behrens, B. J. (2006). *Laboratory manual for physical agents: Theory and practice.* Philadelphia, PA: F. A Davis Company.

Belanger, A.-Y. (2003). *Evidence-based guide to therapeutic physical agents.* Philadelphia, PA: Lippincott Williams & Wilkins.

Bracciano, A. G. (2000). *Physical agent modalities: Theory and application for the occupational therapist.* Thorofare, NJ: SLACK, Inc.

Bracciano, A. G., & Earley, D. (2002). Physical agent modalities. In C. A. Trombly & M. V. Radomski (Eds.), *Occupational therapy for physical dysfunction* (5th ed.). Philadelphia, PA: Lippincott Williams and Wilkins.

Breines, E. B. (2001). Therapeutic occupations and modalities. In L. W. Pedretti & M. B. Early (Eds.), *Occupational therapy: Practice skills for physical dysfunction* (5th ed.). St. Louis, MO: Mosby.

Caillet, R. (1996). *Soft tissue pain and disability* (3rd ed.). Philadelphia, PA: F. A. Davis Co.

Cornish-Painter, C., Peterson, C. Q., et al. (1997). Skill acquisition and competency testing for physical agent modality use. *American Journal of Occupational Therapy, 51*(8), 681-685.

Daniels, L., & Worthingham, C. (1977). *Therapeutic exercise for body alignment and function* (2nd ed.). Philadelphia, PA: W. B. Saunders Co.

Dutton, R. (1989). Guidelines for using both activity and exercise. *American Journal of Occupational Therapy, 43*(9), 573-580.

Glauner, J. H., Ekes, A. M., et al. (1997). A pilot study of the theoretical and technical competence and appropriate education for the use of nine physical agent modalities in occupational therapy practice. *American Journal of Occupational Therapy, 51*(9), 767-774.

Greene, D. P., & Roberts, S. L. (1999). *Kinesiology: Movement in the context of activity.* St. Louis, MO: Mosby.

Joe, B. E. (1996). Can you justify all those treatments? *OT Week, 10*(2), 15-16.

Marrelli, T. M., & Krulish, L. H. (1999). *Home care therapy: Quality, documentation and reimbursement.* Boca Grande, FL: Marrelli and Associates, Inc.

Neistadt, M. E., & Seymour, S. G. (1995). Treatment activity preferences of occupational therapists in adult physical dysfunction settings. *American Journal of Occupational Therapy, 49*(5), 437-443.

Nielson, C., Hansen, R., et al. (1994). A guide for the preparation of occupational therapy practitioners for the use of physical agent modalities. *American Journal of Occupational Therapy.*

Pedretti, L. W., & Wade, I. E. (1996). Therapeutic modalities. In L. W. Pedretti & M. B. Early (Eds.), *Occupational therapy: Practice skills for physical dysfunction* (4th ed.). St. Louis, MO: Mosby.

Post, R. E., Lee, S. L., et al. (Eds.) (1995). *Physical agent modalities. Occupational therapy for physical dysfunction* (4th ed.). Baltimore, MD: Williams and Wilkins.

Shankar, K. (1999). *Exercise prescription.* Philadelphia, PA: Hanley & Belfus, Inc.

Taylor, E., & Humphrey, R. (1991). Survey of physical agent modality use. *American Journal of Occupational Therapy, 45*(10), 924-931.

Turner, A., Foster, M., & Johnson, S. E. (1996). *Occupational therapy and physical dysfunction: Principles, skills and practice* (4th ed.). New York, NY: Churchill Livingstone.

Van Deusen, J., & Brunt, D. (1997). *Assessment in occupational therapy and physical therapy.* Philadelphia, PA: W. B. Saunders Co.

Zelenka, J. P., Floren, A. E., & Jordan, J. J. (1966). Minimal forces to move patients. *American Journal of Occupational Therapy, 50*(5), 354-361.

12

REHABILITATION
ADAPTATION AND COMPENSATION

Your client is unable to wash her face, brush her teeth, apply makeup, and groom and style her hair. This person is limited in the performance of personal hygiene and grooming skills. The meaning of this deficit is determined by the person and includes individual beliefs about hygiene, self-concept concerns, gender issues, and cultural requirements for appropriate grooming. The person also needs sufficient range of motion (ROM), strength, trunk balance, coordination, cognition, vision, sensation, and tone to accomplish the tasks associated with grooming. Characteristics about the task and the level of skill involved also are a factor in the person's ability to engage successfully in hygiene tasks. The steps involved in the process, the time and timing required by the task, the hazards, and the types and characteristics of objects needed for each type of hygiene task may be the cause of the performance limitation. Finally, the environment may not be conducive for task completion. Perhaps the mirror is too high or the sink too low for ease of achievement of hygiene tasks. The role of the occupational therapist is to facilitate this person's performance in areas of occupation (in this case, hygiene) and to consider all aspects of the person, the environment, and the occupation.

Using the Person-Environment-Occupation Model (Law et al., 1996) as a structure to conceptualize the process of occupational therapy, Figure 12-1 illustrates the relationship between the person, environment, and occupation in the ability to engage successfully in occupational performance (hygiene).

The biomechanical treatment approach (discussed in Chapter 11) provides intervention methods to restore or remediate client factors of strength, ROM, tissue integrity,

structural stability, coordination, and endurance. These would be aspects related to the Person in the PEO model. The question this chapter answers is if ROM, strength, endurance, coordination, tissue integrity, and structural stability cannot be regained through remediation intervention, then how can activities (occupation) or the context (environment) be modified to adapt to or compensate for these limitations.

CONCEPTUAL BACKGROUND

The rehabilitation intervention approach is a compensatory and adaptation approach. Often, this approach is used when there is little or no expectation for change or improvement in the performance skills and abilities or when there are residual impairments, and further remediation attempts are unproductive (Holm, Rogers, & James, 1998). When there is limited time for intervention or the client or family prefers a more immediate resolution to functional problems, rehabilitation approaches may also be used.

Compensatory and adaptive strategies are valuable when activity limitations and participation restrictions interfere with occupational performance and when there are problems of safety during occupational performance (Moyers, 1999). While some authors suggest that rehabilitation is a group of techniques rather than a theoretical approach (Dutton, 1995), this intervention approach has been used extensively in occupational therapy since the beginning of the practice of the profession.

Rybski MF.
Kinesiology for Occupational Therapy, Second Edition (pp 355-428)
© 2012 SLACK Incorporated

Figure 12-1. Person-Environment-Occupation Analysis of Hygiene. (Adapted from Law, M., Cooper, B., Strong, S., Stewart, D., Rigby, P., & Letts, L. (1996). Person-Environment-Occupational (PEO) Model: A transactive approach to occupational performance. *Canadian Journal of Occupational Therapy, 63*(1), 9-23.)

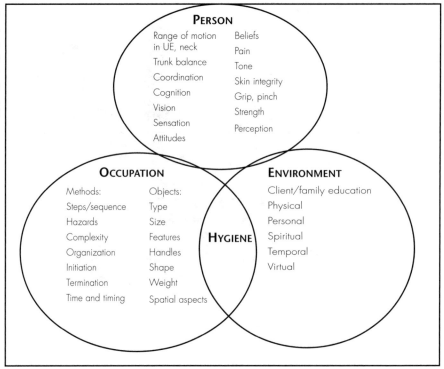

This approach capitalizes on the client's abilities. The focus of this approach is not on the client-level impairment but on the client's ability to participate in areas of occupation. For this reason, this approach is considered a top-down approach where performance is the first consideration, and factors that influence performance are secondary. Rather than trying to restore skills that are impaired, changes are made to the task (adapt the method or task object) or to the environment (environmental modification, caregiver education).

The rehabilitation approach is often used after restorative approaches in treatment of sensorimotor aspects has occurred or can be concurrently used with restorative approaches. Intervention often uses restorative methods initially, especially with neuromuscular diagnoses in which the first 4 to 6 weeks are seen as critical to recovery of function. Compensation and adaptation rehabilitation techniques are seen as abnormal movement patterns and so are viewed as conflicting with remediation efforts (Pal, 2003).

The rehabilitation intervention approach is also used with clients in whom there is a need for immediate success in occupational performance to sustain motivation for rehabilitation or if there are problems of safety during occupational performance (Holm et al., 1998; James, 2008).

The client must have an awareness of the problems interfering with performance and use this information to know how and when to adapt the task or environment or use new devices. The client must be able to make these accommodations in different situations and places and without therapist cues. This requires learning or relearning skills needed for engagement in areas of occupation.

Teaching and Learning

Occupational therapists teach clients, families, and caregivers. Clients and their families learn new ways of performing daily tasks or how to do common activities with new tools, or the client may need to relearn skills that have been lost.

Theories of Learning

How learning occurs and what factors are involved in learning varies based on the theory used to explain it. Behaviorist theory, developed by John B. Watson, includes theorists Edwin Ray Guthrie, Clark L. Hull, Ivan Pavlov, B. F. Skinner, and Edward Lee Thorndike. This model assumes that observable behavior is the focus of learning. It is a stimulus-response theory of learning where the environment shapes behavior (not factors within the individual). Reinforcement and the time between stimulus and response are the major ways that learning occurs. In this approach, it is important to structure the environment to control the learning that occurs (Merriam & Caffarella, 1999).

Cognitivist theory is helpful in that this approach has developed theories about transfer of knowledge (theorists: David Ausubel, Jerome Bruner, R. Gagne, Kurt Koffka, Wolfgang Kohler, Kurt Lewin, Jean Piaget). The focus is on memory and metacognition and how to acquire different types of knowledge, because the locus of control is within the individual learner. The learner's needs and learning style are taken into account; and perception, insight, and meaning are important parts of learning. Learning involves the reorganization of experiences in order to make sense of information from the environment (Merriam & Caffarella, 1999). The therapist using this approach would structure the content of learning activity to facilitate memory and transfer of knowledge.

Carl Rogers and Abraham Maslow considered the affective as well as cognitive dimensions of learning in the humanist theories. Client-centered therapy is often equated with the student-centered learning espoused by this theory. The focus of this approach is on the individual who has potential for growth and self-development. Behavior is not predetermined by either the environment or one's subconscious but is the consequence of choice. Humans can control their own destiny, and people are expected to assume responsibility for their own learning.

Social learning theories combine elements of the behaviorist and cognitivist approaches. People learn from observing others, and these observations take place in a social setting. So, learning involves both imitation and reinforcement (theorists: Albert Bandura and Julian Rotter). Learning is influenced by attention, retention, behavioral rehearsal, and motivation and involves both the individual and the environment. Learning occurs by the process of socialization, social roles, mentoring, and locus of control.

Constructivist theorists assert that learning is a process of constructing meaning, and this occurs at personal and social levels (theorists: John Dewey, Jean Lave, Jean Piaget, and Lev Vygotsky). Learning involves practice, and practice provides a "history" of the learning (Merriam & Caffarella, 1999). Experience provides a resource and stimulus for learning because, due to experiences, cognitive conflict or shared problems or tasks encourage learners to develop new knowledge schemes. Learning is a process of constructing meaning and is how people make sense of their experiences. Experiential learning and self-directed learning is how learning occurs. Reflective practice, associated with the constructivist theories, is postulated to be part of the mechanism in the progression from a novice to expert clinician.

Different theoretic conceptual understandings of learning can be used with different types of clients. A client with traumatic brain injury may require a very structured environment initially to decrease outside distractions to enable occupational performance, which would be consistent with the behaviorist approach. The meaning of the activity is important to all theories except the behaviorist approach, so occupational therapy's emphasis on meaningful and purposeful activity, the role of the environment, and the interaction of person, environment, and occupation seem well-suited to the use of these theories of learning.

Research about learning has debunked three common myths about learning. First, learners are not "blank slates" who are waiting to receive knowledge passively. Our clients come to us with experiences and knowledge. In order to gain new knowledge, the new knowledge must be used and manipulated in such a way that enables the new knowledge to fit with the old knowledge schemata. This requires motivation and practice. Learning takes work. Second, learning is not merely a behavioral process of stimulus-response. In much of the learning we do, there is not a "right" or "wrong" answer. Learning is not a simple process. Third, learning is not independent of the context in which it is learned. The learning process is deeply embedded in the setting, which provides critical cues for performance (Shotwell & Schell, 1999).

Application to Practice

When thinking about teaching new skills to your client to adapt or compensate for performance limitations, you need to consider the client's readiness for learning. What are the specific learning needs of this individual? What is this individual's focus of learning? How does this individual learn, and has this style changed? Consider the client's preferred modes of information processing when presenting new information.

Start by understanding the client's goals, because this enhances learning because it reflects a valued skill. How well does this individual currently perform the skill, and how confident is the client in his or her ability to change skill performance? Self-efficacy is positively linked to performance and persistence in task performance (Gage & Polatajko, 1994; James, 2008), so if the client feels unable to change, that may need to be addressed first. Depression can influence the individual's belief in self-efficacy, and this is an area that is more difficult for cognitively impaired clients or those experiencing intractable or chronic pain. Other issues affecting the teaching-learning process are language and literacy.

Feedback to the client about performance is another variable affecting learning. Extrinsic feedback, provided by the therapist, is useful until the learner understands the movement, skills, or strategy required (Flinn & Radomski, 2008). Providing feedback either as a summary of the performance or gradually by decreasing the timing of the feedback is more effective than immediate or constant feedback. "Bandwidth feedback" (Goodwin & Meeuwsen, 1995) is feedback that is only provided when a skill is not performed within an acceptable range. Intervention should focus on decreasing the amount of extrinsic feedback provided by the therapist and increasing the client's perception, understanding, and realization of his or her own performance and reliance on intrinsic feedback. Having the client perform the skill in the normal context and experience the consequences in a safe way is a way of receiving natural feedback.

Considering the stage of learning of the client may facilitate client-centered motor learning and what intervention will be most successful. Initially, the client may be in the contemplative stage (Fitts & Posner, 1967), in which there is an awareness of the problem. The next stage, determination, is when the client resolves to do something about the problem. For both of these stages, verbal intervention is appropriate, as collaboratively the therapist and client consider the components of the task, task sequence, and critical cues (Flinn & Radomski, 2008). Action is necessary for the next two stages. In the action stage of learning, the client works to address the problem. The final stage, maintenance, is the effort required to maintain the newly learned or relearned skill. To acquire the skill, the client does more than simple repetition; the client needs to be attentive and interested in the task.

Contextual characteristics include the client's length of stay, the environment (physical, social, cultural), and the relationship between context, person, and occupation. The length of stay may suggest different learning strategies that would be more successful given temporal restrictions. An understanding of the expected environment provides information about the critical aspects of the task and cultural meaning to the client. Structuring the environment for optimal learning would

Table 12-1

SUMMARY OF TEACHING STRATEGIES

1. Identify client needs, goals, and preferred learning styles

2. Determine potential barriers to learning

3. Evaluate current skills and potential barriers

4. Use a collaborative approach to enhance learner's participation, trust, and progression from extrinsic to intrinsic feedback

5. Individualize the learning process to the learner's capabilities, and provide the "just right" challenge

6. Provide opportunities for active learning and practice

7. Present learning in real contexts with common objects

8. Arrange practice environments to reflect skill objectives of automaticy, transfer of learning or generalization

9. Test the client's learning by requiring the task to be done independently or in the appropriate time and place

10. Collaboratively discuss progress, and revise learning plans with the client

include having the client perform tasks in a pleasant, authentic, and natural way (MacRae, 2010).

Use real task objects in real task situations (Ma, Trombly, & Robinson-Podolski, 1999; Wu, Trombly, & Lin, 1994). Tasks encountered in the environment must include the typical challenges in a graded way, so the client can verbalize steps, perform the task, and receive feedback about performance. Because you are collaborating with the client in shared goals, this provides a cohesive and safe environment for the client to learn new skills.

Task-specific outcomes are achieved when skills are acquired within the same context in which they will be used. Having a client practice hygiene skills in the bathroom in the morning is a task-specific skill if this is the way the client has previously done the task and will do it in the future. The client can perform this skill in a consistent environment and a consistent sequence to develop routines and habits that can be performed automatically.

Task skills can be learned by performing the skill in a consistent sequence but by varying the environment. This will enhance skill transfer to different contexts. Varying both the task and the environment provides the learner with general strategies that can be used in different situations. Generalization is often the intent of teaching adapted techniques to clients. Teaching a person with quadriplegia how to transfer from the mat to the bed can be generalized to transferring to and from other surfaces. Table 12-1 summarizes the teaching strategies presented.

Assumptions

One assumption of this frame of reference is that the ability to function is essential to well-being and that there are secondary benefits to be gained by improving performance despite physical, cognitive, psychological, or social dysfunctions (Turner, Foster, & Johnson, 1996). Humans are capable

of adapting to their limitations by learning new methods of doing activities, by responding to new teaching processes, and by using adapted objects and environments to their advantage. Clients are capable of capitalizing on their strengths as a healthy means of compensating for their limitations.

Through adaptation and compensation, clients can regain meaning and resumption of roles and a sense of purpose. Motivation for independence is based on the client's values, roles (volitional and habitual subsystems), and context (home environment, resources, financial status, family situation, age). The individual's involvement in choosing methods to improve daily life activities is also seen as an important part of the rehabilitation intervention approach. Use of compensatory strategies can facilitate integration of the client into the family, community, and previous life roles.

The reasoning used in this approach should reflect a top-down approach where successful accomplishment of activities of daily living (ADL), work tasks, and play and leisure pursuits are the focus of intervention rather than specific changes in anatomic, physiologic, or psychological attributes. In a top-down hierarchy, the intervention steps are to first identify the environmental demands and resources of the individual. What aspects of the context (including temporal aspects such as time and disability status) and environment (physical, social, and cultural aspects) are important variables to this person? It is imperative to ask the client and caregiver about the volitional and habitual subsystems. What does the client want to do? What does the client usually do? What does this person need to do to successfully engage in desired roles? How important are specific activities to this person? After gathering this information, an evaluation of the areas of occupation, including work, play, leisure, and self-care is completed to determine functional capabilities of the individual. Prerequisite skills that the patient lacks and task demands are compared with intervention aimed at matching the compensatory or adapted method with the prerequisite skill that the client lacks. Some authors indicate that, in the rehabilitation approach, the client needs to acquire the ability to set his or her own life's direction, control it, and take responsibility for it with the acquisition of an attitude of independence as a basic part of the theoretical base (American Occupational Therapy Association, 1995).

Function/Dysfunction Criteria

The theoretical base of the rehabilitation frame of reference is in medicine, education, and in the physical sciences. Function is the ability to maintain oneself and take care of others and the home; the ability to advance oneself through work, learning, and financial management; and to enhance the self through self-actualizing activities. This would necessitate certain levels of motor, sensory, cognitive, intrapersonal, and interpersonal subskills and role-relevant behaviors. Function is the ability to engage in constructive activity successfully along a continuum of independence.

Dysfunction is the loss of the ability to maintain and care for oneself and others and the home. It is the loss of the ability to advance oneself through work or learning, the loss of financial management, or the loss of participation in self-actualizing activities. While function occurs through normal development, dysfunction occurs through degenerative disorders, disease, or trauma (problems in structure or function). Table 12-2 provides a summary of the focus, assumptions, function/dysfunction, expected outcomes, and methods used in the rehabilitation intervention approach.

Strengths and Limitations

Some merits to this intervention approach are that it is widely documented and extensively used. The foundational concepts are easy to explain to the client and caregiver with intervention often visual, concrete, and with rapid results. A range of options are available and can be easily matched to the needs of the individual. There is no rigid sequencing of intervention steps, and the rehabilitation approach can be used to meet short-term needs as well as to compensate for permanent deficits.

Because the rehabilitation approach is associated with the medical model, there may be the tendency to be reductionistic and use recipe-like thinking rather than clinical reasoning to evaluate the range of intervention options. For example, one may be tempted to provide a long-handled sponge to all clients with total hip replacements without evaluating the actual need for the device or usability and acceptability of the device to that specific individual. The match between the person, occupation, and environment will prevent provision of equipment that is unwanted, unused, or unnecessary.

Often, by providing adapted equipment or by teaching a modified technique, intervention can be relatively inexpensive and rapid. While this is a definite advantage to this approach, it is truly advantageous only if the intervention provided is what the client really needs and not just done because it saved time and money or was at the expense of a more in-depth evaluation of the client and his or her unique situation.

Some clients may refuse to participate in compensatory or adapted techniques or use of special equipment or tools because the use of the device forces recognition of permanent loss of function. An understanding of the client's psychological adjustment to these losses is essential. Depression has been found to be a strong predictor of rehabilitation failure. An attitude of assertiveness is helpful in adjustment, as is a high frustration tolerance and the ability to understand and learn new ways to do things.

Being able to understand abstract concepts, such as joint protection or safety precautions, is important in independent living. Use Allen cognitive levels as a guide to determining cognition required for learning new methods or the use of new devices: Level 4, it is recommended that a caregiver be trained; at Level 5, clients are unable to implement abstract procedures; and at Level 6, it is important to stress problem-solving rather than attempting to train the client in every possible means of compensation via rote learning (Dutton, 1995). The strengths and limitations of the rehabilitation approach are shown in Table 12-3.

The role of the therapist is to work with the client to determine the best method for performing a task and to determine the best teaching process for the client. The therapist will design, construct, recommend, and order the adaptive equipment. Adaptations to the home, work site, or school will be collaboratively decided upon by the therapist, client, family, and others. Identification of community resources is also an important part of the rehabilitation process in returning the client to his or her home environment and in enabling the client to assume responsibility for his or her own health and well-being.

In Figure 12-2, five different strategies are proposed for intervention of loss of grasp needed for hygiene tasks. To decide what approach is most appropriate requires a clear analysis of hygiene tasks (occupation), of the client (person), and of the context in which the tasks usually occur (environment). Understanding the task characteristics (tools used, time requirements, space, etc) is necessary so that the individual's skills can be matched to the task. Table 12-4 provides a partial analysis of hygiene tasks in general and possible adaptations that might be possible. Not all of the suggestions apply to all hygiene tasks, and suggested strategies do not include all functional limitations.

An assessment of the client's skills would involve looking at body structures and functions to ascertain if limitations in client factors are influencing task performance. Safety considerations may be due to client factors (e.g., poor judgment, impulsiveness, or tremors), hazardous tools or methods (e.g., razors), or environmental constraints (e.g., inadequate space or lighting). A task as seemingly simple as brushing teeth or application of make-up, when viewed from the PEO perspective, requires much analysis of the component parts.

The five approaches include remediation if there is expectation of improvement in the limitation and if restoration of the anatomic, physiologic, or psychological attributes would result in improved occupational performance. If remediation is not possible, has plateaued in progression, or if more immediate results are desired, knowledge of available devices or adapted methods are considered as well as accessibility issues in the home or workplace. Caregiver education is the fifth approach represented in Figure 12-1 in which family members are taught ways to enhance client performance and, at times, may even perform the task in lieu of the client.

Table 12-2

SUMMARY OF REHABILITATION APPROACH

Focus	• Top-down approach
	• Evaluation of the performance areas of work, play, and self-care
	• Identify environmental demands and resources
	• Client strengths
	• Client's ability to participate in areas of occupation
	• Little or no expectation for change or improvement in impairments
	• Focus on context, activity demands, performance patterns
	• Need for immediate success in occupational performance to sustain motivation for rehabilitation
	• Activity limitations and participation restrictions interfere with occupational performance
	• Problems of safety during occupational performance
Assumptions	• The ability to function is essential to well-being
	• Motivation is based on the client's values, roles, and context
	• There are secondary benefits to improving performance
	• Humans are capable of adapting to their limitations and capitalizing on their strengths
	• Through adaptation and compensation, clients can regain meaning and resumption of roles and a sense of purpose
Function	• To maintain oneself, take care of others and the home
	• The ability to advance oneself through work, learning, financial management
	• To enhance the self through self-actualizing activities
Expected outcomes	• Learning new skills or use of devices to resume life roles
	• Maintaining or improving quality of life
	• Prevention of disability
	• Enhanced self-efficacy and satisfaction with performance
	• Improved adaptation to occupational challenges
Methods	Changing the task via
	• Adapted task methods or procedures
	• Adapting the task objects, adaptive devices, or orthotics
	Changing the context via
	• Environmental modification
	• Training the caregiver or family
	• Mobility adaptations
	• Disability prevention

Table 12-3

STRENGTHS AND LIMITATIONS OF THE REHABILITATION APPROACH

STRENGTHS	*LIMITATIONS*
Widely documented	Maybe the tendency to be reductionistic
Extensively used	Needs full analysis of need of device or method matched with person, environment, and occupation
Concepts are easy to explain	Not appropriate for client with impaired cognition
Intervention often visual, concrete	Seen as conflicting with other types of intervention
A range of options is available and can be easily matched to the needs of the individual	Need to understand what the changes mean to the client (psychologically, socially, culturally, etc)
Intervention results may be rapid	Transfer and generalization may not occur

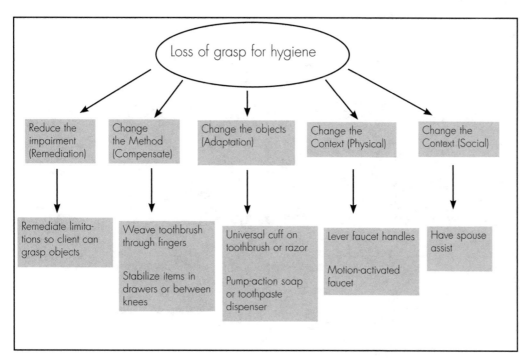

Figure 12-2. Approaches for Intervention of loss of grasp for hygiene tasks. (Adapted from Moyers, P. A. (1999). Guide to occupational therapy practice. *American Journal of Occupational Therapy, 53,* 247-322.)

Table 12-4

ANALYSIS OF HYGIENE TASKS AND RELATED REHABILITATION STRATEGIES

ACTIVITY LIMITATIONS UNABLE TO	IMPAIRMENT IN	REHABILITATION INTERVENTION: HYGIENE				
		SAFETY CONSIDERATIONS	ADAPTED TASK METHOD	ADAPTED TASK OBJECTS	ADAPTED CONTEXT	EDUCATION: CLIENT AND/ OR CAREGIVER WILL
• Reach face, all areas of head • Reach faucet • Pick up/hold/manipulate tools and utensils • Use both hands simultaneously • Attend to activity • Locate needed items • Use items appropriately • Reach sink, mirror • Initiate tasks • Complete tasks • Perform tasks due to pain	• Range of motion in UE, neck • Trunk balance • Incoordination • Cognition • Vision • Sensation • Perception • Weakness in UE, LE • Grasp, pinch • Pain • Abnormal tone • Joint precautions • Wounds or dressings	• Bathroom layout • Caregiver availability • Safety with equipment use • Compliance with precautions • History of falls • Judgment • Medication • Sensory status • Sharp and potentially dangerous tools and utensils	• For ↓ ROM • One arm assists the other if unilateral deficit • Use mouth to open containers • Stabilize items in drawers or between knees • Use nonskid surfaces • Grow a beard! • For ↓ strength and grasp • Use tenodesis grasp • Weave utensils through weak fingers	• For ↓ ROM • Splints for positioning hand • Pump-action containers of soap, toothpaste • Sponge with strap • Long-handled sponge with or without strap or universal cuff • Suction sponge • Built up handles • For ↓ strength and grasp • Splints for positioning hand • Pump-action containers of soap, toothpaste	• Store items in accessible locations • Store items in basket or plastic container to be transported easily • Remove cupboard doors for wheelchair access • Locate mirror for easy viewing • For ↓ strength and grasp • Mount hairdryer or use stand so no need to hold	• Demonstrate adapted task methods • Demonstrate use of adaptive equipment • Have caregiver assist with the tasks

(continued)

Table 12-4 (continued)

ANALYSIS OF HYGIENE TASKS AND RELATED REHABILITATION STRATEGIES

REHABILITATION INTERVENTION: HYGIENE

ACTIVITY LIMITATIONS UNABLE TO	IMPAIRMENT IN	SAFETY CONSIDERATIONS	ADAPTED TASK METHOD	ADAPTED TASK OBJECTS	ADAPTED CONTEXT	EDUCATION: CLIENT AND/ OR CAREGIVER WILL
			• Use of dynamic or tenodesis splints	• Soap-on-a-rope • Flip-top lipsticks • Electric toothbrush • Wash mitt • Adapted nail clipper • Adapted deodorant • Universal cuffs for utensils and tools	• Single lever faucets	
			• For ↓ stability • Stabilize items in drawers or between knees • Use nonskid surfaces • For ↓ coordination • Rest elbow on counter distal stabilization	• For ↓ coordination • Use weighted tools and utensils • Use electric razor, toothbrush	• For ↓ stability • Provide chair to sit to do tasks	

Adapted from Holm, M. B., Rogers, J. C., & James, A. B. (1998). Treatment of occupational performance areas. In M. E. Neistadt & E. B. Crepeau (Eds.), *Willard & Spackman's occupational therapy* (9th ed., pp. 323-363). Philadelphia, PA: Lippincott Williams & Wilkins; and James, A. B. (2008). Restoring the role of independent person. In M. V. Radomski & C. A. T. Latham (Eds.), *Occupational therapy for physical dysfunction* (6th ed., pp. 774-816). Philadelphia, PA: Wolters Kluwer/Lippincott Williams & Wilkins.

Table 12-5

ENERGY CONSERVATION PRINCIPLES

1. Respect pain
2. Rest frequently
3. Prioritize activities
4. Avoid sustained, static positions or isometric muscle contractions
5. Avoid stressful positions
6. Consider the environment

Intervention uses adaptive equipment and ways to adapt the environment to enable change. These adaptations replace normal function or compensate for abnormal function (AOTA, 1995). These rehabilitation methods may include the following:

- Changing the task via
  - Adapted task methods or procedures
  - Adapting the task objects, adaptive devices, or orthotic
- Changing the context via
  - Environmental modification
  - Training the caregiver or family

CHANGING THE TASK

Compensation may be made by altering the task method or procedures or by adapting the task objects in an attempt to match the task to the abilities of the client.

Adapting the Task Method

Altering the task method involves teaching the client new, more efficient, and more effective ways to complete a task using skills that closely correspond with the client's remaining capacities. Use of client skills and abilities in which there are no deficiencies is the aim in the altered task method. Because the client is altering a method and is doing a daily task in a new way, the client must have a sufficient capacity for new learning as well as adequate time for supervised practice of the new skill. Changing the way one puts on a shirt or the way one eats requires motivation by the client and/or caregiver to learn and apply the new methods to the task (Moyers, 1999).

Principle: Adapting task methods substitutes for loss of ROM, strength, use of one side of the body, limited vision, decreased endurance, inadequate stability, or to minimize the effects of spasticity and can simplify work and conserve energy.

Altering the task method may require modifications of techniques, learning new skills, or transferring existing skills to new situations. An example of a modifying technique is having a client with a paralyzed arm put the affected extremity into the shirt first and then dress the unaffected arm. A "trick" movement may be used by a client with C6 quadriplegia who extends and locks the elbow and externally rotates the shoulder as a compensatory movement for the loss of triceps function. Learning to use gravity as an assist rather than as a force to overcome is an adaptive technique, as is changing body mechanics or leverage used in activities so that they work to the client's advantage (e.g., letting gravity pull the forearm down to extend the elbow when there is paralysis of the triceps muscle).

Ways of altering task methods are numerous, flexible, and easily personalized. When the client is successful in performing the task, this enhances self-confidence and self-esteem. Modifying the procedure is usually cost-effective with long-term effectiveness. Another advantage to altering the task method is that the changes made in how the task is done are rarely visible, and so this may be more acceptable to clients (Dutton, 1995). While this is an advantage to some, to others, the lack of external prompts as reminders is a disadvantage. Adapting the method in which daily activities are done can result in successful performance of the task that is safe and efficient. When taught to sit while bathing, 82% of the clients reported that the adapted method permitted independent function that was safe (Chamberlain, Thornley, Stow, & Wright, 1981).

Disadvantages to changing the task by altering the procedure are that this requires new learning and a change of habit. The client and/or caregiver must be motivated and committed to making this change in daily activities. Dutton adds that some clients feel that an analysis of personal habits is seen as a form of criticism and that their privacy has been invaded (Dutton, 1995).

Some specific methods useful to both clinicians and a wide variety of clients are energy conservation, work simplification, and body mechanics. These guidelines are applicable to many different tasks and diagnoses.

Energy Conservation

Energy conservation principles are designed to be used to decrease fatigue and increase participation in activities. Decreased energy can occur due to chronic illness, such as arthritis, asthma, diabetes mellitus, Parkinson's disease, and low back pain. Strokes and spinal cord injuries can make doing daily activities more time-consuming and physically taxing, resulting in increased fatigue.

The major ideas of energy conservation are shown in Table 12-5. Having the client stop an activity when he or she is in pain is an important way to decrease fatigue. Not only might the pain restrict activity participation today but may also limit activities later in the day or the next day. Planning regular rest periods throughout the day also helps to conserve energy. Using a planner or calendar will help to avoid over-scheduling on any one day or week. Alternating difficult tasks with easier tasks helps to maintain vigor throughout the day. Avoiding positions that are particularly stressful, such as overhead postures or sustained positions, also helps to decrease fatigue.

Sitting requires less energy than standing, and if prolonged standing is necessary, shift positions frequently, rest one foot on a footstool or raised surface, and alternate foot placement every few minutes. Consider the environment as an energy source because loud, crowded, poorly lit, smoky rooms are factors that can lead to fatigue.

Bunyog and Griffith (2007) conducted workshops on energy conservation with people aged 66 to 93 years old and found that bathing and toileting activities were the most fatiguing, followed by dressing. After the workshop, the levels of fatigue decreased significantly in all areas, particularly grooming, washing dishes, and shopping.

There is strong evidence in the efficacy of energy conservation programs that looked at changes in fatigue, self-efficacy, and quality of life (Mathiowetz, Matuska, & Murphy, 2001; Vanage, Gilbertson, & Mathiowetz, 2003), reduction of energy expenditure (Ip, Woo, Yue, Kwan, Sum, & Kwok, 2006), and managing fatigue in individuals with multiple sclerosis (Mathiowetz, Finlayson, Matuska, Chen, & Luo, 2005).

Various methods of presenting the energy conservation principles to people have been used. Gerber and colleagues (1987) used a workbook to convey the information, resulting in behavior changes, decreased pain and fatigue, and increased participation and rest. Using phone conference calls did not seem to be an effective way to teach people about these principles (Finlayson, 2005), and the use of an educational-behavioral program did not result in significant differences between the experimental and standard groups, although further study was recommended (Furst, Gerber, Smith, & Fisher, 1987).

Work Simplification

Work simplification techniques are related to energy conservation, and the use of these strategies also results in decreased fatigue. Often, energy conservation and work simplification are used together, and the principles have some overlap. Related to both energy conservation and work simplification are joint protection techniques. These techniques can be considered adapted ways of doing activities and often are used to prevent disability.

Organizing storage by having similar items placed together eliminates unnecessary steps and movement to retrieve items from outlying areas. For example, have canisters of flour and sugar near the mixer in a baking center in the kitchen. Organize heavier items on lower shelves to minimize overhead lifting, which is often a safety issue as well. Keep items used frequently at waist height.

Planning ahead spreads out more strenuous tasks throughout the week so work can be easier and energy can be conserved. Big tasks can be broken down into smaller parts so sufficient time and energy can be devoted to the entire task over time. Making a grocery list, arranged according to item placement in the store, makes meal preparation easier because you will have all ingredients necessary with fewer trips to the store. Shopping when the stores are less crowded and when help is more available will help to avoid crowds and a rushed, noisy environment.

Having an easy flow of work simplifies work. An analogy of easy flow of work can be applied to how your automobile operates. When you drive your car on the highway at a constant rate, there is less wear and tear on the mechanical parts of your car, and you get better gas mileage. Compare this to driving in the city, where there is stop and go driving. This is harder on your car, and you get much less mileage per gallon of gas. A functional task can be done in a smooth flow of work by being organized, planning ahead, and having sufficient time to complete the task.

There are some tasks that the client may not even need to do. Elimination of tasks is a way to simplify work. While this is not always possible, minimizing steps to a task also can make the task easier. Purchasing permanent press shirts and wrinkle-free clothes eliminates ironing, and letting dishes air-dry or using a dishwasher eliminates steps by adapting how these tasks may be done. Leaving a tie knotted and then just pulling it over the head eliminates the need for tying, and using premeasured laundry detergent can eliminate steps and conserve energy.

The use of efficient methods also simplifies work. Sitting to iron or cut vegetables are ways to conserve energy while being more efficient. Use both arms whenever possible, and slide rather than lift objects. A wheeled cart can transport objects in the kitchen or laundry room, and using electric appliances is often more efficient. Making a bed can be done more efficiently by only going around the bed once. Start by smoothing the sheets and covers at the head of the bed on one side. Next, walk to the foot of the bed, and smooth covers there. Finally, walk to the other side of the bed, and smooth the covers toward the head of the bed on that side. Using a ping pong paddle or yardstick to tuck in the sheets can also be done to minimize reaching (Jacobs & Jacobs, 2009).

Of the methods included in energy conservation and work simplification programs, the strategies used most often were changing body position, planning for rest periods, adjusting priorities, and reduced frequency of or simplified work. Least-used strategies were changing the time of day to do an activity and use of adaptive equipment (Matuska, Mathiowetz, & Finlayson, 2007).

A list of suggested ways to conserve energy is presented in Table 12-6, and Table 12-7 lists energy-conservation principles.

The next sections will address adaptations made for clients with specific limitations in physical function. The adaptations presented can be generalized from the specific categories to similar client factor and performance skill deficits that may not have been presented. For example, in clients with weakness in all four extremities due to multiple sclerosis, some of the adaptations for the spinal cord injured (SCI) client may also apply. The specific examples presented are not meant to be an exhaustive list of all possible adaptations that can be made to the ways we do everyday activities. Whole texts are devoted to presenting numerous options for techniques if one has weakness on one side as occurs with a stroke or weakness or paralysis in all four extremities as may occur with a spinal cord injury. These examples are either commonly encountered or are considered representative of adaptations that are encountered in practice. The adaptations made for an individual client are based on the unique characteristics of that person, his or her goals, the expected environment, and the particular tasks that person needs to do. The adaptations

Table 12-6

WORK SIMPLIFICATION EXAMPLES

Self-care	Gather tasks before beginning
	Choose light, loose-fitting clothes or clothes with elastic waistbands
	Velcro closures instead of buttons, hooks, shoelaces
	Sit in firm, straight-backed chair for support when dressing
	Fasten a bra in front of the body then turn to back
	Wear slip-on shoes; elastic shoelaces; long-handled shoe horn
	Belts with magnetic fasteners
	Lightweight wallets or purses, and eliminate all unnecessary items
	Sit while grooming
	Built-up handles on grooming items for easier grip
	Shave with electric razor
	Hair done by professional or family member
	Shower caddy to contain items in shower
	Sit to dress/undress, shower, dry
	Long-handled sponge
	Terrycloth robe to eliminate/minimize thorough drying
Kitchen/meal preparation	Plan menus before shopping
	Plan shopping according to layout of store
	Shop when store is not busy
	Shop at store where employees will unload cart, bag groceries, carry to car
	Ask for help with heavy or high items
	Ask that groceries be bagged lightly so you can lift and carry
	Have family member sort and store groceries if possible
	Use wheeled cart to transport items
	Line up stored canned goods by similar types of items
	Plan menus that require short preparation time and little effort
	Sit when preparing meal
	Use convenient appliances, such as food processor, blender, mixer
	Slide items to transport or use wheeled cart
	Serve directly from baking dish or pan used to cook the food
	Arrange a buffet where guests can serve themselves
	Consider disposable dishes, napkins, silverware
Household/cleaning tasks	Ask family members to make the bed for you
	Allow space on both sides of bed to enable movement on all sides of bed
	Eliminate or reduce knickknacks to decrease dusting
	Use a feather duster and sit to dust when possible
	Use a long-handled dust pan

(continued)

Table 12-6 (continued)

WORK SIMPLIFICATION EXAMPLES

Household/cleaning tasks
Have a wastebasket in every room to eliminate trips
Place pail of water on dolly with wheels when mopping
Use a mop with squeeze control in the handle
Store all cleaning supplies together
Have a duplicate set of cleaning supplies on each floor
Ask for help with heavier cleaning

Laundry
Use paper towels and napkins to reduce amount of laundry
Frequent wash days to avoid large loads
Laundry facilities on main floor if possible
Use wheeled cart to transport clothes to laundry area
If laundry facilities are in the basement, use a laundry chute
Sit while sorting clothes at table
Transferring wet clothes from washer in small amounts at a time
Purchase a laundry basket on wheels
Use tongs or a reacher to remove articles
Consider taking heavy items (blankets, bedspreads) to a laundromat
Sit when removing items from dryer
Sit to fold and sort clean laundry
Avoid line drying, if possible
Purchase permanent-press or wrinkle-free clothing
Do not iron items than can "pass" (sleepwear, sheets, T-shirts)
Select lightweight iron and pad handle if needed
Consider purchase of hand-held steamer unit to rid wrinkles
Sit while ironing, and use an adjustable ironing board
Slide iron rather than lifting it on garment
Rack on wheels for hanger items
Iron in several short sessions rather than all at once

Yardwork
Take frequent rest breaks
Sit when working in garden
Long-handled gardening tools
Consider raised boxes for gardening
Ask for help if activities are too strenuous or cause pain
Take advantage of power tools and labor-saving devices
Ergonomic tools

Recreation
Use devices such as a cardholder (a clean, upturned hairbrush can also hold cards)
Use automatic card shuffler
Substitute weaving for knitting or crocheting
Avoid prolonged flexion of the fingers; let a hoop hold needlework

(continued)

Table 12-6 (continued)

Work Simplification Examples

Recreation

Use elastic scissors that remain open and require little pressure

Prop a book in a bookstand or place on pillow

Lay newspaper flat across a table to read

Use good posture when reading, playing cards, sewing

When fishing, use rod holder to free hands

Use golf cart to conserve energy for the game itself

Reprinted with permission from Jacobs, K., & Jacobs, L. (Eds.). (2009). *Quick reference dictionary for occupational therapy* (5th ed.). Thorofare, NJ: SLACK Incorporated.

Table 12-7

Work Simplification Principles

1. Organize storage
2. Plan ahead
3. Easy flow of work
4. Eliminate steps or tasks
5. Use efficient methods

---

are limited only by the collaborative creativity of the therapist-client relationship.

Adapted Task Methods: Range of Motion

Problems associated with decreased ROM involve being able to reach objects or body parts or being able to have sufficient ROM to grasp objects. For example, decreased shoulder ROM may result in the inability to reach back to put on a coat, and decreased grasp would prevent holding a hairbrush. Specific joint mobility restrictions may also apply, such as after joint replacement surgery.

Precautions for clients with total hip replacement are designed to reduce the possibility of hip dislocation during the first 10 to 14 days following surgery. While individual orthopedic surgeons may have slightly different preferences, the following are guidelines generally accepted for clients following total hip replacement surgery: do not bend trunk more than 90 degrees; do not reach past knees; do not cross ankles or legs when sitting, laying, or getting in/out bed; keep legs apart when lying in bed and toes pointed towards ceiling (Dutton, 1995). There are similar postsurgery restrictions for other joint replacements, which may vary from one facility to another. Activities such as dressing need to be modified to adhere to these precautions.

Dressing is often difficult due to fasteners (especially snaps and buttons). Garments with zippers and fasteners in the front are easier to don than pullover garments. Velcro can be used as a replacement for fasteners, and a loop of webbing can be sewn into socks to enable insertion of a thumb to pull up socks. When making these adaptations, the disadvantage is that this necessitates adaptation of many clothing items. Slip-on shoes or shoes with elastic laces are easier to don/doff for individuals with decreased ROM.

Grooming and toileting can be adapted for people with decreased ROM or grasp by wrapping toilet tissue around the hand so grasping is not necessary. An extra-wide base of support is a helpful position to assume to prevent slacks and underpants from dropping to the floor when readjusting clothes after toileting. Some clients opt to eliminate wearing underwear not only because of the difficulty of donning/doffing but also because of potential skin breakdown in areas that are insensate. An extra broad base of support while seated on the toilet may help to prevent imbalance or falls. Changes in grooming methods might be to simplify the hairstyle and make-up routines, use a beauty salon, or grow a beard (Holm et al., 1998). After a shower or bath, instead of drying off with a towel, have the client don a terrycloth bathrobe.

Child care may be more difficult with decreased ROM, and keeping the child in close proximity will facilitate tasks. Feeding a child in an infant seat or propped on pillows and using clothing with elastic and large closures are some considerations for child care.

Adapted Task Methods: Strength

Adapted procedures and methods that substitute for loss of strength reflect the principles of letting gravity assist, using the mechanical principles of levers, using increased friction to decrease power requirements, and using two hands (Trombly, 1995).

Some general principles of changes in methods due to decreased strength are to teach the client to use the muscles that are remaining to perform tasks. For example, use the strength in the arms and legs when getting on/off the toilet and to push up to get out of a chair. Put one hand on the counter and one on a thigh to push up to a standing position to help support the spine and body weight. Use one hand to support the body while standing at the sink and to avoid bending. In spinal cord injuries, there may be selective tightening for a hook grasp if there is extrinsic tightness, which is useful in hooking fingers on the edge of a transfer board or wedging or "weaving" utensils or tools between tight fingers (Dutton, 1995). Normally, when the wrist is actively extended, the fingers are passively flexed. If the fingers of a client with quadriplegia are allowed to develop some flexion tightness, then by using this tenodesis grasp, the client can pick up light objects and hook the wrist behind the wheelchair upright.

Lessen the force demands of activities to accommodate weaker muscles. Consider storing a 5-lb bag of flour in more than one canister so lifting the canister requires less strength. Use this same principle for items in all areas of the house (kitchen, closets, bathroom, garage, and basement) and for a variety of tasks (e.g., laundry detergent in one-load sizes).

Using both hands to do tasks also is a method to compensate for decreased strength. Also, it is easier to push than to pull objects because pushing uses the muscles in the legs and back and produces better posture.

Weakness can be restricted to only one part of the body (e.g., only one hand) or may affect all four extremities (e.g., spinal cord injuries). The suggestions presented are for those clients with more pervasive loss of strength but apply to clients with less severe losses. Adapted methods for clients with loss of strength on only one side of the body (e.g., stroke) are included in a later section.

The following descriptions of dressing and mobility describe how these activities could be done for a client with C6 spinal cord lesion. A client with a C6 level of spinal cord injury has weakness in wrist flexion, elbow extension, hand and grasp, and paralysis of trunk and lower extremity. Consider the consequences of these patterns of weakness on the ability to dress independently. First, the client will need to learn new ways of moving in the bed, coming to a seated position, and rolling from side to side to don pants. Lower extremity dressing will be done in bed to take advantage of gravity, use of two hands, and momentum. So bed mobility is a necessary prerequisite for lower extremity dressing in bed.

Bed Mobility

Many activities require adaptation in the way that they are done if there is weakness in all four extremities. Some clients will require assistance with bed mobility where the caregiver will roll the client's body at one time with the client assisting as possible (log roll). Bed mobility will require altered methods in order to roll from side to side and to come to a sitting position. Rolling is made easier if the hands can be clasped together and the legs crossed prior to beginning the roll. To return to supine from sidelying, extend the wrist, lock the elbow in extension, and force the left shoulder back toward the bed (Pierson, 1994).

Clients with weakness of all four extremities may need to hold onto a bed rail or use a rope ladder, the side of the bed, a trapeze bar, or the arm of a wheelchair in order to roll from side to side. To use these devices, the client needs good scapular, elbow, and shoulder strength. They may need to grasp or hook

their extended wrist or flexed elbow on these items if there is a decreased grasp.

Coming to a sitting position when there is weakness of all four extremities can be achieved by doing the following:

1. Rolling to one side (e.g., right)
2. Flinging the top arm (left) backward to rest the elbow on the bed, having momentum assist with the body motion
3. Rolling onto the left elbow, and quickly flinging the right arm back to rest the palm of the hand or fist on the bed
4. Roll to the right, and quickly move the left arm to rest the hand on the bed
5. Having achieved a semi-sitting position resting on both hands, with the elbows extended or locked, the client "walks" his or her hand forward to come to a forward-leaning position of greater than 90 degrees of hip flexion to maintain balance

Clients at the C6 spinal cord level can be taught to use their scapular muscles and external rotators to perform the motions of shoulder external rotation and scapular elevation and depression to compensate for the lack of triceps so that the client can move forward in the bed. Using shoulder muscles to substitute for triceps paralysis will enable the client to engage in activities, such as moving up to sit or transfer to a wheelchair.

An alternative way to come to a sitting position might be for the client to do the following:

1. Place hands under hips or in pockets for stabilization.
2. Flex the neck and elbows until weight is on the elbows.
3. Shift elbows backward, one at a time, until weight is on the forearms.
4. Fling one arm backward, laterally rotating and extending the shoulder and elbow until the heel of the hand contacts the mattress (interphalangeal joints should be flexed).
5. Come to sitting by shifting weight onto the extended arm, and repeat this with the other arm.
6. Gain balance with weight on both extended arms.
7. Walk hands toward hips.
8. Using a rope ladder, overhead loops, or a trapeze can also assist with coming to a sitting position.

Once in a seated position, the client needs to learn how to maintain balance while in long sitting position on the bed. Bedrails, an overhead trapeze, and learning to balance on an extended arm will help to compensate for the lack of trunk musculature. The client needs to learn how to regain the upright position after leaning forward, often by leaning over and using the bedrails, flexing the arm, and pulling back up into a seated position. These are additional prerequisite skills to be learned before lower extremity dressing. Adjustable, electric hospital beds make the task of bed mobility easier for clients with weakness of all four extremities.

Lower Extremity Dressing

Minimum criteria for lower extremity dressing would be the following:

1. Muscle strength fair to good in pectorals, rhomboids, supinators, and radial wrist extensors

2. ROM: Knee flexion and extension 0 degrees to 120 degrees to permit sitting with legs extended; hip flexion 0 degrees to 110 degrees

3. Body control, such as the ability to transfer from bed to wheelchair with minimum assistance, the ability to roll from side to side, and balance when sidelying (Pierson, 1994)

In a seated position, the client will position the pants using adaptive equipment or by placing the pants between both hands and extending the wrists (motions available at C6 level and below) to hold and pull the pants into position. By hooking an extended wrist under the distal thigh, the leg can be pulled over, or momentum can carry the legs. The foot is inserted into the pants, and the pants are pulled to knee height. Repeat with the second foot so that both pants legs are at knee height or higher. An alternative method is to have the client cross one leg over the other, insert the foot into pants, and return the leg to the extended position. Repeat on the other side. At this point, the client will need to return to supine.

The client will insert the wrist into a pocket, under the waistband to pull the pants up, rolling from one side and then to the other to pull the pants up over the buttocks. Momentum can be gained by rolling back and forth using proximal musculature and head when trunk muscles are affected. Once the pants are pulled up to the waist, adaptations to clothing (such as Velcro) or adapted devices (zipper pull or string/embroidery floss on zipper) will complete the lower extremity dressing.

Different types of clothing may be more difficult to don than others. While pants with elastic in the cuffs and waistband (such as sweat pants) initially seem easier to put on, sometimes pulling them up over the buttocks proves to be discouraging because the elastic will pull the pants back down if not pulled up far enough. Another disadvantage is that there may not be pockets in which to insert the wrists when pulling the pants up, but the waistband is usually sufficient.

Stiff clothing like blue jeans has some unique challenges. It is difficult to pull jeans up against cotton sheets. Silky material slides easier on sheets. The most difficult aspect about jeans is fastening the metal button. The button is usually larger than most button hook devices, so many clients opt to leave it unbuttoned or remove it completely. If the skin is insensate, the double seams on jeans may cause skin breakdown if the pants are too tight. Clothing that is slightly larger is easier to don and is less constrictive.

Many people choose not to continue wearing undershorts or underwear for several reasons. One, this would be one more piece of clothing to put on in the morning. Dressing is very energy-consuming, and if it takes more than 1 hour to complete, it is not considered functional (James, 2008). A second reason for not wearing undershorts or panties is that this is an additional potential source of skin breakdown. Clients should be cautioned to wear larger clothing that is loose-fitting and that does not bind or impinge the skin. A final reason might be that the undershorts may interfere with the client's bowel and bladder programs.

Upper Extremity Dressing

Adapted techniques for clients with weakness in all four extremities require that certain levels of performance be met as prerequisites for dressing. Following are the minimum criteria for upper extremity dressing:

1. Neck stability is medically cleared.

2. Muscle strength is fair to good in shoulders (deltoid, trapezius, serratus anterior, rotators) and elbows (biceps).

3. Shoulder flexion and abduction 0 degrees to 90 degrees, external and internal shoulder rotation 0 degrees to 30 degrees, and elbow extension and flexion 15 degrees to 140 degrees.

4. Sitting tolerance and balance in bed and/or wheelchair achieved with assistance of bed side rails or wheelchair safety belts (Colenbrander & Fletcher, 1995).

Upper extremity dressing can be done seated in bed or in the wheelchair or chair.

Clients at the C6 spinal cord level lack triceps function and are unable to extend the elbow. The functional implications are that, whenever the arm is brought to heights above the shoulder, the elbow will flex due to the pull of gravity. This use of gravity as an assist is very helpful when dressing and is a good principle of intervention.

For a client with shoulder and scapula muscles, elbow flexion, and wrist extension, the process of putting on a cardigan or button-up shirt is similar to the customary method of dressing. Put one arm into the shirt, using wrist extension under the material or in the armhole to pull it up over the elbow and up to the shoulder.

A very useful muscle action added at the C6 level is that of wrist extension due to the innervation of extensor carpi radialis longus and brevis muscles (radial wrist extensors). The addition of wrist extension is very helpful when trying to dress because the client can use an extended wrist in sleeves and pant legs to assist with putting on the clothing. Two wrists extended and placed with palms together can hold clothing to pull it on or to straighten a shirt or blouse.

Licking the heel of the hand helps to provide friction to aid in pushing the material toward the shoulder. The client can reach around for the armhole and insert the other arm. Balance can be maintained in the wheelchair by hooking the elbow around the wheelchair upright push handle. This positioning permits increased reach without loss of balance. Using both wrists in an extended position, hold the collar and adjust the shirt so the shirt fronts are aligned. Use adapted equipment (buttonhook device) to button the shirt.

Putting on a bra is much the same procedure as putting on a cardigan as described above. It is easier to hook the bra in the front with the bra at waist level and then pull the straps up over the shoulder. Adaptations such as sewn loops or Velcro will enable independent fastening by the client.

An alternative method to putting on a cardigan would be to place the shirt on the lap, collar facing toward the legs. Place one hand into the sleeve, shaking the sleeve to help move it along the arm. Push the heel of the hand along the outside material of the sleeve to push the sleeve up to the elbows. Clients often use their teeth to assist with this, and licking the heel of the hand helps, too, by providing a friction surface. At the axillary border of the sleeve (where the sleeve is sewn to the body of the shirt), have the client extend the wrist and pull the sleeve up over the shoulder. Putting the shirt material between the heels of each extended wrist also helps in pulling the shirt front down. Have the client hook the arm with the shirt sleeve pulled up to the shoulder over the upright of the wheelchair for balance. The client can then lean forward and insert the other arm into the remaining armhole. Pull the arm through the sleeve. Using both extended wrists, straighten the shirt by putting shirt material between the heels of both hands. A summary of adapted dressing methods is shown in Table 12-8.

There are some contraindications relative to UE and LE dressing. Decreased breathing capacity and decreased vital capacity (below 50%) may prevent individuals from performing lower extremity dressing activities (may be able to do upper extremity dressing). Ischial tuberosity decubiti ulcers or high frequency or tendency for skin breakdown during rolling and transfers may be a contraindication for lower extremity dressing. If the client experiences pain in the neck or trunk, then dressing activities should be discontinued (Pierson, 1994). If the client is continually resistant to dressing practice, you need to ascertain if this is a valued goal for this person. If not, then a caregiver or aide needs to learn the most efficient ways to assist with dressing.

Transfer: Dependent Transfers

Prior to beginning any type of transfer, it is important to know the client's strengths and limitations in all areas of function, including physical, cognitive, visual, perceptual, and levels of assistance required with current mobility. Be aware of medical precautions. Be able to identify any line or tubes that may be attached to the person (e.g., IV lines, etc) and be knowledgeable about the purpose of the line and precautions related to movement.

Anticipate any safety issues in the environment, and ensure adequate space for the transfer unencumbered by electric cords, excessive furniture, or equipment. Preview the transfer to preplan the transfer set up, being sure that transfer surfaces are level and even. Explain these planning steps to the client and caregiver so that generalization of learning can occur and transfers can be safely executed in a variety of contexts.

You also need to be aware of your own strengths and limitations and freely ask for help with transfers when needed. Use proper body mechanics when assisting with transfers, and train caregivers to use these as well.

Consideration of the person, the environment, and the task is given prior to the transfer. Activity demands include the objects that will be used in the transfer (such as a sliding board and transfer belt), the space demands, the steps involved in the process (and the cognitive, visual, perceptual skills associated with implementation), required actions (including safety aspects such as locking the wheelchair prior to transfer), and physical requirements and skill prerequisites (strength, ROM, bed mobility, etc) (Merano & Latella, 2008).

There is a range of types of transfers that can be done with clients ranging from dependent transfers (where the client is moved by equipment or one or more people) to independent transfers (client requires neither cues nor assistance). Sliding board transfers are for clients who have weakness and cannot bear weight on the lower extremities. Pivot transfers are for people who can stand and bear some weight on their legs. Clients may need some assistance with pivot or sliding board transfers, or they may be independent.

Levels of assistance are defined to describe the skill of the client and assistance needed. If the client is able to complete

the task including setup with or without adaptive equipment, then the client is independent. Standby assistance is when assistance of one other person is needed to perform activity. Independent with setup is sometimes used to indicate that someone is independent once someone sets up the activity (gets the transfer board, gets grooming supplies, etc). If the client does not require physical contact but does need cueing or coaxing or cannot be left alone due to cognitive deficits, poor balance, or other safety concerns, the client requires supervision. Minimum assistance describes a task in which the client requires assistance with only 25% of the task performance. Similarly, moderate assistance is a task requiring 25% to 49%, and maximum assistance requires 50% to 75% assistance. Total assistance or dependence is when the client is unable to perform any part of the task (75% to 100%). Modified independence is a term that is used to describe a client who is able to perform the task but requires adaptive equipment or an assistive device or takes more than a reasonable amount of time or if the activity involves safety considerations. Contact guard is another term used that denotes that the therapist has a hand on the client at all times (Matthews & Jabri, 2001).

Figure 12-3 provides a flow chart to aid in deciding what type of transfer is appropriate for a particular client.

If the client is unable to assist with the transfer, then a dependent transfer will be done that requires the assistance of one or more people to move the client from one surface to another. One-person transfers can be done with clients who need assistance as well as those with and without the ability to flex at the hip (Table 12-9).

To help a person roll from side to side or to transfer a recumbent person, a draw-sheet, mattress pad, or bed liner can be used, which may also help with transfers. The sheet or pad is rolled and grasped close to the client. One person assists at the head and another one at the hips on one side of the client, and another person assists at the hips on the other side, opposite to the movement. If the client has extraneous movements that cannot be controlled, then wrap the sheet around the patient to control the movements. At a given signal, all three assistants simultaneously lift and move the client toward the surface (Minor & Minor, 2010).

An alternative for dependent clients would be use of a hydraulic lift (i.e., Hoyer lift), which is a good choice for very large or dependent clients and/or small caregivers. Other mechanically assisted transfers have been made from adapted battery-operated winches or garage door openers.

Transfers: Sliding Board

To transfer from the bed or mat to a wheelchair requires the skill of coming to a seated position and requires that the wheelchair be located near the bed with the armrest removed on the side nearest to the bed. One arm is positioned behind to maintain balance, and the other arm is positioned underneath the leg closest to the wheelchair, sliding the leg toward the wheelchair. Once near the wheelchair, position one arm on the wheelchair armrest or seat, and pull the leg toward the chair. Once both legs are on the wheelchair footrests, the client can prepare for a sliding board transfer.

A functional consequence of the loss of triceps strength is that the client will not be able to transfer using a sliding board without learning compensatory movements. Depression transfers can be successfully performed by having the client

---

Table 12-8

ADAPTED DRESSING METHODS: BILATERAL UPPER EXTREMITY WEAKNESS

1. Putting on a cardigan garment.

    The method may be adapted for jackets, blouses, sweaters, shirts, and top portion of dresses that open down the front.

    a.   Patient is sitting in wheelchair. Position shirt on lap with back of shirt up and collar toward knees. The label of the shirt is facing down.

    b.   Put arms under shirt back starting at shirt tail and into sleeve starting at armhole and working toward cuff. Push shirt past elbows.

    c.   Using wrist extension, hook hands under shirt back and gather up material.

    d.   Using shoulder abduction, scapular abduction and adduction, elbow flexion, and slight neck flexion, pass shirt over head.

    e.   By relaxing wrist and shoulders, and with the aid of gravity, the hands may be removed from shirt back and the arms are now completely through the sleeves. Most of the material of the shirt is gathered up at back of patient's neck across shoulders and underarms.

    f.   Shirt is worked into place over shoulders and trunk by alternately shrugging shoulders, leaning forward, with aid of wheelchair arms for balance if necessary, and using elbow flexion and wrist extension.

    g.   Close shirt using buttons, snaps, or Velcro. If the shirt has not been buttoned previously, use a button hook, starting with bottom button, which is easier to see.

    Exceptions to this procedure would be as follows:

    a.   Arrange shirt on table preparatory to putting on.

    b.   When trunk stability is a problem, support elbows on table to assist in flipping shirt over head.

2. Removing a cardigan garment.

    a.   Patient is in wheelchair. Unbutton only the necessary buttons. Use a hook, if necessary.

    b.   Push one shoulder of cardigan at a time off shoulder. Elevate and depress shoulders, rotate trunk, and use gravity so cardigan will slip down arms as far as possible. Use thumbs alternately in armholes to slip sleeves farther down arms.

    c.   Hold one cardigan cuff with opposite thumb and flex elbow to pull arm out of garment. Repeat for other arm. The thumb is used as a "hook" in this step.

Reprinted with permission from Palmer, M. L., & Toms, J. (1992). *Manual for functional training.* Philadelphia: F. A. Davis Co.

---

externally rotate the shoulder while locking the elbows as the weight of the body is shifted to that side. By depressing the scapula to shift the weight off the elbow, the client can inch forward on the sliding board. This same technique can be used in rolling onto the side and pushing up on the elbow during bed mobility activities and for push-ups in the wheelchair for pressure relief.

Getting from a bed into a wheelchair will require a transfer from one surface to another. Use of a transfer or sliding board is the most common transfer type for patients with loss of strength in all four extremities and for clients with C6

injuries and below. In many cases, the client can manipulate the wheelchair parts and position the sliding board to become independent in the transfer. The sliding board acts as a bridge between two surfaces. Because clients can compensate for the lack of triceps by using shoulder external rotation and scapular elevation and depression, a sliding board transfer is also known as a depression transfer.

The use of a sliding board in a depression transfer is a specific example of using an adaptive device that compensates for loss of lower extremity function. Transfers are also adapted methods to achieve mobility. Impediments to successful sliding

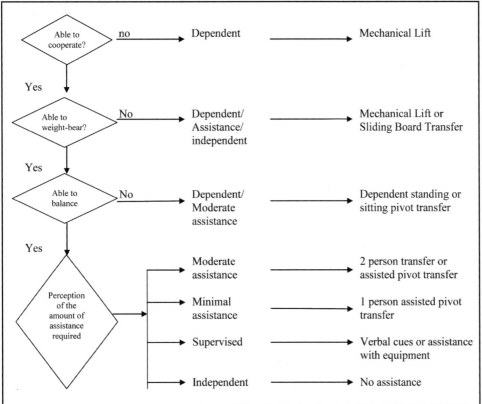

Figure 12-3. Flow chart for transfer decisions. (Adapted from Minor, M. A. D., & Minor, S. D. (2010). *Patient Care Skills.* Upper Saddle River, N.J.: Pearson.)

Table 12-9

ONE-PERSON TRANSFER TECHNIQUES

ONE PERSON TECHNIQUE (CLIENTS WITHOUT LIMITED HIP FLEXION)

- Stand in front of client.
- Place client in forward flexed position with chest lying on the thighs.
- Shifts body weight over knees and ankles rather than on buttocks so allows buttocks to be more easily moved.
- Helper can control movement.

ONE PERSON TECHNIQUE (CLIENTS WITH LIMITED HIP FLEXION)

- Slide buttocks toward edge of wheelchair.
- Place client's knees between the helper's knees.
- Rock client forward slightly and simultaneously pull client's transfer belt and rotate hips to surface to be moved to.

---

board transfers would be poor trunk balance that cannot be compensated, excessive spasticity, excessive body weight, joint tightness, and cognitive deficits.

The choice of a transfer board is dependent upon how much space is available for the transfer, the type of transfer surfaces, and how much strength the client has. Plastic boards are lightweight and good for tub and car transfers. Wooden boards are also available but are heavier to use. Transfer boards can come with cut out areas for the hand to be inserted to help position the board even without grasp, using the hand as a whole to move the board. Webbing straps can be attached to assist with pulling the transfer board out from wheelchair pouches or from storage.

Sliding board transfers can be done independently by the client or with assistance, if needed. Assistance can be provided by holding the client around the ribs, waist, or waistband or by using a transfer belt and maneuvering the client's legs, if necessary. Transfer surfaces optimally should be at the same level for an even transfer.

Positioning the client prior to transfer requires that the trunk be aligned and upright and that there is slight anterior pelvic tilt. The client should be able to shift weight and maneuver the legs without losing balance. If the client is not able to attain these skills, then a caregiver needs to be trained, and the client needs to learn how to tell others the ways in which help is needed. Table 12-10 lists the steps of a depression transfer.

Adapting Task Method: Stability and Coordination

Adapting the task method or procedure to provide stability due to ataxia or incoordination involves using stabilization to counteract the tremors or lack of controlled movement. Teaching the client to stabilize objects being used and to position the body in as stable a position as possible are compensatory ways of providing stability. Ways to position the body may include sitting when possible, bearing weight on the part, and holding the arms close to the body. It is helpful to stabilize proximal parts so the need to control body movement is reduced to just the distal body parts. Stabilizing items in drawers or between the knees if seated makes opening containers easier.

Using larger and less precise fasteners, tools, and objects also helps decrease frustration due to incoordination. For example, using roll-on deodorant is easier and safer for the person with decreased coordination and control. Increasing friction can add stability and can be as simple as placing an object on a wet cloth or towel, or using a nonskid mat. As with other limitations, simplifying tasks (hairstyle, omitting steps, less make-up) are also options that can be used with clients with incoordination.

Dutton recommends using a two-handed proprioceptive neuromuscular facilitation (PNF) technique called "chop and lift" in daily activities, such as getting a glass from a cupboard or washing one's face. Another PNF technique uses surface contact to increase friction and decrease instability. In this case, sliding the hand along the table towards the glass would be a more steady position for the arm. She suggests moving and stopping in the course of an action to improve controlled movement, such as resting on every shelf of a cabinet to lower one's hand (Dutton, 1995).

Adapting Task Methods: Spasticity

While control or inhibition of spasticity is not a remediation goal of either the biomechanical or an adaptation or compensation technique of the rehabilitation intervention approach, some of the techniques advocated by theorists from the neurodevelopmental (NDT)/sensorimotor frame of refer-

ence are useful in controlling spasticity or at least in minimizing the effect spasticity has on the performance of daily tasks. Distal key points of control (NDT technique) can be used with clients who have had a stroke where it is recommended that the unaffected leg be placed under the spastic leg during bed mobility. Another technique is to clasp the hands in reaching activities and transfers. Using proximal points of control, another NDT method, is seen if a client is instructed to lean forward to dangle a spastic arm to use the weight of that arm to protract the scapula and extend the elbow while putting on a shirt (Dutton, 1995). Dutton further recommends using placing, lowering, and weight-bearing (NDT techniques) to control spasticity. An example of this might be seen when the client lowers a spastic arm to the table and uses the arm to hold down a piece of paper (Dutton, 1995).

Adapting Task Methods: Vision

Clients with limited vision use organization as a primary method of compensating. If objects are organized in a specific way, in a particular place, and are routinely replaced after use, then the client with limited vision will always know where to find the object. Instructing caregivers and family members to close drawers and open doors is important to the visually impaired person.

Clothes can be organized by color in a closet. Pinning like items together after wearing them, but before laundering, will eliminate the need for visually sorting after washing and drying. By folding money differently for each denomination, the person with limited vision will be able to differentiate paper money. Food is cut by finding the edge of the food with the fork, moving the fork a bite-sized amount onto the meat, and then cutting, keeping the knife in contact with the food. Pouring liquid is based on the weight of the cup when it is full.

Adapting Task Methods: Cognition

Clear, simple directions presented in a consistent, relaxed manner will help the cognitively impaired client with routines that are reinforced. Each step should be explained, and checklists can be provided to remind the client of necessary steps of the task. There can be checklists for toileting (reminding to flush, wash hands), dressing (bra first, followed by blouse), or cooking tasks (get pan, fill with 1 cup water, etc.). Use of clippers rather than scissors to trim nails is suggested, and awareness of object use and safety is always the primary concern.

Changing eating methods for cognitively impaired clients or those with dysphasia might include reminders via an alarm to eat. Teaching the client to eat only small amounts of food with small bites at one time with reminders to swallow after chewing each bite may also be necessary. Often, the use of soft foods and avoidance of sticky foods makes chewing and swallowing easier (Holm et al., 1998; James, 2008).

Adapting Task Methods: Endurance

Many clients experience decreased endurance, and they need to frequently schedule rests during the day. Energy conservation and work simplification principles are very applicable to those with cardiovascular and respiratory dysfunction, spinal cord injuries, and rheumatoid arthritis (see Tables 12-5, 12-6, and 12-7).

Table 12-10

TECHNIQUE FOR SLIDING BOARD TRANSFER

- Use a wheelchair with removable armrests and footrests
- Provide level transfer surfaces
- Position wheelchair at slight angle to transfer surface
- Lock wheelchair brakes
- Move footrest/leg rest out of the way on the slide towards the destination surface; remove armrest on the destination side
- Client shifts body weight (or is assisted) to the side opposite the direction of movement and places approximately 25% of the board under the buttocks
- Client places one hand on the board and the other on the wheelchair seat and leans forward
- Client then moves the upper body weight in the direction opposite to that in which he or she is going
- Client scoots along the board by flexing/extending elbow or by externally rotating the humerus and locking the elbows to compensate for lack of triceps
- The caregiver may assist with helping the client move along the sliding board by placing his or her hands on the client's hips, waistband, or transfer belt. The caregiver assists by ensuring balance as needed and by managing the client's legs
- After the transfer, the wheelchair armrest and footrests are repositioned

Many of the earlier suggestions for decreased strength and decreased ROM also apply to those clients with poor endurance. Stabilizing items so that they do not need to be held (via nonslip materials or in drawers or between the knees) and task simplification are useful. Dressing in bed may conserve energy, and maintaining good postural alignment throughout the day and during tasks will facilitate task completion.

Adapting Task Methods: Use of One Side of the Body

Consider a client who has paralysis due to a stroke. The limitations will be primarily limited to one side of the body, including the arm, leg, and trunk muscles. There may be tone limitations ranging from total flaccidity to spasticity; there may be no volitional movement or a gradual return of function beginning at the shoulder and scapula and ending with grasp

and release. Greater volitional control and less tone problems will enhance task performance.

Bed Mobility

Clients with loss on one side of the body will be able to roll toward the affected side but will have greater difficulty rolling toward the unaffected side. To roll to the unaffected side from supine, the client places the unaffected foot under the affected leg. Slide the legs to the edge of the bed. Using the unaffected arm, carry the affected arm across the body, and then he or she pulls himself or herself over onto the unaffected side by holding the side of the bed. An alternative method advocated by Bobath is to have the client clasp both hands together with the thumb of the affected hand above the sound one, elbows extended, shoulder flexed above 90 degrees. The client then swings the clasped arms side to side to build momentum, which will carry him or her onto his or her side.

Either method can be carried to the next step or that of coming to a sitting position. Once rolled onto the side, with the unaffected arm, push against the bed while swinging legs over the side of the bed to come to a sitting position. An alternative to crossing the affected leg over the unaffected is to bring the unaffected leg over the edge of the bed while simultaneously pushing against the bed with the unaffected arm. The affected leg will follow, and the client will be in a sitting position. In either case, it is important to roll all of the way onto the side before pushing to sit up, as this will help to position the arm to push up and will create less strain on the back. Once positioned at the side of the bed, the client can prepare to pivot transfer to a wheelchair or use a cane or walker for ambulation.

Transfer: Pivot Transfer

A pivot transfer is a common transfer type for clients with loss of one side of the body. Because the patient can partially weight-bear on one lower extremity, this transfer is a good choice and can be done independently or with assistance. A modified pivot transfer, called a dependent standing or dependent sitting pivot transfer, involves the client remaining in a semi-crouched position, with body over legs, as the therapist or caregiver pivots the buttocks from one surface to the other. If the client is unsafe with a pivot transfer or is unable to perform the transfer even with assistance, sliding board transfers would be the next viable transfer option.

Pivot transfers generally are set up to move the person in a 90-degree angle, although some pivot transfers are in a 180-degree arc if in restricted spaces (like a bathroom). Table 12-11 lists the steps in a pivot transfer. Have the client assist with the transfer as much as possible. Initially, clients are taught to transfer to the uninvolved side ("strong" side), and eventually teaching will involve transfers to the involved side, too, for greater independence. For the client to be considered independent in transfers, he or she should be able to demonstrate adequate safety awareness, have independence in bed mobility, maintain sitting trunk balance, and be able to follow simple directions (Palmer & Toms, 1992). Table 12-12 can be used as a checklist to consider all of the variables related to a safe, efficient transfer.

Dressing

Dressing for a client with loss of function on one side of the body generally follows the pattern that the affected extremity is dressed first and undressed last. In putting on a shirt, the overhead method is seen as less confusing for clients with sensory and perceptual impairments. This overhead method, however, is cumbersome for dresses and not possible for use with coats. Both methods are described in Table 12-13 as are methods for donning trousers, shoes, and socks. The table describes how a person with a left cerebral vascular accident with resultant right-sided weakness or limitation would dress. One-handed shoe tying or slip-on shoes are examples of adaptations for footwear.

Additional Suggestions

Different ways of performing the everyday activities of cutting meat, opening packages, and brushing teeth can be accomplished with little or no extra equipment or devices.

Taping a nail file onto a table for filing nails or placing a jar inside a drawer to stabilize it are two alternative methods for everyday tasks. Using the affected hand to hold or be placed on objects can also be helpful. For example, when writing, the affected hand can be placed on the paper to hold it during the writing process. After toileting, use a wide base of support while standing to pull up pants to prevent pants from sliding down or put the affected hand in the pants pocket.

Clients who have experienced a loss of ROM and strength on one side of the body (e.g., stroke) can use the unaffected arm and leg to independently propel a wheelchair. Alternative ways of transferring, dressing, and moving from one place to another are necessary when the client is able to move only one side of the body.

Child care by one who has one functional upper extremity or impairment of one side of the body would be facilitated by propping an infant on pillows for ease or placing the child in an infant seat. If the client can get up off the floor, dressing the child may be easier done seated on the floor. A Velcro strap is helpful on a dressing table to secure the baby during clothing changes, and suction-based bath seats are useful for bathing.

A final suggestion about adapting methods for all tasks is to determine the tasks that the client is able to do and wants to do and then decide which tasks can be delegated to others.

Adapting Task Methods: Functional Mobility

Being in a wheelchair necessitates learning how to get around in the environment in new ways. Even going through a door in a wheelchair requires a new method. Table 12-14 lists the steps for going through a door with varying levels of strength and functional capacities.

Table 12-15 provides a sample goal statement, method, and the principle behind why you are changing the task. Examples are provided for each goal. The goal for a client who is expected to become independent in grooming would read: "Client will be independent in all grooming tasks using adapted task methods to substitute for loss of ROM in the right upper extremity in 3 days." An alternative way to state this might be, "Using adapted methods to compensate for the loss of ROM in the right upper extremity, the client will be independent in grooming tasks in 3 days."

Adapting Task Objects, Adaptive Devices, and Orthotics

Adapting the task object involves changing or substituting the objects or tools used in activities. Adaptive equipment spans the continuum from low-tech buttonhooks to complex computer equipment and sophisticated environmental control units. Adaptive equipment can be used to enable performance, to compensate for lost function, and to aid in efficient and safe performance of activities.

Provision of adaptive equipment seems deceptively simple, but careful consideration of devices and client need is necessary to avoid costly errors in terms of equipment purchased and devices that are not used or are used incorrectly. Some devices, used incorrectly, may result in additional performance impairments or may prevent proper use of the device (Marrelli & Krulish, 1999). A proper fit between the needs of the individual and the functions and options of the device must

Table 12-11

PROCEDURE FOR PIVOT TRANSFER

- Position wheelchair.
- Lock wheelchair brakes.
- Move footrest/legrest out of the way on the side towards the destination surface.
- Have the patient come forward in the wheelchair; provide assistance so that both hips are even and that the person achieves an anterior pelvic tilt so that body weight is forward.
- Make sure the patient has a sufficiently wide base of support (shoulder distance).
- The therapist blocks the client's knees to provide stability.
- The client is encouraged to assist by pushing up with unaffected parts.
- The patient comes to a controlled stand; the therapist holds onto the client via gait or transfer belt.
- Client and therapist turn/pivot with therapist helping to move the client's affected foot.
- Client is encouraged to reach towards destination surface and ease self onto the surface by using unaffected parts; therapist assists by holding onto transfer belt and easing client onto surface while practicing good body mechanics.

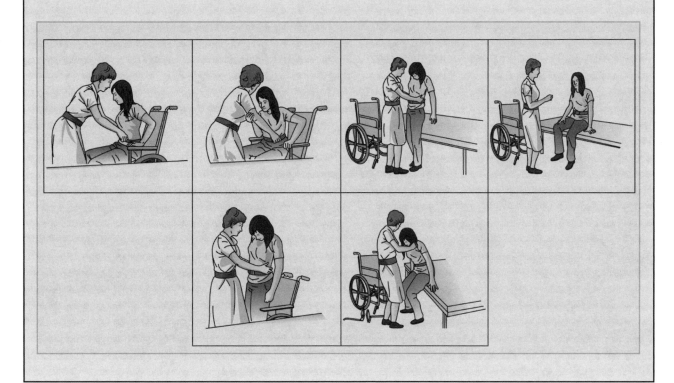

Table 12-12

Pivot Transfer Checklist

Introduction	• Wash hands
	• Introduction to the patient
	• Proper patient identification
Mechanics	Patient and wheelchair position
	• Level surface
	• Wheelchair positioned at 30- to 45-degree angle to transfer surface
	• Lock brakes
	• Remove footrests
	• Feet on floor, shoes on
	Pre-transfer
	• Use nonaffected/stronger arm to push out of chair
	• Move to edge of seat
	• Push up on armrest
	• Come to stand
	• Stable in standing
	• Assessment of client status (dizzy)
	Transfer
	• Reach for transfer surface
	• Management of lower extremity
	• Pivot on strong lower extremity
	• Quarter circle turn
	• Ease into chair
	• Client position after transfer
	• Awareness of therapists' body mechanics
Clinical judgment	Ongoing assessment of client status
	• Consistent eye contact
	• Short, concise, clear directions to patient
	• Periodic verification of patient understanding of task steps

---

be made. Included in the decision is the cost and availability of the device, what adaptations may be needed, and what is involved in maintaining and repairing the device.

> **Principle: Adapting task objects or using adaptive/assistive devices or orthotics compensates for lack of reach, ROM, grasp, strength, vision, use of one side of the body, loss of mobility, and to permit transportation of objects, to overcome architectural barriers, or to ensure proper positioning.**

Adaptive equipment may radically change the way things are done. For example, a long-handled hairbrush may be recommended to a client to enable independent grooming. Use of the long-handled hairbrush minimizes the amount of shoulder ROM that is required to brush the hair, but the device still requires isometric contraction of the wrist and hand muscles with enough strength to hold the brush. In addition, coordination and arm strength is needed to control the longer lever created by the extended handle.

If the equipment or device is complex, the client will need higher levels of cognition to use the devices effectively

Table 12-13

ADAPTED DRESSING METHODS: LOSS OF THE USE OF ONE SIDE OF THE BODY

PUTTING ON A PULLOVER SHIRT

1. Begin sitting with shirt in lap, backside up, neck away from patient (label is facing down).
2. With left hand gather up the back of the shirt to expose the right armhole.
3. Using left hand, lift right hand and place it through the armhole and sleeve.
4. Place left arm through left armhole and sleeve up to the elbow.
5. Using left hand, push right sleeve above right elbow.
6. Gather back of shirt from collar to hemline.
7. Continue holding shirt and work shirt up both arms toward the shoulders.
8. Duck head and pull shirt over it.
9. Pull shirt down in back and front.

PUTTING ON A CARDIGAN GARMENT

1. Begin sitting with shirt in lap, inside up, and collar away from body.
2. Using left arm, place right hand in right armhole (the armhole is diagonally opposite arm).
3. Pull sleeve over hand, grasp collar, and pull sleeve up onto the right shoulder.
4. Toss the rest of the garment behind body.
5. Reach left hand back and place it in armhole.
6. Work sleeve up arm and straighten shirt.
7. Button shirt (easier to start from bottom).

PUTTING ON TROUSERS

1. Begin sitting on side of bed.
2. Using left hand, cross right leg over left.
3. Check to see that trousers are opened completely.
4. Grasp trousers at bottom of front opening and toss down toward right foot.
5. Pull right trouser leg up and over right foot.
6. Place right foot on floor and put left leg in other trouser leg.
7. Pull trousers up over knees.
8. Lie down, bend left hip and knee, push against bed, and raise buttocks.
9. Pull pants over hips; fasten. If patient can stand, omit step 8 and pull trousers on while standing; sit to fasten trousers.

PUTTING ON SOCKS

1. Cross legs and pull on with left hand.

PUTTING ON SHOES OR ORTHOSES

1. Sew tongue to top of shoe at one side to prevent it from doubling over.
2. If brace is attached to shoe, be sure that the leg is in front of brace when putting shoe on.
3. Begin sitting on side of bed with right leg crossed over left.
4. Slip shoe on foot as far as possible.
5. Place a shoehorn in heel of shoe and place foot on floor.
6. Push down on knee, making sure shoehorn stays in place.
7. Fasten shoes (buckles, Velcro, or one-handed tie).

(continued)

Table 12-13 (continued)

ADAPTED DRESSING METHODS: LOSS OF THE USE OF ONE SIDE OF THE BODY

TYING A SHOE WITH ONE HAND

1. Knot one end of the shoe string and lace the shoe, leaving the knotted end at the lowest eyelet.
2. In the top eyelet, feed the end of the shoe string from outside to inside. Throw the end over the top of the laces.
3. Make a loop in the free end of the shoestring and pull it, loop within a loop.
4. Pull the lace tight, being careful not to pull the free end all the way through.
5. To untie, pull the free end.

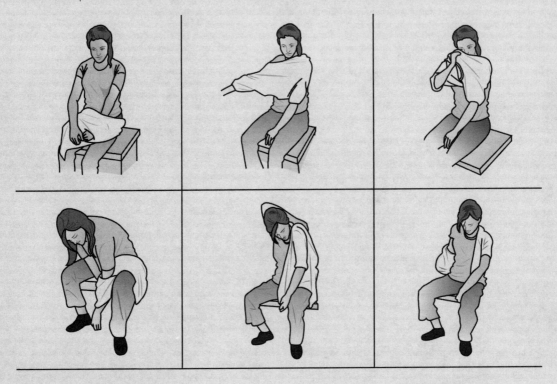

and safely. It is easy to see that a reacher will compensate for decreased ROM, but judgment is needed when getting a heavy soup can down from a high cupboard. Not only might the soup can fall on the person, but there are additional stresses on the wrist and fingers as well as on the cardiovascular system with overhead movements.

Many items of adaptive equipment do not change the way the activity is done, but make the task easier. This is seen with Velcro shoe closures and adding foam to utensils to build up the handles. The fork with a foam handle is used the same way as a fork without foam; the difference is that the person can hold the fork more easily because less hand closure is required.

Another difference is that the foam makes the fork look different, which is one disadvantage to adapting the task

objects for some people. Consideration of the psychological impact of adaptive equipment on the person is important in the consistent use of the devices by the client. By using tools or devices that look different, unwanted attention may be drawn to the person using them, which may be a constant reminder of loss of function. Sometimes, a client would rather have assistance than use adaptive equipment. This is part of the assessment for use of adaptive devices. Some orthotic devices (e.g., mobile arm supports) may be bulky and unattractive, which may be another disadvantage for some clients. Clients have both positive and negative attitudes toward adaptive devices. The device is valued for the function that it affords, but there may be a stigma when using the device (Ali, O'Brien, Hoffmann, Phillips, Garland, Finely, et al., 2008; James, 2008). Collaborate with the client for function and use of the device.

Table 12-14

WHEELCHAIR MANEUVERING THROUGH DOORS

THROUGH DOORS (HIGH QUADRIPLEGIC, C4 OR C5, PULLING DOOR TO OPEN)

1. Starting position: Back chair up to a double or a single door, the handrims just clearing the door that will be opened.
2. Motion:
   a. Place hand in door handle.
   b. Open the door slightly.
   c. Remove hand from door handle.
   d. Using hand against door, push door open until door is past the rim of the wheelchair.
   e. Using the rim to block the door open, turn chair toward the door, keeping the rim against the door.
   f. Propel chair forward with rim against the door until door is fully open.
   g. With the wheel continuing to block the door open, back chair with arm that is opposite the door.
   h. Push off from door and propel through doorway backward.
3. Teaching tip: Because of the tenodesis effect, it is important to place hand in door handle before turning the chair completely backward.

THROUGH DOORS (LOW QUADRIPLEGIC, C-6 TO C-7)

1. Starting position: Back chair up to a double or a single door, the handrims just clearing the door that will be opened.
2. Motion:
   a. Push door open far enough so that the door is blocked by the foot pedals.
   b. With the foot pedals against the door, turn wheelchair toward the open door.
   Caution: Feet will catch the door as the individual turns toward the door. To avoid injury to the feet let the chair roll back a little as the individual turns through the door.
   c. Hold elbow against the door to keep it open.
   d. Propel through.

THROUGH DOOR (PUSHING FORWARD)

1. Starting position: Approach the door to be opened at a slight angle (30 to 40 degrees).
2. Motion:
   a. Push the door open.
   b. Use toes and foot pedals to brace the door open.
   c. Turn wheelchair toward open door.
   Caution: Do not bang into the door with your toes.
3. Teaching tip: The weight of the door will straighten out the chair as it goes through.

THROUGH TWO DOORS (PUSHING DOORS OPEN)

1. Starting position: Center wheelchair between the two doors with one foot pedal against each door.
2. Motion:
   a. Push doors with both feet until wheelchair is through the door past the handrims.
   b. With hands on the doors and elbows flexed, use trunk extension to push doors open.
   c. Propel through the doors.
   Caution: If you do not get far enough through the doors, the weight of the doors will push the chair backwards.

Table 12-15

CHANGING THE TASK: ADAPTED TASK METHODS GOALS

CHANGING THE TASK: ADAPTED TASK METHOD OR PROCEDURE

GOAL	METHOD	PRINCIPLE	EXAMPLE
Client will be (I/modified I) (require maximal/moderate/ minimal/standby/verbal cue/physical cue) to perform tasks (specify which tasks) in _____ weeks/days using…	Adapted task procedures and methods that	Substitute for lost range of motion	• Pivot or sliding board transfers • Affected extremity in first one-sided limitations • Avoid twisting at the knees when transferring • Hip precautions for THR • Garments with zippers and buttons in front • Slip-on shoes • Wrap toilet paper around hand • Wide base of support when pulling pants up in standing position • Eliminate wearing underwear • Wide base of support on toilet to prevent falls • Simplify hairstyle, routines, tasks • Grow a beard • Use terrycloth bathrobe to dry off instead of towel • Large fasteners or Velcro for clothing fasteners
Client will be (I/modified I) (require maximal/moderate/ minimal/standby/verbal cue/physical cue) to perform tasks (specify which tasks) in _____ weeks/days using	Adapted task procedures and methods that	Substitute for loss of strength	• Let gravity assist • Use principles of levers (force arm > resistance arm) • Apply force closer or farther from fulcrum to change length of lever arms

(continued)

Table 12-15 (continued)

CHANGING THE TASK: ADAPTED TASK METHODS GOALS

CHANGING THE TASK: ADAPTED TASK METHOD OR PROCEDURE

GOAL	METHOD	PRINCIPLE	EXAMPLE
			• Increased friction decreases power required for pinch or grasp
			• Use two hands
			• Weave utensils through fingers
			• External rotation and abduction can substitute for supination
			• Hook elbow around wheelchair upright to increase reach
			• Extrinsic tightness for selective tightening for hook grasp
			• Elbow "walk"
			• Lock elbow using external rotation and scapular depression
			• Tenodesis action with wrist extension for grasp
			• Adapted bed mobility
			• Sliding board or pivot transfers
			• Adapted dressing
			• Limit the weight of task objects
			• Let unaffected parts assist
			• Use affected parts to hold objects
			• Careful selection of clothing items (type of material), preferably slightly large for ease in donning
			• Lick the heel of the hand to move shirt material up arm
			• Adapted functional and community mobility
			• Simplify hairstyle, routines, tasks

(continued)

Table 12-15 (continued)

Changing the Task: Adapted Task Methods Goals

CHANGING THE TASK: ADAPTED TASK METHOD OR PROCEDURE

GOAL	METHOD	PRINCIPLE	EXAMPLE
Client will be (I/modified I) (require maximal/ moderate/minimal/standby/verbal cue/physical cue) to perform tasks (specify which tasks) in _____ weeks/days using	Adapted task procedures and methods that	Compensate for lost use of one side of the body	• Adapted dressing by putting affected arm in sleeve first (one-handed limitation) • Let unaffected parts assist • Use affected parts to hold objects • Adapted bed mobility • Pivot transfer • Affected part dressed first and undressed last • One-handed shoe tying, slip-on shoes, Velcro closures • Tape nail file to table to stabilize • Use unaffected arm and leg to propel hemi-wheelchair • Prop infant or use infant seat • Velcro strap on changing table • Adapted functional and community mobility • Simplify hairstyle, routines, tasks
Client will be (I/modified I) (require maximal/ moderate/minimal/standby/verbal cue/physical cue) to perform tasks (specify which tasks) in _____ weeks/days using	Adapted task procedures and methods that	Provide stability (ataxic and uncoordinated movements)	• PNF patterns • Teach client to stabilize objects being used • Stabilize proximal part • Use body in as stable a position as possible • Use larger and/or less precise fasteners, tools, objects • Increase friction

(continued)

Table 12-15 (continued)

CHANGING THE TASK: ADAPTED TASK METHODS GOALS

CHANGING THE TASK: ADAPTED TASK METHOD OR PROCEDURE

GOAL	METHOD	PRINCIPLE	EXAMPLE
Client will be (I/modified I) (require maximal/moderate/minimal/standby/verbal cue/physical cue) to perform tasks (specify which tasks) in _____ weeks/days using	Adapted task procedures and methods that	Minimize the effect of spasticity	• Use "distal key points of control" • Use "proximal points of control" • Use placing, lowering, and weightbearing • Use affected arm as stabilization
Client will be (I/modified I) (require maximal/moderate/minimal/standby/verbal cue/physical cue) to perform tasks (specify which tasks) in _____ weeks/days using	Adapted task procedures and methods that	Substitute for limited vision	• Organize; there is a place for everything • Food on plate organized like a clock • Adapted cutting by means of food placement • Pour liquid based on weight of cup when full • French knots in labels to identify colors • Store all colored clothes in one part of closet • Fold money for each denomination differently
Client will be (I/modified I) (require maximal/moderate/minimal/standby/verbal cue/physical cue) to perform tasks (specify which tasks) in _____ weeks/days using	Adapted task procedures and methods that	Substitute for decreased endurance	• Have grocery list • Frequent rests • Shop during nonpeak times • Work simplification • Energy conservation • Stabilizing items so that they do not need to be held • Dressing in bed • Good posture throughout day

(continued)

Table 12-15, Continued

CHANGING THE TASK: ADAPTED TASK METHODS GOALS

CHANGING THE TASK: ADAPTED TASK METHOD OR PROCEDURE

GOAL	METHOD	PRINCIPLE	EXAMPLE
Client will be (I/modified I) (require maximal/ moderate/minimal/standby assist/verbal or physical cues) to perform (specify) tasks in _____ weeks/days using or Client will consistently use work simplification techniques to _____ (specify task)	Adapted methods that	Simplify work	• Organize storage • Plan ahead • Have an easy flow of work • Eliminate steps and jobs • Use efficient methods • Consider the environment
Client will be (I/modified I) (require maximal/ moderate/minimal/standby assist/verbal or physical cues) to perform (specify) tasks in _____ weeks/days using or Client will consistently use energy conservation techniques to _____ (specify task)	Adapted methods that	Conserve energy	• Respect pain • Rest frequently • Prioritize activities • Avoid sustained, static positions or isometric muscle contractions • Avoid stressful positions • Consider the environment

Another disadvantage to using adaptive equipment is that the device needs to be available every time that activity is done. A person may have adapted utensils or use a mobile arm support or other equipment at home when eating, but would need to take the equipment with him or her when he or she goes out to dinner. The device needs to be used in different contexts, so portability is a consideration or deciding upon alternative ways of achieving the task in different situations.

The use of adaptive devices and orthotics has three advantages. First, adaptive devices have good face validity, and it is easy for clients to correlate improved function with the use of the device. Second, adaptive devices offer a concrete, immediate solution to a functional problem. The client cannot hold a fork, and foam on the utensil or the utensil inserted into a universal cuff resolves the limitation. Third, many adaptive devices are inexpensive (although given the variety of devices available, there is also a range of cost) (Dutton, 1995).

Clients and caregivers need sufficient education about the use of the device and practice in using the device during activities. In a study by Matuska, Mathiowetz, and Finlayson, 13% of the respondents said they did not use adaptive equipment because they were unsure of how to use the device or they could not use it (2007). Even if the clients are satisfied with the level of education about adaptive equipment use, they do not always use the equipment when they get home. In a study by Finlayson and Havixbeck (1992), 97% of the clients reported satisfaction of adaptive equipment education, but 75% of the clients used the equipment. Functional independence and satisfaction with and use of bathing devices improved in clients who received home visits after discharge from the hospital, inferring that contextual generalization of device use is helpful (Chiu & Man, 2004). Assistive devices increase the ability to perform daily activities (Gosman-Hedstrom, Calesson, & Blomstrand, 2002).

Adapting Task Objects: Range of Motion

To compensate for decreased ROM, handles are extended, so reaching the object is possible or built up to make grasping easier. Extensions can be added to the tool itself (adding a long lever to nail clippers) or can be accomplished with the use of a reacher. Built-up eating utensils are available for purchase, or regular forks and spoon handles can be enlarged by using cylindrical foam that fits over the utensil. Making items needed for tasks easily available in close proximity to the task location is a helpful strategy in organization. Use of nonslip materials (Dycem or a damp washcloth) keep items stabilized.

Dressing, Grooming

A dressing stick or reachers can be used to reach the clothing without excessive flexion or external rotation and due to loss of ROM. Sock aids are often used for people with decreased ROM, and this aid is considered valuable to clients after discharge (Finlayson & Havixbeck, 1992). A wall mounted hairdryer would be a helpful tool and adaptation for those who are unable to hold the hairdryer to style their hair. Elastic shoe laces and a long-handled shoe horn are important pieces of equipment for those with limited hip ROM and are often recommended after total hip replacements due to hip precautions after surgery. Buttonhooks can be used to button shirts

(Figure 12-4), and they can be inserted into a universal cuff or enlarged with foam if grasp is limited or may be mounted on a table with a suction cup to stabilize while buttoning.

Bathing, Toileting, Showering

Long-handled sponges making washing the feet and back much easier, and the plastic handle can be heated and bent slightly to provide better contact with skin surfaces. Pump-action containers for soap and shampoo are helpful at the sink and in caddies or wall mounted in the shower. Long-handled toilet aids or tongs can enable acceptable toilet hygiene. A tub bench or shower chair and hand-held shower may enable safe bathing from a seated position. A bedside commode or urinal near the bed may eliminate unsafe trips to the bathroom in the middle of the night. Grab bars near the tub and toilet make transfers to those surfaces safer. Elevated commodes make using the toilet safer and easier.

Eating

Decreased ROM may affect eating by disallowing grasp or reaching the hand to the mouth. For decreased grasp, enlarged handles on eating utensils permit successful self-feeding or inserting utensils into a universal cuff (Figure 12-5). Long straws and adapted cups that don't require grasp enable independent drinking. Opening containers can be done by stabilizing the item in a drawer or between the knees, using an adapted wall-mounted jar opener, or using the teeth to open some packages.

Child Care

For clients with decreased ROM and strength, playing with children and changing a baby can sometimes be done more safely on the floor. For those who lack the ability to get up off of the floor, having a crib that swings open on the side (rather than moving up and down) may make putting the child to bed easier. Clothing for the child is easier to put on if slightly large and with large closures or Velcro.

Additional Suggestions

Keys may need to be adapted so that a person with decreased grasp can use them. This can be done by either buying commercial key adaptations or by adapting the key using splinting material. Lamp switches may need extensions, or inexpensive environmental controls can be used (e.g., clap or motion sensors to activate lights).

Electrical appliances often make tasks easier but not always. For example, some electric can openers require the use of both hands to align the can and force to depress the lever. Review the variety of equipment available online and in catalogs, and try the device with the client before recommending it.

Bowls can be stabilized in a drawer or between the knees, and suction devices attach to the stovetop to stabilize saucepans during cooking. Cutting boards with or without suction devices are available with stainless steel nails so that food can be placed on the nail, stabilizing it, so it can be cut safely. If drawers have small knobs or are difficult to open, loops can be added so that wrist or forearm movements can open the drawer. Silicone spray or bar soap rubbed on drawers will ease the sliding of the drawer.

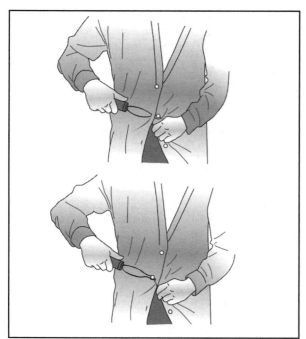

Figure 12-4. Using a buttonhook device.

Adapting Task Objects: Strength and Endurance

Many of the adaptive devices suggested for decreased ROM will also apply to the other functional limitations, including decreased strength. Devices for stabilization eliminate the need for sustained muscle contractions so are helpful for those with decreased strength and endurance. Enlarged handles or items inserted into universal cuffs are helpful for people with decreased hand or arm strength. Loops on clothing and socks facilitate dressing, and loops on drawers permit a person easy access to cupboards, desks, and dressers. Electric appliances, such as electric razors and toothbrushes, make hygiene easier. Plate guards and bowls with raised edges, enlarged handles, or utensils inserted into universal cuffs enable independent self-feeding.

Reachers are also useful but need to be considered carefully to determine which type works best for any individual. Wooden scissors-type reachers require bilateral hand use. Pistol-grip reachers can be used with one hand but require sustained grip and grasping action to open and close the device. Consider the change in forces when an object is picked up with reachers and how this changes lever arm mechanics.

Splints can be used to stabilize the wrist or to provide functional hand positioning. Dynamic splints (such as a flexor-hinge or tenodesis splint) can provide functional hand movements that permit wide gradations of force and skill levels.

Clients on specific bowel and bladder programs need special adaptations. Catheter adaptations are often necessary and may include Velcro straps for ease in removing and attaching legbags, and adapted legbag valves to enable legbag emptying. Suppository inserters are used in bowel programs as are digital stimulators (or rectal stimulation splints). While occupational therapists don't design or teach bowel and bladder programs, we are responsible for assisting in devising methods for performing these programs and for providing adapted equipment used in these procedures.

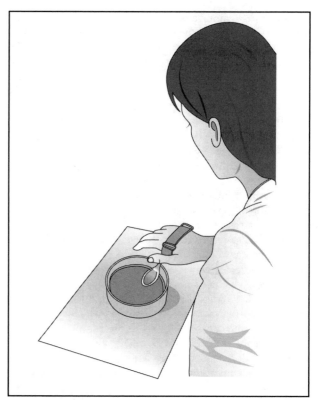

Figure 12-5. Universal cuff.

Lapboards for wheelchair users provide a working surface that is easily reached. It also provides a means to transport items from one place to another, including hot items from a microwave to a table. Wheelchair bags or pouches are often attached to either the side of the wheelchair on the outside of the armrest or between the push handles in the back. Walker bags or a wheeled cart help ambulators transport items.

For clients with severe upper extremity strength losses, a mobile arm support (MAS) or bilateral feeding orthosis (BFO; also known as bilateral forearm orthosis) may permit movement in a frictionless arc. A MAS supports the weight of the arm and assists with shoulder and elbow motion in a gravity-eliminated plane. Suspension slings may also be used by clients with weak or absent shoulder or elbow muscles. A strap is suspended from an overhead rod to support the arm with springs attached to the strap to allow movement in specific directions. Both of these orthotic devices can be used to assist with daily living tasks.

Bedrails, overhead trapeze, or rope ladders are devices that assist in bed mobility and dressing tasks.

Adaptations may need to be made to electrical appliances or electronic aids to daily living (EADLs). These might include adapting remote control devices, VCR and DVD players, iPods, door openers, lights, stereo systems, computer switches, and different types of cell phones (James, 2008).

Writing devices come in a variety of options and are commercially available, or a figure-of-eight writing splint can be made from thermoplastic materials. Splints can be used for writing, and being able to write a legal signature is an important goal. Book holders and automatic page turners are available commercially, or a pencil inserted into a universal cuff with the eraser end pointed out is an easy way to turn pages, dial a phone, or use a keyboard. Books on tape or electronic readers can also be used. Mouthsticks can also be used to turn pages, write, or use a keyboard.

Adapting Task Objects: Coordination

Stabilizing objects and reducing fine motor movements will help in compensating for poor coordination. Suction devices (nail brushes, cutting boards, etc) will hold items steady, and pump-action containers don't require hand function to operate. Using a plate guard or a bowl with raised edges makes eating easier, and a wash mitt or sponge with a strap makes bathing more efficient.

Using elastic shoelaces or slip-on shoes eliminates the need for tying shoes. Having medications with non-childproof lids is helpful for those with poor coordination, decreased grasp, and low vision. Electric devices (such as an electric razor or electric toothbrush) may be good alternatives for the individual with poor coordination. Magnetic jewelry closures make wearing necklaces and bracelets possible.

If tremors are causing the incoordination, weighting the devices used helps to dampen the tremors. Weighted eating utensils have been found to be an effective way to improve feeding performance (McGruder, Cors, Tiernan, & Tomlin, 2003). Heavy cookware helps decrease tremors, and pans, pots, and serving dishes with double handles provide greater stability. A serrated knife is less likely to slip than a straight-edge knife (Holm et al., 1998). Oven controls in the front are easier for people with many types of limitations, including incoordination, loss of ROM, and decreased strength.

Adapting Task Objects: Vision

Contrast is used to differentiate items for the low-vision client, such as using a dark cutting board for cutting a potato and a light cutting board for cutting a tomato. Other contrast examples would be pouring coffee into a light-colored cup and using colored toothpaste, adding contrast to white brush bristles (Beaver & Mann, 1995; Cooper, 1985; Cristarella, 1977; Lempert & Lapolice, 1995). Contrasting colors on thermostats, wall outlets, locks on windows, light switches, oven and stove dials, and on drawstrings on draperies are other examples of color contrast. Contrasting dinnerware is another way that contrast can be used to compensate for decreased vision, and use of separate dishes for different food items is another suggestion for eating. Having the salt shaker in a round container and the pepper in a square one can provide shape contrast.

Use plastic tape of contrasting colors to mark different items. Contrasting colors of colored tape at the end of a cane may make locating the cane easier. Labeling items with large, black print, tactile labels, or Braille can also help discriminate one item from another on a shelf or in a cupboard. Floating a brightly colored object in water can enable judging water height when filling a bathtub (Holm et al., 1998; James, 2008).

Making things easier to see can also be done by changing the lighting. Replacing overhead incandescent lights with fluorescent lights and being aware of glare can help people with low vision. Avoiding shiny surfaces (tabletops, floors) and excessive clutter makes it easier to find objects easily and quickly.

Dressing can be easier if shirts are kept buttoned except at the neck. This makes dressing easier, and the clothing stays on the hanger better. Washable puff paints can be used as a color system of dots or letters on clothing (Holm et al., 1998; James, 2008). Using an electric razor, soap-on-a-rope, and pump soap dispensers are helpful adaptations for grooming.

Magnifying images and types can be done by using a page magnifier, enlarging copies, or using large-print books. The font size on computers can be increased for easier reading, and many devices are "talking" or voice-activated such as computers, watches, kitchen scales, and clocks. Oversized phone dials, calculators, and watches are helpful adaptations for the visually impaired. Computer optical scanning devices help make printed materials easier to read. Prism glasses can be useful while reading in bed.

Adapting Task Objects: Cognition

The primary concern for clients with cognitive deficits is safety. Poor judgment or impulsiveness when using sharp objects or remembering to turn off the stove or iron requires objects that reduce risks to the client. Appliances that turn off automatically (irons, hairdryers) minimize safety threats to clients with decreased memory and attention span. Using alarms to alert the client to perform certain tasks at certain times of the day may help with task performance (for example, an alarm might alert the client when to eat or take medication). Many of the adaptations for cognitively impaired clients are made to the environment by the caregiver or family to ensure safety.

Adapting Task Objects: Use of One Side of the Body

In many tasks, one hand stabilizes the item, and the other hand performs the activity. Consider peeling a potato: one hand holds the potato, while the other hand peels the potato. Consider washing a glass or cutting a sandwich or applying toothpaste to the toothbrush, and the actions of the two hands follow this same pattern. One of the challenges for people who have the use of only one side of the body is that they have lost the stabilizing function of one of the hands, so adapted equipment is needed to provide that. Suction devices are used on many items (nail clippers, bottle brushes, buttonhooks, etc), which provide this stabilizing function. A double-sided suction device (octopus) can stabilize dishes in a sink with one side attaching to the sink and the other suction portion attaching to the plate. Using a suction device on a long-handled sponge may make bathing easier. Nonslip surfaces secure items.

A free-standing toilet paper holder may make it easier for a person with use of one side of the body because the paper is positioned vertically and not horizontally (Ali et al., 2008). Using suspenders or slacks with elastic or hooks may make donning clothing after toileting easier with one hand.

A rocker knife is an additional piece of equipment that makes cutting meat and food easier with one hand. Instead of a back-and-forth method of cutting (or sawing), the rocker knife has a curved cutting edge so the knife is rocked along the edge to cut food.

Adapting Task Objects: Functional Mobility

Wheelchairs, canes, walkers, and other ambulation devices and orthoses permit functional mobility that is otherwise limited by decreased strength, endurance, or ROM. Careful consideration of the client's capacities and limitations, the expected environment, distances covered by the client during daily tasks, and knowledge of mobility devices need to be coordinated. Proper measurement of the wheelchair to ensure a good fit is essential for function and to prevent further disability. For example, a self-reclining wheelchair is very useful for people with low blood pressure, but shear forces may cause skin breakdown in insensate areas. Matching the wheelchair to the person requires collaboration with the client and other health-care professionals. A few of the adaptations to wheelchairs are listed in Table 12-16.

Table 12-16 provides sample goal statements, methods, and the principles behind why you are changing the task by using adapted devices. This is not an exhaustive list of all of the possibilities, and there is redundancy in the types of equipment or devices that are beneficial for a wide variety of limitations.

Changing the Context

Contextual variables are important considerations when implementing intervention. Context includes the physical environment, cultural and social aspects of the individual, virtual or realistic simulation of an environment, and spiritual aspects of tasks and activities.

Cultural and social changes in context may occur as a part of the teaching/learning process experienced by the client and caregivers. This may be reflected in the amount and degree of emphasis on the client's assumption of responsibility for his or her own care. In some cultures, it is the responsibility of family members to care for a disabled person. In other cultures, independence in as many tasks as possible is the valued outcome. Within cultural contexts, there are always individual responses that may or may not include the views of the prevalent cultural group. Assuming that this individual who is your client adheres strictly to all cultural values is not realistic. The importance of learning what the client values and his or her expected outcomes cannot be overstressed.

The contextual areas most obviously focused on in the rehabilitation approach are those that change the physical environment. This is done by changing the environmental structures (such as adding ramps) or by providing mobility devices to enable community or in adapting the environment. Changing the context includes changes to the physical environment and training the family and/or caregivers.

Changing the Physical Environment

Changing the physical environment will enable greater access to public and private facilities and great participation in community activities. Changes in the physical environment will enable greater access to public and private facilities and will promote independence that otherwise may not have been possible.

Enhancing the environment or enhanced therapeutic conditions are also a means to modify cortical maps. Enhanced environments are those that incorporate interesting objects or natural objects and tasks (Wu, 1998), challenging demands (Plautz et al., 2000), intensive or complex movements (Fisher & Sullivan, 2001), and social interactions (Johansson, 2000), which require problem solving and goal attainment (Trombly,

2003). The enhanced environmental conditions result in more time-efficient, more direct, and smoother motion (Wu, Trombly, & Tickle-Degnen, 2000). Engaging clients in these environments can have benefits in physical, social, cultural, and cognitive realms.

> **Principle: Changing the context by modifying the environment will provide access to transportation, housing, public and private facilities, and work and recreational activities.**

Environmental adaptation ranges from architectural design and new construction to slight modifications (Moyers, 1999). Environmental modifications are advantageous when problems cannot be solved any other way. By making these changes, valued roles may be regained.

One disadvantage to environmental adaptations is expense. Ramps, chair glides, elevators, and modified vans are all very expensive solutions to environmental problems. Another disadvantage is that, often, the changes are not portable. If a ramp is installed at one home, it often is not easily transported to another. If the deficit is temporary, then the cost may not seem justified, so the person may be without sufficient modifications to ensure independence.

The socioeconomic status of the person may have some influence on the types and amount of home modification recommendations made by occupational therapists. Publicly insured clients received fewer home modification recommendations compared to privately insured clients and were generally more limited in their ability to pay for equipment and modifications. Unfortunately, publicly insured clients were also discharged from rehabilitation facilities with significantly less functional independence and so may actually have needed the modifications more and for longer periods (Lysack & Neufeld, 2003).

Not all adaptations need to be complex or expensive, though. The President's Committee on Employment of People with Disabilities developed a sample list of adaptations to home and worksites particularly in reference to reasonable accommodation and the ADA. There are often simple solutions to problems of accessibility. A person who is unable to work because the wheelchair will not fit under the desk or a person who cannot transfer to a sofa because the sofa is lower than the wheelchair will benefit from raising these surfaces with wooden blocks. Changing the way a task is done can also provide simple solutions to accessibility problems. Using a tape recorder or transcriber rather than writing is helpful for people who need to write reports, and altering schedules to permit frequent rests is a helpful solution for people with poor endurance.

Home modifications have a positive impact on the client's self-rated ability and perceptions of safety in the home (Petersson, Lilja, Hammel, & Kottorp, 2008), decreased functional limitations and dependence on mobility devices (Fange & Iwarsson, 2005), and increased performance, satisfaction, and occupational performance (Stark, 2004). Caregivers were found to use 81% of the strategies recommended regarding task modification and changes to the social and physical environment (Corcoran & Gitlin, 2001). Environmental intervention and assistive technology use was associated with reduced institutional and nursing/case manager costs (Mann, Ottenbacher, Fraas, Tomita, & Granger, 1999) and engagement in new

Table 12-16

CHANGING THE TASK: ADAPTING TASK OBJECTS, DEVICES, OR ORTHOTICS GOALS

CHANGING THE TASK: ADAPTING THE TASK OBJECTS, DEVICES, OR ORTHOTICS

GOAL	METHOD	PRINCIPLE	EXAMPLE
Client will (be independent/modified I) (require maximal/moderate/minimal/standby assist/verbal or physical cues) to perform ADL (specify) tasks in _____ weeks/days using	Adapted task objects or adaptive devices or orthotics that	Compensate for lack of full reach or range of motion	• Long-handled sponge • Dressing stick • Stocking or sock aid • Elastic shoelaces • Long-handled shoe horn • Long-handled reacher • Long-handled utensils • Long straw with clip • Long-handled comb • Aerosol deodorant adaptation • Wiping tongs • Suppository inserter • Raised toilet seat • Long-handled dustpan • Sponge mop with squeeze handle on side • Long-handled dusters • Adjustable office chairs • Electric or spring-loaded lift for chairs to assist with standing • MAS or suspension sling
Client will (be independent/modified I) (require maximal/moderate/minimal/standby assist/verbal or physical cues) to perform ADL (specify) tasks in _____ weeks/days using	Adaptive device or orthotics that	Compensate for lack of strength	• Mouthstick for typing • Speaker phone • Mobile arm support • Overhead suspension sling • Wheelchair lapboard • Rocker knife

(continued)

Table 12-16 (continued)

CHANGING THE TASK: ADAPTING TASK OBJECTS, DEVICES, OR ORTHOTICS GOALS

CHANGING THE TASK: ADAPTING THE TASK OBJECTS, DEVICES, OR ORTHOTICS

GOAL	METHOD	PRINCIPLE	EXAMPLE
Client will (be independent/modified I) (require maximal/moderate/minimal/standby assist/ verbal or physical cues) to perform ADL (specify) tasks in ____ weeks/days using	Adaptive device or orthotics that	Compensate for lack of strength	• Elevated table to axilla height to support arm and eliminate gravity • Vacuum cleaner wand attachments • Dolly with large casters for carrying a pail of water • Self-propelled vacuum with automatic cord rewinder • Lightweight carpet sweeper • Cordless, handheld vacuum • Lightweight brooms • Rope ladder • Trapeze bars • Sliding board for transfer • Use powered tools and utensils • Wheeled cart to transport items • Wheelchair lapboard • Tub seat • Grab bars • Use of bicycle gloves to push wheelchair
Client will (be independent/modified I) (require maximal/moderate/minimal/standby assist/ verbal or physical cues) to perform ADL (specify) tasks in ____ weeks/days using	Adapting task objects or adaptive devices or orthotics that	Compensate for lack of full hand closure	• Lightweight built-up handles on pencil, utensils, comb, brush, etc. • Universal cuff with utensils, razor • Flexor hinge splint (tenodesis splint) • Corel or nonbreaking dishes

(continued)

Table 12-16 (continued)

CHANGING THE TASK: ADAPTING TASK OBJECTS, DEVICES, OR ORTHOTICS GOALS

CHANGING THE TASK: ADAPTING THE TASK OBJECTS, DEVICES, OR ORTHOTICS

GOAL	METHOD	PRINCIPLE	EXAMPLE
Client will (be independent/modified I) (require maximal/moderate/minimal/standby assist/ verbal or physical cues) to perform ADL (specify) tasks in _____ weeks/days using	Adapting task objects or adaptive devices or orthotics that	Compensate for lack of full hand closure	• Writing devices (use with roller ball pens for less friction). • Phone cuffs • T-shaped handle for tools, wheelchair control • Foam insulator to provide friction or dycem, plastisol • Plexiglas or plastic straws • Wiping tongs • Suppository inserter • Raised toilet seat • Digital stimulators • Adapted or lever faucet handles • Soap on a rope • Liquid soap • Bath mitts • Terrycloth bathrobe • Button hook device • Built-up handles on pots and pans • Built-up handles on brooms, irons • Bottle brush on suction mop with mechanism to wring sponge one-handed • Book holder • Electric page turner • Mouthstick • Levers to EADL equipment

(continued)

Table 12-16 (continued)

CHANGING THE TASK: ADAPTING TASK OBJECTS, DEVICES, OR ORTHOTICS GOALS

CHANGING THE TASK: ADAPTING THE TASK OBJECTS, DEVICES, OR ORTHOTICS

GOAL	METHOD	PRINCIPLE	EXAMPLE
Client will (be independent/modified I) (require maximal/moderate/minimal/standby assist/ verbal or physical cues) to perform ADL (specify) tasks in _____ weeks/days using	Adapting task objects or adaptive devices or orthotics that	Compensate for lack of full hand closure	• Autodialing phones • Use of bicycle gloves to push wheelchair
Client will (be independent/modified I) (require maximal/moderate/minimal/standby assist/verbal or physical cues) to perform ADL (specify) tasks in _____ weeks/days using	Adapting task objects or adaptive devices or orthotics that	Compensate for lack of vision	• Page magnifier • Prism glasses • Long-handled skin inspection mirror • Braille labels • Magnify type or images • Devices to provide auditory, tactile, or kinesthetic feedback • Color contrast • Shape contrast • Computerized optical scanning devices • "Talking" or voice-activated computers, watches, kitchen scales, clocks • Oversized phone dial, watches • Large-print cookbooks, books • Autodialing phones • Corel or nonbreaking dishes
Client will (be independent/modified I) (require maximal/moderate/minimal/standby assist/ verbal or physical cues) to perform ADL (specify) tasks in _____ weeks/days using	Adapting task objects or adaptive devices or orthotics that	Compensate for cognitive deficits	• Appliances that turn off automatically • Alarms or reminder systems

(continued)

Table 12-16 (continued)

CHANGING THE TASK: ADAPTING TASK OBJECTS, DEVICES, OR ORTHOTICS GOALS

CHANGING THE TASK: ADAPTING THE TASK OBJECTS, DEVICES, OR ORTHOTICS

GOAL	METHOD	PRINCIPLE	EXAMPLE
Client will (be independent/modified I) (require maximal/moderate/minimal/standby assist/verbal or physical cues) to perform ADL (specify) tasks in _____ weeks/days using	Adapting task objects or adaptive devices or orthotics that	Compensate for the use of one side of the body	• Suction devices or suction added to tools, utensils (octopus device) • Nonslip materials • Suction device on a long-handled sponge • Long-handled sponge • Dressing stick • Stocking or sock aid • Elastic shoelaces • Long-handled shoe horn • Long-handled reacher • Aerosol deodorant adaptation • Wiping tongs • Suppository inserter • Raised toilet seat • Long-handled dustpan • Sponge mop with squeeze handle on side • Long-handled dusters • Adjustable office chairs • Electric or spring-loaded lift for chairs to assist with standing • Free-standing toilet paper holder • Suspenders or slacks with elastic • Rocker knife • Corel or nonbreaking dishes • Crack eggs one handed • Separate eggs via egg separator, funnel, or large slotted spoon

(continued)

Table 12-16 (continued)

CHANGING THE TASK: ADAPTING TASK OBJECTS, DEVICES, OR ORTHOTICS GOALS

CHANGING THE TASK: ADAPTING THE TASK OBJECTS, DEVICES, OR ORTHOTICS

GOAL	METHOD	PRINCIPLE	EXAMPLE
Client will (be independent/modified I) (require maximal/moderate/minimal/standby assist/verbal or physical cues) to perform ADL (specify) tasks in _____ weeks/days using a	Wheelchair (or cane, walker, etc) that	Compensates for limited mobility by permitting self-propulsion (of the wheelchair) assisted mobility (cane, walker, etc)	Seat: • narrow adult (requires less UE abduction, which is tiring • low hemi-seat (so propelling foot easily reaches floor) • needs to fit to enable propulsion Wheels: • 8" diameter (more stable on rocks, curbs, etc.) • 5" diameter (tighter turns) Rims: • spoke extensions (clients with quadriplegia can push with palm or webspace; often, use bicycle gloves for better traction) • double rims (hemiplegic can push chair with one hand) • flat rim (lighter weight) Electric wheelchair One arm drive wheelchair for some clients with CVA
Client will (be independent/modified I) (require maximal/moderate/minimal/standby assist/verbal or physical cues) to perform ADL (specify) tasks in _____ weeks/days using	Wheelchair (or cane, walker, etc) that	Compensates for limited mobility by permitting transportation of objects	• Pouches attached to wheelchair • Wheelchair laptray/lapboard • Crutch holders attached to wheelchair • Walker bag • Wheeled cart

(continued)

Table 12-16 (continued)

CHANGING THE TASK: ADAPTING TASK OBJECTS, DEVICES, OR ORTHOTICS GOALS

CHANGING THE TASK: ADAPTING THE TASK OBJECTS, DEVICES, OR ORTHOTICS

GOAL	METHOD	PRINCIPLE	EXAMPLE
Client will (be independent/modified I) (require maximal/moderate/minimal/standby assist/verbal or physical cues) to perform ADL (specify) tasks in _____ weeks/days using	Wheelchair (or cane, walker, etc) that	Compensates for limited functional mobility by facilitating proper positioning	Back: • reclining back (better trunk stability) • reclines for low blood pressure • head extension (for poor head control) Arm rest: • adjustable height (to ensure proper arm and trunk support) • arm troughs/lapboards (UE positioning) • offset arms (increases inside width between uprights for wider hips) Leg rest: • elevating (for edema or problems with blood pressure) • calf pad (prevents leg from sliding off footrest) Foot rest: • Heel strap (prevents foot from sliding off footrest) Lateral supports (for better and lateral stability) Seatbelt (keeps hips back so pelvis and trunk can rest against backrest and trunk is erect) Seat: • Hard seat (inhibits LE spasticity; promotes neutral pelvic tilt and symmetrical weightbearing) *(continued)*

Table 12-16 (continued)

CHANGING THE TASK: ADAPTING TASK OBJECTS, DEVICES, OR ORTHOTICS GOALS

CHANGING THE TASK: ADAPTING THE TASK OBJECTS, DEVICES, OR ORTHOTICS

GOAL	METHOD	PRINCIPLE	EXAMPLE
Client will (be independent/modified I) (require maximal/moderate/minimal/standby assist/verbal or physical cues) to perform ADL (specify) tasks in _____ weeks/days using	Wheelchair modifications that	Overcome architectural barriers	Cushion: • ROHO seat cushion (for pressure relief) • Jay seat cushion (for pressure relief and lateral support) • Numerous abductor devices (inhibits leg scissoring and extension, which cause patient to slide out of wheelchair) • High-density foam Seat: narrow adult (for narrow doorways, navigating congested areas) Leg rest: detachable, reduces turning space for wheelchair armrest • Wrap-around armrests are 2" less in overall width • Desk arms permit movement under tables • Detachable armrests to get closer to tables Lapboard when wheelchair won't fit under table, sink, etc.
Client will (be independent/modified I) (require maximal/moderate/minimal/standby assist/verbal or physical cues) to perform ADL (specify) tasks in _____ weeks/days using	Ambulatory aids that	Permit transportation of objects	• Walker pouch or bag • Backpack while crutch-walking • Lapboard • Apron

activities with fewer rests (Niva & Skar, 2006). Occupational therapy assessment and follow-up in home modification is effective (Gitlin, Miller, & Boyce, 1999).

Home Accessibility

The major areas of concern in the home are safety, means of mobility, and client characteristics. Safety includes awareness of sensory deficits, the cognitive abilities to make good decisions and pay attention during task performance, having sufficient physical strength and endurance to complete activities (including grasp), and the ability to use equipment.

Clients may be considered high-risk when considering safety in the home for a variety of reasons. An older person living at home or with an elderly caregiver may put the client in a higher risk category. The physical environment must support basic requirements for safety and health, such as refrigeration availability, safe drinking water, adequate heat, free from pest or infestations (such as fleas, lice), and provide adequate protection from the natural environment. If the client is in an abusive relationship (either the client or the caregiver), has a pattern of problematic interactions with others, refuses family help, or is geographically isolated from others, this, too, can put the client at more risk for unsuccessful community and home re-entry. Pre-existing or co-existing limitations and multiple pathologies make living at home less likely or more difficult.

The home may have limited adaptability, such as only having one bathroom, which is located on the second floor. The client may have used a wood stove but is now on oxygen, or the client used outdoor toilets but is now in a wheelchair. These are examples of limited environmental adaptability. Financial constraints can also limit the type and extent of changes made to the physical environment.

A visit to the client's home allows the therapist to see what tasks the client can do safely, what modifications or adaptations are needed, how much assistance or supervision is required, and what tasks the client cannot do safely. This is particularly valuable because it is the client's own home and the client can demonstrate skills learned in the clinic in the actual environment to which he or she will return. Items in the home that are safety hazards include frayed cords, sharp kitchen implements, use of machinery, repetitive trauma, fires, and exposure to toxic substances. Often, a home visit will identify potential safety hazards that may be overlooked by the family because they see the objects daily and may not see the danger as clearly.

General principles to consider when looking at the client's home are to allow enough space for ease of mobility (whether using a cane, walker, or wheelchair), but not an excessive amount of space, which may be fatiguing. Limiting the travel between tasks completed sequentially will also conserve energy as well as limiting changes in levels.

Often, a client will turn an existing room (often a dining room) into a downstairs bedroom if there is a full bath on the same level to eliminate the need to travel on different levels of the home.

Awareness of lighting, noise, and safety features (such as door peep holes at the level of the user) is important as well.

Entrances

Getting into the home is the first issue of accessibility. This includes having accessible pathways if the client traverses from the street or a garage to the house. When considering building a ramp, the client's strength and balance are factors as is the length and slope of the ramp and the stability of the wheelchair. If a ramp is needed, ramps should follow a 1:12 ratio (i.e., for every rise in the slope, there needs to be 12 inches of ramp). Steeper ramps are possible for those clients in wheelchairs with good upper extremity strength. Because ramps that follow this ratio may become very long, switchback ramps are a good alternative with at least a 5 x 5 foot platform between rises for adequate turning.

Curbs on ramps prevent wheelchair wheels from moving off the ramp, and anti-skid material applied to the ramp surface increases traction for the wheelchair user. This can be commercially available non-skid strips or simply mixing sand in with paint to eliminate slippage. Handrails on both sides of the ramp that extend to the top and bottom of the ramp are useful for people in wheelchairs and for people with low vision as they ascend and descend a ramp.

Outdoor lighting and accessible doorbells and mailboxes are additional features to consider regarding making the outdoor areas accessible. Well-illuminated outdoor features (stairs, sidewalks) are especially useful for people with visual impairments. Painting the first and last steps different colors provides contrast that is helpful when ambulating. Clients with low vision, poor balance, and decreased endurance should avoid open stairwells, low balcony rails, and doors that open directly to steps. Low-hanging objects are also hazards, such as signs and mailboxes.

In general, doorways need to allow a minimum of 36 inches clear opening (32 inches with door open). Adaptations for narrow doorways include removal of the doorframe, removal of the door, offset or fold-back hinges or inset doors, replacing the door with a curtain, or considering the construction of pocket doors (the door slides into the wall). Door thresholds should be 0.5 inches or less with beveled edges. Door sills can be removed, and a thick mat can be placed over the threshold to eliminate the sill bump. Kick plates on the bottom of doors prevent damage to the door when footrests are used to hold a door open. The force needed to open a door should be no more than 5 pounds. Levered door handles or doorknob adaptations permit easy access, and keyless locks are good for clients with limited hand use. Consider whether the client can unlock the door from the outside and also lock it from the inside.

Flooring and Furniture

Wooden floors are easy to traverse for ambulators and for people in wheelchairs as long as the floor is even. Linoleum with a nonslip finish is preferable to a floor with a high gloss. Carpets should have dense loops or be dense with a firm, thin pad. Removal of small area rugs is important to consider for those who are ambulatory, in a wheelchair, have incoordination or ataxia, or have low vision.

Furniture may need to be rearranged to provide enough space for mobility and turning radius. Any unnecessary items can be eliminated, and furniture should be firm and high enough for easy transfer. If a chair or sofa is too low, leg extenders can be used; if the furniture is too high, sofa legs can be removed, and chair legs can be cut to a more appropriate length. Furniture should provide postural support, and the texture of the furniture covering should consider the ease of transfers, risk of skin abrasions, and slipping and shearing forces.

Child Care Suggestions and Modifications

Child care presents several challenges to the disabled client. If the client is ambulatory but bending over is difficult, cribs can be raised by raising the crib legs, adjusting the mattress, or even by using two mattresses. For the client in a wheelchair, the typical drop-side crib can be adapted by cutting the middle of the crib side and attaching two hinges at each end with a latch in the center. Another alternative might be to make the crib side slide along horizontal rather than vertical channels. Bathing an infant may be easiest in a modified kitchen sink for both the wheelchair and ambulatory client. Although a plastic tub or small baby pool can be used to bathe a baby, emptying the tub may be difficult without assistance. Playpens are very useful in confining the child safely.

Raising the legs of a playpen or using a portable crib will minimize bending. Adapting a playpen so that a person in a wheelchair can use it with his or her children would be similar to adapting the crib. While the safest place to handle a baby is on the floor, a changing table at a height of 31 inches is appropriate for most wheelchair users. Carrying a baby can be done by using a bassinet on wheels or a reclining stroller if one is ambulatory. Cloth infant carriers can also be used, but be alert that the front styles require back muscle contraction against a load, while back styles compress the vertebrae. For those in a wheelchair, transporting the child on the lap with a seatbelt is appropriate.

Kitchens

Kitchens that have an "L" or "U" shape are most advantageous because turning and maneuvering a wheelchair is easier. Removable leg rests and detachable armrests enable further reaching into cupboards. Lapboards and wheeled carts help to transport items in the kitchen and to other rooms. For ambulatory clients, a continuous countertop is good, so items can be slid rather than lifted. A drop-leaf kitchen table or a table attached to a wall can provide the proper height work surface and not be in the way. A slide-out board under the counter can add working space at an accessible height.

Cabinets are best if they are 12 inches to 15 inches above the counter or 48 inches above the floor and no more than 10 inches deep. With wheelchair users, there needs to be toe space to clear the wheelchair footrests. Often, removing cabinet and cupboard doors allows easier access. Lazy-Susan devices or turntables and storing frequently used items in cupboards that are between hip and shoulder height for clients who are ambulating prevents undue stretching or bending. Full-extension drawers pull out farther than standard drawers so items in the back of the drawer can be reached. Drawer and cupboard doors may need to have new knobs or magnetic latches installed for easier grasp.

The stove or cook top can be at any level that is usable by the client. Leaving the area below the countertop open will provide knee space for a person in a wheelchair. The danger in that is that spillovers could spill through to the person below. Recessed burners or glass top burners are easier for some to clean, but may be unsafe for visually or cognitively impaired clients who can't see that the burner is on. Front controls prevent reaching over the hot stove to adjust burner temperature unless the person is unsteady and might bump into the burner controls inadvertently or if the client has poor coordination and for those with small children in the home. A mirror can be placed on the stove to help one see into pots and pans on the stove. A disadvantage to the mirror is that it frequently will get splattered, greasy, and fogged by steam.

A built-in wall oven located 30 inches from the floor is very useful. Large oven and stove dials and an audible click are useful, particularly for the visually impaired person. Having an oven with a side-swing door will permit the user to get closer to the oven and handle food more easily. Heat-resistant countertops make placement of hot items easier and safer, and a wheeled cart facilitates transport of items for functional ambulators. A 15 inch flat stick is useful in pushing and pulling the oven rack, and oven mitts are easily obtained. A self-cleaning oven is the optimal choice. Microwave ovens and countertop appliances are easy to place at convenient heights.

A side-by-side refrigerator with adjustable shelves is good because the doors are smaller, easier to open, and require less clearance to access food inside. Turntables on the refrigerator shelves enable easy access to all items. Heaviest items should be placed at lapboard height for clients in wheelchairs. Automatic ice makers and frost-free options save time and are efficient. Adjustable, slide-out shelves make food retrieval more convenient. If a side-by-side refrigerator is not available, a single-door refrigerator with the freezer on the bottom can be used by people who ambulate and by wheelchair users. Newer refrigerator models have split upper doors for the refrigerator and a lower freezer drawer.

Sinks should be between 24 and 48 inches high, 36 inches wide, and 5.5 inches deep. If the sink is too low, the client can develop back strain; if the sink is too high, the elbow may become chafed, and there may be shoulder strain. An overturned dishpan can be placed in the sink with dishes placed on top to raise the level of sink. A mirror angled above the sink will provide visual assistance in seeing objects in the sink (Marrelli & Krulish, 1999). If the area underneath the sink is left open (by removing cupboard doors), a person in a wheelchair can get closer to the sink. Pipes under the sink need to be covered with foam to prevent burns to insensate areas. Two sinks at different heights might accommodate ambulatory and wheelchair users. A spray attachment on the faucet can be used to fill pots so that it will not be necessary to carry a heavy filled pot. Some clients might benefit from foot-activated pedal valves and others by lever faucet handles. If the dishwasher can be raised 8 inches, the user won't have to bend so much to load and unload dishes. Push-button controls are preferable to a knob that is turned on all appliances.

Bathrooms

For ease in maneuverability, a bathroom should have a diameter of 60 inches or a 5 foot square for turning. In tight areas, such as bathrooms, open cupboards under the countertops and sinks enable greater a turning area. Ideally, the toilet should be in a corner opposite from the door. Often, bathroom doors are narrower and ideal dimensions nonexistent. Multilevel sinks, cabinets, and countertops can accommodate all users, and adapted toilet seats and pedestal sinks provide easier access to the sink for wheelchair users (be sure to wrap exposed pipes with foam to prevent burns). If there is sufficient room in the bathroom, a stool can be placed near the sink and counter so that people with decreased endurance can sit while grooming.

Toilets should be 17 to 19 inches high for a level transfer, and standard toilets are 14 to 15 inches from the floor. Elevated toilets or commode chairs can raise the toilet height. Often, the toilet is located next to the bathtub, so tub transfers using a tub seat are difficult because there is minimal space to swing the legs into the tub.

Grab bars are useful in the tub and near the toilet. Grab bars can be wall-mounted or attached via suction, or pivoting grab bars can be moved out of the way, which is helpful in the confined spaces of a bathroom. Grab bars can also be a part of a commode seat, which may assist with movement in the small room. A center pole that goes from floor to ceiling can also be used as a way to provide support and takes up minimal space (Holm et al., 1998). Grab bars usually are 1.25 to 1.5 inches thick and can bear 250 pounds. For visually impaired people, grab bars can be installed in contrasting colors to the background.

Tub benches can be used successfully even if the bathroom dimensions are minimal. They have slats so water will not collect on the chair and are adjustable in height to ensure a level transfer. Hydraulic bathtub chair lifts are available for people who prefer to take baths but are unable to transfer into the tub. Rubber suction-grip tub mats are a possible solution for slippery bathtubs. Shower doors can be removed and replaced with a plastic curtain hung with hooks over a spring-tension bar if shower doors prevent entry into the shower.

Hand-held showers, used with the tub seat, allow a person to sit to bathe safely. The hand-held shower head can replace the existing showerhead, can be provided as an additional showerhead, or can attach directly to the spigot. Hand-held showers can also be mounted to a bar that can allow changes in position. It is best if there is an on/off switch in the handle and that the water shuts off automatically if dropped. Diverter valves can be added so that if other family members do not wish to use the hand-held shower, they can use the regular showerhead. Anti-scald devices are available if the client has decreased sensation, and the temperature on the hot water heater can be adjusted to safe levels to prevent burns. A thermometer can also be used to test water temperature, which should be below 120°F. A tap overflow alarm will alert inattentive or visually impaired people when water levels are too high. An emergency call system is particularly helpful in the bathroom where many falls and accidents occur. Roll-in showers are a nice option for individuals in wheelchairs. Coupled with a stable seat, a roll-in shower can be beneficial for those with poor balance and decreased endurance, because there is no tub to step over.

Stairs, Bedrooms

Stair glides are useful to travel from one level of the home to the next. Generally, there needs to be 22 inches of stair width to install stair glides. For people using a walker or cane, having additional devices on each level may be a solution once the person uses the stairs. Stairs should have double handrails for safer mobility.

Bedrooms should have sufficient space for maneuvering around the bed and for access to other pieces of furniture in the room. Clothing racks in closets may need to be lowered so items can be reached from a wheelchair. Closet doors may need to be removed for easier access. The bed should be at a height that ensures a level transfer, which can be achieved by removing or shortening the legs of the bed or by adding extension blocks to raise the surface. Waterbeds, while helpful to prevent decubiti formation, are difficult surfaces for transfers.

Laundry Facilities

For clients who can ambulate, a top-loading washing machine is best, and a raised dryer would minimize bending. Having a waist-high table for folding clothes nearby would enable a smooth, efficient sequence. For clients in a wheelchair, a front-loading washer is preferred with a dryer with side-hinged doors (which is hard to find). If a top loading washer is used, use of an overhead mirror will enable the person to load the machine. Transporting objects is made easier with a wheeled cart, lapboard, walker bag, wheelchair pouches, or by wearing an apron. Having a clip mounted on the wall will allow one end of a sheet to be held while folding. Mounting an ironing board on a wall at 28 to 32 inches may provide a more appropriate height for this activity. Sitting to sort and fold laundry conserves energy and is a safer method for people with decreased balance and poor endurance.

Safety and Security

When looking at the home overall for accessibility, it may be easy to overlook how easily the client can control the environment and safety features. The client needs to be able to control the lights, open and close the draperies or blinds, and adjust the thermostat. Adaptations may need to be made to burglar alarms or to emergency alarm or alert systems. When evaluating the accessibility of the home, be sure to consider two ways to enter the home and have an evacuation route planned in case of emergency. Clients using a respirator or who are on oxygen need to have a back-up plan in place should there be a power failure. A caregiver or aide will need to perform regular maintenance at the home, including changing smoke alarm batteries periodically and checking breaker boxes. Even if the client is unable to use a fire extinguisher, one should be in the kitchen and other areas of the house.

It is important to reinforce the use of smoke and carbon dioxide detectors, which can be wired into the electrical circuits of the home. This would eliminate the need for batteries. Smoke detectors should be on every level of the home. The pitch of the alarm should be low enough to be heard easily by the elderly.

Cordless utensils (including phones) are excellent devices that diminish the possibility of entanglement in the cord or inadvertently cutting the cord. Appliances that turn off automatically are excellent choices, especially if the client has decreased attention or short-term memory deficits. Devices that turn on automatically (outside lights with timers, coffee pots, etc) are also good choices.

Portable, cellular, or cordless phones are very helpful because a person in a wheelchair may be unable to reach a wall-mounted telephone in an emergency. Checking the location of light switches, thermostats, and door locks may reveal that these are inaccessible for some clients. Buzzers, intercom systems, or environmental control units are commercially available as means of contacting others in the home or in independently controlling the environment.

Community Accessibility

Getting around in the community is possible for the disabled via public and private transportation systems. Information about these options and modifications that can be made to private automobiles and driver's training for the disabled is a valuable resource to share with the client and family.

Parking spaces for the disabled should be clearly marked, located near entrances, and need to be a minimum of 12 foot wide with a 5 foot area serving as a drop-off zone for side-exit vans. The parking space should be connected by an accessible route to an accessible entrance to pathways or buildings. Curb cuts should be textured and have a minimal lip with the street. There should be no more than a 2 degree cross slope, no more than a 5 degree rise, and have 5 feet level surface at the top and bottom for turning. Pathways should be 36 inches wide with a firm, smooth, nonslip continuous surface.

Getting into public buildings can be accomplished either by ramps or by negotiating steps. Helping a client in a wheelchair navigate over curbs and stairs is a valuable skill in which to train a caregiver. To assist with ascending a curb, the easiest and safest method is to ascend the curb with the chair facing forward or toward the curb. This gives the greatest control of the chair and requires the least effort by the caregiver (Pierson, 1994). This technique is discussed in detail in the following section on caregiver education.

Entrances to buildings should have a 36-inch clear opening (32 inch with door open) or power-operated door. The door threshold should be 0.5 inches or less, although fire doors are exempt from these requirements. Doors in a series need adequate space between the doors for door swing. Turnstiles and revolving doors are not considered accessible. Directional signs should be posted indicating accessible entrances and should be 54 to 66 inches off the ground or floor level for easy visibility.

Building interiors need to be 48 inches wide and free of obstructions such as furniture, water fountains, plants, or fire extinguishers. Hard, nonslip surfaces or low-pile carpet are recommended for most types of disabilities. The building interior should not require the person to leave the building or negotiate stairs. Elevator controls should be no more than 60 inches high with doors at least 32 inches wide. Elevator controls should have visual, auditory, and tactile indicators.

Don't forget items like telephone height, light switch location, smoke alarms, elevator call button accessibility, thermostat, and water fountain heights when evaluating accessibility. Remember, too, that accessibility applies not only to those in wheelchairs but also for ambulators and people with low vision, decreased endurance, and many other limitations.

Fast Food Restaurants

Fast food restaurants and bars may present different challenges for the mobility impaired client. High counters make ordering and managing food items more difficult. Getting beverages from refrigerated storage or from soda machines is not always possible without assistance. Built-in booths for seating

may be challenging for transfers, and high stools and tables cannot be used by those in wheelchairs and with poor balance.

Theaters

Theaters also have high counters, and a person in a wheelchair may not be seen beneath the level of the counter and may be overlooked in the crowded, noisy cinema lobby.

Shopping

A person in a wheelchair cannot push a cart so needs to use a basket, which limits the amount they can buy at any one time and may necessitate more frequent shopping. Reachers may be very useful because many items may be out of reach. Store aisles are often narrow and crowded with items making movement through the store more difficult. Clothing stores provide accessible changing rooms, but trying on clothing in the store is difficult due to benches that are too narrow or not long enough to permit laying down to don/doff the items.

Grocery stores present a wide variety of colors, sights, sounds, and smells that can easily overload those people with sensory or perceptual deficits. Shiny, highly waxed floors also provide challenges. The wide array of options requires that many purchase decisions be made, which may be very difficult for those with impaired cognition. Getting groceries from the cart to the car is easily remedied by having grocery employees assist with this task, but the problem recurs when unloading groceries from the car at home.

Barbershop or Beauty Salon, Dentist Office

Getting a haircut or going to a dentist usually occurs by sitting in a special chair that can be raised or lowered to a height that is comfortable to have the task completed. The problem is that some people may be unable to transfer to these chairs, and the haircut or dental work needs to be done with the person in the wheelchair.

Banking

Automatic teller machines and counters inside of banks are designed for people who are standing. Again, the height of the counter or ATM may be limiting for people in a wheelchair. Drive-through banking is designed primarily for cars and may be too low for people in vans.

Assessments of the Environment

Tools to assess the environment are provided in Table 12-17. Some tools listed have demonstrated reliability and validity (Enabler, HOME FAST, SAFER/SAFER-HOME, Craig Hospital Inventory of Environmental Factors [CHIEF], Environmental Functional Independence Measure [Enviro-FIM]), and some are standardized with established norms of performance (Enabler).

Table 12-18 provides a brief summary of the physical adaptations that can be made and possible goal statements for documentation.

Client and Family/Caregiver Education

Family, client, and caregiver education involves teaching the family or caregiver the ways in which the disorder or injury affects the client's performance in work, play, and self-care. The client needs to learn how to care for himself or herself or to instruct others about his or her care needs.

Table 12-17

TOOLS FOR ASSESSMENT OF THE ENVIRONMENT

NAME OF TEST	RESOURCE
Assessment of Home Environments	Yarrow, Rubenstein, & Pedersen (1975)
The Home Observation for Measurement of the Environment (HOME)	Bradley & Caldwell (1997)
Descriptive Home Evaluation	The Rehabilitation Institute, Kansas City, MO
Enabler	Iwarsson & Isacsson (1996)
HOME FAST	Mackenzie, Byles, & Higginbotham (2000)
In-Home Occupational Performance Evaluation (I-HOPE)	Stark, Somerville, & Morris (2010)
SAFER/SAFER-HOME	Oliver, Blathwayt, Brackley, & Tamaki (1993)
Craig Hospital Inventory of Environmental Factors (CHIEF)	Craig Hospital, Denver, CO
Environmental Functional Independence Measure (Enviro-FIM)	Steinfeld & Danforth (1997)
Comprehensive Functional Assessment and Evaluation of the Home	Mann, Ottenbacher, Fraas, Tomita, & Grange (1999)

Often, the ramifications of a chronic disability are not realized by the family who, like the client, is adjusting to the medical and psychosocial aspects of the injury while being only minimally aware of the changes that will be needed in daily activities. The client with chronic disabilities needs to relearn how to move, dress, and perform daily tasks with an altered body and capabilities. The client and the family need help coping with the changes, and the family needs help in learning how to help the client. Provide resources to the client and his or her family to help with the psychological adjustment to the altered physical capacities and refer them to other professionals or support groups if additional counseling is needed.

> **Principle: Changing the context by caregiver and client education will prevent disability and injury by using good body mechanics, ensure joint protection, protect the body from further structural damage, maintain the joints in functional positions, prevent falls, ensure safe mobility and transfers, maintain ROM, prevent hypertrophic scar formation, maintain strength, ensure safety for somatosensory loss, and ensure that precautions are followed.**

Teaching the family the specific ways to provide assistance in dressing, transfers, positioning, as well as provision of the appropriate level and type of cueing is important for successful return to home and for safety. Recommending adaptive equipment or assistive technology and available resources to aid the family/caregiver in providing assistance is an important aspect of caregiver training. Home programs are taught to the family

so that functional gains can be continued or maintained. The family needs to clearly understand the level of assistance or supervision that is required during activities.

Disability Prevention

Much of caregiver training is directed toward safety and preventing further disability or injury for both the caregiver and the client. Disability prevention includes a wide variety of topics, including prevention of deformities, accidents, dependency, the need for institutional care, invalidism, vocational misfits, and misunderstanding and mistreatment (West, 1969).

Caregiver and client education addresses prevention at different levels. For the client, education is at a tertiary level because they already have activity limitations, impairments, or participation restrictions. The goal of education at this level is to maximize function and minimize the detrimental effects of illness or injury. Disability prevention at this level is often related to secondary consequences that may occur because of a physical limitation. For example, clients who have sustained spinal cord injuries are taught skin protection and inspection techniques because insensate skin may develop decubiti ulcers. Clients with unilateral neglect are taught to visually scan the environment to prevent injury due to lack of awareness of objects, in particular visual areas. Avoidance of potentially deforming positions is often recommended for clients with rheumatoid arthritis. This is teaching that is done to prevent further disabilities or to arrest the progression of specific conditions.

Table 12-18

CHANGING THE CONTEXT: CHANGING THE PHYSICAL ENVIRONMENT GOALS

GOAL	METHOD	PRINCIPLE	EXAMPLE
Client will (be independent/modified I) (require maximal/moderate/minimal/standby assist/verbal or physical cues) to perform ADL (specify) tasks in _____ weeks/days using…	Change the context by modifying the environment, which will	Provide access to transportation	Transport system: • Paratransit will not take you out of the house • Kneeling bus • Buses with lifts • Wheelchair tie-downs in buses/trains Private transportation • Modified driver controls • Vans with lifts • Handrails on both sides • Nonslip surface on ramp • Curbs on ramp
Client will (be independent/modified I) (require maximal/moderate/minimal/standby assist/verbal or physical cues) to perform ADL (specify) tasks in _____ weeks/days using…	Change the context by modifying the environment, which will	Provide outside access to housing, public and private facilities, and recreation	Parking spaces • located near entrances, clearly marked • 12 feet wide Curb cuts • Curb cuts should be textured • Minimal lip • No more than 2° cross slope • No more than 5° rise Doorways • Minimum 36" clear opening (32" with door open) • Electric doors with 13-second closure delay • Use of offset or fold-back hinges • Inset doors • Remove door frame

(continued)

Table 12-18 (continued)

CHANGING THE CONTEXT: CHANGING THE PHYSICAL ENVIRONMENT GOALS

GOAL	METHOD	PRINCIPLE	EXAMPLE
Client will (be independent/modified I) (require maximal/moderate/minimal/standby assist/verbal or physical cues) to perform ADL (specify) tasks in _____ weeks/days using…	Change the context by modifying the environment, which	Provide inside access to housing, public and private facilities, and recreation	• Door thresholds should be 0.5 inches or less with beveled edges • Kick plates on the bottom of doors • Five pounds of force or less to open a door • Lever door handles • Adapted keys or keyless locks Elevator • Call button 36" from floor • Door 32" wide • Visual, auditory, tactile buttons Signage • Use of symbols and Braille Ramps • 1:12 ratio • Switchback ramp if ramp is long • 5 x 5 platform for turns Floors • Wooden floors if even • Low pile carpet • Nonslip linoleum Kitchens • "U" or "L" shaped • 12-15" cabinets above counter or 48" from floor, no more than 19" deep • Oven 30" from floor • Side-by-side refrigerator • Sink 24-48" high, 36" wide, 5.5" deep *(continued)*

Table 12-18 (continued)

CHANGING THE CONTEXT: CHANGING THE PHYSICAL ENVIRONMENT GOALS

GOAL	METHOD	PRINCIPLE	EXAMPLE
Client will (be independent/modified I) (require maximal/moderate/minimal/standby assist/verbal or physical cues) to perform ADL (specify) tasks in _____ weeks/days using...	Change the context by modifying the environment, which	Provide inside access to housing, public and private facilities, and recreation	Bathrooms • Diameter of 60" • Toilet 17-19" high • Grab bars • Custom-made shower stall • Roll-in showers

For the caregiver, education is at a secondary prevention level because he or she may be considered at risk for developing disability due to his or her caregiving responsibilities. At-risk populations may include clients just returning home from a recent hospitalization, children born to addicted mothers, or those in whom new behaviors may conflict with strong values or ingrained habituation (Harlowe, 2006). Caregivers are given information that will enable safer ways or use different objects to perform tasks. These methods might include work simplification, energy conservation, and joint protection techniques as well as instruction and practice in using correct body mechanics.

Primary prevention occurs in those people who do not have limitations or impairments. Education about consistent use of seat belts, proper child seat restraints, wearing helmet on bicycles and motorcycles, backpack awareness training, and the hazards of smoking are all directed toward helping people in society at large avoid the onset of unhealthy conditions, diseases, or injuries before a critical event occurs. Primary prevention efforts are directed toward helping people realize the "linkages between occupational behavior and risks for injury" (Harlowe, 2006).

Effective prevention efforts need to include not only an evaluation of capacities of the individual and context, but must include an understanding of the client's beliefs about health risks. The importance and value that the client places on the risk factor may be related to his or her ability to carry out the recommendations. Other factors influencing the success of caregiver and client education are the cost-benefit analysis of using the recommendations in daily activities, one's belief in self-efficacy, and a belief that the recommendations or resources will be helpful (Harlowe, 2006).

The caregiver provides a pivotal role in the family, and it is important to appreciate the complex demands placed on the caregiver as he or she balances the needs of the client with his or her own needs and the needs of the other family members. Often, caregivers work outside of the home so it is reasonable to expect to repeat information more than once. Teaching caregivers and clients includes helping them transfer what they have learned in therapy so that they can apply it at home and in family contexts. Provision of resources and additional support in the community is a useful strategy, particularly for people with degenerative or progressive illnesses or who need ongoing support beyond the provision of the occupational therapy services (Flinn & Radomski, 2008).

Present information to the client and caregiver in a variety of formats to appeal to varied learning styles. Handouts that provide both written instructions and pictures are helpful references. Be sure to include your contact information, local organizations, and websites that may relate to the information provided. Demonstration of techniques need to be done with clear, concise directions and presented in the same way each time. Have the client/family member repeat after the demonstration to demonstrate competence and safety. Providing videotapes and audiotapes is also a useful way to convey information that can be used after the client is discharged. Resource catalogs with recommendations for adaptive equipment marked will make equipment procurement simpler.

The following sections will start with body mechanics and joint protection techniques. These concepts should be part of

Figure 12-6. Example of good body mechanics.

teaching the client and family how to perform all tasks safely. Education regarding mobility and transfers are the next two sections, which provide information for caregivers as to how to ensure that these tasks are done safely. Prevention suggestions for specific limitations and diagnostic categories are provided in the final sections.

Education: Body Mechanics

Occupational therapists often assume a dual responsibility in health-care settings: teaching families good body mechanics and teaching other health-care workers methods to avoid injury (Darragh, Huddleston, & King, 2009). Body mechanics should be taught to the family to ensure safe movement of the client and prevent injuries to the caregiver. Body mechanics should also be consistently practiced by the therapist for the same reasons. Before attempting to move or lift something, be sure that you or the client/caregiver are strong enough to handle the demands of the task. Test the load prior to lifting. Ask for assistance whenever you need it, and do not attempt to lift or move an object that is beyond your capacity.

Sliding, pushing, pulling, or rolling objects rather than lifting uses better body mechanics as well as conserving energy and simplifying the work. Working in a horizontal plane will eliminate or lessen the effects of gravity on the movement of objects. Gravity can be an assist when objects are moved toward the floor. Movements done in an arc also require less force to overcome object inertia and produce movement. Figure 12-6 illustrates the use of good body mechanics when getting an object out of the refrigerator or lifting a garbage can.

Use these mechanical principles to your advantage when moving objects.

Lifting with the legs and not the back is crucial and is done by bending the knees, using the strongest muscles to their best advantage. Keep the back straight, and lift in one smooth

Table 12-19

BODY MECHANICS PRINCIPLES

- Keep objects/people close to body
- Keep wide base of support
- Push before pulling and pull before lifting
- Lift with legs not back by bending knees to squat and then stand
- Use good pacing, lift smoothly
- Pivot whole body en bloc by moving feet
- Maintain your center of gravity close to the object/person's center of gravity
- Use short lever arms for better control

motion. Avoid twisting or over-stretching during the movement. A general rule cited by Gench, Hinson, and Harvey is, "if the head is kept erect while lifting, pushing and pulling, muscular involvement will tend to keep hips low and force the larger, stronger muscles of hip and knee to carry out the task" (1995, p. 184). Use the hip flexor and extensor muscles, and bend the hip and knees rather than bending forward at the waist. Back pain to caregivers and therapists can occur as a result of muscle and ligament strains due to improper lifting or trauma, muscular imbalance, and poor posture coupled with sedentary habits and overeating.

When standing, better body mechanics can be achieved by positioning one foot ahead of other and by bending the knees slightly (Gench et al., 1995). Incorporate pelvic tilt during static sitting or standing to unload the facet joints, aid in pelvic awareness, and decrease muscular tension in low back (Trombly & Radomski, 2002). Keep the back in proper alignment with the ear over the shoulder, the shoulder over the hips, and the hips over the knees. Keep a wide base of support, with feet at least hip distance apart. This will provide a stable base of support over your center of gravity.

Keep objects and people close to your body when moving them. In essence, you are using short lever arms for better control. By doing this, you are avoiding twisting the trunk, and the movement requires less muscular force. Pivot the body as a whole by moving the feet rather than twisting the torso, which can cause stress to ligaments and muscles in the back. Use good pacing, and don't rush the movement. Table 12-19 summarizes the body mechanics principles.

Education: Joint Protection Techniques

Preventing further disability may involve ensuring joint protection. Instruction to the family and client about joint protection techniques is useful for many clients, especially those with fragile joints, as in rheumatoid arthritis. Joint protection techniques involve seven different principles.

The first is to encourage the client to avoid positions of possible deformity. Usually, this would be static, flexed positions. Examples would be to avoid leaning the head on the back of

the hand or to avoid pushing up in a chair with the back of the fingers bent at the knuckles. Instead, push up by leaning on the palm. Avoiding movements in an ulnar direction also would be an example of avoiding a position of possible deformity. Most objects and fasteners are designed for use by right-handed people, using ulnar movements. Consider opening a pickle jar lid: one turns the lid to the ulnar side. An alternative way to open the jar would be to press down on the jar with the palm and use the shoulder, not wrist, to open the lid. Using a knife with a pulling motion is another example. Avoid activities that require a tight grip by enlarging handles or gripping items between the palms of both hands. Avoid wringing out a washcloth; instead, press down to remove excess moisture.

The second major idea behind joint protection techniques is to avoid holding joints or using muscles in one position for a long time. Specific examples include the following:

1. Don't stand too long.
2. Use a book rest or prop it up.
3. Substitute typing for writing.
4. Avoid unnecessary jobs.
5. Activities requiring repetitive actions should be done for short periods only (i.e., wash windows, vacuuming).

Using the strongest joints available helps to protect joints. This third principle is obvious in body mechanics principles that stress using the legs rather than the back to lift. Carrying the purse on the forearm rather than by the hand can decrease stress to joints, and even using the crook of the elbow to stabilize a bowl when stirring rather than using the hands can be protective of the joints. Figure 12-7 illustrates some joint protection suggestions.

This is related to the fourth joint protection principle, that of using each joint to its best mechanical advantage. Using both arms and legs to stand up and maintaining good posture while working at comfortable work heights enables joints to work at their best capacity.

Being able to stop an activity is an important consideration in joint protection and is the fifth principle. Breaking a task into smaller parts might make this easier. For example, if planning a multicourse dinner, prepare different parts of the meal on different days. Prolonging an activity to the point of fatigue may exacerbate the inflammatory process in healing tissues.

Respecting pain, the sixth principle, is important for clients to learn to do because stopping an activity due to pain or avoiding painful activities may prevent further joint damage. As a general guide, if the client experiences pain 1 hour after an activity, the activity was too stressful. The client should conscientiously conserve energy and respect pain when it occurs.

The final joint protection principle is to plan rest periods throughout the day to give joints respite from activity demands. Clients should be encouraged to organize their days so that activities are sequenced based on alternating strenuous activities with rest, using knowledge of fatiguing situations and times of day as guides. Sufficient sleep at night (8 to 10 hours) and frequent small rests during the day would be beneficial. Stretch break software is available that prompts you to take breaks and is available at no cost (http://people.bu.edu/kjacobs).

Figure 12-7. Joint protection suggestions.

Activity	Wrong	Right
Avoid sustained grip	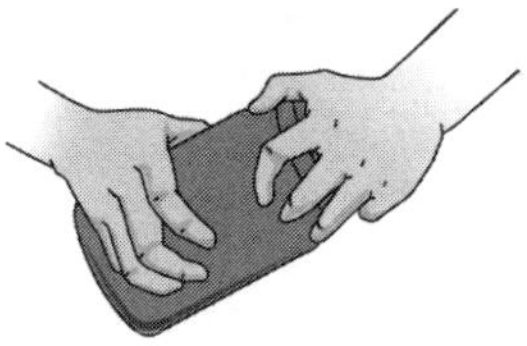 much strain on all fingers	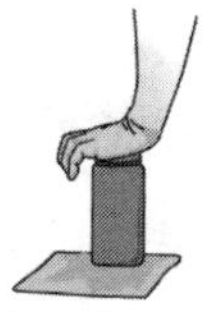 Sponge or towel to prevent sliding and use of palm on jar lid and body weight to open the jar
Cutting meat	 May be painful, or not possible to perform	 Hold knife like a dagger and use a pulling motion
Holding a plate or casserole pan	 Using fingers to lift pan	 Use oven mitts and lift with palms so wrists and elbows do the work
Carry a purse	 Carrying with fingers	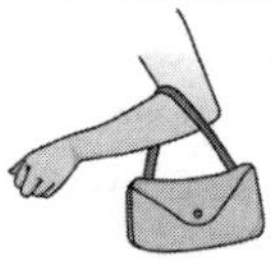 Use elbow or over the shoulder so stronger joints do the work
Backpack	 Over one shoulder or too low or too full of items	 Over both shoulders, properly adjusted
Holding a pot or pan	 Holding with one hand	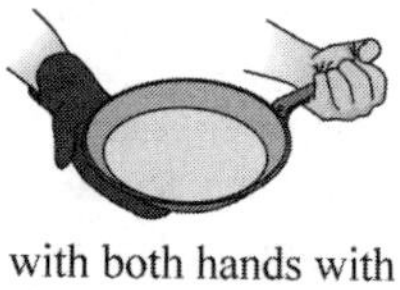 Hold with both hands with one under the pot

Reliable and valid joint protection behavior assessments have been developed to assess how well these techniques are used by clients (Klompenhouwer, Lysack, Dijkers, & Hammond, 2000). There is some support that an educational-behavioral approach is useful in teaching joint protection techniques to clients and families (Hammond & Freeman, 2004), and use of these techniques results in significantly less pain and disability and increased social and health status (Masiero, Boniolo, Wasserman, Machiedo, Volante, & Punzi, 2007). Table 12-20 summarizes the joint protection techniques.

Prevention for the Therapist

Disability prevention applies to therapists, too. Incorporating good body mechanics and joint protection techniques as well as energy conservation and work simplification ideas are as important for the therapist as for the client and caregiver. Sit when possible when working, and when you do stand, stand straight with good posture. Optimally standing on a gel-filled mat will decrease discomfort and facilitate posture changes necessary to avoid static positioning. Raising one foot on a low stool will help to support the lower back for prolonged standing. While not always possible in busy clinics, take a brief break every 20 minutes or change what you are doing (Fecko, Errico, & Jacobs, 2004).

Work-related musculoskeletal injuries and disorders (WRMD) include moderate to severe work-related pain and injury affecting the lower back, neck, shoulder, wrist, and hand (King, Huddleston, & Darragh, 2009). Because the work of occupational therapy often involves heavy lifting, awkward postures, or repetitive motions, therapists are at risk for

<table>
<tr><td>

Table 12-20

Joint Protection Techniques

1. Avoid positions of possible deformity.

2. Avoid holding joints or using muscles in one position for a long time.

3. Use the strongest joints available.

4. Do not start an activity you cannot stop.

5. Use each joint to its mechanical advantage.

6. Respect pain.

7. Regular rest periods.

</td></tr>
</table>

WRMD. Therapists have injury incidence rates that are higher than other professions, such as construction, manufacturing, and agriculture (Darragh et al., 2009; King et al., 2009). This led Darragh and colleagues to conclude that "...work-related injuries and disorders among occupational and physical therapists pose a significant population health problem" (2009, p. 359). Good body mechanics alone will not protect therapists from injury, so inclusion of mechanical lift devices and other minimal-lift equipment can help prevent injury (Darragh et al., 2009).

Education to Ensure Safe Mobility and Transfers

Mobility training and transfers are essential skills to teach the family. If possible, the client should also be taught how to instruct others in his or her own care if unable to perform the tasks independently. Taking responsibility for oneself is an important aspect that may facilitate coping and acceptance of the disability.

Assisting a person with moving from a supine position to sitting involves helping the person roll to the unaffected or stronger side. The client can then slide the feet off the bed, or the caregiver can assist with this. Keeping the client's head and shoulder forward, assist the client in pushing up to sit. Ensure that sitting balance is achieved before removing support. During this process, the caregiver should keep his or her back as straight as possible, bend with the knees, and avoid excessive trunk torsion.

Assisting someone with transfers can be done by one person. For a logroll transfer, start with a locked wheelchair facing the foot of bed, and be sure that the seat is located near the client's hips. Remove the armrest, and place a pillow over the wheel to facilitate the transfer and minimize skin breakdown. The client is rolled onto his or her side with his or her back toward the chair. His or her hips are near the chair seat at the edge of the bed with arms and legs flexed to distribute the weight. Move the hips onto the chair seat, keeping the trunk and upper body on the bed. Then, the upper trunk is moved to an upright position in the chair, and finally the feet are placed on the footrests (Trombly, 1995).

Assisted swivel trapeze transfers can be done by having the client hook an extended wrist around the bar and pulling him or her to a sitting position. With forearms across the bar, the client would then contract his or her elbow flexors while the helper holds onto the legs and pulls off the bed and swings the lower body over to the chair (Trombly, 1995).

If there is weakness in all four extremities, then the assisted depression or sliding board transfer would be a good choice. Assistance is provided by helping to place the transfer board, then by holding onto the client's waist, waistband, or transfer belt to guard against loss of balance. Do not hold onto the belt loops of pants because these rarely remain secure when pulled on strenuously. Further assistance can be provided in helping the person slide across the transfer board and then maneuvering and positioning the person's feet.

If the client can partially weight-bear or has loss of function on one side of the body, an assisted pivot transfer can be used. Assistance can be provided by giving physical and/or verbal cues; by helping the client scoot forward in the chair; by holding the client by a transfer belt, belt loop, or waistband during the transfer; by stabilizing the client in standing; by maneuvering the client's feet during the pivot and when sitting; and by easing the client into the chair.

It may be necessary to assist in repositioning the client once in the wheelchair. The helper stands behind the wheelchair, which is locked and tipped back to point of balance. Gently shake the chair, which assists gravity in sliding the client's hips back into the seat. A precaution is if the client has spasticity, which would necessitate a different method. In that case, have the client flex the elbows so the forearms go across the body. The helper puts his or her arms in between the client's folded arms and the chest wall, under the axilla from behind to grasp each forearm just distal to the forearm. The helper applies an upward force to reposition the client (Trombly, 1995).

Management of wheelchair parts is important to review with the family prior to initiation of transfer skills. Regardless of whether the client can manipulate the wheelchair parts without assistance, the client should be able to instruct others on how to set up the wheelchair prior to and after transfers.

Assisting a wheelchair, navigating curbs can be done either facing toward or away from the curb. This is shown in Figure 12-8. The helper approaches the curb with the front of the wheelchair and depresses the foot projection on the back of the frame. Next, the helper will push down on push handles and the tipping lever to push casters over the curb. The client can assist by leaning forward and pushing on the tire rims as the wheelchair is elevated. Roll the wheelchair forward so all four wheels are on the elevated surface. To ascend a curb backwards, the rear wheels are placed close to the edge of the curb. The helper stands behind the chair and tips the chair so that the casters are elevated. Pull the push handles so that the rear wheels come over the curb. Turn the wheelchair 90 degrees, and gently place the casters down.

Helping a client in a wheelchair descend a curb backwards is the easiest, safest method and requires the least effort by the caregiver. The helper approaches the curb backwards with the back wheels of the wheelchair eased down first. The helper can use his or her hip or thigh to slow the downward movement. Turn the wheelchair 90 degrees, and then ease the casters down. Descending a curb in a forward position is done by tipping

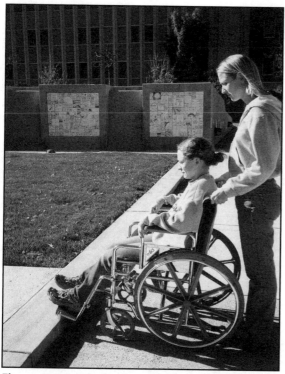

Figure 12-8A. Approach curb with casters facing the curb.

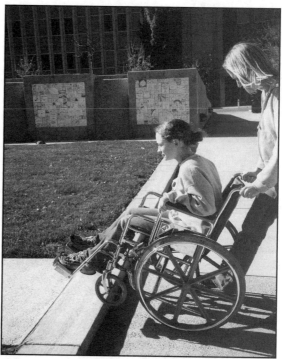

Figure 12-8B. Assistant pushes on handles and tilts wheelchair back, while patient leans forward if possible.

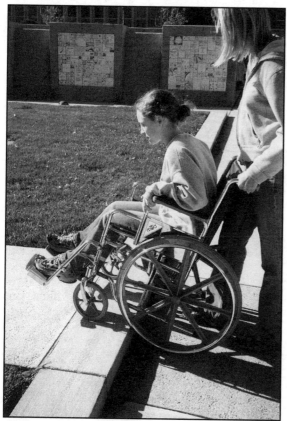

Figure 12-8C. Assistant rolls wheelchair and pushes casters over curb.

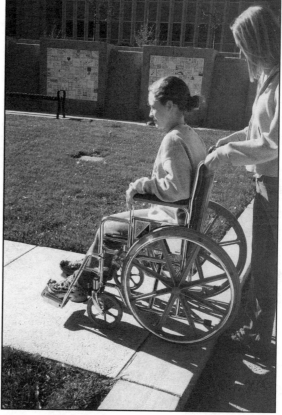

Figure 12-8D. Assistant rolls wheelchair and pushes back wheels over curb.

the wheelchair back to a point of balance with the center of gravity over the rear axle. Then, push the large wheels down gently over the curb (Turner, Foster, & Johnson, 1996).

Two people are needed to assist a client in a wheelchair going up stairs. Approach the stairs backwards with one helper behind the wheelchair. This person tips the wheelchair backward into a balanced position. The second person is in front and holds onto the leg rest upright (onto the frame, not removable parts). The second person helps to maintain the point of balance while pulling the wheelchair up each step, while the person at the foot assists with lifting and maintaining the point of balance (Pierson, 1994). Going down the stairs, the stairs are approached in a forward position. The wheelchair is tipped in balanced positions, and the process above is reversed.

Placing a wheelchair into a car is also a skill that can be taught to the caregivers. The wheelchair can fit either into the trunk or into the back seat of the car, especially if a two-door car is used and the client is loading the wheelchair into the car.

Assisting a person using a walker to ascend a curb is achieved by having the client step close to the curb with the walker and then lifting the walker up. With the stronger leg leading, the client then steps up. The caregiver is behind the client or to the side, holding onto a gait belt or waistband of the pants or onto the client's elbow. Going down a curb is the reverse process: approach close to the curb, lower the walker, and step down with the weaker leg first. The same method would apply to crutch users, too.

Education: Proper Use of Equipment

Teaching the family and caregiver how to use any pieces of adaptive equipment, devices, or orthotics is important for continued and proper use of these items. Education about adapted techniques used in dressing, feeding, and daily living skills as well as adaptations for work and play need to be clearly explained and demonstrated to the family.

Education: Prevention of Structural Damage

Preventing further structural damage is often achieved by orthotic devices or positioning in bed or in the wheelchair. It is important for both the client and the family to know how to apply the devices, minimize potential harm (i.e., skin breakdown, duration of use), and achieve maximal benefit from the devices. Devices may be used to protect a specific vulnerable body part during activities, to relieve pain, or to promote rest for healing structures.

Wrist cock-up splints are useful for clients following a stroke, and arm troughs are placed on wheelchairs to position a dependent extremity. Other examples are seen in prosthetic devices for clients with amputations or with the use of an ankle-foot orthosis (AFO) or knee-ankle-foot orthosis (KAFO). These devices not only position the parts but protect them from further structural damage. Foot drop splints and orthoses maintain functional positions and prevent foot drop deformities, which increase the risk of falls and limit ambulation.

Slings are often used following an upper extremity injury or after a stroke, but their use for subluxed joints is controversial. The use of a sling for the glenohumeral joint that would position the scapula on the ribcage with the glenoid fossa facing upward, forward, and outward would position the joint to compensate for the lack of rotator cuff and superior capsule support. However, no slings assist in realigning the scapula on the ribcage in this way (Gillen). Slings do not reduce subluxation (Zorowitz et al.). Slings may lift the head of the humerus to the level of the glenoid fossa, but scapular and trunk alignments do not occur in slings that are currently available (Gillen). Slings can be used to provide support to prevent overstretching of soft tissue, to decrease traction forces, and to prevent neurovascular injury (Zorowitz et al.). Issues of comfort, cosmetic appeal, and easy donning/doffing are important to consider in recommending a sling. Slings, troughs, and lap trays can be used to support the glenohumeral joint, but care should be taken that their use does not interfere with extremity function.

Positioning a client with loss of movement on one side of the body provides support for weakened parts, which maintains normal symmetry with additional goals of inhibition of abnormal tone, provision of normal sensory input, and increased awareness of the affected side (NDT goals). In supine, pillows are placed under the knees to maintain knee flexion, and a folded sheet is placed under the pelvis on the affected side to maintain symmetrical pelvic alignment. A rolled towel under the head will maintain the head in midline with slight flexion. Specific positioning suggestions may vary by diagnosis.

Education: Maintenance of Function

An advantage to education for the client and family is that the client may be able to live in a less restrictive environment in order to maintain valued roles and interests. Clients and their families may not always see the need for safety education, because, often, safety awareness and precautions require vigilant application of abstract concepts. Clients with cognitive impairments may not be able to visualize safety problems or be able to view their own body while performing a task, and many techniques require changes in habit. In these cases, the obligation of safety will fall to the caregivers.

Maintaining ROM may be indicated for those individuals who are unable to actively move or who are not permitted to move through full, partial, or any amount of motion. This may include a client on extended bed rest (e.g., in a coma) or may be a client with the potential for scar development due to surgery. Passive range of motion (PROM) is performed to prevent deformities, contractures, and shortening of soft tissues, as well as to prevent adhesions and mold collagen into orderly chains. If PROM exercises and activities are not done in healing tissues, collagen fibers would bind into tangles and adhere to adjacent structures (Cooper, 2003; Dutton, 1995).

Maintenance of ROM is motion done through the range that is currently available. No attempt is made to move the part past the currently available ROM. Whether the movement occurs passively, active-assisted, or actively depends upon the capabilities of the client and safety issues. For example, if the client has sensory losses or cognitive/judgment deficits, active motion controlled by the client may progress too far or too fast and harm healing structures. In these cases, therapist intervention in providing passive or active-assisted range via activities or exercise may be indicated.

Some of the benefits of PROM include the following:

- Maintenance of joint and soft tissue integrity.
- Minimize contracture formation.
- Maintain the mechanical elasticity of muscle fibers.

- Assist circulatory and vascular functions.
- Enhance synovial movement for cartilage nutrition.
- May decrease pain.
- Assist in healing after surgery or trauma.
- Helps to maintain an awareness of movement.

PROM will not prevent muscle atrophy or increase strength or endurance because there is no active muscle contraction. While PROM assists circulation and vascular functions, more value is gained by active or active-assisted motion than by passive movement. In addition to the benefits of PROM, when the client actively contracts muscle fibers, additional benefits are gained, such as maintenance of physiologic elasticity and contractility of participating muscles, provision of sensory feedback, increased stimulation for bone integrity, and increased circulation, which also helps prevent thrombosis formation.

The principle of maintaining ROM is that movement needs to go through the full available range in order to be effective. This can be accomplished by activities such as overhead checkers, macramé, putting dishes into a cupboard, placing clothes in a side-loading dryer, or making a bed. ROM can also be maintained by movement in anatomical planes, in muscle range of elongation, or in combined patterns of movement. Self range exercises are particularly beneficial for the client to enable independence and elicit responsibility in this aspect of care. Activities likely to be found in the clinic might be the use of a finger ladder, shoulder wheel, overhead pulley, or a suspension device (i.e., suspension sling). The most valuable activities, though, are those that the client needs to perform in daily life and those performed with active motion.

Active movement can be used to maintain strength by using currently intact muscles and joints and preventing disuse atrophy, pain, and stiffness. Daily activities, such as activities of daily living and wheelchair propulsion, can be used to maintain the strength of the muscles. The goal is conservation and preservation, not increasing the muscle strength.

A special case in maintaining ROM might be the client who has sustained a spinal cord injury and is in a halo device. ROM should not go past 90 degrees (nor will it likely be possible) in shoulder flexion and abduction. Above these levels, the humerus will likely have contact with the halo, restricting movement and causing pain. Other considerations for clients with spinal cord injuries is to avoid full ROM of the finger flexors to permit slight flexion tightening to enable tenodesis grip of the fingers when the wrist is extended.

Assisted motion may be required for the client with burned hands, and gentle active motion can be performed by a client with rheumatoid arthritis, even in an acute flare-up, provided clients respect their pain limits.

Scar prevention devices, which use compression, may slow collagen synthesis. Devices such as pressure dressings, elastomer molds, transparent face masks, and pressure garments apply a uniform pressure gradient to promote healing and minimize heterotrophic scar formation. Caregivers need to learn how to apply these garments and need to be instructed in care to maintain hygiene.

Education: Somatosensory Losses

Clients with somatosensory losses will need to be aware of the areas that are insensate to prevent skin breakdown or damage. Pressure relief is a very important area of disability prevention for clients with lack of sensation. Clients are taught methods to inspect their skin daily. Pressure relief techniques, such as push-ups in the wheelchair or leaning to one side in a wheelchair, are essential to teach to clients with SCI to prevent decubiti ulcers (pressure sores). Clients are taught to relieve pressure at regular intervals, are able to differentiate pressure marks from pressure sores, and are provided with seating devices and special mattresses to minimize skin breakdown due to prolonged pressure.

Awareness of areas lacking sensation is important in compensatory practices. For example, a person with no thermal sensation will need to use alternative methods to prevent burns to the hand when checking water temperature. Decreasing the temperature of the hot water heater and using a thermometer are easily implemented protective adaptations. Clients are taught to compensate for decreased sensation by relying more on vision.

Education: Fall Prevention

Falls in the home occur for many different reasons or combinations of factors. Client-related factors are varied. A client with decreased vision due to cataracts, macular degeneration, glaucoma, homonymous hemianopsia, or low vision will need environmental adaptations, such as contrasting colors on steps and greater illumination.

Clients may have lower extremity dysfunction or weakness due to arthritis, peripheral neuropathies, or foot disorders. Lower extremity weakness is a common contributor to increased incidence of falls. Clients with gait and balance disorders (Parkinson's, stroke, cane, or walker users) are at risk for falls, and people with bladder dysfunction (nocturia, incontinence, frequency of urination) may rush to the bathroom and fall. Fluctuations in blood pressure and cardiovascular disorders (orthostatic hypotension, arrhythmia, syncope) and certain medications (diuretics, antihypertensive, sedatives, psychotropic, NSAIDs) may also be risk factors for falls. Emotional and cognitive deficits also make falling more probable (dementia, depression, anxiety, fear of falling, denial of physical and functional limitations, refusal to use assistive device).

Environmental factors that may cause falls include using unsafe objects like a wheeled cart for support rather than a walker and using objects in an incorrect way. Furniture in poor repair is a danger as are loose scatter rugs. Carpets that are thick or not secured at the edges are hazards to falls. Poor lighting is also a variable to be considered when attempting to prevent falls. This is especially true in congested areas, halls, and entranceways. Lighting intensity may be increased (especially in baths and stairways), using fluorescent lights, and adding tinted windows or Mylar shades to reduce glare may improve lighting throughout the home. Improper footwear is another culprit in falls and tripping hazards, such as excessive clutter or electrical cords also are extrinsic causes of falls in the home.

Falls in the bathroom are common and are serious due to the small space, wet surfaces, lack of maneuverability, and hard surfaces. Common adaptations may include slip-resistant surfaces or non-slip adhesive strips. And falls are not restricted only to the elderly. For children, playground equipment is a potential source of falls, and many playgrounds being built now have reduced the fall impact by being covered with shredded rubber or mulch rather than concrete.

Fall prevention programs are comprehensive and can occur at all three levels of the disability prevention continuum. Primary prevention of falls would consist of controlling environmental hazards, educating people about fall risks, and making a clear link between behavior and falls (Harlowe, 2006). Primary prevention programs in older adults without impairments have resulted in fewer declines in physical health and in improved physical and social functioning, vitality, mental health, and life satisfaction (Clark, Azen, Zemke, Jackson, Carlson, Hay, et al., 1997; Jackson, Carlson, Mandel, Zemke, & Clark, 1998). Secondary prevention training might include a home safety assessment for clients at risk and would include an assessment of attitudes and knowledge about fall risks to ascertain if new behaviors may conflict with strong values and ingrained habituation (Harlowe, 2006). Tertiary prevention efforts are for those clients who have experienced falls on recurrent basis.

Comprehensive programs would provide assessment and information to the family and client about environmental safety, risk taking behavior, assertiveness training, self-efficacy, attitudes about risks, and physical fitness. A complete medical history and fall history would be gathered during an interview with the client and his or her family. Get detailed descriptions about the client's level of physical activity and alcohol use. A complete and updated list of current medications would also be part of the medical information gathered. Physical assessment would include assessment of strength, ROM, coordination, balance problems, gait or mobility deficits, and visual impairments.

The interaction of the person, environment, and occupation must be included in the assessment. Discover what activities the client is doing or wants to do, and match the client and environmental characteristics to the person's capacities. Use the person's outcome expectations and efficacy expectations to determine intervention (Ceranski & Haertlein, 2002). In a study by Cumming and colleagues, 65% of the clients and families were at least partially compliant with 50% or more of the recommendations for home modification for fall prevention. The only factor that differentiated adherers from non-adherers was a belief that home modifications prevent falls and that the clients and families saw no association between home modification and the self-perceived risk of falls or history of falls (Cumming, Thomas, Szonyi, Frampton, Salkeld, & Clemson, 2001). The clients and family need to believe that the modifications will be beneficial and that they need the changes for full participation in activities. In another study, participants in a falls prevention program were very knowledgeable about fall risks, but there was little change in behavior and attitudes (Buri, 1997).

This may require a change in beliefs about levels of activity participation and changes in physical abilities. Strategies that worked before may no longer work, necessitating learning new ways of performing common tasks. Sensory information and coping strategies need to be reinterpreted and revised. Changing efficacy and activity participation will require repetition of tasks in sequence to change habit patterns. Feedback, encouragement, and support from family and near-peer role models may help the client to believe he or she can develop the skills needed for safety in the home (Bandura, 1977; Ceranski & Haertlein, 2002).

Fall prevention programs must be multidimensional. In a study by Campbell and colleagues, a home safety program and an exercise program were used with older adults. Falls and injury were reduced primarily due to the safety education elements of the intervention and not due to the exercise, possibly due to low adherence (Campbell, Robertson, LaGrow, Kerse, Sanderson, & Jacobs, 2005). This may be due to the lack of association between the benefit of exercise in decreasing falls or due to the lack of congruence with valued occupations.

The intervention needs to include the interaction of person-environment-occupation. A cognitive-behavioral program for falls prevention was found to result in a 31% reduction of falls (Clemson, Cumming, Kendig, Swann, Heard, & Taylor, 2004) as compared to usual care alone (Weatherall, 2004), and falls prevention interventions decrease the number and/or severity of falls (Haines, Bennell, Osborne, & Hill, 2004). Multicomponent, multifaceted, and interdisciplinary programs reduce functional difficulties, fear of falling, fall incidence, anxiety about falls, and home hazards, and enhance self-efficacy and adapted coping in older adults with chronic conditions (Close, Ellis, Hooper, Glucksman, Jackson, & Swift, 1999; Gitlin, Winter, Dennis, Corcoran, Schinfeld, & Hauck, 2006; Tolley & Atwal, 2003).

Education: Sternal Precautions

After heart surgery, sternal precautions are in effect. Activities where there is pushing or pulling with the arms, which includes using the arms to push oneself out of bed or a chair, should be avoided. No one should pull on the client's arms during activities or transfers because this applies too much pressure on the sternal area. Precautions for cardiac clients would include monitoring of heart rate and blood pressure and often rely on metabolic equivalent (MET) levels for selected activities. If the heart rate goes up more than 20 beats per minute from the resting pulse or above 120 beats per minute or below 60, a physician should be consulted. If resting systolic blood pressure goes above 150 or below 90 and/or if diastolic blood pressure goes above 90 or below 50, consult a physician. Normal blood pressure is 120/80.

Education: Cognitive Impairment

For clients with cognitive impairments, limiting client choices and monitoring the client for safety are roles that the caregiver may assume. Limiting clothing options by limiting the clothing available and removing extra clothing from closets and drawers can allow the client to make decisions but limits the decisions that are necessary. This strategy can be applied to other tasks, such as meal preparation, housekeeping tasks, or leisure activities. Monitor the comfort of the client throughout the day including how well the clothing fits, the temperature of the room, and medication routines. Help with organization by decreasing clutter and arranging the items for tasks in sequences (i.e., arrange clothing in order to be donned) (Holm et al., 1998). The client may not be able to perform all parts of the task, so the caregiver may initiate the task and the client complete it. Provide only the level of assistance that is needed.

Having the client wear a non-removable bracelet with identifying and contact information would be helpful if the person tends to wander off. Alarms can be installed if doors are opened. Keep hazardous materials and items out of reach such as medications, alcohol, toxic cleaners, sharp objects, matches, and car keys. If safety around the stove is a concern, remove the knobs from appliances and unplug the microwave oven and coffee pot.

Education: Visual Impairment

Caregiver training for family members with visual impairment would depend on the level of disability and vision loss. It may be a change in habit for the family members, such as closing drawers and opening doors, to increase safety for the visually impaired person or returning items to designated places and arranging food in a specific way to make it easier for the person. The help that the caregiver provides may be as simple as checking the appearance of the person to be sure clothing is clean, matched, and appropriate for the weather.

Education: Shoulder Impairments

Clients who have recently had shoulder surgery or have unstable shoulder girdles should not be pulled up in bed or repositioned using their arms. During transfers, support these clients around their trunks rather than under the axilla to avoid structural damage or further instability. This advice is actually true for all clients. Safer transfer mechanisms have been described, and pulling on a client's shoulder girdle is not a recommended practice for any type of transfer.

Table 12-21 provides sample goal statements, the method to achieve the goal, and the principle guiding the intervention with several examples. A sample goal statement for family and client education might be, "Client's mother will demonstrate proper body mechanics and safe techniques consistently while assisting son with stand pivot transfer from bed to wheelchair" (Moyers, 1999).

CASE STUDY

The following case study is provided to demonstrate how theory influences intervention. This case study has elements that are more appropriately considered within a compensatory/adaptation or rehabilitation intervention approach, while other aspects are better served using a remediation or biomechanical construct. This same case study is used in Chapter 11 so that the elements appropriate to each theory can be seen as they relate to this case. Other theories not covered in this text may also be appropriate, and it is important to consider whether all of the approaches selected can be implemented concurrently with others or if the design of the intervention needs to be sequential in nature.

Maggie

Maggie is a 61-year-old woman referred to home health occupational therapy on October 29 with a diagnosis of right Colles' fracture. Maggie fell while shopping at her neighborhood department store on October 27. The physician has ordered occupational therapy to evaluate and treat.

On your first visit to see Maggie, you interview her and find out the following information. Maggie was in excellent health prior to this injury. Maggie's son, Fred, lives in the next town. She works part-time in her son's accounting office. A widow, she lives alone with her cat and her dog. Maggie is concerned that, because she lives alone, she needs to be able to do all of the cooking and cleaning and care for her dog and cat. Her son is only able to come to her home every other day or every 3 days to assist with some of these activities.

Maggie is concerned about her present situation. She is right-handed and states that, although she tries her best, she is not able to do very much for herself. Maggie has been very active in her local church organization. Maggie always heads up the Christmas crafts bazaar. Part of her responsibilities for the bazaar include organizing and teaching at craft nights. Maggie also has been one of the primary cooks for the annual fish dinner held just before Easter. She states that the church especially needs her this year because the other primary cook is moving to another state. She doesn't want to let the church organization down.

In the emergency room on October 27, the emergency room physician performed a closed reduction and applied a plaster cast from mid-humeral level to the metacarpophalangeal (MCP) joints of the right hand. The elbow is casted in approximately 90 degrees of elbow flexion, and the wrist is casted in approximately 30 degrees of wrist flexion. Maggie states that the physician has instructed her to stay home for the next 2 weeks and not to drive or do any housework. This is confirmed when you speak to the physician. A summary of information about Maggie as well as analysis of additional information that is needed and ideas about the next steps are as follows:

Facts

- 61-year-old woman
- Enjoys shopping
- Right Colles' fracture
- Fractures are immobilized and cause pain and edema
- Fractures may limit the use of the UE
- Lives alone
- Was independent in ADLs, IADLs
- Cares for dog and cat
- Active in church craft bazaar and fish dinner
- Anxious about present situation
- Immobilized in plaster case
- Physician restricted activities for 2 weeks

Additional Information Needed

- How did Maggie fall?
- Precautions
- Medical management of Colles' fracture
- Possible complications
- Prognosis to return to prior level of function
- Tasks that Maggie will need to do to resume roles

Actions

- Continue dialogue with Maggie
- Review medical record
- Consult with physician
- Consult orthopedic textbooks and OT texts
- Perform OT assessments

From this information, an evaluation plan can be developed. Several assessments have been listed, but these are not the only assessments that can be used to address the concerns about Maggie.

Evaluation Plan

1. Concerns.
2. Maggie's perception of her strengths and weaknesses and her priorities for therapy.
3. Decreased ADL performance due to right (dominant) UE in cast.
4. Unable to carry out worker and volunteer roles.
5. Unable to participate in leisure pursuits.
6. Sensorimotor concerns (sensory): Possible sensory loss or altered sensation due to pressure on nerves from swelling or tight cast.
7. Sensorimotor concerns (motor): Potential for edema, stiffness, and CRPS (complex regional pain syndrome previously known as RSD or reflex sympathetic dystrophy) in right hand; potential for frozen right shoulder.
8. Psychological and cognitive concerns: Potential for depression due to loss of role performance and feelings of helplessness.

Assessments

1. Canadian Occupational Therapy Performance Measure (COPM) or other client-centered informal interview tool.
2. Observe Maggie attempting to perform ADL tasks; Klein-Bell ADL assessment; self-report; Functional Independence Measure (FIM).
3. Interview Maggie to determine specific worker and volunteer tasks and activities.
4. Interest checklist; formal or informal interview.
5. Sensory evaluation to exposed areas; observe cast for fit, look for areas of redness where cast is pressing against the skin.
6. AROM and PROM of PIP and DIP joints of the right hand; circumferential edema measurements of the digits; pain assessment; AROM and PROM of right shoulder.
7. Formal and informal interview; observe Maggie during evaluation; speak with family members and other team member.

How Assessment Results Will Be Used

1. Set client-centered goals for therapy; establish rapport; gain better understanding of Maggie's occupational context.
2. Assess potential for teaching compensatory techniques or use of adaptive equipment; determine components of ADL task causing difficulty; determine baseline for therapy.
3. Determine readiness for return to work and volunteer duties; assess potential for adapting tasks and activities.
4. Determine realistic leisure interest; assess potential for adapting leisure activities.
5. Set baseline for therapy; make suggestions for alterations of cast if necessary; determine need to teach compensatory techniques or sensory re-education program if sensation is impaired or absent.
6. Set baseline for therapy; develop treatment plan, which includes management of edema and home exercise program (HEP) to maintain ROM and prevent complications.
7. Determine Maggie's motivation for therapy; assess ability to follow directions and carry out HEP, set realistic goals appropriate for Maggie's cognitive and emotional status.

You evaluated Maggie and found the following: Maggie is cognitively intact and motivated to be independent. Maggie is proud as she describes her success at figuring out how to open the foil packet of cat food with her left hand so that she could feed her cat. She is unable to don some types of blouses and dresses over the cast and is not able to don pants or skirts with closures. She requires minimal assistance to don a house dress, underpants, and socks. She is unable to don her bra or panty hose or tie her shoes.

Because of the inability to use her dominant right hand, Maggie is unable to manage buttons, has difficulty feeding herself with her nondominant left hand (spills food on herself frequently), requires minimal assistance with person hygiene after toileting, sponge bathing, and brushing her teeth. She requires maximal assistance in simple meal preparation. Maggie states she is able to get herself some simple food items, such as prepared foods (prepared by her son's wife) from plastic containers. She is dependent in household chores, such as cleaning.

Through further interview, you learn that much of Maggie's work at her son's office involves working with Excel, a spreadsheet computer program. She has a computer at home and is an avid Internet user. She tells you that it is difficult to use the computer now because she cannot use her right hand.

Tasks associated with her responsibilities as chairperson of the Christmas craft bazaar include phoning church members to inform them of meeting times and craft supplies they will be required to bring, and delegation of other responsibilities such as making refreshments, teaching certain crafts, and clean-up duties. She is having difficulty using her rotary dial telephone to keep in touch with other church volunteers.

Table 12-21

CHANGING THE CONTEXT: CLIENT AND CAREGIVER EDUCATION GOALS

CHANGING THE CONTEXT: CLIENT AND FAMILY EDUCATION

GOAL	METHOD	PRINCIPLE	EXAMPLE
Client/client will (be independent/modified I) (Require maximal/moderate/minimal/standby assist/verbal or physical cues) to perform (specify) tasks in _____ weeks/days using	Safety education which	Ensures good body mechanics	• Keep objects/people close to body • Keeps wide base of support; feet wide apart • Slides/pushes/pulls/rolls objects/people rather than lifting • Lifts with legs not back by bending knees to squat and then stand up • Uses good pacing • Lifts smoothly • Pivots whole body by moving feet instead of twisting torso • Maintain your center of gravity close to the object/person's center of gravity. • Use short lever arms for better control.
Client will (be independent/modified I) (Require maximal/moderate/minimal/standby assist/verbal or physical cues) to perform ADL (specify) tasks in _____ weeks/days using	Safety education which	Ensures joint protection	• Avoid positions of possible deformity • Avoid holding joints or using muscles in one position for a long time • Use the strongest joints available • Do not start on activity you cannot stop • Use each joint to it's mechanical advantage • Respect pain • Regular rest periods
Client will (be independent/modified I) (Require maximal/moderate/minimal/standby assist/verbal or physical cues) to perform ADL (specify) tasks in _____ weeks/days using	Orthoses and devices which or	Protects the upper extremity or	• Wheelchair arm trough • Lapboard • Cock up splint for client with hemiplegia • Sling for UE injury

(continued)

Table 12-21 (continued)

Changing the Context: Client and Caregiver Education Goals

CHANGING THE CONTEXT: CLIENT AND FAMILY EDUCATION

GOAL	METHOD	PRINCIPLE	EXAMPLE
	Positioning that	Prevents further structural damage by protecting a specific vulnerable body part during activities or maintains joints and soft tissue in functional positions and prevents deformities	• Foot drop splints and orthosis • Positioning for specific diagnosis
Client/caregiver will (be independent/modified I) (Require maximal/moderate/minimal/standby assist/ verbal or physical cues) to perform (specify) tasks in _____ weeks/days using	Client and caregiver education	Which ensures safe mobility	• Wheelchair parts and maneuverability • Hoyer lift • 2 person carry • Logroll transfer • Assisted swivel transfer • Assisted sliding board • Assisted pivot transfer • Assistance in repositioning • Assistance in ascending/descending curbs and stairs
Client /caregiver will (be independent/modified I) (Require maximal/moderate/minimal/standby assist/ verbal or physical cues) to perform (specify) tasks in _____ weeks/days using	Safety education which	Ensures safe transfers	• Gets wheelchair close as possible • Removes wheelchair parts if necessary • Equalizes heights • Locks wheelchair brakes • Moves client's hips forward to get close to edge of chair/bed • Places feet flat on floor and directly under knees before standing

(continued)

Table 12-21 (continued)

CHANGING THE CONTEXT: CLIENT AND CAREGIVER EDUCATION GOALS

CHANGING THE CONTEXT: CLIENT AND FAMILY EDUCATION

GOAL	METHOD	PRINCIPLE	EXAMPLE
			• Identifies safe landing site for fall before standing up • Moves smoothly
Client/caregiver will (be independent/modified I) (Require maximal/moderate/minimal/standby assist/ verbal or physical cues) to perform (specify) tasks in _____ weeks/days using	Safety education which	Maintains joint range of motion in the current range to prevent adhesions and molds collagen into orderly chains	• Functional patterns in activities and tasks related to interests such as overhead checkers, macramé, putting dishes into a cupboard, placing clothes in a side loading dryer, making a bed • Movement in anatomical planes • Movements in muscle range of elongation • Movement in combined patterns of movement (ex: PNF) • Self range of motion exercises • Range of motion dance • Finger ladder • Shoulder wheel • Overhead pulleys • Suspension devices
Client/caregiver will (be independent/modified I) (Require maximal/moderate/minimal/standby assist/verbal or physical cues) to perform (specify) tasks in _____ weeks/days using	Education to	Prevent scar hypertrophic formation using devices which apply compression, which may slow collagen synthesis	• Pressure dressings • Elastomer molds • Transparent face mask • Pressure garments

(continued)

Table 12-21 (continued)

CHANGING THE CONTEXT: CLIENT AND CAREGIVER EDUCATION GOALS

CHANGING THE CONTEXT: CLIENT AND FAMILY EDUCATION

GOAL	METHOD	PRINCIPLE	EXAMPLE
Client/caregiver will (be independent/modified I) (Require maximal/moderate/minimal/standby assist/verbal or physical cues) to perform (specify) tasks in _____ weeks/days using	Safety education to provide	Maintenance of strength by active use of currently intact muscles and joints which prevents disuse atrophy and stiffness during periods of enforced rest	• Active movement in anatomical planes • Active movement in diagonal PNF patterns • ADLs • Wheelchair propulsion • Leisure activities • Work activities
Client/caregiver will (be independent/modified I) (Require maximal/moderate/minimal/standby assist/verbal or physical cues) to perform ADL (specify) tasks in _____ weeks/days using	Safety education which	Ensures safety for somato-sensory loss	• Inspects skin daily • Differentiates between a pressure mark and pressure area • Relieves pressure at regular intervals • Wears protective splints and orthotics • Uses appropriate positioning • Moves insensate body part carefully using visual feedback
Client/caregiver will (be independent/modified I) (Require maximal/moderate/minimal/standby assist/verbal or physical cues) to perform ADL (specify) tasks in _____ weeks/days using	Safety education which	Ensures cardiac precautions	• Rest if heart rate goes up more than 20 bpm from resting pulse • Rest if heart rate goes above 120 bpm or below 60 bpm • Rest if systolic BP goes above 150 or below 90 (normally SBP = 120) • Rest if diastolic BP goes above 90 or below 50 (normally DBP 80)
Client will (be independent/modified I) (Require maximal/moderate/minimal/standby assist/verbal or physical cues) to perform ADL (specify) tasks in _____ weeks/days using	Safety education which	Ensures that hip precautions are followed	• Avoid hip abduction (e.g., don't cross legs to roll in bed • Avoid hip internal rotation (e.g., don't line up foot with shoe by twisting leg into internal rotation)

(continued)

Table 12-21 (continued)

CHANGING THE CONTEXT: CLIENT AND CAREGIVER EDUCATION GOALS

CHANGING THE CONTEXT: CLIENT AND FAMILY EDUCATION

GOAL	METHOD	PRINCIPLE	EXAMPLE
Client/caregiver will (be independent/modified I) (Require maximal/moderate/minimal/standby assist/ verbal or physical cues) to perform ADL (specify) tasks in _____ weeks/days using	Safety education which	Ensures shoulder precautions	• Never pull client up in bed using his or her arms; use a draw sheet • Do not support a client under his or her arms during transfers; support client around the trunk
			• Avoid hip flexion past 90° (e.g., stand up from chair by leaning back and sliding hips to edge of seat, then extend knee of operated leg, and push off from armrests) • Use elevated commode to maintain hip flexion 90° or less • Do not cross ankles or legs when sitting, laying or getting in/out of bed • When in bed, keep legs apart and toes pointed toward ceiling
Client/caregiver will (be independent/modified I) (Require maximal/moderate/minimal/standby assist/ verbal or physical cues) to perform (specify) tasks in _____ weeks/days using	Safety education which	Ensures sternal precautions (after heart surgery)	• Do not push/pull with arms • Do not use arms to push out of bed or pull on siderails • Do not allow others to pull on your arms

(continued)

Maggie complains of moderate pain in her right upper extremity, mainly the wrist and fingers. She rates her pain an 8 on a scale of 1 to 10. She describes it as a burning, throbbing type of pain. Severe edema is noted in all of the digits of the right hand (approximately two times the size of the left hand). She has a great deal of difficulty flexing her fingers to make a fist (the cast restricts the MP motion). She is unable to oppose her thumb to her index or any other finger. She is unable to hold any objects in her right hand because of the cast. She states that at times her hand feels numb, and other times she has paresthesia. The fingers are discolored, a dusky reddish color.

In your conversation with the home health coordinator and the social worker, you find out that Maggie's insurance will cover home health occupational therapy for up to 3 months for the purpose of improving independence in ADLs. Table 12-9 would be an example of the intervention that would address compensatory, adaptation, or rehabilitation concerns in this case. This is not a complete intervention plan in that many areas that can be addressed in treatment by an occupational therapist are not included. By combining the treatment plan for the biomechanical remediation treatment plan with this, a more complete intervention for Maggie can be realized.

In Table 12-22, deficits as identified previously are related directly to long- and short-term outcomes. Specific principles or rationales are provided so that the reason for the specific intervention is clear. Both the biomechanical and rehabilitation intervention approaches would be useful for intervention with Maggie.

SUMMARY

- The rehabilitation frame of reference is a compensatory and adaptation approach used when there is little or no expectation for change and where there are residual impairments. It may also be used when there is limited time for intervention and when the client or family prefers more immediate resolution of functional problems.

- This intervention approach focuses on the areas of occupation, which includes work, play, leisure, education, instrumental ADLs, and social participation rather than on the underlying performance skills. This is known as a "top-down" approach where intervention starts by identifying the environmental demands and resources with the client.

- Intervention methods include changing the task or changing the context.

- Changing the task can be accomplished by adapting how the task is done or by changing the task objects or using adaptive equipment.

- Changing the context would include environmental modification and training the client and caregivers.

- Disability prevention is accomplished by client and caregiver education. This is especially true of those who already have activity or participation limitations as well as those who may be a member of an at-risk population.

REFERENCES

Ali, N. A., O'Brien, J. M., Hoffmann, S. P., Phillips, G., Garland, A., Finely, J. C. W., et al. (2008). Acquired weakness, handgrip strength, and mortality in critically ill patients. *American Journal of Respiratory Critical Care Medicine.* 178(3), 261-268.

Bandura, A. (1977). Self-efficacy: Toward a unifying theory of behavior change. *Psychology Review, 84,* 181-215.

Bonder, B. R., & Bello-Haas, V. D. (2009). *Functional performance in older adults* (3rd ed.). Philadelphia, PA: F. A. Davis Company.

Bunyog, V. M. & Griffin, C. (2007). Educating the well elderly on work simplification and energy conservation techniques. *OT Practice,* April, 11-12.

Buri, H. (1997). A group programme to prevent falls in elderly hospital patients. *British Journal of Therapy and Rehabilitation, 4,* 550-556.

Campbell, A. J., Robertson, M. C., LaGrow, S. J., Kerse, N. M., Sanderson, G. F., & Jacobs, R. J. (2005). Randomised controlled trial of prevent of falls in people aged 75 with severe visual impairment: The VIP trial. *British Medical Journal, 333,* 817-823.

Ceranski, S., & Haertlein, C. (2002). Helping older adults prevent falls. *OT Practice,* July 22, 12-17.

Chamberlain, M. A., Thornley, G., Stow, J., & Wright, V. (1981). Evaluation of aids and equipment for the bath: II. A possible solution to the problem. *Rheumatology and Rehabilitation, 20,* 38-43.

Chiu, C. W., & Man, D. W. K. (2004). The effect of training older adults with stroke to use home-based assistive devices. *OTJR: Occupation, Participation, and Health, 24,* 113-120.

Clark, F., Azen, S. P., Zemke, R., Jackson, J. M., Carlson, M. E., Hay, J., et al. (1997). Occupational therapy for independent-living older adults: a randomized controlled trial. *Journal of the American Medical Association, 278*(16), 1321-1326.

Clemson, L., Cumming, R. G., Kendig, H., Swann, M., Heard, R., & Taylor, K. (2004). The effectiveness of a community-based program for reducing the incidence of falls in the elderly: A randomized trial. *Journal of American Geriatric Society, 52*(9), 1487-1494.

Close, L., Ellis, M., Hooper, R., Glucksman, E., Jackson, S., & Swift, C. (1999). Prevention of falls in the elderly trial (PROFET): a randomised controlled trial. *Lancet, 353,* 93-97.

Colenbrander, A., & Fletcher, D. C. (1995). Basic concepts and terms for low vision rehabilitation. *American Journal of Occupational Therapy, 49*(9), 865-869.

Cooper, B. A. (1985). A model for implementing color contrast in the environment of the elderly. *American Journal of Occupational Therapy, 39*(4), 253-314.

Cooper, C. (2003). Hand therapy. *Occupational Therapy Practice, 9*(May 10), 17-20.

Corcoran, M., & Gitlin, L. N. (2001). Family caregiver acceptance and use of environmental strategies provided in an occupational therapy intervention. *Physical and Occupational Therapy in Geriatrics, 19,* 1-20.

Cristarella, M. (1977). Visual functions of the elderly. *American Journal of Occupational Therapy, 31*(7), 432-440.

Cumming, R. G., Thomas, M., Szonyi, G., Frampton, G., Salkeld, G., & Clemson, L. (2001). Adherence to occupational therapist recommendations for home modifications and for falls prevention. *American Journal of Occupational Therapy, 53*(6), 64-67.

Darragh, A. R., Huddleston, W., & King, P. (2009). Work-related musculoskeletal injuries and disorders among occupational and physical therapists. *American Journal of Occupational Therapy, 63*(3), 351-362.

Dutton, R. (1995). *Clinical reasoning in physical disabilities.* Baltimore, MD: Williams & Wilkins.

Fange, A., & Iwarsson, S. (2005). Changes in accessibility and usability in housing: An exploration of the house adaptation process. *Occupational Therapy International, 12*(1), 44-59.

Fecko, A., Errico, P., & Jacobs, K. (2004). Everyday ergonomics for therapists. *OT Practice,* August 23, 16-18.

Finlayson, M. (2005). Pilot study of an energy conservation education program delivered by telephone conference call to people with multiple sclerosis. *Neurorehabilitation, 20*(4), 267-277.

Fitts, P. M., & Posner, M. I. (1967). *Human performance.* Belmont, CA: Brooks/Cole.

Table 12-22

TREATMENT PLAN: REHABILITATION/COMPENSATION AND ADAPTATION APPROACH

DEFICITS	STAGE-SPECIFIC CAUSE	SHORT-TERM GOALS	METHODS	PRINCIPLE	SPECIFIC ACTIVITY/MODALITY
1. Spills food when eating with left hand.	1-7. Unable to use right dominant hand as it is nonfunctional due to immobilization and pain. Using left, nondominant hand for activities.	1. Client will use adapted utensil to feed self with nondominant hand without spilling.	1. Provide textured built-up handle to compensate for decreased manipulation of nondominant hand. Provide adaptive aids and educate in use of compensatory techniques for cutting, scooping, and spearing foods with one hand.	1. Textured handle will enable better manipulation of utensil. Compensate for lost RUE function.	1. Client will be provided with textured handle for utensils (plastizote) and instructed how to hold and use them. Client will practice using adapted spoon with left hand.
2. Minimal assist with personal hygiene (toilet hygiene, brushing teeth).		2. Independent personal hygiene using one-handed techniques and adaptive equipment.	2, 3, 4, 5, 6: Educate client in use of compensatory techniques and adaptive aids, which compensate for lost UE function.	2, 3, 4, 5, 6: Compensate for lost RUE function.	2. Client will be educated in and practice one-handed technique of opening toothpaste and applying to brush.
3. Minimal assist with sponge bathing.		3. Independent sponge bathing using one-handed techniques and adaptive aids.			3. Client will be provided with "octopus" suction cup to hold soap and educated how to soap wash cloth using left hand only. Client will learn to wring out wash cloth by pressing against sink and will practice one-handed techniques with assistance and verbal cues from therapist.
4. Minimal assist with donning/doffing house dress, underpants, socks.		4. Independent in donning/doffing house dress, underpants, socks using compensatory techniques.			

(continued)

Table 12-22 (continued)

TREATMENT PLAN: REHABILITATION/COMPENSATION
AND ADAPTATION APPROACH

DEFICITS	STAGE-SPECIFIC CAUSE	SHORT-TERM GOALS	METHODS	PRINCIPLE	SPECIFIC ACTIVITY/MODALITY
					4. Client will be educated in and helped to select type of clothing that will fit easily over cast. Client will be educated in one-handed dressing techniques and will practice these techniques with assistance and verbal cues by the therapist.
5. Dependent in donning bra, panty hose, shoes.		5. Minimal assist donning/doffing bra, pantyhose, shoes using compensatory techniques.			5. Client will be provided with elastic laces and/or taught one handed shoe tying. Client will practice shoe donning/doffing shoes with assistance and verbal cues from therapist. Client will be educated in and will practice one-handed bra donning.
6. Maximal assist with meal preparation		6. Minimal assistance in simple meal preparation using adaptive aids and compensatory techniques.			6. Client will be educated in use of prepackaged food, compensatory techniques for opening packages; cutting board will be adapted with stainless steel nails, etc. Client will be educated in

Table 12-22 (continued)

TREATMENT PLAN: REHABILITATION/COMPENSATION AND ADAPTATION APPROACH

DEFICITS	STAGE-SPECIFIC CAUSE	SHORT-TERM GOALS	METHODS	PRINCIPLE	SPECIFIC ACTIVITY/MODALITY
					energy conservation and work simplification techniques and will practice simple meal preparation with assistance and verbal cues from the therapist.
7. Unable to perform volunteer role of chairperson of craft bazaar.		7. Independent in using phone to contact volunteers to organize bazaar using adaptive aids and compensatory techniques.	7. Educate client in use of compensatory techniques and adaptive aids which compensate for lost RUE function.	7. Compensate for lost RUE function.	7. Client will be assisted in obtaining a speaker phone with large numbers for dialing. Phone will be positioned in easily accessible place and in an area where other volunteer work supplies are placed. Client will be educated in using the phone with one-handed techniques. Client will delegate some duties to other volunteers.
8. Unable to perform worker	8. Unable to use RUE functionally	8. Independent in using home computer	8. Educate client in use of compensatory	8. Compensate for lost function of RUE	8. Home computer will be adapted for left hand use

Flinn, N. A., & Radomski, M. V. (2008). Learning. In M. V. Radomski & C. A. T. Latham (Eds.), *Occupational therapy for physical dysfunction* (6th ed., pp. 382-401).

Furst, G. P., Gerber, L. H., Smith, C. C., & Fisher, S. (1987). A program for improving energy conservation behaviors in adults with rheumatoid arthritis. *American Journal of Occupational Therapy, 41*(2), 101-111.

Gage, M., & Polatajko, H. (1994). Enhancing occupational performance through an understanding of perceived self-efficacy. *American Journal of Occupational Therapy, 48*, 452-461.

Gerber, L., & Furst, G. (1987). Patient education program to teach energy conservation behaviors to patients with rheumatoid arthritis: A pilot study. *Archives of Physical Medicine and Rehabilitation, 68*, 442-445.

Gench, B. E., & Hinson, M. M. (1995). *Anatomical kinesiology.* Dubuque, IA: Eddie Bowers Publishing, Inc.

Gitlin, L. N., Miller, K. S., & Boyce, A. (1999). Bathroom modifications for frail elderly renters: Outcomes of a community-based program. *Technology and Disability, 10*(3), 141-149.

Gitlin, L. N., Winter, L., Dennis, M. P., Corcoran, M., Schinfeld, S., & Hauck, S. W. (2006). A randomized trial of a multi-component home intervention to reduce functional difficulties in older adults. *Journal of the American Geriatric Society, 54*, 809-816.

Goodwin, J. E., & Meeuwsen, H. J. (1995). Using bandwidth knowledge of results to alter relative frequencies during motor skills acquisition. *Research Quarterly for Exercise and Sport, 66*, 99-104.

Gosman-Hedstrom, G., Calesson, L., & Blomstrand, C. (2002). Assistive devices in elderly people after stroke: A longitudinal, randomized study—the Gotegorg 70+ Stroke Study. *Scandinavian Journal of Occupational Therapy, 9*, 109-118.

Haines, T., Bennell, K., Osborne, R., & Hill, K. (2004). Effectiveness of targeted falls prevention in sub-acute hospital setting: Randomised controlled trial. *British Medical Journal, 328*, 676-679.

Hammond, A., & Freeman, K. (2004). The long term outcomes from a randomized controlled trial of an educational behavioural joint protection programme for people with rheumatoid arthritis. *Clinical Rehabilitation, 18*(5), 520-528.

Harlowe, D. (2006). Occupational therapy for prevention of injury and physical dysfunction. In H. M. Pendleton & W. Schultz-Krohn (Eds.), *Pedretti's occupational therapy: Practice skills for physical dysfunction.* St. Louis, MO: Mosby.

Holm, M. B., Rogers, J. C., & James, A. B. (1998). Treatment of occupational performance areas. In M. E. Neistadt & E. B. Crepeau (Eds.), *Willard & Spackman's occupational therapy* (9th ed., pp. 323-363). Philadelphia, PA: Lippincott Williams & Wilkins.

Ip, W. M., Woo, J., Yue, S. Y., Kwan, M., Sum, S. M., & Kwok, T. (2006). Evaluation of the effect of energy conservation techniques in the performance of activity of daily living tasks. *Clinical Rehabilitation, 20*(3), 254-261.

Jackson, J., Carlson, M., Mandel, D., Zemke, R., & Clark, F. A. (1998). Occupation in lifestyle redesign: the well elderly study occupational therapy program. *American Journal of Occupational Therapy, 52*(5), 326-336.

Jacobs, K., & Jacobs, L. (Eds.). (2009). *Quick reference dictionary for occupational therapy* (5th ed.). Thorofare, NJ: SLACK Incorporated.

James, A. B. (2008). Restoring the role of independent person. In M. V. Radomski & C. A. T. Latham (Eds.), *Occupational therapy for physical dysfunction* (6th ed., pp. 774-816). Philadelphia, PA: Wolters Kluwer/Lippincott Williams & Wilkins.

King, P., Huddleston, W., & Darragh, A. R. (2009). Work-related musculoskeletal disorders and injuries: Differences among older and younger occupational and physical therapists. *Journal of Occupational Rehabilitation, 19*, 274-283.

Klompenhouwer, P. L., Lysack, C., Dijkers, M., & Hammond, A. (2000). The joint protection behavior assessment: A reliability study. *American Journal of Occupational Therapy, 54*(5), 516-523.

Law, M., Cooper, B., Strong, S., Rigby, P., & Letts, L. (1996). Person-Environment-Occupational (PEO) Model: A transactive approach to occupational performance. *Canadian Journal of Occupational Therapy, 63*(1), 9-23.

Lempert, J., & Lapolice, D. J. (1995). Functional considerations in evaluation and treatment of the client with low vision. *Am J Occupational Therapy, 49*(9), 885-889.

Lysack, C., & Neufeld, W. S. (2003). Occupational therapist home evaluations: Inequalities but doing the best we can. *American Journal of Occupational Therapy, 57*(4), 369-379.

Ma, H. I., Trombly, C. A., & Robinson-Podolski, C. (1999). The effect of context on skill acquisition and transfer. *American Journal of Occupational Therapy, 58*, 150-158.

MacRae, N. (2010). Training, education, teaching and learning. In K. Sladyk, K. Jacobs, & N. MacRae (Eds.), *Occupational therapy essentials for clinical competence* (pp. 217-228). Thorofare, NJ: SLACK Incorporated.

Mann, W. C., & Ottenbacher, K. J. (1999). Effectiveness of assistive technology and environmental interventions in maintaining independence and reducing home care costs for the frail elderly: A randomized controlled trial. *Archives of Family Medicine, 8*(3), 210-217.

Marrelli, T. M. & Krulish, L. H. (1999). *Home care therapy: Quality, documentation and reimbursement.* Boca Grande, FL: Marrelli & Associates, Inc.

Masiero, S., Boniolo, A., Wassermann, L., Machiedo, H., Volante, D., & Punzi, L. (2007). Effects of an educational-behavioural joint protection program on people with moderate to severe rheumatoid arthritis: a randomized controlled trial. *Clinical Rheumatology, 26*(12), 2043-2050.

Mathiowetz, V., Finlayson, M. L., Matuska, K. M., Chen, H. Y., & Luo, P. (2005). Randomized controlled trial of an energy conservation course for persons with multiple sclerosis. *Multiple Sclerosis, 115*, 592-601.

Mathiowetz, V., Matuska, K. M., & Murphy, M. E. (2001). Efficacy of an energy conservation course for persons with multiple sclerosis. *Archives of Physical Medicine and Rehabilitation, 82*(4), 449-456.

Matthews, M. M., & Jabri, J. L. (2001). Documentation of occupational therapy services. In L. W. Pedretti & M. B. Early (Eds.), *Occupational therapy practice skills for physical dysfunction* (5th ed., pp. 91-100). St. Louis, MO: Mosby.

Matuska, K., & Mathiowetz, V. (2007). Use and perceived effectiveness of energy conservation strategies for managing multiple sclerosis fatigue. *American Journal of Occupational Therapy, 61*(1), 62-69.

McGruder, J., Cors, D., Tiernan, A. M., & Tomlin, G. (2003). Weighted wrist cuffs for tremor reduction during eating in adults with static brain lesions. *American Journal of Occupational Therapy, 57*(5), 507-516.

Merano, C., & Latella, D. (2008). *Occupational therapy interventions: Function and occupations.* Thorofare, NJ: Slack Incorporated.

Merriam, S. B., & Caffarella, R. S. (1999). *Learning in adulthood: A comprehensive guide.* San Francisco, CA: Jossey-Bass.

Minor, M. A. D., & Minor, S. D. (2010). *Patient care skills.* Upper Saddle River, NJ: Pearson.

Moyers, P. A. (1999). Guide to occupational therapy practice. *American Journal of Occupational Therapy, 53*, 247-322.

Niva, B., & Skar, L. (2006). A pilot study of the activity patterns of five elderly persons after a housing adaptation. *Occupational Therapy International, 13*(1), 21-34.

Pal, H. (2003). Movement after stroke: Is compensation the wrong way to go? *Advance for Occupation Therapy Practitioners, 19*(11), 19-30.

Palmer, M. L., & Toms, J. E. (1992). *Manual for functional training.* Philadelphia, PA: F. A. Davis Co.

Petersson, I., Lilja, M., Hammel, J., & Kottorp, A. (2008). Impact of home modification services on ability in everyday life for people ageing with disabilities. *Journal of Rehabilitation Medicine, 40*(4), 253-260.

Pierson, F. (1994). *Principles and techniques of patient care.* Philadelphia, PA: W. B. Saunders Co.

Shotwell, M. P., & Schell, B. (1999). Occupational therapy practitioner as health educator: A framework for active learning. *Physical Disabilities Special Interest Quarterly, 22*(4), 1-3.

Stark, S. (2004). Removing environmental barriers in the homes of older adults with disabilities improves occupational performance. *OTJR: Occupation, Participation, and Health, 24*(1), 32-39.

Tolley, L., & Atwal, A. (2003). Determining the effectiveness of a falls prevention programme to enhance quality of life: An occupational therapy perspective. *British Journal of Occupational Therapy, 66*(6), 269-276.

Trombly, C. A. (1995). Occupation: Purposefulness and meaningfulness as therapeutic mechanisms. *American Journal of Occupational Therapy, 49*(10), 960-970.

Trombly, C. A., & Radomski, M. V. (2002). *Occupational therapy for physical dysfunction.* Philadelphia, PA: Wolters Kluwer/Lippincott, Williams & Wilkins.

Turner, A., Foster, M., & Johnson, S. E. (1996). *Occupational therapy and physical dysfunction: Principles, skills and practice* (4th ed.). New York: Churchill Livingstone.

Vanage, S. M., Gilbertson, K. K., & Mathiowetz, V. (2003). Effects of an energy conservation course on fatigue impact for persons with progressive multiple sclerosis. *American Journal of Occupational Therapy, 57*(3), 315-323.

Weatherall, M. (2004). A targeted falls prevention programme plus usual care significantly reduces falls in elderly people during hospital stays. *Evidence-based Healthcare and Public Health, 8*(5), 273-275.

West, W. (1969). The growing importance of prevention. *American Journal of Occupational Therapy, 23*(3), 226-231.

Wu, C., Trombly, C. A., & Lin, K. (1994). The relationship between occupational form and occupational performance: A kinematic perspective. *American Journal of Occupational Therapy, 48*, 679-687.

Yarrow, L. J., Rubenstein, J. L., & Pedersen, F. A. (1975). *Infant and environment: Early cognitive and motivational development.* Washington, DC: Hemisphere, Walsted, Wiley.

ADDITIONAL RESOURCES

American Occupational Therapy Association. (1994). Uniform terminology for occupational therapy (3rd ed.). *American Journal of Occupational Therapy, 48*(11), 1047-1054.

Andersen, L. T. (2002). *Adult physical disabilities: Case studies for learning.* Thorofare, NJ: SLACK Incorporated.

American Occupational Therapy Association (2003). Fall prevention for people with disabilities and older adults [consumer tip sheet].

Arsdell, L. P. (2005). Following the six Es. *Rehabilitation Management*, June 2005.

Bassett, J. (2002). A strong balance: exercise can eliminate a major cause of falls. *Advance for Occupational Therapy Practitioners*, December 16.

Bowen, J. E. (1999). Health promotion in the new millennium: Opening the lens, adjusting the focus. *OT Practice*, 14-18.

Ceranski, S., & Haertlein, C. (2002). Helping older adults prevent falls. *OT practice*, July 22.

Christiansen, C., & Baum, C. (1997). *Occupational therapy: Enabling function and well-being* (2nd ed.). Thorofare, NJ: SLACK Incorporated.

Clemson, L., Cumming, R. G., & Heard, R. (2003). The development of an assessment to evaluate behavioral factors associated with falling. *American Journal of Occupational Therapy, 57*(4), 380-388.

Cohen, H. S., & Gavia, J. A. (1998). A task for assessing vertigo elicited by repetitive head movements.

Cohen, H., Miller, L. V., Kane-Wineland, M., & Hatfield, C. L. (1995). Vestibular rehabilitation with graded occupations. *American Journal of Occupational Therapy, 49*(4), 362-367.

Collins, L. F. (1996). Understanding visual impairments. *OT Practice*, Jan, 27-33.

Cook, A. M., & Hussey, S. M. (1995). *Assistive technologies: Principles and practice.* St. Louis, MO: C. V. Mosby.

Greene, D. P., & Roberts, S. L. (1999). *Kinesiology: Movement in the context of activity.* St. Louis, MO: C. V. Mosby.

Kielhofner, G. (1992). *Conceptual foundations of occupational therapy.* Philadelphia, PA: F. A. Davis Co.

Kisner, C., & Colby, L. A. (1990). *Therapeutic exercise: Foundations and techniques* (2nd ed.). Philadelphia, PA: F. A. Davis.

Merriam, S. B., & Caffarella, R. S. (1999). *Learning in adulthood: A comprehensive guide.* San Francisco, CA: Jossey-Bass.

Oakley, R. M., Felder, R. B., & Elhajj, I. (2004). Turning student groups into effective teams. *Journal of Student Centered Learning, 2*(1), 9-34.

Pedretti, L. W. (1996). *Occupational therapy: Practice skills for physical dysfunction* (4th ed.). St. Louis, MO: C. V. Mosby.

Radomski, M. V., & Latham, C. A. T. (Eds.) (2008). *Occupational therapy for physical dysfunction.* Philadelphia, PA: Wolters Kluwer/Lippincott Williams & Wilkins.

Shotwell, M. P., & Schell, B. (1999). Occupational therapy practitioner as health educator: A framework for active learning. *Physical Disabilities Special Interest Quarterly, 22*(4), 1-3.

Sine, R., Liss, S. E., Rousch, R. E., Holcomb, J. D., & Wilson, G. (2000). *Basic rehabilitation techniques: A self instructional guide* (4th ed.). Gaithersburg, MD: Aspen Publishers.

Thomas, J. J. (1999). Enhancing patient education: Addressing the issue of literacy. *Physical Disabilities Special Interest Quarterly, 22*(4), 4-6.

Toth-Riddering, A. (1998). Living with age-related macular degeneration. *OT Practice*, Jan, 18-23.

www.aarp.org/universalhome

www.abledata.com

www.bobvila.com/Features/AccessibleDesign

www.design.ncsu.edu/cud/built_env/housing/article_hmod.htm

INDEX